FUNDAMENTAL CONSTANTS

Constant	Symbol	Value	Power of 10	Units
Speed of light	c	2.997 924 58*	10^{8}	$m\ s^{-1}$
Elementary charge	e	1.602 176 565	10^{-19}	C
Planck's constant	h	6.626 069 57	10^{-34}	J s
	$\hbar = h/2\pi$	1.054 571 726	10^{-34}	J s
Boltzmann's constant	k	1.380 6488	10^{-23}	$J\ K^{-1}$
Avogadro's constant	N_A	6.022 141 29	10^{23}	mol^{-1}
Gas constant	$R = N_A k$	8.314 4621		$J\ K^{-1}\ mol^{-1}$
Faraday's constant	$F = N_A e$	9.648 533 65	10^{4}	$C\ mol^{-1}$
Mass				
Electron	m_e	9.109 382 91	10^{-31}	kg
Proton	m_p	1.672 621 777	10^{-27}	kg
Neutron	m_n	1.674 927 351	10^{-27}	kg
Atomic mass constant	m_u	1.660 538 921	10^{-27}	kg
Vacuum permeability	μ_0	4π*	10^{-7}	$J\ s^2\ C^{-2}\ m^{-1}$
Vacuum permittivity	$\varepsilon_0 = 1/\mu_0 c^2$	8.854 187 817	10^{-12}	$J^{-1}\ C^2\ m^{-1}$
	$4\pi\varepsilon_0$	1.112 650 056	10^{-10}	$J^{-1}\ C^2\ m^{-1}$
Bohr magneton	$\mu_B = e\hbar/2m_e$	9.274 009 68	10^{-24}	$J\ T^{-1}$
Nuclear magneton	$\mu_N = e\hbar/2m_p$	5.050 783 53	10^{-27}	$J\ T^{-1}$
Proton magnetic moment	μ_p	1.410 606 743	10^{-26}	$J\ T^{-1}$
g-Value of electron	g_e	2.002 319 304		
Magnetogyric ratio				
Electron	$\gamma_e = -g_e e/2m_e$	−1.001 159 652	10^{10}	$C\ kg^{-1}$
Proton	$\gamma_p = 2\mu_p/\hbar$	2.675 222 004	10^{8}	$C\ kg^{-1}$
Bohr radius	$a_0 = 4\pi\varepsilon_0\hbar^2/e^2 m_e$	5.291 772 109	10^{-11}	m
Rydberg constant	$\tilde{R}_\infty = m_e e^4/8h^3 c\varepsilon_0^2$	1.097 373 157	10^{5}	cm^{-1}
	$hc\tilde{R}_\infty/e$	13.605 692 53		eV
Fine-structure constant	$\alpha = \mu_0 e^2 c/2h$	7.297 352 5698	10^{-3}	
	α^{-1}	1.370 359 990 74	10^{2}	
Second radiation constant	$c_2 = hc/k$	1.438 777 0	10^{-2}	m K
Stefan–Boltzmann constant	$\sigma = 2\pi^5 k^4/15h^3 c^2$	5.670 373	10^{-8}	$W\ m^{-2}\ K^{-4}$
Standard acceleration of free fall	g	9.806 65*		$m\ s^{-2}$
Gravitational constant	G	6.673 84	10^{-11}	$N\ m^2\ kg^{-2}$

* Exact value. For current values of the constants, see the National Institute of Standards and Technology (NIST) website.

PHYSICAL CHEMISTRY

Thermodynamics, Structure, and Change

Volume 1: Thermodynamics and Kinetics

Tenth edition

Peter Atkins
Fellow of Lincoln College,
University of Oxford,
Oxford, UK

Julio de Paula
Professor of Chemistry,
Lewis & Clark College,
Portland, Oregon, USA

W. H. Freeman and Company
New York

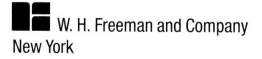

Publisher: Jessica Fiorillo
Associate Director of Marketing: Debbie Clare
Associate Editor: Heidi Bamatter
Media Acquisitions Editor: Dave Quinn
Marketing Assistant: Samantha Zimbler

Library of Congress Control Number: 2014932569

ISBN-13: 978-1-4641-2451-8
ISBN-10: 1-4641-2451-5

Published in Great Britain by Oxford University Press

Distributed under licence in the United States and Canada by
W. H. Freeman and Company.

First printing

W. H. Freeman and Company
41 Madison Avenue
New York, NY 10010
www.whfreeman.com

PREFACE

This new edition is the product of a thorough revision of content and its presentation. Our goal is to make the book even more accessible to students and useful to instructors by enhancing its flexibility. We hope that both categories of user will perceive and enjoy the renewed vitality of the text and the presentation of this demanding but engaging subject.

The text is still divided into three parts, but each chapter is now presented as a series of short and more readily mastered *Topics*. This new structure allows the instructor to tailor the text within the time constraints of the course as omissions will be easier to make, emphases satisfied more readily, and the trajectory through the subject modified more easily. For instance, it is now easier to approach the material either from a 'quantum first' or a 'thermodynamics first' perspective because it is no longer necessary to take a linear path through chapters. Instead, students and instructors can match the choice of Topics to their learning objectives. We have been very careful not to presuppose or impose a particular sequence, except where it is demanded by common sense.

We open with a *Foundations* chapter, which reviews basic concepts of chemistry and physics used through the text. Part 1 now carries the title *Thermodynamics*. New to this edition is coverage of ternary phase diagrams, which are important in applications of physical chemistry to engineering and materials science. Part 2 (*Structure*) continues to cover quantum theory, atomic and molecular structure, spectroscopy, molecular assemblies, and statistical thermodynamics. Part 3 (*Change*) has lost a chapter dedicated to catalysis, but not the material. Enzyme-catalysed reactions are now in Chapter 20, and heterogeneous catalysis is now part of a new Chapter 22 focused on surface structure and processes.

As always, we have paid special attention to helping students navigate and master this material. Each chapter opens with a brief summary of its Topics. Then each Topic begins with three questions: 'Why do you need to know this material?', 'What is the key idea?', and 'What do you need to know already?'. The answers to the third question point to other Topics that we consider appropriate to have studied or at least to refer to as background to the current Topic. The *Checklists* at the end of each Topic are useful distillations of the most important concepts and equations that appear in the exposition.

We continue to develop strategies to make mathematics, which is so central to the development of physical chemistry, accessible to students. In addition to associating *Mathematical background* sections with appropriate chapters, we give more help with the development of equations: we motivate them, justify them, and comment on the steps taken to derive them. We also added a new feature: *The chemist's toolkit*, which offers quick and immediate help on a concept from mathematics or physics.

This edition has more worked *Examples*, which require students to organize their thoughts about how to proceed with complex calculations, and more *Brief illustrations*, which show how to use an equation or deploy a concept in a straightforward way. Both have *Self-tests* to enable students to assess their grasp of the material. We have structured the end-of-chapter *Discussion questions*, *Exercises*, and *Problems* to match the grouping of the Topics, but have added Topic- and Chapter-crossing *Integrated activities* to show that several Topics are often necessary to solve a single problem. The *Resource section* has been restructured and augmented by the addition of a list of integrals that are used (and referred to) throughout the text.

We are, of course, alert to the development of electronic resources and have made a special effort in this edition to encourage the use of web-based tools, which are identified in the *Using the book* section that follows this preface. Important among these tools are *Impact* sections, which provide examples of how the material in the chapters is applied in such diverse areas as biochemistry, medicine, environmental science, and materials science.

Overall, we have taken this opportunity to refresh the text thoroughly, making it even more flexible, helpful, and up to date. As ever, we hope that you will contact us with your suggestions for its continued improvement.

PWA, Oxford
JdeP, Portland

USING THE BOOK

For the tenth edition of *Physical Chemistry: Thermodynamics, Structure, and Change* we have tailored the text even more closely to the needs of students. First, the material within each chapter has been reorganized into discrete topics to improve accessibility, clarity, and flexibility. Second, in addition to the variety of learning features already present, we have significantly enhanced the mathematics support by adding new Chemist's toolkit boxes, and checklists of key concepts at the end of each topic.

Organizing the information

➤ Innovative new structure

Each chapter has been reorganized into short topics, making the text more readable for students and more flexible for instructors. Each topic opens with a comment on why it is important, a statement of the key idea, and a brief summary of the background needed to understand the topic.

> ➤ Why do you need to know this material?
>
> Because chemistry is about matter and the changes that it can undergo, both physically and chemically, the properties of matter underlie the entire discussion in this book.
>
> ➤ What is the key idea?
>
> The bulk properties of matter are related to the identities

➤ Notes on good practice

Our *Notes on good practice* will help you avoid making common mistakes. They encourage conformity to the international language of science by setting out the conventions and procedures adopted by the International Union of Pure and Applied Chemistry (IUPAC).

> applicable only to perfect gases (and other idealized systems) are labelled, as here, with a number in blue.
>
> *A note on good practice* Although the term 'ideal gas' is almost universally used in place of 'perfect gas', there are reasons for preferring the latter term. In an ideal system the interactions between molecules in a mixture are all the same. In a perfect gas not only are the interactions all the same but they are in fact zero. Few, though, make this useful distinction.
>
> Equation A.5, the **perfect gas equation**, is a summary of three empirical conclusions, namely Boyle's law ($p \propto 1/V$ at

➤ Resource section

The comprehensive *Resource section* at the end of the book contains a table of integrals, data tables, a summary of conventions about units, and character tables. Short extracts of these tables often appear in the topics themselves, principally to give an idea of the typical values of the physical quantities we are introducing.

> RESOURCE SECTION
>
> Contents
>
>

➤ Checklist of concepts

A *Checklist of key concepts* is provided at the end of each topic so that you can tick off those concepts which you feel you have mastered.

Checklist of concepts

- ☐ 1. The **entropy** acts as a signpost of spontaneous change.
- ☐ 2. Entropy change is defined in terms of heat transactions (the **Clausius definition**).
- ☐ 3. The **Boltzmann formula** defines absolute entropies in terms of the number of ways of achieving a configuration.

Presenting the mathematics

➤ Justifications

Mathematical development is an intrinsic part of physical chemistry, and to achieve full understanding you need to see how a particular expression is obtained and if any assumptions have been made. The *Justifications* are set off from the text to let you adjust the level of detail to meet your current needs and make it easier to review material.

Justification 3A.1 Heating accompanying reversible adiabatic expansion

This *Justification* is based on two features of the cycle. One feature is that the two temperatures T_h and T_c in eqn 3A.7 lie on the same adiabat in Fig. 3A.7. The second feature is that the energy transferred as heat during the two isothermal stages are

$$q_h = nRT_h \ln\frac{V_B}{V_A} \qquad q_c = nRT_c \ln\frac{V_D}{V_C}$$

We now show that the two volume ratios are related in a very simple way. From the relation between temperature and volume for reversible adiabatic processes (VT^c=constant, Topic 2D):

➤ Chemist's toolkits

New to the tenth edition, the *Chemist's toolkits* are succinct reminders of the mathematical concepts and techniques that you will need in order to understand a particular derivation being described in the main text.

The chemist's toolkit A.1 Quantities and units

The result of a measurement is a **physical quantity** that is reported as a numerical multiple of a unit:

physical quantity = numerical value × unit

It follows that units may be treated like algebraic quantities and may be multiplied, divided, and cancelled. Thus, the expression (physical quantity)/unit is the numerical value (a dimensionless quantity) of the measurement in the specified

➤ Mathematical backgrounds

There are six *Mathematical background* sections dispersed throughout the text. They cover in detail the main mathematical concepts that you need to understand in order to be able to master physical chemistry. Each one is located at the end of the chapter to which it is most relevant.

Mathematical background 1 Differentiat

Two of the most important mathematical techniques in the physical sciences are differentiation and integration. They occur throughout the subject, and it is essential to be aware of the procedures involved.

MB1.1 Differentiation: definitions

Differentiation is concerned with the slopes of functions, such as the rate of change of a variable with time. The formal definition of the **derivative**, df/dx, of a function $f(x)$ is

➤ Annotated equations and equation labels

We have annotated many equations to help you follow how they are developed. An annotation can take you across the equals sign: it is a reminder of the substitution used, an approximation made, the terms that have been assumed constant, the integral used, and so on. An annotation can also be a reminder of the significance of an individual term in an expression. We sometimes color a collection of numbers or symbols to show how they carry from one line to the next. Many of the equations are labelled to highlight their significance.

$$w = -nRT\int_{V_i}^{V_f}\frac{dV}{V} \overset{\text{Integral A.2}}{=} -nRT\ln\frac{V_f}{V_i}$$

Perfect gas, reversible, isothermal Work of expansion (2A.9)

➤ Checklists of equations

You don't have to memorize every equation in the text. A checklist at the end of each topic summarizes the most important equations and the conditions under which they apply.

Checklist of equations

Property	Equation
Compression factor	$Z = V_m/V_m^{\circ}$
Virial equation of state	$pV_m = RT(1+B/V_m+C/V_m^3+\cdots)$
van der Waals equation of state	$p = nRT/(V-nb) - a(n/V)^2$
Reduced variables	$X_r = X_m/X_c$

Setting up and solving problems

➤ Brief illustrations

A *Brief illustration* shows you how to use equations or concepts that have just been introduced in the text. They help you to learn how to use data, manipulate units correctly, and become familiar with the magnitudes of properties. They are all accompanied by a Self-test question which you can use to monitor your progress.

Brief illustration 1C.5 Corresponding states

The critical constants of argon and carbon dioxide are given in Table 1C.2. Suppose argon is at 23 atm and 200 K, its reduced pressure and temperature are then

$$p_r = \frac{23\,\text{atm}}{48.0\,\text{atm}} = 0.48 \qquad T_r = \frac{200\,\text{K}}{150.7\,\text{K}} = 1.33$$

For carbon dioxide to be in a corresponding state, its pressure and temperature would need to be

$$p = 0.48\times(72.9\,\text{atm}) = 35\,\text{atm} \qquad T = 1.33\times 304.2\,\text{K} = 405\,\text{K}$$

Self-test 1C.6 What would be the corresponding state of ammonia?

Answer: 53 atm, 539 K

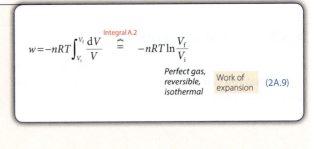

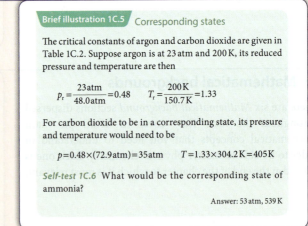

➤ Worked examples

Worked *Examples* are more detailed illustrations of the application of the material, which require you to assemble and develop concepts and equations. We provide a suggested method for solving the problem and then implement it to reach the answer. Worked examples are also accompanied by *Self-test* questions.

Example 3A.2 Calculating the entropy change for a composite process

Calculate the entropy change when argon at 25 °C and 1.00 bar in a container of volume 0.500 dm^3 is allowed to expand to 1.000 dm^3 and is simultaneously heated to 100 °C.

Method As remarked in the text, use reversible isothermal expansion to the final volume, followed by reversible heating at constant volume to the final temperature. The entropy change in the first step is given by eqn 3A.16 and that of the second step, provided C_V is independent of temperature, by eqn 3A.20 (with C_V in place of C_p). In each case we need to

➤ Discussion questions

Discussion questions appear at the end of every chapter, where they are organized by topic. These questions are designed to encourage you to reflect on the material you have just read, and to view it conceptually.

TOPIC 3A Entropy

Discussion questions

3A.1 The evolution of life requires the organization of a very large number of molecules into biological cells. Does the formation of living organisms violate the Second Law of thermodynamics? State your conclusion clearly and present detailed arguments to support it.

3A.2 Discuss the significance of the terms 'dispersal' and 'disorder' in the context of the Second Law.

Exercises

3A.1(a) During a hypothetical process, the entropy of a system increases by 125 J K^{-1} while the entropy of the surroundings decreases by 125 J K^{-1}. Is the process spontaneous?
3A.1(b) During a hypothetical process, the entropy of a system increases by 105 J K^{-1} while the entropy of the surroundings decreases by 95 J K^{-1}. Is the process spontaneous?

3A.2(a) A certain ideal heat engine uses water at the triple point as the hot source and an organic liquid as the cold sink. It withdraws 10.00 kJ of heat from the hot source and generates 3.00 kJ of work. What is the temperature of the organic liquid?
3A.2(b) A certain ideal heat engine uses water at the triple point as the hot source and an organic liquid as the cold sink. It withdraws 2.71 kJ of heat from the hot source and generates 0.71 kJ of work. What is the temperature of the organic liquid?

➤ Exercises and Problems

Exercises and *Problems* are also provided at the end of every chapter, and organized by topic. They prompt you to test your understanding of the topics in that chapter. Exercises are designed as relatively straightforward numerical tests whereas the problems are more challenging. The Exercises come in related pairs, with final numerical answers available on the Book Companion Site for the 'a' questions. Final numerical answers to the odd-numbered problems are also available on the Book Companion Site.

➤ Integrated activities

At the end of most chapters, you will find questions that cross several topics and chapters, and are designed to help you use your knowledge creatively in a variety of ways. Some of the questions refer to the Living Graphs on the Book Companion Site, which you will find helpful for answering them.

➤ Solutions manuals

Two solutions manuals have been written by Charles Trapp, Marshall Cady, and Carmen Giunta to accompany this book.

The *Student Solutions Manual* (ISBN 1-4641-2449-3) provides full solutions to the 'a' exercises and to the odd-numbered problems.

The *Instructor's Solutions Manual* provides full solutions to the 'b' exercises and to the even-numbered problems (available to download from the Book Companion Site for registered adopters of the book only).

BOOK COMPANION SITE

The Book Companion Site to accompany *Physical Chemistry: Thermodynamics, Structure, and Change*, tenth edition provides a number of useful teaching and learning resources for students and instructors.

The site can be accessed at:

http://www.whfreeman.com/pchem10e/

Instructor resources are available only to registered adopters of the textbook. To register, simply visit **http://www.whfreeman.com/pchem10e/** and follow the appropriate links.

Student resources are openly available to all, without registration.

Customers outside North America should visit **www.oxfordtextbooks.co.uk/orc/pchem10e/** to access online resources.

Materials on the Book Companion Site include:

'Impact' sections

'Impact' sections show how physical chemistry is applied in a variety of modern contexts. New for this edition, the Impacts are linked from the text by QR code images. Alternatively, visit the URL displayed next to the QR code image.

Group theory tables

Comprehensive group theory tables are available to download.

Figures and tables from the book

Instructors can find the artwork and tables from the book in ready-to-download format. These may be used for lectures without charge (but not for commercial purposes without specific permission).

Molecular modeling problems

PDFs containing molecular modeling problems can be downloaded, designed for use with the Spartan Student™ software. However they can also be completed using any modeling software that allows Hartree-Fock, density functional, and MP2 calculations.

Living graphs

These interactive graphs can be used to explore how a property changes as various parameters are changed. Living graphs are sometimes referred to in the Integrated activities at the end of a chapter.

ACKNOWLEDGEMENTS

A book as extensive as this could not have been written without significant input from many individuals. We would like to reiterate our thanks to the hundreds of people who contributed to the first nine editions. Many people gave their advice based on the ninth edition, and others, including students, reviewed the draft chapters for the tenth edition as they emerged. We wish to express our gratitude to the following colleagues:

Oleg Antzutkin, *Luleå University of Technology*
Mu-Hyun Baik, *Indiana University — Bloomington*
Maria G. Benavides, *University of Houston — Downtown*
Joseph A. Bentley, *Delta State University*
Maria Bohorquez, *Drake University*
Gary D. Branum, *Friends University*
Gary S. Buckley, *Cameron University*
Eleanor Campbell, *University of Edinburgh*
Lin X. Chen, *Northwestern University*
Gregory Dicinoski, *University of Tasmania*
Niels Engholm Henriksen, *Technical University of Denmark*
Walter C. Ermler, *University of Texas at San Antonio*
Alexander Y. Fadeev, *Seton Hall University*
Beth S. Guiton, *University of Kentucky*
Patrick M. Hare, *Northern Kentucky University*
Grant Hill, *University of Glasgow*
Ann Hopper, *Dublin Institute of Technology*
Garth Jones, *University of East Anglia*
George A. Kaminsky, *Worcester Polytechnic Institute*
Dan Killelea, *Loyola University of Chicago*
Richard Lavrich, *College of Charleston*
Yao Lin, *University of Connecticut*
Tony Masiello, *California State University — East Bay*
Lida Latifzadeh Masoudipour, *California State University — Dominquez Hills*
Christine McCreary, *University of Pittsburgh at Greensburg*
Ricardo B. Metz, *University of Massachusetts Amherst*
Maria Pacheco, *Buffalo State College*
Sid Parrish, Jr., *Newberry College*
Nessima Salhi, *Uppsala University*
Michael Schuder, *Carroll University*
Paul G. Seybold, *Wright State University*
John W. Shriver, *University of Alabama Huntsville*
Jens Spanget-Larsen, *Roskilde University*
Stefan Tsonchev, *Northeastern Illinois University*
A. L. M. van de Ven, *Eindhoven University of Technology*
Darren Walsh, *University of Nottingham*
Nicolas Winter, *Dominican University*
Georgene Wittig, *Carnegie Mellon University*
Daniel Zeroka, *Lehigh University*

Because we prepared this edition at the same time as its sister volume, *Physical Chemistry: Quanta, matter, and change*, it goes without saying that our colleague on that book, Ron Friedman, has had an unconscious but considerable impact on this text too, and we cannot thank him enough for his contribution to this book. Our warm thanks also go to Charles Trapp, Carmen Giunta, and Marshall Cady who once again have produced the *Solutions manuals* that accompany this book and whose comments led us to make a number of improvements. Kerry Karukstis contributed helpfully to the Impacts that are now on the web.

Last, but by no means least, we would also like to thank our two commissioning editors, Jonathan Crowe of Oxford University Press and Jessica Fiorillo of W. H. Freeman & Co., and their teams for their encouragement, patience, advice, and assistance.

FULL CONTENTS

(Contents of Volume 1 are highlighted below)

TABLES

CHEMIST'S TOOLKITS

Foundations

Chemistry is the science of matter and the changes it can undergo. Physical chemistry is the branch of chemistry that establishes and develops the principles of the subject in terms of the underlying concepts of physics and the language of mathematics. It provides the basis for developing new spectroscopic techniques and their interpretation, for understanding the structures of molecules and the details of their electron distributions, and for relating the bulk properties of matter to their constituent atoms. Physical chemistry also provides a window on to the world of chemical reactions, and allows us to understand in detail how they take place.

A Matter

Throughout the text we draw on a number of concepts that should already be familiar from introductory chemistry, such as the 'nuclear model' of the atom, 'Lewis structures' of molecules, and the 'perfect gas equation'. This Topic reviews these and other concepts of chemistry that appear at many stages of the presentation.

B Energy

Because physical chemistry lies at the interface between physics and chemistry, we also need to review some of the concepts from elementary physics that we need to draw on in the text. This Topic begins with a brief summary of 'classical mechanics', our starting point for discussion of the motion and energy of particles. Then it reviews concepts of 'thermodynamics' that should already be part of your chemical vocabulary. Finally, we introduce the 'Boltzmann distribution' and the 'equipartition theorem', which help to establish connections between the bulk and molecular properties of matter.

C Waves

This Topic describes waves, with a focus on 'harmonic waves', which form the basis for the classical description of electromagnetic radiation. The classical ideas of motion, energy, and waves in this Topic and Topic B are expanded with the principles of quantum mechanics (Chapter 7), setting the stage for the treatment of electrons, atoms, and molecules. Quantum mechanics underlies the discussion of chemical structure and chemical change, and is the basis of many techniques of investigation.

A Matter

Contents

➤ **Why do you need to know this material?**

Because chemistry is about matter and the changes that it can undergo, both physically and chemically, the properties of matter underlie the entire discussion in this book.

➤ **What is the key idea?**

The bulk properties of matter are related to the identities and arrangements of atoms and molecules in a sample.

➤ **What do you need to know already?**

This Topic reviews material commonly covered in introductory chemistry.

The presentation of physical chemistry in this text is based on the experimentally verified fact that matter consists of atoms. In this Topic, which is a review of elementary concepts and language widely used in chemistry, we begin to make connections between atomic, molecular, and bulk properties. Most of the material is developed in greater detail later in the text.

A.1 Atoms

The atom of an element is characterized by its **atomic number**, Z, which is the number of protons in its nucleus. The number of neutrons in a nucleus is variable to a small extent, and the **nucleon number** (which is also commonly called the *mass number*), A, is the total number of protons and neutrons in the nucleus. Protons and neutrons are collectively called **nucleons**. Atoms of the same atomic number but different nucleon number are the **isotopes** of the element.

(a) The nuclear model

According to the **nuclear model**, an atom of atomic number Z consists of a nucleus of charge $+Ze$ surrounded by Z electrons each of charge $-e$ (e is the fundamental charge: see inside the front cover for its value and the values of the other fundamental constants). These electrons occupy **atomic orbitals**, which are regions of space where they are most likely to be found, with no more than two electrons in any one orbital. The atomic orbitals are arranged in **shells** around the nucleus, each shell being characterized by the **principal quantum number**, $n=1, 2, \ldots$. A shell consists of n^2 individual orbitals, which are grouped together into n **subshells**; these subshells, and the orbitals they contain, are denoted s, p, d, and f. For all neutral atoms other than hydrogen, the subshells of a given shell have slightly different energies.

(b) The periodic table

The sequential occupation of the orbitals in successive shells results in periodic similarities in the **electronic configurations**, the specification of the occupied orbitals, of atoms when they are arranged in order of their atomic number. This periodicity of structure accounts for the formulation of the **periodic table** (see the inside the back cover). The vertical columns of the periodic table are called **groups** and (in the modern convention) numbered from 1 to 18. Successive rows of the periodic table are called **periods**, the number of the period being equal to the principal quantum number of the **valence shell**, the outermost shell of the atom.

Some of the groups also have familiar names: Group 1 consists of the **alkali metals**, Group 2 (more specifically, calcium, strontium, and barium) of the **alkaline earth metals**, Group 17 of the **halogens**, and Group 18 of the **noble gases.** Broadly speaking, the elements towards the left of the periodic table are **metals** and those towards the right are **non-metals**; the two classes of substance meet at a diagonal line running from boron to polonium, which constitute the **metalloids**, with properties intermediate between those of metals and non-metals.

The periodic table is divided into s, p, d, and f **blocks**, according to the subshell that is last to be occupied in the formulation of the electronic configuration of the atom. The members of the d block (specifically the members of Groups 3–11 in the d block) are also known as the **transition metals**; those of the f block (which is not divided into numbered groups) are sometimes called the **inner transition metals**. The upper row of the f block (Period 6) consists of the **lanthanoids** (still commonly the 'lanthanides') and the lower row (Period 7) consists of the **actinoids** (still commonly the 'actinides').

(c) Ions

A monatomic **ion** is an electrically charged atom. When an atom gains one or more electrons it becomes a negatively charged **anion**; when it loses one or more electrons it becomes a positively charged **cation**. The charge number of an ion is called the **oxidation number** of the element in that state (thus, the oxidation number of magnesium in Mg^{2+} is +2 and that of oxygen in O^{2-} is −2). It is appropriate, but not always done, to distinguish between the oxidation number and the **oxidation state**, the latter being the physical state of the atom with a specified oxidation number. Thus, the oxidation number *of* magnesium is +2 when it is present as Mg^{2+}, and it is present *in* the oxidation state Mg^{2+}.

The elements form ions that are characteristic of their location in the periodic table: metallic elements typically form cations by losing the electrons of their outermost shell and acquiring the electronic configuration of the preceding noble gas atom. Nonmetals typically form anions by gaining electrons and attaining the electronic configuration of the following noble gas atom.

A.2 Molecules

A **chemical bond** is the link between atoms. Compounds that contain a metallic element typically, but far from universally, form **ionic compounds** that consist of cations and anions in a crystalline array. The 'chemical bonds' in an ionic compound are due to the Coulombic interactions between all the ions in the crystal and it is inappropriate to refer to a bond between a specific pair of neighbouring ions. The smallest unit of an ionic compound is called a **formula unit.** Thus $NaNO_3$, consisting of a Na^+ cation and a NO_3^- anion, is the formula unit of sodium nitrate. Compounds that do not contain a metallic element typically form **covalent compounds** consisting of discrete molecules. In this case, the bonds between the atoms of a molecule are **covalent**, meaning that they consist of shared pairs of electrons.

A note on good practice Some chemists use the term 'molecule' to denote the smallest unit of a compound with the composition of the bulk material regardless of whether it is an ionic or covalent compound and thus speak of 'a molecule of NaCl'. We use the term 'molecule' to denote a discrete covalently bonded entity (as in H_2O); for an ionic compound we use 'formula unit'.

(a) Lewis structures

The pattern of bonds between neighbouring atoms is displayed by drawing a **Lewis structure**, in which bonds are shown as lines and **lone pairs** of electrons, pairs of valence electrons that are not used in bonding, are shown as dots. Lewis structures are constructed by allowing each atom to share electrons until it has acquired an **octet** of eight electrons (for hydrogen, a *duplet* of two electrons). A shared pair of electrons is a **single bond**, two shared pairs constitute a **double bond**, and three shared pairs constitute a **triple bond**. Atoms of elements of Period 3 and later can accommodate more than eight electrons in their valence shell and 'expand their octet' to become **hypervalent**, that is, form more bonds than the octet rule would allow (for example, SF_6), or form more bonds to a small number of atoms (see *Brief illustration* A.1). When more than one Lewis structure can be written for a given arrangement of atoms, it is supposed that **resonance**, a blending of the structures, may occur and distribute multiple-bond character over the molecule (for example, the two Kekulé structures of benzene). Examples of these aspects of Lewis structures are shown in Fig. A.1.

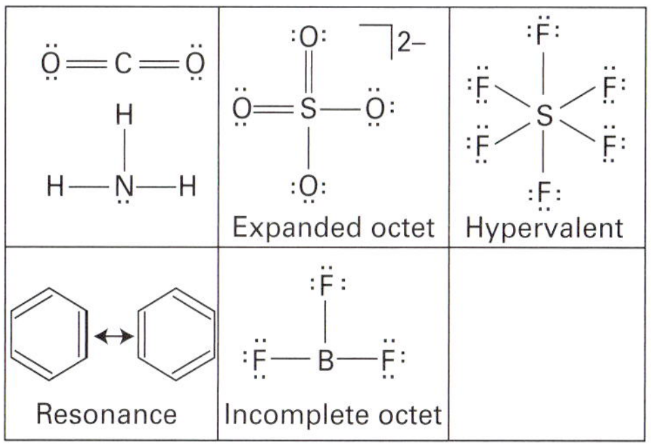

Figure A.1 Examples of Lewis structures.

Brief illustration A.1 Octet expansion

Octet expansion is also encountered in species that do not necessarily require it, but which, if it is permitted, may acquire a lower energy. Thus, of the structures (**1a**) and (**1b**) of the SO_4^{2-} ion, the second has a lower energy than the first. The actual structure of the ion is a resonance hybrid of both structures (together with analogous structures with double bonds in different locations), but the latter structure makes the dominant contribution.

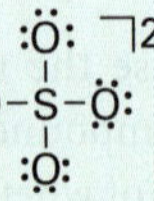

1a

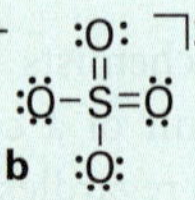

1b

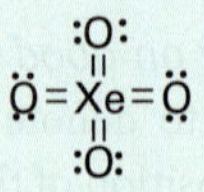

2

Self-test A.1 Draw the Lewis structure for XeO_4.

Answer: See 2

(b) VSEPR theory

Except in the simplest cases, a Lewis structure does not express the three-dimensional structure of a molecule. The simplest approach to the prediction of molecular shape is **valence-shell electron pair repulsion theory** (VSEPR theory). In this approach, the regions of high electron density, as represented by bonds—whether single or multiple—and lone pairs, take up orientations around the central atom that maximize their separations. Then the position of the attached atoms (not the lone pairs) is noted and used to classify the shape of the molecule. Thus, four regions of electron density adopt a tetrahedral arrangement; if an atom is at each of these locations (as in CH_4), then the molecule is tetrahedral; if there is an atom at only three of these locations (as in NH_3), then the molecule is trigonal pyramidal, and so on. The names of the various shapes that are commonly found are shown in Fig. A.2. In a refinement of the theory, lone pairs are assumed to repel bonding pairs more strongly than bonding pairs repel each other. The shape a molecule then adopts, if it is not determined fully by symmetry, is such as to minimize repulsions from lone pairs.

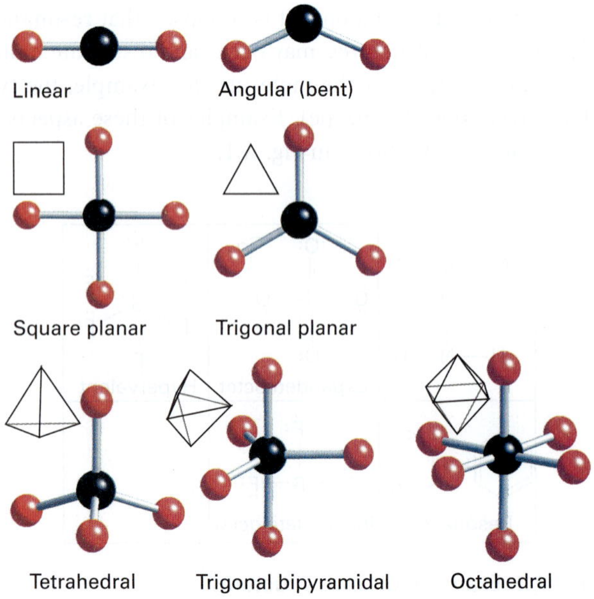

Figure A.2 The shapes of molecules that result from application of VSEPR theory.

Brief illustration A.2 Molecular shapes

In SF_4 the lone pair adopts an equatorial position and the two axial S–F bonds bend away from it slightly, to give a bent see-saw shaped molecule (Fig. A.3).

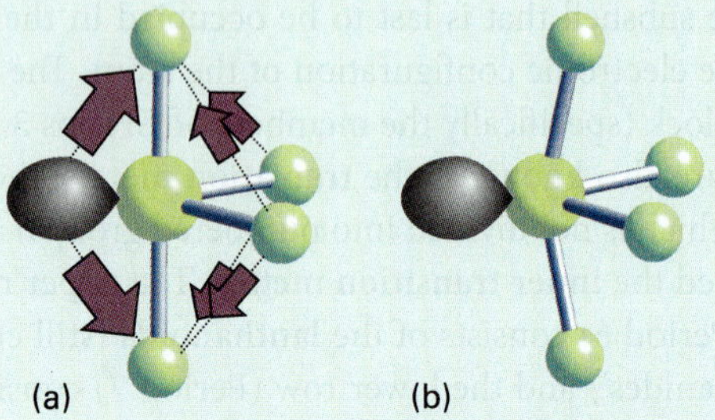

Figure A.3 (a) In SF_4 the lone pair adopts an equatorial position. (b) The two axial S–F bonds bend away from it slightly, to give a bent see-saw shaped molecule.

Self-test A.2 Predict the shape of the SO_3^{2-} ion.

Answer: Trigonal pyramid

(c) Polar bonds

Covalent bonds may be **polar**, or correspond to an unequal sharing of the electron pair, with the result that one atom has a partial positive charge (denoted δ+) and the other a partial negative charge (δ–). The ability of an atom to attract electrons to itself when part of a molecule is measured by the **electronegativity**, χ (chi), of the element. The juxtaposition of equal and opposite partial charges constitutes an **electric dipole**. If those charges are $+Q$ and $-Q$ and they are separated by a distance d, the magnitude of the **electric dipole moment**, μ, is

$$\mu = Qd \quad \text{Definition} \quad \text{Magnitude of the electric dipole moment} \qquad \text{(A.1)}$$

Brief illustration A.3 Nonpolar molecules with polar bonds

Whether or not a molecule as a whole is polar depends on the arrangement of its bonds, for in highly symmetrical molecules there may be no net dipole. Thus, although the linear CO_2 molecule (which is structurally OCO) has polar CO bonds, their effects cancel and the molecule as a whole is nonpolar.

Self-test A.3 Is NH_3 polar?

Answer: Yes

A.3 Bulk matter

Bulk matter consists of large numbers of atoms, molecules, or ions. Its physical state may be solid, liquid, or gas:

A **solid** is a form of matter that adopts and maintains a shape that is independent of the container it occupies.

A **liquid** is a form of matter that adopts the shape of the part of the container it occupies (in a gravitational field, the lower part) and is separated from the unoccupied part of the container by a definite surface.

A **gas** is a form of matter that immediately fills any container it occupies.

A liquid and a solid are examples of a **condensed state** of matter. A liquid and a gas are examples of a **fluid** form of matter: they flow in response to forces (such as gravity) that are applied.

(a) Properties of bulk matter

The state of a bulk sample of matter is defined by specifying the values of various properties. Among them are:

The **mass**, m, a measure of the quantity of matter present (unit: 1 kilogram, 1 kg).

The **volume**, V, a measure of the quantity of space the sample occupies (unit: 1 cubic metre, 1 m^3).

The **amount of substance**, n, a measure of the number of specified entities (atoms, molecules, or formula units) present (unit: 1 mole, 1 mol).

Brief illustration A.4 Volume units

Volume is also expressed as submultiples of 1 m^3, such as cubic decimetres (1 $dm^3 = 10^{-3}$ m^3) and cubic centimetres (1 $cm^3 = 10^{-6}$ m^3). It is also common to encounter the non-SI unit litre (1 L = 1 dm^3) and its submultiple the millilitre (1 mL = 1 cm^3). To carry out simple unit conversions, simply replace the fraction of the unit (such as 1 cm) by its definition (in this case, 10^{-2} m). Thus, to convert 100 cm^3 to cubic decimetres (litres), use 1 cm = 10^{-1} dm, in which case 100 cm^3 = 100 $(10^{-1}\,dm)^3$, which is the same as 0.100 dm^3.

Self-test A.4 Express a volume of 100 mm^3 in units of cm^3.

Answer: 0.100 cm^3

An **extensive property** of bulk matter is a property that depends on the amount of substance present in the sample; an **intensive property** is a property that is independent of the amount of substance. The volume is extensive; the mass density, ρ (rho), with

$$\rho = \frac{m}{V} \qquad \text{Mass density} \qquad (A.2)$$

is intensive.

The **amount of substance**, n (colloquially, 'the number of moles'), is a measure of the number of specified entities present in the sample. 'Amount of substance' is the official name of the quantity; it is commonly simplified to 'chemical amount' or simply 'amount'. The unit 1 mol is currently defined as the number of carbon atoms in exactly 12 g of carbon-12. (In 2011 the decision was taken to replace this definition, but the change has not yet, in 2014, been implemented.) The number of entities per mole is called **Avogadro's constant**, N_A; the currently accepted value is 6.022×10^{23} mol^{-1} (note that N_A is a constant with units, not a pure number).

The **molar mass of a substance**, M (units: formally kilograms per mole but commonly grams per mole, g mol^{-1}) is the mass per mole of its atoms, its molecules, or its formula units. The amount of substance of specified entities in a sample can readily be calculated from its mass, by noting that

$$n = \frac{m}{M} \qquad \text{Amount of substance} \qquad (A.3)$$

A note on good practice Be careful to distinguish atomic or molecular mass (the mass of a single atom or molecule; units kg) from molar mass (the mass per mole of atoms or molecules; units kg mol^{-1}). *Relative* molecular masses of atoms and molecules, $M_r = m/m_u$, where m is the mass of the atom or molecule and m_u is the atomic mass constant (see inside front cover), are still widely called 'atomic weights' and 'molecular weights' even though they are dimensionless quantities and not weights (the gravitational force exerted on an object).

A sample of matter may be subjected to a **pressure**, p (unit: 1 pascal, Pa; 1 Pa = 1 kg m^{-1} s^{-2}), which is defined as the force, F, it is subjected to divided by the area, A, to which that force is applied. A sample of gas exerts a pressure on the walls of its container because the molecules of gas are in ceaseless, random motion, and exert a force when they strike the walls. The frequency of the collisions is normally so great that the force, and therefore the pressure, is perceived as being steady.

Although 1 pascal is the SI unit of pressure (*The chemist's toolkit* A.1), it is also common to express pressure in bar (1 bar = 10^5 Pa) or atmospheres (1 atm = 101 325 Pa exactly), both of which correspond to typical atmospheric pressure. Because many physical properties depend on the pressure acting on a sample, it is appropriate to select a certain value of the pressure to report their values. The **standard pressure** for reporting physical quantities is currently defined as $p^{\ominus} = 1$ bar exactly.

The chemist's toolkit A.1 Quantities and units

The result of a measurement is a **physical quantity** that is reported as a numerical multiple of a unit:

$$\text{physical quantity} = \text{numerical value} \times \text{unit}$$

It follows that units may be treated like algebraic quantities and may be multiplied, divided, and cancelled. Thus, the expression (physical quantity)/unit is the numerical value (a dimensionless quantity) of the measurement in the specified units. For instance, the mass m of an object could be reported as $m = 2.5\,\text{kg}$ or $m/\text{kg} = 2.5$. See Table A.1 in the *Resource section* for a list of units. Although it is good practice to use only SI units, there will be occasions where accepted practice is so deeply rooted that physical quantities are expressed using other, non-SI units. By international convention, all physical quantities are represented by oblique (sloping) symbols; all units are roman (upright).

Units may be modified by a prefix that denotes a factor of a power of 10. Among the most common SI prefixes are those listed in Table A.2 in the *Resource section*. Examples of the use of these prefixes are:

$$1\,\text{nm} = 10^{-9}\,\text{m} \qquad 1\,\text{ps} = 10^{-12}\,\text{s} \qquad 1\,\mu\text{mol} = 10^{-6}\,\text{mol}$$

Powers of units apply to the prefix as well as the unit they modify. For example, $1\,\text{cm}^3 = 1\,(\text{cm})^3$, and $(10^{-2}\,\text{m})^3 = 10^{-6}\,\text{m}^3$. Note that $1\,\text{cm}^3$ does not mean $1\,\text{c}(\text{m}^3)$. When carrying out numerical calculations, it is usually safest to write out the numerical value of an observable in scientific notation (as $n.nnn \times 10^n$).

There are seven SI base units, which are listed in Table A.3 in the *Resource section*. All other physical quantities may be expressed as combinations of these base units (see Table A.4 in the *Resource section*). *Molar concentration* (more formally, but very rarely, *amount of substance concentration*) for example, which is an amount of substance divided by the volume it occupies, can be expressed using the derived units of $\text{mol}\,\text{dm}^{-3}$ as a combination of the base units for amount of substance and length. A number of these derived combinations of units have special names and symbols and we highlight them as they arise.

To specify the state of a sample fully it is also necessary to give its **temperature**, T. The temperature is formally a property that determines in which direction energy will flow as heat when two samples are placed in contact through thermally conducting walls: energy flows from the sample with the higher temperature to the sample with the lower temperature. The symbol T is used to denote the **thermodynamic temperature** which is an absolute scale with $T = 0$ as the lowest point. Temperatures above $T = 0$ are then most commonly expressed by using the **Kelvin scale**, in which the gradations of temperature are expressed as multiples of the unit 1 kelvin (1 K). The Kelvin scale is currently defined by setting the triple point of water (the temperature at which ice, liquid water, and water vapour are in mutual equilibrium) at exactly 273.16 K (as for certain other units, a decision has been taken to revise this definition, but it has not yet, in 2014, been implemented). The freezing point of water (the melting point of ice) at 1 atm is then found experimentally to lie 0.01 K below the triple point, so the freezing point of water is 273.15 K. The Kelvin scale is unsuitable for everyday measurements of temperature, and it is common to use the **Celsius scale**, which is defined in terms of the Kelvin scale as

$$\theta/{}^\circ\text{C} = T/\text{K} - 273.15 \qquad \textit{Definition} \quad \text{Celsius scale} \qquad \text{(A.4)}$$

Thus, the freezing point of water is 0 °C and its boiling point (at 1 atm) is found to be 100 °C (more precisely 99.974 °C). Note that in this text T invariably denotes the thermodynamic (absolute) temperature and that temperatures on the Celsius scale are denoted θ (theta).

A note on good practice Note that we write $T = 0$, not $T = 0\,\text{K}$. General statements in science should be expressed without reference to a specific set of units. Moreover, because T (unlike θ) is absolute, the lowest point is 0 regardless of the scale used to express higher temperatures (such as the Kelvin scale). Similarly, we write $m = 0$, not $m = 0\,\text{kg}$ and $l = 0$, not $l = 0\,\text{m}$.

(b) The perfect gas equation

The properties that define the state of a system are not in general independent of one another. The most important example of a relation between them is provided by the idealized fluid known as a **perfect gas** (also, commonly, an 'ideal gas'):

$$pV = nRT \qquad \text{Perfect gas equation} \qquad \text{(A.5)}$$

Here R is the **gas constant**, a universal constant (in the sense of being independent of the chemical identity of the gas) with the value $8.3145\,\text{J}\,\text{K}^{-1}\,\text{mol}^{-1}$. Throughout this text, equations applicable only to perfect gases (and other idealized systems) are labelled, as here, with a number in blue.

A note on good practice Although the term 'ideal gas' is almost universally used in place of 'perfect gas', there are reasons for preferring the latter term. In an ideal system the interactions between molecules in a mixture are all the same. In a perfect gas not only are the interactions all the same but they are in fact zero. Few, though, make this useful distinction.

Equation A.5, the **perfect gas equation**, is a summary of three empirical conclusions, namely Boyle's law ($p \propto 1/V$ at constant temperature and amount), Charles's law ($p \propto T$ at constant volume and amount), and Avogadro's principle ($V \propto n$ at constant temperature and pressure).

Example A.1 Using the perfect gas equation

Calculate the pressure in kilopascals exerted by 1.25 g of nitrogen gas in a flask of volume 250 cm^3 at 20 °C.

Method To use eqn A.5, we need to know the amount of molecules (in moles) in the sample, which we can obtain from the mass and the molar mass (by using eqn A.3) and to convert the temperature to the Kelvin scale (by using eqn A.4).

Answer The amount of N_2 molecules (of molar mass 28.02 g mol^{-1}) present is

$$n(N_2)=\frac{m}{M(N_2)}=\frac{1.25\,g}{28.02\,g\,mol^{-1}}=\frac{1.25}{28.02}\,mol$$

The temperature of the sample is

$$T/K=20+273.15, \text{ so } T=(20+273.15)\,K$$

Therefore, after rewriting eqn A.5 as $p=nRT/V$,

$$p=\frac{\overbrace{(1.25/28.02)\,mol}^{n}\times\overbrace{(8.3145\,J\,K^{-1}\,mol^{-1})}^{R}\times\overbrace{(20+273.15)\,K}^{T}}{\underbrace{(2.50\times10^{-4})\,m^3}_{V}}$$

$$=\frac{(1.25/28.02)\times(8.3145)\times(20+273.15)}{2.50\times10^{-4}}\,\frac{J}{m^3}$$

$$\overset{1\,J\,m^{-3}=1\,Pa}{=}\;4.35\times10^5\,Pa=435\,kPa$$

A note on good practice It is best to postpone a numerical calculation to the last possible stage, and carry it out in a single step. This procedure avoids rounding errors. When we judge it appropriate to show an intermediate result without committing ourselves to a number of significant figures, we write it as *n.nnn*....

Self-test A.5 Calculate the pressure exerted by 1.22 g of carbon dioxide confined in a flask of volume 500 dm^3 ($5.00\times10^2\,dm^3$) at 37 °C.

Answer: 143 Pa

All gases obey the perfect gas equation ever more closely as the pressure is reduced towards zero. That is, eqn A.5 is an example of a **limiting law**, a law that becomes increasingly valid in a particular limit, in this case as the pressure is reduced to zero. In practice, normal atmospheric pressure at sea level (about 1 atm) is already low enough for most gases to behave almost perfectly, and unless stated otherwise, we assume in this text that the gases we encounter behave perfectly and obey eqn A.5.

A mixture of perfect gases behaves like a single perfect gas. According to **Dalton's law**, the total pressure of such a mixture is the sum of the pressures to which each gas would give rise if it occupied the container alone:

$$p=p_A+p_B+\cdots \qquad \text{Dalton's law} \qquad (A.6)$$

Each pressure, p_J, can be calculated from the perfect gas equation in the form $p_J=n_JRT/V$.

Checklist of concepts

- ☐ 1. In the **nuclear model** of an atom negatively charged electrons occupy atomic orbitals which are arranged in shells around a positively charged nucleus.
- ☐ 2. The **periodic table** highlights similarities in electronic configurations of atoms, which in turn lead to similarities in their physical and chemical properties.
- ☐ 3. **Covalent compounds** consist of discrete molecules in which atoms are linked by covalent bonds.
- ☐ 4. **Ionic compounds** consist of cations and anions in a crystalline array.
- ☐ 5. **Lewis structures** are useful models of the pattern of bonding in molecules.
- ☐ 6. The **valence-shell electron pair repulsion theory** (VSEPR theory) is used to predict the three-dimensional shapes of molecules from their Lewis structures.
- ☐ 7. The electrons in **polar covalent bonds** are shared unequally between the bonded nuclei.
- ☐ 8. The physical states of bulk matter are solid, liquid, or gas.
- ☐ 9. The state of a sample of bulk matter is defined by specifying its properties, such as mass, volume, amount, pressure, and temperature.
- ☐ 10. The **perfect gas equation** is a relation between the pressure, volume, amount, and temperature of an idealized gas.
- ☐ 11. A **limiting law** is a law that becomes increasingly valid in a particular limit.

Checklist of equations

Property	Equation	Comment	Equation number
Electric dipole moment	$\mu = Qd$	μ is the magnitude of the moment	A.1
Mass density	$\rho = m/V$	Intensive property	A.2
Amount of substance	$n = m/M$	Extensive property	A.3
Celsius scale	$\theta/°\text{C} = T/\text{K} - 273.15$	Temperature is an intensive property; 273.15 is exact.	A.4
Perfect gas equation	$pV = nRT$		A.5
Dalton's law	$p = p_A + p_B + \cdots$		A.6

B Energy

Contents

➤ Why do you need to know this material?

Energy is the central unifying concept of physical chemistry, and you need to gain insight into how electrons, atoms, and molecules gain, store, and lose energy.

➤ What is the key idea?

Energy, the capacity to do work, is restricted to discrete values in electrons, atoms, and molecules.

➤ What do you need to know already?

You need to review the laws of motion and principles of electrostatics normally covered in introductory physics and concepts of thermodynamics normally covered in introductory chemistry.

Much of chemistry is concerned with transfers and transformations of energy, and from the outset it is appropriate to define this familiar quantity precisely. We begin here by reviewing **classical mechanics**, which was formulated by Isaac Newton in the seventeenth century, and establishes the vocabulary used to describe the motion and energy of particles. These classical ideas prepare us for **quantum mechanics**, the more fundamental theory formulated in the twentieth century for the study of small particles, such as electrons, atoms, and molecules. We develop the concepts of quantum mechanics throughout the text. Here we begin to see why it is needed as a foundation for understanding atomic and molecular structure.

B.1 Force

Molecules are built from atoms and atoms are built from subatomic particles. To understand their structures we need to know how these bodies move under the influence of the forces they experience.

(a) Momentum

'Translation' is the motion of a particle through space. The **velocity**, $\boldsymbol{v}$, of a particle is the rate of change of its position $\boldsymbol{r}$:

$$\boldsymbol{v}=\frac{\mathrm{d}\boldsymbol{r}}{\mathrm{d}t}$$ *Definition* Velocity (B.1)

For motion confined to a single dimension, we would write $v_x=\mathrm{d}x/\mathrm{d}t$. The velocity and position are vectors, with both direction and magnitude (vectors and their manipulation are treated in detail in *Mathematical background* 5). The magnitude of the velocity is the **speed**, v. The **linear momentum**, $\boldsymbol{p}$, of a particle of mass m is related to its velocity, $\boldsymbol{v}$, by

$$\boldsymbol{p}=m\boldsymbol{v}$$ *Definition* Linear momentum (B.2)

Like the velocity vector, the linear momentum vector points in the direction of travel of the particle (Fig. B.1); its magnitude is denoted p.

The description of rotation is very similar to that of translation. The rotational motion of a particle about a central point is described by its **angular momentum**, $\boldsymbol{J}$. The angular

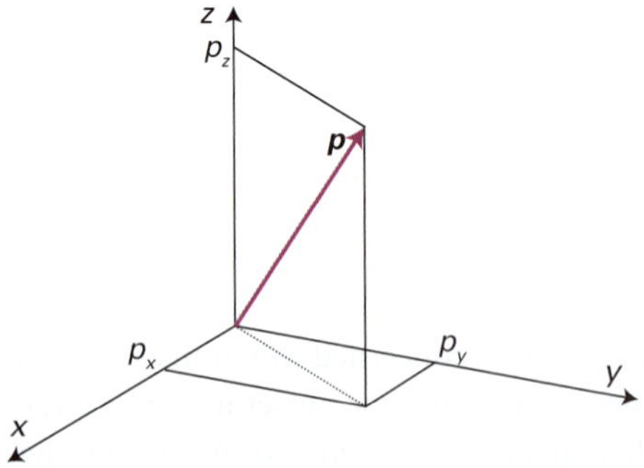

Figure B.1 The linear momentum $\boldsymbol{p}$ is denoted by a vector of magnitude p and an orientation that corresponds to the direction of motion.

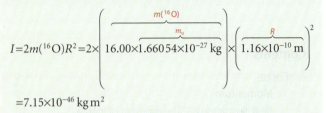

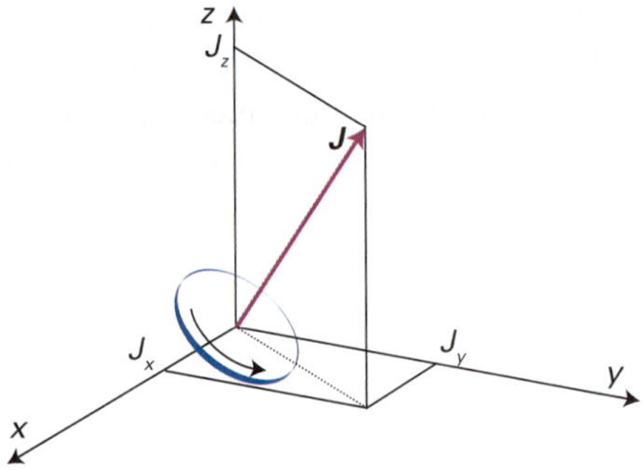

Figure B.2 The angular momentum $\boldsymbol{J}$ of a particle is represented by a vector along the axis of rotation and perpendicular to the plane of rotation. The length of the vector denotes the magnitude J of the angular momentum. The direction of motion is clockwise to an observer looking in the direction of the vector.

momentum is a vector: its magnitude gives the rate at which a particle circulates and its direction indicates the axis of rotation (Fig. B.2). The magnitude of the angular momentum, J, is

$$J = I\omega \qquad \text{Angular momentum} \qquad \text{(B.3)}$$

where ω is the **angular velocity** of the body, its rate of change of angular position (in radians per second), and I is the **moment of inertia**, a measure of its resistance to rotational acceleration. For a point particle of mass m moving in a circle of radius r, the moment of inertia about the axis of rotation is

$$I = mr^2 \qquad \textit{Point particle} \qquad \text{Moment of inertia} \qquad \text{(B.4)}$$

Brief illustration B.1 The moment of inertia

There are two possible axes of rotation in a $C^{16}O_2$ molecule, each passing through the C atom and perpendicular to the axis of the molecule and to each other. Each O atom is at a distance R from the axis of rotation, where R is the length of a CO bond, 116 pm. The mass of each ^{16}O atom is $16.00m_u$, where $m_u = 1.660\,54 \times 10^{-27}$ is the atomic mass constant. The C atom is stationary (it lies on the axis of rotation) and does not contribute to the moment of inertia. Therefore, the moment of inertia of the molecule around the rotation axis is

$$I = 2m(^{16}\text{O})R^2 = 2 \times \left(\overbrace{16.00 \times \overbrace{1.660\,54 \times 10^{-27}\,\text{kg}}^{m_u}}^{m(^{16}\text{O})}\right) \times \left(\overbrace{1.16 \times 10^{-10}\,\text{m}}^{R}\right)^2$$

$$= 7.15 \times 10^{-46}\,\text{kg}\,\text{m}^2$$

Note that the units of moments of inertia are kilograms-metre squared (kg m^2).

Self-test B.1 The moment of inertia for rotation of a hydrogen molecule, $^1H_2$, about an axis perpendicular to its bond is 4.61×10^{-48} kg m^2. What is the bond length of H_2?

Answer: 74.14 pm

(b) Newton's second law of motion

According to **Newton's second law of motion**, *the rate of change of momentum is equal to the force acting on the particle*:

$$\frac{d\boldsymbol{p}}{dt} = \boldsymbol{F} \qquad \text{Newton's second law of motion} \qquad \text{(B.5a)}$$

For motion confined to one dimension, we would write $dp_x/dt = F_x$. Equation B.5a may be taken as the definition of force. The SI units of force are newtons (N), with

$$1\,\text{N} = 1\,\text{kg}\,\text{m}\,\text{s}^{-2}$$

Because $\boldsymbol{p} = m(d\boldsymbol{r}/dt)$, it is sometimes more convenient to write eqn B.5a as

$$m\boldsymbol{a} = \boldsymbol{F} \qquad \boldsymbol{a} = \frac{d^2\boldsymbol{r}}{dt^2} \qquad \textit{Alternative form} \qquad \text{Newton's second law of motion} \qquad \text{(B.5b)}$$

where $\boldsymbol{a}$ is the **acceleration** of the particle, its rate of change of velocity. It follows that if we know the force acting everywhere and at all times, then solving eqn B.5 will give the **trajectory**, the position and momentum of the particle at each instant.

Brief illustration B.2 Newton's second law of motion

A *harmonic oscillator* consists of a particle that experiences a 'Hooke's law' restoring force, one that is proportional to its displacement from equilibrium. An example is a particle of mass m attached to a spring or an atom attached to another by a chemical bond. For a one-dimensional system, $F_x = -k_f x$, where the constant of proportionality k_f is called the *force constant*. Equation B.5b becomes

$$m\frac{d^2x}{dt^2} = -k_f x$$

(Techniques of differentiation are reviewed in *Mathematical background* 1 following Chapter 1.) If $x=0$ at $t=0$, a solution (as may be verified by substitution) is

$$x(t) = A\sin(2\pi\nu t) \qquad \nu = \frac{1}{2\pi}\left(\frac{k_f}{m}\right)^{1/2}$$

This solution shows that the position of the particle varies *harmonically* (that is, as a sine function) with a frequency ν, and that the frequency is high for light particles (m small) attached to stiff springs (k_f large).

Self-test B.2 How does the momentum of the oscillator vary with time?

Answer: $p = 2\pi\nu Am\cos(2\pi\nu t)$

To accelerate a rotation it is necessary to apply a **torque**, T, a twisting force. Newton's equation is then

$$\frac{dJ}{dt} = T \qquad \textit{Definition} \quad \text{Torque} \quad \text{(B.6)}$$

The analogous roles of m and I, of v and ω, and of p and J in the translational and rotational cases respectively should be remembered because they provide a ready way of constructing and recalling equations. These analogies are summarized in Table B.1.

Table B.1 Analogies between translation and rotation

Translation		Rotation	
Property	**Significance**	**Property**	**Significance**
Mass, m	Resistance to the effect of a force	Moment of inertia, I	Resistance to the effect of a torque
Speed, v	Rate of change of position	Angular velocity, ω	Rate of change of angle
Magnitude of linear momentum, p	$p = mv$	Magnitude of angular momentum, J	$J = I\omega$
Translational kinetic energy, E_k	$E_k = \frac{1}{2}mv^2 = p^2/2m$	Rotational kinetic energy, E_k	$E_k = \frac{1}{2}I\omega^2 = J^2/2I$
Equation of motion	$dp/dt = F$	Equation of motion	$dJ/dt = T$

B.2 Energy: a first look

Before defining the term 'energy', we need to develop another familiar concept, that of 'work', more formally. Then we preview the uses of these concepts in chemistry.

(a) Work

Work, w, is done in order to achieve motion against an opposing force. For an infinitesimal displacement through $d\boldsymbol{s}$ (a vector), the work done is

$$dw = -\boldsymbol{F}\cdot d\boldsymbol{s} \qquad \textit{Definition} \quad \text{Work} \quad \text{(B.7a)}$$

where $\boldsymbol{F}\cdot d\boldsymbol{s}$ is the 'scalar product' of the vectors $\boldsymbol{F}$ and $d\boldsymbol{s}$:

$$\boldsymbol{F}\cdot d\boldsymbol{s} = F_x dx + F_y dy + F_z dz \qquad \textit{Definition} \quad \text{Scalar product} \quad \text{(B.7b)}$$

For motion in one dimension, we write $dw = -F_x dx$. The total work done along a path is the integral of this expression, allowing for the possibility that F changes in direction and magnitude at each point of the path. With force in newtons and distance in metres, the units of work are joules (J), with

$$1\,\text{J} = 1\,\text{N m} = 1\,\text{kg m}^2\,\text{s}^{-2}$$

Brief illustration B.3 The work of stretching a bond

The work needed to stretch a chemical bond that behaves like a spring through an infinitesimal distance dx is

$$dw = -F_x dx = -(-k_f x)dx = k_f x dx$$

The total work needed to stretch the bond from zero displacement ($x=0$) at its equilibrium length R_e to a length R, corresponding to a displacement $x = R - R_e$, is

$$w = \int_0^{R-R_e} k_f x dx = k_f \int_0^{R-R_e} x dx = \tfrac{1}{2}k_f(R-R_e)^2$$

We see that the work required increases as the square of the displacement: it takes four times as much work to stretch a bond through 20 pm as it does to stretch the same bond through 10 pm.

Self-test B.3 The force constant of the H–H bond is about $575\,\text{N m}^{-1}$. How much work is needed to stretch this bond by 10 pm?

Answer: 28.8 zJ

(b) The definition of energy

Energy is the capacity to do work. The SI unit of energy is the same as that of work, namely the joule. The rate of supply of energy is called the **power** (P), and is expressed in watts (W):

$$1\,\text{W} = 1\,\text{J s}^{-1}$$

Calories (cal) and kilocalories (kcal) are still encountered in the chemical literature. The calorie is now defined in terms of the joule, with 1 cal=4.184 J (exactly). Caution needs to be exercised as there are several different kinds of calorie. The 'thermochemical calorie', cal_{15}, is the energy required to raise the temperature of 1 g of water at 15 °C by 1 °C and the 'dietary Calorie' is 1 kcal.

A particle may possess two kinds of energy, kinetic energy and potential energy. The **kinetic energy**, E_k, of a body is the energy the body possesses as a result of its motion. For a body of mass m travelling at a speed v,

$$E_k = \tfrac{1}{2}mv^2 \qquad \textit{Definition} \quad \text{Kinetic energy} \qquad \text{(B.8)}$$

It follows from Newton's second law that if a particle of mass m is initially stationary and is subjected to a constant force F for a time τ, then its speed increases from zero to $F\tau/m$ and therefore its kinetic energy increases from zero to

$$E_k = \frac{F^2\tau^2}{2m} \qquad \text{(B.9)}$$

The energy of the particle remains at this value after the force ceases to act. Because the magnitude of the applied force, F, and the time, τ, for which it acts may be varied at will, eqn B.9 implies that the energy of the particle may be increased to any value.

The **potential energy**, E_p or V, of a body is the energy it possesses as a result of its position. Because (in the absence of losses) the work that a particle can do when it is stationary in a given location is equal to the work that had to be done to bring it there, we can use the one-dimensional version of eqn B.7 to write $\mathrm{d}V = -F_x\mathrm{d}x$, and therefore

$$F_x = -\frac{\mathrm{d}V}{\mathrm{d}x} \qquad \textit{Definition} \quad \text{Potential energy} \qquad \text{(B.10)}$$

No universal expression for the potential energy can be given because it depends on the type of force the body experiences. For a particle of mass m at an altitude h close to the surface of the Earth, the gravitational potential energy is

$$V(h) = V(0) + mgh \qquad \text{Gravitational potential energy} \qquad \text{(B.11)}$$

where g is the **acceleration of free fall** (g depends on location, but its 'standard value' is close to 9.81 m s^{-2}). The zero of potential energy is arbitrary. For a particle close to the surface of the Earth, it is common to set $V(0)=0$.

The **total energy** of a particle is the sum of its kinetic and potential energies:

$$E = E_k + E_p, \text{or } E = E_k + V \qquad \textit{Definition} \quad \text{Total energy} \qquad \text{(B.12)}$$

We make use of the apparently universal law of nature that *energy is conserved*; that is, energy can neither be created nor destroyed. Although energy can be transferred from one location to another and transformed from one form to another, the total energy is constant. In terms of the linear momentum, the total energy of a particle is

$$E = \frac{p^2}{2m} + V \qquad \text{(B.13)}$$

This expression may be used in place of Newton's second law to calculate the trajectory of a particle.

Brief illustration B.4 The trajectory of a particle

Consider an argon atom free to move in one direction (along the x-axis) in a region where $V=0$ (so the energy is independent of position). Because $v=\mathrm{d}x/\mathrm{d}t$, it follows from eqns B.1 and B.8 that $\mathrm{d}x/\mathrm{d}t=(2E_k/m)^{1/2}$. As may be verified by substitution, a solution of this differential equation is

$$x(t) = x(0) + \left(\frac{2E_k}{m}\right)^{1/2} t$$

The linear momentum is

$$p(t) = mv(t) = m\frac{\mathrm{d}x}{\mathrm{d}t} = (2mE_k)^{1/2}$$

and is a constant. Hence, if we know the initial position and momentum, we can predict all later positions and momenta exactly.

Self-test B.4 Consider an atom of mass m moving along the x direction with an initial position x_1 and initial speed v_1. If the atom moves for a time interval Δt in a region where the potential energy varies as $V(x)$, what is its speed v_2 at position x_2?

Answer: $v_2 = v_1 \left|\mathrm{d}V(x)/\mathrm{d}x\right|_{x_1} \Delta t/m$

(c) The Coulomb potential energy

One of the most important kinds of potential energy in chemistry is the **Coulomb potential energy** between two electric charges. The Coulomb potential energy is equal to the work that must be done to bring up a charge from infinity to a distance r from a second charge. For a point charge Q_1 at a distance r in a vacuum from another point charge Q_2, their potential energy is

$$V(r)=\frac{Q_1Q_2}{4\pi\varepsilon_0 r} \quad \textit{Definition} \quad \text{Coulomb potential energy} \quad \text{(B.14)}$$

Charge is expressed in coulombs (C), often as a multiple of the fundamental charge, e. Thus, the charge of an electron is $-e$ and that of a proton is $+e$; the charge of an ion is ze, with z the **charge number** (positive for cations, negative for anions). The constant ε_0 (epsilon zero) is the **vacuum permittivity**, a fundamental constant with the value $8.854\times10^{-12}\,C^2\,J^{-1}\,m^{-1}$. It is conventional (as in eqn B.14) to set the potential energy equal to zero at infinite separation of charges. Then two opposite charges have a negative potential energy at finite separations whereas two like charges have a positive potential energy.

Brief illustration B.5 The Coulomb potential energy

The Coulomb potential energy resulting from the electrostatic interaction between a positively charged sodium cation, Na^+, and a negatively charged chloride anion, Cl^-, at a distance of 0.280 nm, which is the separation between ions in the lattice of a sodium chloride crystal, is

$$V=\frac{\overbrace{(-1.602\times10^{-19}\,C)}^{Q(Cl^-)}\times\overbrace{(1.602\times10^{-19}\,C)}^{Q(Na^+)}}{4\pi\times\underbrace{(8.854\times10^{-12}\,C^2\,J^{-1}\,m^{-1})}_{\varepsilon_0}\times\underbrace{(0.280\times10^{-9}\,m)}_{r}}$$

$$=-8.24\times10^{-19}\,J$$

This value is equivalent to a molar energy of

$$V\times N_A=(-8.24\times10^{-19}\,J)\times(6.022\times10^{23}\,mol^{-1})=-496\,kJ\,mol^{-1}$$

A note on good practice: Write units at *every* stage of a calculation and do not simply attach them to a final numerical value. Also, it is often sensible to express all numerical quantities in scientific notation using exponential format rather than SI prefixes to denote powers of ten.

Self-test B.5: The centres of neighbouring cations and anions in magnesium oxide crystals are separated by 0.21 nm. Determine the molar Coulomb potential energy resulting from the electrostatic interaction between a Mg^{2+} and an O^{2-} ion in such a crystal.

Answer: $2600\,kJ\,mol^{-1}$

In a medium other than a vacuum, the potential energy of interaction between two charges is reduced, and the vacuum permittivity is replaced by the **permittivity**, ε, of the medium. The permittivity is commonly expressed as a multiple of the vacuum permittivity:

$$\varepsilon=\varepsilon_r\varepsilon_0 \quad \textit{Definition} \quad \text{Permittivity} \quad \text{(B.15)}$$

with ε_r the dimensionless **relative permittivity** (formerly, the *dielectric constant*). This reduction in potential energy can be substantial: the relative permittivity of water at 25 °C is 80, so the reduction in potential energy for a given pair of charges at a fixed difference (with sufficient space between them for the water molecules to behave as a fluid) is by nearly two orders of magnitude.

Care should be taken to distinguish *potential energy* from *potential*. The potential energy of a charge Q_1 in the presence of another charge Q_2 can be expressed in terms of the **Coulomb potential**, ϕ (phi):

$$V(r)=Q_1\phi(r) \qquad \phi(r)=\frac{Q_2}{4\pi\varepsilon_0 r} \quad \textit{Definition} \quad \text{Coulomb potential} \quad \text{(B.16)}$$

The units of potential are joules per coulomb, $J\,C^{-1}$, so when ϕ is multiplied by a charge in coulombs, the result is in joules. The combination joules per coulomb occurs widely and is called a volt (V):

$$1\,V=1\,J\,C^{-1}$$

If there are several charges Q_2, Q_3, … present in the system, the total potential experienced by the charge Q_1 is the sum of the potential generated by each charge:

$$\phi=\phi_2+\phi_3+\cdots \quad \text{(B.17)}$$

Just as the potential energy of a charge Q_1 can be written $V=Q_1\phi$, so the magnitude of the force on Q_1 can be written $F=Q_1\mathcal{E}$, where $\mathcal{E}$ is the magnitude of the **electric field strength** (units: volts per metre, $V\,m^{-1}$) arising from Q_2 or from some more general charge distribution. The electric field strength (which, like the force, is actually a vector quantity) is the negative gradient of the electric potential. In one dimension, we write the magnitude of the electric field strength as

$$\mathcal{E}=-\frac{d\phi}{dx} \quad \text{Electric field strength} \quad \text{(B.18)}$$

The language we have just developed inspires an important alternative energy unit, the **electronvolt** (eV): 1 eV is defined as the kinetic energy acquired when an electron is accelerated from rest through a potential difference of 1 V. The relation between electronvolts and joules is

$$1\,eV=1.602\times10^{-19}\,J$$

Many processes in chemistry involve energies of a few electronvolts. For example, to remove an electron from a sodium atom requires about 5 eV.

A particularly important way of supplying energy in chemistry (as in the everyday world) is by passing an electric current through a resistance. An **electric current** (I) is defined as the rate of supply of charge, $I = \mathrm{d}Q/\mathrm{d}t$, and is measured in *amperes* (A):

$$1\,\mathrm{A} = 1\,\mathrm{C\,s^{-1}}$$

If a charge Q is transferred from a region of potential ϕ_i, where its potential energy is $Q\phi_i$, to where the potential is ϕ_f and its potential energy is $Q\phi_f$, and therefore through a potential difference $\Delta\phi = \phi_f - \phi_i$, the change in potential energy is $Q\Delta\phi$. The rate at which the energy changes is $(\mathrm{d}Q/\mathrm{d}t)\Delta\phi$, or $I\Delta\phi$. The power is therefore

$$P = I\Delta\phi \qquad \text{Electrical power} \qquad \text{(B.19)}$$

With current in amperes and the potential difference in volts, the power is in watts. The total energy, E, supplied in an interval Δt is the power (the rate of energy supply) multiplied by the duration of the interval:

$$E = P\Delta t = I\Delta\phi\Delta t \qquad \text{(B.20)}$$

The energy is obtained in joules with the current in amperes, the potential difference in volts, and the time in seconds.

(d) Thermodynamics

The systematic discussion of the transfer and transformation of energy in bulk matter is called **thermodynamics**. This subtle subject is treated in detail in the text, but it will be familiar from introductory chemistry that there are two central concepts, the **internal energy**, U (units: joules, J), and the **entropy**, S (units: joules per kelvin, $\mathrm{J\,K^{-1}}$).

The internal energy is the total energy of a system. The **First Law of thermodynamics** states that the internal energy is constant in a system isolated from external influences. The internal energy of a sample of matter increases as its temperature is raised, and we write

$$\Delta U = C\Delta T \qquad \text{Change in internal energy} \qquad \text{(B.21)}$$

where ΔU is the change in internal energy when the temperature of the sample is raised by ΔT. The constant C is called the **heat capacity**, C (units: joules per kelvin, $\mathrm{J\,K^{-1}}$), of the sample. If the heat capacity is large, a small increase in temperature results in a large increase in internal energy. This remark can be expressed in a physically more significant way by inverting it: if the heat capacity is large, then even a large transfer of energy into the system leads to only a small rise in temperature. The heat capacity is an extensive property, and values for a substance are commonly reported as the **molar heat capacity**, $C_m = C/n$ (units: joules per kelvin per mole, $\mathrm{J\,K^{-1}\,mol^{-1}}$) or the **specific heat capacity**, $C_s = C/m$ (units: joules per kelvin per gram, $\mathrm{J\,K^{-1}\,g^{-1}}$), both of which are intensive properties.

Thermodynamic properties are often best discussed in terms of infinitesimal changes, in which case we would write eqn B.21 as $\mathrm{d}U = C\mathrm{d}T$. When this expression is written in the form

$$C = \frac{\mathrm{d}U}{\mathrm{d}T} \qquad \textit{Definition} \quad \text{Heat capacity} \qquad \text{(B.22)}$$

we see that the heat capacity can be interpreted as the slope of the plot of the internal energy of a sample against the temperature.

As will also be familiar from introductory chemistry and will be explained in detail later, for systems maintained at constant pressure it is usually more convenient to modify the internal energy by adding to it the quantity pV, and introducing the **enthalpy**, H (units: joules, J):

$$H = U + pV \qquad \textit{Definition} \quad \text{Enthalpy} \qquad \text{(B.23)}$$

The enthalpy, an extensive property, greatly simplifies the discussion of chemical reactions, in part because changes in enthalpy can be identified with the energy transferred as heat from a system maintained at constant pressure (as in common laboratory experiments).

Brief illustration B.6 The relation between U and H

The internal energy and enthalpy of a perfect gas, for which $pV = nRT$, are related by

$$H = U + nRT$$

Division by n and rearrangement gives

$$H_m - U_m = RT$$

where H_m and U_m are the molar enthalpy and the molar internal energy, respectively. We see that the difference between H_m and U_m increases with temperature.

Self-test B.6 By how much does the molar enthalpy of oxygen gas differ from its molar internal energy at 298 K?

Answer: $2.48\,\mathrm{kJ\,mol^{-1}}$

The **entropy**, S, is a measure of the *quality* of the energy of a system. If the energy is distributed over many modes of motion (for example, the rotational, vibrational, and translational motions for the particles that comprise the system), then the entropy is high. If the energy is distributed over only a small number of modes of motion, then the entropy is low. The **Second Law of thermodynamics** states that any spontaneous (that is, natural) change in an isolated system is accompanied by an increase in the entropy of the system. This tendency is commonly expressed by saying that the natural direction of change is accompanied by dispersal of energy from a localized region or its conversion to a less organized form.

The entropy of a system and its surroundings is of the greatest importance in chemistry because it enables us to identify the spontaneous direction of a chemical reaction and to identify the composition at which the reaction is at **equilibrium**. In a state of *dynamic* equilibrium, which is the character of all chemical equilibria, the forward and reverse reactions are occurring at the same rate and there is no net tendency to change in either direction. However, to use the entropy to identify this state we need to consider both the system and its surroundings. This task can be simplified if the reaction is taking place at constant temperature and pressure, for then it is possible to identify the state of equilibrium as the state at which the **Gibbs energy**, G (units: joules, J), of the system has reached a minimum. The Gibbs energy is defined as

$$G = H - TS$$ *Definition* Gibbs energy (B.24)

and is of the greatest importance in chemical thermodynamics. The Gibbs energy, which informally is called the 'free energy', is a measure of the energy stored in a system that is free to do useful work, such as driving electrons through a circuit or causing a reaction to be driven in its nonspontaneous (unnatural) direction.

B.3 The relation between molecular and bulk properties

The energy of a molecule, atom, or subatomic particle that is confined to a region of space is **quantized**, or restricted to certain discrete values. These permitted energies are called **energy levels**. The values of the permitted energies depend on the characteristics of the particle (for instance, its mass) and the extent of the region to which it is confined. The quantization of energy is most important—in the sense that the allowed energies are widest apart—for particles of small mass confined to small regions of space. Consequently, quantization is very important for electrons in atoms and molecules, but usually unimportant for macroscopic bodies, for which the separation of translational energy levels of particles in containers of macroscopic dimensions is so small that for all practical purposes their translational motion is unquantized and can be varied virtually continuously.

The energy of a molecule other than its unquantized translational motion arises mostly from three modes of motion: rotation of the molecule as a whole, distortion of the molecule through vibration of its atoms, and the motion of electrons around nuclei. Quantization becomes increasingly important as we change focus from rotational to vibrational and then to electronic motion. The separation of rotational energy levels (in small molecules, about 10^{-21} J or 1 zJ, corresponding to about 0.6 kJ mol^{-1}) is smaller than that of vibrational energy levels (about 10−100 zJ, or 6−60 kJ mol^{-1}), which itself is smaller than that of electronic energy levels (about 10^{-18} J or 1 aJ, where a is another uncommon but useful SI prefix, standing for atto, 10^{-18}, corresponding to about 600 kJ mol^{-1}). Figure B.3 depicts these typical energy level separations.

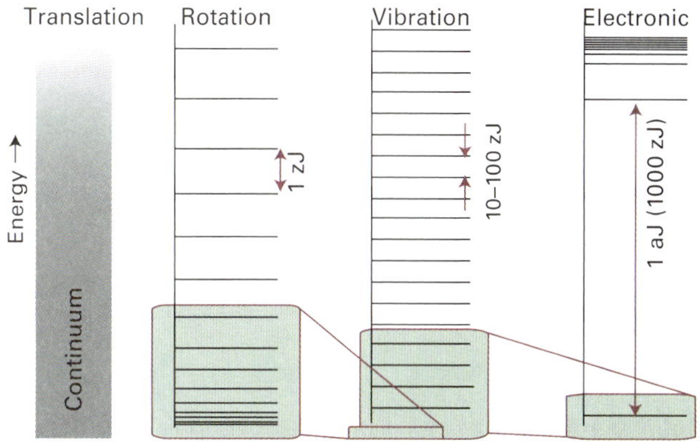

Figure B.3 The energy level separations typical of four types of system. (1 zJ = 10^{-21} J; in molar terms, 1 zJ is equivalent to about 0.6 kJ mol^{-1}.)

(a) The Boltzmann distribution

The continuous thermal agitation that the molecules experience in a sample at $T>0$ ensures that they are distributed over the available energy levels. One particular molecule may be in a state corresponding to a low energy level at one instant, and then be excited into a high energy state a moment later. Although we cannot keep track of the state of a single molecule, we can speak of the *average* numbers of molecules in each state; even though individual molecules may be changing their states as a result of collisions, the average number in each state is constant (provided the temperature remains the same).

The average number of molecules in a state is called the **population** of the state. Only the lowest energy state is occupied at $T=0$. Raising the temperature excites some molecules into higher energy states, and more and more states become accessible as the temperature is raised further (Fig. B.4). The formula for calculating the relative populations of states of various energies is called the **Boltzmann distribution** and was derived by the Austrian scientist Ludwig Boltzmann towards the end of the nineteenth century. This formula gives the ratio of the numbers of particles in states with energies ε_i and ε_j as

$$\frac{N_i}{N_j} = e^{-(\varepsilon_i - \varepsilon_j)/kT}$$ Boltzmann distribution (B.25a)

where k is **Boltzmann's constant**, a fundamental constant with the value $k = 1.381 \times 10^{-23}$ J K^{-1}. In chemical applications it is common to use not the individual energies but energies per mole of molecules, E_i, with $E_i = N_A\varepsilon_i$, where N_A is Avogadro's

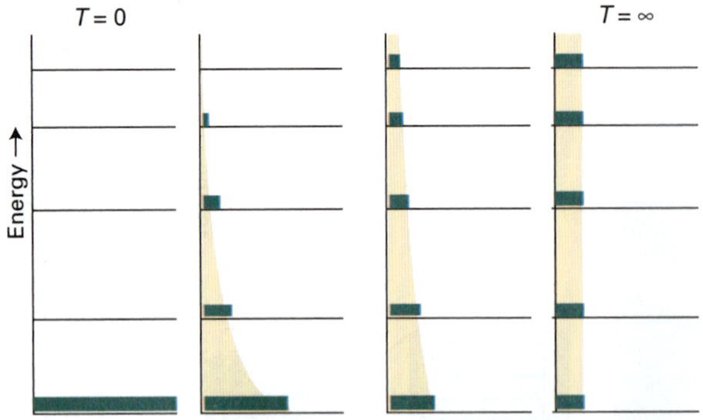

Figure B.4 The Boltzmann distribution of populations for a system of five energy levels as the temperature is raised from zero to infinity.

constant. When both the numerator and denominator in the exponential are multiplied by N_A, eqn B.25a becomes

$$\frac{N_i}{N_j}=e^{-(E_i-E_j)/RT}$$ *Alternative form* Boltzmann distribution (B.25b)

where $R=N_Ak$. We see that k is often disguised in 'molar' form as the gas constant. The Boltzmann distribution provides the crucial link for expressing the macroscopic properties of matter in terms of microscopic behaviour.

Brief illustration B.7 Relative populations

Methyl cyclohexane molecules may exist in one of two conformations, with the methyl group in either an equatorial or axial position. The equatorial form is lower in energy with the axial form being 6.0 kJ mol^{-1} higher in energy. At a temperature of 300 K, this difference in energy implies that the relative populations of molecules in the axial and equatorial states is

$$\frac{N_a}{N_e}=e^{-(E_a-E_e)/RT}=e^{-(6.0\times10^3\,\mathrm{J\,mol^{-1}})/(8.3145\,\mathrm{J\,K^{-1}\,mol^{-1}}\times300\,\mathrm{K})}=0.090$$

where E_a and E_e are molar energies. The number of molecules in an axial conformation is therefore just 9 per cent of those in the equatorial conformation.

Self-test B.7 Determine the temperature at which the relative proportion of molecules in axial and equatorial conformations in a sample of methyl cyclohexane is 0.30 or 30 per cent.

Answer: 600 K

The important features of the Boltzmann distribution to bear in mind are:

Physical interpretation

- The distribution of populations is an exponential function of energy and temperature.
- At a high temperature more energy levels are occupied than at a low temperature.
- More levels are significantly populated if they are close together in comparison with kT (like rotational and translational states), than if they are far apart (like vibrational and electronic states).

Figure B.5 summarizes the form of the Boltzmann distribution for some typical sets of energy levels. The peculiar shape of the population of rotational levels stems from the fact that eqn B.25 applies to *individual states*, and for molecular rotation quantum theory shows that the number of rotational states corresponding to a given energy level—broadly speaking, the number of planes of rotation—increases with energy; therefore, although the population of each *state* decreases with energy, the population of the *levels* goes through a maximum.

One of the simplest examples of the relation between microscopic and bulk properties is provided by **kinetic molecular theory**, a model of a perfect gas. In this model, it is assumed that the molecules, imagined as particles of negligible size, are in ceaseless, random motion and do not interact except during their brief collisions. Different speeds correspond to different energies, so the Boltzmann formula can be used to predict the proportions of molecules having a specific speed at a particular temperature. The expression giving the fraction of molecules that have a particular speed is called the **Maxwell–Boltzmann distribution** and has the features summarized in Fig. B.6. The Maxwell–Boltzmann distribution can be used to show that the average speed, v_{mean}, of the molecules depends on the temperature and their molar mass as

$$v_{\text{mean}}=\left(\frac{8RT}{\pi M}\right)^{1/2}$$ *Perfect gas* Average speed of molecules (B.26)

Thus, the average speed is high for light molecules at high temperatures. The distribution itself gives more information. For instance, the tail towards high speeds is longer at high temperatures than at low, which indicates that at high temperatures more molecules in a sample have speeds much higher than average.

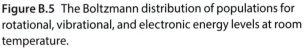

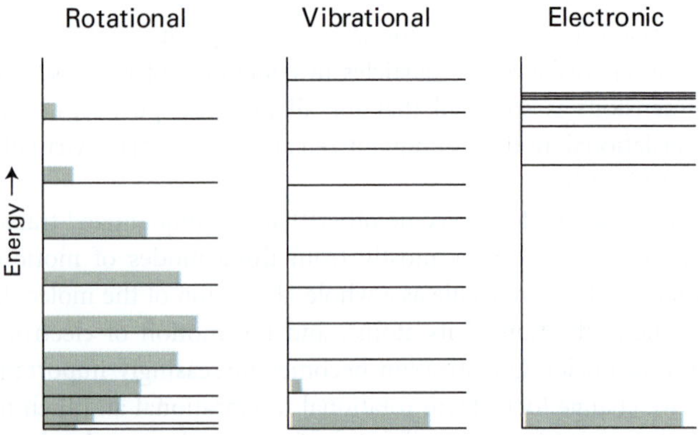

Figure B.5 The Boltzmann distribution of populations for rotational, vibrational, and electronic energy levels at room temperature.

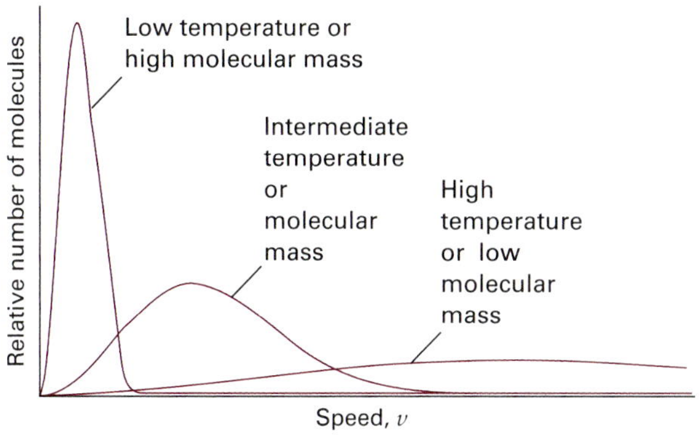

Figure B.6 The (Maxwell–Boltzmann) distribution of molecular speeds with temperature and molar mass. Note that the most probable speed (corresponding to the peak of the distribution) increases with temperature and with decreasing molar mass, and simultaneously the distribution becomes broader.

(b) Equipartition

Although the Boltzmann distribution can be used to calculate the average energy associated with each mode of motion of an atom or molecule in a sample at a given temperature, there is a much simpler shortcut. When the temperature is so high that many energy levels are occupied, we can use the **equipartition theorem**:

> For a sample at thermal equilibrium the average value of each quadratic contribution to the energy is $\frac{1}{2}kT$.

By a 'quadratic contribution' we mean a term that is proportional to the square of the momentum (as in the expression for the kinetic energy, $E_k = p^2/2m$) or the displacement from an equilibrium position (as for the potential energy of a harmonic oscillator, $E_p = \frac{1}{2}k_f x^2$). The theorem is strictly valid only at high temperatures or if the separation between energy levels is small because under these conditions many states are populated. The equipartition theorem is most reliable for translational and rotational modes of motion. The separation between vibrational and electronic states is typically greater than for rotation or translation, and so the equipartition theorem is unreliable for these types of motion.

Brief illustration B.8 Average molecular energies

An atom or molecule may move in three dimensions and its translational kinetic energy is therefore the sum of three quadratic contributions

$$E_{\text{trans}} = \tfrac{1}{2}mv_x^2 + \tfrac{1}{2}mv_y^2 + \tfrac{1}{2}mv_z^2$$

The equipartition theorem predicts that the average energy for each of these quadratic contributions is $\frac{1}{2}kT$. Thus, the average kinetic energy is $E_{\text{trans}} = 3\times\frac{1}{2}kT = \frac{3}{2}kT$. The molar translational energy is thus $E_{\text{trans,m}} = \frac{3}{2}kT \times N_A = \frac{3}{2}RT$. At 300 K

$$E_{\text{trans,m}} = \tfrac{3}{2}\times(8.3145\,\text{J K}^{-1}\,\text{mol}^{-1})\times(300\,\text{K}) = 3700\,\text{J mol}^{-1}$$
$$= 3.7\,\text{kJ mol}^{-1}$$

Self-test B.8 A linear molecule may rotate about two axes in space, each of which counts as a quadratic contribution. Calculate the rotational contribution to the molar energy of a collection of linear molecules at 500 K.

Answer: $4.2\,\text{kJ mol}^{-1}$

Checklist of concepts

- ☐ 1. **Newton's second law of motion** states that the rate of change of momentum is equal to the force acting on the particle.
- ☐ 2. **Work** is done in order to achieve motion against an opposing force.
- ☐ 3. **Energy** is the capacity to do work.
- ☐ 4. The **kinetic energy** of a particle is the energy it possesses as a result of its motion.
- ☐ 5. The **potential energy** of a particle is the energy it possesses as a result of its position.
- ☐ 6. The total energy of a particle is the sum of its kinetic and potential energies.
- ☐ 7. The **Coulomb potential energy** between two charges separated by a distance r varies as $1/r$.
- ☐ 8. The **First Law of thermodynamics** states that the internal energy is constant in a system isolated from external influences.
- ☐ 9. The **Second Law of thermodynamics** states that any spontaneous change in an isolated system is accompanied by an increase in the entropy of the system.
- ☐ 10. **Equilibrium** is the state at which the **Gibbs energy** of the system has reached a minimum.
- ☐ 11. The energy levels of confined particles are quantized.
- ☐ 12. The **Boltzmann distribution** is a formula for calculating the relative populations of states of various energies.
- ☐ 13. The **equipartition theorem** states that for a sample at thermal equilibrium the average value of each quadratic contribution to the energy is $\frac{1}{2}kT$.

Checklist of equations

Property	Equation	Comment	Equation number
Velocity	$\boldsymbol{v}=\mathrm{d}\boldsymbol{r}/\mathrm{d}t$	Definition	B.1
Linear momentum	$\boldsymbol{p}=m\boldsymbol{v}$	Definition	B.2
Angular momentum	$J=I\omega, I=mr^2$	Point particle	B.3–B.4
Force	$\boldsymbol{F}=m\boldsymbol{a}=\mathrm{d}\boldsymbol{p}/\mathrm{d}t$	Definition	B.5
Torque	$\boldsymbol{T}=\mathrm{d}\boldsymbol{J}/\mathrm{d}t$	Definition	B.6
Work	$\mathrm{d}w=-\boldsymbol{F}\cdot\mathrm{d}\boldsymbol{s}$	Definition	B.7
Kinetic energy	$E_\mathrm{k}=\frac{1}{2}mv^2$	Definition	B.8
Potential energy and force	$F_x=-\mathrm{d}V/\mathrm{d}x$	One dimension	B.10
Coulomb potential energy	$V(r)=Q_1Q_2/4\pi\varepsilon_0 r$	Vacuum	B.14
Coulomb potential	$\phi=Q_2/4\pi\varepsilon_0 r$	Vacuum	B.16
Electric field strength	$\mathcal{E}=-\mathrm{d}\phi/\mathrm{d}x$	One dimension	B.18
Electrical power	$P=I\Delta\phi$	I is the current	B.19
Heat capacity	$C=\mathrm{d}U/\mathrm{d}T$	U is the internal energy	B.22
Enthalpy	$H=U+pV$	Definition	B.23
Gibbs energy	$G=H-TS$	Definition	B.24
Boltzmann distribution	$N_i/N_j=\mathrm{e}^{-(\varepsilon_i-\varepsilon_j)/kT}$		B.25a
Average speed of molecules	$v_\mathrm{mean}=(8RT/\pi M)^{1/2}$	Perfect gas	B.26

C Waves

Contents

➤ Why do you need to know this material?

Several important investigative techniques in physical chemistry, such as spectroscopy and X-ray diffraction, involve electromagnetic radiation, a wavelike electromagnetic disturbance. We shall also see that the properties of waves are central to the quantum mechanical description of electrons in atoms and molecules. To prepare for those discussions, we need to understand the mathematical description of waves.

➤ What is the key idea?

A wave is a disturbance that propagates through space with a displacement that can be expressed as a harmonic function.

➤ What do you need to know already?

You need to be familiar with the properties of harmonic (sine and cosine) functions.

A **wave** is an oscillatory disturbance that travels through space. Examples of such disturbances include the collective motion of water molecules in ocean waves and of gas particles in sound waves. A **harmonic wave** is a wave with a displacement that can be expressed as a sine or cosine function.

C.1 Harmonic waves

A harmonic wave is characterized by a **wavelength**, λ (lambda), the distance between the neighbouring peaks of the wave, and its **frequency**, ν (nu), the number of times per second at which its displacement at a fixed point returns to its original value (Fig. C.1). The frequency is measured in *hertz*, where $1\,\text{Hz}=1\,\text{s}^{-1}$. The wavelength and frequency are related by

$$\lambda\nu=v \qquad \text{Relation between frequency and wavelength} \qquad \text{(C.1)}$$

where v is the speed of propagation of the wave.

First, consider the snapshot of a harmonic wave at $t=0$. The displacement $\psi(x,t)$ varies with position x as

$$\psi(x,0)=A\cos\{(2\pi/\lambda)x+\phi\} \qquad \text{Harmonic wave at } t=0 \qquad \text{(C.2a)}$$

where A is the **amplitude** of the wave, the maximum height of the wave, and ϕ is the **phase** of the wave, the shift in the location of the peak from $x=0$ and which may lie between $-\pi$ and π (Fig. C.2). As time advances, the peaks migrate along the x-axis (the direction of propagation), and at any later instant the displacement is

$$\psi(x,t)=A\cos\{(2\pi/\lambda)x-2\pi\nu t+\phi\} \qquad \text{Harmonic wave at } t>0 \qquad \text{(C.2b)}$$

A given wave can also be expressed as a sine function with the same argument but with ϕ replaced by $\phi+\frac{1}{2}\pi$.

If two waves, in the same region of space, with the same wavelength, have different phases then the resultant wave, the sum of the two, will have either enhanced or diminished amplitude. If the phases differ by $\pm\pi$ (so the peaks of one wave coincide with the troughs of the other), then the resultant wave, the sum of the two, will have a diminished amplitude. This effect is called **destructive interference.** If the phases of the two waves

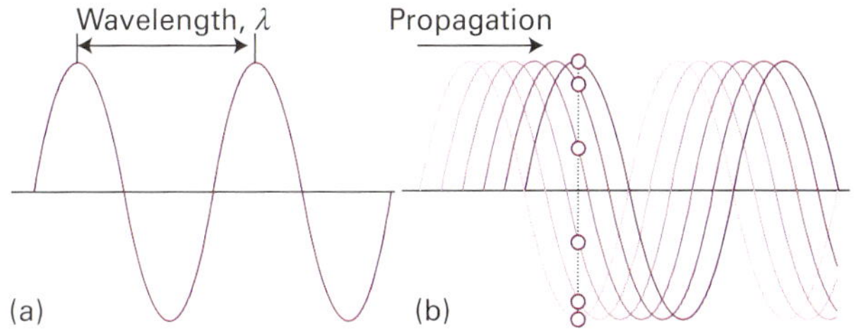

Figure C.1 (a) The wavelength, λ, of a wave is the peak-to-peak distance. (b) The wave is shown travelling to the right at a speed v. At a given location, the instantaneous amplitude of the wave changes through a complete cycle (the six dots show half a cycle) as it passes a given point. The frequency, ν, is the number of cycles per second that occur at a given point. Wavelength and frequency are related by $\lambda\nu=v$.

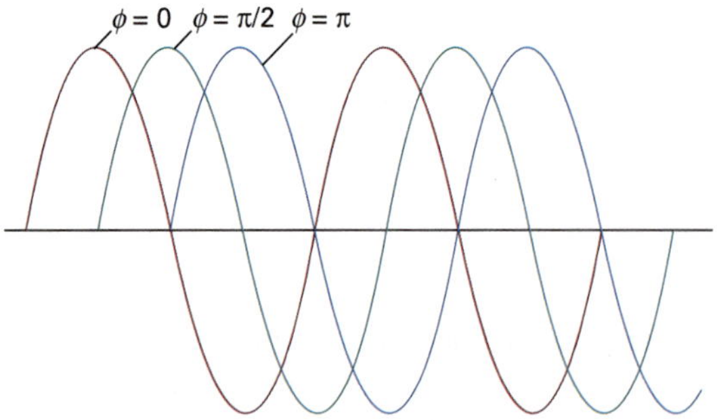

Figure C.2 The phase ϕ of a wave specifies the relative location of its peaks.

are the same (coincident peaks), the resultant has an enhanced amplitude. This effect is called **constructive interference.**

Brief illustration C.1 Resultant waves

To gain insight into cases in which the phase difference is a value other than $\pm\pi$, consider the addition of the waves $f(x)=\cos(2\pi x/\lambda)$ and $g(x)=\cos\{(2\pi x/\lambda)+\phi\}$. Figure C.3 shows plots of $f(x)$, $g(x)$, and $f(x)+g(x)$ against x/λ for $\phi=\pi/3$. The resultant wave has a greater amplitude than either $f(x)$ or $g(x)$, and has peaks between the peaks of $f(x)$ and $g(x)$.

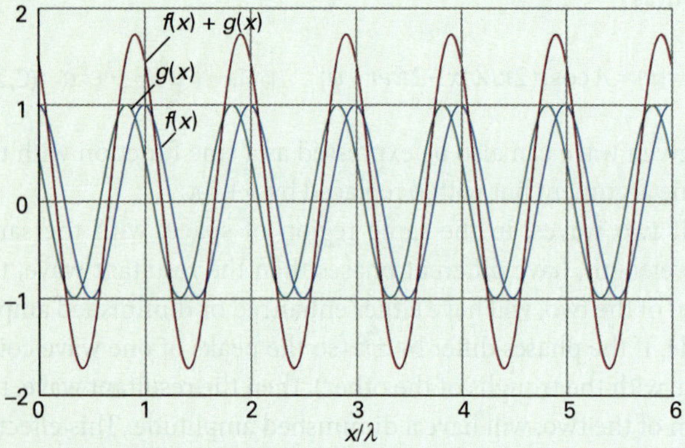

Figure C.3 Interference between the waves discussed in *Brief illustration* C.1.

Self-test C.1 Consider the same waves, but with $\phi=3\pi/4$. Does the resultant wave have diminished or enhanced amplitude?

Answer: Diminished amplitude

C.2 The electromagnetic field

Light is a form of electromagnetic radiation. In classical physics, electromagnetic radiation is understood in terms of the **electromagnetic field**, an oscillating electric and magnetic disturbance that spreads as a harmonic wave through space. An **electric field** acts on charged particles (whether stationary or moving) and a **magnetic field** acts only on moving charged particles.

The wavelength and frequency of an electromagnetic wave in a vacuum are related by

$$\lambda\nu=c \qquad \textit{Electromagnetic wave in a vacuum} \qquad \text{Relation between frequency and wavelength} \qquad \text{(C.3)}$$

where $c=2.997\,924\,58\times10^{8}\,\mathrm{m\,s^{-1}}$ (which we shall normally quote as $2.998\times10^{8}\,\mathrm{m\,s^{-1}}$) is the speed of light in a vacuum. When the wave is passing through a medium (even air), its speed is reduced to c' and, although the frequency remains unchanged, its wavelength is reduced accordingly. The reduced speed of light in a medium is normally expressed in terms of the **refractive index**, n_r, of the medium, where

$$n_r=\frac{c}{c'} \qquad \text{Refractive index} \qquad \text{(C.4)}$$

The refractive index depends on the frequency of the light, and for visible light typically increases with frequency. It also depends on the physical state of the medium. For yellow light in water at 25 °C, $n_r=1.3$, so the wavelength is reduced by 30 per cent.

The classification of the electromagnetic field according to its frequency and wavelength is summarized in Fig. C.4. It is often desirable to express the characteristics of an electromagnetic wave by giving its **wavenumber**, $\tilde{\nu}$ (nu tilde), where

$$\tilde{\nu}=\frac{\nu}{c}=\frac{1}{\lambda} \qquad \textit{Electromagnetic radiation} \qquad \text{Wavenumber} \qquad \text{(C.5)}$$

A wavenumber can be interpreted as the number of complete wavelengths in a given length (of vacuum). Wavenumbers are normally reported in reciprocal centimetres ($\mathrm{cm^{-1}}$), so a wavenumber of $5\,\mathrm{cm^{-1}}$ indicates that there are 5 complete wavelengths in 1 cm.

Brief illustration C.2 Wavenumbers

The wavenumber of electromagnetic radiation of wavelength 660 nm is

$$\tilde{\nu}=\frac{1}{\lambda}=\frac{1}{660\times10^{-9}\,\mathrm{m}}=1.5\times10^{6}\,\mathrm{m^{-1}}=15\,000\,\mathrm{cm^{-1}}$$

You can avoid errors in converting between units of $\mathrm{m^{-1}}$ and $\mathrm{cm^{-1}}$ by remembering that wavenumber represents the number of wavelengths in a given distance. Thus, a wavenumber expressed as the number of waves per centimetre and hence in units of $\mathrm{cm^{-1}}$ must be 100 times less than the equivalent quantity expressed per metre in units of $\mathrm{m^{-1}}$.

Self-test C.2 Calculate the wavenumber and frequency of red light, of wavelength 710 nm.

Answer: $\tilde{\nu}=1.41\times10^{6}\,\mathrm{m^{-1}}=1.41\times10^{4}\,\mathrm{cm^{-1}}$, $\nu=422\,\mathrm{THz}$ ($1\,\mathrm{THz}=10^{12}\,\mathrm{s^{-1}}$)

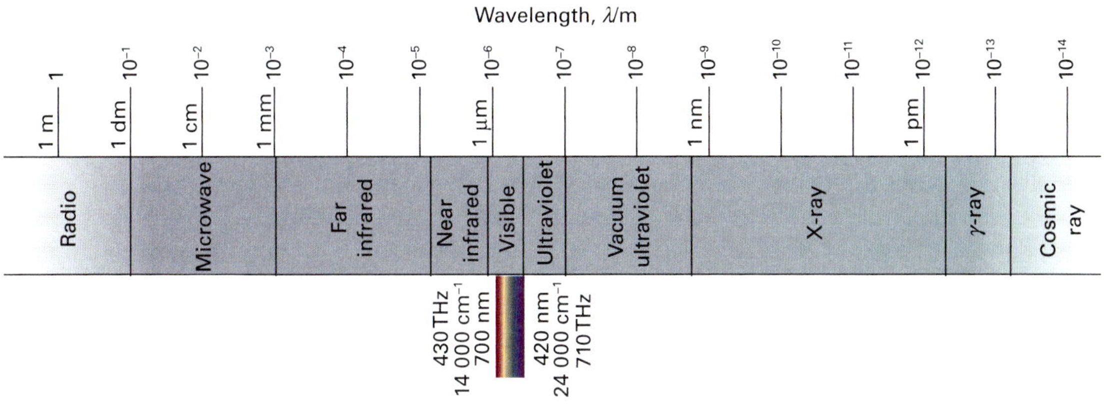

Figure C.4 The electromagnetic spectrum and its classification into regions (the boundaries are not precise).

The functions that describe the oscillating electric field, $\mathcal{E}(x,t)$, and magnetic field, $\mathcal{B}(x,t)$, travelling along the x-direction with wavelength λ and frequency ν are

$$\mathcal{E}(x,t) = \mathcal{E}_0 \cos\{(2\pi/\lambda)x - 2\pi\nu t + \phi\}$$ *Electromagnetic radiation* Electric field (C.6a)

$$\mathcal{B}(x,t) = \mathcal{B}_0 \cos\{(2\pi/\lambda)x - 2\pi\nu t + \phi\}$$ *Electromagnetic radiation* Magnetic field (C.6b)

where $\mathcal{E}_0$ and $\mathcal{B}_0$ are the amplitudes of the electric and magnetic fields, respectively, and ϕ is the phase of the wave. In this case the amplitude is a vector quantity, because the electric and magnetic fields have direction as well as amplitude. The magnetic field is perpendicular to the electric field and both are perpendicular to the propagation direction (Fig. C.5). According to classical electromagnetic theory, the **intensity** of electromagnetic radiation, a measure of the energy associated with the wave, is proportional to the square of the amplitude of the wave.

Equation C.6 describes electromagnetic radiation that is **plane polarized**; it is so called because the electric and magnetic fields each oscillate in a single plane. The plane of polarization may be orientated in any direction around the direction of propagation. An alternative mode of polarization is **circular polarization**, in which the electric and magnetic fields rotate around the direction of propagation in either a clockwise or an anticlockwise sense but remain perpendicular to it and to each other (Fig. C.6).

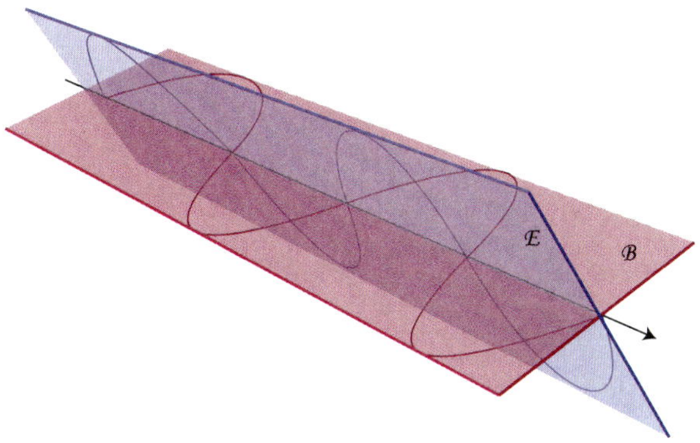

Figure C.5 In a plane polarized wave, the electric and magnetic fields oscillate in orthogonal planes and are perpendicular to the direction of propagation.

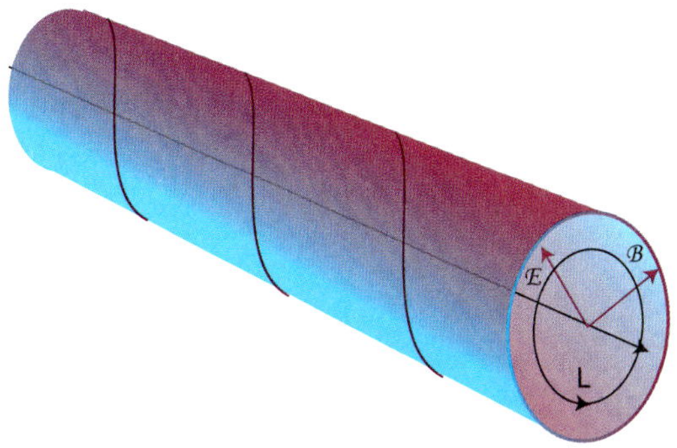

Figure C.6 In a circularly polarized wave, the electric and magnetic fields rotate around the direction of propagation but remain perpendicular to one another. The illustration also defines 'right' and 'left-handed' polarizations ('left-handed' polarization is shown as L).

Checklist of concepts

☐ 1. A **wave** is an oscillatory disturbance that travels through space.

☐ 2. A **harmonic wave** is a wave with a displacement that can be expressed as a sine or cosine function.

- ☐ 3. A harmonic wave is characterized by a **wavelength**, **frequency**, **phase**, and **amplitude**.
- ☐ 4. **Destructive interference** between two waves of the same wavelength but different phases leads to a resultant wave with diminished amplitude.
- ☐ 5. **Constructive interference** between two waves of the same wavelength and phase leads to a resultant wave with enhanced amplitude.
- ☐ 6. The **electromagnetic field** is an oscillating electric and magnetic disturbance that spreads as a harmonic wave through space.
- ☐ 7. An **electric field** acts on charged particles (whether stationary or moving).
- ☐ 8. A **magnetic field** acts only on moving charged particles.
- ☐ 9. In **plane polarized** electromagnetic radiation, the electric and magnetic fields each oscillate in a single plane and are mutually perpendicular.
- ☐ 10. In **circular polarization**, the electric and magnetic fields rotate around the direction of propagation in either a clockwise or an anticlockwise sense but remain perpendicular to it and each other.

Checklist of equations

Property	Equation	Comment	Equation number
Relation between the frequency and wavelength	$\lambda\nu=v$	For electromagnetic radiation in a vacuum, $v=c$	C.1
Refractive index	$n_r=c/c'$	Definition; $n_r\geq 1$	C.4
Wavenumber	$\tilde{\nu}=\nu/c=1/\lambda$	Electromagnetic radiation	C.5

FOUNDATIONS

TOPIC A Matter

Discussion questions

A.1 Summarize the features of the nuclear model of the atom. Define the terms atomic number, nucleon number, and mass number.

A.2 Where in the periodic table are metals, non-metals, transition metals, lanthanoids, and actinoids found?

A.3 Summarize what is meant by a single bond and a multiple bond.

A.4 Summarize the principal concepts of the VSEPR theory of molecular shape.

A.5 Compare and contrast the properties of the solid, liquid, and gas states of matter.

Exercises

A.1(a) Express the typical ground-state electron configuration of an atom of an element in (i) Group 2, (ii) Group 7, (iii) Group 15 of the periodic table.
A.1(b) Express the typical ground-state electron configuration of an atom of an element in (i) Group 3, (ii) Group 5, (iii) Group 13 of the periodic table.

A.2(a) Identify the oxidation numbers of the elements in (i) $MgCl_2$, (ii) FeO, (iii) Hg_2Cl_2.
A.2(b) Identify the oxidation numbers of the elements in (i) CaH_2, (ii) CaC_2, (iii) LiN_3.

A.3(a) Identify a molecule with a (i) single, (ii) double, (iii) triple bond between a carbon and a nitrogen atom.
A.3(b) Identify a molecule with (i) one, (i) two, (iii) three lone pairs on the central atom.

A.4(a) Draw the Lewis (electron dot) structures of (i) SO_3^{2-}, (ii) XeF_4, (iii) P_4.
A.4(b) Draw the Lewis (electron dot) structures of (i) O_3, (ii) ClF_3^+, (iii) N_3^-.

A.5(a) Identify three compounds with an incomplete octet.
A.5(b) Identify four hypervalent compounds.

A.6(a) Use VSEPR theory to predict the structures of (i) PCl_3, (ii) PCl_5, (iii) XeF_2, (iv) XeF_4.
A.6(b) Use VSEPR theory to predict the structures of (i) H_2O_2, (ii) FSO_3^-, (iii) KrF_2, (iv) PCl_4^+.

A.7(a) Identify the polarities (by attaching partial charges $\delta+$ and $\delta-$) of the bonds (i) C–Cl, (ii) P–H, (iii) N–O.
A.7(b) Identify the polarities (by attaching partial charges $\delta+$ and $\delta-$) of the bonds (i) C–H, (ii) P–S, (iii) N–Cl.

A.8(a) State whether you expect the following molecules to be polar or nonpolar: (i) CO_2, (ii) SO_2, (iii) N_2O, (iv) SF_4.
A.8(b) State whether you expect the following molecules to be polar or nonpolar: (i) O_3, (ii) XeF_2, (iii) NO_2, (iv) C_6H_{14}.

A.9(a) Arrange the molecules in Exercise A.8(a) by increasing dipole moment.
A.9(b) Arrange the molecules in Exercise A.8(b) by increasing dipole moment.

A.10(a) Classify the following properties as extensive or intensive: (i) mass, (ii) mass density, (iii) temperature, (iv) number density.
A.10(b) Classify the following properties as extensive or intensive: (i) pressure, (ii) specific heat capacity, (iii) weight, (iv) molality.

A.11(a) Calculate (i) the amount of C_2H_5OH (in moles) and (ii) the number of molecules present in 25.0 g of ethanol.
A.11(b) Calculate (i) the amount of $C_6H_{12}O_6$ (in moles) and (ii) the number of molecules present in 5.0 g of glucose.

A.12(a) Calculate (i) the mass, (ii) the weight on the surface of the Earth (where $g = 9.81\ m\ s^{-2}$) of 10.0 mol $H_2O(l)$.
A.12(b) Calculate (i) the mass, (ii) the weight on the surface of Mars (where $g = 3.72\ m\ s^{-2}$) of 10.0 mol $C_6H_6(l)$.

A.13(a) Calculate the pressure exerted by a person of mass 65 kg standing (on the surface of the Earth) on shoes with soles of area 150 cm^2.
A.13(b) Calculate the pressure exerted by a person of mass 60 kg standing (on the surface of the Earth) on shoes with stiletto heels of area 2 cm^2 (assume that the weight is entirely on the heels).

A.14(a) Express the pressure calculated in Exercise A.13(a) in atmospheres.
A.14(b) Express the pressure calculated in Exercise A.13(b) in atmospheres.

A.15(a) Express a pressure of 1.45 atm in (i) pascal, (ii) bar.
A.15(b) Express a pressure of 222 atm in (i) pascal, (ii) bar.

A.16(a) Convert blood temperature, 37.0 °C, to the Kelvin scale.
A.16(b) Convert the boiling point of oxygen, 90.18 K, to the Celsius scale.

A.17(a) Equation A.4 is a relation between the Kelvin and Celsius scales. Devise the corresponding equation relating the Fahrenheit and Celsius scales and use it to express the boiling point of ethanol (78.5 °C) in degrees Fahrenheit.
A.17(b) The Rankine scale is a version of the thermodynamic temperature scale in which the degrees (°R) are the same size as degrees Fahrenheit. Derive an expression relating the Rankine and Kelvin scales and express the freezing point of water in degrees Rankine.

A.18(a) A sample of hydrogen gas was found to have a pressure of 110 kPa when the temperature was 20.0 °C. What can its pressure be expected to be when the temperature is 7.0 °C?
A.18(b) A sample of 325 mg of neon occupies 2.00 dm^3 at 20.0 °C. Use the perfect gas law to calculate the pressure of the gas.

A.19(a) At 500 °C and 93.2 kPa, the mass density of sulfur vapour is 3.710 kg m^{-3}. What is the molecular formula of sulfur under these conditions?
A.19(b) At 100 °C and 16.0 kPa, the mass density of phosphorus vapour is 0.6388 kg m^{-3}. What is the molecular formula of phosphorus under these conditions?

A.20(a) Calculate the pressure exerted by 22 g of ethane behaving as a perfect gas when confined to 1000 cm^3 at 25.0 °C.
A.20(b) Calculate the pressure exerted by 7.05 g of oxygen behaving as a perfect gas when confined to 100 cm^3 at 100.0 °C.

A.21(a) A vessel of volume 10.0 dm^3 contains 2.0 mol H_2 and 1.0 mol N_2 at 5.0 °C. Calculate the partial pressure of each component and their total pressure.
A.21(b) A vessel of volume 100 cm^3 contains 0.25 mol O_2 and 0.034 mol CO_2 at 10.0 °C. Calculate the partial pressure of each component and their total pressure.

TOPIC B Energy

Discussion questions

B.1 What is energy?

B.2 Distinguish between kinetic and potential energy.

B.3 State the Second Law of thermodynamics. Can the entropy of the system that is not isolated from its surroundings decrease during a spontaneous process?

B.4 What is meant by quantization of energy? In what circumstances are the effects of quantization most important for microscopic systems?

B.5 What are the assumptions of the kinetic molecular theory?

B.6 What are the main features of the Maxwell–Boltzmann distribution of speeds?

Exercises

B.1(a) A particle of mass 1.0 g is released near the surface of the Earth, where the acceleration of free fall is $g = 9.81\,\mathrm{m\,s^{-2}}$. What will be its speed and kinetic energy after (i) 1.0 s, (ii) 3.0 s. Ignore air resistance.
B.1(b) The same particle in Exercise B.1(a) is released near the surface of Mars, where the acceleration of free fall is $g = 3.72\,\mathrm{m\,s^{-2}}$. What will be its speed and kinetic energy after (i) 1.0 s, (ii) 3.0 s. Ignore air resistance.

B.2(a) An ion of charge ze moving through water is subject to an electric field of strength $\mathcal{E}$ which exerts a force $ze\mathcal{E}$, but it also experiences a frictional drag proportional to its speed s and equal to $6\pi\eta Rs$, where R is its radius and η (eta) is the viscosity of the medium. What will be its terminal velocity?
B.2(b) A particle descending through a viscous medium experiences a frictional drag proportional to its speed s and equal to $6\pi\eta Rs$, where R is its radius and η (eta) is the viscosity of the medium. If the acceleration of free fall is denoted g, what will be the terminal velocity of a sphere of radius R and mass density ρ?

B.3(a) Confirm that the general solution of the harmonic oscillator equation of motion ($m\mathrm{d}^2x/\mathrm{d}t^2 = -k_f x$) is $x(t) = A\sin\omega t + B\cos\omega t$ with $\omega = (k_f/m)^{1/2}$.
B.3(b) Consider a harmonic oscillator with $B = 0$ (in the notation of Exercise B.3(a)); relate the total energy at any instant to its maximum displacement amplitude.

B.4(a) The force constant of a C–H bond is about $450\,\mathrm{N\,m^{-1}}$. How much work is needed to stretch the bond by (i) 10 pm, (ii) 20 pm?
B.4(b) The force constant of the H–H bond is about $510\,\mathrm{N\,m^{-1}}$. How much work is needed to stretch the bond by 20 pm?

B.5(a) An electron is accelerated in an electron microscope from rest through a potential difference $\Delta\phi = 100\,\mathrm{kV}$ and acquires an energy of $e\Delta\phi$. What is its final speed? What is its energy in electronvolts (eV)?
B.5(b) A $C_6H_4^{2+}$ ion is accelerated in a mass spectrometer from rest through a potential difference $\Delta\phi = 20\,\mathrm{kV}$ and acquires an energy of $e\Delta\phi$. What is its final speed? What is its energy in electronvolts (eV)?

B.6(a) Calculate the work that must be done in order to remove a Na^+ ion from 200 pm away from a Cl^- ion to infinity (in a vacuum). What work would be needed if the separation took place in water?
B.6(b) Calculate the work that must be done in order to remove an Mg^{2+} ion from 250 pm away from an O^{2-} ion to infinity (in a vacuum). What work would be needed if the separation took place in water?

B.7(a) Calculate the electric potential due to the nuclei at a point in a LiH molecule located at 200 pm from the Li nucleus and 150 pm from the H nucleus.
B.7(b) Plot the electric potential due to the nuclei at a point in a Na^+Cl^- ion pair located on a line half way between the nuclei (the internuclear separation is 283 pm) as the point approaches from infinity and ends at the mid-point between the nuclei.

B.8(a) An electric heater is immersed in a flask containing 200 g of water, and a current of 2.23 A from a 15.0 V supply is passed for 12.0 minutes. How much energy is supplied to the water? Estimate the rise in temperature (for water, $C = 75.3\,\mathrm{J\,K^{-1}\,mol^{-1}}$).

B.8(b) An electric heater is immersed in a flask containing 150 g of ethanol, and a current of 1.12 A from a 12.5 V supply is passed for 172 s. How much energy is supplied to the ethanol? Estimate the rise in temperature (for ethanol, $C = 111.5\,\mathrm{J\,K^{-1}\,mol^{-1}}$).

B.9(a) The heat capacity of a sample of iron was $3.67\,\mathrm{J\,K^{-1}}$. By how much would its temperature rise if 100 J of energy were transferred to it as heat?
B.9(b) The heat capacity of a sample of water was $5.77\,\mathrm{J\,K^{-1}}$. By how much would its temperature rise if 50.0 kJ of energy were transferred to it as heat?

B.10(a) The molar heat capacity of lead is $26.44\,\mathrm{J\,K^{-1}\,mol^{-1}}$. How much energy must be supplied (by heating) to 100 g of lead to increase its temperature by 10.0 °C?
B.10(b) The molar heat capacity of water is $75.2\,\mathrm{J\,K^{-1}\,mol^{-1}}$. How much energy must be supplied by heating to 10.0 g of water to increase its temperature by 10.0 °C?

B.11(a) The molar heat capacity of ethanol is $111.46\,\mathrm{J\,K^{-1}\,mol^{-1}}$. What is its specific heat capacity?
B.11(b) The molar heat capacity of sodium is $28.24\,\mathrm{J\,K^{-1}\,mol^{-1}}$. What is its specific heat capacity?

B.12(a) The specific heat capacity of water is $4.18\,\mathrm{J\,K^{-1}\,g^{-1}}$. What is its molar heat capacity?
B.12(b) The specific heat capacity of copper is $0.384\,\mathrm{J\,K^{-1}\,g^{-1}}$. What is its molar heat capacity?

B.13(a) By how much does the molar enthalpy of hydrogen gas differ from its molar internal energy at 1000 °C? Assume perfect gas behaviour.
B.13(b) The mass density of water is $0.997\,\mathrm{g\,cm^{-3}}$. By how much does the molar enthalpy of water differ from its molar internal energy at 298 K?

B.14(a) Which do you expect to have the greater entropy at 298 K and 1 bar, liquid water or water vapour?
B.14(b) Which do you expect to have the greater entropy at 0 °C and 1 atm, liquid water or ice?

B.15(a) Which do you expect to have the greater entropy, 100 g of iron at 300 K or 3000 K?
B.15(b) Which do you expect to have the greater entropy, 100 g of water at 0 °C or 100 °C?

B.16(a) Give three examples of a system that is in dynamic equilibrium.
B.16(b) Give three examples of a system that is in static equilibrium.

B.17(a) Suppose two states differ in energy by 1.0 eV (electronvolts, see inside the front cover); what is the ratio of their populations at (a) 300 K, (b) 3000 K?
B.17(b) Suppose two states differ in energy by 2.0 eV (electronvolts, see inside the front cover); what is the ratio of their populations at (a) 200 K, (b) 2000 K?

B.18(a) Suppose two states differ in energy by 1.0 eV, what can be said about their populations when $T = 0$?
B.18(b) Suppose two states differ in energy by 1.0 eV, what can be said about their populations when the temperature is infinite?

B.19(a) A typical vibrational excitation energy of a molecule corresponds to a wavenumber of 2500 cm^{-1} (convert to an energy separation by multiplying by hc; see *Foundations* C). Would you expect to find molecules in excited vibrational states at room temperature (20 °C)?
B.19(b) A typical rotational excitation energy of a molecule corresponds to a frequency of about 10 GHz (convert to an energy separation by multiplying by h; see *Foundations* C). Would you expect to find gas-phase molecules in excited rotational states at room temperature (20 °C)?

B.20(a) Suggest a reason why most molecules survive for long periods at room temperature.
B.20(b) Suggest a reason why the rates of chemical reactions typically increase with increasing temperature.

B.21(a) Calculate the relative mean speeds of N_2 molecules in air at 0 °C and 40 °C.
B.21(b) Calculate the relative mean speeds of CO_2 molecules in air at 20 °C and 30 °C.

B.22(a) Calculate the relative mean speeds of N_2 and CO_2 molecules in air.
B.22(b) Calculate the relative mean speeds of Hg_2 and H_2 molecules in a gaseous mixture.

B.23(a) Use the equipartition theorem to calculate the contribution of translational motion to the internal energy of 5.0 g of argon at 25 °C.
B.23(b) Use the equipartition theorem to calculate the contribution of translational motion to the internal energy of 10.0 g of helium at 30 °C.

B.24(a) Use the equipartition theorem to calculate the contribution to the total internal energy of a sample of 10.0 g of (i) carbon dioxide, (ii) methane at 20 °C; take into account translation and rotation but not vibration.
B.24(b) Use the equipartition theorem to calculate the contribution to the total internal energy of a sample of 10.0 g of lead at 20 °C, taking into account the vibrations of the atoms.

B.25(a) Use the equipartition theorem to compute the molar heat capacity of argon.
B.25(b) Use the equipartition theorem to compute the molar heat capacity of helium.

B.26(a) Use the equipartition theorem to estimate the heat capacity of (i) carbon dioxide, (ii) methane.
B.26(b) Use the equipartition theorem to estimate the heat capacity of (i) water vapour, (ii) lead.

TOPIC C Waves

Discussion questions

C.1 How many types of wave motion can you identify?

C.2 What is the wave nature of the sound of a sudden 'bang'?

Exercises

C.1(a) What is the speed of light in water if the refractive index of the latter is 1.33?
C.1(b) What is the speed of light in benzene if the refractive index of the latter is 1.52?

C.2(a) The wavenumber of a typical vibrational transition of a hydrocarbon is 2500 cm^{-1}. Calculate the corresponding wavelength and frequency.
C.2(b) The wavenumber of a typical vibrational transition of an O–H bond is 3600 cm^{-1}. Calculate the corresponding wavelength and frequency.

Integrated activities

F.1 In Topic 1B we show that for a perfect gas the fraction of molecules that have a speed in the range v to $v+dv$ is $f(v)dv$, where

$$f(v)=4\pi\left(\frac{M}{2\pi RT}\right)^{3/2} v^2 e^{-Mv^2/2RT}$$

is the Maxwell–Boltzmann distribution (eqn 1B.4). Use this expression and mathematical software, a spreadsheet, or the *Living graphs* on the web site of this book for the following exercises:

(a) Refer to the graph in Fig. B.6. Plot different distributions by keeping the molar mass constant at 100 g mol^{-1} and varying the temperature of the sample between 200 K and 2000 K.
(b) Evaluate numerically the fraction of molecules with speeds in the range 100 m s^{-1} to 200 m s^{-1} at 300 K and 1000 K.

F.2 Based on your observations from Problem F.1, provide a molecular interpretation of temperature.

PART ONE

Thermodynamics

Part 1 of the text develops the concepts of thermodynamics, the science of the transformations of energy. Thermodynamics provides a powerful way to discuss equilibria and the direction of natural change in chemistry. Its concepts apply to both physical change, such as fusion and vaporization, and chemical change, including electrochemistry. We see that through the concepts of energy, enthalpy, entropy, Gibbs energy, and the chemical potential it is possible to obtain a unified view of these core features of chemistry and to treat equilibria quantitatively.

The chapters in Part 1 deal with the bulk properties of matter; those of Part 2 show how these properties stem from the behaviour of individual atoms.

CHAPTER 1

The properties of gases

A **gas** is a form of matter that fills whatever container it occupies. This chapter establishes the properties of gases that will be used throughout the text.

1A The perfect gas

The chapter begins with an account of an idealized version of a gas, a 'perfect gas', and shows how its equation of state may be assembled from the experimental observations summarized by Boyle's law, Charles's law, and Avogadro's principle.

1B The kinetic model

One central feature of physical chemistry is its role in building models of molecular behaviour that seek to explain observed phenomena. A prime example of this procedure is the development of a molecular model of a perfect gas in terms of a collection of molecules (or atoms) in ceaseless, essentially random motion. This model is the basis of 'kinetic molecular theory'. As well as accounting for the gas laws, this theory can be used to predict the average speed at which molecules move in a gas, and that speed's dependence on temperature. In combination with the Boltzmann distribution (*Foundations* B), the kinetic theory can also be used to predict the spread of molecular speeds and its dependence on molecular mass and temperature.

1C Real gases

The perfect gas is an excellent starting point for the discussion of properties of all gases, and its properties are invoked throughout the chapters on thermodynamics that follow this chapter. However, actual gases, 'real gases', have properties that differ from those of perfect gases, and we need to be able to interpret these deviations and build the effects of molecular attractions and repulsions into our model. The discussion of real gases is another example of how initially primitive models in physical chemistry are elaborated to take into account more detailed observations.

What is the impact of this material?

The perfect gas law and the kinetic theory can be applied to the study of phenomena confined to a reaction vessel or encompassing an entire planet or star. We have identified two applications. In *Impact* I1.1 we see how the gas laws are used in the discussion of meteorological phenomena—the weather. In *Impact* I1.2 we examine how the kinetic model of gases has a surprising application: to the discussion of dense stellar media, such as the interior of the Sun.

To read more about the impact of this material, scan the QR code, or go to bcs.whfreeman.com/webpub/chemistry/pchem10e/impact/pchem-1-1.html

1A The perfect gas

Contents

➤ Why do you need to know this material?

Equations related to perfect gases provide the basis for the development of many equations in thermodynamics. The perfect gas law is also a good first approximation for accounting for the properties of real gases.

➤ What is the key idea?

The perfect gas law, which is based on a series of empirical observations, is a limiting law that is obeyed increasingly well as the pressure of a gas tends to zero.

➤ What do you need to know already?

You need to be aware of the concepts of pressure and temperature introduced in *Foundations* A.

In molecular terms, a gas consists of a collection of molecules that are in ceaseless motion and which interact significantly with one another only when they collide. The properties of gases were among the first to be established quantitatively (largely during the seventeenth and eighteenth centuries) when the technological requirements of travel in balloons stimulated their investigation.

1A.1 Variables of state

The **physical state** of a sample of a substance, its physical condition, is defined by its physical properties. Two samples of the same substance that have the same physical properties are in the same state. The variables needed to specify the state of a system are the amount of substance it contains, n, the volume it occupies, V, the pressure, p, and the temperature, T.

(a) Pressure

The origin of the force exerted by a gas is the incessant battering of the molecules on the walls of its container. The collisions are so numerous that they exert an effectively steady force, which is experienced as a steady pressure. The SI unit of pressure, the *pascal* (Pa, $1\,\text{Pa}=1\,\text{N m}^{-2}$) is introduced in *Foundations* A. As discussed there, several other units are still widely used (Table 1A.1). A pressure of 1 bar is the **standard pressure** for reporting data; we denote it $p^{\ominus}$.

If two gases are in separate containers that share a common movable wall (a 'piston', Fig. 1A.1), the gas that has the higher pressure will tend to compress (reduce the volume of) the gas that has lower pressure. The pressure of the high-pressure gas will fall as it expands and that of the low-pressure gas will rise as it is compressed. There will come a stage when the two pressures are equal and the wall has no further tendency to move. This condition of equality of pressure on either side of a movable wall is a state of **mechanical equilibrium** between the two gases. The pressure of a gas is therefore an indication of whether a container that contains the gas will be in mechanical equilibrium with another gas with which it shares a movable wall.

Table 1A.1 Pressure units*

Name	Symbol	Value
pascal	1 Pa	$\mathbf{1\,N\,m^{-2}, 1\,kg\,m^{-1}\,s^{-2}}$
bar	1 bar	$\mathbf{10^5\,Pa}$
atmosphere	1 atm	**101.325 kPa**
torr	1 Torr	**(101 325/760) Pa** = 133.32… Pa
millimetres of mercury	1 mmHg	133.322… Pa
pounds per square inch	1 psi	6.894 757… kPa

* Values in bold are exact.

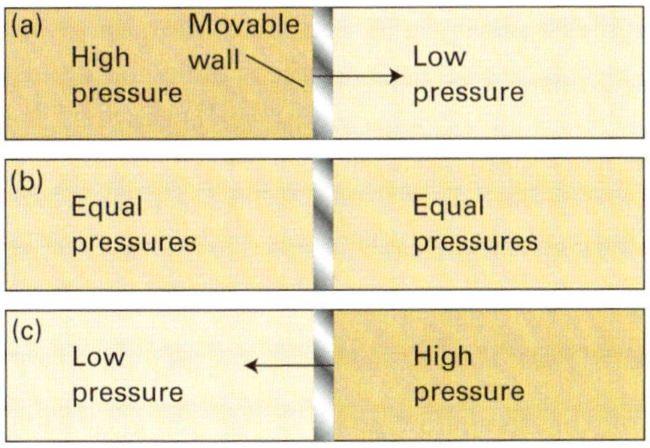

Figure 1A.1 When a region of high pressure is separated from a region of low pressure by a movable wall, the wall will be pushed into one region or the other, as in (a) and (c). However, if the two pressures are identical, the wall will not move (b). The latter condition is one of mechanical equilibrium between the two regions.

The pressure exerted by the atmosphere is measured with a *barometer*. The original version of a barometer (which was invented by Torricelli, a student of Galileo) was an inverted tube of mercury sealed at the upper end. When the column of mercury is in mechanical equilibrium with the atmosphere, the pressure at its base is equal to that exerted by the atmosphere. It follows that the height of the mercury column is proportional to the external pressure.

Example 1A.1 Calculating the pressure exerted by a column of liquid

Derive an equation for the pressure at the base of a column of liquid of mass density ρ (rho) and height h at the surface of the Earth. The pressure exerted by a column of liquid is commonly called the 'hydrostatic pressure'.

Method According to *Foundations* A, the pressure is the force, F, divided by the area, A, to which the force is applied: $p=F/A$. For a mass m subject to a gravitational field at the surface of the earth, $F=mg$, where g is the acceleration of free fall. To calculate F we need to know the mass m of the column of liquid, which is its mass density, ρ, multiplied by its volume, V: $m=\rho V$. The first step, therefore, is to calculate the volume of a cylindrical column of liquid.

Answer Let the column have cross-sectional area A, then its volume is Ah and its mass is $m=\rho Ah$. The force the column of this mass exerts at its base is

$$F=mg=\rho Ahg$$

The pressure at the base of the column is therefore

$$p=\frac{F}{A}=\frac{\rho Agh}{A}=\rho gh \qquad \text{Hydrostatic pressure} \qquad (1A.1)$$

Note that the hydrostatic pressure is independent of the shape and cross-sectional area of the column. The mass of the column of a given height increases as the area, but so does the area on which the force acts, so the two cancel.

Self-test 1A.1 Derive an expression for the pressure at the base of a column of liquid of length l held at an angle θ (theta) to the vertical (**1**).

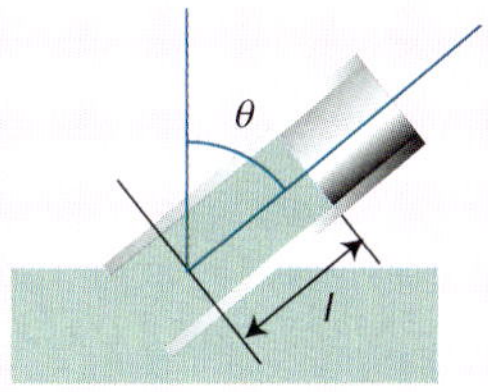

1

Answer: $p=\rho gl\cos\theta$

The pressure of a sample of gas inside a container is measured by using a pressure gauge, which is a device with properties that respond to the pressure. For instance, a *Bayard–Alpert pressure gauge* is based on the ionization of the molecules present in the gas and the resulting current of ions is interpreted in terms of the pressure. In a *capacitance manometer*, the deflection of a diaphragm relative to a fixed electrode is monitored through its effect on the capacitance of the arrangement. Certain semiconductors also respond to pressure and are used as transducers in solid-state pressure gauges.

(b) Temperature

The concept of temperature is introduced in *Foundations* A. In the early days of thermometry (and still in laboratory practice today), temperatures were related to the length of a column of liquid, and the difference in lengths shown when the thermometer was first in contact with melting ice and then with boiling water was divided into 100 steps called 'degrees', the lower point being labelled 0. This procedure led to the **Celsius scale** of temperature. In this text, temperatures on the Celsius scale are denoted θ (theta) and expressed in *degrees Celsius* (°C). However, because different liquids expand to different extents, and do not always expand uniformly over a given range, thermometers constructed from different materials showed different numerical values of the temperature between their fixed points. The pressure of a gas, however, can be used to construct a **perfect-gas temperature scale** that is independent of the identity of the gas. The perfect-gas scale turns out to be identical to the **thermodynamic temperature scale** introduced in Topic 3A, so we shall use the latter term from now on to avoid a proliferation of names.

On the thermodynamic temperature scale, temperatures are denoted T and are normally reported in *kelvins* (K; not °K). Thermodynamic and Celsius temperatures are related by the exact expression

$$T/\text{K} = \theta/°\text{C} + 273.15$$ Definition of Celsius scale (1A.2)

This relation is the current definition of the Celsius scale in terms of the more fundamental Kelvin scale. It implies that a difference in temperature of 1 °C is equivalent to a difference of 1 K.

A note on good practice We write $T=0$, not $T=0\,\text{K}$ for the zero temperature on the thermodynamic temperature scale. This scale is absolute, and the lowest temperature is 0 regardless of the size of the divisions on the scale (just as we write $p=0$ for zero pressure, regardless of the size of the units we adopt, such as bar or pascal). However, we write 0 °C because the Celsius scale is not absolute.

Brief illustration 1A.1 Temperature conversion

To express 25.00 °C as a temperature in kelvins, we use eqn 1A.4 to write

$$T/\text{K} = (25.00\,°\text{C})/°\text{C} + 273.15 = 25.00 + 273.15 = 298.15$$

Note how the units (in this case, °C) are cancelled like numbers. This is the procedure called 'quantity calculus' in which a physical quantity (such as the temperature) is the product of a numerical value (25.00) and a unit (1 °C); see *The chemist's toolkit* A.1 of *Foundations*. Multiplication of both sides by the unit K then gives $T=298.15\,\text{K}$.

A note on good practice When the units need to be specified in an equation, the approved procedure, which avoids any ambiguity, is to write (physical quantity)/units, which is a dimensionless number, just as (25.00 °C)/°C = 25.00 in this illustration. Units may be multiplied and cancelled just like numbers.

1A.2 Equations of state

Although in principle the state of a pure substance is specified by giving the values of n, V, p, and T, it has been established experimentally that it is sufficient to specify only three of these variables, for then the fourth variable is fixed. That is, it is an experimental fact that each substance is described by an **equation of state**, an equation that interrelates these four variables.

The general form of an equation of state is

$$p = f(T,V,n)$$ General form of an equation of state (1A.3)

This equation tells us that if we know the values of n, T, and V for a particular substance, then the pressure has a fixed value. Each substance is described by its own equation of state, but we know the explicit form of the equation in only a few special cases. One very important example is the equation of state of a 'perfect gas', which has the form $p=nRT/V$, where R is a constant independent of the identity of the gas.

The equation of state of a perfect gas was established by combining a series of empirical laws.

(a) The empirical basis

We assume that the following individual gas laws are familiar:

Boyle's law: $pV = \text{constant}$, at constant n, T (1A.4a)

Charles's law: $V = \text{constant} \times T$, at constant n, p (1A.4b)

$p = \text{constant} \times T$, at constant n, V (1A.4c)

Avogadro's principle:

$V = \text{constant} \times n$ at constant p, T (1A.4d)

Boyle's and Charles's laws are examples of a **limiting law**, a law that is strictly true only in a certain limit, in this case $p \to 0$. For example, if it is found empirically that the volume of a substance fits an expression $V = aT + bp + cp^2$, then in the limit of $p \to 0$, $V = aT$. Throughout this text, equations valid in this limiting sense are labelled with a blue equation number, as in these expressions. Although these relations are strictly true only at $p=0$, they are reasonably reliable at normal pressures ($p \approx 1$ bar) and are used widely throughout chemistry.

Avogadro's principle is commonly expressed in the form 'equal volumes of gases at the same temperature and pressure contain the same numbers of molecules'. It is a principle rather than a law (a summary of experience) because it depends on the validity of a model, in this case the existence of molecules. Despite there now being no doubt about the existence of molecules, it is still a model-based principle rather than a law.

Figure 1A.2 depicts the variation of the pressure of a sample of gas as the volume is changed. Each of the curves in the

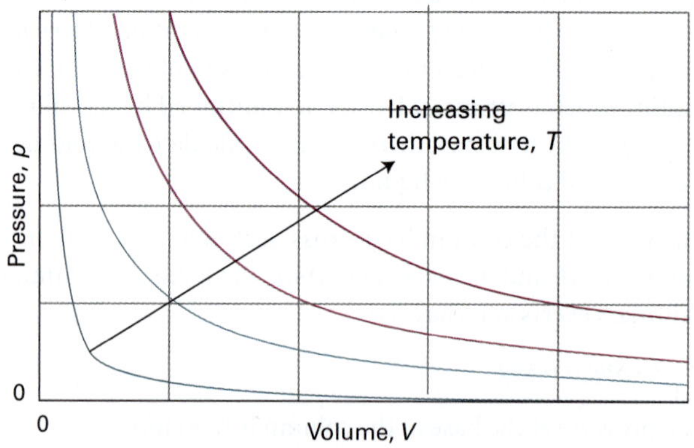

Figure 1A.2 The pressure–volume dependence of a fixed amount of perfect gas at different temperatures. Each curve is a hyperbola (pV=constant) and is called an isotherm.

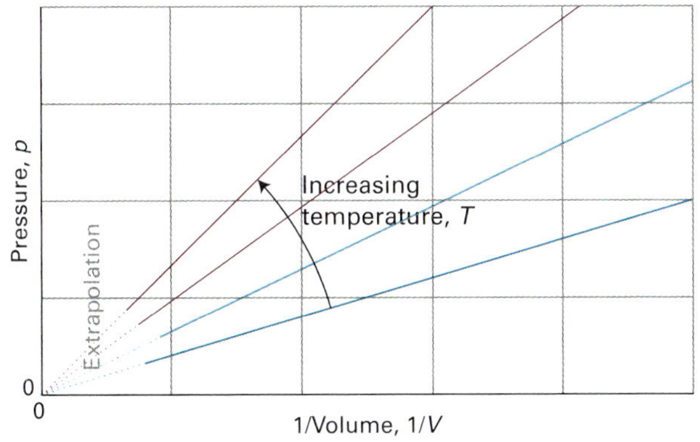

Figure 1A.3 Straight lines are obtained when the pressure is plotted against $1/V$ at constant temperature.

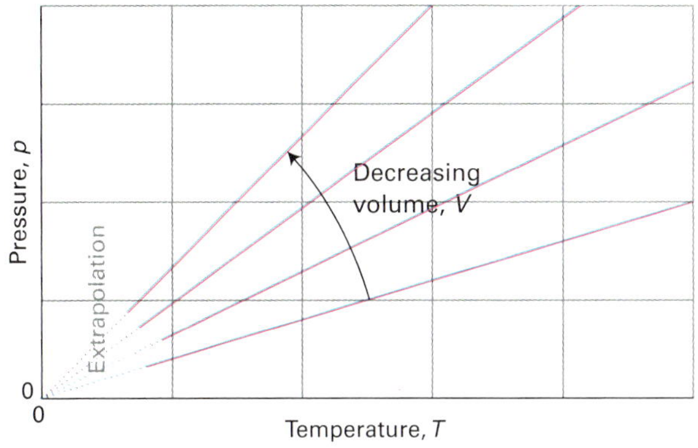

Figure 1A.5 The pressure also varies linearly with the temperature at constant volume, and extrapolates to zero at $T=0$ (−273 °C).

graph corresponds to a single temperature and hence is called an **isotherm**. According to Boyle's law, the isotherms of gases are hyperbolas (a curve obtained by plotting y against x with $xy=$constant, or $y=$constant$/x$). An alternative depiction, a plot of pressure against 1/volume, is shown in Fig. 1A.3. The linear variation of volume with temperature summarized by Charles's law is illustrated in Fig. 1A.4. The lines in this illustration are examples of **isobars**, or lines showing the variation of properties at constant pressure. Figure 1A.5 illustrates the linear variation of pressure with temperature. The lines in this diagram are **isochores**, or lines showing the variation of properties at constant volume.

A note on good practice To test the validity of a relation between two quantities, it is best to plot them in such a way that they should give a straight line, for deviations from a straight line are much easier to detect than deviations from a curve. The development of expressions that, when plotted, give a straight line is a very important and common procedure in physical chemistry.

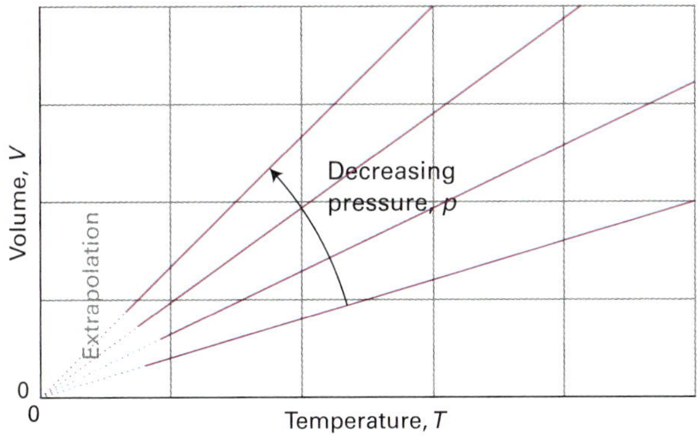

Figure 1A.4 The variation of the volume of a fixed amount of gas with the temperature at constant pressure. Note that in each case the isobars extrapolate to zero volume at $T=0$, or $\theta=-273$ °C.

The empirical observations summarized by eqn 1A.5 can be combined into a single expression:

$$pV = \text{constant} \times nT$$

This expression is consistent with Boyle's law ($pV=$constant) when n and T are constant, with both forms of Charles's law ($p \propto T$, $V \propto T$) when n and either V or p are held constant, and with Avogadro's principle ($V \propto n$) when p and T are constant. The constant of proportionality, which is found experimentally to be the same for all gases, is denoted R and called the (molar) **gas constant**. The resulting expression

$$pV = nRT \qquad \text{Perfect gas law} \qquad (1A.5)$$

is the **perfect gas law** (or *perfect gas equation of state*). It is the approximate equation of state of any gas, and becomes increasingly exact as the pressure of the gas approaches zero. A gas that obeys eqn 1A.5 exactly under all conditions is called a **perfect gas** (or *ideal gas*). A **real gas**, an actual gas, behaves more like a perfect gas the lower the pressure, and is described exactly by eqn 1A.5 in the limit of $p \to 0$. The gas constant R can be determined by evaluating $R=pV/nT$ for a gas in the limit of zero pressure (to guarantee that it is behaving perfectly). However, a more accurate value can be obtained by measuring the speed of sound in a low-pressure gas (argon is used in practice), for the speed of sound depends on the value of R and extrapolating its value to zero pressure. Another route to its value is to recognize (as explained in *Foundations* B) that it is related to Boltzmann's constant, k, by

$$R = N_A k \qquad \text{The (molar) gas constant} \qquad (1A.6)$$

where N_A is Avogadro's constant. There are currently (in 2014) plans to use this relation as the sole route to R, with defined values of N_A and k. Table 1A.2 lists the values of R in a variety of units.

A note on good practice Despite 'ideal gas' being the more common term, we prefer 'perfect gas'. As explained in Topic 5A, in an 'ideal mixture' of A and B, the AA, BB, and AB

Table 1A.2 The gas constant ($R = N_A k$)

R	
8.314 47	J K^{-1} mol^{-1}
8.205 74 × 10^{-2}	dm^3 atm K^{-1} mol^{-1}
8.314 47 × 10^{-2}	dm^3 bar K^{-1} mol^{-1}
8.314 47	Pa m^3 K^{-1} mol^{-1}
62.364	dm^3 Torr K^{-1} mol^{-1}
1.987 21	cal K^{-1} mol^{-1}

interactions are all the same but not necessarily zero. In a perfect gas, not only are the interactions all the same, they are also zero.

The surface in Fig. 1A.6 is a plot of the pressure of a fixed amount of perfect gas against its volume and thermodynamic temperature as given by eqn 1A.5. The surface depicts the only possible states of a perfect gas: the gas cannot exist in states that do not correspond to points on the surface. The graphs in Figs. 1A.2 and 1A.4 correspond to the sections through the surface (Fig. 1A.7).

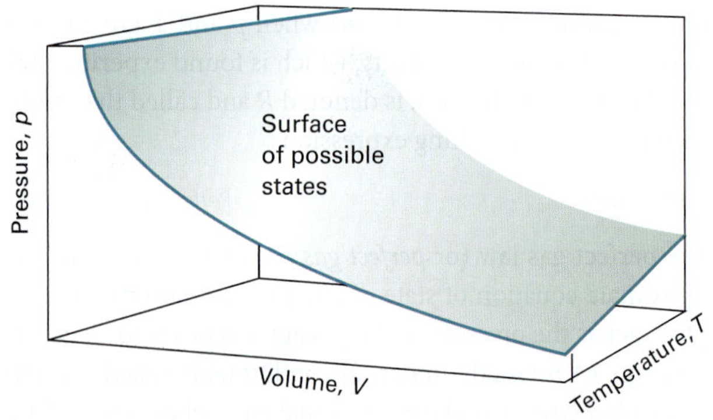

Figure 1A.6 A region of the p,V,T surface of a fixed amount of perfect gas. The points forming the surface represent the only states of the gas that can exist.

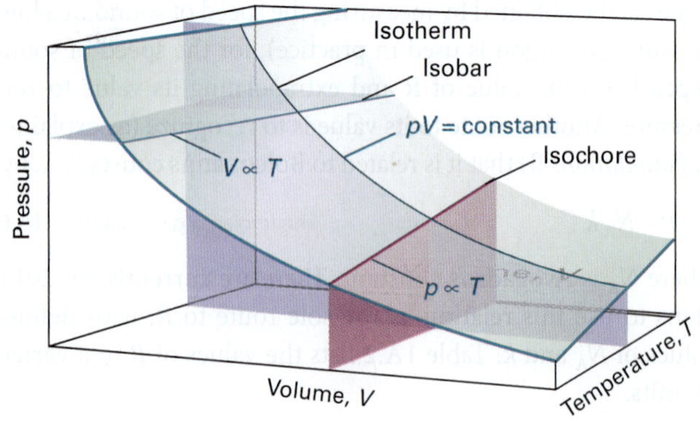

Figure 1A.7 Sections through the surface shown in Fig. 1A.6 at constant temperature give the isotherms shown in Fig. 1A.2, the isobars shown in Fig. 1A.4, and the isochores shown in Fig. 1A.5.

Example 1A.2 Using the perfect gas law

In an industrial process, nitrogen is heated to 500 K in a vessel of constant volume. If it enters the vessel at 100 atm and 300 K, what pressure would it exert at the working temperature if it behaved as a perfect gas?

Method We expect the pressure to be greater on account of the increase in temperature. The perfect gas law in the form $pV/nT = R$ implies that if the conditions are changed from one set of values to another, then because pV/nT is equal to a constant, the two sets of values are related by the 'combined gas law'

$$\frac{p_1V_1}{n_1T_1} = \frac{p_2V_2}{n_2T_2} \quad \text{Combined gas law} \quad (1A.7)$$

This expression is easily rearranged to give the unknown quantity (in this case p_2) in terms of the known. The known and unknown data are summarized as follows:

	n	p	V	T
Initial	Same	100	Same	300
Final	Same	?	Same	500

Answer Cancellation of the volumes (because $V_1 = V_2$) and amounts (because $n_1 = n_2$) on each side of the combined gas law results in

$$\frac{p_1}{T_1} = \frac{p_2}{T_2}$$

which can be rearranged into

$$p_2 = \frac{T_2}{T_1} \times p_1$$

Substitution of the data then gives

$$p_2 = \frac{500\text{ K}}{300\text{ K}} \times (100\text{ atm}) = 167\text{ atm}$$

Experiment shows that the pressure is actually 183 atm under these conditions, so the assumption that the gas is perfect leads to a 10 per cent error.

Self-test 1A.2 What temperature would result in the same sample exerting a pressure of 300 atm?

Answer: 900 K

The perfect gas law is of the greatest importance in physical chemistry because it is used to derive a wide range of relations that are used throughout thermodynamics. However, it is also of considerable practical utility for calculating the properties of a gas under a variety of conditions. For instance, the molar volume, $V_m = V/n$, of a perfect gas under the conditions called **standard ambient temperature and pressure** (SATP), which means

298.15 K and 1 bar (that is, exactly 10^5 Pa), is easily calculated from $V_m = RT/p$ to be 24.789 dm^3 mol^{-1}. An earlier definition, **standard temperature and pressure** (STP), was 0 °C and 1 atm; at STP, the molar volume of a perfect gas is 22.414 dm^3 mol^{-1}.

The molecular explanation of Boyle's law is that if a sample of gas is compressed to half its volume, then twice as many molecules strike the walls in a given period of time than before it was compressed. As a result, the average force exerted on the walls is doubled. Hence, when the volume is halved the pressure of the gas is doubled, and pV is a constant. Boyle's law applies to all gases regardless of their chemical identity (provided the pressure is low) because at low pressures the average separation of molecules is so great that they exert no influence on one another and hence travel independently. The molecular explanation of Charles's law lies in the fact that raising the temperature of a gas increases the average speed of its molecules. The molecules collide with the walls more frequently and with greater impact. Therefore they exert a greater pressure on the walls of the container. For a quantitative account of these relations, see Topic 1B.

(b) Mixtures of gases

When dealing with gaseous mixtures, we often need to know the contribution that each component makes to the total pressure of the sample. The **partial pressure**, p_J, of a gas J in a mixture (any gas, not just a perfect gas), is defined as

$$p_J = x_J p \qquad \textit{Definition} \quad \text{Partial pressure} \quad (1A.8)$$

where x_J is the **mole fraction** of the component J, the amount of J expressed as a fraction of the total amount of molecules, n, in the sample:

$$x_J = \frac{n_J}{n} \qquad n = n_A + n_B + \cdots \qquad \textit{Definition} \quad \text{Mole fraction} \quad (1A.9)$$

When no J molecules are present, $x_J = 0$; when only J molecules are present, $x_J = 1$. It follows from the definition of x_J that, whatever the composition of the mixture, $x_A + x_B + \ldots = 1$ and therefore that the sum of the partial pressures is equal to the total pressure:

$$p_A + p_B + \cdots = (x_A + x_B + \cdots)p = p \qquad (1A.10)$$

This relation is true for both real and perfect gases.

When all the gases are perfect, the partial pressure as defined in eqn 1A.9 is also the pressure that each gas would exert if it occupied the same container alone at the same temperature. The latter is the original meaning of 'partial pressure'. That identification was the basis of the original formulation of **Dalton's law**:

> The pressure exerted by a mixture of gases is the sum of the pressures that each one would exert if it occupied the container alone. — Dalton's law

Now, however, the relation between partial pressure (as defined in eqn 1A.8) and total pressure (as given by eqn 1A.10) is true for all gases and the identification of partial pressure with the pressure that the gas would exert on its own is valid only for a perfect gas.

Example 1A.3 Calculating partial pressures

The mass percentage composition of dry air at sea level is approximately N_2: 75.5; O_2: 23.2; Ar: 1.3. What is the partial pressure of each component when the total pressure is 1.20 atm?

Method We expect species with a high mole fraction to have a proportionally high partial pressure. Partial pressures are defined by eqn 1A.8. To use the equation, we need the mole fractions of the components. To calculate mole fractions, which are defined by eqn 1A.9, we use the fact that the amount of molecules J of molar mass M_J in a sample of mass m_J is $n_J = m_J/M_J$. The mole fractions are independent of the total mass of the sample, so we can choose the latter to be exactly 100 g (which makes the conversion from mass percentages very easy). Thus, the mass of N_2 present is 75.5 per cent of 100 g, which is 75.5 g.

Answer The amounts of each type of molecule present in 100 g of air, in which the masses of N_2, O_2, and Ar are 75.5 g, 23.2 g, and 1.3 g, respectively, are

$$n(N_2) = \frac{75.5\,\text{g}}{28.02\,\text{g mol}^{-1}} = \frac{75.5}{28.02}\,\text{mol} = 2.69\,\text{mol}$$

$$n(O_2) = \frac{23.2\,\text{g}}{32.00\,\text{g mol}^{-1}} = \frac{23.2}{32.00}\,\text{mol} = 0.725\,\text{mol}$$

$$n(\text{Ar}) = \frac{1.3\,\text{g}}{39.95\,\text{g mol}^{-1}} = \frac{1.3}{39.95}\,\text{mol} = 0.033\,\text{mol}$$

The total is 3.45 mol. The mole fractions are obtained by dividing each of the above amounts by 3.45 mol and the partial pressures are then obtained by multiplying the mole fraction by the total pressure (1.20 atm):

	N_2	O_2	Ar
Mole fraction:	0.780	0.210	0.0096
Partial pressure/atm:	0.936	0.252	0.012

We have not had to assume that the gases are perfect: partial pressures are defined as $p_J = x_J p$ for any kind of gas.

Self-test 1A.3 When carbon dioxide is taken into account, the mass percentages are 75.52 (N_2), 23.15 (O_2), 1.28 (Ar), and 0.046 (CO_2). What are the partial pressures when the total pressure is 0.900 atm?

Answer: 0.703, 0.189, 0.0084, 0.00027 atm

Checklist of concepts

- ☐ 1. The **physical state** of a sample of a substance, its physical condition, is defined by its physical properties.
- ☐ 2. **Mechanical equilibrium** is the condition of equality of pressure on either side of a shared movable wall.
- ☐ 3. An **equation of state** is an equation that interrelates the variables that define the state of a substance.
- ☐ 4. Boyle's and Charles's laws are examples of a **limiting law**, a law that is strictly true only in a certain limit, in this case $p \to 0$.
- ☐ 5. An **isotherm** is a line in a graph that corresponds to a single temperature.
- ☐ 6. An **isobar** is a line in a graph that corresponds to a single pressure.
- ☐ 7. An **isochore** is a line in a graph that corresponds to a single volume.
- ☐ 8. A **perfect gas** is a gas that obeys the perfect gas law under all conditions.
- ☐ 9. **Dalton's law** states that the pressure exerted by a mixture of (perfect) gases is the sum of the pressures that each one would exert if it occupied the container alone.

Checklist of equations

Property	Equation	Comment	Equation number
Relation between temperature scales	$T/\mathrm{K} = \theta/{}^\circ\mathrm{C} + 273.15$	273.15 is exact	1A.2
Equation of state	$p = f(n,V,T)$		1A.3
Perfect gas law	$pV = nRT$	Valid for real gases in the limit $p \to 0$	1A.5
Partial pressure	$p_J = x_J p$	Valid for all gases	1A.8

1B The kinetic model

Contents

➤ **Why do you need to know this material?**

This material illustrates an important skill in science: the ability to extract quantitative information from a qualitative model. Moreover, the model is used in the discussion of the transport properties of gases (Topic 19A), reaction rates in gases (Topic 20F), and catalysis (Topic 22C).

➤ **What is the key idea?**

A gas consists of molecules of negligible size in ceaseless random motion and obeying the laws of classical mechanics in their collisions.

➤ **What do you need to know already?**

You need to be aware of Newton's second law of motion, that the acceleration of a body is proportional to the force acting on it, and the conservation of linear momentum.

In the **kinetic theory** of gases (which is sometimes called the *kinetic-molecular theory*, KMT) it is assumed that the only contribution to the energy of the gas is from the kinetic energies of the molecules. The kinetic model is one of the most remarkable—and arguably most beautiful—models in physical chemistry, for from a set of very slender assumptions, powerful quantitative conclusions can be reached.

1B.1 The model

The kinetic model is based on three assumptions:

1. The gas consists of molecules of mass m in ceaseless random motion obeying the laws of classical mechanics.
2. The size of the molecules is negligible, in the sense that their diameters are much smaller than the average distance travelled between collisions.
3. The molecules interact only through brief elastic collisions.

An **elastic collision** is a collision in which the total translational kinetic energy of the molecules is conserved.

(a) Pressure and molecular speeds

From the very economical assumptions of the kinetic model, we show in the following *Justification* that the pressure and volume of the gas are related by

$$pV = \tfrac{1}{3}nMv_{\rm rms}^2 \qquad \textit{Perfect gas} \quad \text{Pressure} \quad (1B.1)$$

where $M = mN_A$, the molar mass of the molecules of mass m, and $v_{\rm rms}$ is the square root of the mean of the squares of the speeds, v, of the molecules:

$$v_{\rm rms} = \langle v^2 \rangle^{1/2} \qquad \textit{Definition} \quad \text{Root-mean-square speed} \quad (1B.2)$$

Justification 1.1B The pressure of a gas according to the kinetic model

Consider the arrangement in Fig. 1B.1. When a particle of mass m that is travelling with a component of velocity v_x parallel to the x-axis collides with the wall on the right and is reflected, its linear momentum changes from mv_x before the collision to $-mv_x$ after the collision (when it is travelling in the opposite direction). The x-component of momentum therefore changes by $2mv_x$ on each collision (the y- and z-components are unchanged). Many molecules collide with the wall in an interval Δt, and the total change of momentum is the product of the change in momentum of each molecule multiplied

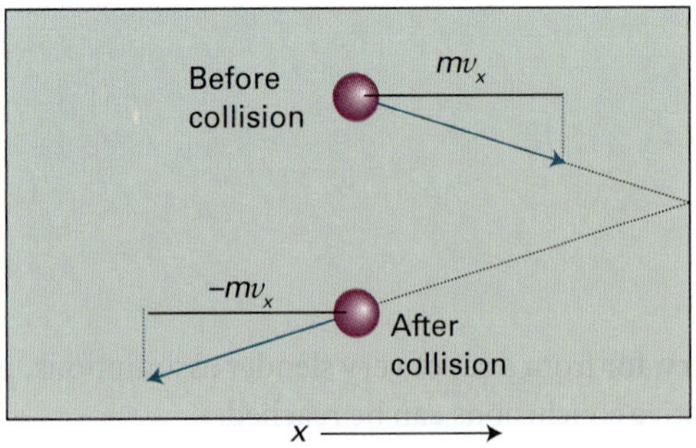

Figure 1B.1 The pressure of a gas arises from the impact of its molecules on the walls. In an elastic collision of a molecule with a wall perpendicular to the *x*-axis, the *x*-component of velocity is reversed but the *y*- and *z*-components are unchanged.

by the number of molecules that reach the wall during the interval.

Because a molecule with velocity component v_x can travel a distance $v_x\Delta t$ along the *x*-axis in an interval Δt, all the molecules within a distance $v_x\Delta t$ of the wall will strike it if they are travelling towards it (Fig. 1B.2). It follows that if the wall has area A, then all the particles in a volume $A\times v_x\Delta t$ will reach the wall (if they are travelling towards it). The number density of particles is nN_A/V, where n is the total amount of molecules in the container of volume V and N_A is Avogadro's constant, so the number of molecules in the volume $Av_x\Delta t$ is $(nN_A/V)\times Av_x\Delta t$.

At any instant, half the particles are moving to the right and half are moving to the left. Therefore, the average number of collisions with the wall during the interval Δt is $\frac{1}{2}nN_A Av_x\Delta t/V$. The total momentum change in that interval is the product of this number and the change $2mv_x$:

$$\text{Momentum change} = \frac{nN_A Av_x\Delta t}{2V}\times 2mv_x$$

$$= \frac{n\overbrace{mN_A}^{M} Av_x^2\Delta t}{V} = \frac{nMAv_x^2\Delta t}{V}$$

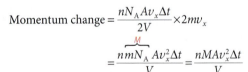

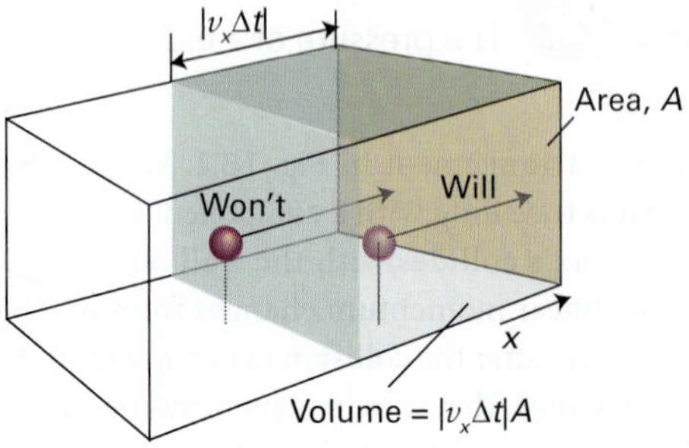

Figure 1B.2 A molecule will reach the wall on the right within an interval Δt if it is within a distance $v_x\Delta t$ of the wall and travelling to the right.

Next, to find the force, we calculate the rate of change of momentum, which is this change of momentum divided by the interval Δt during which it occurs:

$$\text{Rate of change of momentum} = \frac{nMAv_x^2}{V}$$

This rate of change of momentum is equal to the force (by Newton's second law of motion). It follows that the pressure, the force divided by the area, is

$$\text{Pressure} = \frac{nMv_x^2}{V}$$

Not all the molecules travel with the same velocity, so the detected pressure, p, is the average (denoted $\langle\ldots\rangle$) of the quantity just calculated:

$$p = \frac{nM\langle v_x^2\rangle}{V}$$

This expression already resembles the perfect gas equation of state.

To write an expression for the pressure in terms of the root-mean-square speed, $v_{\rm rms}$, we begin by writing the speed of a single molecule, v, as $v^2 = v_x^2 + v_y^2 + v_z^2$. Because the root-mean-square speed is defined as $v_{\rm rms} = \langle v^2\rangle^{1/2}$, it follows that

$$v_{\rm rms}^2 = \langle v^2\rangle = \langle v_x^2\rangle + \langle v_y^2\rangle + \langle v_z^2\rangle$$

However, because the molecules are moving randomly, all three averages are the same. It follows that $v_{\rm rms}^2 = \langle 3v_x^2\rangle$. Equation 1B.1 follows immediately by substituting $\langle v_x^2\rangle = \frac{1}{3}\langle v_{\rm rms}^2\rangle$ into $p = nM\langle v_x^2\rangle/V$.

Equation 1B.1 is one of the key results of the kinetic model. We see that, if the root-mean-square speed of the molecules depends only on the temperature, then at constant temperature

$$pV = \text{constant}$$

which is the content of Boyle's law. Moreover, for eqn 1B.1 to be the equation of state of a perfect gas, its right-hand side must be equal to nRT. It follows that the root-mean-square speed of the molecules in a gas at a temperature T must be

$$v_{\rm rms} = \left(\frac{3RT}{M}\right)^{1/2} \qquad \textit{Perfect gas} \quad \text{RMS speed} \quad (1B.3)$$

Brief illustration 1B.1 Molecular speeds

For N_2 molecules at 25 °C, we use $M = 28.02\ \text{g mol}^{-1}$, then

$$v_{\rm rms} = \left\{\frac{3\times(8.3145\ \text{J K}^{-1}\ \text{mol}^{-1})\times(298\ \text{K})}{0.02802\ \text{kg mol}^{-1}}\right\}^{1/2} = 515\ \text{m s}^{-1}$$

Shortly we shall encounter the mean speed, v_{mean}, and the most probable speed v_{mp}; they are, respectively,

$$v_{mean} = \left(\frac{8}{3\pi}\right)^{1/2} v_{rms} = 0.921\ldots \times (515\,\mathrm{m\,s^{-1}}) = 475\,\mathrm{m\,s^{-1}}$$

$$v_{mp} = \left(\frac{2}{3}\right)^{1/2} v_{rms} = 0.816\ldots \times (515\,\mathrm{m\,s^{-1}}) = 420\,\mathrm{m\,s^{-1}}$$

Self-test 1B.1 Evaluate the root-mean-square speed of H_2 molecules at 25 °C.

Answer: 1.92 km s⁻¹

(b) The Maxwell–Boltzmann distribution of speeds

Equation 1B.2 is an expression for the mean square speed of molecules. However, in an actual gas the speeds of individual molecules span a wide range, and the collisions in the gas continually redistribute the speeds among the molecules. Before a collision, a molecule may be travelling rapidly, but after a collision it may be accelerated to a very high speed, only to be slowed again by the next collision. The fraction of molecules that have speeds in the range v to $v+\mathrm{d}v$ is proportional to the width of the range, and is written $f(v)\mathrm{d}v$, where $f(v)$ is called the **distribution of speeds**. Note that, in common with other distribution functions, $f(v)$ acquires physical significance only after it is multiplied by the range of speeds of interest. In the following *Justification* we show that the fraction of molecules that have a speed in the range v to $v+\mathrm{d}v$ is $f(v)\mathrm{d}v$, where

$$f(v) = 4\pi\left(\frac{M}{2\pi RT}\right)^{3/2} v^2 e^{-Mv^2/2RT} \quad \text{Perfect gas} \quad \text{Maxwell–Boltzmann distribution} \quad (1B.4)$$

The function $f(v)$ is called the **Maxwell–Boltzmann distribution of speeds**.

Justification 1B.2 The Maxwell–Boltzmann distribution of speeds

The Boltzmann distribution (*Foundations* B) implies that the fraction of molecules with velocity components v_x, v_y, and v_z is proportional to an exponential function of their kinetic energy: $f(v) = Ke^{-\varepsilon/kT}$, where K is a constant of proportionality. The kinetic energy is

$$\varepsilon = \tfrac{1}{2}mv_x^2 + \tfrac{1}{2}mv_y^2 + \tfrac{1}{2}mv_z^2$$

Therefore, we can use the relation $a^{x+y+z} = a^x a^y a^z$ to write

$$f(v) = Ke^{-(mv_x^2+mv_y^2+mv_z^2)/2kT} = Ke^{-mv_x^2/2kT}e^{-mv_y^2/2kT}e^{-mv_z^2/2kT}$$

The distribution factorizes into three terms, and we can write $f(v) = f(v_x)f(v_y)f(v_z)$ and $K = K_xK_yK_z$, with

$$f(v_x) = K_x e^{-mv_x^2/2kT}$$

and likewise for the other two axes.

To determine the constant K_x, we note that a molecule must have a velocity component somewhere in the range $-\infty < v_x < \infty$, so

$$\int_{-\infty}^{\infty} f(v_x)\mathrm{d}v_x = 1$$

Substitution of the expression for $f(v_x)$ then gives

$$1 = K_x \int_{-\infty}^{\infty} e^{-mv_x^2/2kT}\mathrm{d}v_x \overset{\text{Integral G.1}}{=} K_x\left(\frac{2\pi kT}{m}\right)^{1/2}$$

Therefore, $K_x = (m/2\pi kT)^{1/2}$ and at this stage we can write

$$f(v_x) = \left(\frac{m}{2\pi kT}\right)^{1/2} e^{-mv_x^2/2kT} \quad (1B.5)$$

The probability that a molecule has a velocity in the range v_x to $v_x+\mathrm{d}v_x$, v_y to $v_y+\mathrm{d}v_y$, v_z to $v_z+\mathrm{d}v_z$, is therefore

$$f(v_x)f(v_y)f(v_z) = \left(\frac{m}{2\pi kT}\right)^{3/2} e^{-mv_x^2/2kT}e^{-mv_y^2/2kT}e^{-mv_z^2/2kT} \times \mathrm{d}v_x\mathrm{d}v_y\mathrm{d}v_z$$

$$= \left(\frac{m}{2\pi kT}\right)^{3/2} e^{-mv^2/2kT}\mathrm{d}v_x\mathrm{d}v_y\mathrm{d}v_z$$

where $v^2 = v_x^2 + v_y^2 + v_z^2$.

To evaluate the probability that the molecules have a speed in the range v to $v+\mathrm{d}v$ regardless of direction we think of the three velocity components as defining three coordinates in 'velocity space', with the same properties as ordinary space except that the coordinates are labelled (v_x, v_y, v_z) instead of (x,y,z). Just as the volume element in ordinary space is $\mathrm{d}x\mathrm{d}y\mathrm{d}z$, so the volume element in velocity space is $\mathrm{d}v_x\mathrm{d}v_y\mathrm{d}v_z$. The sum of all the volume elements in ordinary space that lie at a distance r from the centre is the volume of a spherical shell of radius r and thickness $\mathrm{d}r$. That volume is the product of its surface area, $4\pi r^2$, and its thickness $\mathrm{d}r$, and is therefore $4\pi r^2\mathrm{d}r$. Similarly, the analogous volume in velocity space is the volume of a shell of radius v and thickness $\mathrm{d}v$, namely $4\pi v^2\mathrm{d}v$ (Fig. 1B.3). Now, because $f(v_x)f(v_y)f(v_z)$, the term in blue in the last equation, depends only on v^2, and has the same value everywhere in a shell of radius v, the total probability of the molecules possessing a speed in the range v to $v+\mathrm{d}v$ is the product of the term in blue and the volume of the

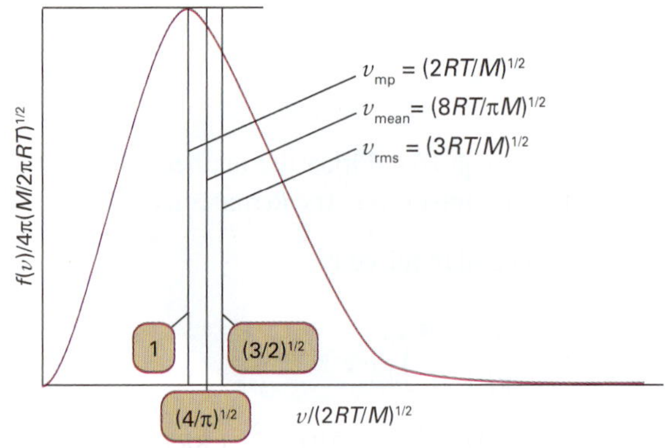

Figure 1B.7 A summary of the conclusions that can be deduced form the Maxwell distribution for molecules of molar mass M at a temperature T: v_{mp} is the most probable speed, v_{mean} is the mean speed, and v_{rms} is the root-mean-square speed.

This result is much harder to derive, but the diagram in Fig. 1B.8 should help to show that it is plausible. For the relative mean speed of two dissimilar molecules of masses m_A and m_B:

$$v_{rel} = \left(\frac{8kT}{\pi\mu}\right)^{1/2} \quad \mu = \frac{m_A m_B}{m_A + m_B}$$ *Perfect gas* Mean relative speed (1B.10b)

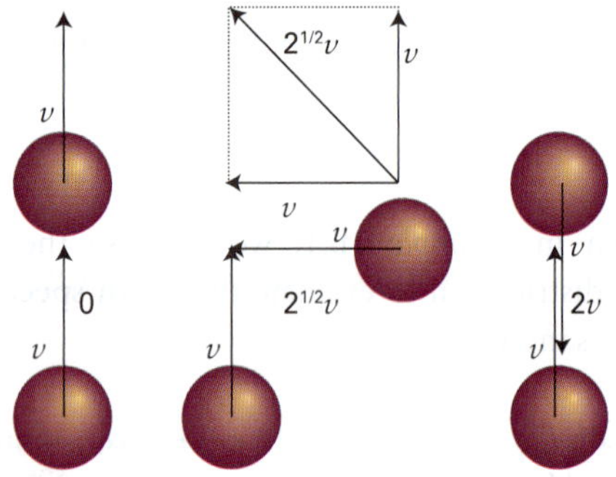

Figure 1B.8 A simplified version of the argument to show that the mean relative speed of molecules in a gas is related to the mean speed. When the molecules are moving in the same direction, the mean relative speed is zero; it is $2v$ when the molecules are approaching each other. A typical mean direction of approach is from the side, and the mean speed of approach is then $2^{1/2}v$. The last direction of approach is the most characteristic, so the mean speed of approach can be expected to be about $2^{1/2}v$. This value is confirmed by more detailed calculation.

Brief illustration 1B.2 Relative molecular speeds

We have already seen (in *Brief illustration* 1B.1) that the rms speed of N_2 molecules at 25 °C is 515 m s^{-1}. It follows from eqn 1B.10a that their relative mean speed is

$$v_{rel} = 2^{1/2} \times (515\,\text{m s}^{-1}) = 728\,\text{m s}^{-1}$$

Self-test 1B.3 What is the relative mean speed of N_2 and H_2 molecules in a gas at 25 °C?

Answer: 1.83 km s^{-1}

1B.2 Collisions

The kinetic model enables us to make the qualitative picture of a gas as a collection of ceaselessly moving, colliding molecules more quantitative. In particular, it enables us to calculate the frequency with which molecular collisions occur and the distance a molecule travels on average between collisions.

(a) The collision frequency

Although the kinetic-molecular theory assumes that the molecules are point-like, we can count a 'hit' whenever the centres of two molecules come within a distance d of each other, where d, the collision diameter, is of the order of the actual diameters of the molecules (for impenetrable hard spheres d is the diameter). As we show in the following *Justification*, we can use the kinetic model to deduce that the **collision frequency**, z, the number of collisions made by one molecule divided by the time interval during which the collisions are counted, when there are N molecules in a volume V is

$$z = \sigma v_{rel} \mathcal{N}$$ *Perfect gas* Collision frequency (1B.11a)

with $\mathcal{N} = N/V$, the number density, and v_{rel} given by eqn 1B.10. The area $\sigma = \pi d^2$ is called the **collision cross-section** of the molecules. Some typical collision cross-sections are given in Table 1B.1. In terms of the pressure (as is also shown in the following *Justification*),

$$z = \frac{\sigma v_{rel} p}{kT}$$ *Perfect gas* Collision frequency (1B.11b)

Table 1B.1* Collision cross-sections, σ/nm^2

	σ/nm^2
C_6H_6	0.88
CO_2	0.52
He	0.21
N_2	0.43

* More values are given in the *Resource section*.

Justification 1B.3 The collision frequency according to the kinetic model

Consider the positions of all the molecules except one to be frozen. Then note what happens as one mobile molecule travels through the gas with a mean relative speed v_{rel} for a

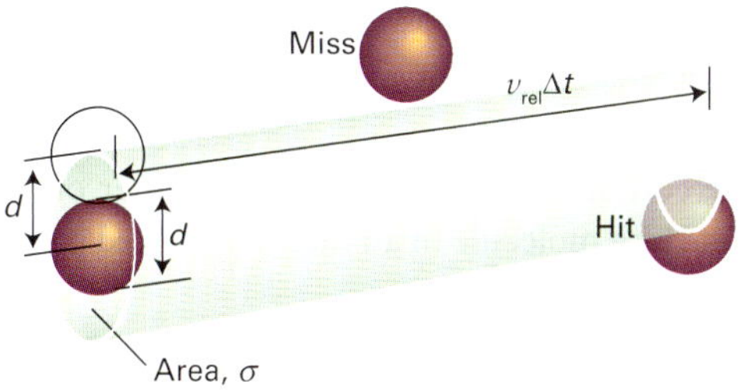

Figure 1B.9 The calculation of the collision frequency and the mean free path in the kinetic theory of gases.

time Δt. In doing so it sweeps out a 'collision tube' of cross-sectional area $\sigma=\pi d^2$ and length $v_{rel}\Delta t$ and therefore of volume $\sigma v_{rel}\Delta t$ (Fig. 1B.9). The number of stationary molecules with centres inside the collision tube is given by the volume of the tube multiplied by the number density $\mathcal{N}=N/V$, and is $\mathcal{N}\sigma v_{rel}\Delta t$. The number of hits scored in the interval Δt is equal to this number, so the number of collisions divided by the time interval is $\mathcal{N}\sigma v_{rel}$, which is eqn 1B.11a. The expression in terms of the pressure of the gas is obtained by using the perfect gas equation to write

$$\mathcal{N}=\frac{N}{V}=\frac{nN_A}{V}=\frac{nN_A}{nRT/p}=\frac{p}{kT}$$

Equation 1B.11a shows that, at constant volume, the collision frequency increases with increasing temperature. Equation 1B.11b shows that, at constant temperature, the collision frequency is proportional to the pressure. Such a proportionality is plausible because the greater the pressure, the greater the number density of molecules in the sample, and the rate at which they encounter one another is greater even though their average speed remains the same.

Brief illustration 1B.3 Molecular collisions

For an N_2 molecule in a sample at 1.00 atm (101 kPa) and 25 °C, from *Brief illustration* 1B.2 we know that $v_{rel}=728\,m\,s^{-1}$. Therefore, from eqn 1B.11b, and taking $\sigma=0.45\,nm^2$ (corresponding to $0.45\times10^{-18}\,m^2$) from Table 1B.1,

$$z=\frac{(0.43\times10^{-18}\,m^2)\times(728\,m\,s^{-1})\times(1.01\times10^5\,Pa)}{(1.381\times10^{-23}\,J\,K^{-1})\times(298\,K)}$$
$$=7.7\times10^9\,s^{-1}$$

so a given molecule collides about 8×10^9 times each second. We are beginning to appreciate the timescale of events in gases.

Self-test 1B.4 Evaluate the collision frequency between H_2 molecules in a gas under the same conditions.

Answer: $4.1\times10^9\,s^{-1}$

(b) The mean free path

Once we have the collision frequency, we can calculate the **mean free path**, λ (lambda), the average distance a molecule travels between collisions. If a molecule collides with a frequency z, it spends a time $1/z$ in free flight between collisions, and therefore travels a distance $(1/z)v_{rel}$. It follows that the mean free path is

$$\lambda=\frac{v_{rel}}{z} \qquad \textit{Perfect gas} \quad \text{Mean free path} \quad (1B.12)$$

Substitution of the expression for z in eqn 1B.11b gives

$$\lambda=\frac{kT}{\sigma p} \qquad \textit{Perfect gas} \quad \text{Mean free path} \quad (1B.13)$$

Doubling the pressure reduces the mean free path by half.

Brief illustration 1B.4 The mean free path

In *Brief illustration* 1B.2 we noted that $v_{rel}=728\,m\,s^{-1}$ for N_2 molecules at 25 °C, and in *Brief illustration* 1B.3 that $z=7.7\times10^9\,s^{-1}$ when the pressure is 1.00 atm. Under these circumstances, the mean free path of N_2 molecules is

$$\lambda=\frac{728\,m\,s^{-1}}{7.7\times10^9\,s^{-1}}=9.5\times10^{-8}\,m$$

or 95 nm, about 10^3 molecular diameters.

Self-test 1B.5 Evaluate the mean free path of benzene molecules at 25 °C in a sample where the pressure is 0.10 atm.

Answer: 460 nm

Although the temperature appears in eqn 1B.13, in a sample of constant volume, the pressure is proportional to T, so T/p remains constant when the temperature is increased. Therefore, the mean free path is independent of the temperature in a sample of gas in a container of fixed volume: the distance between collisions is determined by the number of molecules present in the given volume, not by the speed at which they travel.

In summary, a typical gas (N_2 or O_2) at 1 atm and 25 °C can be thought of as a collection of molecules travelling with a mean speed of about $500\,m\,s^{-1}$. Each molecule makes a collision within about 1 ns, and between collisions it travels about 10^3 molecular diameters. The kinetic model of gases is valid and the gas behaves nearly perfectly if the diameter of the molecules is much smaller than the mean free path ($d\ll\lambda$), for then the molecules spend most of their time far from one another.

Attractive forces are ineffective when the molecules are far apart (well to the right in Fig. 1C.1). Intermolecular forces are also important when the temperature is so low that the molecules travel with such low mean speeds that they can be captured by one another.

The consequences of these interactions are shown by shapes of experimental isotherms (Fig. 1C.2). At low pressures, when the sample occupies a large volume, the molecules are so far apart for most of the time that the intermolecular forces play no significant role, and the gas behaves virtually perfectly. At moderate pressures, when the average separation of the molecules is only a few molecular diameters, the attractive forces dominate the repulsive forces. In this case, the gas can be expected to be more compressible than a perfect gas because the forces help to draw the molecules together. At high pressures, when the average separation of the molecules is small, the repulsive forces dominate and the gas can be expected to be less compressible because now the forces help to drive the molecules apart.

Consider what happens when we compress (reduce the volume of) a sample of gas initially in the state marked A in Fig. 1C.2 at constant temperature by pushing in a piston. Near A, the pressure of the gas rises in approximate agreement with Boyle's law. Serious deviations from that law begin to appear when the volume has been reduced to B.

At C (which corresponds to about 60 atm for carbon dioxide), all similarity to perfect behaviour is lost, for suddenly the piston slides in without any further rise in pressure: this stage is represented by the horizontal line CDE. Examination of the contents of the vessel shows that just to the left of C a liquid appears, and there are two phases separated by a sharply defined surface. As the volume is decreased from C through D to E, the amount of liquid increases. There is no additional resistance to the piston because the gas can respond by condensing. The pressure corresponding to the line CDE, when both liquid and vapour are present in equilibrium, is called the **vapour pressure** of the liquid at the temperature of the experiment.

At E, the sample is entirely liquid and the piston rests on its surface. Any further reduction of volume requires the exertion of considerable pressure, as is indicated by the sharply rising line to the left of E. Even a small reduction of volume from E to F requires a great increase in pressure.

(a) The compression factor

As a first step in making these observations quantitative we introduce the **compression factor**, Z, the ratio of the measured molar volume of a gas, $V_m = V/n$, to the molar volume of a perfect gas, V_m°, at the same pressure and temperature:

$$Z = \frac{V_m}{V_m^\circ} \qquad \textit{Definition} \quad \text{Compression factor} \qquad (1C.1)$$

Because the molar volume of a perfect gas is equal to RT/p, an equivalent expression is $Z = RT/pV_m^\circ$, which we can write as

$$pV_m = RTZ \qquad (1C.2)$$

Because for a perfect gas $Z=1$ under all conditions, deviation of Z from 1 is a measure of departure from perfect behaviour.

Some experimental values of Z are plotted in Fig. 1C.3. At very low pressures, all the gases shown have $Z \approx 1$ and behave nearly perfectly. At high pressures, all the gases have $Z > 1$, signifying that they have a larger molar volume than a perfect gas. Repulsive forces are now dominant. At intermediate pressures, most gases have $Z < 1$, indicating that the attractive forces are reducing the molar volume relative to that of a perfect gas.

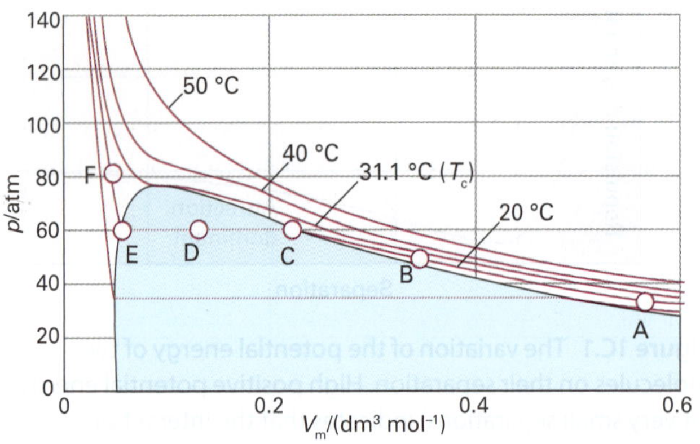

Figure 1C.2 Experimental isotherms of carbon dioxide at several temperatures. The 'critical isotherm', the isotherm at the critical temperature, is at 31.1 °C.

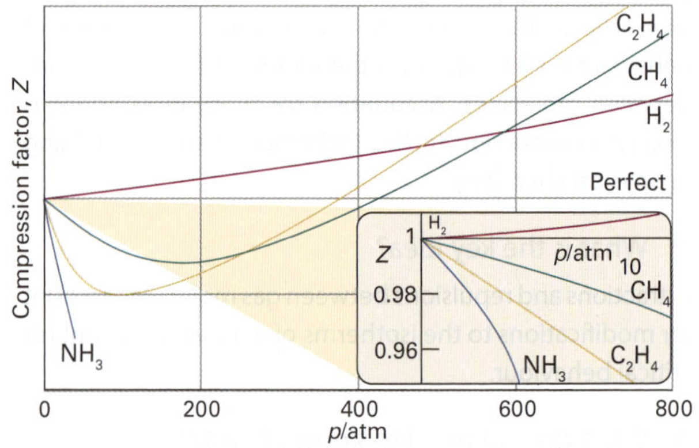

Figure 1C.3 The variation of the compression factor, Z, with pressure for several gases at 0 °C. A perfect gas has $Z=1$ at all pressures. Notice that, although the curves approach 1 as $p \to 0$, they do so with different slopes.

Brief illustration 1C.1 The compression factor

The molar volume of a perfect gas at 500 K and 100 bar is $V_m^\circ = 0.416\,\text{dm}^3\,\text{mol}^{-1}$. The molar volume of carbon dioxide under the same conditions is $V_m = 0.366\,\text{dm}^3\,\text{mol}^{-1}$. It follows that at 500 K

$$Z = \frac{0.366\,\text{dm}^3\,\text{mol}^{-1}}{0.416\,\text{dm}^3\,\text{mol}^{-1}} = 0.880$$

The fact that $Z < 1$ indicates that attractive forces dominate repulsive forces under these conditions.

Self-test 1C.1 The mean molar volume of air at 60 bar and 400 K is $0.9474\,\text{dm}^3\,\text{mol}^{-1}$. Are attractions or repulsions dominant?

Answer: Repulsions

(b) Virial coefficients

Now we relate Z to the experimental isotherms in Fig. 1C.2. At large molar volumes and high temperatures the real-gas isotherms do not differ greatly from perfect-gas isotherms. The small differences suggest that the perfect gas law $pV_m = RT$ is in fact the first term in an expression of the form

$$pV_m = RT(1 + B'p + C'p^2 + \cdots) \tag{1C.3a}$$

This expression is an example of a common procedure in physical chemistry, in which a simple law that is known to be a good first approximation (in this case $pV_m = RT$) is treated as the first term in a series in powers of a variable (in this case p). A more convenient expansion for many applications is

$$pV_m = RT\left(1 + \frac{B}{V_m} + \frac{C}{V_m^2} + \cdots\right) \quad \text{Virial equation of state} \tag{1C.3b}$$

These two expressions are two versions of the **virial equation of state.**[1] By comparing the expression with eqn 1C.2 we see that the term in parentheses in eqn 1C.3b is just the compression factor, Z.

The coefficients B, C, ..., which depend on the temperature, are the second, third, ... **virial coefficients** (Table 1C.1); the first virial coefficient is 1. The third virial coefficient, C, is usually less important than the second coefficient, B, in the sense that at typical molar volumes $C/V_m^2 \ll B/V_m$. The values of the virial coefficients of a gas are determined from measurements of its compression factor.

[1] The name comes from the Latin word for force. The coefficients are sometimes denoted B_2, B_3, ...

Table 1C.1* Second virial coefficients, $B/(\text{cm}^3\,\text{mol}^{-1})$

	Temperature	
	273 K	600 K
Ar	−21.7	11.9
CO_2	−149.7	−12.4
N_2	−10.5	21.7
Xe	−153.7	−19.6

* More values are given in the *Resource section*.

Brief illustration 1C.2 The virial equation of state

To use eqn 1C.3b (up to the B term), to calculate the pressure exerted at 100 K by 0.104 mol O_2(g) in a vessel of volume 0.225 dm^3, we begin by calculating the molar volume:

$$V_m = \frac{V}{n_{O_2}} = \frac{0.225\,\text{dm}^3}{0.104\,\text{mol}} = 2.16\,\text{dm}^3\,\text{mol}^{-1} = 2.16\times10^{-3}\,\text{m}^3\,\text{mol}^{-1}$$

Then, by using the value of B found in Table 1C.1 of the *Resource section*,

$$\begin{aligned} p &= \frac{RT}{V_m}\left(1 + \frac{B}{V_m}\right) \\ &= \frac{(8.3145\,\text{J}\,\text{mol}^{-1}\,\text{K}^{-1})\times(100\,\text{K})}{2.16\times10^{-3}\,\text{m}^3\,\text{mol}^{-1}}\left(1 - \frac{1.975\times10^{-4}\,\text{m}^3\,\text{mol}^{-1}}{2.16\times10^{-3}\,\text{m}^3\,\text{mol}^{-1}}\right) \\ &= 3.50\times10^5\,\text{Pa, or } 350\,\text{kPa} \end{aligned}$$

where we have used $1\,\text{Pa} = 1\,\text{J}\,\text{m}^{-3}$. The perfect gas equation of state would give the calculated pressure as 385 kPa, or 10 per cent higher than the value calculated by using the virial equation of state. The deviation is significant because under these conditions $B/V_m \approx 0.1$ which is not negligible relative to 1.

Self-test 1C.2 What pressure would 4.56 g of nitrogen gas in a vessel of volume 2.25 dm^3 exert at 273 K if it obeyed the virial equation of state?

Answer: 104 kPa

An important point is that although the equation of state of a real gas may coincide with the perfect gas law as $p \to 0$, not all its properties necessarily coincide with those of a perfect gas in that limit. Consider, for example, the value of dZ/dp, the slope of the graph of compression factor against pressure. For a perfect gas $dZ/dp = 0$ (because $Z = 1$ at all pressures), but for a real gas from eqn 1C.3a we obtain

$$\frac{dZ}{dp} = B' + 2pC' + \cdots \to B' \text{ as } p \to 0 \tag{1C.4a}$$

However, B' is not necessarily zero, so the slope of Z with respect to p does not necessarily approach 0 (the perfect gas

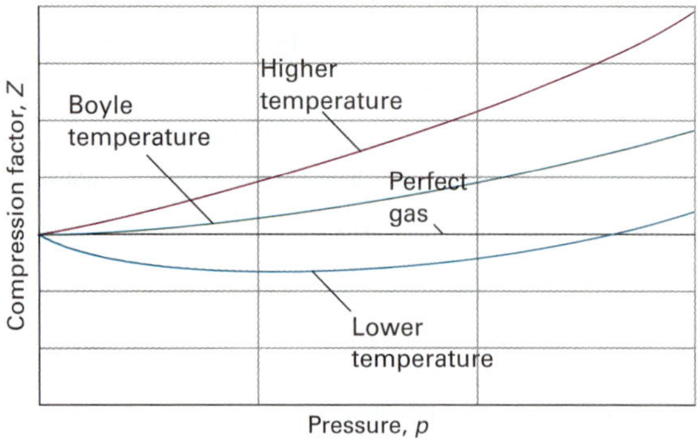

Figure 1C.4 The compression factor, Z, approaches 1 at low pressures, but does so with different slopes. For a perfect gas, the slope is zero, but real gases may have either positive or negative slopes, and the slope may vary with temperature. At the Boyle temperature, the slope is zero and the gas behaves perfectly over a wider range of conditions than at other temperatures.

value), as we can see in Fig. 1C.4. Because several physical properties of gases depend on derivatives, the properties of real gases do not always coincide with the perfect gas values at low pressures. By a similar argument,

$$\frac{dZ}{d(1/V_m)} \to B \text{ as } V_m \to \infty \qquad (1C.4b)$$

Because the virial coefficients depend on the temperature, there may be a temperature at which $Z \to 1$ with zero slope at low pressure or high molar volume (as in Fig. 1C.4). At this temperature, which is called the **Boyle temperature**, T_B, the properties of the real gas do coincide with those of a perfect gas as $p \to 0$. According to eqn 1C.4b, Z has zero slope as $p \to 0$ if $B=0$, so we can conclude that $B=0$ at the Boyle temperature. It then follows from eqn 1C.3 that $pV_m \approx RT_B$ over a more extended range of pressures than at other temperatures because the first term after 1 (that is, B/V_m) in the virial equation is zero and C/V_m^2 and higher terms are negligibly small. For helium $T_B = 22.64$ K; for air $T_B = 346.8$ K; more values are given in Table 1C.2.

(d) Critical constants

The isotherm at the temperature T_c (304.19 K, or 31.04 °C for CO_2) plays a special role in the theory of the states of matter.

Table 1C.2* Critical constants of gases

	p_c/atm	V_c/(cm^3 mol^{-1})	T_c/K	Z_c	T_B/K
Ar	48.0	75.3	150.7	0.292	411.5
CO_2	72.9	94.0	304.2	0.274	714.8
He	2.26	57.8	5.2	0.305	22.64
O_2	50.14	78.0	154.8	0.308	405.9

* More values are given in the *Resource section*.

An isotherm slightly below T_c behaves as we have already described: at a certain pressure, a liquid condenses from the gas and is distinguishable from it by the presence of a visible surface. If, however, the compression takes place at T_c itself, then a surface separating two phases does not appear and the volumes at each end of the horizontal part of the isotherm have merged to a single point, the **critical point** of the gas. The temperature, pressure, and molar volume at the critical point are called the **critical temperature**, T_c, **critical pressure**, p_c, and **critical molar volume**, V_c, of the substance. Collectively, p_c, V_c, and T_c are the **critical constants** of a substance (Table 1C.2).

At and above T_c, the sample has a single phase which occupies the entire volume of the container. Such a phase is, by definition, a gas. Hence, the liquid phase of a substance does not form above the critical temperature. The single phase that fills the entire volume when $T > T_c$ may be much denser that we normally consider typical of gases, and the name **supercritical fluid** is preferred.

Brief illustration 1C.3 The critical temperature

The critical temperature of oxygen signifies that it is impossible to produce liquid oxygen by compression alone if its temperature is greater than 155 K. To liquefy oxygen—to obtain a fluid phase that does not occupy the entire volume—the temperature must first be lowered to below 155 K, and then the gas compressed isothermally.

Self-test 1C.3 Under which conditions can liquid nitrogen be formed by the application of pressure?

Answer: At $T < 126$ K

1C.2 The van der Waals equation

We can draw conclusions from the virial equations of state only by inserting specific values of the coefficients. It is often useful to have a broader, if less precise, view of all gases. Therefore, we introduce the approximate equation of state suggested by J.D. van der Waals in 1873. This equation is an excellent example of an expression that can be obtained by thinking scientifically about a mathematically complicated but physically simple problem; that is, it is a good example of 'model building'.

(a) Formulation of the equation

The **van der Waals equation** is

$$p = \frac{nRT}{V - nb} - a\frac{n^2}{V^2} \qquad \text{Van der Waals equation of state} \qquad (1C.5a)$$

and a derivation is given in the following *Justification*. The equation is often written in terms of the molar volume $V_m = V/n$ as

$$p = \frac{RT}{V_m - b} - \frac{a}{V_m^2} \qquad (1C.5b)$$

The constants a and b are called the **van der Waals coefficients**. As can be understood from the following *Justification*, a represents the strength of attractive interactions and b that of the repulsive interactions between the molecules. They are characteristic of each gas but independent of the temperature (Table 1C.3). Although a and b are not precisely defined molecular properties, they correlate with physical properties such as critical temperature, vapour pressure, and enthalpy of vaporization that reflect the strength of intermolecular interactions. Correlations have also been sought where intermolecular forces might play a role. For example, the potency of certain general anaesthetics shows a correlation in the sense that a higher activity is observed with lower values of a (Fig. 1C.5).

Table 1C.3* van der Waals coefficients

	$a/(\text{atm dm}^6\,\text{mol}^{-2})$	$b/(10^{-2}\,\text{dm}^3\,\text{mol}^{-1})$
Ar	1.337	3.20
CO_2	3.610	4.29
He	0.0341	2.38
Xe	4.137	5.16

* More values are given in the *Resource section*.

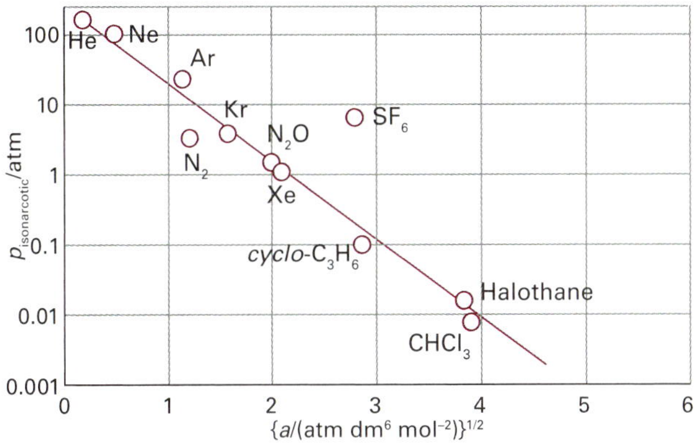

Figure 1C.5 The correlation of the effectiveness of a gas as an anaesthetic and the van der Waals parameter a. (Based on R.J. Wulf and R.M. Featherstone, *Anesthesiology* **18**, 97 (1957).) The isonarcotic pressure is the pressure required to bring about the same degree of anaesthesia.

Justification 1C.1 The van der Waals equation of state

The repulsive interactions between molecules are taken into account by supposing that they cause the molecules to behave as small but impenetrable spheres. The non-zero volume of the molecules implies that instead of moving in a volume V they are restricted to a smaller volume $V - nb$, where nb is approximately the total volume taken up by the molecules themselves. This argument suggests that the perfect gas law $p = nRT/V$ should be replaced by

$$p = \frac{nRT}{V - nb}$$

when repulsions are significant. To calculate the excluded volume we note that the closest distance of two hard-sphere molecules of radius r, and volume $V_{\text{molecule}} = \frac{4}{3}\pi r^3$, is $2r$, so the volume excluded is $\frac{4}{3}\pi(2r)^3$ or $8V_{\text{molecule}}$. The volume excluded per molecule is one-half this volume, or $4V_{\text{molecule}}$, so $b \approx 4V_{\text{molecule}}N_A$.

The pressure depends on both the frequency of collisions with the walls and the force of each collision. Both the frequency of the collisions and their force are reduced by the attractive interaction, which act with a strength proportional to the molar concentration, n/V, of molecules in the sample. Therefore, because both the frequency and the force of the collisions are reduced by the attractive interactions, the pressure is reduced in proportion to the square of this concentration. If the reduction of pressure is written as $a(n/V)^2$, where a is a positive constant characteristic of each gas, the combined effect of the repulsive and attractive forces is the van der Waals equation of state as expressed in eqn 1C.5.

In this *Justification* we have built the van der Waals equation using vague arguments about the volumes of molecules and the effects of forces. The equation can be derived in other ways, but the present method has the advantage that it shows how to derive the form of an equation out of general ideas. The derivation also has the advantage of keeping imprecise the significance of the coefficients a and b: they are much better regarded as empirical parameters that represent attractions and repulsions, respectively, rather than as precisely defined molecular properties.

Example 1C.1 Using the van der Waals equation to estimate a molar volume

Estimate the molar volume of CO_2 at 500 K and 100 atm by treating it as a van der Waals gas.

Method We need to find an expression for the molar volume by solving the van der Waals equation, eqn 1C.5b. To do so, we multiply both sides of the equation by $(V_m - b)V_m^2$, to obtain

$$(V_m - b)V_m^2 p = RTV_m^2 - (V_m - b)a$$

Then, after division by p, collect powers of V_m to obtain

$$V_m^3 - \left(b + \frac{RT}{p}\right)V_m^2 + \left(\frac{a}{p}\right)V_m = 0$$

Although closed expressions for the roots of a cubic equation can be given, they are very complicated. Unless analytical solutions are essential, it is usually more expedient to solve such equations with commercial software; graphing calculators can also be used to help identify the acceptable root.

Answer According to Table 1C.3, $a = 3.592\,\mathrm{dm^6\,atm\,mol^{-2}}$ and $b = 4.267\times10^{-2}\,\mathrm{dm^3\,mol^{-1}}$. Under the stated conditions, $RT/p = 0.410\,\mathrm{dm^3\,mol^{-1}}$. The coefficients in the equation for V_m are therefore

$$b + RT/p = 0.453\,\mathrm{dm^3\,mol^{-1}}$$
$$a/p = 3.61\times10^{-2}(\mathrm{dm^3\,mol^{-1}})^2$$
$$ab/p = 1.55\times10^{-3}(\mathrm{dm^3\,mol^{-1}})^3$$

Therefore, on writing $x = V_m/(\mathrm{dm^3\,mol^{-1}})$, the equation to solve is

$$x^3 - 0.453x^2 + (3.61\times10^{-2})x - (1.55\times10^{-3}) = 0$$

The acceptable root is $x = 0.366$ (Fig. 1C.6), which implies that $V_m = 0.366\,\mathrm{dm^3\,mol^{-1}}$. For a perfect gas under these conditions, the molar volume is $0.410\,\mathrm{dm^3\,mol^{-1}}$.

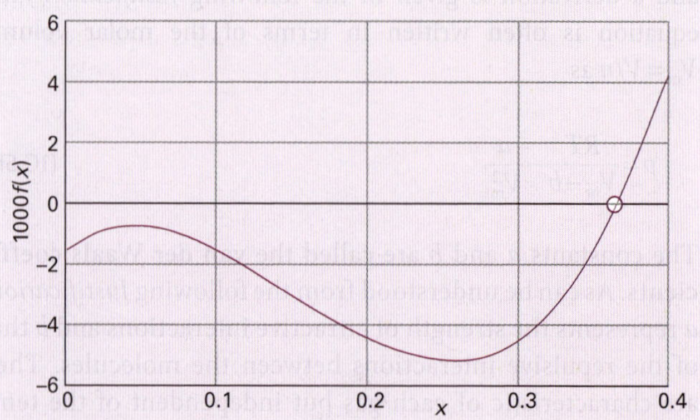

Figure 1C.6 The graphical solution of the cubic equation for V in *Example* 1C.1.

Self-test 1C.4 Calculate the molar volume of argon at 100 °C and 100 atm on the assumption that it is a van der Waals gas.

Answer: $0.298\,\mathrm{dm^3\,mol^{-1}}$

(b) The features of the equation

We now examine to what extent the van der Waals equation predicts the behaviour of real gases. It is too optimistic to expect a single, simple expression to be the true equation of state of all substances, and accurate work on gases must resort to the virial equation, use tabulated values of the coefficients at various temperatures, and analyse the systems numerically. The advantage of the van der Waals equation, however, is that it is analytical (that is, expressed symbolically) and allows us to draw some general conclusions about real gases. When the equation fails we must use one of the other equations of state that have been proposed (some are listed in Table 1C.4), invent a new one, or go back to the virial equation.

Table 1C.4 Selected equations of state

	Equation	Reduced form*	Critical constants		
			p_c	V_c	T_c
Perfect gas	$p=\dfrac{nRT}{V}$				
van der Waals	$p=\dfrac{nRT}{V-nb}-\dfrac{n^2a}{V^2}$	$p_r=\dfrac{8T_r}{3V_r-1}-\dfrac{3}{V_r^2}$	$\dfrac{a}{27b^2}$	$3b$	$\dfrac{8a}{27bR}$
Berthelot	$p=\dfrac{nRT}{V-nb}-\dfrac{n^2a}{TV^2}$	$p_r=\dfrac{8T_r}{3V_r-1}-\dfrac{3}{T_rV_r^2}$	$\dfrac{1}{12}\left(\dfrac{2aR}{3b^3}\right)^{1/2}$	$3b$	$\dfrac{2}{3}\left(\dfrac{2a}{3bR}\right)^{1/2}$
Dieterici	$p=\dfrac{nRTe^{-aRTV/n}}{V-nb}$	$p_r=\dfrac{T_re^{2(1-1/T_rV_r)}}{2V_r-1}$	$\dfrac{a}{4e^2b^2}$	$2b$	$\dfrac{a}{4bR}$
Virial	$p=\dfrac{nRT}{V}\left\{1+\dfrac{nB(T)}{V}+\dfrac{n^2C(T)}{V^2}+\cdots\right\}$				

* Reduced variables are defined in Section 1C.2(c). Equations of state are sometimes expressed in terms of the molar volume, $V_m = V/n$.

That having been said, we can begin to judge the reliability of the equation by comparing the isotherms it predicts with the experimental isotherms in Fig. 1C.2. Some calculated isotherms are shown in Fig. 1C.7 and Fig. 1C.8. Apart from the oscillations below the critical temperature, they do resemble experimental isotherms quite well. The oscillations, the **van der Waals' loops**, are unrealistic because they suggest that under some conditions an increase of pressure results in an increase of volume. Therefore they are replaced by horizontal lines drawn so the loops define equal areas above and below the lines: this procedure is called the **Maxwell construction** (**1**). The van der Waals coefficients, such as those in Table 1C.3, are found by fitting the calculated curves to the experimental curves.

1

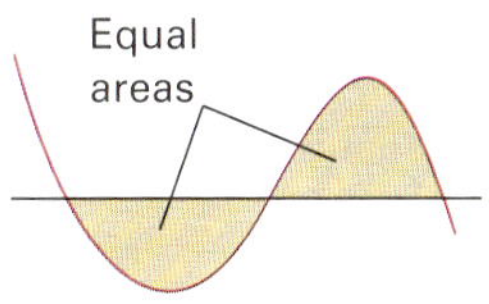

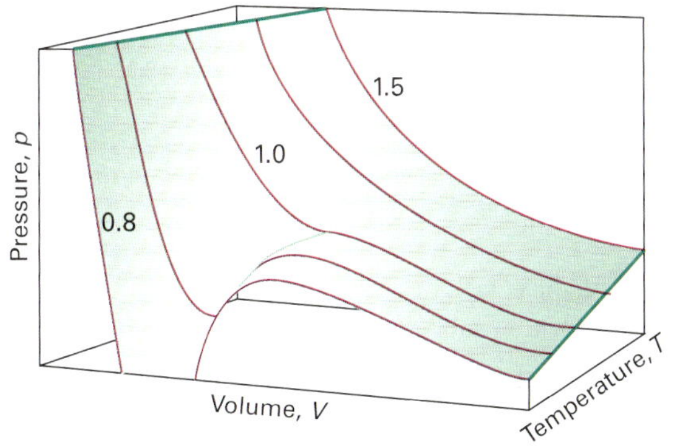

Figure 1C.7 The surface of possible states allowed by the van der Waals equation. Compare this surface with that shown in Fig. 1C.8.

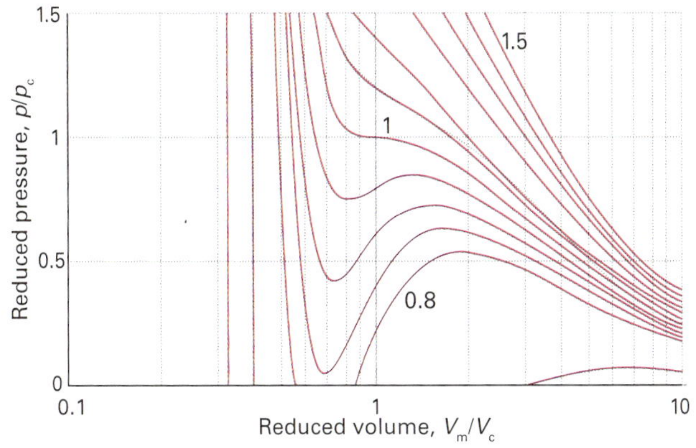

Figure 1C.8 van der Waals isotherms at several values of T/T_c. Compare these curves with those in Fig. 1C.2. The van der Waals loops are normally replaced by horizontal straight lines. The critical isotherm is the isotherm for $T/T_c = 1$.

The principal features of the van der Waals equation can be summarized as follows.

1. Perfect gas isotherms are obtained at high temperatures and large molar volumes.

When the temperature is high, RT may be so large that the first term in eqn 1C.5b greatly exceeds the second. Furthermore, if the molar volume is large in the sense $V_m \gg b$, then the denominator $V_m - b \approx V_m$. Under these conditions, the equation reduces to $p = RT/V_m$, the perfect gas equation.

2. Liquids and gases coexist when the attractive and repulsive effects are in balance.

The van der Waals loops occur when both terms in eqn 1C.5b have similar magnitudes. The first term arises from the kinetic energy of the molecules and their repulsive interactions; the second represents the effect of the attractive interactions.

3. The critical constants are related to the van der Waals coefficients.

For $T < T_c$, the calculated isotherms oscillate, and each one passes through a minimum followed by a maximum. These extrema converge as $T \to T_c$ and coincide at $T = T_c$; at the critical point the curve has a flat inflexion (**2**). From the properties of curves, we know that an inflexion of this type occurs when both the first and second derivatives are zero. Hence, we can find the critical constants by calculating these derivatives and setting them equal to zero at the critical point:

2

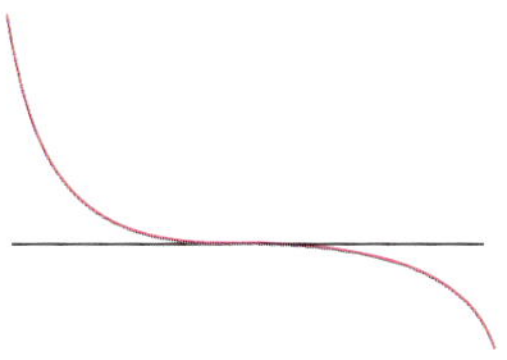

$$\frac{\mathrm{d}p}{\mathrm{d}V_m} = -\frac{RT}{(V_m - b)^2} + \frac{2a}{V_m^3} = 0$$

$$\frac{\mathrm{d}^2 p}{\mathrm{d}V_m^2} = \frac{2RT}{(V_m - b)^3} - \frac{6a}{V_m^4} = 0$$

The solutions of these two equations (and using eqn 1C.5b to calculate p_c from V_c and T_c) are

$$V_c = 3b \qquad p_c = \frac{a}{27b^2} \qquad T_c = \frac{8a}{27Rb} \tag{1C.6}$$

These relations provide an alternative route to the determination of a and b from the values of the critical constants. They can be tested by noting that the **critical compression factor**, Z_c, is predicted to be equal to

$$Z_c = \frac{p_c V_c}{RT_c} = \frac{3}{8} \tag{1C.7}$$

for all gases that are described by the van der Waals equation near the critical point. We see from Table 1C.2 that although $Z_c < \frac{3}{8} = 0.375$, it is approximately constant (at 0.3) and the discrepancy is reasonably small.

Brief illustration 1C.4 Criteria for perfect gas behaviour

For benzene $a = 18.57$ atm dm^6 mol^{-2} (1.882 Pa m^6 mol^{-2}) and $b = 0.1193$ dm^3 mol^{-1} (1.193×10^{-4} m^3 mol^{-1}); its normal boiling point is 353 K. Treated as a perfect gas at $T = 400$ K and $p = 1.0$ atm, benzene vapour has a molar volume of $V_m = RT/p = 33$ dm mol^{-1}, so the criterion $V_m \gg b$ for perfect gas behaviour is satisfied. It follows that $a/V_m^2 \approx 0.017$ atm, which is 1.7 per cent of 1.0 atm. Therefore, we can expect benzene vapour to deviate only slightly from perfect gas behaviour at this temperature and pressure.

Self-test 1C.5 Can argon gas be treated as a perfect gas at 400 K and 3.0 atm?

Answer: Yes

(c) The principle of corresponding states

An important general technique in science for comparing the properties of objects is to choose a related fundamental property of the same kind and to set up a relative scale on that basis. We have seen that the critical constants are characteristic properties of gases, so it may be that a scale can be set up by using them as yardsticks. We therefore introduce the dimensionless **reduced variables** of a gas by dividing the actual variable by the corresponding critical constant:

$$V_r = \frac{V_m}{V_c} \quad p_r = \frac{p}{p_c} \quad T_r = \frac{T}{T_c} \qquad \text{Definition} \quad \text{Reduced variables} \qquad (1C.8)$$

If the reduced pressure of a gas is given, we can easily calculate its actual pressure by using $p = p_r p_c$, and likewise for the volume and temperature. van der Waals, who first tried this procedure, hoped that gases confined to the same reduced volume, V_r, at the same reduced temperature, T_r, would exert the same reduced pressure, p_r. The hope was largely fulfilled (Fig. 1C.9). The illustration shows the dependence of the compression factor on the reduced pressure for a variety of gases at various reduced temperatures. The success of the procedure is strikingly clear: compare this graph with Fig. 1C.3, where similar data are plotted without using reduced variables. The observation that real gases at the same reduced volume and reduced temperature exert the same reduced pressure is called the **principle of corresponding states**. The principle is only an approximation. It works best for gases composed of spherical molecules; it fails, sometimes badly, when the molecules are non-spherical or polar.

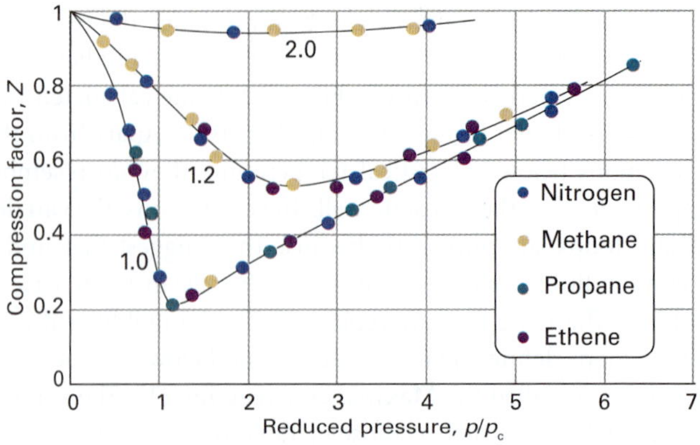

Figure 1C.9 The compression factors of four of the gases shown in Fig. 1C.3 plotted using reduced variables. The curves are labelled with the reduced temperature $T_r = T/T_c$. The use of reduced variables organizes the data on to single curves.

Brief illustration 1C.5 Corresponding states

The critical constants of argon and carbon dioxide are given in Table 1C.2. Suppose argon is at 23 atm and 200 K, its reduced pressure and temperature are then

$$p_r = \frac{23\,\text{atm}}{48.0\,\text{atm}} = 0.48 \qquad T_r = \frac{200\,\text{K}}{150.7\,\text{K}} = 1.33$$

For carbon dioxide to be in a corresponding state, its pressure and temperature would need to be

$$p = 0.48\times(72.9\,\text{atm}) = 35\,\text{atm} \qquad T = 1.33\times304.2\,\text{K} = 405\,\text{K}$$

Self-test 1C.6 What would be the corresponding state of ammonia?

Answer: 53 atm, 539 K

The van der Waals equation sheds some light on the principle. First, we express eqn 1C.5b in terms of the reduced variables, which gives

$$p_r p_c = \frac{RT_rT_c}{V_rV_c - b} - \frac{a}{V_r^2V_c^2}$$

Then we express the critical constants in terms of a and b by using eqn 1C.8:

$$\frac{ap_r}{27b^2} = \frac{8aT_r/27b}{3bpV_r - b} - \frac{a}{9b^2V_r^2}$$

which can be reorganized into

$$p_r = \frac{8T_r}{3V_r - 1} - \frac{3}{V_r^2} \qquad (1C.9)$$

This equation has the same form as the original, but the coefficients a and b, which differ from gas to gas, have disappeared. It follows that if the isotherms are plotted in terms of the reduced variables (as we did in fact in Fig. 1C.8 without drawing attention to the fact), then the same curves are obtained whatever the gas. This is precisely the content of the principle of corresponding states, so the van der Waals equation is compatible with it.

Looking for too much significance in this apparent triumph is mistaken, because other equations of state also accommodate the principle (like those in Table 1C.4). In fact, all we need are two parameters playing the roles of a and b, for then the equation can always be manipulated into reduced form. The observation that real gases obey the principle approximately amounts to saying that the effects of the attractive and repulsive interactions can each be approximated in terms of a single parameter. The importance of the principle is then not so much its theoretical interpretation but the way that it enables the properties of a range of gases to be coordinated on to a single diagram (for example, Fig. 1C.9 instead of Fig. 1C.3).

Checklist of concepts

☐ 1. The extent of deviations from perfect behaviour is summarized by introducing the **compression factor**.

☐ 2. The **virial equation** is an empirical extension of the perfect gas equation that summarizes the behaviour of real gases over a range of conditions.

☐ 3. The isotherms of a real gas introduce the concepts of **vapour pressure** and **critical behaviour**.

☐ 4. A gas can be liquefied by pressure alone only if its temperature is at or below its **critical temperature**.

☐ 5. The **van der Waals equation** is a model equation of state for a real gas expressed in terms of two parameters, one (a) corresponding to molecular attractions and the other (b) to molecular repulsions.

☐ 6. The van der Waals equation captures the general features of the behaviour of real gases, including their critical behaviour.

☐ 7. The properties of real gases are coordinated by expressing their equations of state in terms of **reduced variables**.

Checklist of equations

Property	Equation	Comment	Equation number
Compression factor	$Z = V_m / V_m^\circ$	Definition	1C.1
Virial equation of state	$pV_m = RT(1 + B/V_m + C/V_m^2 + \cdots)$	B, C depend on temperature	1C.3
van der Waals equation of state	$p = nRT/(V - nb) - a(n/V)^2$	a parameterizes attractions, b parameterizes repulsions	1C.5
Reduced variables	$X_r = X/X_c$	$X = p$, V_m, or T	1C.8

CHAPTER 1 The properties of gases

TOPIC 1A The perfect gas

Discussion questions

1A.1 Explain how the perfect gas equation of state arises by combination of Boyle's law, Charles's law, and Avogadro's principle.

1A.2 Explain the term 'partial pressure' and explain why Dalton's law is a limiting law.

Exercises

1A.1(a) Could 131 g of xenon gas in a vessel of volume 1.0 dm³ exert a pressure of 20 atm at 25 °C if it behaved as a perfect gas? If not, what pressure would it exert?

1A.1(b) Could 25 g of argon gas in a vessel of volume 1.5 dm³ exert a pressure of 2.0 bar at 30 °C if it behaved as a perfect gas? If not, what pressure would it exert?

1A.2(a) A perfect gas undergoes isothermal compression, which reduces its volume by 2.20 dm³. The final pressure and volume of the gas are 5.04 bar and 4.65 dm³, respectively. Calculate the original pressure of the gas in (i) bar, (ii) atm.

1A.2(b) A perfect gas undergoes isothermal compression, which reduces its volume by 1.80 dm³. The final pressure and volume of the gas are 1.97 bar and 2.14 dm³, respectively. Calculate the original pressure of the gas in (i) bar, (ii) torr.

1A.3(a) A car tyre (i.e. an automobile tire) was inflated to a pressure of 24 lb in^{-2} (1.00 atm = 14.7 lb in^{-2}) on a winter's day when the temperature was −5 °C. What pressure will be found, assuming no leaks have occurred and that the volume is constant, on a subsequent summer's day when the temperature is 35 °C? What complications should be taken into account in practice?

1A.3(b) A sample of hydrogen gas was found to have a pressure of 125 kPa when the temperature was 23 °C. What can its pressure be expected to be when the temperature is 11 °C?

1A.4(a) A sample of 255 mg of neon occupies 3.00 dm³ at 122 K. Use the perfect gas law to calculate the pressure of the gas.

1A.4(b) A homeowner uses 4.00×10^3 m³ of natural gas in a year to heat a home. Assume that natural gas is all methane, CH_4, and that methane is a perfect gas for the conditions of this problem, which are 1.00 atm and 20 °C. What is the mass of gas used?

1A.5(a) A diving bell has an air space of 3.0 m³ when on the deck of a boat. What is the volume of the air space when the bell has been lowered to a depth of 50 m? Take the mean density of sea water to be 1.025 g cm^{-3} and assume that the temperature is the same as on the surface.

1A.5(b) What pressure difference must be generated across the length of a 15 cm vertical drinking straw in order to drink a water-like liquid of density 1.0 g cm^{-3}?

1A.6(a) A manometer consists of a U-shaped tube containing a liquid. One side is connected to the apparatus and the other is open to the atmosphere. The pressure inside the apparatus is then determined from the difference in heights of the liquid. Suppose the liquid is water, the external pressure is 770 Torr, and the open side is 10.0 cm lower than the side connected to the apparatus. What is the pressure in the apparatus? (The density of water at 25 °C is 0.997 07 g cm^{-3}.)

1.A6(b) A manometer like that described in Exercise 1.6(a) contained mercury in place of water. Suppose the external pressure is 760 Torr, and the open side is 10.0 cm higher than the side connected to the apparatus. What is the pressure in the apparatus? (The density of mercury at 25 °C is 13.55 g cm^{-3}.)

1A.7(a) In an attempt to determine an accurate value of the gas constant, R, a student heated a container of volume 20.000 dm³ filled with 0.251 32 g of helium gas to 500 °C and measured the pressure as 206.402 cm of water in a manometer at 25 °C. Calculate the value of R from these data. (The density of water at 25 °C is 0.997 07 g cm^{-3}; the construction of a manometer is described in Exercise 1.6(a).)

1A.7(b) The following data have been obtained for oxygen gas at 273.15 K. Calculate the best value of the gas constant R from them and the best value of the molar mass of O_2.

p/atm	0.750 000	0.500 000	0.250 000
V_m/(dm³ mol^{-1})	29.9649	44.8090	89.6384

1A.8(a) At 500 °C and 93.2 kPa, the mass density of sulfur vapour is 3.710 kg m^{-3}. What is the molecular formula of sulfur under these conditions?

1A.8(b) At 100 °C and 16.0 kPa, the mass density of phosphorus vapour is 0.6388 kg m^{-3}. What is the molecular formula of phosphorus under these conditions?

1A.9(a) Calculate the mass of water vapour present in a room of volume 400 m³ that contains air at 27 °C on a day when the relative humidity is 60 per cent.

1A.9(b) Calculate the mass of water vapour present in a room of volume 250 m³ that contains air at 23 °C on a day when the relative humidity is 53 per cent.

1A.10(a) Given that the density of air at 0.987 bar and 27 °C is 1.146 kg m^{-3}, calculate the mole fraction and partial pressure of nitrogen and oxygen assuming that (i) air consists only of these two gases, (ii) air also contains 1.0 mole per cent Ar.

1A.10(b) A gas mixture consists of 320 mg of methane, 175 mg of argon, and 225 mg of neon. The partial pressure of neon at 300 K is 8.87 kPa. Calculate (i) the volume and (ii) the total pressure of the mixture.

1A.11(a) The density of a gaseous compound was found to be 1.23 kg m^{-3} at 330 K and 20 kPa. What is the molar mass of the compound?

1A.11(b) In an experiment to measure the molar mass of a gas, 250 cm³ of the gas was confined in a glass vessel. The pressure was 152 Torr at 298 K, and after correcting for buoyancy effects, the mass of the gas was 33.5 mg. What is the molar mass of the gas?

1A.12(a) The densities of air at −85 °C, 0 °C, and 100 °C are 1.877 g dm^{-3}, 1.294 g dm^{-3}, and 0.946 g dm^{-3}, respectively. From these data, and assuming that air obeys Charles's law, determine a value for the absolute zero of temperature in degrees Celsius.

1A.12(b) A certain sample of a gas has a volume of 20.00 dm³ at 0 °C and 1.000 atm. A plot of the experimental data of its volume against the Celsius temperature, θ, at constant p, gives a straight line of slope 0.0741 dm³ °C^{-1}. From these data determine the absolute zero of temperature in degrees Celsius.

1A.13(a) A vessel of volume 22.4 dm^3 contains 2.0 mol H_2 and 1.0 mol N_2 at 273.15 K. Calculate (i) the mole fractions of each component, (ii) their partial pressures, and (iii) their total pressure.

1A.13(b) A vessel of volume 22.4 dm^3 contains 1.5 mol H_2 and 2.5 mol N_2 at 273.15 K. Calculate (i) the mole fractions of each component, (ii) their partial pressures, and (iii) their total pressure.

Problems

1A.1 Recent communication with the inhabitants of Neptune have revealed that they have a Celsius-type temperature scale, but based on the melting point (0 °N) and boiling point (100 °N) of their most common substance, hydrogen. Further communications have revealed that the Neptunians know about perfect gas behaviour and they find that in the limit of zero pressure, the value of pV is 28 dm^3 atm at 0 °N and 40 dm^3 atm at 100 °N. What is the value of the absolute zero of temperature on their temperature scale?

1A.2 Deduce the relation between the pressure and mass density, ρ, of a perfect gas of molar mass M. Confirm graphically, using the following data on dimethyl ether at 25 °C, that perfect behaviour is reached at low pressures and find the molar mass of the gas.

p/kPa	12.223	25.20	36.97	60.37	85.23	101.3
$\rho/(\text{kg m}^{-3})$	0.225	0.456	0.664	1.062	1.468	1.734

1A.3 Charles's law is sometimes expressed in the form $V = V_0(1+\alpha\theta)$, where θ is the Celsius temperature, is a constant, and V_0 is the volume of the sample at 0 °C. The following values for have been reported for nitrogen at 0 °C:

p/Torr	749.7	599.6	333.1	98.6
$10^3\alpha/°\text{C}^{-1}$	3.6717	3.6697	3.6665	3.6643

For these data calculate the best value for the absolute zero of temperature on the Celsius scale.

1A.4 The molar mass of a newly synthesized fluorocarbon was measured in a gas microbalance. This device consists of a glass bulb forming one end of a beam, the whole surrounded by a closed container. The beam is pivoted, and the balance point is attained by raising the pressure of gas in the container, so increasing the buoyancy of the enclosed bulb. In one experiment, the balance point was reached when the fluorocarbon pressure was 327.10 Torr; for the same setting of the pivot, a balance was reached when CHF_3 (M = 70.014 g mol^{-1}) was introduced at 423.22 Torr. A repeat of the experiment with a different setting of the pivot required a pressure of 293.22 Torr of the fluorocarbon and 427.22 Torr of the CHF_3. What is the molar mass of the fluorocarbon? Suggest a molecular formula.

1A.5 A constant-volume perfect gas thermometer indicates a pressure of 6.69 kPa at the triple point temperature of water (273.16 K). (a) What change of pressure indicates a change of 1.00 K at this temperature? (b) What pressure indicates a temperature of 100.00 °C? (c) What change of pressure indicates a change of 1.00 K at the latter temperature?

1A.6 A vessel of volume 22.4 dm^3 contains 2.0 mol H_2 and 1.0 mol N_2 at 273.15 K initially. All the H_2 reacted with sufficient N_2 to form NH_3. Calculate the partial pressures and the total pressure of the final mixture.

1A.7 Atmospheric pollution is a problem that has received much attention. Not all pollution, however, is from industrial sources. Volcanic eruptions can be a significant source of air pollution. The Kilauea volcano in Hawaii emits 200–300 t of SO_2 per day. If this gas is emitted at 800 °C and 1.0 atm, what volume of gas is emitted?

1A.8 Ozone is a trace atmospheric gas which plays an important role in screening the Earth from harmful ultraviolet radiation, and the abundance of ozone is commonly reported in *Dobson units*. One Dobson unit is the thickness, in thousandths of a centimetre, of a column of gas if it were collected as a pure gas at 1.00 atm and 0 °C. What amount of O_3 (in moles) is found in a column of atmosphere with a cross-sectional area of 1.00 dm^2 if the abundance is 250 Dobson units (a typical mid-latitude value)? In the seasonal Antarctic ozone hole, the column abundance drops below 100 Dobson units; how many moles of O_3 are found in such a column of air above a 1.00 dm^2 area? Most atmospheric ozone is found between 10 and 50 km above the surface of the earth. If that ozone is spread uniformly through this portion of the atmosphere, what is the average molar concentration corresponding to (a) 250 Dobson units, (b) 100 Dobson units?

1A.9 The barometric formula (see *Impact* 1.1) relates the pressure of a gas of molar mass M at an altitude h to its pressure p_0 at sea level. Derive this relation by showing that the change in pressure dp for an infinitesimal change in altitude dh where the density is ρ is $dp = -\rho g dh$. Remember that ρ depends on the pressure. Evaluate (a) the pressure difference between the top and bottom of a laboratory vessel of height 15 cm, and (b) the external atmospheric pressure at a typical cruising altitude of an aircraft (11 km) when the pressure at ground level is 1.0 atm.

1A.10 Balloons are still used to deploy sensors that monitor meteorological phenomena and the chemistry of the atmosphere. It is possible to investigate some of the technicalities of ballooning by using the perfect gas law. Suppose your balloon has a radius of 3.0 m and that it is spherical. (a) What amount of H_2 (in moles) is needed to inflate it to 1.0 atm in an ambient temperature of 25 °C at sea level? (b) What mass can the balloon lift at sea level, where the density of air is 1.22 kg m^{-3}? (c) What would be the payload if He were used instead of H_2?

1A.11‡ The preceding problem is most readily solved with the use of Archimedes principle, which states that the lifting force is equal to the difference between the weight of the displaced air and the weight of the balloon. Prove Archimedes principle for the atmosphere from the barometric formula. *Hint*: Assume a simple shape for the balloon, perhaps a right circular cylinder of cross-sectional area A and height h.

1A.12‡ Chlorofluorocarbons such as CCl_3F and CCl_2F_2 have been linked to ozone depletion in Antarctica. As of 1994, these gases were found in quantities of 261 and 509 parts per trillion (10^{12}) by volume (World Resources Institute, *World resources* 1996 –97). Compute the molar concentration of these gases under conditions typical of (a) the mid-latitude troposphere (10 °C and 1.0 atm) and (b) the Antarctic stratosphere (200 K and 0.050 atm).

1A.13‡ The composition of the atmosphere is approximately 80 per cent nitrogen and 20 per cent oxygen by mass. At what height above the surface of the Earth would the atmosphere become 90 per cent nitrogen and 10 per cent oxygen by mass? Assume that the temperature of the atmosphere is constant at 25 °C. What is the pressure of the atmosphere at that height?

‡ These problems were supplied by Charles Trapp and Carmen Giunta.

TOPIC 1B The kinetic model

Discussion questions

1B.1 Specify and analyse critically the assumptions that underlie the kinetic model of gases.

1B.2 Provide molecular interpretations for the dependencies of the mean free path on the temperature, pressure, and size of gas molecules.

Exercises

1B.1(a) Determine the ratios of (i) the mean speeds, (ii) the mean translational kinetic energies of H_2 molecules and Hg atoms at 20 °C.
1B.1(b) Determine the ratios of (i) the mean speeds, (ii) the mean kinetic energies of He atoms and Hg atoms at 25 °C.

1B.2(a) Calculate the root mean square speeds of H_2 and O_2 molecules at 20 °C.
1B.2(b) Calculate the root mean square speeds of CO_2 molecules and He atoms at 20 °C.

1B.3(a) Use the Maxwell–Boltzmann distribution of speeds to estimate the fraction of N_2 molecules at 400 K that have speeds in the range 200 to 210 m s^{-1}.
1B.3(b) Use the Maxwell–Boltzmann distribution of speeds to estimate the fraction of CO_2 molecules at 400 K that have speeds in the range 400 to 405 m s^{-1}.

1B.4(a) Calculate the most probable speed, the mean speed, and the mean relative speed of CO_2 molecules in air at 20 °C.
1B.4(b) Calculate the most probable speed, the mean speed, and the mean relative speed of H_2 molecules in air at 20 °C.

1B.5(a) Assume that air consists of N_2 molecules with a collision diameter of 395 pm. Calculate (i) the mean speed of the molecules, (ii) the mean free path, (iii) the collision frequency in air at 1.0 atm and 25 °C.
1B.5(b) The best laboratory vacuum pump can generate a vacuum of about 1 nTorr. At 25 °C and assuming that air consists of N_2 molecules with a collision diameter of 395 pm, calculate (i) the mean speed of the molecules, (ii) the mean free path, (iii) the collision frequency in the gas.

1B.6(a) At what pressure does the mean free path of argon at 20 °C become comparable to the diameter of a 100 cm^3 vessel that contains it? Take $\sigma = 0.36\,\text{nm}^2$.
1B.6(b) At what pressure does the mean free path of argon at 20 °C become comparable to 10 times the diameters of the atoms themselves?

1B.7(a) At an altitude of 20 km the temperature is 217 K and the pressure 0.050 atm. What is the mean free path of N_2 molecules? ($\sigma = 0.43\,\text{nm}^2$).
1B.7(b) At an altitude of 15 km the temperature is 217 K and the pressure 12.1 kPa. What is the mean free path of N_2 molecules? ($\sigma = 0.43\,\text{nm}^2$).

Problems

1B.1 A rotating slotted-disc apparatus like that in Fig. 1B.5 consists of five coaxial 5.0 cm diameter disks separated by 1.0 cm, the slots in their rims being displaced by 2.0° between neighbours. The relative intensities, I, of the detected beam of Kr atoms for two different temperatures and at a series of rotation rates were as follows:

ν/Hz	20	40	80	100	120
I (40 K)	0.846	0.513	0.069	0.015	0.002
I (100 K)	0.592	0.485	0.217	0.119	0.057

Find the distributions of molecular velocities, $f(v_x)$, at these temperatures, and check that they conform to the theoretical prediction for a one-dimensional system.

1B.2 A Knudsen cell was used to determine the vapour pressure of germanium at 1000 °C. During an interval of 7200 s the mass loss through a hole of radius 0.50 mm amounted to 43 µg. What is the vapour pressure of germanium at 1000 °C? Assume the gas to be monatomic.

1B.3 Start from the Maxwell–Boltzmann distribution and derive an expression for the most probable speed of a gas of molecules at a temperature T. Go on to demonstrate the validity of the equipartition conclusion that the average translational kinetic energy of molecules free to move in three dimensions is $\frac{3}{2}kT$.

1B.4 Consider molecules that are confined to move in a plane (a two-dimensional gas). Calculate the distribution of speeds and determine the mean speed of the molecules at a temperature T.

1B.5 A specially constructed velocity-selector accepts a beam of molecules from an oven at a temperature T but blocks the passage of molecules with a speed greater than the mean. What is the mean speed of the emerging beam, relative to the initial value, treated as a one-dimensional problem?

1B.6 What, according to the Maxwell–Boltzmann distribution, is the proportion of gas molecules having (a) more than, (b) less than the root mean square speed? (c) What are the proportions having speeds greater and smaller than the mean speed?

1B.7 Calculate the fractions of molecules in a gas that have a speed in a range Δv at the speed nv_{mp} relative to those in the same range at v_m itself? This calculation can be used to estimate the fraction of very energetic molecules (which is important for reactions). Evaluate the ratio for $n=3$ and $n=4$.

1B.8 Derive an expression for $\langle v^n \rangle^{1/n}$ from the Maxwell–Boltzmann distribution of speeds. You will need standard integrals given in the *Resource section*.

1B.9 Calculate the escape velocity (the minimum initial velocity that will take an object to infinity) from the surface of a planet of radius R. What is the value for (a) the Earth, $R = 6.37 \times 10^6$ m, $g = 9.81$ m s^{-2}, (b) Mars, $R = 3.38 \times 10^6$ m, $m_{Mars}/m_{Earth} = 0.108$. At what temperatures do H_2, He, and O_2 molecules have mean speeds equal to their escape speeds? What proportion of the molecules have enough speed to escape when the temperature is (a) 240 K, (b) 1500 K? Calculations of this kind are very important in considering the composition of planetary atmospheres.

1B.10 The principal components of the atmosphere of the Earth are diatomic molecules, which can rotate as well as translate. Given that the translational kinetic energy density of the atmosphere is 0.15 J cm^{-3}, what is the total kinetic energy density, including rotation? The average rotational energy of a linear molecule is kT.

1B.11 Plot different Maxwell–Boltzmann speed distributions by keeping the molar mass constant at 100 g mol^{-1} and varying the temperature of the sample between 200 K and 2000 K.

1B.12 Evaluate numerically the fraction of molecules with speeds in the range 100 m s^{-1} to 200 m s^{-1} at 300 K and 1000 K.

TOPIC 1C Real gases

Discussion questions

1C.1 Explain how the compression factor varies with pressure and temperature and describe how it reveals information about intermolecular interactions in real gases.

1C.2 What is the significance of the critical constants?

1C.3 Describe the formulation of the van der Waals equation and suggest a rationale for one other equation of state in Table 1C.6.

1C.4 Explain how the van der Waals equation accounts for critical behaviour.

Exercises

1C.1(a) Calculate the pressure exerted by 1.0 mol C_2H_6 behaving as a van der Waals gas when it is confined under the following conditions: (i) at 273.15 K in 22.414 dm^3, (ii) at 1000 K in 100 cm^3. Use the data in Table 1C.3.
1C.1(b) Calculate the pressure exerted by 1.0 mol H_2S behaving as a van der Waals gas when it is confined under the following conditions: (i) at 273.15 K in 22.414 dm^3, (ii) at 500 K in 150 cm^3. Use the data in Table 1C.3.

1C.2(a) Express the van der Waals parameters $a = 0.751$ atm dm^6 mol^{-2} and $b = 0.0226$ dm^3 mol^{-1} in SI base units.
1C.2(b) Express the van der Waals parameters $a = 1.32$ atm dm^6 mol^{-2} and $b = 0.0436$ dm^3 mol^{-1} in SI base units.

1C.3(a) A gas at 250 K and 15 atm has a molar volume 12 per cent smaller than that calculated from the perfect gas law. Calculate (i) the compression factor under these conditions and (ii) the molar volume of the gas. Which are dominating in the sample, the attractive or the repulsive forces?
1C.3(b) A gas at 350 K and 12 atm has a molar volume 12 per cent larger than that calculated from the perfect gas law. Calculate (i) the compression factor under these conditions and (ii) the molar volume of the gas. Which are dominating in the sample, the attractive or the repulsive forces?

1C.4(a) In an industrial process, nitrogen is heated to 500 K at a constant volume of 1.000 m^3. The gas enters the container at 300 K and 100 atm. The mass of the gas is 92.4 kg. Use the van der Waals equation to determine the approximate pressure of the gas at its working temperature of 500 K. For nitrogen, $a = 1.352$ dm^6 atm mol^{-2}, $b = 0.0387$ dm^3 mol^{-1}.
1C.4(b) Cylinders of compressed gas are typically filled to a pressure of 200 bar. For oxygen, what would be the molar volume at this pressure and 25 °C based on (i) the perfect gas equation, (ii) the van der Waals equation? For oxygen, $a = 1.364$ dm^6 atm mol^{-2}, $b = 3.19\times10^{-2}$ dm^3 mol^{-1}.

1C.5(a) Suppose that 10.0 mol C_2H_6(g) is confined to 4.860 dm^3 at 27 °C. Predict the pressure exerted by the ethane from (i) the perfect gas and (ii) the van der Waals equations of state. Calculate the compression factor based on these calculations. For ethane, $a = 5.507$ dm^6 atm mol^{-2}, $b = 0.0651$ dm^3 mol^{-1}.
1C.5(b) At 300 K and 20 atm, the compression factor of a gas is 0.86. Calculate (i) the volume occupied by 8.2 mmol of the gas under these conditions and (ii) an approximate value of the second virial coefficient B at 300 K.

1C.6(a) The critical constants of methane are $p_c = 45.6$ atm, $V_c = 98.7$ cm^3 mol^{-1}, and $T_c = 190.6$ K. Calculate the van der Waals parameters of the gas and estimate the radius of the molecules.
1C.6(b) The critical constants of ethane are $p_c = 48.20$ atm, $V_c = 148$ cm^3 mol^{-1}, and $T_c = 305.4$ K. Calculate the van der Waals parameters of the gas and estimate the radius of the molecules.

1C.7(a) Use the van der Waals parameters for chlorine in Table 1C.3 of the *Resource section* to calculate approximate values of (i) the Boyle temperature of chlorine and (ii) the radius of a Cl_2 molecule regarded as a sphere.
1C.7(b) Use the van der Waals parameters for hydrogen sulfide in Table 1C.3 of the *Resource section* to calculate approximate values of (i) the Boyle temperature of the gas and (ii) the radius of a H_2S molecule regarded as a sphere.

1C.8(a) Suggest the pressure and temperature at which 1.0 mol of (i) NH_3, (ii) Xe, (iii) He will be in states that correspond to 1.0 mol H_2 at 1.0 atm and 25 °C.
1C.8(b) Suggest the pressure and temperature at which 1.0 mol of (i) H_2S, (ii) CO_2, (iii) Ar will be in states that correspond to 1.0 mol N_2 at 1.0 atm and 25 °C.

1C.9(a) A certain gas obeys the van der Waals equation with $a = 0.50$ m^6 Pa mol^{-2}. Its volume is found to be 5.00×10^{-4} m^3 mol^{-1} at 273 K and 3.0 MPa. From this information calculate the van der Waals constant b. What is the compression factor for this gas at the prevailing temperature and pressure?
1C.9(b) A certain gas obeys the van der Waals equation with $a = 0.76$ m^6 Pa mol^{-2}. Its volume is found to be 4.00×10^{-4} m^3 mol^{-1} at 288 K and 4.0 MPa. From this information calculate the van der Waals constant b. What is the compression factor for this gas at the prevailing temperature and pressure?

Problems

1C.1 Calculate the molar volume of chlorine gas at 350 K and 2.30 atm using (a) the perfect gas law and (b) the van der Waals equation. Use the answer to (a) to calculate a first approximation to the correction term for attraction and then use successive approximations to obtain a numerical answer for part (b).

1C.2 At 273 K measurements on argon gave $B = -21.7$ cm^3 mol^{-1} and $C = 1200$ cm^6 mol^{-2}, where B and C are the second and third virial coefficients in the expansion of Z in powers of $1/V_m$. Assuming that the perfect gas law holds sufficiently well for the estimation of the second and third terms of the expansion, calculate the compression factor of argon at 100 atm and 273 K. From your result, estimate the molar volume of argon under these conditions.

1C.3 Calculate the volume occupied by 1.00 mol N_2 using the van der Waals equation in the form of a virial expansion at (a) its critical temperature, (b) its Boyle temperature, and (c) its inversion temperature. Assume that the pressure is 10 atm throughout. At what temperature is the gas most perfect? Use the following data: $T_c = 126.3$ K, $a = 1.352$ dm^6 atm mol^{-2}, $b = 0.0387$ dm^3 mol^{-1}.

1C.4‡ The second virial coefficient of methane can be approximated by the empirical equation $B(T) = a + be^{-c/T^2}$, where $a = -0.1993$ bar^{-1}, $b = 0.2002$ bar^{-1}, and $c = 1131$ K^2 with 300 K < T < 600 K. What is the Boyle temperature of methane?

1C.5 The mass density of water vapour at 327.6 atm and 776.4 K is 133.2 kg m^{-3}. Given that for water $T_c = 647.4$ K, $p_c = 21.3$ atm, $a = 5.464$ dm^6 atm mol^{-2}, $b = 0.03049$ dm^3 mol^{-1}, and $M = 18.02$ g mol^{-1}, calculate (a) the molar volume. Then calculate the compression factor (b) from the data, (c) from the virial expansion of the van der Waals equation.

1C.6 The critical volume and critical pressure of a certain gas are 160 cm^3 mol^{-1} and 40 atm, respectively. Estimate the critical temperature by assuming that the gas obeys the Berthelot equation of state. Estimate the radii of the gas molecules on the assumption that they are spheres.

1C.7 Estimate the coefficients a and b in the Dieterici equation of state from the critical constants of xenon. Calculate the pressure exerted by 1.0 mol Xe when it is confined to $1.0\,\text{dm}^3$ at 25 °C.

1C.8 Show that the van der Waals equation leads to values of $Z<1$ and $Z>1$, and identify the conditions for which these values are obtained.

1C.9 Express the van der Waals equation of state as a virial expansion in powers of $1/V_m$ and obtain expressions for B and C in terms of the parameters a and b. The expansion you will need is $(1-x)^{-1}=1+x+x^2+\ldots$. Measurements on argon gave $B=-21.7\,\text{cm}^3\,\text{mol}^{-1}$ and $C=1200\,\text{cm}^6\,\text{mol}^{-2}$ for the virial coefficients at 273 K. What are the values of a and b in the corresponding van der Waals equation of state?

1C.10‡ Derive the relation between the critical constants and the Dieterici equation parameters. Show that $Z_c=2e^{-2}$ and derive the reduced form of the Dieterici equation of state. Compare the van der Waals and Dieterici predictions of the critical compression factor. Which is closer to typical experimental values?

1C.11 A scientist proposed the following equation of state:

$$p=\frac{RT}{V_m}-\frac{B}{V_m^2}+\frac{C}{V_m^3}$$

Show that the equation leads to critical behaviour. Find the critical constants of the gas in terms of B and C and an expression for the critical compression factor.

1C.12 Equations 1C.3a and 1C.3b are expansions in p and $1/V_m$, respectively. Find the relation between B, C and B', C'.

1C.13 The second virial coefficient B' can be obtained from measurements of the density ρ of a gas at a series of pressures. Show that the graph of p/ρ against p should be a straight line with slope proportional to B'. Use the data on dimethyl ether in Problem 1A.2 to find the values of B' and B at 25 °C.

1C.14 The equation of state of a certain gas is given by $p=RT/V_m+(a+bT)/V_m^2$, where a and b are constants. Find $(\partial V/\partial T)_p$.

1C.15 The following equations of state are occasionally used for approximate calculations on gases: (gas A) $pV_m=RT(1+b/V_m)$, (gas B) $p(V_m-b)=RT$. Assuming that there were gases that actually obeyed these equations of state, would it be possible to liquefy either gas A or B? Would they have a critical temperature? Explain your answer.

1C.16 Derive an expression for the compression factor of a gas that obeys the equation of state $p(V-nb)=nRT$, where b and R are constants. If the pressure and temperature are such that $V_m=10b$, what is the numerical value of the compression factor?

1C.17‡ The discovery of the element argon by Lord Rayleigh and Sir William Ramsay had its origins in Rayleigh's measurements of the density of nitrogen with an eye toward accurate determination of its molar mass. Rayleigh prepared some samples of nitrogen by chemical reaction of nitrogen-containing compounds; under his standard conditions, a glass globe filled with this 'chemical nitrogen' had a mass of 2.2990 g. He prepared other samples by removing oxygen, carbon dioxide, and water vapor from atmospheric air; under the same conditions, this 'atmospheric nitrogen' had a mass of 2.3102 g (Lord Rayleigh, *Royal Institution Proceedings* **14**, 524 (1895)). With the hindsight of knowing accurate values for the molar masses of nitrogen and argon, compute the mole fraction of argon in the latter sample on the assumption that the former was pure nitrogen and the latter a mixture of nitrogen and argon.

1C.18‡ A substance as elementary and well known as argon still receives research attention. Stewart and Jacobsen have published a review of thermodynamic properties of argon (R.B. Stewart and R.T. Jacobsen, *J. Phys. Chem. Ref. Data* **18**, 639 (1989)) which included the following 300 K isotherm.

p/MPa	0.4000	0.5000	0.6000	0.8000	1.000
$V_m/(\text{dm}^3\,\text{mol}^{-1})$	6.2208	4.9736	4.1423	3.1031	2.4795
p/MPa	1.500	2.000	2.500	3.000	4.000
$V_m/(\text{dm}^3\,\text{mol}^{-1})$	1.6483	1.2328	0.98357	0.81746	0.60998

(a) Compute the second virial coefficient, B, at this temperature. (b) Use non-linear curve-fitting software to compute the third virial coefficient, C, at this temperature.

1C.19 Use mathematical software, a spreadsheet, or the *Living graphs* on the web site for this book to: (a) Explore how the pressure of 1.5 mol CO_2(g) varies with volume as it is compressed at (a) 273 K, (b) 373 K from $30\,\text{dm}^3$ to $15\,\text{dm}^3$. (c) Plot the data as p against $1/V$.

1C.20 Calculate the molar volume of chlorine gas on the basis of the van der Waals equation of state at 250 K and 150 kPa and calculate the percentage difference from the value predicted by the perfect gas equation.

1C.21 Is there a set of conditions at which the compression factor of a van der Waals gas passes through a minimum? If so, how does the location and value of the minimum value of Z depend on the coefficients a and b?

Mathematical background 1 Differentiation and integration

Two of the most important mathematical techniques in the physical sciences are differentiation and integration. They occur throughout the subject, and it is essential to be aware of the procedures involved.

MB1.1 Differentiation: definitions

Differentiation is concerned with the slopes of functions, such as the rate of change of a variable with time. The formal definition of the **derivative**, df/dx, of a function $f(x)$ is

$$\frac{df}{dx}=\lim_{\delta x\to 0}\frac{f(x+\delta x)-f(x)}{\delta x}$$ *Definition* First derivative (MB1.1)

As shown in Fig. MB1.1, the derivative can be interpreted as the slope of the tangent to the graph of $f(x)$. A positive first derivative indicates that the function slopes upwards (as x increases), and a negative first derivative indicates the opposite. It is sometimes convenient to denote the first derivative as $f'(x)$. The **second derivative**, d^2f/dx^2, of a function is the derivative of the first derivative (here denoted f'):

$$\frac{d^2f}{dx^2}=\lim_{\delta x\to 0}\frac{f'(x+\delta x)-f'(x)}{\delta x}$$ *Definition* Second derivative (MB1.2)

It is sometimes convenient to denote the second derivative f''. As shown in Fig. MB1.1, the second derivative of a function can be interpreted as an indication of the sharpness of the curvature of the function. A positive second derivative indicates that the function is ∪ shaped, and a negative second derivative indicates that it is ∩ shaped.

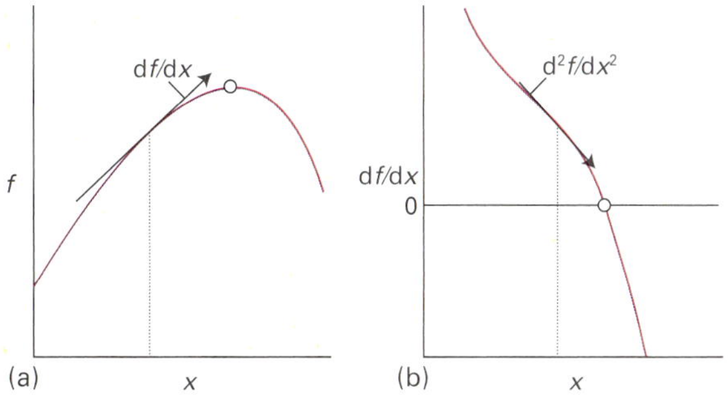

Figure MB1.1 (a) The first derivative of a function is equal to the slope of the tangent to the graph of the function at that point. The small circle indicates the extremum (in this case, maximum) of the function, where the slope is zero. (b) The second derivative of the same function is the slope of the tangent to a graph of the first derivative of the function. It can be interpreted as an indication of the curvature of the function at that point.

The derivatives of some common functions are as follows:

$$\frac{d}{dx}x^n=nx^{n-1}$$ (MB1.3a)

$$\frac{d}{dx}e^{ax}=ae^{ax}$$ (MB1.3b)

$$\frac{d}{dx}\sin ax=a\cos ax \qquad \frac{d}{dx}\cos ax=-a\sin ax$$ (MB1.3c)

$$\frac{d}{dx}\ln ax=\frac{1}{x}$$ (MB1.3d)

When a function depends on more than one variable, we need the concept of a **partial derivative**, $\partial f/\partial x$. Note the change from d to ∂: partial derivatives are dealt with at length in *Mathematical background* 2; all we need know at this stage is that they signify that all variables other than the stated variable are regarded as constant when evaluating the derivative.

Brief illustration MB1.1 Partial derivatives

Suppose we are told that f is a function of two variables, and specifically $f=4x^2y^3$. Then, to evaluate the partial derivative of f with respect to x, we regard y as a constant (just like the 4), and obtain

$$\frac{\partial f}{\partial x}=\frac{\partial}{\partial x}(4x^2y^3)=4y^3\frac{\partial}{\partial x}x^2=8xy^3$$

Similarly, to evaluate the partial derivative of f with respect to y, we regard x as a constant (again, like the 4), and obtain

$$\frac{\partial f}{\partial y}=\frac{\partial}{\partial y}(4x^2y^3)=4x^2\frac{\partial}{\partial y}y^3=12x^2y^2$$

MB1.2 Differentiation: manipulations

It follows from the definition of the derivative that a variety of combinations of functions can be differentiated by using the following rules:

$$\frac{d}{dx}(u+v)=\frac{du}{dx}+\frac{dv}{dx}$$ (MB1.4a)

$$\frac{d}{dx}uv=u\frac{dv}{dx}+v\frac{du}{dx}$$ (MB1.4b)

$$\frac{d}{dx}\frac{u}{v}=\frac{1}{v}\frac{du}{dx}-\frac{u}{v^2}\frac{dv}{dx}$$ (MB1.4c)

Brief illustration MB1.2 Derivatives

To differentiate the function $f = \sin^2 ax/x^2$ use eqn MB1.4 to write

$$\begin{aligned}\frac{d}{dx}\frac{\sin^2 ax}{x^2} &= \frac{d}{dx}\left(\frac{\sin ax}{x}\right)\left(\frac{\sin ax}{x}\right) = 2\left(\frac{\sin ax}{x}\right)\frac{d}{dx}\left(\frac{\sin ax}{x}\right)\\ &= 2\left(\frac{\sin ax}{x}\right)\left\{\frac{1}{x}\frac{d}{dx}\sin ax + \sin ax \frac{d}{dx}\frac{1}{x}\right\}\\ &= 2\left\{\frac{a}{x^2}\sin ax \cos ax - \frac{\sin^2 ax}{x^3}\right\}\end{aligned}$$

The function and this first derivative are plotted in Fig. MB1.2.

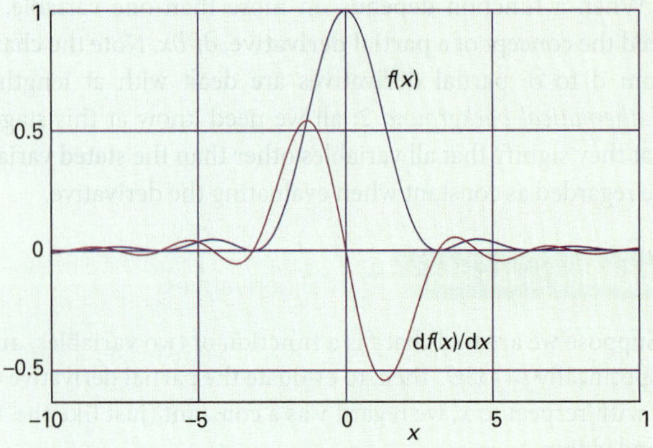

Figure MB1.2 The function considered in *Brief illustration* MB1.2 and its first derivative.

MB1.3 Series expansions

One application of differentiation is to the development of power series for functions. The **Taylor series** for a function $f(x)$ in the vicinity of $x = a$ is

$$\begin{aligned}f(x) &= f(a) + \left(\frac{df}{dx}\right)_a (x-a) + \frac{1}{2!}\left(\frac{d^2 f}{dx^2}\right)_a (x-a)^2 + \cdots\\ &= \sum_{n=0}^{\infty} \frac{1}{n!}\left(\frac{d^n f}{dx^n}\right)_a (x-a)^n \end{aligned} \qquad \text{Taylor series} \quad \text{(MB1.5)}$$

where the notation $(\ldots)_a$ means that the derivative is evaluated at $x = a$ and $n!$ denotes a **factorial** given by

$$n! = n(n-1)(n-2)\ldots 1, \quad 0! = 1 \qquad \text{Factorial} \quad \text{(MB1.6)}$$

The **Maclaurin series** for a function is a special case of the Taylor series in which $a = 0$.

Brief illustration MB1.3 Series expansion

To evaluate the expansion of $\cos x$ around $x = 0$ we note that

$$\left(\frac{d}{dx}\cos x\right)_0 = (-\sin x)_0 = 0 \qquad \left(\frac{d^2}{dx^2}\cos x\right)_0 = (-\cos x)_0 = -1$$

and in general

$$\left(\frac{d^n}{dx^n}\cos x\right)_0 = \begin{cases} 0 \text{ for } n \text{ odd} \\ (-1)^{n/2} \text{ for } n \text{ even}\end{cases}$$

Therefore,

$$\cos x = \sum_{n\text{ even}}^{\infty} \frac{(-1)^{n/2}}{n!} x^n = 1 - \tfrac{1}{2}x^2 + \tfrac{1}{24}x^4 - \cdots$$

The following Taylor series (specifically, Maclaurin series) are used at various stages in the text:

$$(1+x)^{-1} = 1 - x + x^2 - \cdots = \sum_{n=0}^{\infty} (-1)^n x^n \qquad \text{(MB1.7a)}$$

$$e^x = 1 + x + \tfrac{1}{2}x^2 + \cdots = \sum_{n=0}^{\infty} \frac{x^n}{n!} \qquad \text{(MB1.7b)}$$

$$\ln(1+x) = x - \tfrac{1}{2}x^2 + \tfrac{1}{3}x^3 - \cdots = \sum_{n=1}^{\infty} (-1)^{n+1}\frac{x^n}{n} \qquad \text{(MB1.7c)}$$

Taylor series are used to simplify calculations, for when $x \ll 1$ it is possible, to a good approximation, to terminate the series after one or two terms. Thus, provided $x \ll 1$ we can write

$$(1+x)^{-1} \approx 1 - x \qquad \text{(MB1.8a)}$$

$$e^x \approx 1 + x \qquad \text{(MB1.8b)}$$

$$\ln(1+x) \approx x \qquad \text{(MB1.8c)}$$

A series is said to **converge** if the sum approaches a finite, definite value as n approaches infinity. If the sum does not approach a finite, definite value, then the series is said to **diverge**. Thus, the series in eqn MB1.7a converges for $x < 1$ and diverges for $x \geq 1$. There are a variety of tests for convergence, which are explained in mathematics texts.

MB1.4 Integration: definitions

Integration (which formally is the inverse of differentiation) is concerned with the areas under curves. The **integral** of a

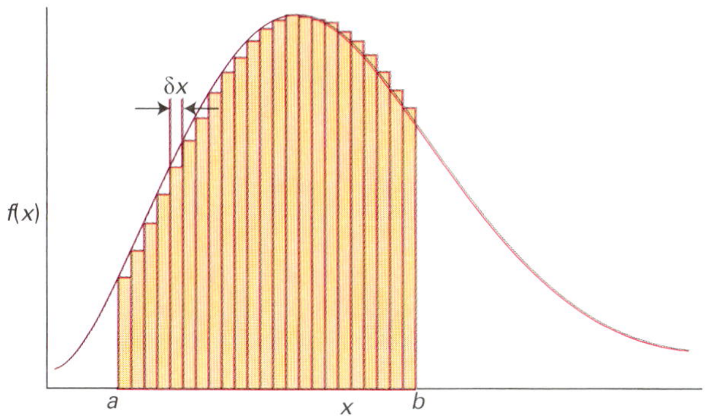

Figure MB1.3 A definite integral is evaluated by forming the product of the value of the function at each point and the increment δx, with $\delta x \to 0$, and then summing the products $f(x)\delta x$ for all values of x between the limits a and b. It follows that the value of the integral is the area under the curve between the two limits.

function $f(x)$, which is denoted $\int f\,dx$ (the symbol $\int$ is an elongated S denoting a sum), between the two values $x=a$ and $x=b$ is defined by imagining the x axis as divided into strips of width δx and evaluating the following sum:

$$\int_a^b f(x)dx = \lim_{\delta x \to 0} \sum_i f(x_i)\delta x \qquad \textit{Definition} \quad \text{Integration} \quad \text{(MB1.9)}$$

As can be appreciated from Fig. MB1.3, the integral is the area under the curve between the limits a and b. The function to be integrated is called the **integrand**. It is an astonishing mathematical fact that the integral of a function is the inverse of the differential of that function in the sense that if we differentiate f and then integrate the resulting function, then we obtain the original function f (to within a constant). The function in eqn MB1.9 with the limits specified is called a **definite integral**. If it is written without the limits specified, then we have an **indefinite integral**. If the result of carrying out an indefinite integration is $g(x)+C$, where C is a constant, the following notation is used to evaluate the corresponding definite integral:

$$I = \int_a^b f(x)dx = \{g(x)+C\}\Big|_a^b = \{g(b)+C\} - \{g(a)+C\}$$
$$= g(b) - g(a) \qquad \text{Definite integral} \quad \text{(MB1.10)}$$

Note that the constant of integration disappears. The definite and indefinite integrals encountered in this text are listed in the *Resource section*.

MB1.5 Integration: manipulations

When an indefinite integral is not in the form of one of those listed in the *Resource section* it is sometimes possible to transform it into one of the forms by using integration techniques such as:

Substitution. Introduce a variable u related to the independent variable x (for example, an algebraic relation such as $u=x^2-1$ or a trigonometric relation such as $u=\sin x$). Express the differential dx in terms of du (for these substitutions, $du=2x\,dx$ and $du=\cos x\,dx$, respectively). Then transform the original integral written in terms of x into an integral in terms of u upon which, in some cases, a standard form such as one of those listed in the *Resource section* can be used.

Brief illustration MB1.4 Integration by substitution

To evaluate the indefinite integral $\int \cos^2 x \sin x\,dx$ we make the substitution $u=\cos x$. It follows that $du/dx=-\sin x$, and therefore that $\sin x\,dx=-du$. The integral is therefore

$$\int \cos^2 x \sin x\,dx = -\int u^2\,du = -\tfrac{1}{3}u^3 + C = -\tfrac{1}{3}\cos^3 x + C$$

To evaluate the corresponding definite integral, we have to convert the limits on x into limits on u. Thus, if the limits are $x=0$ and $x=\pi$, the limits become $u=\cos 0=1$ and $u=\cos \pi=-1$:

$$\int_0^{\pi} \cos^2 x \sin x\,dx = -\int_1^{-1} u^2\,du = \{-\tfrac{1}{3}u^3 + C\}\Big|_1^{-1} = \frac{2}{3}$$

Integration by parts. For two functions $f(x)$ and $g(x)$:

$$\int f\frac{dg}{dx}dx = fg - \int g\frac{df}{dx}dx \qquad \text{Integration by parts} \quad \text{(MB1.11a)}$$

which may be abbreviated as:

$$\int f dg = fg - \int g df \qquad \text{(MB1.11b)}$$

Brief illustration MB1.5 Integration by parts

Integrals over xe^{-ax} and their analogues occur commonly in the discussion of atomic structure and spectra. They may be integrated by parts, as in the following:

$$\int_0^{\infty} \overbrace{x}^{f}\, \overbrace{e^{-ax}}^{dg/dx}\,dx = \overbrace{x}^{f}\,\overbrace{\frac{e^{-ax}}{-a}}^{g}\Bigg|_0^{\infty} - \int_0^{\infty} \overbrace{\frac{e^{-ax}}{-a}}^{g}\;\overbrace{1}^{df/dx}\,dx$$
$$= -\frac{xe^{-ax}}{a}\Bigg|_0^{\infty} + \frac{1}{a}\int_0^{\infty} e^{-ax}\,dx = 0 - \frac{e^{-ax}}{a^2}\Bigg|_0^{\infty}$$
$$= \frac{1}{a^2}$$

MB1.6 Multiple integrals

A function may depend on more than one variable, in which case we may need to integrate over both the variables:

$$I=\int_a^b\int_c^d f(x,y)\mathrm{d}x\mathrm{d}y \qquad \text{(MB1.12)}$$

We (but not everyone) adopt the convention that a and b are the limits of the variable x and c and d are the limits for y (as depicted by the colours in this instance). This procedure is simple if the function is a product of functions of each variable and of the form $f(x,y)=X(x)Y(y)$. In this case, the double integral is just a product of each integral:

$$I=\int_a^b\int_c^d X(x)Y(y)\mathrm{d}x\mathrm{d}y=\int_a^b X(x)\mathrm{d}x\int_c^d Y(y)\mathrm{d}y \qquad \text{(MB1.13)}$$

Brief illustration MB1.6 A double integral

Double integrals of the form

$$I=\int_0^{L_1}\int_0^{L_2}\sin^2(\pi x/L_1)\sin^2(\pi y/L_2)\mathrm{d}x\mathrm{d}y$$

occur in the discussion of the translational motion of a particle in two dimensions, where L_1 and L_2 are the maximum extents of travel along the x- and y-axes, respectively. To evaluate I we use eqn MB1.13 and an integral listed in the *Resource section* to write

$$\begin{aligned}I &\overset{\text{Integral T.2}}{=} \int_0^{L_1}\sin^2(\pi x/L_1)\mathrm{d}x\int_0^{L_2}\sin^2(\pi y/L_2)\mathrm{d}y\\ &=\left\{\tfrac{1}{2}x-\frac{\sin(2\pi x/L_1)}{4\pi/L_1}+C\right\}\Bigg|_0^{L_1}\left\{\tfrac{1}{2}y-\frac{\sin(2\pi y/L_2)}{4\pi/L_2}+C\right\}\Bigg|_0^{L_2}\\ &=\tfrac{1}{4}L_1L_2\end{aligned}$$

CHAPTER 2

The First Law

The release of energy can be used to provide heat when a fuel burns in a furnace, to produce mechanical work when a fuel burns in an engine, and to generate electrical work when a chemical reaction pumps electrons through a circuit. In chemistry, we encounter reactions that can be harnessed to provide heat and work, reactions that liberate energy that is unused but which give products we require, and reactions that constitute the processes of life. **Thermodynamics**, the study of the transformations of energy, enables us to discuss all these matters quantitatively and to make useful predictions.

2A Internal energy

First, we examine the ways in which a system can exchange energy with its surroundings in terms of the work it may do or have done on it or the heat that it may produce or absorb. These considerations lead to the definition of the 'internal energy', the total energy of a system, and the formulation of the 'First Law' of thermodynamics, which states that the internal energy of an isolated system is constant.

2B Enthalpy

The second major concept of the chapter is 'enthalpy', which is a very useful book-keeping property for keeping track of the heat output (or requirements) of physical processes and chemical reactions that take place at constant pressure. Experimentally, changes in internal energy or enthalpy may be measured by techniques known collectively as 'calorimetry'.

2C Thermochemistry

'Thermochemistry' is the study of heat transactions during chemical reactions. We describe both computational and experimental methods for the determination of enthalpy changes associated with both physical and chemical changes.

2D State functions and exact differentials

We also begin to unfold some of the power of thermodynamics by showing how to establish relations between different properties of a system. We see that one very useful aspect of thermodynamics is that a property can be measured indirectly by measuring others and then combining their values. The relations we derive also enable us to discuss the liquefaction of gases and to establish the relation between the heat capacities of a substance under different conditions.

2E Adiabatic changes

'Adiabatic' processes occur without transfer of energy as heat. We focus on adiabatic changes involving perfect gases because they figure prominently in our presentation of thermodynamics.

What is the impact of this material?

Concepts of thermochemistry apply to the chemical reactions associated with the conversion of food into energy in organisms, and so form a basis for the discussion of bioenergetics. In *Impact* I2.1, we explore some of the thermochemical calculations related to the metabolism of fats, carbohydrates, and proteins.

To read more about the impact of this material, scan the QR code, or go to bcs.whfreeman.com/webpub/chemistry/pchem10e/impact/pchem-2-1.html

(b) The molecular interpretation of heat and work

In molecular terms, heating is the transfer of energy that makes use of disorderly, apparently random, molecular motion in the surroundings. The disorderly motion of molecules is called **thermal motion**. The thermal motion of the molecules in the hot surroundings stimulates the molecules in the cooler system to move more vigorously and, as a result, the energy of the system is increased. When a system heats its surroundings, molecules of the system stimulate the thermal motion of the molecules in the surroundings (Fig. 2A.3).

In contrast, work is the transfer of energy that makes use of organized motion in the surroundings (Fig. 2A.4). When a weight is raised or lowered, its atoms move in an organized way (up or down). The atoms in a spring move in an orderly way when it is wound; the electrons in an electric current move in the same direction. When a system does work it causes atoms or electrons in its surroundings to move in an organized way. Likewise, when work is done on a system, molecules in the surroundings are used to transfer energy to it in an organized way, as the atoms in a weight are lowered or a current of electrons is passed.

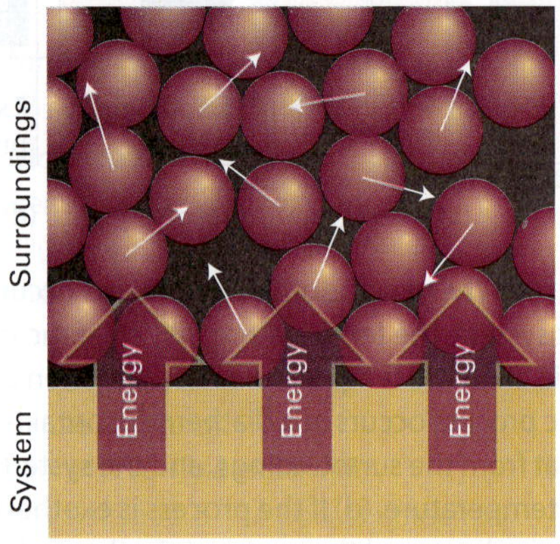

Figure 2A.3 When energy is transferred to the surroundings as heat, the transfer stimulates random motion of the atoms in the surroundings. Transfer of energy from the surroundings to the system makes use of random motion (thermal motion) in the surroundings.

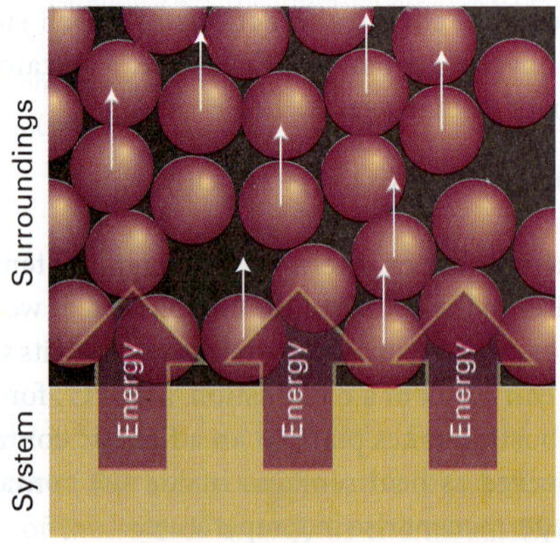

Figure 2A.4 When a system does work, it stimulates orderly motion in the surroundings. For instance, the atoms shown here may be part of a weight that is being raised. The ordered motion of the atoms in a falling weight does work on the system.

The distinction between work and heat is made in the surroundings. The fact that a falling weight may stimulate thermal motion in the system is irrelevant to the distinction between heat and work: work is identified as energy transfer making use of the organized motion of atoms in the surroundings, and heat is identified as energy transfer making use of thermal motion in the surroundings. In the adiabatic compression of a gas, for instance, work is done on the system as the atoms of the compressing weight descend in an orderly way, but the effect of the incoming piston is to accelerate the gas molecules to higher average speeds. Because collisions between molecules quickly randomize their directions, the orderly motion of the atoms of the weight is in effect stimulating thermal motion in the gas. We observe the falling weight, the orderly descent of its atoms, and report that work is being done even though it is stimulating thermal motion.

2A.2 The definition of internal energy

In thermodynamics, the total energy of a system is called its **internal energy**, U. The internal energy is the total kinetic and potential energy of the constituents (the atoms, ions, or molecules) of the system. It does not include the kinetic energy arising from the motion of the system as a whole, such as its kinetic energy as it accompanies the Earth on its orbit round the Sun. That is, the internal energy is the energy 'internal' to the system. We denote by ΔU the change in internal energy when a system changes from an initial state i with internal energy U_i to a final state f of internal energy U_f:

$$\Delta U = U_f - U_i \qquad (2A.1)$$

Throughout thermodynamics, we use the convention that $\Delta X = X_f - X_i$, where X is a property (a 'state function') of the system.

The internal energy is a **state function** in the sense that its value depends only on the current state of the system and is independent of how that state has been prepared. In other words, internal energy is a function of the properties that determine the current state of the system. Changing any one of the state variables, such as the pressure, results in a change in internal energy. That the internal energy is a state function has consequences of the greatest importance, as we shall start to unfold in Topic 2D.

The internal energy is an extensive property of a system (a property that depends on the amount of substance present, *Foundations* A) and is measures in joules ($1\,\mathrm{J} = 1\,\mathrm{kg\,m^2\,s^{-2}}$). The molar internal energy, U_m, is the internal energy divided by the amount of substance in a system, $U_m = U/n$; it is an intensive property (a property independent of the amount of substance) and commonly reported in kilojoules per mole ($\mathrm{kJ\,mol^{-1}}$).

(a) Molecular interpretation of internal energy

A molecule has a certain number of motional degrees of freedom, such as the ability to translate (the motion of its centre of mass through space), rotate around its centre of mass, or vibrate (as its bond lengths and angles change, leaving its centre of mass unmoved). Many physical and chemical properties depend on the energy associated with each of these modes of motion. For example, a chemical bond might break if a lot of energy becomes concentrated in it, for instance as vigorous vibration.

The 'equipartition theorem' of classical mechanics introduced in *Foundations* B can be used to predict the contributions of each mode of motion of a molecule to the total energy of a collection of non-interacting molecules (that is, of a perfect gas, and providing quantum effects can be ignored). For translation and rotational modes the contribution of a mode is proportional to the temperature, so the internal energy of a sample increases as the temperature is raised.

Brief illustration 2A.2 The internal energy of a perfect gas

In *Foundations* B it is shown that the mean energy of a molecule due to its translational motion is $\frac{3}{2}kT$ and therefore to the molar energy of a collection the contribution is $\frac{3}{2}RT$. Therefore, considering only the translational contribution to internal energy,

$$U_m(T) = U_m(0) + \tfrac{3}{2}N_A kT = U_m(0) + \tfrac{3}{2}RT$$

where $U_m(0)$, the internal energy at $T = 0$, can be greater than zero (see, for example, Chapter 8). At 25 °C, $RT = 2.48\,\mathrm{kJ\,mol^{-1}}$, so the translational motion contributes $3.72\,\mathrm{kJ\,mol^{-1}}$ to the molar internal energy of gases.

Self-test 2A.2 Calculate the molar internal energy of carbon dioxide at 25 °C, taking into account its translational and rotational degrees of freedom.

Answer: $U_m(T) = U_m(0) + \frac{5}{2}RT$

The contribution to the internal energy of a collection of perfect gas molecules is independent of the volume occupied by the molecules: there are no intermolecular interactions in a perfect gas, so the distance between the molecules has no effect on the energy. That is, *the internal energy of a perfect gas is independent of the volume it occupies.*

The internal energy of interacting molecules in condensed phases also has a contribution from the potential energy of their interaction, but no simple expressions can be written down in general. Nevertheless, it remains true that as the temperature of a system is raised, the internal energy increases as the various modes of motion become more highly excited.

(b) The formulation of the First Law

It has been found experimentally that the internal energy of a system may be changed either by doing work on the system or by heating it. Whereas we may know how the energy transfer has occurred (because we can see if a weight has been raised or lowered in the surroundings, indicating transfer of energy by doing work, or if ice has melted in the surroundings, indicating transfer of energy as heat), the system is blind to the mode employed. *Heat and work are equivalent ways of changing a system's internal energy.* A system is like a bank: it accepts deposits in either currency, but stores its reserves as internal energy. It is also found experimentally that if a system is isolated from its surroundings, then no change in internal energy takes place. This summary of observations is now known as the **First Law of thermodynamics** and is expressed as follows:

The internal energy of an isolated system is constant.

First Law of thermodynamics

We cannot use a system to do work, leave it isolated, and then come back expecting to find it restored to its original state with the same capacity for doing work. The experimental evidence for this observation is that no 'perpetual motion machine', a machine that does work without consuming fuel or using some other source of energy, has ever been built.

These remarks may be summarized as follows. If we write w for the work done on a system, q for the energy transferred as heat to a system, and ΔU for the resulting change in internal energy, then it follows that

$$\Delta U = q + w \qquad \text{Mathematical statement of the First Law} \qquad (2A.2)$$

Equation 2A.2 summarizes the equivalence of heat and work and the fact that the internal energy is constant in an isolated system (for which $q = 0$ and $w = 0$). The equation states that the change in internal energy of a closed system is equal to the energy that passes through its boundary as heat or work. It employs the 'acquisitive convention', in which w and q are positive if energy is transferred to the system as work or heat and are negative if energy is lost from the system. In other words, we view the flow of energy as work or heat from the system's perspective.

Brief illustration 2A.3 Changes in internal energy

If an electric motor produced 15 kJ of energy each second as mechanical work and lost 2 kJ as heat to the surroundings, then the change in the internal energy of the motor each second is $\Delta U = -2\,\text{kJ} - 15\,\text{kJ} = -17\,\text{kJ}$. Suppose that, when a spring was wound, 100 J of work was done on it but 15 J escaped to the surroundings as heat. The change in internal energy of the spring is $\Delta U = 100\,\text{J} - 15\,\text{J} = +85\,\text{J}$.

A note on good practice Always include the sign of ΔU (and of ΔX in general), even if it is positive.

Self-test 2A.3 A generator does work on an electric heater by forcing an electric current through it. Suppose 1 kJ of work is done on the heater and it heats its surroundings by 1 kJ. What is the change in internal energy of the heater?

Answer: 0

2A.3 Expansion work

The way is opened to powerful methods of calculation by switching attention to infinitesimal changes of state (such as infinitesimal change in temperature) and infinitesimal changes in the internal energy $\mathrm{d}U$. Then, if the work done on a system is $\mathrm{d}w$ and the energy supplied to it as heat is $\mathrm{d}q$, in place of eqn 2A.2 we have

$$\mathrm{d}U = \mathrm{d}q + \mathrm{d}w \qquad (2A.3)$$

To use this expression we must be able to relate $\mathrm{d}q$ and $\mathrm{d}w$ to events taking place in the surroundings.

We begin by discussing **expansion work**, the work arising from a change in volume. This type of work includes the work done by a gas as it expands and drives back the atmosphere. Many chemical reactions result in the generation of gases (for instance, the thermal decomposition of calcium carbonate or the combustion of octane), and the thermodynamic characteristics of the reaction depend on the work that must be done to make room for the gas it has produced. The term 'expansion work' also includes work associated with negative changes of volume, that is, compression.

(a) The general expression for work

The calculation of expansion work starts from the definition used in physics, which states that the work required to move an object a distance $\mathrm{d}z$ against an opposing force of magnitude $|F|$ is

$$\mathrm{d}w = -|F|\mathrm{d}z \qquad \textit{Definition} \quad \text{Work done} \quad (2A.4)$$

The negative sign tells us that, when the system moves an object against an opposing force of magnitude $|F|$, and there are no other changes, then the internal energy of the system doing the work will decrease. That is, if $\mathrm{d}z$ is positive (motion to positive z), $\mathrm{d}w$ is negative, and the internal energy decreases ($\mathrm{d}U$ in eqn 2A.3 is negative provided that $\mathrm{d}q=0$).

Now consider the arrangement shown in Fig. 2A.5, in which one wall of a system is a massless, frictionless, rigid, perfectly fitting piston of area A. If the external pressure is p_{ex}, the magnitude of the force acting on the outer face of the piston is $|F| = p_{ex}A$. When the system expands through a distance $\mathrm{d}z$ against an external pressure p_{ex}, it follows that the work done is $\mathrm{d}w = -p_{ex}A\mathrm{d}z$. The quantity $A\mathrm{d}z$ is the change in volume, $\mathrm{d}V$, in the course of the expansion. Therefore, the work done when the system expands by $\mathrm{d}V$ against a pressure p_{ex} is

$$\mathrm{d}w = -p_{ex}\mathrm{d}V \qquad \text{Expansion work} \quad (2A.5a)$$

To obtain the total work done when the volume changes from an initial value V_i to a final value V_f we integrate this expression between the initial and final volumes:

$$w = -\int_{V_i}^{V_f} p_{ex}\mathrm{d}V \qquad (2A.5b)$$

The force acting on the piston, $p_{ex}A$, is equivalent to the force arising from a weight that is raised as the system expands. If the system is compressed instead, then the same weight is lowered in the surroundings and eqn 2A.5b can still be used, but now $V_f < V_i$. It is important to note that it is still the external pressure that determines the magnitude of the work. This somewhat perplexing conclusion seems to be inconsistent with the fact that the gas *inside* the container is opposing the compression. However, when a gas is compressed, the ability of the *surroundings* to do work is diminished by an amount determined by the weight that is lowered, and it is this energy that is transferred into the system.

Other types of work (for example, electrical work), which we shall call either **non-expansion work** or **additional work**, have

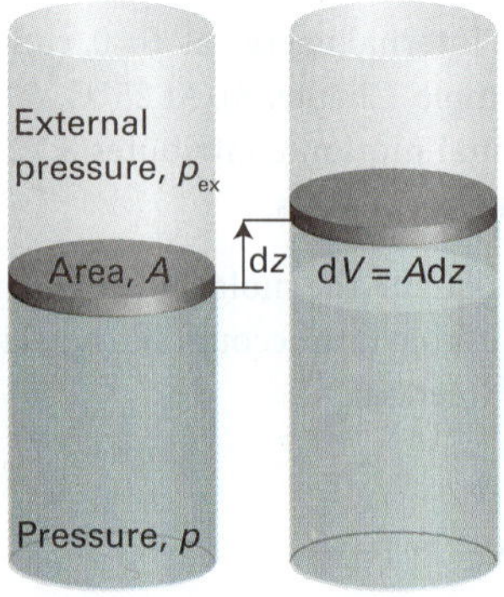

Figure 2A.5 When a piston of area A moves out through a distance $\mathrm{d}z$, it sweeps out a volume $\mathrm{d}V = A\mathrm{d}z$. The external pressure p_{ex} is equivalent to a weight pressing on the piston, and the magnitude of the force opposing expansion is $|F| = p_{ex}A$.

Table 2A.1 Varieties of work*

Type of work	dw	Comments	Units†
Expansion	$-p_{ex}dV$	p_{ex} is the external pressure dV is the change in volume	Pa m^3
Surface expansion	$\gamma d\sigma$	γ is the surface tension $d\sigma$ is the change in area	N m^{-1} m^2
Extension	fdl	f is the tension dl is the change in length	N m
Electrical	ϕdQ	ϕ is the electric potential dQ is the change in charge	V C
	$Qd\phi$	$d\phi$ is the potential difference Q is the charge transferred	V C

* In general, the work done on a system can be expressed in the form $dw=-|F|dz$, where $|F|$ is the magnitude of a 'generalized force' and dz is a 'generalized displacement'.

† For work in joules (J). Note that 1 N m = 1 J and 1 V C = 1 J.

analogous expressions, with each one the product of an intensive factor (the pressure, for instance) and an extensive factor (the change in volume). Some are collected in Table 2A.1. For the present we continue with the work associated with changing the volume, the expansion work, and see what we can extract from eqn 2A.5b.

Brief illustration 2A.4 The work of extension

To establish an expression for the work of stretching an elastomer, a polymer that can stretch and contract, to an extension l given that the force opposing extension is proportional to the displacement from the resting state of the elastomer we write $|F|=k_f x$, where k_f is a constant and x is the displacement. It then follows from eqn 2A.4 that for an infinitesimal displacement from x to $x+dx$, $dw=-k_f x dx$. For the overall work of displacement from $x=0$ to the final extension l,

$$w=-\int_0^l k_f x\, dx=-\frac{1}{2}k_f l^2$$

Self-test 2A.4 Suppose the restoring force weakens as the elastomer is stretched, and $k_f(x)=a-bx^{1/2}$. Evaluate the work of extension to l.

Answer: $w=-\frac{1}{2}al^2+\frac{2}{5}bl^{5/2}$

(b) Expansion against constant pressure

Suppose that the external pressure is constant throughout the expansion. For example, the piston may be pressed on by the atmosphere, which exerts the same pressure throughout the expansion. A chemical example of this condition is the expansion of a gas formed in a chemical reaction in a container that can expand. We can evaluate eqn 2A.5b by taking the constant p_{ex} outside the integral:

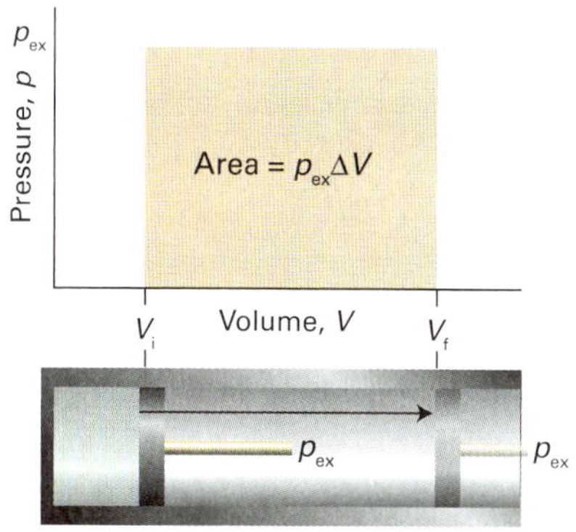

Figure 2A.6 The work done by a gas when it expands against a constant external pressure, p_{ex}, is equal to the shaded area in this example of an indicator diagram.

$$w=-p_{ex}\int_{V_i}^{V_f} dV=-p_{ex}(V_f-V_i)$$

Therefore, if we write the change in volume as $\Delta V=V_f-V_i$,

$$w=-p_{ex}\Delta V \quad \textit{Constant external pressure} \quad \text{Expansion work} \quad (2A.6)$$

This result is illustrated graphically in Fig. 2A.6, which makes use of the fact that an integral can be interpreted as an area. The magnitude of w, denoted $|w|$, is equal to the area beneath the horizontal line at $p=p_{ex}$ lying between the initial and final volumes. A pV-graph used to illustrate expansion work is called an **indicator diagram**; James Watt first used one to indicate aspects of the operation of his steam engine.

Free expansion is expansion against zero opposing force. It occurs when $p_{ex}=0$. According to eqn 2A.6,

$$w=0 \quad \text{Work of free expansion} \quad (2A.7)$$

That is, no work is done when a system expands freely. Expansion of this kind occurs when a gas expands into a vacuum.

Example 2A.1 Calculating the work of gas production

Calculate the work done when 50 g of iron reacts with hydrochloric acid to produce $FeCl_2(aq)$ and hydrogen in (a) a closed vessel of fixed volume, (b) an open beaker at 25 °C.

Method We need to judge the magnitude of the volume change and then to decide how the process occurs. If there is no change in volume, there is no expansion work however the process takes place. If the system expands against a constant external pressure, the work can be calculated from eqn 2A.6. A general feature of processes in which a condensed phase changes into a gas is that the volume of the former may usually be neglected relative to that of the gas it forms.

Answer In (a) the volume cannot change, so no expansion work is done and $w=0$. In (b) the gas drives back the atmosphere and therefore $w=-p_{ex}\Delta V$. We can neglect the initial volume because the final volume (after the production of gas) is so much larger and $\Delta V = V_f - V_i \approx V_f = nRT/p_{ex}$, where n is the amount of H_2 produced. Therefore,

$$w = -p_{ex}\Delta V \approx -p_{ex} \times \frac{nRT}{p_{ex}} = -nRT$$

Because the reaction is $Fe(s) + 2\,HCl(aq) \rightarrow FeCl_2(aq) + H_2(g)$, we know that 1 mol H_2 is generated when 1 mol Fe is consumed, and n can be taken as the amount of Fe atoms that react. Because the molar mass of Fe is 55.85 g mol^{-1}, it follows that

$$w = -\frac{50\,\text{g}}{55.85\,\text{g mol}^{-1}} \times (8.3145\,\text{J K}^{-1}\,\text{mol}^{-1}) \times (298\,\text{K})$$
$$\approx -2.2\,\text{kJ}$$

The system (the reaction mixture) does 2.2 kJ of work driving back the atmosphere. Note that (for this perfect gas system) the magnitude of the external pressure does not affect the final result: the lower the pressure, the larger the volume occupied by the gas, so the effects cancel.

Self-test 2A.5 Calculate the expansion work done when 50 g of water is electrolysed under constant pressure at 25 °C.

Answer: −10 kJ

(c) Reversible expansion

A **reversible change** in thermodynamics is a change that can be reversed by an infinitesimal modification of a variable. The key word 'infinitesimal' sharpens the everyday meaning of the word 'reversible' as something that can change direction. One example of reversibility that we have encountered already is the thermal equilibrium of two systems with the same temperature. The transfer of energy as heat between the two is reversible because, if the temperature of either system is lowered infinitesimally, then energy flows into the system with the lower temperature. If the temperature of either system at thermal equilibrium is raised infinitesimally, then energy flows out of the hotter system. There is obviously a very close relationship between reversibility and equilibrium: systems at equilibrium are poised to undergo reversible change.

Suppose a gas is confined by a piston and that the external pressure, p_{ex}, is set equal to the pressure, p, of the confined gas. Such a system is in mechanical equilibrium with its surroundings because an infinitesimal change in the external pressure in either direction causes changes in volume in opposite directions. If the external pressure is reduced infinitesimally, the gas expands slightly. If the external pressure is increased infinitesimally, the gas contracts slightly. In either case the change is reversible in the thermodynamic sense. If, on the other hand, the external pressure differs measurably from the internal pressure, then changing p_{ex} infinitesimally will not decrease it below the pressure of the gas, so will not change the direction of the process. Such a system is not in mechanical equilibrium with its surroundings and the expansion is thermodynamically irreversible.

To achieve reversible expansion we set p_{ex} equal to p at each stage of the expansion. In practice, this equalization could be achieved by gradually removing weights from the piston so that the downward force due to the weights always matches the changing upward force due to the pressure of the gas. When we set $p_{ex} = p$, eqn 2A.5a becomes

$$dw = -p_{ex}dV = -pdV \qquad \text{Reversible expansion work} \qquad (2A.8a)$$

Although the pressure inside the system appears in this expression for the work, it does so only because p_{ex} has been set equal to p to ensure reversibility. The total work of reversible expansion from an initial volume V_i to a final volume V_f is therefore

$$w = -\int_{V_i}^{V_f} p\,dV \qquad (2A.8b)$$

The integral can be evaluated once we know how the pressure of the confined gas depends on its volume. Equation 2A.8b is the link with the material covered in the Topics of Chapter 1 for, if we know the equation of state of the gas, then we can express p in terms of V and evaluate the integral.

(d) Isothermal reversible expansion

Consider the isothermal, reversible expansion of a perfect gas. The expansion is made isothermal by keeping the system in thermal contact with its surroundings (which may be a constant-temperature bath). Because the equation of state is $pV = nRT$, we know that at each stage $p = nRT/V$, with V the volume at that stage of the expansion. The temperature T is constant in an isothermal expansion, so (together with n and R) it may be taken outside the integral. It follows that the work of reversible isothermal expansion of a perfect gas from V_i to V_f at a temperature T is

$$w = -nRT\int_{V_i}^{V_f}\frac{dV}{V} \overset{\text{Integral A.2}}{=} -nRT\ln\frac{V_f}{V_i} \qquad \text{Perfect gas, reversible, isothermal} \qquad \text{Work of expansion} \qquad (2A.9)$$

Brief illustration 2A.5 The work of isothermal reversible expansion

When a sample of 1.00 mol Ar, regarded here as a perfect gas, undergoes an isothermal reversible expansion at 20.0 °C from 10.0 dm^3 to 30.0 dm^3 the work done is

$$w = -(1.00\,\text{mol}) \times (8.3145\,\text{J K}^{-1}\,\text{mol}^{-1}) \times (293.2\,\text{K}) \ln\frac{30.0\,\text{dm}^3}{10.0\,\text{dm}^3}$$
$$= -2.68\,\text{kJ}$$

Self-test 2A.6 Suppose that attractions are important between gas molecules, and the equation of state is $p=nRT/V-n^2a/V^2$. Derive an expression for the reversible, isothermal expansion of this gas. Is more or less work done *on the surroundings* when it expands (compared with a perfect gas)?

Answer: $w=-nRT\ln(V_f/V_i)-n^2a(1/V_f-1/V_i)$; less

When the final volume is greater than the initial volume, as in an expansion, the logarithm in eqn 2A.9 is positive and hence $w<0$. In this case, the system has done work on the surroundings and there is a corresponding negative contribution to its internal energy. (Note the cautious language: we shall see later that there is a compensating influx of energy as heat, so overall the internal energy is constant for the isothermal expansion of a perfect gas.) The equations also show that more work is done for a given change of volume when the temperature is increased: at a higher temperature the greater pressure of the confined gas needs a higher opposing pressure to ensure reversibility and the work done is correspondingly greater.

We can express the result of the calculation as an indicator diagram, for the magnitude of the work done is equal to the area under the isotherm $p=nRT/V$ (Fig. 2A.7). Superimposed on the diagram is the rectangular area obtained for irreversible expansion against constant external pressure fixed at the same final value as that reached in the reversible expansion. More work is obtained when the expansion is reversible (the area is greater) because matching the external pressure to the internal pressure at each stage of the process ensures that none of the system's pushing power is wasted. We cannot obtain more work than for the reversible process because increasing the external pressure even infinitesimally at any stage results in compression. We may infer from this discussion that, because some pushing power is wasted when $p>p_{ex}$, the maximum work available from a system operating between specified initial and final states and passing along a specified path is obtained when the change takes place reversibly.

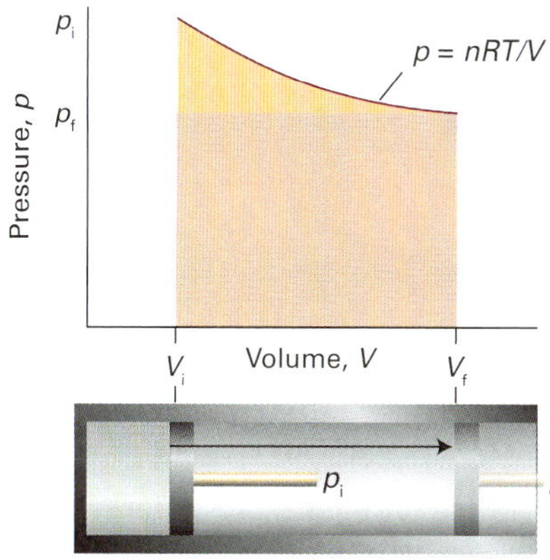

Figure 2A.7 The work done by a perfect gas when it expands reversibly and isothermally is equal to the area under the isotherm $p=nRT/V$. The work done during the irreversible expansion against the same final pressure is equal to the rectangular area shown slightly darker. Note that the reversible work is greater than the irreversible work.

We have introduced the connection between reversibility and maximum work for the special case of a perfect gas undergoing expansion. In Topic 3A we see that it applies to all substances and to all kinds of work.

2A.4 Heat transactions

In general, the change in internal energy of a system is

$$dU=dq+dw_{exp}+dw_e \qquad (2A.10)$$

where dw_e is work in addition (e for 'extra') to the expansion work, dw_{exp}. For instance, dw_e might be the electrical work of driving a current through a circuit. A system kept at constant volume can do no expansion work, so $dw_{exp}=0$. If the system is also incapable of doing any other kind of work (if it is not, for instance, an electrochemical cell connected to an electric motor), then $dw_e=0$ too. Under these circumstances:

$$dU=dq \qquad \text{Heat transferred at constant volume} \qquad (2A.11a)$$

We express this relation by writing $dU=dq_V$, where the subscript implies a change at constant volume. For a measurable change between states i and f along a path at constant volume,

$$\overbrace{\int_i^f dU}^{U_f-U_i}=\overbrace{\int_i^f dq}^{q_V}$$

which we summarize as

$$\Delta U=q_V \qquad (2A.11b)$$

Note that we do not write the integral over dq as Δq because q, unlike U, is not a state function. It follows that, by measuring the energy supplied to a constant-volume system as heat ($q_V>0$) or released from it as heat ($q_V<0$) when it undergoes a change of state, we are in fact measuring the change in its internal energy.

(a) Calorimetry

Calorimetry is the study of the transfer of energy as heat during physical and chemical processes. A **calorimeter** is a device for measuring energy transferred as heat. The most common device for measuring q_V (and therefore ΔU) is an **adiabatic bomb calorimeter** (Fig. 2A.8). The process we wish to study—which may be a chemical reaction—is initiated inside a constant-volume container, the 'bomb'. The bomb is immersed in

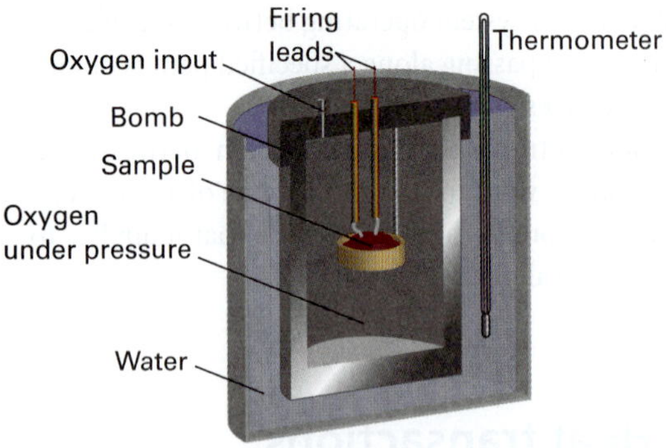

Figure 2A.8 A constant-volume bomb calorimeter. The 'bomb' is the central vessel, which is strong enough to withstand high pressures. The calorimeter (for which the heat capacity must be known) is the entire assembly shown here. To ensure adiabaticity, the calorimeter is immersed in a water bath with a temperature continuously readjusted to that of the calorimeter at each stage of the combustion.

a stirred water bath, and the whole device is the calorimeter. The calorimeter is also immersed in an outer water bath. The water in the calorimeter and of the outer bath are both monitored and adjusted to the same temperature. This arrangement ensures that there is no net loss of heat from the calorimeter to the surroundings (the bath) and hence that the calorimeter is adiabatic.

The change in temperature, ΔT, of the calorimeter is proportional to the energy that the reaction releases or absorbs as heat. Therefore, by measuring ΔT we can determine q_V and hence find ΔU. The conversion of ΔT to q_V is best achieved by calibrating the calorimeter using a process of known energy output and determining the **calorimeter constant**, the constant C in the relation

$$q = C\Delta T \qquad (2A.12)$$

The calorimeter constant may be measured electrically by passing a constant current, I, from a source of known potential difference, $\Delta\phi$, through a heater for a known period of time, t, for then

$$q = It\Delta\phi \qquad (2A.13)$$

Electrical charge is measured in *coulombs*, C. The motion of charge gives rise to an electric current, I, measured in coulombs per second, or *amperes*, A, where $1\ \text{A} = 1\ \text{C s}^{-1}$. If a constant current I flows through a potential difference $\Delta\phi$ (measured in volts, V), the total energy supplied in an interval t is $It\Delta\phi$. Because $1\ \text{A V s} = 1\ (\text{C s}^{-1})\ \text{V s} = 1\ \text{C V} = 1\ \text{J}$, the energy is obtained in joules with the current in amperes, the potential difference in volts, and the time in seconds.

Brief illustration 2A.6 Electrical heating

If a current of 10.0 A from a 12 V supply is passed for 300 s, then from eqn 2A.13 the energy supplied as heat is

$$q = (10.0\ \text{A}) \times (12\ \text{V}) \times (300\ \text{s}) = 3.6 \times 10^4\ \text{A V s} = 36\ \text{kJ}$$

because 1 A V s = 1 J. If the observed rise in temperature is 5.5 K, then the calorimeter constant is $C = (36\ \text{kJ})/(5.5\ \text{K}) = 6.5\ \text{kJ K}^{-1}$.

Self-test 2A.7 What is the value of the calorimeter constant if the temperature rises by 4.8 °C when a current of 8.6 A from an 11 V supply is passed for 280 s?

Answer: $5.5\ \text{kJ K}^{-1}$

Alternatively, C may be determined by burning a known mass of substance (benzoic acid is often used) that has a known heat output. With C known, it is simple to interpret an observed temperature rise as a release of heat.

(b) Heat capacity

The internal energy of a system increases when its temperature is raised. The increase depends on the conditions under which the heating takes place and for the present we suppose that the system has a constant volume. For example, it may be a gas in a container of fixed volume. If the internal energy is plotted against temperature, then a curve like that in Fig. 2A.9 may be obtained. The slope of the tangent to the curve at any temperature is called the **heat capacity** of the system at that temperature. The **heat capacity at constant volume** is denoted C_V and is defined formally as

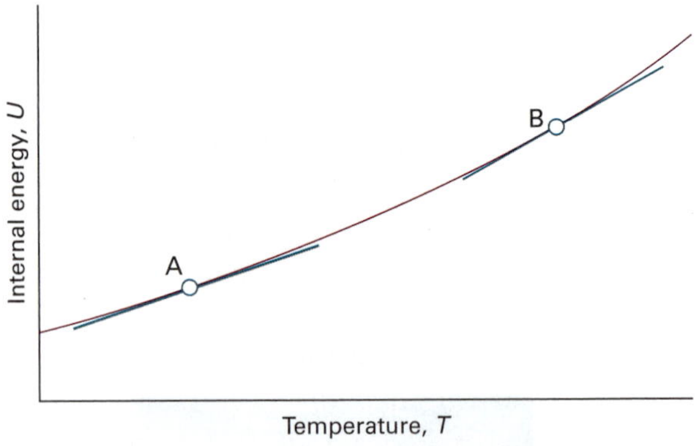

Figure 2A.9 The internal energy of a system increases as the temperature is raised; this graph shows its variation as the system is heated at constant volume. The slope of the tangent to the curve at any temperature is the heat capacity at constant volume at that temperature. Note that, for the system illustrated, the heat capacity is greater at B than at A.

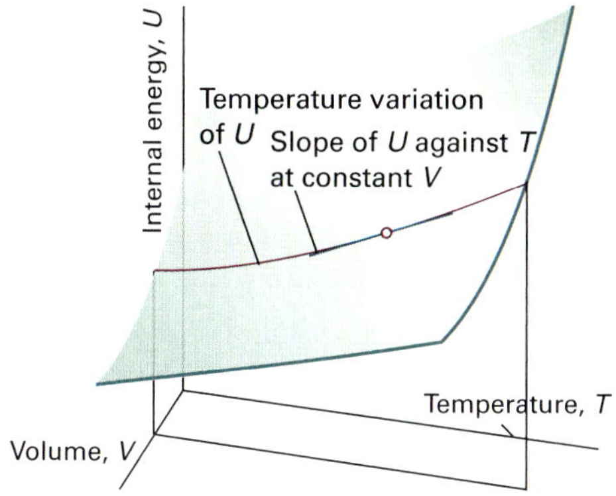

Figure 2A.10 The internal energy of a system varies with volume and temperature, perhaps as shown here by the surface. The variation of the internal energy with temperature at one particular constant volume is illustrated by the curve drawn parallel to T. The slope of this curve at any point is the partial derivative $(\partial U/\partial T)_V$.

$$C_V = \left(\frac{\partial U}{\partial T}\right)_V \quad \textit{Definition} \quad \text{Heat capacity at constant volume} \quad (2A.14)$$

Partial derivatives are reviewed in *Mathematical background* 2 following this chapter. The internal energy varies with the temperature and the volume of the sample, but here we are interested only in its variation with the temperature, the volume being held constant (Fig. 2A.10).

Brief illustration 2A.7 Heat capacity

The heat capacity of a monatomic perfect gas can be calculated by inserting the expression for the internal energy derived in *Brief illustration* 2A.2 where we saw that $U_m(T)=U_m(0)+\frac{3}{2}RT$, so from eqn 2A.14

$$C_{V,m} = \frac{\partial}{\partial T}\{U_m(0)+\tfrac{3}{2}RT\}=\tfrac{3}{2}R$$

The numerical value is 12.47 J K^{-1} mol^{-1}.

Self-test 2A.8 Estimate the molar constant-volume heat capacity of carbon dioxide.

Answer: $\frac{5}{2}R=21\,\text{J K}^{-1}\,\text{mol}^{-1}$

Heat capacities are extensive properties: 100 g of water, for instance, has 100 times the heat capacity of 1 g of water (and therefore requires 100 times the energy as heat to bring about the same rise in temperature). The **molar heat capacity at constant volume**, $C_{V,m}=C_V/n$, is the heat capacity per mole of substance, and is an intensive property (all molar quantities are intensive). Typical values of $C_{V,m}$ for polyatomic gases are close to 25 J K^{-1} mol^{-1}. For certain applications it is useful to know the **specific heat capacity** (more informally, the 'specific heat') of a substance, which is the heat capacity of the sample divided by the mass, usually in grams: $C_{V,s}=C_V/m$. The specific heat capacity of water at room temperature is close to 4.2 J K^{-1} g^{-1}. In general, heat capacities depend on the temperature and decrease at low temperatures. However, over small ranges of temperature at and above room temperature, the variation is quite small and for approximate calculations heat capacities can be treated as almost independent of temperature.

The heat capacity is used to relate a change in internal energy to a change in temperature of a constant-volume system. It follows from eqn 2A.14 that

$$dU=C_V dT \quad \textit{Constant volume} \quad (2A.15a)$$

That is, at constant volume, an infinitesimal change in temperature brings about an infinitesimal change in internal energy, and the constant of proportionality is C_V. If the heat capacity is independent of temperature over the range of temperatures of interest, then

$$\Delta U=\int_{T_1}^{T_2} C_V dT = C_V\int_{T_1}^{T_2} dT = C_V \overbrace{(T_2-T_1)}^{\Delta T}$$

and a measurable change of temperature, ΔT, brings about a measurable change in internal energy, ΔU, where

$$\Delta U=C_V\Delta T \quad \textit{Constant volume} \quad (2A.15b)$$

Because a change in internal energy can be identified with the heat supplied at constant volume (eqn 2A.11b), the last equation can also be written

$$q_V=C_V\Delta T \quad (2A.16)$$

This relation provides a simple way of measuring the heat capacity of a sample: a measured quantity of energy is transferred as heat to the sample (electrically, for example), and the resulting increase in temperature is monitored. The ratio of the energy transferred as heat to the temperature rise it causes ($q_V/\Delta T$) is the constant-volume heat capacity of the sample.

Brief illustration 2A.8 The determination of a heat capacity

Suppose a 55 W electric heater immersed in a gas in a constant-volume adiabatic container was on for 120 s and it was found that the temperature of the gas rose by 5.0 °C (an increase equivalent to 5.0 K). The heat supplied is (55 W)×(120 s) = 6.6 kJ (we have used 1 J = 1 W s), so the heat capacity of the sample is

$$C_V=\frac{6.6\text{ kJ}}{5.0\text{ K}}=1.3\text{ kJ K}^{-1}$$

Self-test 2A.9 When 229 J of energy is supplied as heat to 3.0 mol of a gas at constant volume, the temperature of the gas increases by 2.55 °C. Calculate C_V and the molar heat capacity at constant volume.

Answer: $89.8\,\mathrm{J\,K^{-1}}$, $29.9\,\mathrm{J\,K^{-1}\,mol^{-1}}$

A large heat capacity implies that, for a given quantity of energy transferred as heat, there will be only a small increase in temperature (the sample has a large capacity for heat). An infinite heat capacity implies that there will be no increase in temperature however much energy is supplied as heat. At a phase transition, such as at the boiling point of water, the temperature of a substance does not rise as energy is supplied as heat: the energy is used to drive the endothermic transition, in this case to vaporize the water, rather than to increase its temperature. Therefore, at the temperature of a phase transition, the heat capacity of a sample is infinite. The properties of heat capacities close to phase transitions are treated more fully in Topic 4B.

Checklist of concepts

- ☐ 1. **Work** is done to achieve motion against an opposing force.
- ☐ 2. **Energy** is the capacity to do work.
- ☐ 3. **Heating** is the transfer of energy that makes use of disorderly molecular motion.
- ☐ 4. Work is the transfer of energy that makes use of organized motion.
- ☐ 5. **Internal energy**, the total energy of a system, is a state function.
- ☐ 6. The **equipartition theorem** can be used to estimate the contribution to the internal energy of classical modes of motion.
- ☐ 7. The **First Law** states that the internal energy of an isolated system is constant.
- ☐ 8. Free expansion (expansion against zero pressure) does no work.
- ☐ 9. To achieve **reversible expansion**, the external pressure is matched at every stage to the pressure of the system.
- ☐ 10. The energy transferred as heat at constant volume is equal to the change in internal energy of the system.
- ☐ 11. **Calorimetry** is the measurement of heat transactions.

Checklist of equations

Property	Equation	Comment	Equation number
First Law of thermodynamics	$\Delta U = q + w$	Acquisitive convention	2A.2
Work of expansion	$\mathrm{d}w = -p_{\mathrm{ex}}\mathrm{d}V$		2A.5a
Work of expansion against a constant external pressure	$w = -p_{\mathrm{ex}}\Delta V$	$p_{\mathrm{ex}} = 0$ corresponds to free expansion	2A.6
Reversible work of expansion of a gas	$w = -nRT\ln(V_f/V_i)$	Isothermal, perfect gas	2A.9
Internal energy change	$\Delta U = q_V$	Constant volume, no other forms of work	2A.11b
Electrical heating	$q = It\Delta\phi$		2A.13
Heat capacity at constant volume	$C_V = (\partial U/\partial T)_V$	Definition	2A.14

2B Enthalpy

Contents

➤ Why do you need to know this material?

The concept of enthalpy is central to many thermodynamic discussions about processes taking place under conditions of constant pressure, such as the discussion of the heat requirements or output of physical transformations and chemical reactions.

➤ What is the key idea?

A change in enthalpy is equal to the energy transferred as heat at constant pressure.

➤ What do you need to know already?

This Topic makes use of the discussion of internal energy (Topic 2A) and draws on some aspects of perfect gases (Topic 1A).

The change in internal energy is not equal to the energy transferred as heat when the system is free to change its volume, such as when it is able to expand or contract under conditions of constant pressure. Under these circumstances some of the energy supplied as heat to the system is returned to the surroundings as expansion work (Fig. 2B.1), so dU is less than dq. In this case the energy supplied as heat at constant pressure is equal to the change in another thermodynamic property of the system, the enthalpy.

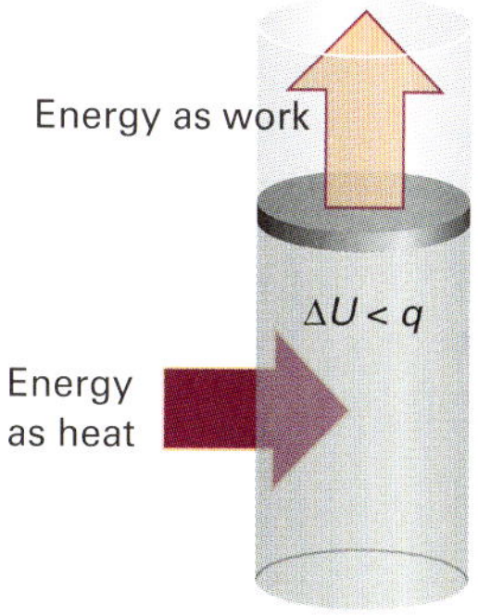

Figure 2B.1 When a system is subjected to constant pressure and is free to change its volume, some of the energy supplied as heat may escape back into the surroundings as work. In such a case, the change in internal energy is smaller than the energy supplied as heat.

2B.1 The definition of enthalpy

The **enthalpy**, H, is defined as

$$H = U + pV \qquad \text{Definition} \quad \text{Enthalpy} \qquad (2B.1)$$

where p is the pressure of the system and V is its volume. Because U, p, and V are all state functions, the enthalpy is a state function too. As is true of any state function, the change in enthalpy, ΔH, between any pair of initial and final states is independent of the path between them.

(a) Enthalpy change and heat transfer

Although the definition of enthalpy may appear arbitrary, it has important implications for thermochemistry. For instance, we show in the following *Justification* that eqn 2B.1 implies that *the change in enthalpy is equal to the energy supplied as heat at constant pressure* (provided the system does no additional work):

$$dH = dq_p \qquad \text{Heat transferred at constant pressure} \qquad (2B.2a)$$

For a measurable change between states i and f along a path at constant pressure, we write

$$\overbrace{\int_i^f dH}^{H_f - H_i} = \overbrace{\int_i^f dq_p}^{q_p}$$

and summarize the result as

$$\Delta H = q_p \qquad (2B.2b)$$

Note that we do not write the integral over dq as Δq because q, unlike H, is not a state function.

Brief illustration 2B.1 A change in enthalpy

Water is heated to boiling under a pressure of 1.0 atm. When an electric current of 0.50 A from a 12 V supply is passed for 300 s through a resistance in thermal contact with it, it is found that 0.798 g of water is vaporized. The enthalpy change is

$$\Delta H = q_p = It\Delta\phi = (0.50\,\text{A})\times(12\,\text{V})\times(300\,\text{s}) = (0.50\times 12\times 300)\,\text{J}$$

Here we have used 1 A V s = 1 J. Because 0.798 g of water is (0.798 g)/(18.02 g mol^{-1}) = (0.798/18.02) mol H_2O, the enthalpy of vaporization per mole of H_2O is

$$\Delta H_\text{m} = \frac{(0.50\times 12\times 300)\,\text{J}}{(0.798/18.02)\,\text{mol}} = +41\,\text{kJ mol}^{-1}$$

Self-test 2B.1 The molar enthalpy of vaporization of benzene at its boiling point (353.25 K) is 30.8 kJ mol^{-1}. For how long would the same 12 V source need to supply a 0.50 A current in order to vaporize a 10 g sample?

Answer: 6.6×10^2 s

Justification 2B.1 The relation $\Delta H = q_p$

For a general infinitesimal change in the state of the system, U changes to $U+\text{d}U$, p changes to $p+\text{d}p$, and V changes to $V+\text{d}V$, so from the definition in eqn 2B.1, H changes from $U+pV$ to

$$\begin{aligned} H+\text{d}H &= (U+\text{d}U)+(p+\text{d}p)(V+\text{d}V) \\ &= U+\text{d}U+pV+p\text{d}V+V\text{d}p+\text{d}p\text{d}V \end{aligned}$$

The last term is the product of two infinitesimally small quantities and can therefore be neglected. As a result, after recognizing $U+pV=H$ on the right (in blue), we find that H changes to

$$H+\text{d}H = H+\text{d}U+p\text{d}V+V\text{d}p$$

and hence that

$$\text{d}H = \text{d}U+p\text{d}V+V\text{d}p$$

If we now substitute dU=dq+dw into this expression, we get

$$\text{d}H = \text{d}q+\text{d}w+p\text{d}V+V\text{d}p$$

If the system is in mechanical equilibrium with its surroundings at a pressure p and does only expansion work, we can write dw=−pdV and obtain

$$\text{d}H = \text{d}q+V\text{d}p$$

Now we impose the condition that the heating occurs at constant pressure by writing dp=0. Then

$$\text{d}H = \text{d}q \text{ (at constant pressure, no additional work)}$$

as in eqn 2B.2a. Equation 2B.2b then follows, as explained in the text.

(b) Calorimetry

The process of measuring heat transactions between a system and its surroundings is called **calorimetry**. An enthalpy change can be measured calorimetrically by monitoring the temperature change that accompanies a physical or chemical change occurring at constant pressure. A calorimeter for studying processes at constant pressure is called an **isobaric calorimeter**. A simple example is a thermally insulated vessel open to the atmosphere: the heat released in the reaction is monitored by measuring the change in temperature of the contents. For a combustion reaction an **adiabatic flame calorimeter** may be used to measure ΔT when a given amount of substance burns in a supply of oxygen (Fig. 2B.2).

Another route to ΔH is to measure the internal energy change by using a bomb calorimeter, and then to convert ΔU to ΔH. Because solids and liquids have small molar volumes, for them pV_m is so small that the molar enthalpy and molar internal energy are almost identical ($H_\text{m}=U_\text{m}+pV_\text{m}\approx U_\text{m}$). Consequently, if a process involves only solids or liquids, the values of ΔH and ΔU are almost identical. Physically, such processes are accompanied by a very small change in volume; the system does negligible work on the surroundings when the process occurs, so the energy supplied as heat stays entirely

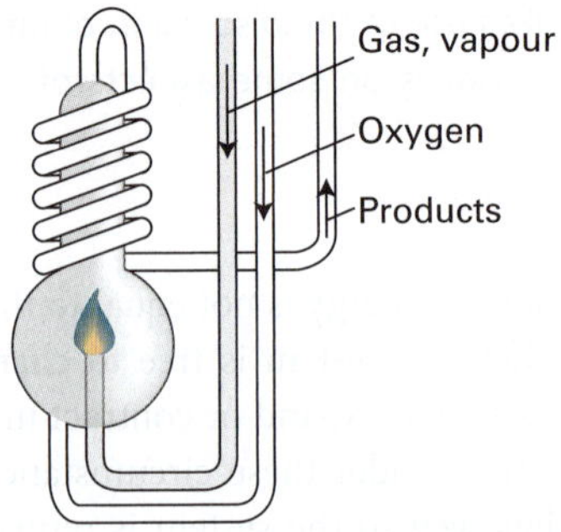

Figure 2B.2 A constant-pressure flame calorimeter consists of this component immersed in a stirred water bath. Combustion occurs as a known amount of reactant is passed through to fuel the flame, and the rise of temperature is monitored.

within the system. The most sophisticated way to measure enthalpy changes, however, is to use a *differential scanning calorimeter* (DSC), as explained in Topic 2C. Changes in enthalpy and internal energy may also be measured by non-calorimetric methods (see Topic 6C).

Example 2B.1 Relating ΔH and ΔU

The change in molar internal energy when $CaCO_3$(s) as calcite converts to another form, aragonite, is +0.21 kJ mol^{-1}. Calculate the difference between the molar enthalpy and internal energy changes when the pressure is 1.0 bar given that the densities of the polymorphs are 2.71 g cm^{-3} (calcite) and 2.93 g cm^{-3} (aragonite).

Method The starting point for the calculation is the relation between the enthalpy of a substance and its internal energy (eqn 2B.1). The difference between the two quantities can be expressed in terms of the pressure and the difference of their molar volumes, and the latter can be calculated from their molar masses, M, and their mass densities, ρ, by using $\rho = M/V_m$.

Answer The change in enthalpy when the transition occurs is

$$\begin{aligned}\Delta H_m &= H_m(\text{aragonite}) - H_m(\text{calcite})\\ &= \{U_m(\text{a}) + pV_m(\text{a})\} - \{U_m(\text{c}) + pV_m(\text{c})\}\\ &= \Delta U_m + p\{V_m(\text{a}) - V_m(\text{c})\}\end{aligned}$$

where a denotes aragonite and c calcite. It follows by substituting $V_m = M/\rho$ that

$$\Delta H_m - \Delta U_m = pM\left(\frac{1}{\rho(\text{a})} - \frac{1}{\rho(\text{c})}\right)$$

Substitution of the data, using $M = 100.09$ g mol^{-1}, gives

$$\begin{aligned}\Delta H_m - \Delta U_m &= (1.0\times10^5\,\text{Pa})\times(100.09\,\text{g mol}^{-1})\\ &\quad\times\left(\frac{1}{2.93\,\text{g cm}^{-3}} - \frac{1}{2.71\,\text{g cm}^{-3}}\right)\\ &= -2.8\times10^5\,\text{Pa cm}^3\,\text{mol}^{-1} = -0.28\,\text{Pa m}^3\,\text{mol}^{-1}\end{aligned}$$

Hence (because 1 Pa m^3 = 1 J), $\Delta H_m - \Delta U_m = -0.28$ J mol^{-1}, which is only 0.1 per cent of the value of ΔU_m. We see that it is usually justifiable to ignore the difference between the molar enthalpy and internal energy of condensed phases, except at very high pressures, when $p\Delta V_m$ is no longer negligible.

Self-test 2B.2 Calculate the difference between ΔH and ΔU when 1.0 mol Sn(s, grey) of density 5.75 g cm^{-3} changes to Sn(s, white) of density 7.31 g cm^{-3} at 10.0 bar. At 298 K, $\Delta H = +2.1$ kJ.

Answer: $\Delta H - \Delta U = -4.4$ J

In contrast to processes involving condensed phases, the values of the changes in internal energy and enthalpy may differ significantly for processes involving gases. Thus, the enthalpy of a perfect gas is related to its internal energy by using $pV = nRT$ in the definition of H:

$$H = U + pV = U + nRT \qquad (2B.3)$$

This relation implies that the change of enthalpy in a reaction that produces or consumes gas under isothermal conditions is

$$\Delta H = \Delta U + \Delta n_g RT \qquad \textit{Perfect gas, isothermal} \qquad \text{Relation between } \Delta H \text{ and } \Delta U \qquad (2B.4)$$

where Δn_g is the change in the amount of gas molecules in the reaction.

Brief illustration 2B.2 Processes involving gases

In the reaction $2\,H_2(g) + O_2(g) \rightarrow 2\,H_2O(l)$, 3 mol of gas-phase molecules are replaced by 2 mol of liquid-phase molecules, so $\Delta n_g = -3$ mol. Therefore, at 298 K, when $RT = 2.5$ kJ mol^{-1}, the enthalpy and internal energy changes taking place in the system are related by

$$\Delta H_m - \Delta U_m = (-3\,\text{mol})\times RT \approx -7.4\,\text{kJ mol}^{-1}$$

Note that the difference is expressed in kilojoules, not joules as in *Example* 2B.2. The enthalpy change is smaller (in this case, less negative) than the change in internal energy because, although heat escapes from the system when the reaction occurs, the system contracts when the liquid is formed, so energy is restored to it from the surroundings.

Self-test 2B.3 Calculate the value of $\Delta H_m - \Delta U_m$ for the reaction $N_2(g) + 3\,H_2(g) \rightarrow 2\,NH_3(g)$.

Answer: -5.0 kJ mol^{-1}

2B.2 The variation of enthalpy with temperature

The enthalpy of a substance increases as its temperature is raised. The relation between the increase in enthalpy and the increase in temperature depends on the conditions (for example, constant pressure or constant volume).

(a) Heat capacity at constant pressure

The most important condition is constant pressure, and the slope of the tangent to a plot of enthalpy against temperature at constant pressure is called the **heat capacity at constant**

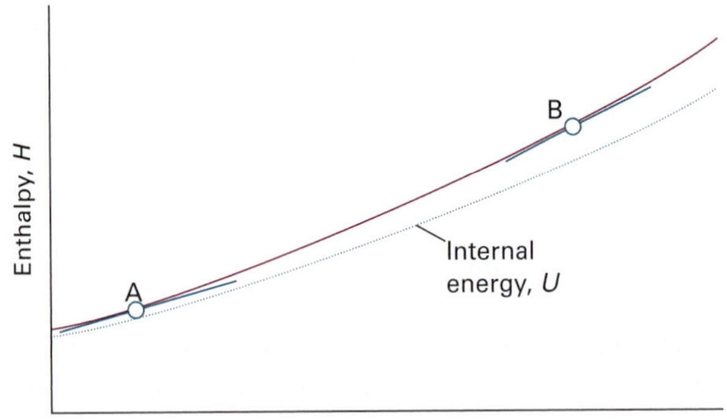

Figure 2B.3 The constant-pressure heat capacity at a particular temperature is the slope of the tangent to a curve of the enthalpy of a system plotted against temperature (at constant pressure). For gases, at a given temperature the slope of enthalpy versus temperature is steeper than that of internal energy versus temperature, and $C_{p,m}$ is larger than $C_{V,m}$.

pressure (or *isobaric heat capacity*), C_p, at a given temperature (Fig. 2B.3). More formally:

$$C_p = \left(\frac{\partial H}{\partial T}\right)_p \quad \textit{Definition} \quad \text{Heat capacity at constant pressure} \qquad (2B.5)$$

The heat capacity at constant pressure is the analogue of the heat capacity at constant volume (Topic 1A) and is an extensive property. The **molar heat capacity at constant pressure**, $C_{p,m}$, is the heat capacity per mole of substance; it is an intensive property.

The heat capacity at constant pressure is used to relate the change in enthalpy to a change in temperature. For infinitesimal changes of temperature,

$$dH = C_p dT \quad \text{(at constant pressure)} \qquad (2B.6a)$$

If the heat capacity is constant over the range of temperatures of interest, then for a measurable increase in temperature

$$\Delta H = \int_{T_1}^{T_2} C_p dT = C_p \int_{T_1}^{T_2} dT = C_p \overbrace{(T_2 - T_1)}^{\Delta T}$$

which we can summarize as

$$\Delta H = C_p \Delta T \quad \text{(at constant pressure)} \qquad (2B.6b)$$

Because a change in enthalpy can be equated with the energy supplied as heat at constant pressure, the practical form of the latter equation is

$$q_p = C_p \Delta T \qquad (2B.7)$$

Table 2B.1* Temperature variation of molar heat capacities, $C_{p,m}/(\text{J K}^{-1}\text{ mol}^{-1}) = a + bT + c/T^2$

	a	$b/(10^{-3}\text{ K}^{-1})$	$c/(10^5\text{ K}^2)$
C(s, graphite)	16.86	4.77	−8.54
CO_2(g)	44.22	8.79	−8.62
H_2O(l)	75.29	0	0
N_2(g)	28.58	3.77	−0.50

* More values are given in the *Resource section*.

This expression shows us how to measure the heat capacity of a sample: a measured quantity of energy is supplied as heat under conditions of constant pressure (as in a sample exposed to the atmosphere and free to expand), and the temperature rise is monitored.

The variation of heat capacity with temperature can sometimes be ignored if the temperature range is small; this approximation is highly accurate for a monatomic perfect gas (for instance, one of the noble gases at low pressure). However, when it is necessary to take the variation into account, a convenient approximate empirical expression is

$$C_{p,m} = a + bT + \frac{c}{T^2} \qquad (2B.8)$$

The empirical parameters a, b, and c are independent of temperature (Table 2B.1) and are found by fitting this expression to experimental data.

Example 2B.2 Evaluating an increase in enthalpy with temperature

What is the change in molar enthalpy of N_2 when it is heated from 25 °C to 100 °C? Use the heat capacity information in Table 2B.1.

Method The heat capacity of N_2 changes with temperature, so we cannot use eqn 2B.6b (which assumes that the heat capacity of the substance is constant). Therefore, we must use eqn 2B.6a, substitute eqn 2B.8 for the temperature dependence of the heat capacity, and integrate the resulting expression from 25 °C (298 K) to 100 °C (373 K).

Answer For convenience, we denote the two temperatures T_1 (298 K) and T_2 (373 K). The relation we require is

$$\int_{H_m(T_1)}^{H_m(T_2)} dH_m = \int_{T_1}^{T_2} \left(a + bT + \frac{c}{T^2}\right) dT$$

After using Integral A.1 in the *Resource section*, it follows that

$$H_m(T_2) - H_m(T_1) = a(T_2 - T_1) + \tfrac{1}{2}b(T_2^2 - T_1^2) - c\left(\frac{1}{T_2} - \frac{1}{T_1}\right)$$

Substitution of the numerical data results in

$$H_m(373\,K) = H_m(298\,K) + 2.20\,kJ\,mol^{-1}$$

If we had assumed a constant heat capacity of 29.14 J K^{-1} mol^{-1} (the value given by eqn 2B.8 for $T = 298$ K), we would have found that the two enthalpies differed by 2.19 kJ mol^{-1}.

Self-test 2B.4 At very low temperatures the heat capacity of a solid is proportional to T^3, and we can write $C_{p,m} = aT^3$. What is the change in enthalpy of such a substance when it is heated from 0 to a temperature T (with T close to 0)?

Answer: $\Delta H_m = \frac{1}{4}aT^4$

(b) The relation between heat capacities

Most systems expand when heated at constant pressure. Such systems do work on the surroundings and therefore some of the energy supplied to them as heat escapes back to the surroundings. As a result, the temperature of the system rises less than when the heating occurs at constant volume. A smaller increase in temperature implies a larger heat capacity, so we conclude that in most cases the heat capacity at constant pressure of a system is larger than its heat capacity at constant volume. We show in Topic 2D that there is a simple relation between the two heat capacities of a perfect gas:

$$C_p - C_V = nR \quad \textit{Perfect gas} \quad \text{Relation between heat capacities} \quad (2B.9)$$

It follows that the molar heat capacity of a perfect gas is about 8 J K^{-1} mol^{-1} larger at constant pressure than at constant volume. Because the molar constant-volume heat capacity of a monatomic gas is about $\frac{3}{2}R = 12$ J K^{-1} mol^{-1}, the difference is highly significant and must be taken into account.

Checklist of concepts

- ☐ 1. Energy transferred as heat at constant pressure is equal to the change in **enthalpy** of a system.
- ☐ 2. Enthalpy changes are measured in a constant-pressure calorimeter.
- ☐ 3. The **heat capacity at constant pressure** is equal to the slope of enthalpy with temperature.

Checklist of equations

Property	Equation	Comment	Equation number
Enthalpy	$H = U + pV$	Definition	2B.1
Heat transfer at constant pressure	$dH = dq_p$, $\Delta H = q_p$	No additional work	2B.2
Relation between ΔH and ΔU	$\Delta H = \Delta U + \Delta n_g RT$	Molar volumes of the participating condensed phases are negligible; isothermal process	2B.4
Heat capacity at constant pressure	$C_p = (\partial H/\partial T)_p$	Definition	2B.5
Relation between heat capacities	$C_p - C_V = nR$	Perfect gas	2B.9

An immediate conclusion is that, because all enthalpies of fusion are positive, the enthalpy of sublimation of a substance is greater than its enthalpy of vaporization (at a given temperature).

1

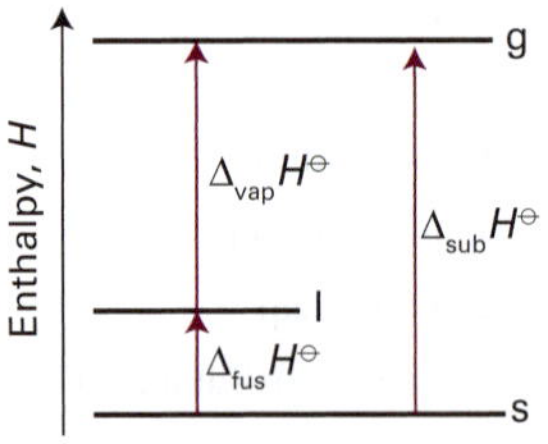

Another consequence of H being a state function is that the standard enthalpy changes of a forward process and its reverse differ in sign (2):

$$\Delta H^{\ominus}(A \to B) = -\Delta H^{\ominus}(B \to A) \qquad (2C.2)$$

For instance, because the enthalpy of vaporization of water is +44 kJ mol^{-1} at 298 K, its enthalpy of condensation at that temperature is −44 kJ mol^{-1}.

2

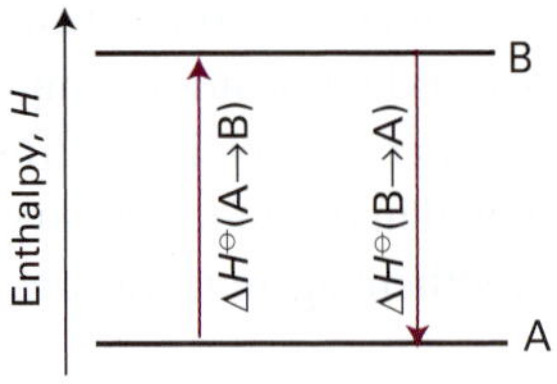

The vaporization of a solid often involves a large increase in energy, especially when the solid is ionic and the strong Coulombic interaction of the ions must be overcome in a process such as

$$MX(s) \to M^+(g) + X^-(g)$$

The **lattice enthalpy**, ΔH_L, is the change in standard molar enthalpy for this process. The lattice enthalpy is equal to the lattice internal energy at $T=0$; at normal temperatures they differ by only a few kilojoules per mole, and the difference is normally neglected.

Experimental values of the lattice enthalpy are obtained by using a **Born–Haber cycle**, a closed path of transformations starting and ending at the same point, one step of which is the formation of the solid compound from a gas of widely separated ions.

Brief illustration 2C.1 A Born–Haber cycle

A typical Born–Haber cycle, for potassium chloride, is shown in Fig. 2C.1.

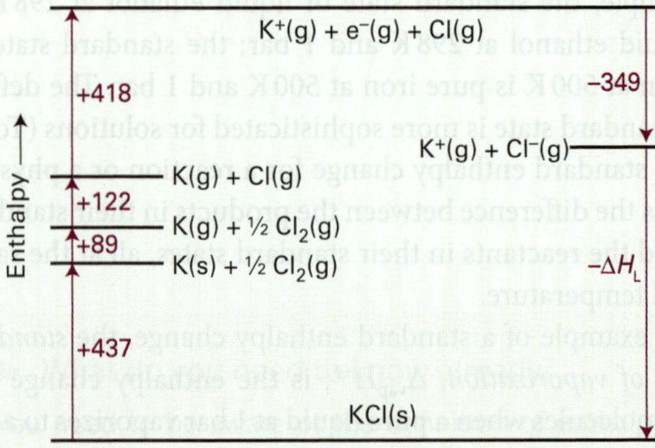

Figure 2C.1 The Born–Haber cycle for KCl at 298 K. Enthalpy changes are in kilojoules per mole.

It consists of the following steps (for convenience, starting at the elements):

	$\Delta H^{\ominus}$/(kJ mol^{-1})	
1. Sublimation of K(s)	+89	[dissociation enthalpy of K(s)]
2. Dissociation of $\frac{1}{2}Cl_2(g)$	+122	[$\frac{1}{2}$ × dissociation enthalpy of $Cl_2(g)$]
3. Ionization of K(g)	+418	[ionization enthalpy of K(g)]
4. Electron attachment to Cl(g)	−349	[electron gain enthalpy of Cl(g)]
5. Formation of solid from gas	$-\Delta H_L$/(kJ mol^{-1})	
6. Decomposition of compound	+437	[negative of enthalpy of formation of KCl(s)]

Because the sum of these enthalpy changes is equal to zero, we can infer from

$$89 + 122 + 418 - 349 - \Delta H_L/(\text{kJ mol}^{-1}) + 437 = 0$$

that $\Delta H_L = +717$ kJ mol^{-1}.

Self-test 2C.1 Assemble a similar cycle for the lattice enthalpy of magnesium chloride.

Answer: 2523 kJ mol^{-1}

Lattice enthalpies obtained in the same way as in *Brief illustration* 2C.1 are listed in Table 2C.3. They are large when the ions are highly charged and small, for then they are close together and attract each other strongly. We examine the quantitative relation between lattice enthalpy and structure in Topic 18B.

(b) Enthalpies of chemical change

Now we consider enthalpy changes that accompany chemical reactions. There are two ways of reporting the change in enthalpy that accompanies a chemical reaction. One is to write

Table 2C.3* Lattice enthalpies at 298 K, $\Delta H_L/(\text{kJ mol}^{-1})$

NaF	787
NaBr	751
MgO	3850
MgS	3406

* More values are given in the *Resource section*.

the **thermochemical equation**, a combination of a chemical equation and the corresponding change in standard enthalpy:

$$CH_4(g)+2\,O_2(g)\rightarrow CO_2(g)+2\,H_2O(g) \quad \Delta H^{\ominus}=-890\,\text{kJ}$$

$\Delta H^{\ominus}$ is the change in enthalpy when reactants in their standard states change to products in their standard states:

Pure, separate reactants in their standard states
→ pure, separate products in their standard states

Except in the case of ionic reactions in solution, the enthalpy changes accompanying mixing and separation are insignificant in comparison with the contribution from the reaction itself. For the combustion of methane, the standard value refers to the reaction in which 1 mol CH_4 in the form of pure methane gas at 1 bar reacts completely with 2 mol O_2 in the form of pure oxygen gas to produce 1 mol CO_2 as pure carbon dioxide at 1 bar and 2 mol H_2O as pure liquid water at 1 bar; the numerical value is for the reaction at 298.15 K.

Alternatively, we write the chemical equation and then report the **standard reaction enthalpy**, $\Delta_r H^{\ominus}$ (or 'standard enthalpy of reaction'). Thus, for the combustion of methane, we write

$$CH_4(g)+2\,O_2(g)\rightarrow CO_2(g)+2\,H_2O(g)$$
$$\Delta_r H^{\ominus}=-890\,\text{kJ mol}^{-1}$$

For a reaction of the form 2 A+B→3 C+D the standard reaction enthalpy would be

$$\Delta_r H^{\ominus}=\{3H_m^{\ominus}(C)+H_m^{\ominus}(D)\}-\{2H_m^{\ominus}(A)+H_m^{\ominus}(B)\}$$

where $H_m^{\ominus}(J)$ is the standard molar enthalpy of species J at the temperature of interest. Note how the 'per mole' of $\Delta_r H^{\ominus}$ comes directly from the fact that molar enthalpies appear in this expression. We interpret the 'per mole' by noting the stoichiometric coefficients in the chemical equation. In this case, 'per mole' in $\Delta_r H^{\ominus}$ means 'per 2 mol A', 'per mole B', 'per 3 mol C', or 'per mol D'. In general,

$$\Delta_r H^{\ominus}=\sum_{\text{Products}}\nu H_m^{\ominus}-\sum_{\text{Reactants}}\nu H_m^{\ominus} \quad \textit{Definition} \quad \text{Standard reaction enthalpy} \quad (2C.3)$$

where in each case the molar enthalpies of the species are multiplied by their (dimensionless and positive) stoichiometric coefficients, ν. This formal definition is of little practical value because the absolute values of the standard molar enthalpies are unknown: we see in Section 2C.2a how that problem is overcome.

Some standard reaction enthalpies have special names and a particular significance. For instance, the **standard enthalpy of combustion**, $\Delta_c H^{\ominus}$, is the standard reaction enthalpy for the complete oxidation of an organic compound to CO_2 gas and liquid H_2O if the compound contains C, H, and O, and to N_2 gas if N is also present.

Brief illustration 2C.2 Enthalpy of combustion

The combustion of glucose is

$$C_6H_{12}O_6(s)+6\,O_2(g)\rightarrow 6\,CO_2(g)+6\,H_2O(l)$$
$$\Delta_c H^{\ominus}=-2808\,\text{kJ mol}^{-1}$$

The value quoted shows that 2808 kJ of heat is released when 1 mol $C_6H_{12}O_6$ burns under standard conditions (at 298 K). More values are given in Table 2C.4.

Self-test 2C.2 Predict the heat output of the combustion of 1.0 dm³ of octane at 298 K. Its mass density is 0.703 g cm⁻³.

Answer: 34 MJ

Table 2C.4* Standard enthalpies of formation ($\Delta_f H^{\ominus}$) and combustion ($\Delta_c H^{\ominus}$) of organic compounds at 298 K

	$\Delta_f H^{\ominus}/(\text{kJ mol}^{-1})$	$\Delta_c H^{\ominus}/(\text{kJ mol}^{-1})$
Benzene, $C_6H_6(l)$	+49.0	−3268
Ethane, $C_2H_6(g)$	−84.7	−1560
Glucose, $C_6H_{12}O_6(s)$	−1274	−2808
Methane, $CH_4(g)$	−74.8	−890
Methanol, $CH_3OH(l)$	−238.7	−721

* More values are given in the *Resource section*.

(c) Hess's law

Standard enthalpies of individual reactions can be combined to obtain the enthalpy of another reaction. This application of the First Law is called **Hess's law**:

> The standard enthalpy of an overall reaction is the sum of the standard enthalpies of the individual reactions into which a reaction may be divided. *Hess's law*

The individual steps need not be realizable in practice: they may be hypothetical reactions, the only requirement being that their chemical equations should balance. The thermodynamic basis of the law is the path-independence of the value of $\Delta_r H^{\ominus}$ and the implication that we may take the specified reactants, pass through any (possibly hypothetical) set of reactions to the specified products, and overall obtain the same change of enthalpy. The importance of Hess's law is that series of related compounds. Nor does the approach distinguish between geometrical isomers, where the same atoms and bonds may be present but experimentally the enthalpies of formation might be significantly different.

Computer-aided molecular modelling has largely displaced this more primitive approach. Commercial software packages use the principles developed in Topic 10E to calculate the standard enthalpy of formation of a molecule drawn on the computer screen. These techniques can be applied to different conformations of the same molecule. In the case of methylcyclohexane, for instance, the calculated conformational energy difference ranges from 5.9 to 7.9 kJ mol^{-1}, with the equatorial conformer having the lower standard enthalpy of formation. These estimates compare favourably with the experimental value of 7.5 kJ mol^{-1}. However, good agreement between calculated and experimental values is relatively rare. Computational methods almost always predict correctly which conformer is more stable but do not always predict the correct magnitude of the conformational energy difference. The most reliable technique for the determination of enthalpies of formation remains calorimetry, typically by using enthalpies of combustion.

Brief illustration 2C.5 Molecular modelling

Each software package has its own procedures; the general approach, though, is the same in most cases: the structure of the molecule is specified and the nature of the calculation selected. When the procedure is applied to the axial and equatorial isomers of methylcyclohexane, a typical value for the standard enthalpy of formation of equatorial isomer in the gas phase is −183 kJ mol^{-1} (using the AM1 semi-empirical procedure) whereas that for the axial isomer is −177 kJ mol^{-1}, a difference of 6 kJ mol^{-1}. The experimental difference is 7.5 kJ mol^{-1}.

Self-test 2C.6 If you have access to modelling software, repeat this calculation for the two isomers of cyclohexanol.

Answer: Using AM1: eq: −345 kJ mol^{-1}; ax: −349 kJ mol^{-1}

2C.3 The temperature dependence of reaction enthalpies

The standard enthalpies of many important reactions have been measured at different temperatures. However, in the absence of this information, standard reaction enthalpies at different temperatures may be calculated from heat capacities and the reaction enthalpy at some other temperature (Fig. 2C.2). In many cases heat capacity data are more accurate than reaction enthalpies. Therefore, providing the information is available, the procedure we are about to describe is more accurate than the direct measurement of a reaction enthalpy at an elevated temperature.

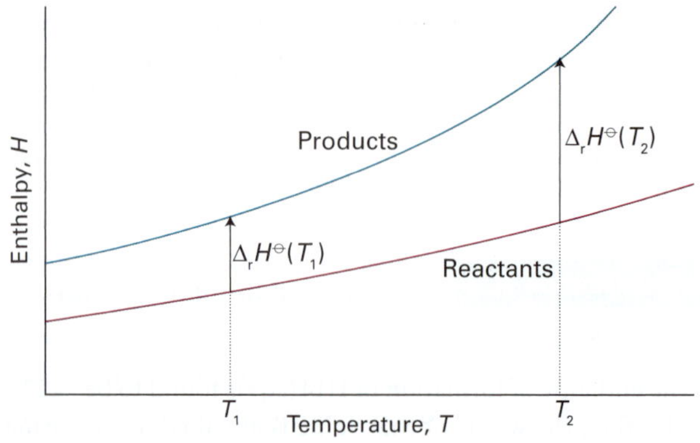

Figure 2C.2 When the temperature is increased, the enthalpy of the products and the reactants both increase, but may do so to different extents. In each case, the change in enthalpy depends on the heat capacities of the substances. The change in reaction enthalpy reflects the difference in the changes of the enthalpies.

It follows from eqn 2B.6a ($dH=C_p dT$) that, when a substance is heated from T_1 to T_2, its enthalpy changes from $H(T_1)$ to

$$H(T_2)=H(T_1)+\int_{T_1}^{T_2} C_p \mathrm{d}T \quad (2C.6)$$

(We have assumed that no phase transition takes place in the temperature range of interest.) Because this equation applies to each substance in the reaction, the standard reaction enthalpy changes from $\Delta_r H^\ominus(T_1)$ to

$$\Delta_r H^\ominus(T_2)=\Delta_r H^\ominus(T_1)+\int_{T_1}^{T_2} \Delta_r C_p^\ominus \mathrm{d}T \quad \text{Kirchhoff's law} \quad (2C.7a)$$

where $\Delta_r C_p^\ominus$ is the difference of the molar heat capacities of products and reactants under standard conditions weighted by the stoichiometric coefficients that appear in the chemical equation:

$$\Delta_r C_{p,\mathrm{m}}^\ominus = \sum_{\text{Products}} \nu C_{p,\mathrm{m}}^\ominus - \sum_{\text{Reactants}} \nu C_{p,\mathrm{m}}^\ominus \quad (2C.7b)$$

or, in the notation of eqn 2C.5b,

$$\Delta_r C_{p,\mathrm{m}}^\ominus = \sum_{\mathrm{J}} \nu_{\mathrm{J}} C_{p,\mathrm{m}}^\ominus(\mathrm{J}) \quad (2C.7c)$$

Equation 2C.7a is known as **Kirchhoff's law**. It is normally a good approximation to assume that $\Delta_r C_p^\ominus$ is independent of the temperature, at least over reasonably limited ranges. Although the individual heat capacities may vary, their difference varies less significantly. In some cases the temperature dependence of heat capacities is taken into account by using eqn 2B.8.

Example 2C.2 Using Kirchhoff's law

The standard enthalpy of formation of $H_2O(g)$ at 298 K is −241.82 kJ mol^{-1}. Estimate its value at 100 °C given the following values of the molar heat capacities at constant pressure: $H_2O(g)$: 33.58 J K^{-1} mol^{-1}; $H_2(g)$: 28.84 J K^{-1} mol^{-1}; $O_2(g)$: 29.37 J^{-1} mol^{-1}. Assume that the heat capacities are independent of temperature.

Method When $\Delta_r C_p^{\ominus}$ is independent of temperature in the range T_1 to T_2, the integral in eqn 2C.7a evaluates to $(T_2-T_1)\Delta_r C_p^{\ominus}$. Therefore,

$$\Delta_r H^{\ominus}(T_2)=\Delta_r H^{\ominus}(T_1)+(T_2-T_1)\Delta_r C_p^{\ominus}$$

To proceed, write the chemical equation, identify the stoichiometric coefficients, and calculate $\Delta_r C_p^{\ominus}$ from the data.

Answer The reaction is $H_2(g)+\frac{1}{2}O_2(g)\rightarrow H_2O(g)$, so

$$\begin{aligned}\Delta_r C_p^{\ominus} &= C_{p,m}^{\ominus}(H_2O,g)-\{C_{p,m}^{\ominus}(H_2,g)+\tfrac{1}{2}C_{p,m}^{\ominus}(O_2,g)\}\\ &=-9.94\,\mathrm{J\,K^{-1}\,mol^{-1}}\end{aligned}$$

It then follows that

$$\begin{aligned}\Delta_r H^{\ominus}(373\,\mathrm{K}) &= -241.82\,\mathrm{kJ\,mol^{-1}}+(75\,\mathrm{K})\\ &\quad\times(-9.94\,\mathrm{J\,K^{-1}\,mol^{-1}})=-242.6\,\mathrm{kJ\,mol^{-1}}\end{aligned}$$

Self-test 2C.7 Estimate the standard enthalpy of formation of cyclohexane, $C_6H_{12}(l)$, at 400 K from the data in Table 2C.6.

Answer: −163 kJ mol^{-1}

2C.4 Experimental techniques

The classic tool of thermochemistry is the calorimeter, as summarized in Topic 2B. However, technological advances have been made that allow measurements to be made on samples with mass as little as a few milligrams. We describe two of them here.

(a) Differential scanning calorimetry

A **differential scanning calorimeter** (DSC) measures the energy transferred as heat to or from a sample at constant pressure during a physical or chemical change. The term 'differential' refers to the fact that the behaviour of the sample is compared to that of a reference material that does not undergo a physical or chemical change during the analysis. The term 'scanning' refers to the fact that the temperatures of the sample and reference material are increased, or scanned, during the analysis.

A DSC consists of two small compartments that are heated electrically at a constant rate. The temperature, T, at time t during a linear scan is $T=T_0+\alpha t$, where T_0 is the initial temperature

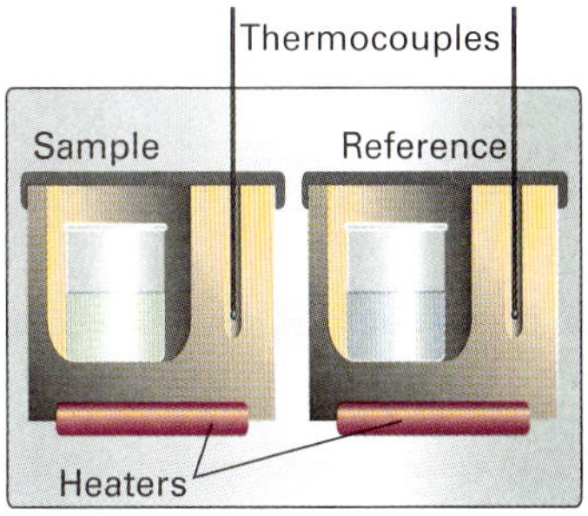

Figure 2C.3 A differential scanning calorimeter. The sample and a reference material are heated in separate but identical metal heat sinks. The output is the difference in power needed to maintain the heat sinks at equal temperatures as the temperature rises.

and α is the scan rate. A computer controls the electrical power supply that maintains the same temperature in the sample and reference compartments throughout the analysis (Fig. 2C.3).

If no physical or chemical change occurs in the sample at temperature T, we write the heat transferred to the sample as $q_p=C_p\Delta T$, where $\Delta T=T-T_0$ and we have assumed that C_p is independent of temperature. Because $T=T_0+\alpha t$, $\Delta T=\alpha t$. The chemical or physical process requires the transfer of $q_p+q_{p,\mathrm{ex}}$, where $q_{p,\mathrm{ex}}$ is the excess energy transferred as heat needed to attain the same change in temperature of the sample as the control. The quantity $q_{p,\mathrm{ex}}$ is interpreted in terms of an apparent change in the heat capacity at constant pressure of the sample, C_p, during the temperature scan:

$$C_{p,\mathrm{ex}}=\frac{q_{p,\mathrm{ex}}}{\Delta T}=\frac{q_{p,\mathrm{ex}}}{\alpha t}=\frac{P_{\mathrm{ex}}}{\alpha} \quad (2C.8)$$

where $P_{\mathrm{ex}}=q_{p,\mathrm{ex}}/t$ is the excess electrical power necessary to equalize the temperature of the sample and reference compartments. A DSC trace, also called a **thermogram**, consists of a plot of $C_{p,\mathrm{ex}}$ against T (Fig. 2C.4). The enthalpy change associated with the process is

$$\Delta H=\int_{T_1}^{T_2} C_{p,\mathrm{ex}}\mathrm{d}T \quad (2C.9)$$

where T_1 and T_2 are, respectively, the temperatures at which the process begins and ends. This relation shows that the enthalpy change is equal to the area under the plot of $C_{p,\mathrm{ex}}$ against T.

The technique is used, for instance, to assess the stability of proteins, nucleic acids, and membranes. The thermogram shown in Fig. 2C.4 indicates that the protein ubiquitin undergoes an endothermic conformational change in which a large number of non-covalent interactions (such as hydrogen bonds) are broken simultaneously and result in denaturation, the loss of the protein's three-dimensional structure. The area under the curve represents the heat absorbed in this process and

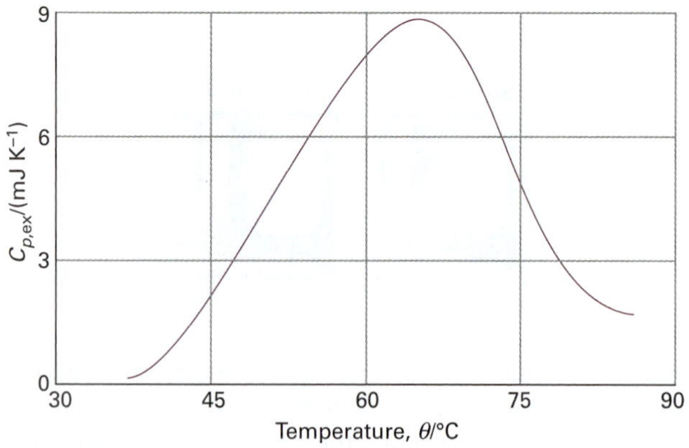

Figure 2C.4 A thermogram for the protein ubiquitin at pH = 2.45. The protein retains its native structure up to about 45 °C and then undergoes an endothermic conformational change. (Adapted from B. Chowdhry and S. LeHarne, *J. Chem. Educ.* **74**, 236 (1997).)

can be identified with the enthalpy change. The thermogram also reveals the formation of new intermolecular interactions in the denatured form. The increase in heat capacity accompanying the native → denatured transition reflects the change from a more compact native conformation to one in which the more exposed amino acid side chains in the denatured form have more extensive interactions with the surrounding water molecules.

(b) Isothermal titration calorimetry

Isothermal titration calorimetry (ITC) is also a 'differential' technique in which the thermal behaviour of a sample is compared with that of a reference. The apparatus is shown in Fig. 2C.5. One of the thermally conducting vessels, which have a volume of a few millilitres ($10^{-6}\,m^3$), contains the reference (water for instance) and a heater rated at a few milliwatts. The second vessel contains one of the reagents, such as a solution of a macromolecule with binding sites; it also contains a heater. At the start of the experiment, both heaters are activated, and then precisely determined amounts (of volume of about a microlitre, $10^{-9}\,m^3$) of the second reagent are added to the reaction cell. The power required to maintain the same temperature differential with the reference cell is monitored. If the reaction is exothermic, less power is needed; if it is endothermic, then more power must be supplied.

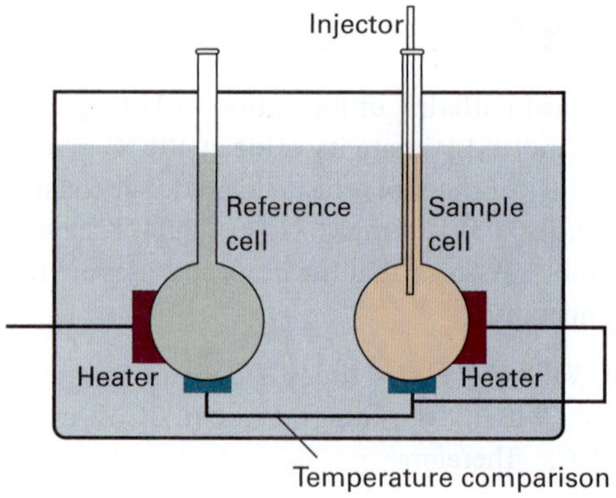

Figure 2C.5 A schematic diagram of the apparatus used for isothermal titration calorimetry.

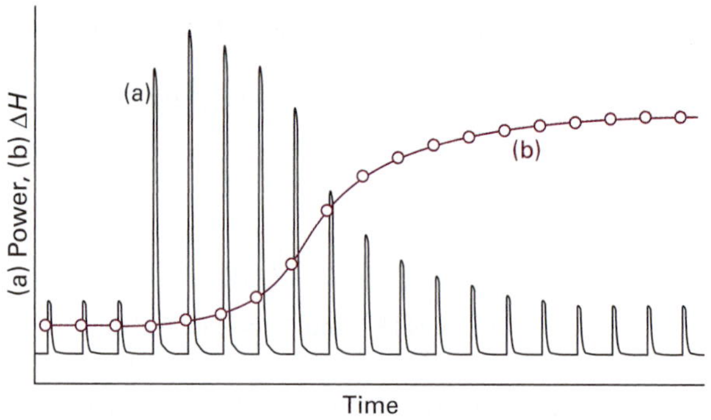

Figure 2C.6 (a) The record of the power applied as each injection is made, and (b) the sum of successive enthalpy changes in the course of the titration.

A typical result is shown in Fig. 2C.6, which shows the power needed to maintain the temperature differential: from the power and the length of time, δt, for which it is supplied, the heat supplied, δq_i, for the injection i can be calculated from $\delta q_i = P_i \delta t$. If the volume of solution is V and the molar concentration of unreacted reagent A is c_i at the time of the ith injection, then the change in its concentration at that injection is δc_i and the heat generated (or absorbed) by the reaction is $V\Delta_r H \delta c_i = \delta q_i$. The sum of all such quantities, given that the sum of δc_i is the known initial concentration of the reactant, can then be interpreted as the value of $\Delta_r H$ for the reaction.

Checklist of concepts

- ☐ 1. The **standard enthalpy of transition** is equal to the energy transferred as heat at constant pressure in the transition.
- ☐ 2. A **thermochemical equation** is a chemical equation and its associated change in enthalpy.
- ☐ 3. **Hess's law** states that the standard enthalpy of an overall reaction is the sum of the standard enthalpies of the individual reactions into which a reaction may be divided.

☐ 4. **Standard enthalpies of formation** are defined in terms of the reference states of elements.

☐ 5. The **standard reaction enthalpy** is expressed as the difference of the standard enthalpies of formation of products and reactants.

☐ 6. Computer modelling is used to estimate standard enthalpies of formation.

☐ 7. The temperature dependence of a reaction enthalpy is expressed by **Kirchhoff's law.**

Checklist of equations

Property	Equation	Comment	Equation number
The standard reaction enthalpy	$\Delta_r H^\ominus = \sum_{\text{Products}} \nu\Delta_f H^\ominus - \sum_{\text{Reactants}} \nu\Delta_f H^\ominus$ $\Delta_r H^\ominus = \sum_J \nu_J \Delta_f H^\ominus(J)$	ν: stoichiometric coefficients; ν_J: (signed) stoichiometric numbers	2C.5
Kirchhoff's law	$\Delta_r H^\ominus(T_2) = \Delta_r H^\ominus(T_2) + \int_{T_1}^{T_2} \Delta_r C_p^\ominus \mathrm{d}T$		2C.7a
	$\Delta_r C_{p,\mathrm{m}}^\ominus = \sum_J \nu_J C_{p,\mathrm{m}}^\ominus(J)$		2C.7c
	$\Delta_r H^\ominus(T_2) = \Delta_r H^\ominus(T_1) + (T_2 - T_1)\Delta_r C_p^\ominus$	If $\Delta_r C_p^\ominus$ independent of temperature	

2D State functions and exact differentials

Contents

➤ Why do you need to know this material?

Thermodynamics gives us the power to derive relations between a variety of properties: this Topic is a first introduction to the manipulation of equations involving state functions. In the process, we obtain such important relations as that between heat capacities. An important technological consequence is the Joule–Thomson effect for cooling gases, which is derived here.

➤ What is the key idea?

The fact that internal energy and enthalpy are state functions leads to relations between thermodynamic properties.

➤ What do you need to know already?

You need to be aware that internal energy and enthalpy are state functions (Topics 2B and 2C) and be familiar with heat capacity. You need to be able to make use of several simple relations involving partial derivatives (*Mathematical background* 2).

A **state function** is a property that depends only on the current state of a system and is independent of its history. The internal energy and enthalpy are two examples of state functions. Physical quantities that do depend on the path between two states are called **path functions**. Examples of path functions are the work and the heating that are done when preparing a state. We do not speak of a system in a particular state as possessing work or heat. In each case, the energy transferred as work or heat relates to the path being taken between states, not the current state itself.

A part of the richness of thermodynamics is that it uses the mathematical properties of state functions to draw far-reaching conclusions about the relations between physical properties and thereby establish connections that may be completely unexpected. The practical importance of this ability is that we can combine measurements of different properties to obtain the value of a property we require.

2D.1 Exact and inexact differentials

Consider a system undergoing the changes depicted in Fig. 2D.1. The initial state of the system is i and in this state the internal energy is U_i. Work is done by the system as it expands adiabatically to a state f. In this state the system has an internal energy U_f and the work done on the system as it changes along Path 1 from i to f is w. Notice our use of language: U is a property of the state; w is a property of the path. Now consider another process, Path 2, in which the initial and final states are the same as those in Path 1 but in which the expansion is not adiabatic. The internal energy of both the initial and the final states are the same as before (because U is a state function). However, in the second path an energy q' enters the system as heat and the work w' is not the same as w. The work and the heat are path functions.

If a system is taken along a path (for example, by heating it), U changes from U_i to U_f, and the overall change is the sum (integral) of all the infinitesimal changes along the path:

$$\Delta U = \int_i^f \mathrm{d}U \qquad (2D.1)$$

The value of ΔU depends on the initial and final states of the system but is independent of the path between them. This path-independence of the integral is expressed by saying that $\mathrm{d}U$

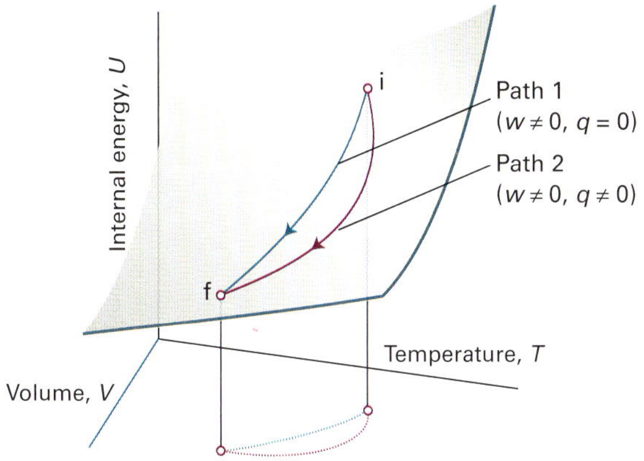

Figure 2D.1 As the volume and temperature of a system are changed, the internal energy changes. An adiabatic and a non-adiabatic path are shown as Path 1 and Path 2, respectively: they correspond to different values of q and w but to the same value of ΔU.

is an 'exact differential'. In general, an **exact differential** is an infinitesimal quantity that, when integrated, gives a result that is independent of the path between the initial and final states.

When a system is heated, the total energy transferred as heat is the sum of all individual contributions at each point of the path:

$$q = \int_{\text{i,path}}^{\text{f}} \mathrm{d}q \qquad (2D.2)$$

Notice the differences between this equation and eqn 2D.1. First, we do not write Δq, because q is not a state function and the energy supplied as heat cannot be expressed as $q_f - q_i$. Secondly, we must specify the path of integration because q depends on the path selected (for example, an adiabatic path has $q=0$, whereas a non-adiabatic path between the same two states would have $q \neq 0$). This path-dependence is expressed by saying that dq is an 'inexact differential'. In general, an **inexact differential** is an infinitesimal quantity that, when integrated, gives a result that depends on the path between the initial and final states. Often dq is written đq to emphasize that it is inexact and requires the specification of a path.

The work done on a system to change it from one state to another depends on the path taken between the two specified states; for example, in general the work is different if the change takes place adiabatically and non-adiabatically. It follows that dw is an inexact differential. It is often written đw.

Example 2D.1 Calculating work, heat, and change in internal energy

Consider a perfect gas inside a cylinder fitted with a piston. Let the initial state be T,V_i and the final state be T,V_f. The change of state can be brought about in many ways, of which the two simplest are the following: Path 1, in which there is free expansion against zero external pressure; Path 2, in which there is reversible, isothermal expansion. Calculate w, q, and ΔU for each process.

Method To find a starting point for a calculation in thermodynamics, it is often a good idea to go back to first principles and to look for a way of expressing the quantity we are asked to calculate in terms of other quantities that are easier to calculate. It is argued in Topic 2B that the internal energy of a perfect gas depends only on the temperature and is independent of the volume those molecules occupy, so for any isothermal change, $\Delta U=0$. We also know that in general $\Delta U=q+w$. The question depends on being able to combine the two expressions. Topic 2A presents a number of expressions for the work done in a variety of processes, and here we need to select the appropriate ones.

Answer Because $\Delta U=0$ for both paths and $\Delta U=q+w$, in each case $q=-w$. The work of free expansion is zero (eqn 2A.7 of Topic 2A, $w=0$); so in Path 1, $w=0$ and therefore $q=0$ too. For Path 2, the work is given by eqn 2A.9 of Topic 2A ($w=-nRT\ln(V_f/V_i)$) and consequently $q=nRT\ln(V_f/V_i)$.

Self-test 2D.1 Calculate the values of q, w, and ΔU for an irreversible isothermal expansion of a perfect gas against a constant nonzero external pressure.

Answer: $q=p_{ex}\Delta V$, $w=-p_{ex}\Delta V$, $\Delta U=0$

2D.2 Changes in internal energy

We begin to unfold the consequences of dU being an exact differential by exploring a closed system of constant composition (the only type of system considered in the rest of this Topic). The internal energy U can be regarded as a function of V, T, and p, but, because there is an equation of state (Topic 1A), stating the values of two of the variables fixes the value of the third. Therefore, it is possible to write U in terms of just two independent variables: V and T, p and T, or p and V. Expressing U as a function of volume and temperature fits the purpose of our discussion.

(a) General considerations

Because the internal energy is a function of the volume and the temperature, when these two quantities change, the internal energy changes by

$$\mathrm{d}U = \left(\frac{\partial U}{\partial V}\right)_T \mathrm{d}V + \left(\frac{\partial U}{\partial T}\right)_V \mathrm{d}T \qquad (2D.3)$$

General expression for a change in U with T and V

The interpretation of this equation is that, in a closed system of constant composition, any infinitesimal change in the internal

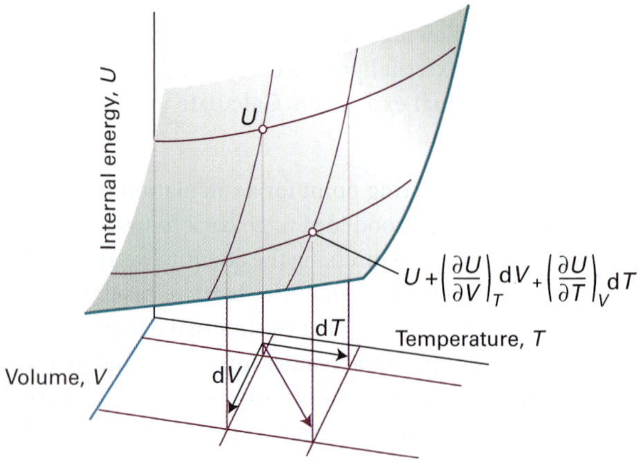

Figure 2D.2 An overall change in U, which is denoted dU, arises when both V and T are allowed to change. If second-order infinitesimals are ignored, the overall change is the sum of changes for each variable separately.

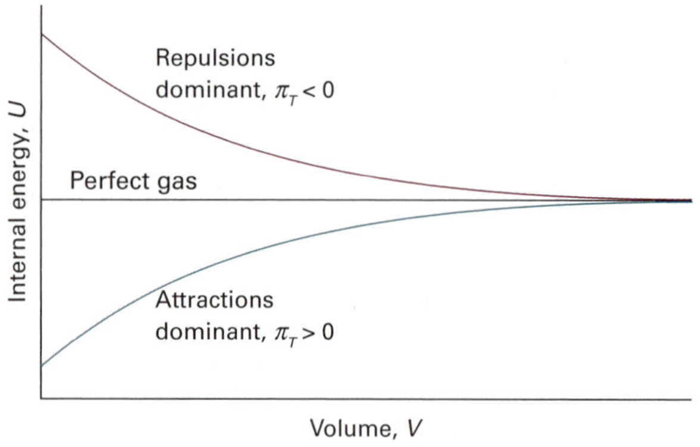

Figure 2D.4 For a perfect gas, the internal energy is independent of the volume (at constant temperature). If attractions are dominant in a real gas, the internal energy increases with volume because the molecules become farther apart on average. If repulsions are dominant, the internal energy decreases as the gas expands.

energy is proportional to the infinitesimal changes of volume and temperature, the coefficients of proportionality being the two partial derivatives (Fig. 2D.2).

In many cases partial derivatives have a straightforward physical interpretation, and thermodynamics gets shapeless and difficult only when that interpretation is not kept in sight. In the present case, we have already met $(\partial U/\partial T)_V$ in Topic 2A, where we saw that it is the constant-volume heat capacity, C_V. The other coefficient, $(\partial U/\partial V)_T$, plays a major role in thermodynamics because it is a measure of the variation of the internal energy of a substance as its volume is changed at constant temperature (Fig. 2D.3). We shall denote it π_T and, because it has the same dimensions as pressure but arises from the interactions between the molecules within the sample, call it the **internal pressure**:

$$\pi_T=\left(\frac{\partial U}{\partial V}\right)_T \qquad \textit{Definition} \quad \text{Internal pressure} \qquad (2D.4)$$

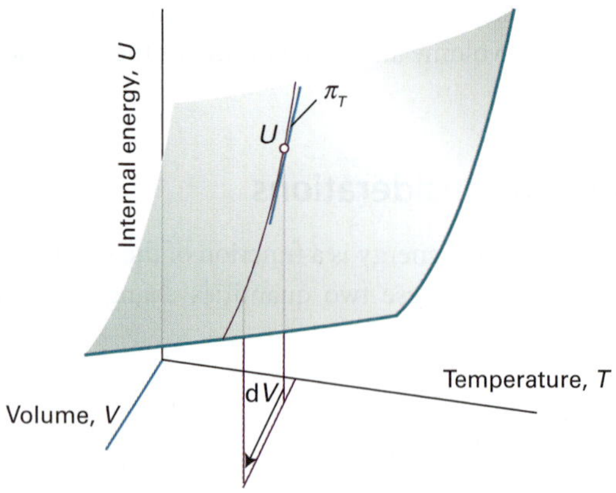

Figure 2D.3 The internal pressure, π_T, is the slope of U with respect to V with the temperature T held constant.

In terms of the notation C_V and π_T, eqn 2D.3 can now be written

$$dU=\pi_T dV+C_V dT \qquad (2D.5)$$

When there are no interactions between the molecules, the internal energy is independent of their separation and hence independent of the volume of the sample. Therefore, for a perfect gas we can write $\pi_T=0$. If the gas is described by the van der Waals equation with a, the parameter corresponding to attractive interactions, dominant, then an increase in volume increases the average separation of the molecules and therefore raises the internal energy. In this case, we expect $\pi_T>0$ (Fig. 2D.4).

The statement $\pi_T=0$ (that is, the internal energy is independent of the volume occupied by the sample) can be taken to be the definition of a perfect gas, for in Topic 3D we see that it implies the equation of state $pV\propto T$.

James Joule thought that he could measure π_T by observing the change in temperature of a gas when it is allowed to expand into a vacuum. He used two metal vessels immersed in a water bath (Fig. 2D.5). One was filled with air at about 22 atm and the other was evacuated. He then tried to measure the change

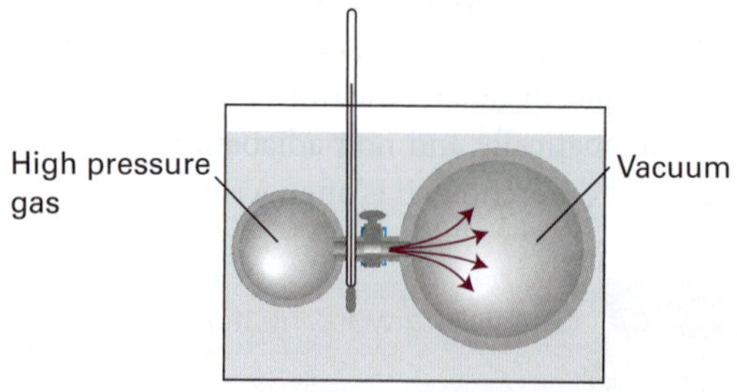

Figure 2D.5 A schematic diagram of the apparatus used by Joule in an attempt to measure the change in internal energy when a gas expands isothermally. The heat absorbed by the gas is proportional to the change in temperature of the bath.

in temperature of the water of the bath when a stopcock was opened and the air expanded into a vacuum. He observed no change in temperature.

The thermodynamic implications of the experiment are as follows. No work was done in the expansion into a vacuum, so $w=0$. No energy entered or left the system (the gas) as heat because the temperature of the bath did not change, so $q=0$. Consequently, within the accuracy of the experiment, $\Delta U=0$. Joule concluded that U does not change when a gas expands isothermally and therefore that $\pi_T=0$. His experiment, however, was crude. In particular, the heat capacity of the apparatus was so large that the temperature change that gases do in fact cause was too small to measure. Nevertheless, from his experiment Joule had extracted an essential limiting property of a gas, a property of a perfect gas, without detecting the small deviations characteristic of real gases.

(b) Changes in internal energy at constant pressure

Partial derivatives have many useful properties and some that we shall draw on frequently are reviewed in *Mathematical background* 2. Skilful use of them can often turn some unfamiliar quantity into a quantity that can be recognized, interpreted, or measured.

As an example, suppose we want to find out how the internal energy varies with temperature when the pressure rather than the volume of the system is kept constant. If we divide both sides of eqn 2D.5 by $\mathrm{d}T$ and impose the condition of constant pressure on the resulting differentials, so that $\mathrm{d}U/\mathrm{d}T$ on the left becomes $(\partial U/\partial T)_p$, we obtain

$$\left(\frac{\partial U}{\partial T}\right)_p=\pi_T\left(\frac{\partial V}{\partial T}\right)_p+C_V$$

It is usually sensible in thermodynamics to inspect the output of a manipulation like this to see if it contains any recognizable physical quantity. The partial derivative on the right in this expression is the slope of the plot of volume against temperature (at constant pressure). This property is normally tabulated as the **expansion coefficient**, α, of a substance, which is defined as

$$\alpha=\frac{1}{V}\left(\frac{\partial V}{\partial T}\right)_p \quad \textit{Definition} \quad \text{Expansion coefficient} \quad (2D.6)$$

and physically is the fractional change in volume that accompanies a rise in temperature. A large value of α means that the volume of the sample responds strongly to changes in temperature. Table 2D.1 lists some experimental values of α. For future reference, it also lists the **isothermal compressibility**, κ_T (kappa), which is defined as

Table 2D.1* Expansion coefficients (α) and isothermal compressibilities (κ_T) at 298 K

	$\alpha/(10^{-4}\ \mathrm{K}^{-1})$	$\kappa_T/(10^{-6}\ \mathrm{bar}^{-1})$
Benzene	12.4	90.9
Diamond	0.030	0.185
Lead	0.861	2.18
Water	2.1	49.0

* More values are given in the *Resource section*.

$$\kappa_T=-\frac{1}{V}\left(\frac{\partial V}{\partial p}\right)_T \quad \textit{Definition} \quad \text{Isothermal compressibility} \quad (2D.7)$$

The isothermal compressibility is a measure of the fractional change in volume when the pressure is increased by a small amount; the negative sign in the definition ensures that the compressibility is a positive quantity, because an increase of pressure, implying a positive $\mathrm{d}p$, brings about a reduction of volume, a negative $\mathrm{d}V$.

Example 2D.2 Calculating the expansion coefficient of a gas

Derive an expression for the expansion coefficient of a perfect gas.

Method The expansion coefficient is defined in eqn 2D.6. To use this expression, substitute the expression for V in terms of T obtained from the equation of state for the gas. As implied by the subscript in eqn 2D.6, the pressure, p, is treated as a constant.

Answer Because $pV=nRT$, we can write

$$\alpha=\frac{1}{V}\left(\frac{\partial(nRT/p)}{\partial T}\right)_p=\frac{1}{V}\times\frac{nR}{p}\frac{\mathrm{d}T}{\mathrm{d}T}=\frac{nR}{pV}=\frac{1}{T}$$

The higher the temperature, the less responsive is the volume of a perfect gas to a change in temperature.

Self-test 2D.2 Derive an expression for the isothermal compressibility of a perfect gas.

Answer: $\kappa_T=1/p$

When we introduce the definition of α into the equation for $(\partial U/\partial T)_p$, we obtain

$$\left(\frac{\partial U}{\partial T}\right)_p=\alpha\pi_T V+C_V \quad (2D.8)$$

This equation is entirely general (provided the system is closed and its composition is constant). It expresses the dependence of the internal energy on the temperature at constant pressure in terms of C_V, which can be measured in one experiment, in terms of α, which can be measured in another, and in terms of the quantity π_T. For a perfect gas, $\pi_T=0$, so then

$$\left(\frac{\partial U}{\partial T}\right)_p = C_V \qquad \text{(2D.9)}$$

That is, although the constant-volume heat capacity of a perfect gas is defined as the slope of a plot of internal energy against temperature at constant volume, for a perfect gas C_V is also the slope at constant pressure.

Equation 2D.9 provides an easy way to derive the relation between C_p and C_V for a perfect gas. Thus, we can use it to express both heat capacities in terms of derivatives at constant pressure:

$$C_p - C_V = \overbrace{\left(\frac{\partial H}{\partial T}\right)_p}^{\text{Definition of } C_p} - \overbrace{\left(\frac{\partial U}{\partial T}\right)_p}^{\text{eqn 2D.9}}$$

Then we introduce $H = U + pV = U + nRT$ into the first term, which results in

$$C_p - C_V = \left(\frac{\partial (U+nRT)}{\partial T}\right)_p - \left(\frac{\partial U}{\partial T}\right)_p = nR \qquad \text{(2D.10)}$$

We show in the following *Justification* that in general

$$C_p - C_V = \frac{\alpha^2 TV}{\kappa_T} \qquad \text{(2D.11)}$$

Equation 2D.11 applies to any substance (that is, it is 'universally true'). It reduces to eqn 2D.10 for a perfect gas when we set $\alpha = 1/T$ and $\kappa_T = 1/p$. Because expansion coefficients α of liquids and solids are small, it is tempting to deduce from eqn 2D.11 that for them $C_p \approx C_V$. But this is not always so, because the compressibility κ_T might also be small, so α^2/κ_T might be large. That is, although only a little work need be done to push back the atmosphere, a great deal of work may have to be done to pull atoms apart from one another as the solid expands.

Brief illustration 2D.1 The relation between heat capacities

The expansion coefficient and isothermal compressibility of water at 25 °C are given in Table 2D.1 as $2.1\times10^{-4}\,\mathrm{K^{-1}}$ and $4.96\times10^{-5}\,\mathrm{atm^{-1}}$ ($4.90\times10^{-10}\,\mathrm{Pa^{-1}}$), respectively. The molar volume of water at that temperature, $V_\mathrm{m} = M/\rho$ (where ρ is the mass density) is $18.1\,\mathrm{cm^3\,mol^{-1}}$ ($1.81\times10^{-5}\,\mathrm{m^3\,mol^{-1}}$). Therefore, from eqn 2D.11, the difference in molar heat capacities (which is given by V_m in place of V) is

$$C_{p,\mathrm{m}} - C_{V,\mathrm{m}} = \frac{(2.1\times10^{-4}\,\mathrm{K^{-1}})^2\times(298\,\mathrm{K})\times(1.81\times10^{-5}\,\mathrm{m^3\,mol^{-1}})}{4.90\times10^{-10}\,\mathrm{Pa^{-1}}}$$
$$= 0.485\,\mathrm{Pa\,m^3\,mol^{-1}} = 0.485\,\mathrm{J\,mol^{-1}}$$

For water, $C_{p,\mathrm{m}} = 75.3\,\mathrm{J\,K^{-1}\,mol^{-1}}$, so $C_{V,\mathrm{m}} = 74.8\,\mathrm{J\,K^{-1}\,mol^{-1}}$. In some cases, the two heat capacities differ by as much as 30 per cent.

Self-test 2D.3 Evaluate the difference in molar heat capacities for benzene; use data from the *Resource section*.

Answer: $45\,\mathrm{J\,K^{-1}\,mol^{-1}}$

Justification 2D.1 The relation between heat capacities

A useful rule when doing a problem in thermodynamics is to go back to first principles. In the present problem we do this twice, first by expressing C_p and C_V in terms of their definitions and then by inserting the definition $H = U + pV$:

$$C_p - C_V = \left(\frac{\partial H}{\partial T}\right)_p - C_V$$
$$= \left(\frac{\partial U}{\partial T}\right)_p + \left(\frac{\partial (pV)}{\partial T}\right)_p - C_V$$

Equation 2D.8, $(\partial U/\partial T)_p = \alpha\pi_T V + C_V$, lets us write the difference of the first and third terms as $\alpha\pi_T V$. We can simplify the remaining term by noting that, because p is constant,

$$\left(\frac{\partial (pV)}{\partial T}\right)_p = p\left(\frac{\partial V}{\partial T}\right)_p = \alpha pV$$

Collecting the two contributions gives

$$C_p - C_V = \alpha(p + \pi_T)V$$

The first term on the right, αpV, is a measure of the work needed to push back the atmosphere; the second term on the right, $\alpha\pi_T V$, is the work required to separate the molecules composing the system.

At this point we can go further by using the (Second Law) result proved in Topic 3D that

$$\pi_T = T\left(\frac{\partial p}{\partial T}\right)_V - p$$

When this expression is inserted in the last equation we obtain

$$C_p - C_V = \alpha TV\left(\frac{\partial p}{\partial T}\right)_V$$

We now transform the remaining partial derivative. With V regarded as a function of p and T, when these two quantities change the resulting change in V is

$$\mathrm{d}V=\left(\frac{\partial V}{\partial T}\right)_p \mathrm{d}T+\left(\frac{\partial V}{\partial p}\right)_T \mathrm{d}p$$

For the volume to be constant, $\mathrm{d}V=0$ implies that

$$\left(\frac{\partial V}{\partial T}\right)_p \mathrm{d}T=-\left(\frac{\partial V}{\partial p}\right)_T \mathrm{d}p \text{ at constant volume}$$

On division by $\mathrm{d}T$, this relation becomes

$$\left(\frac{\partial V}{\partial T}\right)_p=-\left(\frac{\partial V}{\partial p}\right)_T\left(\frac{\partial p}{\partial T}\right)_V$$

and therefore

$$\left(\frac{\partial p}{\partial T}\right)_V=-\frac{(\partial V/\partial T)_p}{(\partial V/\partial p)_T}=\frac{\alpha}{\kappa_T}$$

Insertion of this relation into the expression above for C_p-C_V produces eqn 2D.11.

2D.3 The Joule–Thomson effect

We can carry out a similar set of operations on the enthalpy, $H=U+pV$. The quantities U, p, and V are all state functions; therefore H is also a state function and $\mathrm{d}H$ is an exact differential. It turns out that H is a useful thermodynamic function when the pressure is under our control: we saw a sign of that in the relation $\Delta H=q_p$ (this is eqn 2B.2b of Topic 2B). We shall therefore regard H as a function of p and T, and adapt the argument in Section 2D.2 for the variation of U to find an expression for the variation of H with temperature at constant volume. As explained in the following *Justification*, we find that for a closed system of constant composition,

$$\mathrm{d}H=-\mu C_p\mathrm{d}p+C_p\mathrm{d}T \qquad (2D.12)$$

where the **Joule–Thomson coefficient**, μ (mu), is defined as

$$\mu=\left(\frac{\partial T}{\partial p}\right)_H \qquad \textit{Definition} \quad \text{Joule–Thomson coefficient} \qquad (2D.13)$$

This relation will prove useful for relating the heat capacities at constant pressure and volume and for a discussion of the liquefaction of gases.

Justification 2D.2 The variation of enthalpy with pressure and temperature

Because H is a function of p and T we can write, when these two quantities change by an infinitesimal amount, the enthalpy changes by

$$\mathrm{d}H=\left(\frac{\partial H}{\partial p}\right)_T \mathrm{d}p+\left(\frac{\partial H}{\partial T}\right)_p \mathrm{d}T$$

The second partial derivative is C_p; our task here is to express $(\partial H/\partial p)_T$ in terms of recognizable quantities. If the enthalpy is constant, $\mathrm{d}H=0$ and this expression then requires that

$$\left(\frac{\partial H}{\partial p}\right)_T \mathrm{d}p=-C_p\mathrm{d}T \text{ at constant } H$$

Division of both sides by $\mathrm{d}p$ then gives

$$\left(\frac{\partial H}{\partial p}\right)_T=-C_p\left(\frac{\partial T}{\partial p}\right)_H=-C_p\mu$$

Equation 2D.13 now follows directly.

(a) Observation of the Joule–Thomson effect

The analysis of the Joule–Thomson coefficient is central to the technological problems associated with the liquefaction of gases. We need to be able to interpret it physically and to measure it. As shown in the following *Justification*, the cunning required to impose the constraint of constant enthalpy, so that the process is **isenthalpic**, was supplied by Joule and William Thomson (later Lord Kelvin). They let a gas expand through a porous barrier from one constant pressure to another and monitored the difference of temperature that arose from the expansion (Fig. 2D.6). The whole apparatus was insulated so that the process was adiabatic. They observed a lower temperature on the low pressure side, the difference in temperature being proportional to the pressure difference they maintained.

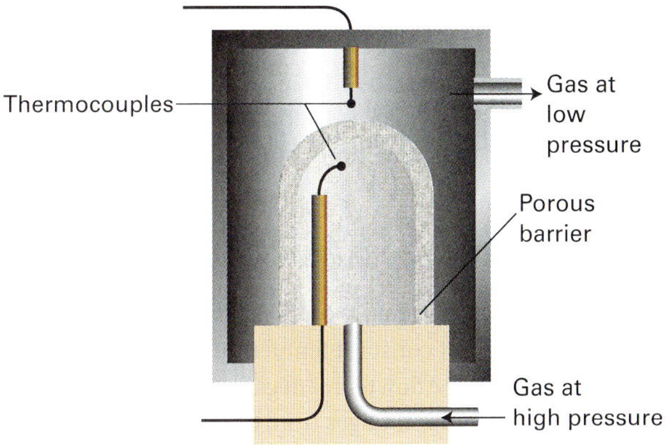

Figure 2D.6 The apparatus used for measuring the Joule–Thomson effect. The gas expands through the porous barrier, which acts as a throttle, and the whole apparatus is thermally insulated. As explained in the text, this arrangement corresponds to an isenthalpic expansion (expansion at constant enthalpy). Whether the expansion results in a heating or a cooling of the gas depends on the conditions.

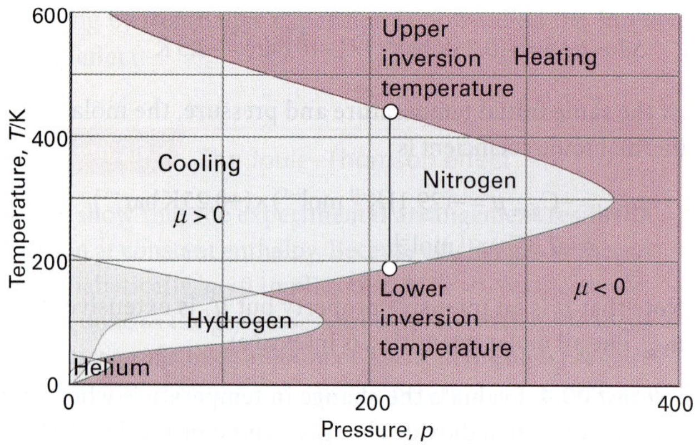

Figure 2D.11 The inversion temperatures for three real gases, nitrogen, hydrogen, and helium.

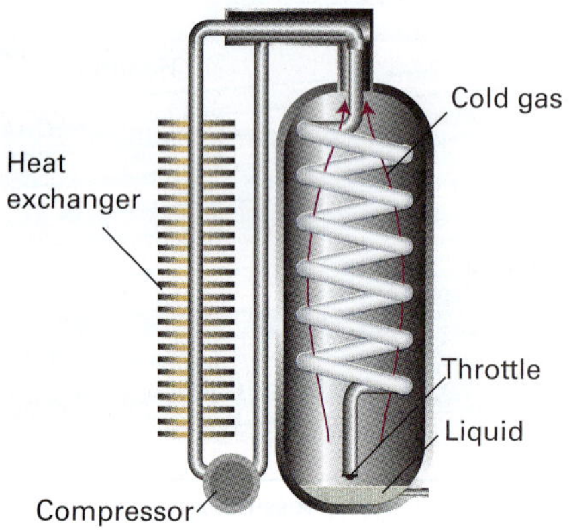

Figure 2D.12 The principle of the Linde refrigerator is shown in this diagram. The gas is recirculated, and so long as it is beneath its inversion temperature it cools on expansion through the throttle. The cooled gas cools the high-pressure gas, which cools still further as it expands. Eventually liquefied gas drips from the throttle.

does not necessarily approach zero as the pressure is reduced even though the equation of state of the gas approaches that of a perfect gas. The coefficient behaves like the properties discussed in Topic 1C in the sense that it depends on derivatives and not on p, V, and T themselves.

(b) The molecular interpretation of the Joule–Thomson effect

The kinetic model of gases (Topic 1B) and the equipartition theorem (*Foundations* B) jointly imply that the mean kinetic energy of molecules in a gas is proportional to the temperature. It follows that reducing the average speed of the molecules is equivalent to cooling the gas. If the speed of the molecules can be reduced to the point that neighbours can capture each other by their intermolecular attractions, then the cooled gas will condense to a liquid.

To slow the gas molecules, we make use of an effect similar to that seen when a ball is thrown into the air: as it rises it slows in response to the gravitational attraction of the Earth and its kinetic energy is converted into potential energy. We saw in Topic 1C that molecules in a real gas attract each other (the attraction is not gravitational, but the effect is the same). It follows that, if we can cause the molecules to move apart from each other, like a ball rising from a planet, then they should slow. It is very easy to move molecules apart from each other: we simply allow the gas to expand, which increases the average separation of the molecules. To cool a gas, therefore, we allow it to expand without allowing any energy to enter from outside as heat. As the gas expands, the molecules move apart to fill the available volume, struggling as they do so against the attraction of their neighbours. Because some kinetic energy must be converted into potential energy to reach greater separations, the molecules travel more slowly as their separation increases. This sequence of molecular events explains the Joule–Thomson effect: the cooling of a real gas by adiabatic expansion. The cooling effect, which corresponds to $\mu>0$, is observed under conditions when attractive interactions are dominant ($Z<1$, where Z is the compression factor defined in eqn 1C.1, $Z=V_m/V_m^\circ$), because the molecules have to climb apart against the attractive force in order for them to travel more slowly. For molecules under conditions when repulsions are dominant ($Z>1$), the Joule–Thomson effect results in the gas becoming warmer, or $\mu<0$.

Checklist of concepts

- ☐ 1. The quantity $\mathrm{d}U$ is an exact differential; $\mathrm{d}w$ and $\mathrm{d}q$ are not.
- ☐ 2. The change in internal energy may be expressed in terms of changes in temperature and pressure.
- ☐ 3. The **internal pressure** is the variation of internal energy with volume at constant temperature.
- ☐ 4. **Joule's experiment** showed that the internal pressure of a perfect gas is zero.
- ☐ 5. The change in internal energy with pressure and temperature is expressed in terms of the internal pressure and the heat capacity and leads to a general expression for the relation between heat capacities.
- ☐ 6. The **Joule–Thomson effect** is the change in temperature of a gas when it undergoes isenthalpic expansion.

Checklist of equations

Property	Equation	Comment	Equation number
Change in $U(V,T)$	$dU = (\partial U/\partial V)_T dV + (\partial U/\partial T)_V dT$	Constant composition	2D.3
Internal pressure	$\pi_T = (\partial U/\partial V)_T$	Definition; for a perfect gas, $\pi_T = 0$	2D.4
Change in $U(V,T)$	$dU = \pi_T dV + C_V dT$	Constant composition	2D.5
Expansion coefficient	$\alpha = (1/V)(\partial V/\partial T)_p$	Definition	2D.6
Isothermal compressibility	$\kappa_T = -(1/V)(\partial V/\partial p)_T$	Definition	2D.7
Relation between heat capacities	$C_p - C_V = nR$	Perfect gas	2D.10
	$C_p - C_V = \alpha^2 TV/\kappa_T$		2D.11
Change in $H(p,T)$	$dH = -\mu C_p dp + C_p dT$	Constant composition	2D.12
Joule–Thomson coefficient	$\mu = (\partial T/\partial p)_H$	For a perfect gas, $\mu = 0$	2D.13
Isothermal Joule–Thomson coefficient	$\mu_T = (\partial H/\partial p)_T$	For a perfect gas, $\mu_T = 0$	2D.14
Relation between coefficients	$\mu_T = -C_p\mu$		2D.15

2E Adiabatic changes

Contents

➤ **Why do you need to know this material?**

Adiabatic processes complement isothermal processes, and are used in the discussion of the Second Law of thermodynamics.

➤ **What is the key idea?**

The temperature of a perfect gas falls when it does work in an adiabatic expansion.

➤ **What do you need to know already?**

This Topic makes use of the discussion of the properties of gases (Topic 1A), particularly the perfect gas law. It also uses the definitions of heat capacity at constant volume (Topic 1B) and constant pressure (Topic 2B), and the relation between them (Topic 2D).

The temperature falls when a gas expands adiabatically (in a thermally insulated container). Work is done, but as no heat enters the system, the internal energy falls, and therefore the temperature of the working gas also falls. In molecular terms, the kinetic energy of the molecules falls as work is done, so their average speed decreases, and hence the temperature falls too.

2E.1 The change in temperature

To calculate the change in temperature that results from a process we focus first on the change in internal energy. The change in internal energy of a perfect gas when the temperature is changed from T_i to T_f and the volume is changed from V_i to V_f can be expressed as the sum of two steps (Fig. 2E.1). In the first step, only the volume changes and the temperature is held constant at its initial value. However, because the internal

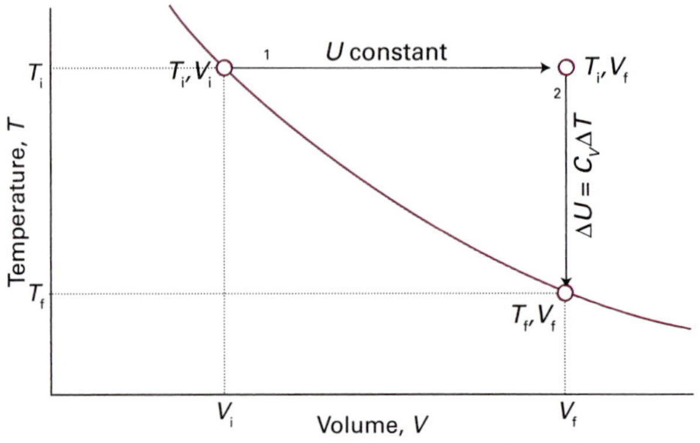

Figure 2E.1 To achieve a change of state from one temperature and volume to another temperature and volume, we may consider the overall change as composed of two steps. In the first step, the system expands at constant temperature; there is no change in internal energy if the system consists of a perfect gas. In the second step, the temperature of the system is reduced at constant volume. The overall change in internal energy is the sum of the changes for the two steps.

energy of a perfect gas is independent of the volume the molecules occupy (Topic 2A), the overall change in internal energy arises solely from the second step, the change in temperature at constant volume. Provided the heat capacity is independent of temperature, this change is

$$\Delta U = (T_f - T_i)C_V = C_V\Delta T$$

Because the expansion is adiabatic, we know that $q=0$; then because $\Delta U = q + w$, it follows that $\Delta U = w_{ad}$. The subscript 'ad' denotes an adiabatic process. Therefore, by equating the two expressions for ΔU, we obtain

$$w_{ad} = C_V\Delta T \qquad \textit{Perfect gas} \qquad \text{Work of adiabatic change} \qquad (2E.1)$$

That is, the work done during an adiabatic expansion of a perfect gas is proportional to the temperature difference between the initial and final states. That is exactly what we expect on molecular grounds, because the mean kinetic energy is proportional to T, so a change in internal energy arising from temperature alone is also expected to be proportional to ΔT.

In the following *Justification* we show that, based on this result, the initial and final temperatures of a perfect gas that undergoes reversible adiabatic expansion (reversible expansion in a thermally insulated container) can be calculated from

$$T_f = T_i\left(\frac{V_i}{V_f}\right)^{1/c} \qquad c = C_{V,m}/R \qquad \textit{Adiabatic, reversible, perfect gas} \qquad \text{Final temperature} \qquad (2E.2a)$$

By raising each side of this expression to the power c, an equivalent expression is

$$V_i T_i^c = V_f T_f^c \qquad c = C_{V,m}/R$$

Adiabatic, reversible, perfect gas — Final temperature (2E.2b)

This result is often summarized in the form $VT^c = \text{constant}$.

Brief illustration 2E.1 The change in temperature

Consider the adiabatic, reversible expansion of 0.020 mol Ar, initially at 25 °C, from 0.50 dm³ to 1.00 dm³. The molar heat capacity of argon at constant volume is 12.47 J K⁻¹ mol⁻¹, so $c = 1.501$. Therefore, from eqn 2B.2a,

$$T_f = (298\,\text{K})\left(\frac{0.50\,\text{dm}^3}{1.00\,\text{dm}^3}\right)^{1/1.501} = 188\,\text{K}$$

It follows that $\Delta T = -110\,\text{K}$, and therefore, from eqn 2E.1, that

$$w = \{(0.020\,\text{mol}) \times (12.48\,\text{J K}^{-1}\,\text{mol}^{-1})\} \times (-110\,\text{K}) = -27\,\text{J}$$

Note that temperature change is independent of the amount of gas but the work is not.

Self-test 2E.1 Calculate the final temperature, the work done, and the change of internal energy when ammonia is used in a reversible adiabatic expansion from 0.50 dm³ to 2.00 dm³, the other initial conditions being the same.

Answer: 194 K, −56 J, −56 J

Justification 2E.1 Changes in temperature

Consider a stage in a reversible adiabatic expansion when the pressure inside and out is p. The work done when the gas expands by dV is $dw = -p dV$; however, for a perfect gas, $dU = C_V dT$. Therefore, because for an adiabatic change ($dq = 0$) $dU = dw + dq = dw$, we can equate these two expressions for dU and write

$$C_V dT = -p dV$$

We are dealing with a perfect gas, so we can replace p by nRT/V and obtain

$$\frac{C_V dT}{T} = -\frac{nR dV}{V}$$

To integrate this expression we note that T is equal to T_i when V is equal to V_i, and is equal to T_f when V is equal to V_f at the end of the expansion. Therefore,

$$C_V \int_{T_i}^{T_f} \frac{dT}{T} = -nR \int_{V_i}^{V_f} \frac{dV}{V}$$

(We are taking C_V to be independent of temperature.) Then, because $\int dx/x = \ln x + \text{constant}$, we obtain

$$C_V \ln\frac{T_f}{T_i} = -nR \ln\frac{V_f}{V_i}$$

Because $\ln(x/y) = -\ln(y/x)$, this expression rearranges to

$$\frac{C_V}{nR} \ln\frac{T_f}{T_i} = \ln\frac{V_i}{V_f}$$

With $c = C_V/nR$ we obtain (because $\ln x^a = a \ln x$)

$$\ln\left(\frac{T_f}{T_i}\right)^c = \ln\frac{V_i}{V_f}$$

which implies that $(T_f/T_i)^c = (V_i/V_f)$ and, upon rearrangement, eqn 2E.2.

2E.2 The change in pressure

We show in the following *Justification* that the pressure of a perfect gas that undergoes reversible adiabatic expansion from a volume V_i to a volume V_f is related to its initial pressure by

$$p_f V_f^\gamma = p_i V_i^\gamma$$

Perfect gas — Reversible adiabatic expansion (2E.3)

where $\gamma = C_{p,m}/C_{V,m}$. This result is commonly summarized in the form $pV^\gamma = \text{constant}$.

Justification 2E.2 The relation between pressure and volume

The initial and final states of a perfect gas satisfy the perfect gas law regardless of how the change of state takes place, so we can use $pV = nRT$ to write

$$\frac{p_i V_i}{p_f V_f} = \frac{T_i}{T_f}$$

However, from eqn 2E.2 we know that $T_i/T_f = (V_f/V_i)^{1/c}$. Therefore,

$$\frac{p_i V_i}{p_f V_f} = \left(\frac{V_f}{V_i}\right)^{1/c}, \text{ so } \frac{p_i}{p_f}\left(\frac{V_i}{V_f}\right)^{1/c+1} = 1$$

We now use the result from Topic 2B that $C_{p,m} - C_{V,m} = R$ to note that

$$\frac{1}{c} + 1 = \frac{1+c}{c} = \frac{R + C_{V,m}}{C_{V,m}} = \frac{C_{p,m}}{C_{V,m}} = \gamma$$

It follows that

$$\frac{p_i}{p_f}\left(\frac{V_i}{V_f}\right)^{\gamma}=1$$

which rearranges to eqn 2E.3.

For a monatomic perfect gas, $C_{V,m}=\frac{3}{2}R$ (Topic 2A) and $C_{p,m}=\frac{5}{2}R$ (from $C_{p,m}-C_{V,m}=R$), so $\gamma=\frac{5}{3}$. For a gas of nonlinear polyatomic molecules (which can rotate as well as translate; vibrations make little contribution at normal temperatures), $C_{V,m}=3R$ and $C_{p,m}=4R$, so $\gamma=\frac{4}{3}$. The curves of pressure versus volume for adiabatic change are known as **adiabats**, and one for a reversible path is illustrated in Fig. 2E.2. Because $\gamma>1$, an adiabat falls more steeply ($p\propto 1/V^{\gamma}$) than the corresponding isotherm ($p\propto 1/V$). The physical reason for the difference is that, in an isothermal expansion, energy flows into the system as heat and maintains the temperature; as a result, the pressure does not fall as much as in an adiabatic expansion.

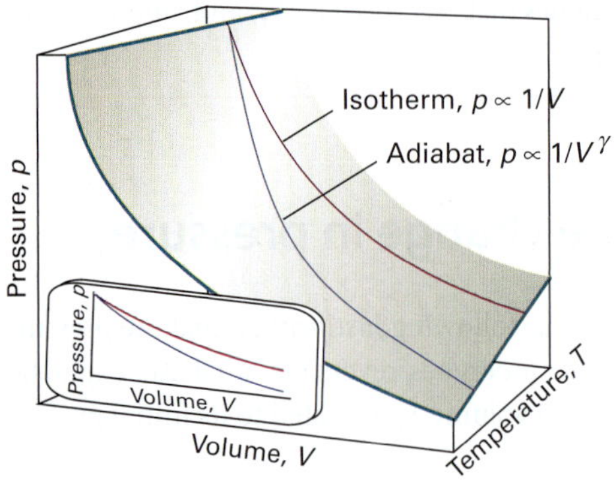

Figure 2E.2 An adiabat depicts the variation of pressure with volume when a gas expands adiabatically. Note that the pressure declines more steeply for an adiabat than it does for an isotherm because the temperature decreases in the former.

Brief illustration 2E.2 Adiabatic expansion

When a sample of argon (for which $\gamma=\frac{5}{3}$) at 100 kPa expands reversibly and adiabatically to twice its initial volume the final pressure will be

$$p_f=\left(\frac{V_i}{V_f}\right)^{\gamma}p_i=\left(\frac{1}{2}\right)^{5/3}\times(100\,\text{kPa})=32\,\text{kPa}$$

For an isothermal doubling of volume, the final pressure would be 50 kPa.

Self-test 2E.2 What is the final pressure when a sample of carbon dioxide at 100 kPa expands reversibly and adiabatically to five times its initial volume?

Answer: 13 kPa

Checklist of concepts

☐ 1. The temperature of a gas falls when it undergoes adiabatic expansion (and does work).

☐ 2. An **adiabat** is a curve showing how pressure varies with volume in an adiabatic process.

Checklist of equations

Property	Equation	Comment	Equation number
Work of adiabatic expansion	$w_{ad}=C_V\Delta T$	Perfect gas	2E.1
Final temperature	$T_f=T_i(V_i/V_f)^{1/c}$ $c=C_{V,m}/R$	Perfect gas, reversible expansion	2E.2a
	$V_iT_i^c=V_fT_f^c$		2E.2b
Adiabats	$p_fV_f^{\gamma}=p_iV_i^{\gamma}$, $\gamma=C_{p,m}/C_{V,m}$		2E.3

CHAPTER 2 The First Law

Assume all gases are perfect unless stated otherwise. Unless otherwise stated, thermochemical data are for 298.15 K.

TOPIC 2A Internal energy

Discussion questions

2A.1 Describe and distinguish the various uses of the words 'system' and 'state' in physical chemistry.

2A.2 Describe the distinction between heat and work in thermodynamic and molecular terms, the latter in terms of populations and energy levels.

2A.3 Identify varieties of additional work.

Exercises

2A.1(a) Use the equipartition theorem to estimate the molar internal energy relative to $U(0)$ of (i) I_2, (ii) CH_4, (iii) C_6H_6 in the gas phase at 25 °C.
2A.1(b) Use the equipartition theorem to estimate the molar internal energy relative to $U(0)$ of (i) O_3, (ii) C_2H_6, (iii) SO_2 in the gas phase at 25 °C.

2A.2(a) Which of (i) pressure, (ii) temperature, (iii) work, (iv) enthalpy are state functions?
2A.2(b) Which of (i) volume, (ii) heat, (iii) internal energy, (iv) density are state functions?

2A.3(a) A chemical reaction takes place in a container of cross-sectional area 50 cm². As a result of the reaction, a piston is pushed out through 15 cm against an external pressure of 1.0 atm. Calculate the work done by the system.
2A.3(b) A chemical reaction takes place in a container of cross-sectional area 75.0 cm². As a result of the reaction, a piston is pushed out through 25.0 cm against an external pressure of 150 kPa. Calculate the work done by the system.

2A.4(a) A sample consisting of 1.00 mol Ar is expanded isothermally at 20 °C from 10.0 dm³ to 30.0 dm³ (i) reversibly, (ii) against a constant external pressure equal to the final pressure of the gas, and (iii) freely (against zero external pressure). For the three processes calculate q, w, and ΔU.
2A.4(b) A sample consisting of 2.00 mol He is expanded isothermally at 0 °C from 5.0 dm³ to 20.0 dm³ (i) reversibly, (ii) against a constant external pressure equal to the final pressure of the gas, and (iii) freely (against zero external pressure). For the three processes calculate q, w, and ΔU.

2A.5(a) A sample consisting of 1.00 mol of perfect gas atoms, for which $C_{V,m} = \frac{3}{2}R$, initially at $p_1 = 1.00$ atm and $T_1 = 300$ K, is heated reversibly to 400 K at constant volume. Calculate the final pressure, ΔU, q, and w.
2A.5(b) A sample consisting of 2.00 mol of perfect gas molecules, for which $C_{V,m} = \frac{5}{2}R$, initially at $p_1 = 111$ kPa and $T_1 = 277$ K, is heated reversibly to 356 K at constant volume. Calculate the final pressure, ΔU, q, and w.

2A.6(a) A sample of 4.50 g of methane occupies 12.7 dm³ at 310 K. (i) Calculate the work done when the gas expands isothermally against a constant external pressure of 200 Torr until its volume has increased by 3.3 dm³. (ii) Calculate the work that would be done if the same expansion occurred reversibly.
2A.6(b) A sample of argon of mass 6.56 g occupies 18.5 dm³ at 305 K. (i) Calculate the work done when the gas expands isothermally against a constant external pressure of 7.7 kPa until its volume has increased by 2.5 dm³. (ii) Calculate the work that would be done if the same expansion occurred reversibly.

Problems

2A.1 Calculate the work done during the isothermal reversible expansion of a van der Waals gas (Topic 1C). Plot on the same graph the indicator diagrams (graphs of pressure against volume) for the isothermal reversible expansion of (a) a perfect gas, (b) a van der Waals gas in which $a = 0$ and $b = 5.11 \times 10^{-2}$ dm³ mol⁻¹, and (c) $a = 4.2$ dm⁶ atm mol⁻² and $b = 0$. The values selected exaggerate the imperfections but give rise to significant effects on the indicator diagrams. Take $V_i = 1.0$ dm³, $n = 1.0$ mol, and $T = 298$ K.

2A.2 A sample consisting of 1.0 mol $CaCO_3$(s) was heated to 800 °C, when it decomposed. The heating was carried out in a container fitted with a piston that was initially resting on the solid. Calculate the work done during complete decomposition at 1.0 atm. What work would be done if instead of having a piston the container was open to the atmosphere?

2A.3 Calculate the work done during the isothermal reversible expansion of a gas that satisfies the virial equation of state, eqn 1C.3. Evaluate (a) the work for 1.0 mol Ar at 273 K (for data, see Table 1C.1) and (b) the same amount of a perfect gas. Let the expansion be from 500 cm³ to 1000 cm³ in each case.

2A.4 Express the work of isothermal reversible expansion of a van der Waals gas in reduced variables (Topic 1C) and find a definition of reduced work that makes the overall expression independent of the identity of the gas. Calculate the work of isothermal reversible expansion along the critical isotherm from V_c to xV_c.

2A.5 Suppose that a DNA molecule resists being extended from an equilibrium, more compact conformation with a restoring force $F = -k_f x$, where x is the difference in the end-to-end distance of the chain from an equilibrium value and k_f is the force constant. Use this model to write an expression for the work that must be done to extend a DNA molecule by a distance x. Draw a graph of your conclusion.

2A.6 A better model of a DNA molecule is the 'one-dimensional freely jointed chain', in which a rigid unit of length l can only make an angle of 0° or 180° with an adjacent unit. In this case, the restoring force of a chain extended by $x=nl$ is given by

$$F=\frac{kT}{2l}\ln\left(\frac{1+\nu}{1-\nu}\right) \qquad \nu=\frac{n}{N}$$

where k is Boltzmann's constant. (a) What is the magnitude of the force that must be applied to extend a DNA molecule with $N=200$ by 90 nm? (b) Plot the restoring force against ν, noting that ν can be either positive or negative. How is the variation of the restoring force with end-to-end distance different from that predicted by Hooke's law? (c) Keep in mind that the difference in end-to-end distance from an equilibrium value is $x=nl$ and, consequently, $dx=ldn=Nld\nu$, and write an expression for the work of extending a DNA molecule. (d) Calculate the work of extending a DNA molecule from $\nu=0$ to $\nu=1.0$. *Hint*: You must integrate the expression for w. The task can be accomplished easily with mathematical software.

2A.7 As a continuation of Problem 2A.6, (a) show that for small extensions of the chain, when $\nu \ll 1$, the restoring force is given by

$$F\approx\frac{\nu kT}{l}=\frac{nkT}{Nl}$$

(b) Is the variation of the restoring force with extension of the chain given in part (a) different from that predicted by Hooke's law? Explain your answer.

TOPIC 2B Enthalpy

Discussion questions

2B.1 Explain the difference between the change in internal energy and the change in enthalpy accompanying a process.

2B.2 Why is the heat capacity at constant pressure of a substance normally greater than its heat capacity at constant volume?

Exercises

2B.1(a) When 229 J of energy is supplied as heat to 3.0 mol Ar(g), the temperature of the sample increases by 2.55 K. Calculate the molar heat capacities at constant volume and constant pressure of the gas.

2B.1(b) When 178 J of energy is supplied as heat to 1.9 mol of gas molecules, the temperature of the sample increases by 1.78 K. Calculate the molar heat capacities at constant volume and constant pressure of the gas.

2B.2(a) The constant-pressure heat capacity of a sample of a perfect gas was found to vary with temperature according to the expression $C_p/(\text{J K}^{-1})=20.17+0.3665(T/\text{K})$. Calculate q, w, and ΔH when the temperature is raised from 25 °C to 100 °C (i) at constant pressure, (ii) at constant volume.

2B.2(b) The constant-pressure heat capacity of a sample of a perfect gas was found to vary with temperature according to the expression $C_p/(\text{J K}^{-1})=20.17+0.4001(T/\text{K})$. Calculate q, w, and ΔH when the temperature is raised from 25 °C to 100 °C (i) at constant pressure, (ii) at constant volume.

2B.3(a) When 3.0 mol O_2 is heated at a constant pressure of 3.25 atm, its temperature increases from 260 K to 285 K. Given that the molar heat capacity of O_2 at constant pressure is 29.4 J K^{-1} mol^{-1}, calculate q, ΔH, and ΔU.

2B.3(b) When 2.0 mol CO_2 is heated at a constant pressure of 1.25 atm, its temperature increases from 250 K to 277 K. Given that the molar heat capacity of CO_2 at constant pressure is 37.11 J K^{-1} mol^{-1}, calculate q, ΔH, and ΔU.

Problems

2B.1 The following data show how the standard molar constant-pressure heat capacity of sulfur dioxide varies with temperature. By how much does the standard molar enthalpy of SO_2(g) increase when the temperature is raised from 298.15 K to 1500 K?

T/K	300	500	700	900	1100	1300	1500
$C^{\ominus}_{p,\text{m}}/(\text{J K}^{-1}\text{mol}^{-1})$	39.909	46.490	50.829	53.407	54.993	56.033	56.759

2B.2 The following data show how the standard molar constant-pressure heat capacity of ammonia depends on the temperature. Use mathematical software to fit an expression of the form of eqn 2B.8 to the data and determine the values of a, b, and c. Explore whether it would be better to express the data as $C_{p,\text{m}}=\alpha+\beta T+\gamma T^2$, and determine the values of these coefficients.

T/K	300	400	500	600	700	800	900	1000
$C^{\ominus}_{p,\text{m}}/(\text{J K}^{-1}\text{mol}^{-1})$	35.678	38.674	41.994	45.229	48.269	51.112	53.769	56.244

2B.3 A sample consisting of 2.0 mol CO_2 occupies a fixed volume of 15.0 dm^3 at 300 K. When it is supplied with 2.35 kJ of energy as heat its temperature increases to 341 K. Assume that CO_2 is described by the van der Waals equation of state (Topic 1C) and calculate w, ΔU, and ΔH.

2B.4 (a) Express $(\partial C_V/\partial V)_T$ as a second derivative of U and find its relation to $(\partial U/\partial V)_T$ and $(\partial C_p/\partial p)_T$ as a second derivative of H and find its relation to $(\partial H/\partial p)_T$. (b) From these relations show that $(\partial C_V/\partial V)_T=0$ and $(\partial C_p/\partial p)_T=0$ for a perfect gas.

TOPIC 2C Thermochemistry

Discussion questions

2C.1 Describe two calorimetric methods for the determination of enthalpy changes that accompany chemical processes.

2C.2 Distinguish between 'standard state' and 'reference state', and indicate their applications.

Exercises

2C.1(a) For tetrachloromethane, $\Delta_{vap}H^{\ominus} = 30.0\,kJ\,mol^{-1}$. Calculate q, w, ΔH, and ΔU when 0.75 mol $CCl_4(l)$ is vaporized at 250 K and 750 Torr.
2C.1(b) For ethanol, $\Delta_{vap}H^{\ominus} = 43.5\,kJ\,mol^{-1}$. Calculate q, w, ΔH, and ΔU when 1.75 mol $C_2H_5OH(l)$ is vaporized at 260 K and 765 Torr.

2C.2(a) The standard enthalpy of formation of ethylbenzene is $-12.5\,kJ\,mol^{-1}$. Calculate its standard enthalpy of combustion.
2C.2(b) The standard enthalpy of formation of phenol is $-165.0\,kJ\,mol^{-1}$. Calculate its standard enthalpy of combustion.

2C.3(a) The standard enthalpy of combustion of cyclopropane is $-2091\,kJ\,mol^{-1}$ at 25 °C. From this information and enthalpy of formation data for $CO_2(g)$ and $H_2O(g)$, calculate the enthalpy of formation of cyclopropane. The enthalpy of formation of propene is $+20.42\,kJ\,mol^{-1}$. Calculate the enthalpy of isomerization of cyclopropane to propene.
2C.3(b) From the following data, determine $\Delta_f H^{\ominus}$ for diborane, $B_2H_6(g)$, at 298 K:

(1)	$B_2H_6(g) + 3\,O_2(g) \rightarrow B_2O_3(s) + 3\,H_2O(g)$	$\Delta_r H^{\ominus} = -1941\,kJ\,mol^{-1}$
(2)	$2\,B(s) + \frac{3}{2}\,O_2(g) \rightarrow B_2O_3(s)$	$\Delta_r H^{\ominus} = -2368\,kJ\,mol^{-1}$
(3)	$H_2(g) + \frac{1}{2}\,O_2(g) \rightarrow H_2O(g)$	$\Delta_r H^{\ominus} = -241.8\,kJ\,mol^{-1}$

2C.4(a) Given that the standard enthalpy of formation of HCl(aq) is $-167\,kJ\,mol^{-1}$, what is the value of $\Delta_f H^{\ominus}(Cl^-, aq)$?
2C.4(b) Given that the standard enthalpy of formation of HI(aq) is $-55\,kJ\,mol^{-1}$, what is the value of $\Delta_f H^{\ominus}(I^-, aq)$?

2C.5(a) When 120 mg of naphthalene, $C_{10}H_8(s)$, was burned in a bomb calorimeter the temperature rose by 3.05 K. Calculate the calorimeter constant. By how much will the temperature rise when 150 mg of phenol, $C_6H_5OH(s)$, is burned in the calorimeter under the same conditions?
2C.5(b) When 225 mg of anthracene, $C_{14}H_{10}(s)$, was burned in a bomb calorimeter the temperature rose by 1.75 K. Calculate the calorimeter constant. By how much will the temperature rise when 125 mg of phenol, $C_6H_5OH(s)$, is burned in the calorimeter under the same conditions? ($\Delta_c H^{\ominus}(C_{14}H_{10},s) = -7061\,kJ\,mol^{-1}$.)

2C.6(a) Given the reactions (1) and (2) below, determine (i) $\Delta_r H^{\ominus}$ and $\Delta_r U^{\ominus}$ for reaction (3), (ii) $\Delta_f H^{\ominus}$ for both HCl(g) and $H_2O(g)$ all at 298 K.

(1)	$H_2(g) + Cl_2(g) \rightarrow 2\,HCl(g)$	$\Delta_r H^{\ominus} = -184.62\,kJ\,mol^{-1}$
(2)	$H_2(g) + O_2(g) \rightarrow 2\,H_2O(g)$	$\Delta_r H^{\ominus} = -483.64\,kJ\,mol^{-1}$
(3)	$4\,HCl(g) + O_2(g) \rightarrow 2\,Cl_2(g) + 2\,H_2O(g)$	

2C.6(b) Given the reactions (1) and (2) below, determine (i) $\Delta_r H^{\ominus}$ and $\Delta_r U^{\ominus}$ for reaction (3), (ii) $\Delta_f H^{\ominus}$ for both HI(g) and $H_2O(g)$ all at 298 K.

(1)	$H_2(g) + I_2(s) \rightarrow 2\,HI(g)$	$\Delta_r H^{\ominus} = +52.96\,kJ\,mol^{-1}$
(2)	$2\,H_2(g) + O_2(g) \rightarrow 2\,H_2O(g)$	$\Delta_r H^{\ominus} = -483.64\,kJ\,mol^{-1}$
(3)	$4\,HI(g) + O_2(g) \rightarrow 2\,I_2(s) + 2\,H_2O(g)$	

2C.7(a) For the reaction $C_2H_5OH(l) + 3\,O_2(g) \rightarrow 2\,CO_2(g) + 3\,H_2O(g)$, $\Delta_r U^{\ominus} = -1373\,kJ\,mol^{-1}$ at 298 K. Calculate $\Delta_r H^{\ominus}$.
2C.7(b) For the reaction $2\,C_6H_5COOH(s) + 15\,O_2(g) \rightarrow 14\,CO_2(g) + 6\,H_2O(g)$, $\Delta_r U^{\ominus} = -772.7\,kJ\,mol^{-1}$ at 298 K. Calculate $\Delta_r H^{\ominus}$.

2C.8(a) From the data in Tables 2C.2 and 2C.3, calculate $\Delta_r H^{\ominus}$ and $\Delta_r U^{\ominus}$ at (i) 298 K, (ii) 478 K for the reaction $C(graphite) + H_2O(g) \rightarrow CO(g) + H_2(g)$. Assume all heat capacities to be constant over the temperature range of interest.
2C.8(b) Calculate $\Delta_r H^{\ominus}$ and $\Delta_r U^{\ominus}$ at 298 K and $\Delta_r H^{\ominus}$ at 427 K for the hydrogenation of ethyne (acetylene) to ethene (ethylene) from the enthalpy of combustion and heat capacity data in Tables 2C.5 and 2C.6. Assume the heat capacities to be constant over the temperature range involved.

2C.9(a) Estimate $\Delta_r H^{\ominus}$(500 K) for the combustion of methane, $CH_4(g) + 2\,O_2(g) \rightarrow CO_2(g) + 2\,H_2O(g)$ by using the data on the temperature dependence of heat capacities in Table 2B.1.
2C.9(b) Estimate $\Delta_r H^{\ominus}$(478 K) for the combustion of naphthalene, $C_{10}H_8(l) + 12\,O_2(g) \rightarrow 10\,CO_2(g) + 4\,H_2O(g)$ by using the data on the temperature dependence of heat capacities in Table 2B.1.

2C.10(a) Set up a thermodynamic cycle for determining the enthalpy of hydration of Mg^{2+} ions using the following data: enthalpy of sublimation of Mg(s), $+167.2\,kJ\,mol^{-1}$; first and second ionization enthalpies of Mg(g), 7.646 eV and 15.035 eV; dissociation enthalpy of $Cl_2(g)$, $+241.6\,kJ\,mol^{-1}$; electron gain enthalpy of Cl(g), -3.78 eV; enthalpy of solution of $MgCl_2(s)$, $-150.5\,kJ\,mol^{-1}$; enthalpy of hydration of $Cl^-(g)$, $-383.7\,kJ\,mol^{-1}$.
2C.10(b) Set up a thermodynamic cycle for determining the enthalpy of hydration of Ca^{2+} ions using the following data: enthalpy of sublimation of Ca(s), $+178.2\,kJ\,mol^{-1}$; first and second ionization enthalpies of Ca(g), $589.7\,kJ\,mol^{-1}$ and $1145\,kJ\,mol^{-1}$; enthalpy of vaporization of bromine, $+30.91\,kJ\,mol^{-1}$; dissociation enthalpy of $Br_2(g)$, $+192.9\,kJ\,mol^{-1}$; electron gain enthalpy of Br(g), $-331.0\,kJ\,mol^{-1}$; enthalpy of solution of $CaBr_2(s)$, $-103.1\,kJ\,mol^{-1}$; enthalpy of hydration of $Br^-(g)$, $-289\,kJ\,mol^{-1}$.

Problems

2C.1 A sample of the sugar D-ribose ($C_5H_{10}O_5$) of mass 0.727 g was placed in a constant-volume bomb calorimeter and then ignited in the presence of excess oxygen. The temperature rose by 0.910 K. In a separate experiment in the same calorimeter, the combustion of 0.825 g of benzoic acid, for which the internal energy of combustion is $-3251\,kJ\,mol^{-1}$, gave a temperature rise of 1.940 K. Calculate the enthalpy of formation of D-ribose.

2C.2 The standard enthalpy of formation of bis(benzene)chromium was measured in a calorimeter. It was found for the reaction $Cr(C_6H_6)_2(s) \rightarrow Cr(s) + 2\,C_6H_6(g)$ that $\Delta_r U^{\ominus}(583\,K) = +8.0\,kJ\,mol^{-1}$. Find the corresponding reaction enthalpy and estimate the standard enthalpy of formation of the compound at 583 K. The constant-pressure molar heat capacity of benzene is $136.1\,J\,K^{-1}\,mol^{-1}$ in its liquid range and $81.67\,J\,K^{-1}\,mol^{-1}$ as a gas.

2C.3‡ From the enthalpy of combustion data in Table 2C.1 for the alkanes methane through octane, test the extent to which the relation $\Delta_c H^\ominus = k\{(M/(\text{g mol}^{-1})\}^n$ holds and find the numerical values for k and n. Predict $\Delta_c H^\ominus$ for decane and compare to the known value.

2C.4‡ Kolesov et al. reported the standard enthalpy of combustion and of formation of crystalline C_{60} based on calorimetric measurements (V.P. Kolesov et al., *J. Chem. Thermodynamics* **28**, 1121 (1996)). In one of their runs, they found the standard specific internal energy of combustion to be −36.0334 kJ g^{-1} at 298.15 K. Compute $\Delta_c H^\ominus$ and $\Delta_f H^\ominus$ of C_{60}.

2C.5‡ A thermodynamic study of $DyCl_3$ (E.H.P. Cordfunke et al., *J. Chem. Thermodynamics* **28**, 1387 (1996)) determined its standard enthalpy of formation from the following information

(1) $DyCl_3(s) \rightarrow DyCl_3(aq, \text{in } 4.0\ \text{M}\ HCl)$ $\Delta_r H^\ominus = -180.06\ \text{kJ mol}^{-1}$

(2) $Dy(s) + 3\,HCl(aq, 4.0\ \text{M}) \rightarrow DyCl_3(aq, \text{in } 4.0\ \text{M}\ HCl(aq)) + \frac{3}{2}H_2(g)$ $\Delta_r H^\ominus = -699.43\ \text{kJ mol}^{-1}$

(3) $\frac{1}{2}H_2(g) + \frac{1}{2}Cl_2(g) \rightarrow HCl(aq, 4.0\ \text{M})$ $\Delta_r H^\ominus = -158.31\ \text{kJ mol}^{-1}$

Determine $\Delta_f H^\ominus(DyCl_3, s)$ from these data.

2C.6‡ Silylene (SiH_2) is a key intermediate in the thermal decomposition of silicon hydrides such as silane (SiH_4) and disilane (Si_2H_6). H.K. Moffat et al. (*J. Phys. Chem.* **95**, 145 (1991)) report $\Delta_f H^\ominus(SiH_2) = +274\ \text{kJ mol}^{-1}$. If $\Delta_f H^\ominus(SiH_4) = +34.3\ \text{kJ mol}^{-1}$ and $\Delta_f H^\ominus(Si_2H_6) = +80.3\ \text{kJ mol}^{-1}$, compute the standard enthalpies of the following reactions:

(a) $SiH_4(g) \rightarrow SiH_2(g) + H_2(g)$
(b) $Si_2H_6(g) \rightarrow SiH_2(g) + SiH_4(g)$

2C.7 As remarked in Problem 2B.2, it is sometimes appropriate to express the temperature dependence of the heat capacity by the empirical expression $C_{p,m} = \alpha + \beta T + \gamma T^2$. Use this expression to estimate the standard enthalpy of combustion of methane at 350 K. Use the following data:

	$\alpha/(\text{J K}^{-1}\ \text{mol}^{-1})$	$\beta/(\text{mJ K}^{-2}\ \text{mol}^{-1})$	$\gamma/(\mu\text{J K}^{-3}\ \text{mol}^{-1})$
$CH_4(g)$	14.16	75.5	−17.99
$CO_2(g)$	26.86	6.97	−0.82
$O_2(g)$	25.72	12.98	−3.862
$H_2O(g)$	30.36	9.61	1.184

2C.8 Figure 2.1 shows the experimental DSC scan of hen white lysozyme (G. Privalov et al., *Anal. Biochem.* **79**, 232 (1995)) converted to joules (from calories). Determine the enthalpy of unfolding of this protein by integration of the curve and the change in heat capacity accompanying the transition.

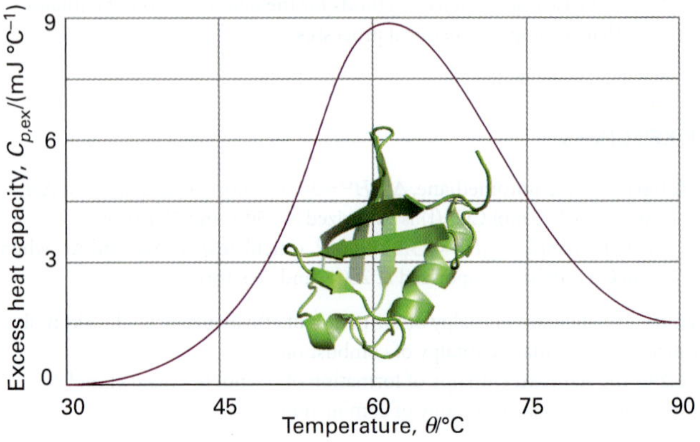

Figure 2.1 The experimental DSC scan of hen white lysozyme.

2C.9 An average human produces about 10 MJ of heat each day through metabolic activity. If a human body were an isolated system of mass 65 kg with the heat capacity of water, what temperature rise would the body experience? Human bodies are actually open systems, and the main mechanism of heat loss is through the evaporation of water. What mass of water should be evaporated each day to maintain constant temperature?

2C.10 In biological cells that have a plentiful supply of oxygen, glucose is oxidized completely to CO_2 and H_2O by a process called *aerobic oxidation*. Muscle cells may be deprived of O_2 during vigorous exercise and, in that case, one molecule of glucose is converted to two molecules of lactic acid ($CH_3CH(OH)COOH$) by a process called *anaerobic glycolysis*. (a) When 0.3212 g of glucose was burned in a bomb calorimeter of calorimeter constant 641 J K^{-1} the temperature rose by 7.793 K. Calculate (i) the standard molar enthalpy of combustion, (ii) the standard internal energy of combustion, and (iii) the standard enthalpy of formation of glucose.
(b) What is the biological advantage (in kilojoules per mole of energy released as heat) of complete aerobic oxidation compared with anaerobic glycolysis to lactic acid?

TOPIC 2D State functions and exact differentials

Discussion questions

2D.1 Suggest (with explanation) how the internal energy of a van der Waals gas should vary with volume at constant temperature.

2D.2 Explain why a perfect gas does not have an inversion temperature.

Exercises

2D.1(a) Estimate the internal pressure, π_T, of water vapour at 1.00 bar and 400 K, treating it as a van der Waals gas. *Hint:* Simplify the approach by estimating the molar volume by treating the gas as perfect.
2D.1(b) Estimate the internal pressure, π_T, of sulfur dioxide at 1.00 bar and 298 K, treating it as a van der Waals gas. *Hint:* Simplify the approach by estimating the molar volume by treating the gas as perfect.

‡ These problems were provided by Charles Trapp and Carmen Giunta.

2D.2(a) For a van der Waals gas, $\pi_T = a/V_m^2$. Calculate ΔU_m for the isothermal expansion of nitrogen gas from an initial volume of 1.00 dm^3 to 20.00 dm^3 at 298 K. What are the values of q and w?
2D.2(b) Repeat Exercise 2D.2(a) for argon, from an initial volume of 1.00 dm^3 to 30.00 dm^3 at 298 K.

2D.3(a) The volume of a certain liquid varies with temperature as

$$V=V^{\ominus}\{0.75+3.9\times10^{-4}(T/\mathrm{K})+1.48\times10^{-6}(T/\mathrm{K})^2\}$$

where $V^{\ominus}$ is its volume at 300 K. Calculate its expansion coefficient, α, at 320 K.

2D.3(b) The volume of a certain liquid varies with temperature as

$$V=V^{\ominus}\{0.77+3.7\times10^{-4}(T/\mathrm{K})+1.52\times10^{-6}(T/\mathrm{K})^2\}$$

where $V^{\ominus}$ is its volume at 298 K. Calculate its expansion coefficient, α, at 310 K.

2D.4(a) The isothermal compressibility of water at 293 K is $4.96\times10^{-5}\,\mathrm{atm^{-1}}$. Calculate the pressure that must be applied in order to increase its density by 0.10 per cent.

2D.4(b) The isothermal compressibility of lead at 293 K is $2.21\times10^{-6}\,\mathrm{atm^{-1}}$. Calculate the pressure that must be applied in order to increase its density by 0.10 per cent.

2D.5(a) Given that $\mu=0.25\,\mathrm{K\,atm^{-1}}$ for nitrogen, calculate the value of its isothermal Joule–Thomson coefficient. Calculate the energy that must be supplied as heat to maintain constant temperature when 10.0 mol N_2 flows through a throttle in an isothermal Joule–Thomson experiment and the pressure drop is 85 atm.

2D.5(b) Given that $\mu=1.11\,\mathrm{K\,atm^{-1}}$ for carbon dioxide, calculate the value of its isothermal Joule–Thomson coefficient. Calculate the energy that must be supplied as heat to maintain constant temperature when 10.0 mol CO_2 flows through a throttle in an isothermal Joule–Thomson experiment and the pressure drop is 75 atm.

Problems

2D.1‡ In 2006, the Intergovernmental Panel on Climate Change (IPCC) considered a global average temperature rise of 1.0–3.5 °C likely by the year 2100, with 2.0 °C its best estimate. Predict the average rise in sea level due to thermal expansion of sea water based on temperature rises of 1.0 °C, 2.0 °C, and 3.5 °C given that the volume of the Earth's oceans is $1.37\times10^9\,\mathrm{km^3}$ and their surface area is $361\times10^6\,\mathrm{km^2}$, and state the approximations which go into the estimates.

2D.2 The heat capacity ratio of a gas determines the speed of sound in it through the formula $c_s=(\gamma RT/M)^{1/2}$, where $\gamma=C_p/C_V$ and M is the molar mass of the gas. Deduce an expression for the speed of sound in a perfect gas of (a) diatomic, (b) linear triatomic, (c) nonlinear triatomic molecules at high temperatures (with translation and rotation active). Estimate the speed of sound in air at 25 °C.

2D.3 Starting from the expression $C_p-C_V=T(\partial p/\partial T)_V(\partial V/\partial T)_p$, use the appropriate relations between partial derivatives to show that

$$C_p-C_V=\frac{T(\partial V/\partial T)_p^2}{(\partial V/\partial p)_T}$$

Evaluate C_p-C_V for a perfect gas.

2D.4 (a) Write expressions for $\mathrm{d}V$ and $\mathrm{d}p$ given that V is a function of p and T and p is a function of V and T. (b) Deduce expressions for $\mathrm{d}\ln V$ and $\mathrm{d}\ln p$ in terms of the expansion coefficient and the isothermal compressibility.

2D.5 Rearrange the van der Waals equation of state, $p=nRT/(V-nb)-n^2a/V^2$, to give an expression for T as a function of p and V (with n constant). Calculate $(\partial T/\partial p)_V$ and confirm that $(\partial T/\partial p)_V=1/(\partial p/\partial T)_V$. Go on to confirm Euler's chain relation (*Mathematical background* 2).

2D.6 Calculate the isothermal compressibility and the expansion coefficient of a van der Waals gas (see Problem 2D.5). Show, using Euler's chain relation (*Mathematical background* 2), that $\kappa_T R=\alpha(V_m-b)$.

2D.7 The speed of sound, c_s, in a gas of molar mass M is related to the ratio of heat capacities γ by $c_s=(\gamma RT/M)^{1/2}$. Show that $c_s=(\gamma p/\rho)^{1/2}$, where ρ is the mass density of the gas. Calculate the speed of sound in argon at 25 °C.

2D.8‡ A gas obeying the equation of state $p(V-nb)=nRT$ is subjected to a Joule–Thomson expansion. Will the temperature increase, decrease, or remain the same?

2D.9 Use the fact that $(\partial U/\partial V)_T=a/V_m^2$ for a van der Waals gas (Topic 1C) to show that $\mu C_{p,m}\approx(2a/RT)-b$ by using the definition of μ and appropriate relations between partial derivatives. *Hint*: Use the approximation $pV_m\approx RT$ when it is justifiable to do so.

2D.10‡ Concerns over the harmful effects of chlorofluorocarbons on stratospheric ozone have motivated a search for new refrigerants. One such alternative is 2,2-dichloro-1,1,1-trifluoroethane (refrigerant 123). Younglove and McLinden published a compendium of thermophysical properties of this substance (B.A. Younglove and M. McLinden, *J. Phys. Chem. Ref. Data* **23**, 7 (1994)), from which properties such as the Joule–Thomson coefficient μ can be computed. (a) Compute μ at 1.00 bar and 50 °C given that $(\partial H/\partial p)_T=-3.29\times10^3\,\mathrm{J\,MPa^{-1}\,mol^{-1}}$ and $C_{p,m}=110.0\,\mathrm{J\,K^{-1}\,mol^{-1}}$. (b) Compute the temperature change which would accompany adiabatic expansion of 2.0 mol of this refrigerant from 1.5 bar to 0.5 bar at 50 °C.

2D.11‡ Another alternative refrigerant (see preceding problem) is 1,1,1,2-tetrafluoroethane (refrigerant HFC-134a). A compendium of thermophysical properties of this substance has been published (R. Tillner-Roth and H.D. Baehr, *J. Phys. Chem. Ref. Data* **23**, 657 (1994)) from which properties such as the Joule–Thomson coefficient μ can be computed. (a) Compute μ at 0.100 MPa and 300 K from the following data (all referring to 300 K):

p/MPa	0.080	0.100	0.12
Specific enthalpy/(kJ kg^{-1})	426.48	426.12	425.76

(The specific constant-pressure heat capacity is $0.7649\,\mathrm{kJ\,K^{-1}\,kg^{-1}}$.) (b) Compute μ at 1.00 MPa and 350 K from the following data (all referring to 350 K):

p/MPa	0.80	1.00	1.2
Specific enthalpy/(kJ kg^{-1})	461.93	459.12	42B.15

(The specific constant-pressure heat capacity is $1.0392\,\mathrm{kJ\,K^{-1}\,kg^{-1}}$.)

TOPIC 2E Adiabatic changes

Discussion questions

2E.1 Why are adiabats steeper than isotherms?

2E.2 Why do heat capacities play a role in the expressions for adiabatic expansion?

Exercises

2E.1(a) Use the equipartition principle to estimate the values of $\gamma = C_p/C_V$ for gaseous ammonia and methane. Do this calculation with and without the vibrational contribution to the energy. Which is closer to the expected experimental value at 25 °C?

2E.1(b) Use the equipartition principle to estimate the value of $\gamma = C_p/C_V$ for carbon dioxide. Do this calculation with and without the vibrational contribution to the energy. Which is closer to the expected experimental value at 25 °C?

2E.2(a) Calculate the final temperature of a sample of argon of mass 12.0 g that is expanded reversibly and adiabatically from 1.0 dm^3 at 273.15 K to 3.0 dm^3.

2E.2(b) Calculate the final temperature of a sample of carbon dioxide of mass 16.0 g that is expanded reversibly and adiabatically from 500 cm^3 at 298.15 K to 2.00 dm^3.

2E.3(a) A sample consisting of 1.0 mol of perfect gas molecules with $C_V = 20.8\ \mathrm{J\ K^{-1}}$ is initially at 4.25 atm and 300 K. It undergoes reversible adiabatic expansion until its pressure reaches 2.50 atm. Calculate the final volume and temperature and the work done.

2E.3(b) A sample consisting of 2.5 mol of perfect gas molecules with $C_{p,\mathrm{m}} = 20.8\ \mathrm{J\ K^{-1}\ mol^{-1}}$ is initially at 240 kPa and 325 K. It undergoes reversible adiabatic expansion until its pressure reaches 150 kPa. Calculate the final volume and temperature and the work done.

2E.4(a) A sample of carbon dioxide of mass 2.45 g at 27.0 °C is allowed to expand reversibly and adiabatically from 500 cm^3 to 3.00 dm^3. What is the work done by the gas?

2E.4(b) A sample of nitrogen of mass 3.12 g at 23.0 °C is allowed to expand reversibly and adiabatically from 400 cm^3 to 2.00 dm^3. What is the work done by the gas?

2E.5(a) Calculate the final pressure of a sample of carbon dioxide that expands reversibly and adiabatically from 67.4 kPa and 0.50 dm^3 to a final volume of 2.00 dm^3. Take $\gamma = 1.4$.

2E.5(b) Calculate the final pressure of a sample of water vapour that expands reversibly and adiabatically from 97.3 Torr and 400 cm^3 to a final volume of 5.0 dm^3. Take $\gamma = 1.3$.

Problem

2E.1 The constant-volume heat capacity of a gas can be measured by observing the decrease in temperature when it expands adiabatically and reversibly. The value of $\gamma = C_p/C_V$ can be inferred if the decrease in pressure is also measured and the constant-pressure heat capacity deduced by combining the two values. A fluorocarbon gas was allowed to expand reversibly and adiabatically to twice its volume; as a result, the temperature fell from 298.15 K to 248.44 K and its pressure fell from 202.94 kPa to 81.840 kPa. Evaluate C_p.

Integrated activities

2.1 Give examples of state functions and discuss why they play a critical role in thermodynamics.

2.2 The thermochemical properties of hydrocarbons are commonly investigated by using molecular modelling methods. (a) Use software to predict $\Delta_c H^\ominus$ values for the alkanes methane through pentane. To calculate $\Delta_c H^\ominus$ values, estimate the standard enthalpy of formation of $C_nH_{2n+2}(g)$ by performing semi-empirical calculations (for example, AM1 or PM3 methods) and use experimental standard enthalpy of formation values for $CO_2(g)$ and $H_2O(l)$. (b) Compare your estimated values with the experimental values of $\Delta_c H^\ominus$ (Table 2C.4) and comment on the reliability of the molecular modelling method. (c) Test the extent to which the relation $\Delta_c H^\ominus = \mathrm{constant} \times \{(M/(\mathrm{g\ mol^{-1}})\}^n$ holds and determine the numerical values of the constant and n.

2.3 Use mathematical software, a spreadsheet, or the *Living graphs* on the web site for this book to:

(a) Calculate the work of isothermal reversible expansion of 1.0 mol $CO_2(g)$ at 298 K from 1.0 m^3 to 3.0 m^3 on the basis that it obeys the van der Waals equation of state.

(b) Explore how the parameter γ affects the dependence of the pressure on the volume. Does the pressure–volume dependence become stronger or weaker with increasing volume?

Mathematical background 2 Multivariate calculus

A thermodynamic property of a system typically depends on a number of variables, such as the internal energy depending on the amount, volume, and temperature. To understand how these properties vary with the conditions we need to understand how to manipulate their derivatives. This is the field of **multivariate calculus**, the calculus of several variables.

MB2.1 Partial derivatives

A **partial derivative** of a function of more than one variable, such as $f(x,y)$, is the slope of the function with respect to one of the variables, all the other variables being held constant (Fig. MB2.1). Although a partial derivative shows how a function changes when one variable changes, it may be used to determine how the function changes when more than one variable changes by an infinitesimal amount. Thus, if f is a function of x and y, then when x and y change by dx and dy, respectively, f changes by

$$\mathrm{d}f = \left(\frac{\partial f}{\partial x}\right)_y \mathrm{d}x + \left(\frac{\partial f}{\partial y}\right)_x \mathrm{d}y \qquad \text{(MB2.1)}$$

where the symbol ∂ ('curly d') is used (instead of d) to denote a partial derivative and the subscript on the parentheses indicates which variable is being held constant. The quantity $\mathrm{d}f$ is also called the **differential** of f. Successive partial derivatives may be taken in any order:

$$\left(\frac{\partial}{\partial y}\left(\frac{\partial f}{\partial x}\right)_y\right)_x = \left(\frac{\partial}{\partial x}\left(\frac{\partial f}{\partial y}\right)_x\right)_y \qquad \text{(MB2.2)}$$

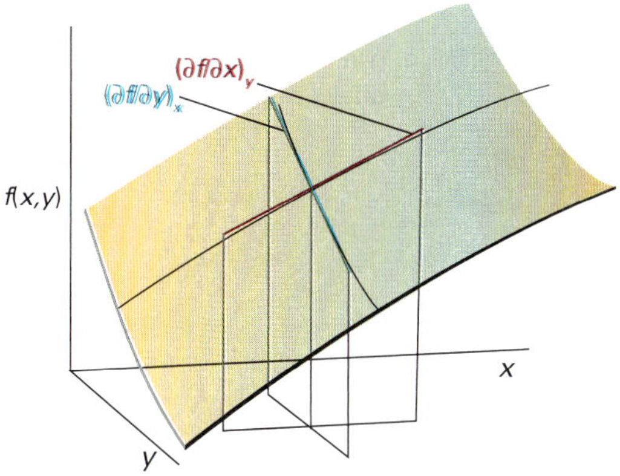

Figure MB2.1 A function of two variables, $f(x,y)$, as depicted by the coloured surface and the two partial derivatives, $(\partial f/\partial x)_y$ and $(\partial f/\partial y)_x$, the slope of the function parallel to the x- and y-axes, respectively. The function plotted here is $f(x,y) = ax^3y + by^2$ with $a = 1$ and $b = -2$.

Brief illustration MB2.1 Partial derivatives

Suppose that $f(x,y) = ax^3y + by^2$ (the function plotted in Fig. MB2.1) then

$$\left(\frac{\partial f}{\partial x}\right)_y = 3ax^2y \qquad \left(\frac{\partial f}{\partial y}\right)_x = ax^3 + 2by$$

Then, when x and y undergo infinitesimal changes, f changes by

$$\mathrm{d}f = 3ax^2y\,\mathrm{d}x + (ax^3 + 2by)\mathrm{d}y$$

To verify that the order of taking the second partial derivative is irrelevant, we form

$$\left(\frac{\partial}{\partial y}\left(\frac{\partial f}{\partial x}\right)_y\right)_x = \left(\frac{\partial(3ax^2y)}{\partial y}\right)_x = 3ax^2$$

$$\left(\frac{\partial}{\partial x}\left(\frac{\partial f}{\partial y}\right)_x\right)_y = \left(\frac{\partial(ax^3+2by)}{\partial x}\right)_y = 3ax^2$$

In the following, z is a variable on which x and y depend (for example, x, y, and z might correspond to p, V, and T).

Relation 1. When x is changed at constant z:

$$\left(\frac{\partial f}{\partial x}\right)_z = \left(\frac{\partial f}{\partial x}\right)_y + \left(\frac{\partial f}{\partial y}\right)_x\left(\frac{\partial y}{\partial x}\right)_z \qquad \text{(MB2.3a)}$$

Relation 2

$$\left(\frac{\partial y}{\partial x}\right)_z = \frac{1}{(\partial x/\partial y)_z} \qquad \text{(MB2.3b)}$$

Relation 3

$$\left(\frac{\partial x}{\partial y}\right)_z = -\left(\frac{\partial x}{\partial z}\right)_y\left(\frac{\partial z}{\partial y}\right)_x \qquad \text{(MB2.3c)}$$

By combining Relations 2 and 3 we obtain the **Euler chain relation**:

$$\left(\frac{\partial y}{\partial x}\right)_z\left(\frac{\partial x}{\partial z}\right)_y\left(\frac{\partial z}{\partial y}\right)_x = -1 \qquad \text{Euler chain relation} \quad \text{(MB2.4)}$$

MB2.2 Exact differentials

The relation in eqn MB2.2 is the basis of a test for an **exact differential**; that is, the test of whether

$$\mathrm{d}f = g(x,y)\mathrm{d}x + h(x,y)\mathrm{d}y \qquad \text{(MB2.5)}$$

has the form in eqn MB2.1. If it has that form, then g can be identified with $(\partial f/\partial x)_y$ and h can be identified with $(\partial f/\partial y)_x$. Then eqn MB2.2 becomes

$$\left(\frac{\partial g}{\partial y}\right)_x = \left(\frac{\partial h}{\partial x}\right)_y \qquad \text{Test for exact differential} \qquad \text{(MB2.6)}$$

Brief illustration MB2.2 Exact differentials

Suppose, instead of the form $df = 3ax^2y\,dx + (ax^3 + 2by)dy$ in the previous *Brief illustration*, we were presented with the expression

$$df = \overbrace{3ax^2y}^{g(x,y)}\,dx + \overbrace{(ax^2 + 2by)}^{h(x,y)}\,dy$$

with ax^2 in place of ax^3 inside the second parentheses. To test whether this is an exact differential, we form

$$\left(\frac{\partial g}{\partial y}\right)_x = \left(\frac{\partial(3ax^2y)}{\partial y}\right)_x = 3ax^2$$

$$\left(\frac{\partial h}{\partial x}\right)_y = \left(\frac{\partial(ax^2+2by)}{\partial x}\right)_y = 2ax$$

These two expressions are not equal, so this form of df is not an exact differential and there is not a corresponding integrated function of the form $f(x,y)$.

If df is exact, then we can do two things:

- From a knowledge of the functions g and h we can reconstruct the function f.
- Be confident that the integral of df between specified limits is independent of the path between those limits.

The first conclusion is best demonstrated with a specific example.

Brief illustration MB2.3 The reconstruction of an equation

We consider the differential $df = 3ax^2y\,dx + (ax^3 + 2by)dy$, which we know to be exact. Because $(\partial f/\partial x)_y = 3ax^2y$, we can integrate with respect to x with y held constant, to obtain

$$f = \int df = \int 3ax^2y\,dx = 3ay\int x^2\,dx = ax^3y + k$$

where the 'constant' of integration k may depend on y (which has been treated as a constant in the integration), but not on x. To find $k(y)$, we note that $(\partial f/\partial y)_x = ax^3 + 2by$, and therefore

$$\left(\frac{\partial f}{\partial y}\right)_x = \left(\frac{\partial(ax^3y+k)}{\partial y}\right)_x = ax^3 + \frac{dk}{dy} = ax^3 + 2by$$

Therefore

$$\frac{dk}{dy} = 2by$$

from which it follows that $k = by^2 + \text{constant}$. We have found, therefore, that

$$f(x,y) = ax^3y + by^2 + \text{constant}$$

which, apart from the constant, is the original function in the *Brief illustration* MB2.1. The value of the constant is pinned down by stating the boundary conditions; thus, if it is known that $f(0,0) = 0$, then the constant is zero.

To demonstrate that the integral of df is independent of the path is now straight forward. Because df is a differential, its integral between the limits a and b is

$$\int_a^b df = f(b) - f(a)$$

The value of the integral depends only on the values at the end points and is independent of the path between them. If df is not an exact differential, the function f does not exist, and this argument no longer holds. In such cases, the integral of df does depend on the path.

Brief illustration MB2.4 Path-dependent integration

Consider the inexact differential (the expression with ax^2 in place of ax^3 inside the second parentheses):

$$df = 3ax^2y\,dx + (ax^2 + 2by)dy$$

Suppose we integrate df from (0,0) to (2,2) along the two paths shown in Fig. MB2.2. Along Path 1,

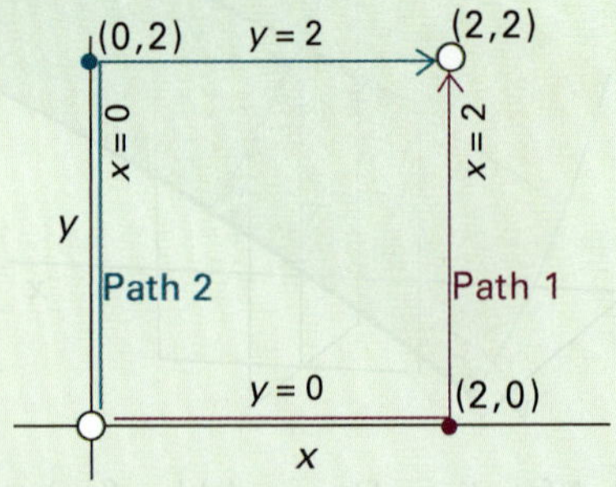

Figure MB2.2 The two integration paths referred to in *Brief illustration* MB2.4.

$$\int_{\text{Path 1}} \mathrm{d}f = \int_{0,0}^{2,0} 3ax^2y\,\mathrm{d}x + \int_{2,0}^{2,2} (ax^2+2by)\,\mathrm{d}y$$
$$= 0 + 4a\int_0^2 \mathrm{d}y + 2b\int_0^2 y\,\mathrm{d}y = 8a+4b$$

whereas along Path 2,

$$\int_{\text{Path 2}} \mathrm{d}f = \int_{0,2}^{2,2} 3ax^2y\,\mathrm{d}x + \int_{0,0}^{0,2} (ax^2+2by)\,\mathrm{d}y$$
$$= 6a\int_0^2 x^2\,\mathrm{d}x + 0 + 2b\int_0^2 y\,\mathrm{d}y = 16a+4b$$

The two integrals are not the same.

An inexact differential may sometimes be converted into an exact differential by multiplication by a factor known as an *integrating factor*. A physical example is the integrating factor $1/T$ that converts the inexact differential $\mathrm{d}q_{\text{rev}}$ into the exact differential $\mathrm{d}S$ in thermodynamics (Topic 3A).

Brief illustration MB2.5 An integrating factor

We have seen that the differential $\mathrm{d}f = 3ax^2y\mathrm{d}x + (ax^2+2by)\mathrm{d}y$ is inexact; the same is true when we set $b=0$ and consider $\mathrm{d}f = 3ax^2y\mathrm{d}x + ax^2\mathrm{d}y$ instead. Suppose we multiply this $\mathrm{d}f$ by x^my^n and write $x^my^n\mathrm{d}f = \mathrm{d}f'$, then we obtain

$$\mathrm{d}f' = \overbrace{3ax^{m+2}y^{n+1}}^{g(x,y)}\,\mathrm{d}x + \overbrace{ax^{m+2}y^n}^{h(x,y)}\,\mathrm{d}y$$

We evaluate the following two partial derivatives:

$$\left(\frac{\partial g}{\partial y}\right)_x = \left(\frac{\partial(3ax^{m+2}y^{n+1})}{\partial y}\right)_x = 3a(n+1)x^{m+2}y^n$$
$$\left(\frac{\partial h}{\partial x}\right)_y = \left(\frac{\partial(ax^{m+2}y^n)}{\partial x}\right)_y = a(m+2)x^{m+1}y^n$$

For the new differential to be exact, these two partial derivatives must be equal, so we write

$$3a(n+1)x^{m+2}y^n = a(m+2)x^{m+1}y^n$$

which simplifies to

$$3(n+1)x = m+2$$

The only solution that is independent of x is $n=-1$ and $m=-2$. It follows that

$$\mathrm{d}f' = 3a\mathrm{d}x + (a/y)\mathrm{d}y$$

is an exact differential. By the procedure already illustrated, its integrated form is $f'(x,y) = 3ax + a\ln y + \text{constant}$.

CHAPTER 3

The Second and Third Laws

Some things happen naturally, some things don't. Some aspect of the world determines the **spontaneous** direction of change, the direction of change that does not require work to bring it about. An important point, though, is that throughout this text 'spontaneous' must be interpreted as a natural *tendency* that may or may not be realized in practice. Thermodynamics is silent on the rate at which a spontaneous change in fact occurs, and some spontaneous processes (such as the conversion of diamond to graphite) may be so slow that the tendency is never realized in practice whereas others (such as the expansion of a gas into a vacuum) are almost instantaneous.

3A Entropy

The direction of change is related to the *distribution of energy and matter*, and spontaneous changes are always accompanied by a dispersal of energy or matter. To quantify this concept we introduce the property called 'entropy', which is central to the formulation of the 'Second Law of thermodynamics'. That law governs all spontaneous change.

3B The measurement of entropy

To make the Second Law quantitative, it is necessary to measure the entropy of a substance. We see that measurement, perhaps with calorimetric methods, of the energy transferred as heat during a physical process or chemical reaction leads to determination of the entropy change and, consequently, the direction of spontaneous change. The discussion in this Topic also leads to the 'Third Law of thermodynamics', which helps us to understand the properties of matter at very low temperatures and to set up an absolute measure of the entropy of a substance.

3C Concentrating on the system

One problem with dealing with the entropy is that it requires separate calculations of the changes taking place in the system and the surroundings. Providing we are willing to impose certain restrictions on the system, that problem can be overcome by introducing the 'Gibbs energy'. Indeed, most thermodynamic calculations in chemistry focus on the change in Gibbs energy, not the direct measurement of the entropy change.

3D Combining the First and Second Laws

Finally, we bring the First and Second Laws together and begin to see the considerable power of thermodynamics for accounting for the properties of matter.

What is the impact of this material?

The Second Law is at the heart of the operation of engines of all types, including devices resembling engines that are used to cool objects. See *Impact* I3.1 for an application to the technology of refrigeration. Entropy considerations are also important in modern electronic materials for it permits a quantitative discussion of the concentration of impurities. See *Impact* I3.2 for a note about how measurement of the entropy at low temperatures gives insight into the purity of materials used as superconductors.

To read more about the impact of this material, scan the QR code, or go to bcs.whfreeman.com/webpub/chemistry/pchem10e/impact/pchem-3-1.html

3A Entropy

Contents

➤ Why do you need to know this material?

Entropy is the concept on which almost all applications of thermodynamics in chemistry are based: it explains why some reactions take place and others do not.

➤ What is the key idea?

The change in entropy of a system can be calculated from the heat transferred to it reversibly.

➤ What do you need to know already?

You need to be familiar with the First-Law concepts of work, heat, and internal energy (Topic 2A). The Topic draws on the expression for work of expansion of a perfect gas (Topic 2A) and on the changes in volume and temperature that accompany the reversible adiabatic expansion of a perfect gas (Topic 2D).

What determines the direction of spontaneous change? It is not the total energy of the isolated system. The First Law of thermodynamics states that energy is conserved in any process, and we cannot disregard that law now and say that everything tends towards a state of lower energy. When a change occurs, the total energy of an isolated system remains constant but it is parcelled out in different ways. Can it be, therefore, that the direction of change is related to the *distribution* of energy? We shall see that this idea is the key, and that spontaneous changes are always accompanied by a dispersal of energy or matter.

3A.1 The Second Law

We can begin to understand the role of the dispersal of energy and matter by thinking about a ball (the system) bouncing on a floor (the surroundings). The ball does not rise as high after each bounce because there are inelastic losses in the materials of the ball and floor. The kinetic energy of the ball's overall motion is spread out into the energy of thermal motion of its particles and those of the floor that it hits. The direction of spontaneous change is towards a state in which the ball is at rest with all its energy dispersed into disorderly thermal motion of molecules in the air and of the atoms of the virtually infinite floor (Fig. 3A.1).

A ball resting on a warm floor has never been observed to start bouncing. For bouncing to begin, something rather special would need to happen. In the first place, some of the thermal motion of the atoms in the floor would have to accumulate in a single, small object, the ball. This accumulation requires a spontaneous localization of energy from the myriad vibrations of the atoms of the floor into the much smaller number of atoms that constitute the ball (Fig. 3A.2). Furthermore, whereas the thermal motion is random, for the ball to move upwards its

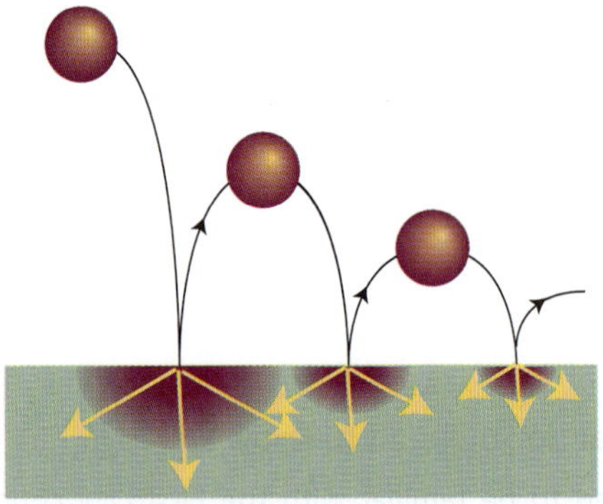

Figure 3A.1 The direction of spontaneous change for a ball bouncing on a floor. On each bounce some of its energy is degraded into the thermal motion of the atoms of the floor, and that energy disperses. The reverse has never been observed to take place on a macroscopic scale.

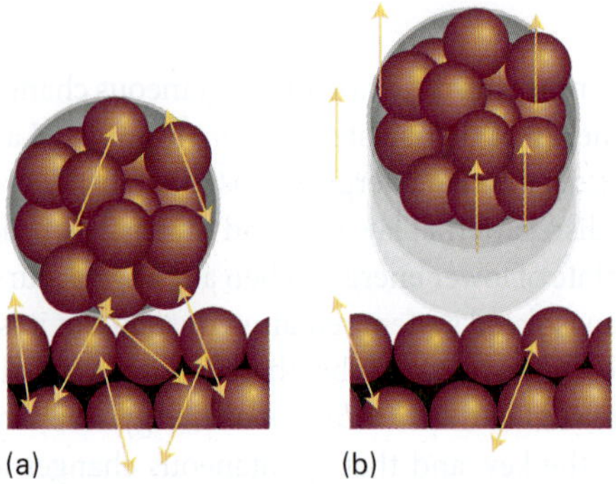

Figure 3A.2 The molecular interpretation of the irreversibility expressed by the Second Law. (a) A ball resting on a warm surface; the atoms are undergoing thermal motion (vibration, in this instance), as indicated by the arrows. (b) For the ball to fly upwards, some of the random vibrational motion would have to change into coordinated, directed motion. Such a conversion is highly improbable.

atoms must all move in the same direction. The localization of random, disorderly motion as concerted, ordered motion is so unlikely that we can dismiss it as virtually impossible.[1]

We appear to have found the signpost of spontaneous change: *we look for the direction of change that leads to dispersal of the total energy of the isolated system.* This principle accounts for the direction of change of the bouncing ball, because its energy is spread out as thermal motion of the atoms of the floor. The reverse process is not spontaneous because it is highly improbable that energy will become localized, leading to uniform motion of the ball's atoms.

Matter also has a tendency to disperse in disorder. A gas does not contract spontaneously because to do so the random motion of its molecules, which spreads out the distribution of molecules throughout the container, would have to take them all into the same region of the container. The opposite change, spontaneous expansion, is a natural consequence of matter becoming more dispersed as the gas molecules occupy a larger volume.

The recognition of two classes of process, spontaneous and non-spontaneous, is summarized by the **Second Law of thermodynamics**. This law may be expressed in a variety of equivalent ways. One statement was formulated by Kelvin:

> No process is possible in which the sole result is the absorption of heat from a reservoir and its complete conversion into work.

For example, it has proved impossible to construct an engine like that shown in Fig. 3A.3, in which heat is drawn from a hot reservoir and completely converted into work. All real heat engines have both a hot source and a cold sink; some energy is always discarded into the cold sink as heat and not converted into work. The Kelvin statement is a generalization of the everyday observation that we have already discussed, that a ball at rest on a surface has never been observed to leap spontaneously upwards. An upward leap of the ball would be equivalent to the conversion of heat from the surface into work. Another statement of the Second Law is due to Rudolf Clausius (Fig. 3A.4):

> Heat does not flow spontaneously from a cool body to a hotter body.

To achieve the transfer of heat to a hotter body, it is necessary to do work on the system, as in a refrigerator.

These two empirical observations turn out to be aspects of a single statement in which the Second Law is expressed in terms of a new state function, the **entropy**, S. We shall see that the entropy (which we shall define shortly, but is a measure of the energy and matter dispersed in a process) lets us assess whether one state is accessible from another by a spontaneous change:

> The entropy of an isolated system increases in the course of a spontaneous change: $\Delta S_{tot} > 0$

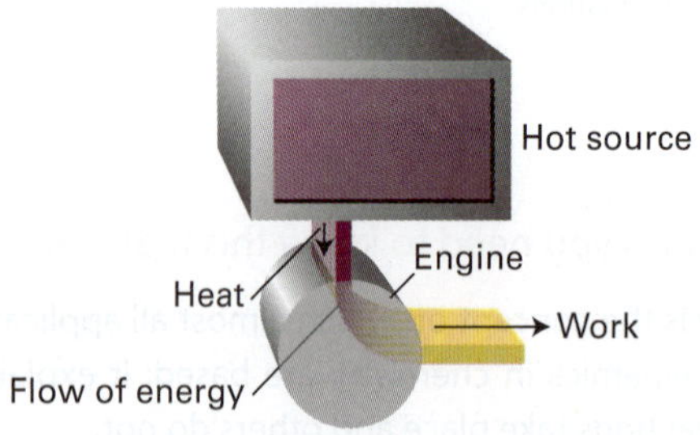

Figure 3A.3 The Kelvin statement of the Second Law denies the possibility of the process illustrated here, in which heat is changed completely into work, there being no other change. The process is not in conflict with the First Law because energy is conserved.

[1] Concerted motion, but on a much smaller scale, is observed as *Brownian motion*, the jittering motion of small particles suspended in a liquid or gas.

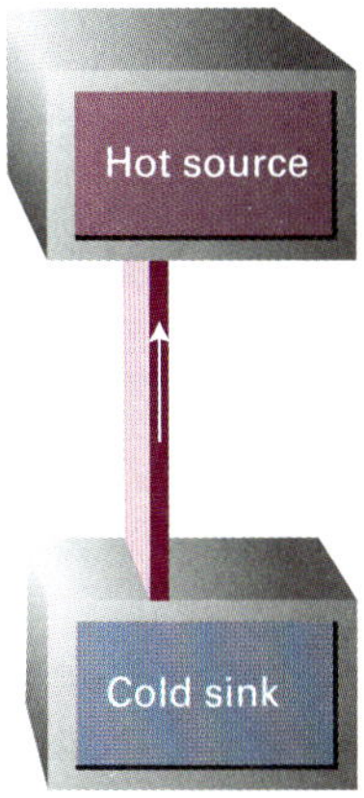

Figure 3A.4 The Clausius statement of the Second Law denies the possibility of the process illustrated here, in which energy as heat migrates from a cool source to a hot sink, there being no other change. The process is not in conflict with the First Law because energy is conserved.

where S_{tot} is the total entropy of the system and its surroundings. Thermodynamically irreversible processes (like cooling to the temperature of the surroundings and the free expansion of gases) are spontaneous processes, and hence must be accompanied by an increase in total entropy.

In summary, the First Law uses the internal energy to identify *permissible* changes; the Second Law uses the entropy to identify the *spontaneous changes* among those permissible changes.

3A.2 The definition of entropy

To make progress, and to turn the Second Law into a quantitatively useful expression, we need to define and then calculate the entropy change accompanying various processes. There are two approaches, one classical and one molecular. They turn out to be equivalent, but each one enriches the other.

(a) The thermodynamic definition of entropy

The thermodynamic definition of entropy concentrates on the change in entropy, dS, that occurs as a result of a physical or chemical change (in general, as a result of a 'process'). The definition is motivated by the idea that a change in the extent to which energy is dispersed depends on how much energy is transferred as heat. As explained in Topic 2A, heat stimulates random motion in the surroundings. On the other hand, work stimulates uniform motion of atoms in the surroundings and so does not change their entropy.

The thermodynamic definition of entropy is based on the expression

$$\mathrm{d}S=\frac{\mathrm{d}q_{\mathrm{rev}}}{T} \qquad \textit{Definition} \quad \text{Entropy change} \qquad (3A.1)$$

For a measurable change between two states i and f,

$$\Delta S=\int_{\mathrm{i}}^{\mathrm{f}}\frac{\mathrm{d}q_{\mathrm{rev}}}{T} \qquad (3A.2)$$

That is, to calculate the difference in entropy between any two states of a system, we find a *reversible* path between them, and integrate the energy supplied as heat at each stage of the path divided by the temperature at which heating occurs.

A note on good practice According to eqn 3A.1, when the energy transferred as heat is expressed in joules and the temperature is in kelvins, the units of entropy are joules per kelvin ($\mathrm{J\,K^{-1}}$). Entropy is an extensive property. Molar entropy, the entropy divided by the amount of substance, $S_m=S/n$, is expressed in joules per kelvin per mole ($\mathrm{J\,K^{-1}\,mol^{-1}}$). The units of entropy are the same as those of the gas constant, R, and molar heat capacities. Molar entropy is an intensive property.

Example 3A.1 Calculating the entropy change for the isothermal expansion of a perfect gas

Calculate the entropy change of a sample of perfect gas when it expands isothermally from a volume V_i to a volume V_f.

Method The definition of entropy instructs us to find the energy supplied as heat for a reversible path between the stated initial and final states regardless of the actual manner in which the process takes place. A simplification is that the expansion is isothermal, so the temperature is a constant and may be taken outside the integral in eqn 3A.2. The energy absorbed as heat during a reversible isothermal expansion of a perfect gas can be calculated from $\Delta U=q+w$ and $\Delta U=0$, which implies that $q=-w$ in general and therefore that $q_{rev}=-w_{rev}$ for a reversible change. The work of reversible isothermal expansion is calculated in Topic 2A. The change in molar entropy is calculated from $\Delta S_m=\Delta S/n$.

Answer Because the temperature is constant, eqn 3A.2 becomes

$$\Delta S=\frac{1}{T}\int_{\mathrm{i}}^{\mathrm{f}}\mathrm{d}q_{\mathrm{rev}}=\frac{q_{\mathrm{rev}}}{T}$$

From Topic 2A we know that

$$q_{\mathrm{rev}}=-w_{\mathrm{rev}}=nRT\ln\frac{V_{\mathrm{f}}}{V_{\mathrm{i}}}$$

It follows that

$$\Delta S=nR\ln\frac{V_{\mathrm{f}}}{V_{\mathrm{i}}} \quad \text{and} \quad \Delta S_{\mathrm{m}}=R\ln\frac{V_{\mathrm{f}}}{V_{\mathrm{i}}}$$

Self-test 3A.1 Calculate the change in entropy when the pressure of a fixed amount of perfect gas is changed isothermally from p_i to p_f. What is this change due to?

Answer: $\Delta S=nR\ln(p_i/p_f)$; the change in volume when the gas is compressed or expands

The definition in eqn 3A.1 is used to formulate an expression for the change in entropy of the surroundings, ΔS_{sur}. Consider an infinitesimal transfer of heat dq_{sur} to the surroundings. The surroundings consist of a reservoir of constant volume, so the energy supplied to them by heating can be identified with the change in the internal energy of the surroundings, dU_{sur}.[2] The internal energy is a state function, and dU_{sur} is an exact differential. These properties imply that dU_{sur} is independent of how the change is brought about and in particular is independent of whether the process is reversible or irreversible. The same remarks therefore apply to dq_{sur}, to which dU_{sur} is equal. Therefore, we can adapt the definition in eqn 3A.1, delete the constraint 'reversible', and write

$$dS = \frac{dq_{rev,sur}}{T_{sur}} = \frac{dq_{sur}}{T_{sur}} \qquad \text{Entropy change of the surroundings} \qquad (3A.3a)$$

Furthermore, because the temperature of the surroundings is constant whatever the change, for a measurable change

$$\Delta S_{sur} = \frac{q_{sur}}{T_{sur}} \qquad (3A.3b)$$

That is, regardless of how the change is brought about in the system, reversibly or irreversibly, we can calculate the change of entropy of the surroundings by dividing the heat transferred by the temperature at which the transfer takes place.

Equation 3A.3 makes it very simple to calculate the changes in entropy of the surroundings that accompany any process. For instance, for any adiabatic change, $q_{sur}=0$, so

$$\Delta S_{sur} = 0 \qquad \text{Adiabatic change} \qquad (3A.4)$$

This expression is true however the change takes place, reversibly or irreversibly, provided no local hot spots are formed in the surroundings. That is, it is true so long as the surroundings remain in internal equilibrium. If hot spots do form, then the localized energy may subsequently disperse spontaneously and hence generate more entropy.

Brief illustration 3A.1 The entropy change of the surroundings

To calculate the entropy change in the surroundings when 1.00 mol $H_2O(l)$ is formed from its elements under standard conditions at 298 K, we use $\Delta H^{\ominus} = -286$ kJ from Table 2C.2. The energy released as heat is supplied to the surroundings, now regarded as being at constant pressure, so $q_{sur} = +286$ kJ. Therefore,

$$\Delta S_{sur} = \frac{2.86\times10^5\,\text{J mol}^{-1}}{298\,\text{K}} = +960\,\text{J K}^{-1}$$

This strongly exothermic reaction results in an increase in the entropy of the surroundings as energy is released as heat into them.

Self-test 3A.2 Calculate the entropy change in the surroundings when 1.00 mol $N_2O_4(g)$ is formed from 2.00 mol $NO_2(g)$ under standard conditions at 298 K.

Answer: $-192\,\text{J K}^{-1}$

[2] Alternatively, the surroundings can be regarded as being at constant pressure, in which case we could equate dq_{sur} to dH_{sur}.

We are now in a position to see how the definition of entropy is consistent with Kelvin's and Clausius's statements of the Second Law. In the arrangement shown in Fig. 3A.3, the entropy of the hot source is reduced as energy leaves it as heat, but no other change in entropy occurs (the transfer of energy as work does not result in the production of entropy); consequently the arrangement does not produce work. In Clausius version, the entropy of the cold source in Fig 3A.4 decreases when a certain quantity of energy leaves it as heat, but when that heat enters the hot sink the rise in entropy is not as great. Therefore, overall there is a decrease in entropy: the process is not spontaneous.

(b) The statistical definition of entropy

The entry point into the molecular interpretation of the Second Law of thermodynamics is Boltzmann's insight, first mentioned in *Foundations* B, that an atom or molecule can possess only certain values of the energy, called its 'energy levels'. The continuous thermal agitation that molecules experience at $T>0$ ensures that they are distributed over the available energy levels. Boltzmann also made the link between the distribution of molecules over energy levels and the entropy. He proposed that the entropy of a system is given by

$$S = k \ln \mathcal{W} \qquad \text{Boltzmann formula for the entropy} \qquad (3A.5)$$

where $k = 1.381\times10^{-23}\,\text{J K}^{-1}$ and $\mathcal{W}$ is the number of **microstates**, the number of ways in which the molecules of a system can be arranged while keeping the total energy constant. Each microstate lasts only for an instant and corresponds to a certain distribution of molecules over the available energy levels. When we measure the properties of a system, we are measuring an average taken over the many microstates the system can occupy under the conditions of the experiment. The concept of the number of microstates makes quantitative the ill-defined qualitative concepts of 'disorder' and 'the dispersal of matter and energy' that are used widely to introduce the concept of entropy: a more disorderly distribution of matter and a greater dispersal of energy corresponds to a greater number of microstates associated with the same total energy. This point is discussed in much greater detail in Topic 15E.

Equation 3A.5 is known as the **Boltzmann formula** and the entropy calculated from it is sometimes called the **statistical**

entropy. We see that if $\mathcal{W}=1$, which corresponds to one microstate (only one way of achieving a given energy, all molecules in exactly the same state), then $S=0$ because $\ln 1=0$. However, if the system can exist in more than one microstate, then $\mathcal{W}>1$ and $S>0$. If the molecules in the system have access to a greater number of energy levels, then there may be more ways of achieving a given total energy; that is, there are more microstates for a given total energy, $\mathcal{W}$ is greater, and the entropy is greater than when fewer states are accessible. Therefore, the statistical view of entropy summarized by the Boltzmann formula is consistent with our previous statement that the entropy is related to the dispersal of energy and matter. In particular, for a gas of particles in a container, the energy levels become closer together as the container expands (Fig. 3A.5; this is a conclusion from quantum theory that is verified in Topic 8A). As a result, more microstates become possible, $\mathcal{W}$ increases, and the entropy increases, exactly as we inferred from the thermodynamic definition of entropy.

Brief illustration 3A.2 The Boltzmann formula

Suppose that each diatomic molecule in a solid sample can be arranged in either of two orientations and that there are $N=6.022\times10^{23}$ molecules in the sample (that is, 1 mol of molecules). Then $\mathcal{W}=2^N$ and the entropy of the sample is

$$S=k\ln 2^N=Nk\ln 2=(6.022\times10^{23})\times(1.381\times10^{-23}\,\text{J K}^{-1})\ln 2$$
$$=5.76\,\text{J K}^{-1}$$

Self-test 3A.3 What is the molar entropy of a similar system in which each molecule can be arranged in four different orientations?

Answer: $11.5\,\text{J K}^{-1}\,\text{mol}^{-1}$

The molecular interpretation of entropy advanced by Boltzmann also suggests the thermodynamic definition given by eqn 3A.1. To appreciate this point, consider that molecules in a system at high temperature can occupy a large number of the available energy levels, so a small additional transfer of energy as heat will lead to a relatively small change in the number of accessible energy levels. Consequently, the number of microstates does not increase appreciably and neither does the entropy of the system. In contrast, the molecules in a system at low temperature have access to far fewer energy levels (at $T=0$, only the lowest level is accessible), and the transfer of the same quantity of energy by heating will increase the number of accessible energy levels and the number of microstates significantly. Hence, the change in entropy upon heating will be greater when the energy is transferred to a cold body than when it is transferred to a hot body. This argument suggests that the change in entropy for a given transfer of energy as heat should be greater at low temperatures than at high, as in eqn 3A.1.

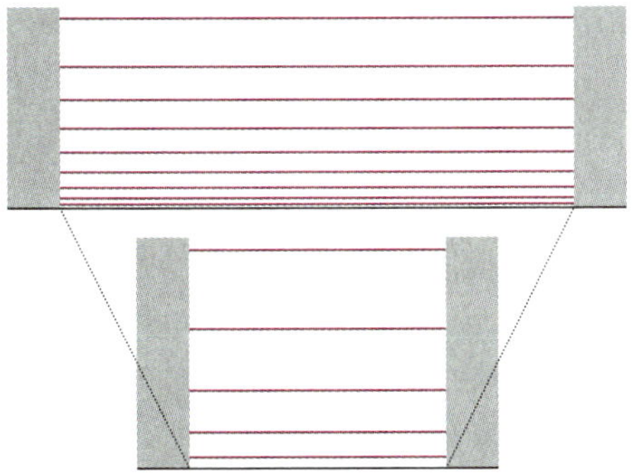

Figure 3A.5 When a box expands, the energy levels move closer together and more become accessible to the molecules. As a result the number of ways of achieving the same energy (the value of $\mathcal{W}$) increases, and so therefore does the entropy.

3A.3 The entropy as a state function

Entropy is a state function. To prove this assertion, we need to show that the integral of dS is independent of path. To do so, it is sufficient to prove that the integral of eqn 3A.1 around an arbitrary cycle is zero, for that guarantees that the entropy is the same at the initial and final states of the system regardless of the path taken between them (Fig. 3A.6). That is, we need to show that

$$\oint \mathrm{d}S=\oint\frac{\mathrm{d}q_{\text{rev}}}{T}=0 \qquad (3A.6)$$

where the symbol $\oint$ denotes integration around a closed path. There are three steps in the argument:

1. First, to show that eqn 3A.6 is true for a special cycle (a 'Carnot cycle') involving a perfect gas.

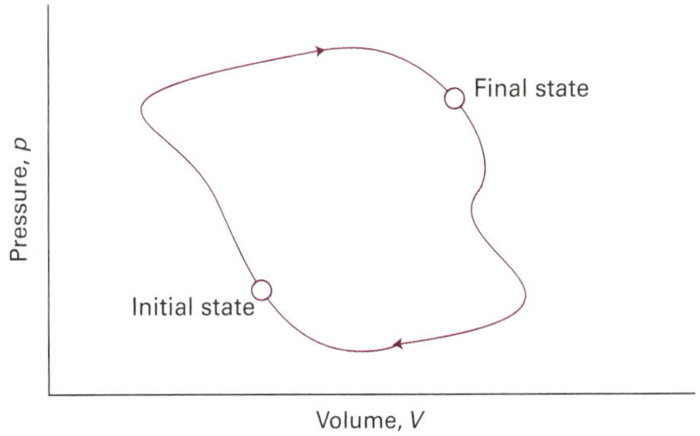

Figure 3A.6 In a thermodynamic cycle, the overall change in a state function (from the initial state to the final state and then back to the initial state again) is zero.

2. Then to show that the result is true whatever the working substance.
3. Finally, to show that the result is true for any cycle.

(a) The Carnot cycle

A **Carnot cycle**, which is named after the French engineer Sadi Carnot, consists of four reversible stages (Fig. 3A.7):

1. Reversible isothermal expansion from A to B at T_h; the entropy change is q_h/T_h, where q_h is the energy supplied to the system as heat from the hot source.
2. Reversible adiabatic expansion from B to C. No energy leaves the system as heat, so the change in entropy is zero. In the course of this expansion, the temperature falls from T_h to T_c, the temperature of the cold sink.
3. Reversible isothermal compression from C to D at T_c. Energy is released as heat to the cold sink; the change in entropy of the system is q_c/T_c; in this expression q_c is negative.
4. Reversible adiabatic compression from D to A. No energy enters the system as heat, so the change in entropy is zero. The temperature rises from T_c to T_h.

The total change in entropy around the cycle is the sum of the changes in each of these four steps:

$$\oint \mathrm{d}S = \frac{q_h}{T_h} + \frac{q_c}{T_c}$$

However, we show in the following *Justification* that for a perfect gas

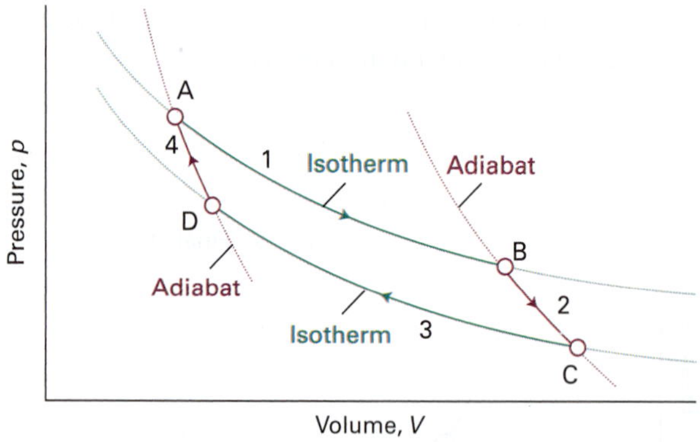

Figure 3A.7 The basic structure of a Carnot cycle. In Step 1, there is isothermal reversible expansion at the temperature T_h. Step 2 is a reversible adiabatic expansion in which the temperature falls from T_h to T_c. In Step 3 there is an isothermal reversible compression at T_c, and that isothermal step is followed by an adiabatic reversible compression, which restores the system to its initial state.

$$\frac{q_h}{q_c} = -\frac{T_h}{T_c} \qquad (3A.7)$$

Substitution of this relation into the preceding equation gives zero on the right, which is what we wanted to prove.

Justification 3A.1 Heating accompanying reversible adiabatic expansion

This *Justification* is based on two features of the cycle. One feature is that the two temperatures T_h and T_c in eqn 3A.7 lie on the same adiabat in Fig. 3A.7. The second feature is that the energy transferred as heat during the two isothermal stages are

$$q_h = nRT_h \ln\frac{V_B}{V_A} \qquad q_c = nRT_c \ln\frac{V_D}{V_C}$$

We now show that the two volume ratios are related in a very simple way. From the relation between temperature and volume for reversible adiabatic processes (VT^c=constant, Topic 2D):

$$V_A T_h^c = V_D T_c^c \qquad V_C T_c^c = V_B T_h^c$$

Multiplication of the first of these expressions by the second gives

$$V_A V_C T_h^c T_c^c = V_D V_B T_h^c T_c^c$$

which, on cancellation of the temperatures, simplifies to

$$\frac{V_D}{V_C} = \frac{V_A}{V_B}$$

With this relation established, we can write

$$q_c = nRT_c \ln\frac{V_D}{V_C} = nRT_c \ln\frac{V_A}{V_B} = -nRT_c \ln\frac{V_B}{V_A}$$

and therefore

$$\frac{q_h}{q_c} = \frac{nRT_h \ln(V_B/V_A)}{-nRT_c \ln(V_B/V_A)} = -\frac{T_h}{T_c}$$

as in eqn 3A.7. For clarification, note that q_h is negative (heat is withdrawn from the hot source) and q_c is positive (heat is deposited in the cold sink), so their ratio is negative.

Brief illustration 3A.3 The Carnot cycle

The Carnot cycle can be regarded as a representation of the changes taking place in an actual idealized engine, where heat is converted into work. (However, other cycles are closer approximations to real engines.) In an engine running in accord with the Carnot cycle, 100 J of energy is withdrawn from the hot source ($q_h=-100$ J) at 500 K and some is used to do work, with the remainder deposited in the cold sink at 300 K. According to eqn 3A.7, the amount of heat deposited is

$$q_c=-q_h\times\frac{T_c}{T_h}=-(-100\text{ J})\times\frac{300\text{ K}}{500\text{ K}}=+60\text{ J}$$

That means that 40 J was used to do work.

Self-test 3A.4 How much work can be extracted when the temperature of the hot source is increased to 800 K?

Answer: 62 J

In the second step we need to show that eqn 3A.6 applies to any material, not just a perfect gas (which is why, in anticipation, we have not labelled it in blue). We begin this step of the argument by introducing the **efficiency**, η (eta), of a heat engine:

$$\eta=\frac{\text{work performed}}{\text{heat absorbed from hot source}}=\frac{|w|}{|q_h|} \quad \text{Definition of efficiency} \quad (3A.8)$$

We are using modulus signs to avoid complications with signs: all efficiencies are positive numbers. The definition implies that the greater the work output for a given supply of heat from the hot reservoir, the greater is the efficiency of the engine. We can express the definition in terms of the heat transactions alone, because (as shown in Fig. 3A.8), the energy supplied as work by the engine is the difference between the energy supplied as heat by the hot reservoir and returned to the cold reservoir:

$$\eta=\frac{|q_h|-|q_c|}{|q_h|}=1-\frac{|q_c|}{|q_h|} \quad (3A.9)$$

It then follows from eqn 3A.7 written as $|q_c|/|q_h|=T_c/T_h$ (see the concluding remark in *Justification* 3A.1) that

$$\eta=1-\frac{T_c}{T_h} \quad \text{Carnot efficiency} \quad (3A.10)$$

Brief illustration 3A.4 Thermal efficiency

A certain power station operates with superheated steam at 300 °C ($T_h=573$ K) and discharges the waste heat into the environment at 20 °C ($T_c=293$ K). The theoretical efficiency is therefore

$$\eta=1-\frac{293\text{ K}}{573\text{ K}}=0.489, \text{ or } 48.9 \text{ per cent}$$

In practice, there are other losses due to mechanical friction and the fact that the turbines do not operate reversibly.

Self-test 3A.5 At what temperature of the hot source would the theoretical efficiency reach 80 per cent?

Answer: 1465 K

Now we are ready to generalize this conclusion. The Second Law of thermodynamics implies that *all reversible engines have the same efficiency regardless of their construction*. To see the truth of this statement, suppose two reversible engines are coupled together and run between the same two reservoirs (Fig. 3A.9). The working substances and details of construction of the two engines are entirely arbitrary. Initially, suppose that engine A is more efficient than engine B, and that we choose a setting of the controls that causes engine B to acquire energy as heat q_c from the cold reservoir and to release a certain

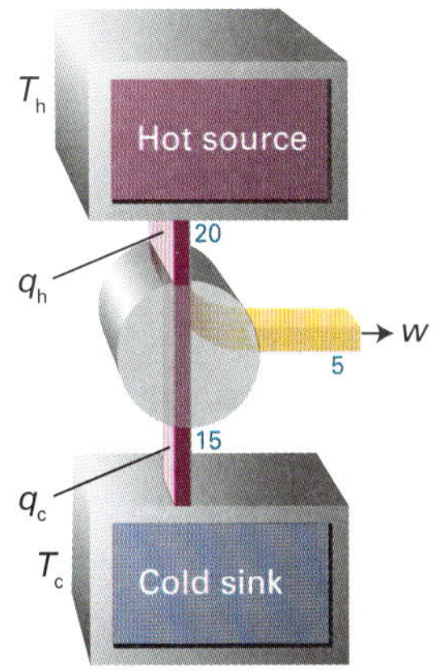

Figure 3A.8 Suppose an energy q_h (for example, 20 kJ) is supplied to the engine and q_c is lost from the engine (for example, $q_c=-15$ kJ) and discarded into the cold reservoir. The work done by the engine is equal to q_h+q_c (for example, 20 kJ + (−15 kJ) = 5 kJ). The efficiency is the work done divided by the energy supplied as heat from the hot source.

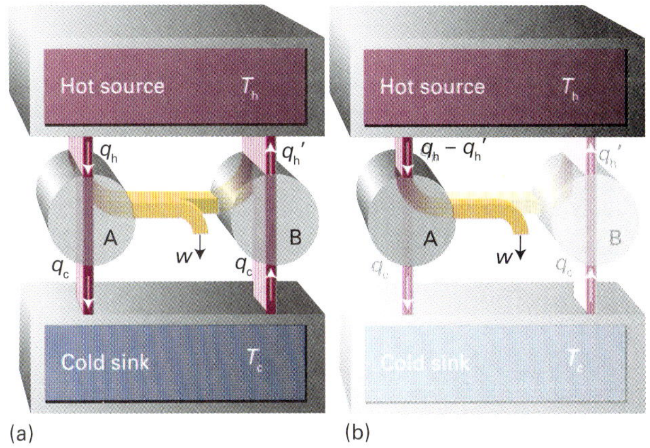

Figure 3A.9 (a) The demonstration of the equivalence of the efficiencies of all reversible engines working between the same thermal reservoirs is based on the flow of energy represented in this diagram. (b) The net effect of the processes is the conversion of heat into work without there being a need for a cold sink: this is contrary to the Kelvin statement of the Second Law.

quantity of energy as heat into the hot reservoir. However, because engine A is more efficient than engine B, not all the work that A produces is needed for this process, and the difference can be used to do work. The net result is that the cold reservoir is unchanged, work has been done, and the hot reservoir has lost a certain amount of energy. This outcome is contrary to the Kelvin statement of the Second Law, because some heat has been converted directly into work. In molecular terms, the random thermal motion of the hot reservoir has been converted into ordered motion characteristic of work. Because the conclusion is contrary to experience, the initial assumption that engines A and B can have different efficiencies must be false. It follows that the relation between the heat transfers and the temperatures must also be independent of the working material, and therefore that eqn 3A.10 is always true for any substance involved in a Carnot cycle.

For the final step in the argument, we note that any reversible cycle can be approximated as a collection of Carnot cycles and the integral around an arbitrary path is the sum of the integrals around each of the Carnot cycles (Fig. 3A.10). This approximation becomes exact as the individual cycles are allowed to become infinitesimal. The entropy change around each individual cycle is zero (as demonstrated above), so the sum of entropy changes for all the cycles is zero. However, in the sum, the entropy change along any individual path is cancelled by the entropy change along the path it shares with the neighbouring cycle. Therefore, all the entropy changes cancel except for those along the perimeter of the overall cycle. That is,

$$\sum_{\text{all}} \frac{q_{\text{rev}}}{T} = \sum_{\text{perimeter}} \frac{q_{\text{rev}}}{T} = 0$$

In the limit of infinitesimal cycles, the non-cancelling edges of the Carnot cycles match the overall cycle exactly, and the sum

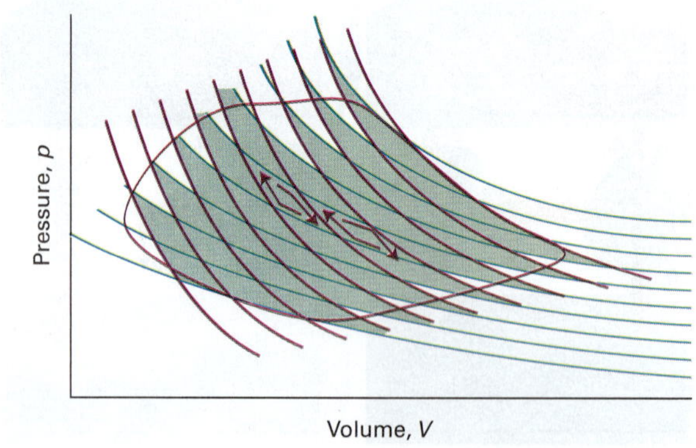

Figure 3A.10 A general cycle can be divided into small Carnot cycles. The match is exact in the limit of infinitesimally small cycles. Paths cancel in the interior of the collection, and only the perimeter, an increasingly good approximation to the true cycle as the number of cycles increases, survives. Because the entropy change around every individual cycle is zero, the integral of the entropy around the perimeter is zero too.

becomes an integral. Equation 3A.6 then follows immediately. This result implies that dS is an exact differential and therefore that S is a state function.

(b) The thermodynamic temperature

Suppose we have an engine that is working reversibly between a hot source at a temperature T_{h} and a cold sink at a temperature T, then we know from eqn 3A.10 that

$$T = (1-\eta)T_{\text{h}} \tag{3A.11}$$

This expression enabled Kelvin to define the **thermodynamic temperature scale** in terms of the efficiency of a heat engine: we construct an engine in which the hot source is at a known temperature and the cold sink is the object of interest. The temperature of the latter can then be inferred from the measured efficiency of the engine. The **Kelvin scale** (which is a special case of the thermodynamic temperature scale) is currently defined by using water at its triple point as the notional hot source and defining that temperature as 273.16 K exactly.[3]

Brief illustration 3A.5 The thermodynamic temperature

A heat engine was constructed that used a hot source at the triple point temperature of water and used as a cold source a cooled liquid. The efficiency of the engine was measured as 0.400. The temperature of the liquid is therefore

$$T = (1-0.400)\times(273.16\,\text{K}) = 164\,\text{K}$$

Self-test 3A.6 What temperature would be reported for the hot source if a thermodynamic efficiency of 0.500 was measured when the cold sink was at 273.16 K?

Answer: 546 K

(c) The Clausius inequality

We now show that the definition of entropy is consistent with the Second Law. To begin, we recall that more work is done when a change is reversible than when it is irreversible. That is, $|\text{d}w_{\text{rev}}| \geq |\text{d}w|$. Because d$w$ and dw_{rev} are negative when energy leaves the system as work, this expression is the same as $-\text{d}w_{\text{rev}} \geq -\text{d}w$, and hence $\text{d}w - \text{d}w_{\text{rev}} \geq 0$. Because the internal energy is a state function, its change is the same for irreversible and reversible paths between the same two states, so we can also write:

$$\text{d}U = \text{d}q + \text{d}w = \text{d}q_{\text{rev}} + \text{d}w_{\text{rev}}$$

[3] Discussions are in progress to replace this definition by another that is independent of the specification of a particular substance.

It follows that $dq_{rev} - dq = dw - dw_{rev} \geq 0$, or $dq_{rev} \geq dq$, and therefore that $dq_{rev}/T \geq dq/T$. Now we use the thermodynamic definition of the entropy (eqn 3A.1; $dS = dq_{rev}/T$) to write

$$dS \geq \frac{dq}{T} \quad \text{Clausius inequality} \quad (3A.12)$$

This expression is the **Clausius inequality**. It proves to be of great importance for the discussion of the spontaneity of chemical reactions, as is shown in Topic 3C.

Brief illustration 3A.6 The Clausius inequality

Consider the transfer of energy as heat from one system—the hot source—at a temperature T_h to another system—the cold sink—at a temperature T_c (Fig. 3A.11).

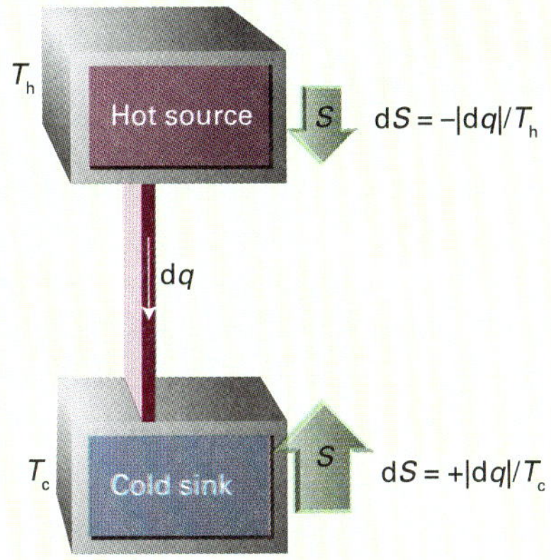

Figure 3A.11 When energy leaves a hot reservoir as heat, the entropy of the reservoir decreases. When the same quantity of energy enters a cooler reservoir, the entropy increases by a larger amount. Hence, overall there is an increase in entropy and the process is spontaneous. Relative changes in entropy are indicated by the sizes of the arrows.

When $|dq|$ leaves the hot source (so $dq_h < 0$), the Clausius inequality implies that $dS \geq dq_h/T_h$. When $|dq|$ enters the cold sink the Clausius inequality implies that $dS \geq dq_c/T_c$ (with $dq_c > 0$). Overall, therefore,

$$dS \geq \frac{dq_h}{T_h} + \frac{dq_c}{T_c}$$

However, $dq_h = -dq_c$, so

$$dS \geq -\frac{dq_c}{T_h} + \frac{dq_c}{T_c} = \left(\frac{1}{T_c} - \frac{1}{T_h}\right) dq_c$$

which is positive (because $dq_c > 0$ and $T_h \geq T_c$). Hence, cooling (the transfer of heat from hot to cold) is spontaneous, as we know from experience.

Self-test 3A.7 What is the change in entropy when 1.0 J of energy as heat transfers from a large block of iron at 30 °C to another large block at 20 °C?

Answer: $+0.1\ \text{mJ K}^{-1}$

We now suppose that the system is isolated from its surroundings, so that $dq = 0$. The Clausius inequality implies that

$$dS \geq 0 \quad (3A.13)$$

and we conclude that *in an isolated system the entropy cannot decrease when a spontaneous change occurs*. This statement captures the content of the Second Law.

3A.4 Entropy changes accompanying specific processes

We now see how to calculate the entropy changes that accompany a variety of basic processes.

(a) Expansion

We established in *Example* 3A.1 that the change in entropy of a perfect gas that expands isothermally from V_i to V_f is

$$\Delta S = nR \ln \frac{V_f}{V_i} \quad \text{Entropy change for the isothermal expansion of a perfect gas} \quad (3A.14)$$

Because S is a state function, the value of ΔS *of the system* is independent of the path between the initial and final states, so this expression applies whether the change of state occurs reversibly or irreversibly. The logarithmic dependence of entropy on volume is illustrated in Fig. 3A.12.

The *total* change in entropy, however, does depend on how the expansion takes place. For any process the energy lost as heat from the system is acquired by the surroundings, so $dq_{sur} = -dq$. For a reversible change we use the expression in *Example* 3A.1 ($q_{rev} = nRT \ln(V_f/V_i)$); consequently, from eqn 3A.3b

$$\Delta S_{sur} = \frac{q_{sur}}{T} = -\frac{q_{rev}}{T} = -nR \ln \frac{V_f}{V_i} \quad (3A.15)$$

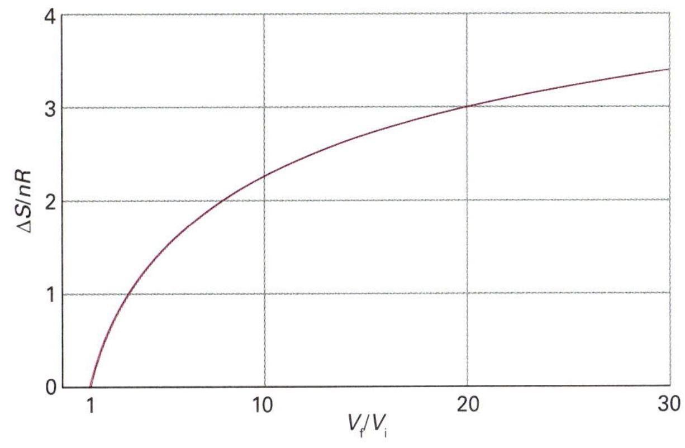

Figure 3A.12 The logarithmic increase in entropy of a perfect gas as it expands isothermally.

This change is the negative of the change in the system, so we can conclude that $\Delta S_{tot}=0$, which is what we should expect for a reversible process. If, on the other hand, the isothermal expansion occurs freely ($w=0$), then $q=0$ (because $\Delta U=0$). Consequently, $\Delta S_{sur}=0$, and the total entropy change is given by eqn 3A.17 itself:

$$\Delta S_{tot} = nR \ln \frac{V_f}{V_i} \qquad (3A.16)$$

In this case, $\Delta S_{tot}>0$, as we expect for an irreversible process.

Brief illustration 3A.7 Entropy of expansion

When the volume of any perfect gas is doubled at any constant temperature, $V_f/V_i=2$ and the change in molar entropy of the system is

$$\Delta S_m = (8.3145\,J\,K^{-1}\,mol^{-1})\times \ln 2 = +5.76\,J\,K^{-1}\,mol^{-1}$$

If the change is carried out reversibly, the change in entropy of the surroundings is $-5.76\,J\,K^{-1}\,mol^{-1}$ (the 'per mole' meaning per mole of gas molecules in the sample). The total change in entropy is 0. If the expansion is free, the change in molar entropy of the gas is still $+5.76\,J\,K^{-1}\,mol^{-1}$, but that of the surroundings is 0, and the total change is $+5.76\,J\,K^{-1}\,mol^{-1}$.

Self-test 3A.8 Calculate the change in entropy when a perfect gas expands isothermally to 10 times its initial volume (a) reversibly, (b) irreversibly against zero pressure.

Answer: (a) $\Delta S_m=+19\,J\,K^{-1}\,mol^{-1}$, $\Delta S_{surr}=-19\,J\,K^{-1}\,mol^{-1}$, $\Delta S_{tot}=0$; (b) $\Delta S_m=+19\,J\,K^{-1}\,mol^{-1}$, $\Delta S_{surr}=0$, $\Delta S_{tot}=+19\,J\,K^{-1}\,mol^{-1}$

(b) Phase transitions

The degree of dispersal of matter and energy changes when a substance freezes or boils as a result of changes in the order with which the molecules pack together and the extent to which the energy is localized or dispersed. Therefore, we should expect the transition to be accompanied by a change in entropy. For example, when a substance vaporizes, a compact condensed phase changes into a widely dispersed gas and we can expect the entropy of the substance to increase considerably. The entropy of a solid also increases when it melts to a liquid and when that liquid turns into a gas.

Consider a system and its surroundings at the **normal transition temperature**, T_{trs}, the temperature at which two phases are in equilibrium at 1 atm. This temperature is 0 °C (273 K) for ice in equilibrium with liquid water at 1 atm, and 100 °C (373 K) for water in equilibrium with its vapour at 1 atm. At the transition temperature, any transfer of energy as heat between the system and its surroundings is reversible because the two phases in the system are in equilibrium. Because at constant pressure $q=\Delta_{trs}H$, the change in molar entropy *of the system* is[4]

$$\Delta_{trs}S = \frac{\Delta_{trs}H}{T_{trs}} \qquad \text{At the transition temperature} \qquad \text{Entropy of phase transition} \qquad (3A.17)$$

If the phase transition is exothermic ($\Delta_{trs}H<0$, as in freezing or condensing), then the entropy change of the system is negative. This decrease in entropy is consistent with the increased order of a solid compared with a liquid and with the increased order of a liquid compared with a gas. The change in entropy of the surroundings, however, is positive because energy is released as heat into them, and at the transition temperature the total change in entropy is zero. If the transition is endothermic ($\Delta_{trs}H>0$, as in melting and vaporization), then the entropy change of the system is positive, which is consistent with dispersal of matter in the system. The entropy of the surroundings decreases by the same amount, and overall the total change in entropy is zero.

Table 3A.1 lists some experimental entropies of transition. Table 3A.2 lists in more detail the standard entropies of vaporization of several liquids at their boiling points. An interesting feature of the data is that a wide range of liquids give approximately the same standard entropy of vaporization (about $85\,J\,K^{-1}\,mol^{-1}$): this empirical observation is called

Table 3A.1* Standard entropies (and temperatures) of phase transitions, $\Delta_{trs}S^{\ominus}/(J\,K^{-1}\,mol^{-1})$

	Fusion (at T_f)	Vaporization (at T_b)
Argon, Ar	14.17 (at 83.8 K)	74.53 (at 87.3 K)
Benzene, C_6H_6	38.00 (at 279 K)	87.19 (at 353 K)
Water, H_2O	22.00 (at 273.15 K)	109.0 (at 373.15 K)
Helium, He	4.8 (at 8 K and 30 bar)	19.9 (at 4.22 K)

* More values are given in the *Resource section*.

Table 3A.2* The standard enthalpies and entropies of vaporization of liquids at their normal boiling points

	$\Delta_{vap}H^{\ominus}/(kJ\,mol^{-1})$	$\theta_b/°C$	$\Delta_{vap}S^{\ominus}/(J\,K^{-1}\,mol^{-1})$
Benzene	30.8	80.1	87.2
Carbon tetrachloride	30	76.7	85.8
Cyclohexane	30.1	80.7	85.1
Hydrogen sulfide	18.7	−60.4	87.9
Methane	8.18	−161.5	73.2
Water	40.7	100.0	109.1

* More values are given in the *Resource section*.

[4] According to Topic 2C, $\Delta_{trs}H$ is an enthalpy change per mole of substance; so $\Delta_{trs}S$ is also a molar quantity.

Trouton's rule. The explanation of Trouton's rule is that a comparable change in volume occurs when any liquid evaporates and becomes a gas. Hence, all liquids can be expected to have similar standard entropies of vaporization. Liquids that show significant deviations from Trouton's rule do so on account of strong molecular interactions that result in a partial ordering of their molecules. As a result, there is a greater change in disorder when the liquid turns into a vapour than for a fully disordered liquid. An example is water, where the large entropy of vaporization reflects the presence of structure arising from hydrogen-bonding in the liquid. Hydrogen bonds tend to organize the molecules in the liquid so that they are less random than, for example, the molecules in liquid hydrogen sulfide (in which there is no hydrogen bonding). Methane has an unusually low entropy of vaporization. A part of the reason is that the entropy of the gas itself is slightly low ($186\,\mathrm{J\,K^{-1}\,mol^{-1}}$ at 298 K); the entropy of N_2 under the same conditions is $192\,\mathrm{J\,K^{-1}\,mol^{-1}}$. As explained in Topic 12B, fewer rotational states are accessible at room temperature for molecules with low moments of inertia (like CH_4) than for molecules with relatively high moments of inertia (like N_2), so their molar entropy is slightly lower.

Brief illustration 3A.8 Trouton's rule

There is no hydrogen bonding in liquid bromine and Br_2 is a heavy molecule that is unlikely to display unusual behaviour in the gas phase, so it is safe to use Trouton's rule. To predict the standard molar enthalpy of vaporization of bromine given that it boils at 59.2 °C, we use the rule in the form

$$\Delta_{\mathrm{vap}}H^{\ominus}=T_{\mathrm{b}}\times(85\,\mathrm{J\,K^{-1}mol^{-1}})$$

Substitution of the data then gives

$$\Delta_{\mathrm{vap}}H^{\ominus}=(332.4\,\mathrm{K})\times(85\,\mathrm{J\,K^{-1}mol^{-1}})=+2.8\times10^{3}\,\mathrm{J\,mol^{-1}}$$
$$=+28\,\mathrm{kJ\,mol^{-1}}$$

The experimental value is $+29.45\,\mathrm{kJ\,mol^{-1}}$.

Self-test 3A.9 Predict the enthalpy of vaporization of ethane from its boiling point, −88.6 °C.

Answer: $16\,\mathrm{kJ\,mol^{-1}}$

(c) Heating

Equation 3A.2 can be used to calculate the entropy of a system at a temperature T_{f} from a knowledge of its entropy at another temperature T_{i} and the heat supplied to change its temperature from one value to the other:

$$S(T_{\mathrm{f}})=S(T_{\mathrm{i}})+\int_{T_{\mathrm{i}}}^{T_{\mathrm{f}}}\frac{\mathrm{d}q_{\mathrm{rev}}}{T} \qquad (3A.18)$$

We shall be particularly interested in the entropy change when the system is subjected to constant pressure (such as from the atmosphere) during the heating. Then, from the definition of constant-pressure heat capacity (eqn 2B.5, $C_p=(\partial H/\partial T)_p$, written as $\mathrm{d}q_{\mathrm{rev}}=C_p\mathrm{d}T$):

$$S(T_{\mathrm{f}})=S(T_{\mathrm{i}})+\int_{T_{\mathrm{i}}}^{T_{\mathrm{f}}}\frac{C_p\mathrm{d}T}{T} \qquad (3A.19)$$

Constant pressure — Entropy variation with temperature

The same expression applies at constant volume, but with C_p replaced by C_V. When C_p is independent of temperature in the temperature range of interest, it can be taken outside the integral and we obtain

$$S(T_{\mathrm{f}})=S(T_{\mathrm{i}})+C_p\int_{T_{\mathrm{i}}}^{T_{\mathrm{f}}}\frac{\mathrm{d}T}{T}=S(T_{\mathrm{i}})+C_p\ln\frac{T_{\mathrm{f}}}{T_{\mathrm{i}}} \qquad (3A.20)$$

with a similar expression for heating at constant volume. The logarithmic dependence of entropy on temperature is illustrated in Fig. 3A.13.

Brief illustration 3A.9 Entropy change on heating

The molar constant-volume heat capacity of water at 298 K is $75.3\,\mathrm{J\,K^{-1}\,mol^{-1}}$. The change in molar entropy when it is heated from 20 °C (293 K) to 50 °C (323 K), supposing the heat capacity to be constant in that range, is therefore

$$\Delta S_{\mathrm{m}}=S_{\mathrm{m}}(323\,\mathrm{K})-S_{\mathrm{m}}(293\,\mathrm{K})=(75.3\,\mathrm{J\,K^{-1}\,mol^{-1}})\times\ln\frac{323\,\mathrm{K}}{293\,\mathrm{K}}$$
$$=+7.34\,\mathrm{J\,K^{-1}\,mol^{-1}}$$

Self-test 3A.10 What is the change when further heating takes the temperature from 50 °C to 80 °C?

Answer: $+5.99\,\mathrm{J\,K^{-1}\,mol^{-1}}$

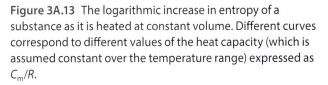

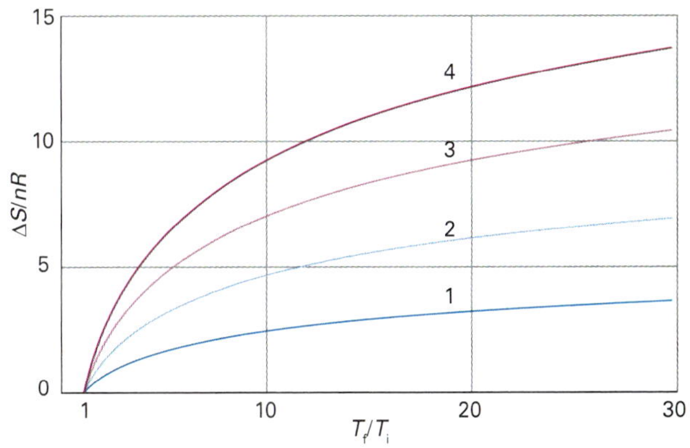

Figure 3A.13 The logarithmic increase in entropy of a substance as it is heated at constant volume. Different curves correspond to different values of the heat capacity (which is assumed constant over the temperature range) expressed as C_{m}/R.

(d) Composite processes

In many cases, more than one parameter changes. For instance, it might be the case that both the volume and the temperature of a gas are different in the initial and final states. Because S is a state function, we are free to choose the most convenient path from the initial state to the final state, such as reversible isothermal expansion to the final volume, followed by reversible heating at constant volume to the final temperature. Then the total entropy change is the sum of the two contributions.

Example 3A.2 Calculating the entropy change for a composite process

Calculate the entropy change when argon at 25 °C and 1.00 bar in a container of volume 0.500 dm^3 is allowed to expand to 1.000 dm^3 and is simultaneously heated to 100 °C.

Method As remarked in the text, use reversible isothermal expansion to the final volume, followed by reversible heating at constant volume to the final temperature. The entropy change in the first step is given by eqn 3A.16 and that of the second step, provided C_V is independent of temperature, by eqn 3A.20 (with C_V in place of C_p). In each case we need to know n, the amount of gas molecules, and can calculate it from the perfect gas equation and the data for the initial state from $n = p_iV_i/RT_i$. The molar heat capacity at constant volume is given by the equipartition theorem as $\frac{3}{2}R$. (The equipartition theorem is reliable for monatomic gases: for others and in general use experimental data like that in Tables 2C.1 and 2C.2 of the *Resource section*, converting to the value at constant volume by using the relation $C_{p,m} - C_{V,m} = R$.)

Answer From eqn 3A.16 the entropy change in the isothermal expansion from V_i to V_f is

$$\Delta S(\text{Step 1}) = nR\ln\frac{V_f}{V_i}$$

From eqn 3A.20, the entropy change in the second step, from T_i to T_f at constant volume, is

$$\Delta S(\text{Step 2}) = nC_{V,m}\ln\frac{T_f}{T_i} = \tfrac{3}{2}nR\ln\frac{T_f}{T_i} = nR\ln\left(\frac{T_f}{T_i}\right)^{3/2}$$

The overall entropy change of the system, the sum of these two changes, is

$$\Delta S = nR\ln\frac{V_f}{V_i} + nR\ln\left(\frac{T_f}{T_i}\right)^{3/2} = nR\ln\frac{V_f}{V_i}\left(\frac{T_f}{T_i}\right)^{3/2}$$

(We have used $\ln x + \ln y = \ln xy$.) Now we substitute $n = p_iV_i/RT_i$ and obtain

$$\Delta S = \frac{p_iV_i}{T_i}\ln\frac{V_f}{V_i}\left(\frac{T_f}{T_i}\right)^{3/2}$$

At this point we substitute the data:

$$\Delta S = \frac{(1.00\times10^5\,\text{Pa})\times(0.500\times10^{-3}\,\text{m}^3)}{298\,\text{K}}\times\ln\frac{1.000}{0.500}\left(\frac{373}{298}\right)^{3/2}$$

$$= +0.173\,\text{J K}^{-1}$$

A note on good practice It is sensible to proceed as generally as possible before inserting numerical data so that, if required, the formula can be used for other data and to avoid rounding errors.

Self-test 3A.11 Calculate the entropy change when the same initial sample is compressed to 0.0500 dm^3 and cooled to −25 °C.

Answer: −0.44 J K^{-1}

Checklist of concepts

- ☐ 1. The **entropy** acts as a signpost of spontaneous change.
- ☐ 2. Entropy change is defined in terms of heat transactions (the **Clausius definition**).
- ☐ 3. The **Boltzmann formula** defines absolute entropies in terms of the number of ways of achieving a configuration.
- ☐ 4. The **Carnot cycle** is used to prove that entropy is a state function.
- ☐ 5. The **efficiency** of a heat engine is the basis of the definition of the thermodynamic temperature scale and one realization, the Kelvin scale.
- ☐ 6. The **Clausius inequality** is used to show that the entropy increases in a spontaneous change and therefore that the Clausius definition is consistent with the Second Law.
- ☐ 7. The entropy of a perfect gas increases when it expands isothermally.
- ☐ 8. The change in entropy of a substance accompanying a change of state at its transition temperature is calculated from its enthalpy of transition.
- ☐ 9. The increase in entropy when a substance is heated is expressed in terms of its heat capacity.

Checklist of equations

Property	Equation	Comment	Equation number
Thermodynamic entropy	$dS = dq_{rev}/T$	Definition	3A.1
Entropy change of surroundings	$\Delta S_{sur} = q_{sur}/T_{sur}$		3A.3b
Boltzmann formula	$S = k \ln \mathcal{W}$	Definition	3A.5
Carnot efficiency	$\eta = 1 - T_c/T_h$	Reversible processes	3A.10
Thermodynamic temperature	$T = (1-\eta)T_h$		3A.11
Clausius inequality	$dS \geq dq/T$		3A.12
Entropy of isothermal expansion	$\Delta S = nR \ln(V_f/V_i)$	Perfect gas	3A.14
Entropy of transition	$\Delta_{trs}S = \Delta_{trs}H/T_{trs}$	At the transition temperature	3A.17
Variation of the entropy with temperature	$S(T_f) = S(T_i) + C \ln(T_f/T_i)$	The heat capacity, C, is independent of temperature and no phase transitions occur	3A.20

3B The measurement of entropy

Contents

➤ Why do you need to know this material?

For entropy to be a quantitatively useful concept it is important to be able to measure it: the calorimetric procedure is described here. The discussion also introduces the Third Law of thermodynamics, which has important implications for the measurement of entropies and (as shown in later Topics) the attainment of absolute zero.

➤ What is the key idea?

The entropy of a perfectly crystalline solid is zero at $T=0$.

➤ What do you need to know already?

You need to be familiar with the expression for the temperature dependence of entropy and how entropies of transition are calculated (Topic 3A). The discussion of residual entropy draws on the Boltzmann formula for the entropy (Topic 3A).

The entropy of a substance can be determined in two ways. One, which is the subject of this Topic, is to make calorimetric measurements of the heat required to raise the temperature of a sample from $T=0$ to the temperature of interest. The other, which is described in Topic 15E, is to use calculated parameters or spectroscopic data and to calculate the entropy by using Boltzmann's statistical definition.

3B.1 The calorimetric measurement of entropy

It is established in Topic 3A that the entropy of a system at a temperature T is related to its entropy at $T=0$ by measuring its heat capacity C_p at different temperatures and evaluating the integral in eqn 3A.19 ($S(T_f)=S(T_i)+\int_{T_i}^{T_f} C_p\mathrm{d}T/T$). The entropy of transition ($\Delta_{trs}H/T_{trs}$) for each phase transition between $T=0$ and the temperature of interest must then be included in the overall sum. For example, if a substance melts at T_f and boils at T_b, then its molar entropy above its boiling temperature is given by

$$S_m(T)=S_m(0)+\overbrace{\int_0^{T_f}\frac{C_{p,m}(s,T)}{T}\mathrm{d}T}^{\text{Heat solid to its melting point}}+\overbrace{\frac{\Delta_{fus}H}{T_f}}^{\text{Entropy of fusion}}+\overbrace{\int_{T_f}^{T_b}\frac{C_{p,m}(l,T)}{T}\mathrm{d}T}^{\text{Heat liquid to its boiling point}}+\overbrace{\frac{\Delta_{vap}H}{T_b}}^{\text{Entropy of vaporization}}+\overbrace{\int_{T_b}^{T}\frac{C_{p,m}(g,T)}{T}\mathrm{d}T}^{\text{Heat vapour to the final temperature}} \quad (3B.1)$$

All the properties required, except $S_m(0)$, can be measured calorimetrically, and the integrals can be evaluated either graphically or, as is now more usual, by fitting a polynomial to the data and integrating the polynomial analytically. The former procedure is illustrated in Fig. 3B.1: the area under the curve of $C_{p,m}/T$ against T is the integral required. Provided all measurements are made at 1 bar on a pure material, the final value is the **standard entropy**, $S^{\ominus}(T)$ and, on division by the amount of substance n, its **standard molar entropy**, $S_m^{\ominus}(T)=S^{\ominus}(T)/n$. Because $\mathrm{d}T/T=\mathrm{d}\ln T$, an alternative procedure is to evaluate the area under a plot of $C_{p,m}$ against $\ln T$.

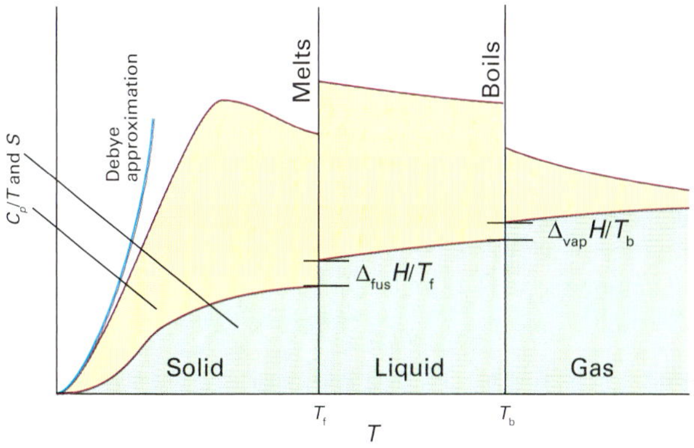

Figure 3B.1 The variation of C_p/T with the temperature for a sample is used to evaluate the entropy, which is equal to the area beneath the upper curve up to the corresponding temperature, plus the entropy of each phase transition passed.

Brief illustration 3B.1 The standard molar entropy

The standard molar entropy of nitrogen gas at 25 °C has been calculated from the following data:

	$S_m^\ominus/(J\,K^{-1}\,mol^{-1})$
Debye extrapolation*	1.92
Integration, from 10 K to 35.61 K	25.25
Phase transition at 35.61 K	6.43
Integration, from 35.61 K to 63.14 K	23.38
Fusion at 63.14 K	11.42
Integration, from 63.14 K to 77.32 K	11.41
Vaporization at 77.32 K	72.13
Integration, from 77.32 K to 298.15 K	39.20
Correction for gas imperfection	0.92
Total	192.06

Therefore, $S_m^\ominus(298.15\,K) = S_m(0) + 192.1\,J\,K^{-1}\,mol^{-1}$.

*This extrapolation is explained immediately following.

One problem with the determination of entropy is the difficulty of measuring heat capacities near $T=0$. There are good theoretical grounds for assuming that the heat capacity of a non-metallic solid is proportional to T^3 when T is low (see Topic 7A), and this dependence is the basis of the **Debye extrapolation**. In this method, C_p is measured down to as low a temperature as possible and a curve of the form aT^3 is fitted to the data. That fit determines the value of a, and the expression $C_{p,m} = aT^3$ is assumed valid down to $T=0$.

Example 3B.1 Calculating the entropy at low temperatures

The molar constant-pressure heat capacity of a certain solid at 4.2 K is 0.43 J K^{-1} mol^{-1}. What is its molar entropy at that temperature?

Method Because the temperature is so low, we can assume that the heat capacity varies with temperature as aT^3, in which case we can use eqn 3A.19 (quoted in the opening paragraph of 3B.1) to calculate the entropy at a temperature T in terms of the entropy at $T=0$ and the constant a. When the integration is carried out, it turns out that the result can be expressed in terms of the heat capacity at the temperature T, so the data can be used directly to calculate the entropy.

Answer The integration required is

$$S_m(T) = S_m(0) + \int_0^T \frac{aT^3}{T}\,dT = S_m(0) + a\int_0^T T^2\,dT$$
$$= S_m(0) + \tfrac{1}{3}aT^3 = S_m(0) + \tfrac{1}{3}C_{p,m}(T)$$

from which it follows that

$$S_m(4.2\,K) = S_m(0) + 0.14\,J\,K^{-1}\,mol^{-1}$$

Self-test 3B.1 For metals, there is also a contribution to the heat capacity from the electrons which is linearly proportional to T when the temperature is low. Find its contribution to the entropy at low temperatures.

Answer: $S(T) = S(0) + C_p(T)$

3B.2 The Third Law

We now address the problem of the value of $S(0)$. At $T=0$, all energy of thermal motion has been quenched, and in a perfect crystal all the atoms or ions are in a regular, uniform array. The localization of matter and the absence of thermal motion suggest that such materials also have zero entropy. This conclusion is consistent with the molecular interpretation of entropy, because $S=0$ if there is only one way of arranging the molecules and only one microstate is accessible (all molecules occupy the ground state, $\mathcal{W}=1$).

(a) The Nernst heat theorem

The experimental observation that turns out to be consistent with the view that the entropy of a regular array of molecules is zero at $T=0$ is summarized by the **Nernst heat theorem**:

> The entropy change accompanying any physical or chemical transformation approaches zero as the temperature approaches zero: $\Delta S \to 0$ as $T \to 0$ provided all the substances involved are perfectly ordered.

Nernst heat theorem

Brief illustration 3B.2 The Nernst heat theorem

Consider the entropy of the transition between orthorhombic sulfur, α, and monoclinic sulfur, β, which can be calculated from the transition enthalpy (−402 J mol⁻¹) at the transition temperature (369 K):

$$\Delta_{trs}S = S_m(\beta) - S_m(\alpha) = \frac{-402\,\text{J mol}^{-1}}{369\,\text{K}} = -1.09\,\text{J K}^{-1}\,\text{mol}^{-1}$$

The two individual entropies can also be determined by measuring the heat capacities from $T=0$ up to $T=369$ K. It is found that $S_m(\alpha) = S_m(\alpha,0) + 37\,\text{J K}^{-1}\,\text{mol}^{-1}$ and $S_m(\beta) = S_m(\beta,0) + 38\,\text{J K}^{-1}\,\text{mol}^{-1}$. These two values imply that at the transition temperature

$$\Delta_{trs}S = S_m(\alpha,0) - S_m(\beta,0) = -1\,\text{J K}^{-1}\,\text{mol}^{-1}$$

On comparing this value with the one above, we conclude that $S_m(\alpha,0) - S_m(\beta,0) \approx 0$, in accord with the theorem.

Self-test 3B.2 Two forms of a metallic solid (see *Self-test* 3B.1) undergo a phase transition at T_{trs}, which is close to $T=0$. What is the enthalpy of transition at T_{trs} in terms of the heat capacities of the two polymorphs?

Answer: $\Delta_{trs}H(T_{trs}) = T_{trs}\Delta C_p(T_{trs})$

It follows from the Nernst theorem, that if we arbitrarily ascribe the value zero to the entropies of elements in their perfect crystalline form at $T=0$, then all perfect crystalline compounds also have zero entropy at $T=0$ (because the change in entropy that accompanies the formation of the compounds, like the entropy of all transformations at that temperature, is zero). This conclusion is summarized by the **Third Law of thermodynamics**:

The entropy of all perfect crystalline substances is zero at $T=0$. Third Law of thermodynamics

As far as thermodynamics is concerned, choosing this common value as zero is a matter of convenience. The molecular interpretation of entropy, however, justifies the value $S=0$ at $T=0$ because then, as we have remarked, $\mathcal{W}=1$.

In certain cases $\mathcal{W}>1$ at $T=0$ and therefore $S(0)>0$. This is the case if there is no energy advantage in adopting a particular orientation even at absolute zero. For instance, for a diatomic molecule AB there may be almost no energy difference between the arrangements …AB AB AB… and …BA AB BA…, so $\mathcal{W}>1$ even at $T=0$. If $S(0)>0$ we say that the substance has a **residual entropy**. Ice has a residual entropy of 3.4 J K⁻¹ mol⁻¹. It stems from the arrangement of the hydrogen bonds between neighbouring water molecules: a given O atom has two short O—H bonds and two long O···H bonds to its neighbours, but there is a degree of randomness in which two bonds are short and which two are long.

Example 3B.2 Estimating a residual entropy

Estimate the residual entropy of ice by taking into account the distribution of hydrogen bonds and chemical bonds about the oxygen atom of one H_2O molecule. The experimental value is 3.4 J K⁻¹ mol⁻¹.

Method Focus on the O atom, and consider the number of ways that that O atom can have two short (chemical) bonds and two long hydrogen bonds to its four neighbours. Refer to Fig. 3B.2.

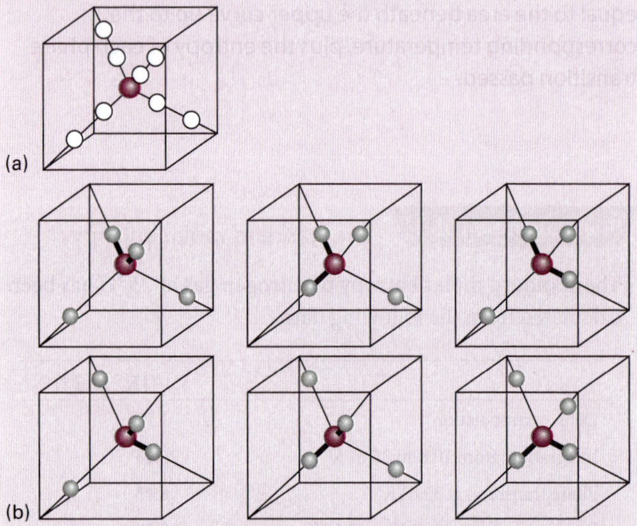

Figure 3B.2 The model of ice showing (a) the local structure of an oxygen atom and (b) the array of chemical and hydrogen bonds used to calculate the residual entropy of ice.

Answer Suppose each H atom can lie either close to or far from its 'parent' O atom, as depicted in Fig. 3B.2. The total number of these conceivable arrangements in a sample that contains N H_2O molecules and therefore $2N$ H atoms is 2^{2N}. Now consider a single central O atom. The total number of possible arrangements of locations of H atoms around the central O atom of one H_2O molecule is $2^4=16$. Of these 16 possibilities, only 6 correspond to two short and two long bonds. That is, only $\frac{6}{16}=\frac{3}{8}$ of all possible arrangements are possible, and for N such molecules only $(3/8)^N$ of all possible arrangements are possible. Therefore, the total number of allowed arrangements in the crystal is $2^{2N}(3/8)^N = 4^N(3/8)^N = (3/2)^N$. If we suppose that all these arrangements are energetically identical, the residual entropy is

$$S(0) = k\ln\left(\tfrac{3}{2}\right)^N = Nk\ln\tfrac{3}{2} = nN_Ak\ln\tfrac{3}{2} = nR\ln\tfrac{3}{2}$$

and the residual molar entropy would be

$$S_m(0) = R \ln \tfrac{3}{2} = 3.4\,\mathrm{J\,K^{-1}\,mol^{-1}}$$

in accord with the experimental value.

Self-test 3B.3 What would be the residual molar entropy of HCF_3 on the assumption that each molecule could take up one of four tetrahedral orientations in a crystal?

Answer: $11.5\,\mathrm{J\,K^{-1}\,mol^{-1}}$

(b) Third-Law entropies

Entropies reported on the basis that $S(0)=0$ are called **Third-Law entropies** (and commonly just 'entropies'). When the substance is in its standard state at the temperature T, the **standard (Third-Law) entropy** is denoted $S^\ominus(T)$. A list of values at 298 K is given in Table 3B.1.

The **standard reaction entropy**, $\Delta_r S^\ominus$, is defined, like the standard reaction enthalpy in Topic 2C, as the difference between the molar entropies of the pure, separated products and the pure, separated reactants, all substances being in their standard states at the specified temperature:

$$\Delta_r S^\ominus = \sum_{\text{Products}} \nu S_m^\ominus - \sum_{\text{Reactants}} \nu S_m^\ominus \qquad \textit{Definition} \quad \text{Standard reaction entropy} \qquad (3B.2a)$$

In this expression, each term is weighted by the appropriate stoichiometric coefficient. A more sophisticated approach is to adopt the notation introduced in Topic 2C and to write

$$\Delta_r S^\ominus = \sum_J \nu_J S_m^\ominus(J) \qquad (3B.2b)$$

where the ν_J are signed (+ for products, − for reactants) stoichiometric numbers. Standard reaction entropies are likely to be positive if there is a net formation of gas in a reaction, and are likely to be negative if there is a net consumption of gas.

Just as in the discussion of enthalpies in Topic 2C, where it is acknowledged that solutions of cations cannot be prepared in the absence of anions, the standard molar entropies of ions in solution are reported on a scale in which the standard entropy of the H^+ ions in water is taken as zero at all temperatures:

$$S^\ominus(H^+, aq) = 0 \qquad \textit{Convention} \quad \text{Ions in solution} \qquad (3B.3)$$

The values based on this choice are listed in Table 2C.5 in the *Resource section*.[1] Because the entropies of ions in water are values relative to the hydrogen ion in water, they may be either positive or negative. A positive entropy means that an ion has a higher molar entropy than H^+ in water and a negative entropy means that the ion has a lower molar entropy than H^+ in water. Ion entropies vary as expected on the basis that they are related to the degree to which the ions order the water molecules around them in the solution. Small, highly charged ions induce local structure in the surrounding water, and the disorder of

Table 3B.1* Standard Third-Law entropies at 298 K, $S_m^\ominus/(\mathrm{J\,K^{-1}\,mol^{-1}})$

	$S_m^\ominus/(\mathrm{J\,K^{-1}\,mol^{-1}})$
Solids	
Graphite, C(s)	5.7
Diamond, C(s)	2.4
Sucrose, $C_{12}H_{22}O_{11}$(s)	360.2
Iodine, I_2(s)	116.1
Liquids	
Benzene, C_6H_6(l)	173.3
Water, H_2O(l)	69.9
Mercury, Hg(l)	76.0
Gases	
Methane, CH_4(g)	186.3
Carbon dioxide, CO_2(g)	213.7
Hydrogen, H_2(g)	130.7
Helium, He(g)	126.2
Ammonia, NH_3(g)	192.4

* More values are given in the *Resource section*.

Brief illustration 3B.3 The standard reaction entropy

To calculate the standard reaction entropy of $H_2(g)+\frac{1}{2}O_2(g)\rightarrow H_2O(l)$ at 298 K, we use the data in Table 2C.5 of the *Resource section* to write

$$\begin{aligned}\Delta_r S^\ominus &= S_m^\ominus(H_2O,l) - \{S_m^\ominus(H_2,g) + \tfrac{1}{2}S_m^\ominus(O_2,g)\}\\ &= 69.9\,\mathrm{J\,K^{-1}\,mol^{-1}} - \{130.7 + \tfrac{1}{2}(205.1)\}\,\mathrm{J\,K^{-1}\,mol^{-1}}\\ &= -163.4\,\mathrm{J\,K^{-1}\,mol^{-1}}\end{aligned}$$

The negative value is consistent with the conversion of two gases to a compact liquid.

A note on good practice Do not make the mistake of setting the standard molar entropies of elements equal to zero: they have nonzero values (provided $T>0$), as we have already discussed.

Self-test 3B.4 Calculate the standard reaction entropy for the combustion of methane to carbon dioxide and liquid water at 298 K.

Answer: $-243\,\mathrm{J\,K^{-1}\,mol^{-1}}$

[1] In terms of the language introduced in Topic 5A, the entropies of ions in solution are actually *partial molar entropies*, for their values include the consequences of their presence on the organization of the solvent molecules around them.

the solution is decreased more than in the case of large, singly charged ions. The absolute, Third-Law standard molar entropy of the proton in water can be estimated by proposing a model of the structure it induces, and there is some agreement on the value $-21\,\mathrm{J\,K^{-1}\,mol^{-1}}$. The negative value indicates that the proton induces order in the solvent.

Brief illustration 3B.4 Absolute and relative ion entropies

The standard molar entropy of $Cl^-(aq)$ is $+57\,\mathrm{J\,K^{-1}\,mol^{-1}}$ and that of $Mg^{2+}(aq)$ is $-128\,\mathrm{J\,K^{-1}\,mol^{-1}}$. That is, the partial molar entropy of $Cl^-(aq)$ is $57\,\mathrm{J\,K^{-1}\,mol^{-1}}$ higher than that of the proton in water (presumably because it induces less local structure in the surrounding water), whereas that of $Mg^{2+}(aq)$ is $128\,\mathrm{J\,K^{-1}\,mol^{-1}}$ lower (presumably because its higher charge induces more local structure in the surrounding water).

Self-test 3B.5 Estimate the absolute values of the partial molar entropies of these ions.

Answer: $+36\,\mathrm{J\,K^{-1}\,mol^{-1}}$, $-149\,\mathrm{J\,K^{-1}\,mol^{-1}}$

Checklist of concepts

- ☐ 1. Entropies are determined calorimetrically by measuring the heat capacity of a substance from low temperatures up to the temperature of interest.
- ☐ 2. The **Debye-T^3 law** is used to estimate heat capacities of non-metallic solids close to $T=0$.
- ☐ 3. The **Nernst heat theorem** states that the entropy change accompanying any physical or chemical transformation approaches zero as the temperature approaches zero: $\Delta S \to 0$ as $T \to 0$ provided all the substances involved are perfectly ordered.
- ☐ 4. The **Third Law of thermodynamics** states that the entropy of all perfect crystalline substances is zero at $T=0$.
- ☐ 5. The **residual entropy** of a solid is the entropy arising from disorder that persists at $T=0$.
- ☐ 6. **Third-Law entropies** are entropies based on $S(0)=0$.
- ☐ 7. The **standard entropies of ions in solution** are based on setting $S^{\ominus}(H^+,aq)=0$ at all temperatures.
- ☐ 8. The **standard reaction entropy**, $\Delta_r S^{\ominus}$, is the difference between the molar entropies of the pure, separated products and the pure, separated reactants, all substances being in their standard states.

Checklist of equations

Property	Equation	Comment	Equation number
Standard molar entropy from calorimetry	See eqn 3B.1	Sum of contributions from $T=0$ to temperature of interest	3B.1
Standard reaction entropy	$\Delta_r S^{\ominus} = \sum_{\text{Products}} \nu S_m^{\ominus} - \sum_{\text{Reactants}} \nu S_m^{\ominus}$ $\Delta_r S^{\ominus} = \sum_J \nu_J S_m^{\ominus}(J)$	ν: (positive) stoichiometric coefficients; ν_J: (signed) stoichiometric numbers	3B.2

3C Concentrating on the system

Contents

➤ Why do you need to know this material?

Most processes of interest in chemistry occur at constant temperature and pressure. Under these conditions, thermodynamic processes are discussed in terms of the Gibbs energy, which is introduced in this Topic. The Gibbs energy is the foundation of the discussion of phase equilibria, chemical equilibrium, and bioenergetics.

➤ What is the key idea?

The Gibbs energy is a signpost of spontaneous change at constant temperature and pressure, and is equal to the maximum non-expansion work that a system can do.

➤ What do you need to know already?

This Topic develops the Clausius inequality (Topic 3A) and draws on information about standard states and reaction enthalpy introduced in Topic 2C. The derivation of the Born equation uses information about the energy of one electric charge in the field of another (*Foundations* B).

Entropy is the basic concept for discussing the direction of natural change, but to use it we have to analyse changes in both the system and its surroundings. In Topic 3A it is shown that it is always very simple to calculate the entropy change in the surroundings (from $\Delta S_{sur}=q_{sur}/T_{sur}$); here we see that it is possible to devise a simple method for taking that contribution into account automatically. This approach focuses our attention on the system and simplifies discussions. Moreover, it is the foundation of all the applications of chemical thermodynamics that follow.

3C.1 The Helmholtz and Gibbs energies

Consider a system in thermal equilibrium with its surroundings at a temperature T. When a change in the system occurs and there is a transfer of energy as heat between the system and the surroundings, the Clausius inequality (eqn 3A.12, $dS \geq dq/T$) reads

$$dS - \frac{dq}{T} \geq 0 \tag{3C.1}$$

We can develop this inequality in two ways according to the conditions (of constant volume or constant pressure) under which the process occurs.

(a) Criteria of spontaneity

First, consider heating at constant volume. Then, in the absence of additional (non-expansion) work, we can write $dq_V = dU$; consequently

$$dS - \frac{dU}{T} \geq 0 \tag{3C.2}$$

The importance of the inequality in this form is that it expresses the criterion for spontaneous change solely in terms of the state functions of the system. The inequality is easily rearranged into

$$T\mathrm{d}S \geq \mathrm{d}U \quad \text{(constant } V\text{, no additional work)} \tag{3C.3}$$

Because $T>0$, at either constant internal energy ($\mathrm{d}U=0$) or constant entropy ($\mathrm{d}S=0$) this expression becomes, respectively,

$$\mathrm{d}S_{U,V} \geq 0 \qquad \mathrm{d}U_{S,V} \leq 0 \tag{3C.4}$$

where the subscripts indicate the constant conditions.

Equation 3C.4 expresses the criteria for spontaneous change in terms of properties relating to the system. The first inequality states that, in a system at constant volume and constant internal energy (such as an isolated system), the entropy increases in a spontaneous change. That statement is essentially the content of the Second Law. The second inequality is less obvious, for it says that if the entropy and volume of the system are constant, then the internal energy must decrease in a spontaneous change. Do not interpret this criterion as a tendency of the system to sink to lower energy. It is a disguised statement about entropy and should be interpreted as implying that if the entropy of the system is unchanged, then there must be an increase in entropy of the surroundings, which can be achieved only if the energy of the system decreases as energy flows out as heat.

When energy is transferred as heat at constant pressure, and there is no work other than expansion work, we can write $\mathrm{d}q_p=\mathrm{d}H$ and obtain

$$T\mathrm{d}S \geq \mathrm{d}H \quad \text{(constant } p\text{, no additional work)} \tag{3C.5}$$

At either constant enthalpy or constant entropy this inequality becomes, respectively,

$$\mathrm{d}S_{H,p} \geq 0 \quad \mathrm{d}H_{S,p} \leq 0 \tag{3C.6}$$

The interpretations of these inequalities are similar to those of eqn 3C.4. The entropy of the system at constant pressure must increase if its enthalpy remains constant (for there can then be no change in entropy of the surroundings). Alternatively, the enthalpy must decrease if the entropy of the system is constant, for then it is essential to have an increase in entropy of the surroundings.

Brief illustration 3C.1 Spontaneous changes at constant volume

A concrete example of the criterion $\mathrm{d}S_{U,V}\geq 0$ is the diffusion of a solute B through a solvent A that form an ideal solution (in the sense of Topic 5B, in which AA, BB, and AB interactions are identical). There is no change in internal energy or volume of the system or the surroundings as B spreads into A, but the process is spontaneous.

Self-test 3C.1 Invent an example of the criterion $\mathrm{d}U_{S,V}\leq 0$.

Answer: A phase change in which one perfectly ordered phase changes into another of lower energy and equal density at $T=0$

Because eqns 3C.4 and 3C.6 have the forms $\mathrm{d}U-T\mathrm{d}S\leq 0$ and $\mathrm{d}H-T\mathrm{d}S\leq 0$, respectively, they can be expressed more simply by introducing two more thermodynamic quantities. One is the **Helmholtz energy**, A, which is defined as

$$A=U-TS \qquad \textit{Definition} \quad \text{Helmholtz energy} \tag{3C.7}$$

The other is the **Gibbs energy**, G:

$$G=H-TS \qquad \textit{Definition} \quad \text{Gibbs energy} \tag{3C.8}$$

All the symbols in these two definitions refer to the system.

When the state of the system changes at constant temperature, the two properties change as follows:

$$\text{(a) } \mathrm{d}A=\mathrm{d}U-T\mathrm{d}S \quad \text{(b) } \mathrm{d}G=\mathrm{d}H-T\mathrm{d}S \tag{3C.9}$$

When we introduce eqns 3C.4 and 3C.6, respectively, we obtain the criteria of spontaneous change as

$$\text{(a) } \mathrm{d}A_{T,V} \leq 0 \quad \text{(b) } \mathrm{d}G_{T,p} \leq 0 \qquad \text{Criteria of spontaneous change} \tag{3C.10}$$

These inequalities, especially the second, are the most important conclusions from thermodynamics for chemistry. They are developed in subsequent sections, Topics, and chapters.

Brief illustration 3C.2 The spontaneity of endothermic reactions

The existence of spontaneous endothermic reactions provides an illustration of the role of G. In such reactions, H increases, the system rises spontaneously to states of higher enthalpy, and $\mathrm{d}H>0$. Because the reaction is spontaneous we know that $\mathrm{d}G<0$ despite $\mathrm{d}H>0$; it follows that the entropy of the system increases so much that $T\mathrm{d}S$ outweighs $\mathrm{d}H$ in $\mathrm{d}G=\mathrm{d}H-T\mathrm{d}S$. Endothermic reactions are therefore driven by the increase of entropy of the system, and this entropy change overcomes the reduction of entropy brought about in the surroundings by the inflow of heat into the system ($\mathrm{d}S_{\mathrm{sur}}=-\mathrm{d}H/T$ at constant pressure).

Self-test 3C.2 Why are so many exothermic reactions spontaneous?

Answer: With $\mathrm{d}H<0$, it is common for $\mathrm{d}G<0$ unless $T\mathrm{d}S$ is strongly negative.

(b) Some remarks on the Helmholtz energy

A change in a system at constant temperature and volume is spontaneous if $\mathrm{d}A_{T,V} \le 0$. That is, a change under these conditions is spontaneous if it corresponds to a decrease in the Helmholtz energy. Such systems move spontaneously towards states of lower A if a path is available. The criterion of equilibrium, when neither the forward nor reverse process has a tendency to occur, is

$$\mathrm{d}A_{T,V} = 0 \qquad (3C.11)$$

The expressions $\mathrm{d}A = \mathrm{d}U - T\mathrm{d}S$ and $\mathrm{d}A < 0$ are sometimes interpreted as follows. A negative value of $\mathrm{d}A$ is favoured by a negative value of $\mathrm{d}U$ and a positive value of $T\mathrm{d}S$. This observation suggests that the tendency of a system to move to lower A is due to its tendency to move towards states of lower internal energy and higher entropy. However, this interpretation is false because the tendency to lower A is solely a tendency towards states of greater overall entropy. *Systems change spontaneously if in doing so the total entropy of the system and its surroundings increases, not because they tend to lower internal energy.* The form of $\mathrm{d}A$ may give the impression that systems favour lower energy, but that is misleading: $\mathrm{d}S$ is the entropy change of the system, $-\mathrm{d}U/T$ is the entropy change of the surroundings (when the volume of the system is constant), and their total tends to a maximum.

Brief illustration 3C.3 Spontaneous change at constant volume

A bouncing ball comes to rest. The spontaneous direction of change is one in which the energy of the ball (potential at the top of its bounce, kinetic when it strikes the floor) spreads out into the surroundings on each bounce. When the ball is still, the energy of the universe is the same as initially, but the energy of the ball is dispersed over the surroundings.

Self-test 3C.3 What other spontaneous similar mechanical processes have a similar explanation?

Answer: One example: A pendulum coming to rest through friction.

(c) Maximum work

It turns out, as we show in the following *Justification*, that A carries a greater significance than being simply a signpost of spontaneous change: *the change in the Helmholtz function is equal to the maximum work accompanying a process at constant temperature*:

$$\mathrm{d}w_{\max} = \mathrm{d}A \qquad \textit{Constant temperature} \qquad \text{Maximum work} \qquad (3C.12)$$

As a result, A is sometimes called the 'maximum work function', or the 'work function'.[1]

Justification 3C.1 Maximum work

To demonstrate that maximum work can be expressed in terms of the changes in Helmholtz energy, we combine the Clausius inequality $\mathrm{d}S \ge \mathrm{d}q/T$ in the form $T\mathrm{d}S \ge \mathrm{d}q$ with the First Law, $\mathrm{d}U = \mathrm{d}q + \mathrm{d}w$, and obtain

$$\mathrm{d}U \le T\mathrm{d}S + \mathrm{d}w$$

$\mathrm{d}U$ is smaller than the term of the right because $\mathrm{d}q$ has been replaced by $T\mathrm{d}S$, which in general is larger than $\mathrm{d}q$. This expression rearranges to

$$\mathrm{d}w \ge \mathrm{d}U - T\mathrm{d}S$$

It follows that the most negative value of $\mathrm{d}w$, and therefore the maximum energy that can be obtained from the system as work, is given by

$$\mathrm{d}w_{\max} = \mathrm{d}U - T\mathrm{d}S$$

and that this work is done only when the path is traversed reversibly (because then the equality applies). Because at constant temperature $\mathrm{d}A = \mathrm{d}U - T\mathrm{d}S$, we conclude that $\mathrm{d}w_{\max} = \mathrm{d}A$.

When a macroscopic isothermal change takes place in the system, eqn 3C.12 becomes

$$w_{\max} = \Delta A \qquad \textit{Constant temperature} \qquad \text{Maximum work} \qquad (3C.13)$$

with

$$\Delta A = \Delta U - T\Delta S \qquad \text{Constant temperature} \qquad (3C.14)$$

This expression shows that, depending on the sign of $T\Delta S$, not all the change in internal energy may be available for doing work. If the change occurs with a decrease in entropy (of the system), in which case $T\Delta S < 0$, then the right-hand side of this equation is not as negative as ΔU itself, and consequently the maximum work is less than ΔU. For the change to be spontaneous, some of the energy must escape as heat in order to generate enough entropy in the surroundings to overcome the reduction in entropy in the system (Fig. 3C.1). In this case, Nature is demanding a tax on the internal energy as it is converted into work. This is the origin of the alternative name 'Helmholtz free energy' for A, because ΔA is that part of the change in internal energy that we are free to use to do work.

Further insight into the relation between the work that a system can do and the Helmholtz energy is to recall that work is

[1] *Arbeit* is the German word for work; hence the symbol A.

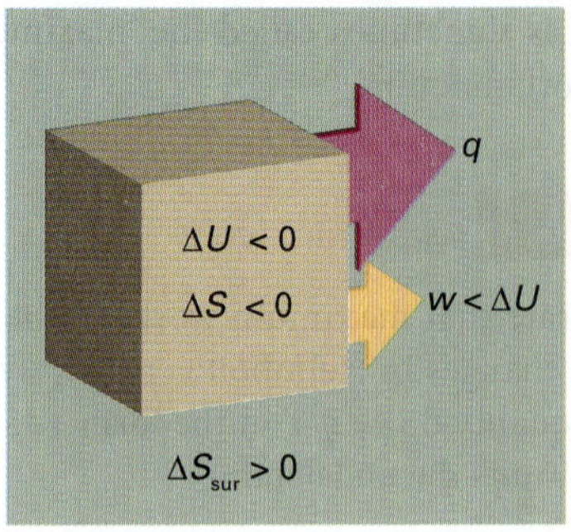

Figure 3C.1 In a system not isolated from its surroundings, the work done may be different from the change in internal energy. Moreover, the process is spontaneous if overall the entropy of the global, isolated system increases. In the process depicted here, the entropy of the system decreases, so that of the surroundings must increase in order for the process to be spontaneous, which means that energy must pass from the system to the surroundings as heat. Therefore, less work than ΔU can be obtained.

energy transferred to the surroundings as the uniform motion of atoms. We can interpret the expression $A = U - TS$ as showing that A is the total internal energy of the system, U, less a contribution that is stored as energy of thermal motion (the quantity TS). Because energy stored in random thermal motion cannot be used to achieve uniform motion in the surroundings, only the part of U that is not stored in that way, the quantity $U - TS$, is available for conversion into work.

If the change occurs with an increase of entropy of the system (in which case $T\Delta S > 0$), the right-hand side of the equation is more negative than ΔU. In this case, the maximum work that can be obtained from the system is greater than ΔU. The explanation of this apparent paradox is that the system is not isolated and energy may flow in as heat as work is done. Because the entropy of the system increases, we can afford a reduction of the entropy of the surroundings yet still have, overall, a spontaneous process. Therefore, some energy (no more than the value of $T\Delta S$) may leave the surroundings as heat and contribute to the work the change is generating (Fig. 3C.2). Nature is now providing a tax refund.

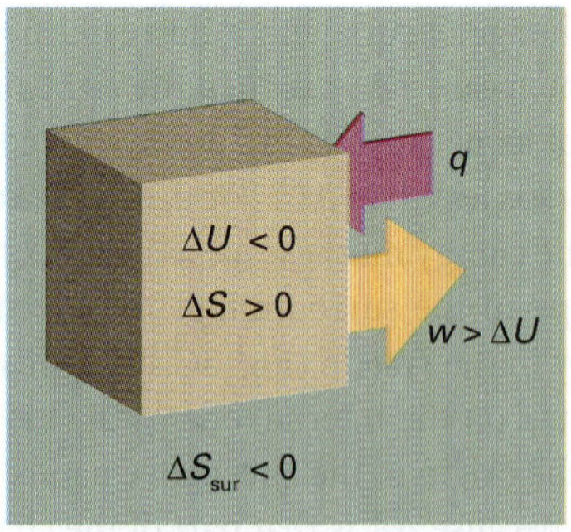

Figure 3C.2 In this process, the entropy of the system increases; hence we can afford to lose some entropy of the surroundings. That is, some of their energy may be lost as heat to the system. This energy can be returned to them as work. Hence the work done can exceed ΔU.

Example 3C.1 Calculating the maximum available work

When 1.000 mol $C_6H_{12}O_6$ (glucose) is oxidized to carbon dioxide and water at 25 °C according to the equation $C_6H_{12}O_6(s) + 6\,O_2(g) \rightarrow 6\,CO_2(g) + 6\,H_2O(l)$ calorimetric measurements give $\Delta_r U^{\ominus} = -2808\,\text{kJ mol}^{-1}$ and $\Delta_r S^{\ominus} = +182.4\,\text{J K}^{-1}\,\text{mol}^{-1}$ at 25 °C. How much of this energy change can be extracted as (a) heat at constant pressure, (b) work?

Method We know that the heat released at constant pressure is equal to the value of ΔH, so we need to relate $\Delta_r H^{\ominus}$ to $\Delta_r U^{\ominus}$, which is given. To do so, we suppose that all the gases involved are perfect, and use eqn 2B.4 ($\Delta H = \Delta U + \Delta n_g RT$) in the form $\Delta_r H = \Delta_r U + \Delta\nu_g RT$. For the maximum work available from the process we use eqn 3C.13.

Answer (a) Because $\Delta\nu_g = 0$, we know that $\Delta_r H^{\ominus} = \Delta_r U^{\ominus} = -2808\,\text{kJ mol}^{-1}$. Therefore, at constant pressure, the energy available as heat is $2808\,\text{kJ mol}^{-1}$. (b) Because $T = 298$ K, the value of $\Delta_r A^{\ominus}$ is

$$\Delta_r A^{\ominus} = \Delta_r U^{\ominus} - T\Delta_r S^{\ominus} = -2862\,\text{kJ mol}^{-1}$$

Therefore, the combustion of 1.000 mol $C_6H_{12}O_6$ can be used to produce up to 2862 kJ of work. The maximum work available is greater than the change in internal energy on account of the positive entropy of reaction (which is partly due to the generation of a large number of small molecules from one big one). The system can therefore draw in energy from the surroundings (so reducing their entropy) and make it available for doing work.

Self-test 3C.4 Repeat the calculation for the combustion of 1.000 mol $CH_4(g)$ under the same conditions, using data from Table 2C.4.

Answer: $|q_p| = 890\,\text{kJ}$, $|w_{max}| = 813\,\text{kJ}$

(d) Some remarks on the Gibbs energy

The Gibbs energy (the 'free energy') is more common in chemistry than the Helmholtz energy because, at least in laboratory chemistry, we are usually more interested in changes occurring at constant pressure than at constant volume. The criterion $dG_{T,p} \leq 0$ carries over into chemistry as the observation that, *at constant temperature and pressure, chemical reactions are spontaneous in the direction of decreasing Gibbs energy*. Therefore, if we want to know whether a reaction is spontaneous, the pressure and temperature being constant, we assess the change in the Gibbs energy. If G decreases as the reaction proceeds, then the reaction has a spontaneous tendency to convert the reactants into products. If G increases, then the reverse reaction is spontaneous. The criterion for equilibrium, when neither the forward nor reverse process is spontaneous, under conditions of constant temperature and pressure is

$$dG_{T,p} = 0 \qquad (3C.15)$$

The existence of spontaneous endothermic reactions provides an illustration of the role of G. In such reactions, H increases, the system rises spontaneously to states of higher enthalpy, and $dH > 0$. Because the reaction is spontaneous we know that $dG < 0$ despite $dH > 0$; it follows that the entropy of the system increases so much that TdS outweighs dH in $dG = dH - TdS$. Endothermic reactions are therefore driven by the increase of entropy of the system, and this entropy change overcomes the reduction of entropy brought about in the surroundings by the inflow of heat into the system ($dS_{sur} = -dH/T$ at constant pressure).

(e) Maximum non-expansion work

The analogue of the maximum work interpretation of ΔA, and the origin of the name 'free energy', can be found for ΔG. In the following *Justification*, we show that at constant temperature and pressure, the maximum additional (non-expansion) work, $w_{add,max}$, is given by the change in Gibbs energy:

$$dw_{add,max} = dG \qquad \textit{Constant temperature and pressure} \qquad \text{Maximum non-expansion work} \qquad (3C.16a)$$

The corresponding expression for a measurable change is

$$w_{add,max} = \Delta G \qquad \textit{Constant temperature and pressure} \qquad \text{Maximum non-expansion work} \qquad (3C.16b)$$

This expression is particularly useful for assessing the electrical work that may be produced by fuel cells and electrochemical cells, and we shall see many applications of it.

Justification 3C.2 Maximum non-expansion work

Because $H = U + pV$, the change in enthalpy for a general change in conditions is

$$dH = dq + dw + d(pV)$$

The corresponding change in Gibbs energy ($G = H - TS$) is

$$dG = dH - TdS - SdT = dq + dw + d(pV) - TdS - SdT$$

When the change is isothermal we can set $dT = 0$; then

$$dG = dq + dw + d(pV) - TdS$$

When the change is reversible, $dw = dw_{rev}$ and $dq = dq_{rev} = TdS$, so for a reversible, isothermal process

$$dG = TdS + dw_{rev} + d(pV) - TdS = dw_{rev} + d(pV)$$

The work consists of expansion work, which for a reversible change is given by $-pdV$, and possibly some other kind of work (for instance, the electrical work of pushing electrons through a circuit or of raising a column of liquid); this additional work we denote dw_{add}. Therefore, with $d(pV) = pdV + Vdp$,

$$dG = (-pdV + dw_{add,rev}) + pdV + Vdp = dw_{add,rev} + Vdp$$

If the change occurs at constant pressure (as well as constant temperature), we can set $dp = 0$ and obtain $dG = dw_{add,rev}$. Therefore, at constant temperature and pressure, $dw_{add,rev} = dG$. However, because the process is reversible, the work done must now have its maximum value, so eqn 3C.16 follows.

Example 3C.2 Calculating the maximum non-expansion work of a reaction

How much energy is available for sustaining muscular and nervous activity from the combustion of 1.00 mol of glucose molecules under standard conditions at 37 °C (blood temperature)? The standard entropy of reaction is +182.4 J K^{-1} mol^{-1}.

Method The non-expansion work available from the reaction is equal to the change in standard Gibbs energy for the reaction ($\Delta_r G^\ominus$, a quantity defined more fully below). To calculate this quantity, it is legitimate to ignore the temperature-dependence of the reaction enthalpy, to obtain $\Delta_r H^\ominus$ from Table 2C.5, and to substitute the data into $\Delta_r G^\ominus = \Delta_r H^\ominus - T\Delta_r S^\ominus$.

Answer Because the standard reaction enthalpy is -2808 kJ mol^{-1}, it follows that the standard reaction Gibbs energy is

$$\Delta_r G^\ominus = -2808\,\text{kJ mol}^{-1} - (310\,\text{K}) \times (182.4\,\text{J K}^{-1}\,\text{mol}^{-1})$$
$$= -2865\,\text{kJ mol}^{-1}$$

Therefore, $w_{add,max} = -2865$ kJ for the combustion of 1 mol glucose molecules, and the reaction can be used to do up to 2865 kJ of non-expansion work. To place this result in perspective, consider that a person of mass 70 kg needs to do 2.1 kJ of work to climb vertically through 3.0 m; therefore, at least 0.13 g of glucose is needed to complete the task (and in practice significantly more).

Self-test 3C.5 How much non-expansion work can be obtained from the combustion of 1.00 mol CH_4(g) under standard conditions at 298 K? Use $\Delta_r S^\ominus = -243$ J K^{-1} mol^{-1}.

Answer: 818 kJ

a point charge at its centre, so we can use the last expression and write

$$\phi(r_i)=\frac{Q}{4\pi\varepsilon r_i}$$

The work of bringing up a charge dQ to the sphere is $\phi(r_i)$dQ. Therefore, the total work of charging the sphere from 0 to z_ie is

$$w=\int_0^{z_ie}\phi(r_i)\mathrm{d}Q=\frac{1}{4\pi\varepsilon r_i}\int_0^{z_ie}Q\mathrm{d}Q=\frac{z_i^2e^2}{8\pi\varepsilon r_i}$$

This electrical work of charging, when multiplied by Avogadro's constant, is the molar Gibbs energy for charging the ions.

The work of charging an ion in a vacuum is obtained by setting $\varepsilon=\varepsilon_0$, the vacuum permittivity. The corresponding value for charging the ion in a medium is obtained by setting $\varepsilon=\varepsilon_r\varepsilon_0$, where ε_r is the relative permittivity of the medium. It follows that the change in molar Gibbs energy that accompanies the transfer of ions from a vacuum to a solvent is the difference of these two quantities:

$$\Delta_{\mathrm{solv}}G^{\ominus}=\frac{z_i^2e^2N_A}{8\pi\varepsilon r_i}-\frac{z_i^2e^2N_A}{8\pi\varepsilon_0 r_i}=\frac{z_i^2e^2N_A}{8\pi\varepsilon_r\varepsilon_0 r_i}-\frac{z_i^2e^2N_A}{8\pi\varepsilon_0 r_i}$$
$$=-\frac{z_i^2e^2N_A}{8\pi\varepsilon_0 r_i}\left(1-\frac{1}{\varepsilon_r}\right)$$

which is eqn 3B.20.

Calorimetry (for ΔH directly, and for S via heat capacities) is only one of the ways of determining Gibbs energies. They may also be obtained from equilibrium constants (Topic 6A) and electrochemical measurements (Topic 6D), and for gases they may be calculated using data from spectroscopic observations (Topic 15F).

Checklist of concepts

☐ 1. The **Clausius inequality** implies a number of criteria for spontaneous change under a variety of conditions that may be expressed in terms of the properties of the system alone; they are summarized by introducing the Helmholtz and Gibbs energies.

☐ 2. A **spontaneous process** at constant temperature and volume is accompanied by a decrease in the Helmholtz energy.

☐ 3. The change in the Helmholtz function is equal to the **maximum work** accompanying a process at constant temperature.

☐ 4. A spontaneous process at constant temperature and pressure is accompanied by a decrease in the Gibbs energy.

☐ 5. The change in the Gibbs function is equal to the **maximum non-expansion work** accompanying a process at constant temperature and pressure.

☐ 6. **Standard Gibbs energies of formation** are used to calculate the standard Gibbs energies of reactions.

☐ 7. The standard Gibbs energies of formation of ions may be estimated from a thermodynamic cycle and the **Born equation.**

Checklist of equations

Property	Equation	Comment	Equation number
Criteria of spontaneity	(a) $\mathrm{d}S_{U,V}\geq 0$ (b) $\mathrm{d}U_{S,V}\leq 0$	Constant volume (etc.)*	3C.4
	(a) $\mathrm{d}S_{H,p}\geq 0$ (b) $\mathrm{d}H_{S,p}\leq 0$	Constant pressure (etc.)	3C.6
Helmholtz energy	$A=U-TS$	Definition	3C.7
Gibbs energy	$G=H-TS$	Definition	3C.8
	(a) $\mathrm{d}A_{T,V}\leq 0$ (b) $\mathrm{d}G_{T,p}\leq 0$	Constant temperature (etc.)	3C.10
Equilibrium	$\mathrm{d}A_{T,V}=0$	Constant volume (etc.)	3C.11
Maximum work	$\mathrm{d}w_{\max}=\mathrm{d}A$, $w_{\max}=\Delta A$	Constant temperature	3C.12

Property	Equation	Comment	Equation number
Equilibrium	$\mathrm{d}G_{T,p}=0$	Constant pressure (etc.)	3C.15
Maximum non-expansion work	$\mathrm{d}w_{\mathrm{add,max}}=\mathrm{d}G$, $w_{\mathrm{add,max}}=\Delta G$	Constant temperature and pressure	3C.16
Standard Gibbs energy of reaction	$\Delta_r G^{\ominus}=\Delta_r H^{\ominus}-T\Delta_r S^{\ominus}$	Definition	3C.17
	$\Delta_r G^{\ominus}=\sum_J \nu_J \Delta_f G_m^{\ominus}(J)$	Practical implementation	3C.18
Ions in solution	$\Delta_f G^{\ominus}(H^+,aq)=0$	Convention	3C.19
Born equation	$\Delta_{\mathrm{solv}} G^{\ominus}=-(z_i^2 e^2 N_A/8\pi\varepsilon_0 r_i)(1-1/\varepsilon_r)$	Solvent a continuum	3C.20

* 'etc.' indicates that the conditions are as expressed by the subscripts.

This relation is called a **thermodynamic equation of state** because it is an expression for pressure in terms of a variety of thermodynamic properties of the system. We are now ready to derive it by using a Maxwell relation.

Justification 3D.1 The thermodynamic equation of state

We obtain an expression for the coefficient π_T by dividing both sides of eqn 3D.1 by dV, imposing the constraint of constant temperature, which gives

$$\overbrace{\left(\frac{\partial U}{\partial V}\right)_T}^{\pi_T}=\left(\frac{\partial U}{\partial S}\right)_V\overbrace{\left(\frac{\partial S}{\partial V}\right)_T}^{}+\overbrace{\left(\frac{\partial U}{\partial V}\right)_S}^{}$$

with the annotations π_T over $(\partial U/\partial V)_T$, T over $(\partial U/\partial S)_V$, and p over $(\partial U/\partial V)_S$.

Next, we introduce the two relations in eqn 3D.3 (as indicated by the annotations) and the definition of π_T to obtain

$$\pi_T=T\left(\frac{\partial S}{\partial V}\right)_T-p$$

The third Maxwell relation in Table 3D.1 turns $(\partial S/\partial V)_T$ into $(\partial p/\partial T)_V$, which completes the derivation of eqn 3D.6.

Example 3D.2 Deriving a thermodynamic relation

Show thermodynamically that $\pi_T=0$ for a perfect gas, and compute its value for a van der Waals gas.

Method Proving a result 'thermodynamically' means basing it entirely on general thermodynamic relations and equations of state, without drawing on molecular arguments (such as the existence of intermolecular forces). We know that for a perfect gas, $p=nRT/V$, so this relation should be used in eqn 3D.6. Similarly, the van der Waals equation is given in Table 1C.3, and for the second part of the question it should be used in eqn 3D.6.

Answer For a perfect gas we write

$$\left(\frac{\partial p}{\partial T}\right)_V=\left(\frac{\partial nRT/V}{\partial T}\right)_V=\frac{nR}{V}$$

Then, eqn 3D.6 becomes

$$\pi_T=\frac{nRT}{V}-p=0$$

The equation of state of a van der Waals gas is

$$p=\frac{nRT}{V-nb}-a\frac{n^2}{V^2}$$

Because a and b are independent of temperature,

$$\left(\frac{\partial p}{\partial T}\right)_V=\left(\frac{\partial nRT/(V-nb)}{\partial T}\right)_V=\frac{nR}{V-nb}$$

Therefore, from eqn 3D.6,

$$\pi_T=\frac{nRT}{V-nb}-p=\frac{nRT}{V-nb}-\left(\frac{nRT}{V-nb}-a\frac{n^2}{V^2}\right)=a\frac{n^2}{V^2}$$

This result for π_T implies that the internal energy of a van der Waals gas increases when it expands isothermally (that is, $(\partial U/\partial V)_T>0$), and that the increase is related to the parameter a, which models the attractive interactions between the particles. A larger molar volume, corresponding to a greater average separation between molecules, implies weaker mean intermolecular attractions, so the total energy is greater.

Self-test 3D.2 Calculate π_T for a gas that obeys the virial equation of state (Table 1C.3).

Answer: $\pi_T=RT^2(\partial B/\partial T)_V/V_m^2+\cdots$

3D.2 Properties of the Gibbs energy

The same arguments that we have used for U can be used for the Gibbs energy $G=H-TS$. They lead to expressions showing how G varies with pressure and temperature that are important for discussing phase transitions and chemical reactions.

(a) General considerations

When the system undergoes a change of state, G may change because H, T, and S all change and

$$dG=dH-d(TS)=dH-TdS-SdT$$

Because $H=U+pV$, we know that

$$dH=dU+d(pV)=dU+pdV+Vdp$$

and therefore

$$dG=dU+pdV+Vdp-TdS-SdT$$

For a closed system doing no non-expansion work, we can replace dU by the fundamental equation $dU=TdS-pdV$ and obtain

$$dG=TdS-pdV+pdV+Vdp-TdS-SdT$$

Four terms now cancel on the right, and we conclude that for a closed system in the absence of non-expansion work and at constant composition

$$dG=Vdp-SdT \qquad (3D.7)$$

The fundamental equation of chemical thermodynamics

This expression, which shows that a change in G is proportional to a change in p or T, suggests that G may be best regarded as a function of p and T. It may be regarded as the **fundamental equation of chemical thermodynamics** as it is so central to the application of thermodynamics to chemistry: it suggests that G is an important quantity in chemistry because the pressure and temperature are usually the variables under our control. In other words, G carries around the combined consequences of the First and Second Laws in a way that makes it particularly suitable for chemical applications.

The same argument that led to eqn 3D.3, when applied to the exact differential $\mathrm{d}G = V\mathrm{d}p - S\mathrm{d}T$, now gives

$$\left(\frac{\partial G}{\partial T}\right)_p = -S \qquad \left(\frac{\partial G}{\partial p}\right)_T = V \qquad \text{The variation of } G \text{ with } T \text{ and } p \qquad (3D.8)$$

These relations show how the Gibbs energy varies with temperature and pressure (Fig. 3D.1). The first implies that:

- Because $S > 0$ for all substances, G always *decreases* when the temperature is raised (at constant pressure and composition).
- Because $(\partial G/\partial T)_p$ becomes more negative as S increases, G decreases most sharply with increasing temperature when the entropy of the system is large.

Physical interpretation

Therefore, the Gibbs energy of the gaseous phase of a substance, which has a high molar entropy, is more sensitive to temperature than its liquid and solid phases (Fig. 3D.2). Similarly, the second relation implies that:

- Because $V > 0$ for all substances, G always *increases* when the pressure of the system is increased (at constant temperature and composition).

Physical interpretation

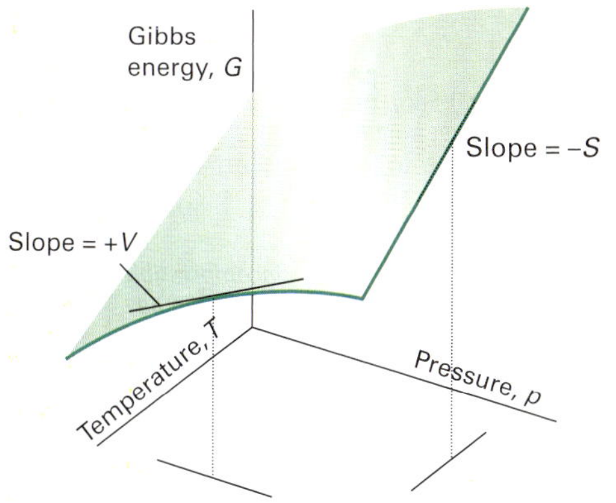

Figure 3D.1 The variation of the Gibbs energy of a system with (a) temperature at constant pressure and (b) pressure at constant temperature. The slope of the former is equal to the negative of the entropy of the system and that of the latter is equal to the volume.

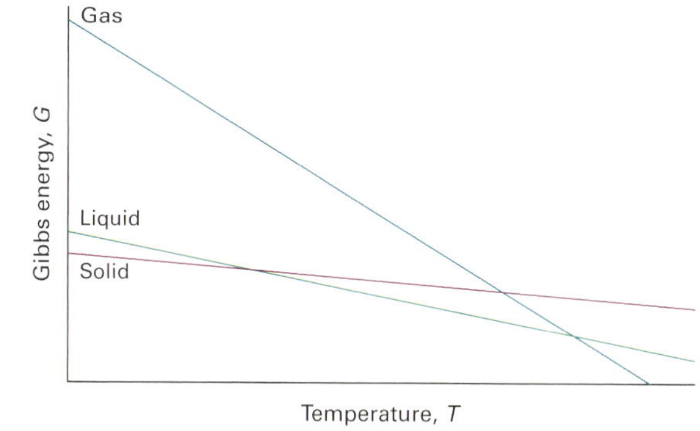

Figure 3D.2 The variation of the Gibbs energy with the temperature is determined by the entropy. Because the entropy of the gaseous phase of a substance is greater than that of the liquid phase, and the entropy of the solid phase is smallest, the Gibbs energy changes most steeply for the gas phase, followed by the liquid phase, and then the solid phase of the substance.

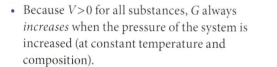

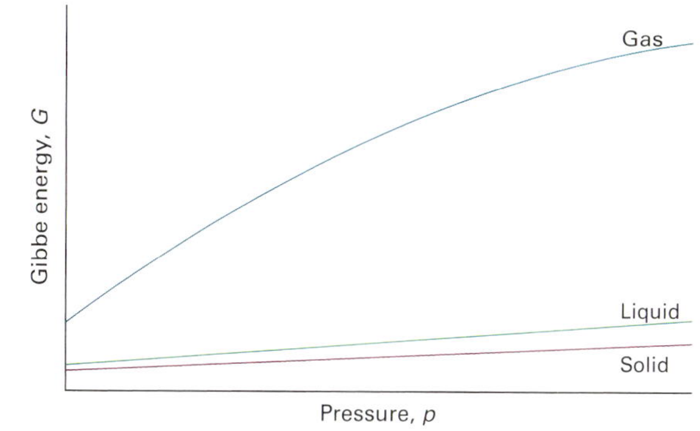

Figure 3D.3 The variation of the Gibbs energy with the pressure is determined by the volume of the sample. Because the volume of the gaseous phase of a substance is greater than that of the same amount of liquid phase, and the entropy of the solid phase is smallest (for most substances), the Gibbs energy changes most steeply for the gas phase, followed by the liquid phase, and then the solid phase of the substance. Because the volumes of the solid and liquid phases of a substance are similar, their molar Gibbs energies vary by similar amounts as the pressure is changed.

- Because $(\partial G/\partial p)_T$ increases with V, G is more sensitive to pressure when the volume of the system is large.

Because the molar volume of the gaseous phase of a substance is greater than that of its condensed phases, the molar Gibbs energy of a gas is more sensitive to pressure than its liquid and solid phases (Fig. 3D.3).

Brief illustration 3D.1 The variation of molar Gibbs energy

The standard molar entropy of liquid water at 298 K is 69.91 J K^{-1} mol^{-1}. It follows that when the temperature is increased by 5.0 K, the molar Gibbs energy changes by

Note that whereas the change in molar Gibbs energy for a condensed phase is a few joules per mole, for a gas the change is of the order of kilojoules per mole

Self-test 3D.5 By how much does the molar Gibbs energy of a perfect gas differ from its standard value at 298 K when its pressure is 0.10 bar?

Answer: $-5.7\,\text{kJ mol}^{-1}$

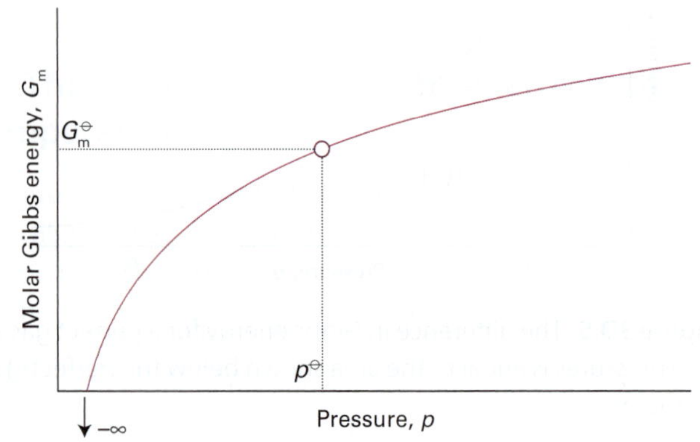

Figure 3D.6 The molar Gibbs energy of a perfect gas varies as $\ln p$, and the standard state is reached at $p^{\ominus}$. Note that, as $p \to 0$, the molar Gibbs energy becomes negatively infinite.

The logarithmic dependence of the molar Gibbs energy on the pressure predicted by eqn 3D.15 is illustrated in Fig. 3D.6. This very important expression, the consequences of which we unfold in the following chapters, applies to perfect gases (which is usually a good enough approximation). The following section shows how to accommodate imperfections.

(d) The fugacity

At various stages in the development of physical chemistry it is necessary to switch from a consideration of idealized systems to real systems. In many cases it is desirable to preserve the form of the expressions that have been derived for an idealized system. Then deviations from the idealized behaviour can be expressed most simply. For instance, the pressure-dependence of the molar Gibbs energy of a real gas might resemble that shown in Fig. 3D.7. To adapt eqn 3D.14 to this case, we replace the true pressure, p, by an effective pressure, called the **fugacity**,[1] f, and write

$$G_m = G_m^{\ominus} + RT\ln(f/p^{\ominus}) \qquad \textit{Definition} \quad \text{Fugacity} \qquad (3D.16)$$

The fugacity, a function of the pressure and temperature, is defined so that this relation is exactly true. A very similar approach is taken in the discussion of real solutions (Topic 5E),

[1] The name 'fugacity' comes from the Latin for 'fleetness' in the sense of 'escaping tendency'; fugacity has the same dimensions as pressure.

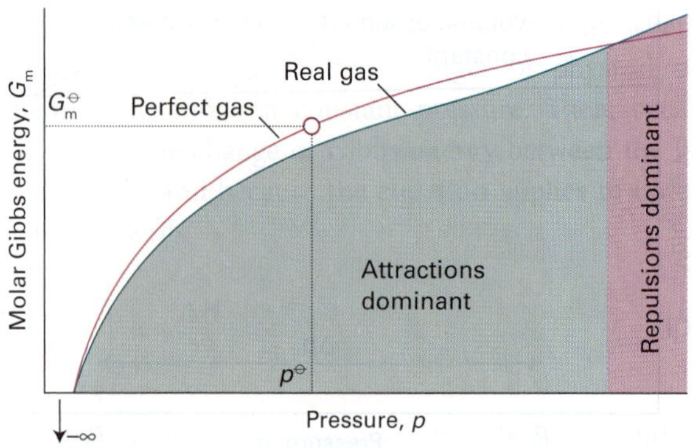

Figure 3D.7 The molar Gibbs energy of a real gas. As $p \to 0$, the molar Gibbs energy coincides with the value for a perfect gas (shown by the purple line). When attractive forces are dominant (at intermediate pressures), the molar Gibbs energy is less than that of a perfect gas and the molecules have a lower 'escaping tendency'. At high pressures, when repulsive forces are dominant, the molar Gibbs energy of a real gas is greater than that of a perfect gas. Then the 'escaping tendency' is increased.

where 'activities' are effective concentrations. Indeed, $f/p^{\ominus}$ may be regarded as a gas-phase activity.

Although thermodynamic expressions in terms of fugacities derived from this expression are exact, they are useful only if we know how to interpret fugacities in terms of actual pressures. To develop this relation we write the fugacity as

$$f = \phi p \qquad \textit{Definition} \quad \text{Fugacity coefficient} \qquad (3D.17)$$

where ϕ is the dimensionless **fugacity coefficient**, which in general depends on the temperature, the pressure, and the identity of the gas. We show in the following *Justification* that the fugacity coefficient is related to the compression factor, Z, of a gas (Topic 1C) by

$$\ln\phi = \int_0^p \frac{Z-1}{p}\,dp \qquad (3D.18)$$

Provided we know how Z varies with pressure up to the pressure of interest, this expression enable us to determine the fugacity coefficient and hence, through eqn 3D.17, to relate the fugacity to the pressure of the gas.

Justification 3D.3 The fugacity coefficient

Equation 3D.12a is true for all gases whether real or perfect. Expressing it in terms of the fugacity by using eqn 3D.16 turns it into

$$\int_{p'}^{p} V_m dp = G_m(p) - G_m(p') = \left\{G_m^{\ominus} + RT\ln\frac{f}{p^{\ominus}}\right\} - \left\{G_m^{\ominus} + RT\ln\frac{f'}{p^{\ominus}}\right\} = RT\ln\frac{f}{f'}$$

In this expression, f is the fugacity when the pressure is p and f' is the fugacity when the pressure is p'. If the gas were perfect, we would write

$$\int_{p'}^{p} \overbrace{V_{\text{perfect,m}}}^{RT/p}\mathrm{d}p = RT\int_{p'}^{p}\frac{1}{p}\mathrm{d}p = RT\ \ln\frac{p}{p'}$$

The difference between the two equations is

$$\int_{p'}^{p}(V_{\text{m}} - V_{\text{perfect,m}})\mathrm{d}p = RT\left(\ln\frac{f}{f'} - \ln\frac{p}{p'}\right) = RT\ln\frac{f/f'}{p/p'} = RT\ln\frac{f/p}{f'/p'}$$

which can be rearranged into

$$\ln\frac{f/p}{f'/p'} = \frac{1}{RT}\int_{p'}^{p}(V_{\text{m}} - V_{\text{perfect,m}})\mathrm{d}p$$

When $p' \to 0$, the gas behaves perfectly and f' becomes equal to the pressure, p'. Therefore, $f'/p' \to 1$ as $p' \to 0$. If we take this limit, which means setting $f'/p' = 1$ on the left and $p' = 0$ on the right, the last equation becomes

$$\ln\frac{f}{p} = \frac{1}{RT}\int_{0}^{p}(V_{\text{m}} - V_{\text{perfect,m}})\mathrm{d}p$$

Then, with $\phi = f/p$,

$$\ln\phi = \frac{1}{RT}\int_{0}^{p}(V_{\text{m}} - V_{\text{perfect,m}})\mathrm{d}p$$

For a perfect gas, $V_{\text{perfect,m}} = RT/p$. For a real gas, $V_{\text{m}} = RTZ/p$, where Z is the compression factor of the gas (Topic 1C). With these two substitutions, we obtain eqn 3D.18.

For a perfect gas, $\phi = 1$ at all pressures and temperatures. We know from Fig. 1C.9 that for most gases $Z < 1$ up to moderate pressures, but that $Z > 1$ at higher pressures. If $Z < 1$ throughout the range of integration, then the integrand in eqn 3D.18 is negative and $\phi < 1$. This value implies that $f < p$ (the molecules tend to stick together) and that the molar Gibbs energy of the gas is less than that of a perfect gas. At higher pressures, the range over which $Z > 1$ may dominate the range over which $Z < 1$. The integral is then positive, $\phi > 1$, and $f > p$ (the repulsive interactions are dominant and tend to drive the particles apart). Now the molar Gibbs energy of the gas is greater than that of the perfect gas at the same pressure.

Figure 3D.8, which has been calculated using the full van der Waals equation of state, shows how the fugacity coefficient depends on the pressure in terms of the reduced variables (Topic 1C). Because critical constants are available in Table 1C.2, the graphs can be used for quick estimates of the fugacities of a wide range of gases. Table 3D.2 gives some explicit values for nitrogen.

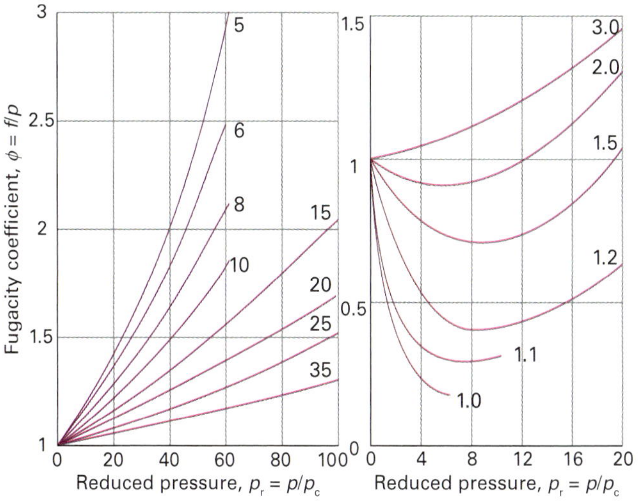

Figure 3D.8 The fugacity coefficient of a van der Waals gas plotted using the reduced variables of the gas. The curves are labelled with the reduced temperature $T_r = T/T_c$.

Table 3D.2* The fugacity of nitrogen at 273 K, f/atm

p/atm	f/atm
1	0.999 55
10	9.9560
100	97.03
1000	1839

* More values are given in the *Resource section*.

Brief illustration 3D.3 The fugacity of a real gas

To use Fig. 3D.8 to estimate the fugacity of carbon dioxide at 400 K and 400 atm, we note from Table 1C.2 that its critical constants are $p_c = 72.85$ atm and $T_c = 304.2$ K. In terms of reduced variables, the gas has $p_r = (400\ \text{atm})/(72.85\ \text{atm}) = 5.5$ and $T_r = (400\ \text{K})/(304.2\ \text{K}) = 1.31$. From Fig. 3D.8 (interpolating by eye), these conditions correspond to $\phi \approx 0.4$ and therefore to $f \approx 160$ atm.

Self-test 3D.6 At what temperature would carbon dioxide have a fugacity of 400 atm when its pressure is 400 atm?

Answer: At about $T_r = 2.5$, corresponding to $T = 760$ K

Checklist of concepts

- ☐ 1. The **fundamental equation**, a combination of the First and Second Laws, is an expression for the change in internal energy that accompanies changes in the volume and entropy of a system.
- ☐ 2. Relations between thermodynamic properties are generated by combining thermodynamic and mathematical expressions for changes in their values.
- ☐ 3. The **Maxwell relations** are a series of relations between derivatives of thermodynamic properties based on criteria for changes in the properties being exact differentials.
- ☐ 4. The Maxwell relations are used to derive the **thermodynamic equation of state** and to determine how the internal energy of a substance varies with volume.
- ☐ 5. The variation of the Gibbs energy of a system suggests that it is best regarded as a function of pressure and temperature.
- ☐ 6. The Gibbs energy of a substance decreases with temperature and increases with pressure.
- ☐ 7. The variation of Gibbs energy with temperature is related to the enthalpy by the **Gibbs–Helmholtz equation**.
- ☐ 8. The Gibbs energies of solids and liquids are almost independent of pressure; those of gases vary linearly with the logarithm of the pressure.
- ☐ 9. The **fugacity** is a kind of effective pressure of a real gas.

Checklist of equations

Property	Equation	Comment	Equation number
Fundamental equation	$\mathrm{d}U = T\mathrm{d}S - p\mathrm{d}V$	No additional work	3D.1
Fundamental equation of chemical thermodynamics	$\mathrm{d}G = V\mathrm{d}p - S\mathrm{d}T$	No additional work	3D.7
Variation of G	$(\partial G/\partial p)_T = V$ and $(\partial G/\partial T)_p = -S$	Composition constant	3D.8
Gibbs–Helmholtz equation	$(\partial (G/T)/\partial T)_p = -H/T^2$	Composition constant	3D.10
Pressure dependence of G	$G_m(p_f) = G_m(p_i) + (p_f - p_i)V_m$	Incompressible substance	3D.13
	$G_m(p_f) = G_m(p_i) + RT\ln(p_f/p_i)$	Perfect gas	3D.14
	$G_m = G_m^{\ominus} + RT\ln(p/p^{\ominus})$	Perfect gas	3D.15
Fugacity	$G_m = G_m^{\ominus} + RT\ln(f/p^{\ominus})$	Definition	3D.16
Fugacity coefficient	$f = \phi p$	Definition	3D.17
	$\ln\phi = \int_0^p \{(Z-1)/p\}\mathrm{d}p$	Determination	3D.18

CHAPTER 3 The Second and Third Laws

Assume that all gases are perfect and that data refer to 298.15 K unless otherwise stated.

TOPIC 3A Entropy

Discussion questions

3A.1 The evolution of life requires the organization of a very large number of molecules into biological cells. Does the formation of living organisms violate the Second Law of thermodynamics? State your conclusion clearly and present detailed arguments to support it.

3A.2 Discuss the significance of the terms 'dispersal' and 'disorder' in the context of the Second Law.

3A.3 Discuss the relationships between the various formulations of the Second Law of thermodynamics.

3A.4 Account for deviations from Trouton's rule for liquids such as water and ethanol. Is their entropy of vaporization larger or smaller than $85\,J\,K^{-1}\,mol^{-1}$? Why?

Exercises

3A.1(a) During a hypothetical process, the entropy of a system increases by $125\,J\,K^{-1}$ while the entropy of the surroundings decreases by $125\,J\,K^{-1}$. Is the process spontaneous?

3A.1(b) During a hypothetical process, the entropy of a system increases by $105\,J\,K^{-1}$ while the entropy of the surroundings decreases by $95\,J\,K^{-1}$. Is the process spontaneous?

3A.2(a) A certain ideal heat engine uses water at the triple point as the hot source and an organic liquid as the cold sink. It withdraws 10.00 kJ of heat from the hot source and generates 3.00 kJ of work. What is the temperature of the organic liquid?

3A.2(b) A certain ideal heat engine uses water at the triple point as the hot source and an organic liquid as the cold sink. It withdraws 2.71 kJ of heat from the hot source and generates 0.71 kJ of work. What is the temperature of the organic liquid?

3A.3(a) Calculate the change in entropy when 100 kJ of energy is transferred reversibly and isothermally as heat to a large block of copper at (i) 0 °C, (ii) 50 °C.

3A.3(b) Calculate the change in entropy when 250 kJ of energy is transferred reversibly and isothermally as heat to a large block of lead at (i) 20 °C, (ii) 100 °C.

3A.4(a) Which of F_2(g) and I_2(g) is likely to have the higher standard molar entropy at 298 K?

3A.4(b) Which of H_2O(g) and CO_2(g) is likely to have the higher standard molar entropy at 298 K?

3A.5(a) Calculate the change in entropy when 15 g of carbon dioxide gas is allowed to expand from $1.0\,dm^3$ to $3.0\,dm^3$ at 300 K.

3A.5(b) Calculate the change in entropy when 4.00 g of nitrogen is allowed to expand from $500\,cm^3$ to $750\,cm^3$ at 300 K.

3A.6(a) Predict the enthalpy of vaporization of benzene from its normal boiling point, 80.1 °C.

3A.6(b) Predict the enthalpy of vaporization of cyclohexane from its normal boiling point, 80.7 °C.

3A.7(a) Calculate the molar entropy of a constant-volume sample of neon at 500 K given that it is $146.22\,J\,K^{-1}\,mol^{-1}$ at 298 K.

3A.7(b) Calculate the molar entropy of a constant-volume sample of argon at 250 K given that it is $154.84\,J\,K^{-1}\,mol^{-1}$ at 298 K.

3A.8(a) Calculate ΔS (for the system) when the state of 3.00 mol of perfect gas atoms, for which $C_{p,m} = \frac{5}{2}R$, is changed from 25 °C and 1.00 atm to 125 °C and 5.00 atm. How do you rationalize the sign of ΔS?

3A.8(b) Calculate ΔS (for the system) when the state of 2.00 mol diatomic perfect gas molecules, for which $C_{p,m} = \frac{5}{2}R$, is changed from 25 °C and 1.50 atm to 135 °C and 7.00 atm. How do you rationalize the sign of ΔS?

3A.9(a) Calculate ΔS_{tot} when two copper blocks, each of mass 1.00 kg, one at 50 °C and the other at 0 °C are placed in contact in an isolated container. The specific heat capacity of copper is $0.385\,J\,K^{-1}\,g^{-1}$ and may be assumed constant over the temperature range involved.

3A.9(b) Calculate ΔS_{tot} when two iron blocks, each of mass 10.0 kg , one at 100 °C and the other at 25 °C, are placed in contact in an isolated container. The specific heat capacity of iron is $0.449\,J\,K^{-1}\,g^{-1}$ and may be assumed constant over the temperature range involved.

3A.10(a) Calculate the change in the entropies of the system and the surroundings, and the total change in entropy, when a sample of nitrogen gas of mass 14 g at 298 K and 1.00 bar doubles its volume in (i) an isothermal reversible expansion, (ii) an isothermal irreversible expansion against $p_{ex}=0$, and (iii) an adiabatic reversible expansion.

3A.10(b) Calculate the change in the entropies of the system and the surroundings, and the total change in entropy, when the volume of a sample of argon gas of mass 21 g at 298 K and 1.50 bar increases from $1.20\,dm^3$ to $4.60\,dm^3$ in (i) an isothermal reversible expansion, (ii) an isothermal irreversible expansion against $p_{ex}=0$, and (iii) an adiabatic reversible expansion.

3A.11(a) The enthalpy of vaporization of chloroform ($CHCl_3$) is $29.4\,kJ\,mol^{-1}$ at its normal boiling point of 334.88 K. Calculate (i) the entropy of vaporization of chloroform at this temperature and (ii) the entropy change of the surroundings.

3A.11(b) The enthalpy of vaporization of methanol is $35.27\,kJ\,mol^{-1}$ at its normal boiling point of 64.1 °C. Calculate (i) the entropy of vaporization of methanol at this temperature and (ii) the entropy change of the surroundings.

3A.12(a) Calculate the change in entropy of the system when 10.0 g of ice at −10.0 °C is converted into water vapour at 115.0 °C and at a constant pressure of 1 bar. The constant-pressure molar heat capacity of H_2O(s) and H_2O(l) is $75.291\,J\,K^{-1}\,mol^{-1}$ and that of H_2O(g) is $33.58\,J\,K^{-1}\,mol^{-1}$.

3A.12(b) Calculate the change in entropy of the system when 15.0 g of ice at −12.0 °C is converted to water vapour at 105.0 °C at a constant pressure of 1 bar. For data, see the preceding exercise.

Problems

3A.1 Represent the Carnot cycle on a temperature–entropy diagram and show that the area enclosed by the cycle is equal to the work done.

3A.2 The cycle involved in the operation of an internal combustion engine is called the *Otto cycle*. Air can be considered to be the working substance and can be assumed to be a perfect gas. The cycle consists of the following steps: (1) Reversible adiabatic compression from A to B, (2) reversible constant-volume pressure increase from B to C due to the combustion of a small amount of fuel, (3) reversible adiabatic expansion from C to D, and (4) reversible and constant-volume pressure decrease back to state A. Determine the change in entropy (of the system and of the surroundings) for each step of the cycle and determine an expression for the efficiency of the cycle, assuming that the heat is supplied in Step 2. Evaluate the efficiency for a compression ratio of 10:1. Assume that in state A, $V=4.00\,\mathrm{dm^3}$, $p=1.00\,\mathrm{atm}$, and $T=300\,\mathrm{K}$, that $V_A=10V_B$, $p_C/p_B=5$, and that $C_{p,\mathrm{m}}=\frac{7}{2}R$.

3A.3 Prove that two reversible adiabatic paths can never cross. Assume that the energy of the system under consideration is a function of temperature only. (*Hint*: Suppose that two such paths can intersect, and complete a cycle with the two paths plus one isothermal path. Consider the changes accompanying each stage of the cycle and show that they conflict with the Kelvin statement of the Second Law.)

3A.4 To calculate the work required to lower the temperature of an object, we need to consider how the coefficient of performance c (see *Impact* I3.1) changes with the temperature of the object. (a) Find an expression for the work of cooling an object from T_i to T_f when the refrigerator is in a room at a temperature T_h. *Hint*: Write $\mathrm{d}w=\mathrm{d}q/c(T)$, relate $\mathrm{d}q$ to $\mathrm{d}T$ through the heat capacity C_p, and integrate the resulting expression. Assume that the heat capacity is independent of temperature in the range of interest. (b) Use the result in part (a) to calculate the work needed to freeze 250 g of water in a refrigerator at 293 K. How long will it take when the refrigerator operates at 100 W?

3A.5 The expressions that apply to the treatment of refrigerators (Problem 3A.4) also describe the behaviour of heat pumps, where warmth is obtained from the back of a refrigerator while its front is being used to cool the outside world. Heat pumps are popular home heating devices because they are very efficient. Compare heating of a room at 295 K by each of two methods: (a) direct conversion of 1.00 kJ of electrical energy in an electrical heater, and (b) use of 1.00 kJ of electrical energy to run a reversible heat pump with the outside at 260 K. Discuss the origin of the difference in the energy delivered to the interior of the house by the two methods.

3A.6 Calculate the difference in molar entropy (a) between liquid water and ice at −5 °C, (b) between liquid water and its vapour at 95 °C and 1.00 atm. The differences in heat capacities on melting and on vaporization are $37.3\,\mathrm{J\,K^{-1}\,mol^{-1}}$ and $-41.9\,\mathrm{J\,K^{-1}\,mol^{-1}}$, respectively. Distinguish between the entropy changes of the sample, the surroundings, and the total system, and discuss the spontaneity of the transitions at the two temperatures.

3A.7 The molar heat capacity of chloroform (trichloromethane, $CHCl_3$) in the range 240 K to 330 K is given by $C_{p,\mathrm{m}}/(\mathrm{J\,K^{-1}\,mol^{-1}})=91.47+7.5\times10^{-2}\,(T/\mathrm{K})$. In a particular experiment, 1.00 mol $CHCl_3$ is heated from 273 K to 300 K. Calculate the change in molar entropy of the sample.

3A.8 A block of copper of mass 2.00 kg ($C_{p,\mathrm{m}}=24.44\,\mathrm{J\,K^{-1}\,mol^{-1}}$) and temperature 0 °C is introduced into an insulated container in which there is 1.00 mol $H_2O(g)$ at 100 °C and 1.00 atm. (a) Assuming all the steam is condensed to water, what will be the final temperature of the system, the heat transferred from water to copper, and the entropy change of the water, copper, and the total system? (b) In fact, some water vapour is present at equilibrium. From the vapour pressure of water at the temperature calculated in (a), and assuming that the heat capacities of both gaseous and liquid water are constant and given by their values at that temperature, obtain an improved value of the final temperature, the heat transferred, and the various entropies. (*Hint*: You will need to make plausible approximations.)

3A.9 A sample consisting of 1.00 mol of perfect gas molecules at 27 °C is expanded isothermally from an initial pressure of 3.00 atm to a final pressure of 1.00 atm in two ways: (a) reversibly, and (b) against a constant external pressure of 1.00 atm. Determine the values of q, w, ΔU, ΔH, ΔS, ΔS_{surr}, and ΔS_{tot} for each path.

3A.10 A block of copper of mass 500 g and initially at 293 K is in thermal contact with an electric heater of resistance 1.00 kΩ and negligible mass. A current of 1.00 A is passed for 15.0 s. Calculate the change in entropy of the copper, taking $C_{p,\mathrm{m}}=24.4\,\mathrm{J\,K^{-1}\,mol^{-1}}$. The experiment is then repeated with the copper immersed in a stream of water that maintains its temperature at 293 K. Calculate the change in entropy of the copper and the water in this case.

3A.11 Find an expression for the change in entropy when two blocks of the same substance and of equal mass, one at the temperature T_h and the other at T_c, are brought into thermal contact and allowed to reach equilibrium. Evaluate the change for two blocks of copper, each of mass 500 g, with $C_{p,\mathrm{m}}=24.4\,\mathrm{J\,K^{-1}\,mol^{-1}}$, taking $T_h=500\,\mathrm{K}$ and $T_c=250\,\mathrm{K}$.

3A.12 According to Newton's law of cooling, the rate of change of temperature is proportional to the temperature difference between the system and its surroundings. Given that $S(T)-S(T_i)=C\ln(T/T_i)$, where T_i is the initial temperature and C the heat capacity, deduce an expression for the rate of change of entropy of the system as it cools.

3A.13 The protein lysozyme unfolds at a transition temperature of 75.5 °C and the standard enthalpy of transition is $509\,\mathrm{kJ\,mol^{-1}}$. Calculate the entropy of unfolding of lysozyme at 25.0 °C, given that the difference in the constant-pressure heat capacities upon unfolding is $6.28\,\mathrm{kJ\,K^{-1}\,mol^{-1}}$ and can be assumed to be independent of temperature. *Hint*: Imagine that the transition at 25.0 °C occurs in three steps: (i) heating of the folded protein from 25.0 °C to the transition temperature, (ii) unfolding at the transition temperature, and (iii) cooling of the unfolded protein to 25.0 °C. Because the entropy is a state function, the entropy change at 25.0 °C is equal to the sum of the entropy changes of the steps.

TOPIC 3B The measurement of entropy

Discussion question

3B.1 Discuss why the standard entropies of ions in solution may be positive, negative, or zero.

Exercises

3B.1(a) Calculate the residual molar entropy of a solid in which the molecules can adopt (i) three, (ii) five, (iii) six orientations of equal energy at $T=0$.

3B.1(b) Suppose that the hexagonal molecule $C_6H_nF_{6-n}$ has a residual entropy on account of the similarity of the H and F atoms. Calculate the residual for each value of n.

3B.2(a) Calculate the standard reaction entropy at 298 K of

(i) $2CH_3CHO(g)+O_2(g) \rightarrow 2CH_3COOH(l)$

(ii) $2AgCl(s)+Br_2(l) \rightarrow 2AgBr(s)+Cl_2(g)$

(iii) $Hg(l)+Cl_2(g) \rightarrow HgCl_2(s)$

3B.2(b) Calculate the standard reaction entropy at 298 K of

(i) $Zn(s)+Cu^{2+}(aq) \rightarrow Zn^{2+}(aq)+Cu(s)$

(ii) $C_{12}H_{22}O_{11}(s)+12O_2(g) \rightarrow 12CO_2(g)+11H_2O(l)$

Problems

3B.1 The standard molar entropy of $NH_3(g)$ is 192.45 J K^{-1} mol^{-1} at 298 K, and its heat capacity is given by eqn 2B.8 with the coefficients given in Table 2B.1. Calculate the standard molar entropy at (a) 100 °C and (b) 500 °C.

3B.2 The molar heat capacity of lead varies with temperature as follows:

T/K	10	15	20	25	30	50
$C_{p,m}$/(J K^{-1} mol^{-1})	2.8	7.0	10.8	14.1	16.5	21.4
T/K	70	100	150	200	250	298
$C_{p,m}$/(J K^{-1} mol^{-1})	23.3	24.5	25.3	25.8	26.2	26.6

Calculate the standard Third-Law entropy of lead at (a) 0 °C and (b) 25 °C.

3B.3 From standard enthalpies of formation, standard entropies, and standard heat capacities available from tables in the *Resource section*, calculate:
(a) the standard enthalpies and entropies at 298 K and 398 K for the reaction $CO_2(g)+H_2(g) \rightarrow CO(g)+H_2O(g)$. Assume that the heat capacities are constant over the temperature range involved.

3B.4 The molar heat capacity of anhydrous potassium hexacyanoferrate(II) varies with temperature as follows:

T/K	10	20	30	40	50	60
$C_{p,m}$/(J K^{-1} mol^{-1})	2.09	14.43	36.44	62.55	87.03	111.0
T/K	70	80	90	100	110	150
$C_{p,m}$/(J K^{-1} mol^{-1})	131.4	149.4	165.3	179.6	192.8	237.6
T/K	160	170	180	190	200	
$C_{p,m}$/(J K^{-1} mol^{-1})	247.3	256.5	265.1	273.0	280.3	

Calculate the molar enthalpy relative to its value at $T=0$ and the Third-Law entropy at each of these temperatures.

3B.5 The compound 1,3,5-trichloro-2,4,6-trifluorobenzene is an intermediate in the conversion of hexachlorobenzene to hexafluorobenzene, and its thermodynamic properties have been examined by measuring its heat capacity over a wide temperature range (R.L. Andon and J.F. Martin, *J. Chem. Soc. Faraday Trans. I*, 871 (1973)). Some of the data are as follows:

T/K	14.14	16.33	20.03	31.15	44.08	64.81
$C_{p,m}$/(J K^{-1} mol^{-1})	9.492	12.70	18.18	32.54	46.86	66.36
T/K	100.90	140.86	183.59	225.10	262.99	298.06
$C_{p,m}$/(J K^{-1} mol^{-1})	95.05	121.3	144.4	163.7	180.2	196.4

Calculate the molar enthalpy relative to its value at $T=0$ and the Third-Law molar entropy of the compound at these temperatures.

3B.6‡ Given that $S_m^{\ominus}=29.79$ J K^{-1} mol^{-1} for bismuth at 100 K and the following tabulated heat capacity data (D.G. Archer, *J. Chem. Eng. Data* **40**, 1015 (1995)), compute the standard molar entropy of bismuth at 200 K.

T/K	100	120	140	150	160	180	200
$C_{p,m}$/(J K^{-1} mol^{-1})	23.00	23.74	24.25	24.44	24.61	24.89	25.11

Compare the value to the value that would be obtained by taking the heat capacity to be constant at 24.44 J K^{-1} mol^{-1} over this range.

3B.7 Derive an expression for the molar entropy of a monatomic solid on the basis of the Einstein and Debye models and plot the molar entropy against the temperature (use T/θ in each case, with θ the Einstein or Debye temperature). Use the following expressions for the temperature-dependence of the heat capacities:

$$\text{Einstein: } C_{V,m}(T)=3Rf^{E}(T) \quad f^{E}(T)=\left(\frac{\theta^{E}}{T}\right)^{2}\left(\frac{e^{\theta^{E}/2T}}{e^{\theta^{E}/T}-1}\right)^{2}$$

$$\text{Debye: } C_{V,m}(T)=3Rf^{D}(T) \quad f^{D}(T)=3\left(\frac{T}{\theta^{D}}\right)^{2}\int_{0}^{\theta^{D}/T}\frac{x^{4}e^{x}}{(e^{x}-1)^{2}}\,dx$$

Use mathematical software to evaluate the appropriate expressions.

3B.8 An average human DNA molecule has 5×10^{8} binucleotides (rungs on the DNA ladder) of four different kinds. If each rung were a random choice of one of these four possibilities, what would be the residual entropy associated with this typical DNA molecule?

TOPIC 3C Concentrating on the system

Discussion questions

3C.1 The following expressions have been used to establish criteria for spontaneous change: $dA_{T,V}<0$ and $dG_{T,p}<0$. Discuss the origin, significance, and applicability of each criterion.

3C.2 Under what circumstances, and why, can the spontaneity of a process be discussed in terms of the properties of the system alone?

‡ These problems were provided by Charles Trapp and Carmen Giunta.

Exercises

3C.1(a) Combine the reaction entropies calculated in Exercise 3B.2(a) with the reaction enthalpies, and calculate the standard reaction Gibbs energies at 298 K.

3C.1(b) Combine the reaction entropies calculated in Exercise 3B.2(b) with the reaction enthalpies, and calculate the standard reaction Gibbs energies at 298 K.

3C.2(a) Calculate the standard Gibbs energy of the reaction $4\,HI(g)+O_2(g)\rightarrow 2\,I_2(s)+2\,H_2O(l)$ at 298 K, from the standard entropies and enthalpies of formation given in the *Resource section*.

3C.2(b) Calculate the standard Gibbs energy of the reaction $CO(g)+CH_3CH_2OH(l)\rightarrow CH_3CH_2COOH(l)$ at 298 K, from the standard entropies and enthalpies of formation given in the *Resource section*.

3C.3(a) Calculate the maximum non-expansion work per mole that may be obtained from a fuel cell in which the chemical reaction is the combustion of methane at 298 K.

3C.3(b) Calculate the maximum non-expansion work per mole that may be obtained from a fuel cell in which the chemical reaction is the combustion of propane at 298 K.

3C.4(a) Use standard Gibbs energies of formation to calculate the standard reaction Gibbs energies at 298 K of the reactions

(i) $2\,CH_3CHO(g)+O_2(g)\rightarrow 2\,CH_3COOH(l)$
(ii) $2\,AgCl(s)+Br_2(l)\rightarrow 2\,AgBr(s)+Cl_2(g)$
(iii) $Hg(l)+Cl_2(g)\rightarrow HgCl_2(s)$

3C.4(b) Use standard Gibbs energies of formation to calculate the standard reaction Gibbs energies at 298 K of the reactions

(i) $Zn(s)+Cu^{2+}(aq)\rightarrow Zn^{2+}(aq)+Cu(s)$
(ii) $C_{12}H_{22}O_{11}(s)+12\,O_2(g)\rightarrow 12\,CO_2(g)+11\,H_2O(l)$

3C.5(a) The standard enthalpy of combustion of ethyl acetate ($CH_3COOC_2H_5$) is $-2231\,kJ\,mol^{-1}$ at 298 K and its standard molar entropy is $259.4\,J\,K^{-1}\,mol^{-1}$. Calculate the standard Gibbs energy of formation of the compound at 298 K.

3C.5(b) The standard enthalpy of combustion of the amino acid glycine (NH_2CH_2COOH) is $-969\,kJ\,mol^{-1}$ at 298 K and its standard molar entropy is $103.5\,J\,K^{-1}\,mol^{-1}$. Calculate the standard Gibbs energy of formation of glycine at 298 K.

Problems

3C.1 Consider a perfect gas contained in a cylinder and separated by a frictionless adiabatic piston into two sections A and B. All changes in B are isothermal; that is, a thermostat surrounds B to keep its temperature constant. There is 2.00 mol of the gas molecules in each section. Initially $T_A=T_B=300\,K$, $V_A=V_B=2.00\,dm^3$. Energy is supplied as heat to Section A and the piston moves to the right reversibly until the final volume of Section B is $1.00\,dm^3$. Calculate (a) ΔS_A and ΔS_B, (b) ΔA_A and ΔA_B, (c) ΔG_A and ΔG_B, (d) ΔS of the total system and its surroundings. If numerical values cannot be obtained, indicate whether the values should be positive, negative, or zero or are indeterminate from the information given. (Assume $C_{V,m}=20\,J\,K^{-1}\,mol^{-1}$.)

3C.2 Calculate the molar internal energy, molar entropy, and molar Helmholtz energy of a collection of harmonic oscillators and plot your expressions as a function of T/θ^V, where $\theta^V=h\nu/k$.

3C.3 In biological cells, the energy released by the oxidation of foods is stored in adenosine triphosphate (ATP or ATP^{4-}). The essence of ATP's action is its ability to lose its terminal phosphate group by hydrolysis and to form adenosine diphosphate (ADP or ADP^{3-}):

$$ATP^{4-}(aq)+H_2O(l)\rightarrow ADP^{3-}(aq)+HPO_4^{2-}(aq)+H_3O^+(aq)$$

At pH = 7.0 and 37 °C (310 K, blood temperature) the enthalpy and Gibbs energy of hydrolysis are $\Delta_rH=-20\,kJ\,mol^{-1}$ and $\Delta_rG=-31\,kJ\,mol^{-1}$, respectively. Under these conditions, the hydrolysis of 1 mol $ATP^{4-}(aq)$ results in the extraction of up to 31 kJ of energy that can be used to do non-expansion work, such as the synthesis of proteins from amino acids, muscular contraction, and the activation of neuronal circuits in our brains. (a) Calculate and account for the sign of the entropy of hydrolysis of ATP at pH = 7.0 and 310 K. (b) Suppose that the radius of a typical biological cell is 10 μm and that inside it 1×10^6 ATP molecules are hydrolysed each second. What is the power density of the cell in watts per cubic metre ($1\,W=1\,J\,s^{-1}$)? A computer battery delivers about 15 W and has a volume of $100\,cm^3$. Which has the greater power density, the cell or the battery? (c) The formation of glutamine from glutamate and ammonium ions requires $14.2\,kJ\,mol^{-1}$ of energy input. It is driven by the hydrolysis of ATP to ADP mediated by the enzyme glutamine synthetase. How many moles of ATP must be hydrolysed to form 1 mol glutamine?

TOPIC 3D Combining the First and Second Laws

Discussion questions

3D.1 Suggest a physical interpretation of the dependence of the Gibbs energy on the temperature.

3D.2 Suggest a physical interpretation of the dependence of the Gibbs energy on the pressure.

Exercises

3D.1(a) Suppose that 2.5 mmol $N_2(g)$ occupies $42\,cm^3$ at 300 K and expands isothermally to $600\,cm^3$. Calculate ΔG for the process.

3D.1(b) Suppose that 6.0 mmol Ar(g) occupies $52\,cm^3$ at 298 K and expands isothermally to $122\,cm^3$. Calculate ΔG for the process.

3D.2(a) The change in the Gibbs energy of a certain constant-pressure process was found to fit the expression $\Delta G/J=-85.40+36.5(T/K)$. Calculate the value of ΔS for the process.

3D.2(b) The change in the Gibbs energy of a certain constant-pressure process was found to fit the expression $\Delta G/\text{J} = -73.1 + 42.8(T/\text{K})$. Calculate the value of ΔS for the process.

3D.3(a) Estimate the change in the Gibbs energy and molar Gibbs energy of 1.0 dm^3 of octane when the pressure acting on it is increased from 1.0 atm to 100 atm. The mass density of octane is 0.703 g cm^{-3}.

3D.3(b) Estimate the change in the Gibbs energy and molar Gibbs energy of 100 cm^3 of water when the pressure acting on it is increased from 100 kPa to 500 kPa. The mass density of water is 0.997 g cm^{-3}.

3D.4(a) Calculate the change in the molar Gibbs energy of hydrogen gas when its pressure is increased isothermally from 1.0 atm to 100.0 atm at 298 K.
3D.4(b) Calculate the change in the molar Gibbs energy of oxygen when its pressure is increased isothermally from 50.0 kPa to 100.0 kPa at 500 K.

Problems

3D.1 Calculate $\Delta_r G^\ominus$ (375 K) for the reaction $2\,CO(g) + O_2(g) \rightarrow 2\,CO_2(g)$ from the value of $\Delta_r G^\ominus$ (298 K), $\Delta_r H^\ominus$ (298 K), and the Gibbs–Helmholtz equation.

3D.2 Estimate the standard reaction Gibbs energy of $N_2(g) + 3\,H_2(g) \rightarrow 2\,NH_3(g)$ at (a) 500 K, (b) 1000 K from their values at 298 K.

3D.3 At 298 K the standard enthalpy of combustion of sucrose is −5797 kJ mol^{-1} and the standard Gibbs energy of the reaction is −6333 kJ mol^{-1}. Estimate the additional non-expansion work that may be obtained by raising the temperature to blood temperature, 37 °C.

3D.4 Two empirical equations of state of a real gas are as follows:

$$\text{van der Waals}: p = \frac{RT}{V_m - b} - \frac{a}{V_m^2}$$

$$\text{Dieterici}: p = \frac{RT e^{-a/RTV_m}}{V_m - b}$$

Evaluate $(\partial S/\partial V)_T$ for each gas. For an isothermal expansion, for which kind of gas (also consider a perfect gas) will ΔS be greatest? Explain your conclusion.

3D.5 Two of the four Maxwell relations were derived in the text, but two were not. Complete their derivation by showing that $(\partial S/\partial V)_T = (\partial p/\partial T)_V$ and $(\partial T/\partial p)_S = (\partial V/\partial S)_p$.

3D.6 (a) Use the Maxwell relations to express the derivatives $(\partial S/\partial V)_T$, $(\partial V/\partial S)_p$, $(\partial p/\partial S)_V$, and $(\partial V/\partial S)_p$ in terms of the heat capacities, the expansion coefficient $\alpha = (1/V)(\partial V/\partial T)_p$, and the isothermal compressibility, $\kappa_T = -(1/V)(\partial V/\partial p)_T$. (b) The Joule coefficient, μ_J, is defined as $\mu_J = (\partial T/\partial V)_U$. Show that $\mu_J C_V = p - \alpha T/\kappa_T$.

3D.7 Suppose that S is regarded as a function of p and T. Show that $T\mathrm{d}S = C_p\mathrm{d}T - \alpha TV\mathrm{d}p$. Hence, show that the energy transferred as heat when the pressure on an incompressible liquid or solid is increased by Δp is equal to $-\alpha TV\Delta p$, where $\alpha = (1/V)(\partial V/\partial T)_p$. Evaluate q when the pressure acting on 100 cm^3 of mercury at 0 °C is increased by 1.0 kbar. ($\alpha = 1.82 \times 10^{-4}\,\text{K}^{-1}$.)

3D.8 Equation 3D.6 ($\pi_T = T(\partial p/\partial T)_V - p$) expresses the internal pressure π_T in terms of the pressure and its derivative with respect to temperature. Express π_T in terms of the molecular partition function.

3D.9 Explore the consequences of replacing the equation of state of a perfect gas by the van der Waals equation of state for the pressure-dependence of the molar Gibbs energy. Proceed in three steps. First, consider the case when $a = 0$ and only repulsions are significant. Then consider the case when $b = 0$ and only attractions are significant. For the latter, you should consider making the approximation that the attractions are weak. Finally, explore the full expression by using mathematical software. In each case plot your results graphically and account physically for the deviations from the perfect gas expression.

3D.10‡ Nitric acid hydrates have received much attention as possible catalysts for heterogeneous reactions which bring about the Antarctic ozone hole. Worsnop et al. (*Science* **259**, 71 (1993)) investigated the thermodynamic stability of these hydrates under conditions typical of the polar winter stratosphere. They report thermodynamic data for the sublimation of mono-, di-, and trihydrates to nitric acid and water vapour, $HNO_3 \cdot nH_2O(s) \rightarrow HNO_3(g) + nH_2O(g)$, for $n = 1$, 2, and 3. Given $\Delta_r G^\ominus$ and $\Delta_r H^\ominus$ for these reactions at 220 K, use the Gibbs–Helmholtz equation to compute $\Delta_r G^\ominus$ at 190 K.

n	1	2	3
$\Delta_r G^\ominus/(\text{kJ mol}^{-1})$	46.2	69.4	93.2
$\Delta_r H^\ominus/(\text{kJ mol}^{-1})$	127	188	237

Integrated activities

3.1 A gaseous sample consisting of 1.00 mol molecules is described by the equation of state $pV_m = RT(1 + Bp)$. Initially at 373 K, it undergoes Joule–Thomson expansion from 100 atm to 1.00 atm. Given that $C_{p,m} = \frac{5}{2}R$, $\mu = 0.21\,\text{K atm}^{-1}$, $B = -0.525(\text{K}/T)\,\text{atm}^{-1}$ and that these are constant over the temperature range involved, calculate ΔT and ΔS for the gas.

3.2 Discuss the relationship between the thermodynamic and statistical definitions of entropy.

3.3 Use mathematical software, a spreadsheet, or the *Living graphs* on the web site for this book to:

(a) Evaluate the change in entropy of 1.00 mol CO_2(g) on expansion from 0.001 m^3 to 0.010 m^3 at 298 K, treated as a van der Waals gas.
(b) Allow for the temperature dependence of the heat capacity by writing $C = a + bT + c/T^2$, and plot the change in entropy for different values of the three coefficients (including negative values of c).
(c) Show how the first derivative of G, $(\partial G/\partial p)_T$, varies with pressure, and plot the resulting expression over a pressure range. What is the physical significance of $(\partial G/\partial p)_T$?
(d) Evaluate the fugacity coefficient as a function of the reduced volume of a van der Waals gas and plot the outcome for a selection of reduced temperatures over the range $0.8 \leq V_r \leq 3$.

CHAPTER 4

Physical transformations of pure substances

Vaporization, melting (fusion), and the conversion of graphite to diamond are all examples of changes of phase without change of chemical composition. The discussion of the phase transitions of pure substances is among the simplest applications of thermodynamics to chemistry, and is guided by the principle that the tendency of systems at constant temperature and pressure is to minimize their Gibbs energy.

4A Phase diagrams of pure substances

First, we see that one type of phase diagram is a map of the pressures and temperatures at which each phase of a substance is the most stable. The thermodynamic criterion of phase stability enables us to deduce a very general result, the 'phase rule', which summarizes the constraints on the equilibria between phases. In preparation for later chapters, we express the rule in a general way that can be applied to systems of more than one component. Then, we describe the interpretation of empirically determined phase diagrams for a selection of substances.

4B Thermodynamic aspects of phase transitions

Here we consider the factors that determine the positions and shapes of the boundaries between the regions on a phase diagram. The practical importance of the expressions we derive is that they show how the vapour pressure of a substance varies with temperature and how the melting point varies with pressure. Transitions between phases are classified by noting how various thermodynamic functions change when the transition occurs. This chapter also introduces the 'chemical potential', a property that will be at the centre of our discussions of mixtures and chemical reactions.

What is the impact of this material?

The properties of carbon dioxide in its supercritical fluid phase can form the basis for novel and useful chemical separation methods, and have considerable promise for 'green' chemistry synthetic procedures. Its properties and applications are discussed in *Impact* I4.1.

To read more about the impact of this material, scan the QR code, or go to bcs.whfreeman.com/webpub/chemistry/pchem10e/impact/pchem-4-1.html

4A Phase diagrams of pure substances

Contents

➤ Why do you need to know this material?

Phase diagrams summarize the behaviour of substances under different conditions. In metallurgy, the ability to control the microstructure resulting from phase equilibria makes it possible to tailor the mechanical properties of the materials to a particular application.

➤ What is the key idea?

A pure substance tends to adopt the phase with the lowest chemical potential.

➤ What do you need to know already?

This Topic builds on the fact that the Gibbs energy is a signpost of spontaneous change under conditions of constant temperature and pressure (Topic 3C).

One of the most succinct ways of presenting the physical changes of state that a substance can undergo is in terms of its 'phase diagram'. This material is also the basis of the discussion of mixtures in Chapter 5.

4A.1 The stabilities of phases

Thermodynamics provides a powerful language for describing and understanding the stabilities and transformations of phases, but to apply it we need to employ definitions carefully.

(a) The number of phases

A **phase** is a form of matter that is uniform throughout in chemical composition and physical state. Thus, we speak of solid, liquid, and gas phases of a substance, and of its various solid phases, such as the white and black allotropes of phosphorus or the aragonite and calcite polymorphs of calcium carbonate.

A note on good practice An *allotrope* is a particular molecular form of an element (such as O_2 and O_3) and may be solid, liquid, or gas. A *polymorph* is one of a number of solid phases of an element or compound.

The number of phases in a system is denoted P. A gas, or a gaseous mixture, is a single phase ($P=1$), a crystal of a substance is a single phase, and two fully miscible liquids form a single phase.

Brief illustration 4A.1 The number of phases

A solution of sodium chloride in water is a single phase ($P=1$). Ice is a single phase even though it might be chipped into small fragments. A slurry of ice and water is a two-phase system ($P=2$) even though it is difficult to map the physical boundaries between the phases. A system in which calcium carbonate undergoes the thermal decomposition $CaCO_3(s) \rightarrow CaO(s) + CO_2(g)$ consists of two solid phases (one consisting of calcium carbonate and the other of calcium oxide) and one gaseous phase (consisting of carbon dioxide), so $P=3$.

Self-test 4A.1 How many phases are present in a sealed, half-full vessel containing water?

Answer: 2

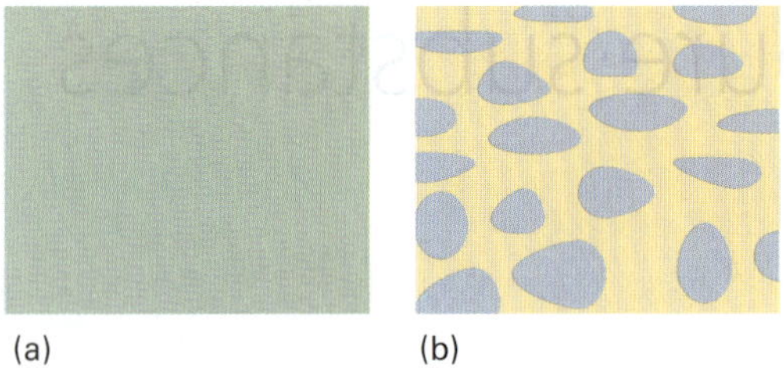

Figure 4A.1 The difference between (a) a single-phase solution, in which the composition is uniform on a microscopic scale, and (b) a dispersion, in which regions of one component are embedded in a matrix of a second component.

Two metals form a two-phase system ($P=2$) if they are immiscible, but a single-phase system ($P=1$), an alloy, if they are miscible. This example shows that it is not always easy to decide whether a system consists of one phase or of two. A solution of solid B in solid A—a homogeneous mixture of the two substances—is uniform on a molecular scale. In a solution, atoms of A are surrounded by atoms of A and B, and any sample cut from the sample, even microscopically small, is representative of the composition of the whole.

A dispersion is uniform on a macroscopic scale but not on a microscopic scale, for it consists of grains or droplets of one substance in a matrix of the other. A small sample could come entirely from one of the minute grains of pure A and would not be representative of the whole (Fig. 4A.1). Dispersions are important because, in many advanced materials (including steels), heat treatment cycles are used to achieve the precipitation of a fine dispersion of particles of one phase (such as a carbide phase) within a matrix formed by a saturated solid solution phase.

(b) Phase transitions

A **phase transition**, the spontaneous conversion of one phase into another phase, occurs at a characteristic temperature for a given pressure. The **transition temperature**, T_{trs}, is the temperature at which the two phases are in equilibrium and the Gibbs energy of the system is minimized at the prevailing pressure.

Brief illustration 4A.2 Phase transitions

At 1 atm, ice is the stable phase of water below 0 °C, but above 0 °C liquid water is more stable. This difference indicates that below 0 °C the Gibbs energy decreases as liquid water changes into ice and that above 0 °C the Gibbs energy decreases as ice changes into liquid water. The numerical values of the Gibbs energies are considered in the next *Brief illustration*.

Self-test 4A.2 Which has the higher standard molar Gibbs energy at 105 °C, liquid water or its vapour?

Answer: Liquid water

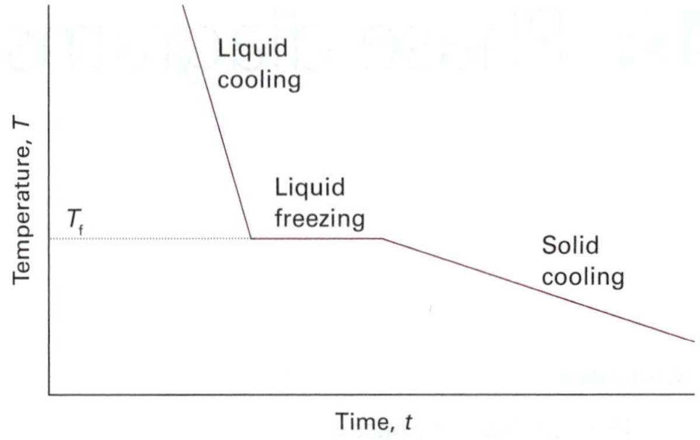

Figure 4A.2 A cooling curve at constant pressure. The flat section corresponds to the pause in the fall of temperature while the first-order exothermic transition (freezing) occurs. This pause enables T_f to be located even if the transition cannot be observed visually.

Detecting a phase transition is not always as simple as seeing water boil in a kettle, so special techniques have been developed. One technique is **thermal analysis**, which takes advantage of the heat that is evolved or absorbed during any transition. The transition is detected by noting that the temperature does not change even though heat is being supplied or removed from the sample (Fig. 4A.2). Differential scanning calorimetry (Topic 2C) is also used. Thermal techniques are useful for solid–solid transitions, where simple visual inspection of the sample may be inadequate. X-ray diffraction (Topic 18A) also reveals the occurrence of a phase transition in a solid, for different structures are found on either side of the transition temperature.

As always, it is important to distinguish between the thermodynamic description of a process and the rate at which the process occurs. A phase transition that is predicted from thermodynamics to be spontaneous may occur too slowly to be significant in practice. For instance, at normal temperatures and pressures the molar Gibbs energy of graphite is lower than that of diamond, so there is a thermodynamic tendency for diamond to change into graphite. However, for this transition to take place, the C atoms must change their locations, which is an immeasurably slow process in a solid except at high temperatures. The discussion of the rate of attainment of equilibrium is a kinetic problem and is outside the range of thermodynamics. In gases and liquids the mobilities of the molecules allow phase transitions to occur rapidly, but in solids thermodynamic instability may be frozen in. Thermodynamically unstable phases that persist because the transition is kinetically hindered are called **metastable phases**. Diamond is a metastable but persistent phase of carbon under normal conditions.

(c) Thermodynamic criteria of phase stability

All our considerations will be based on the Gibbs energy of a substance, and in particular on its molar Gibbs energy, G_m. In

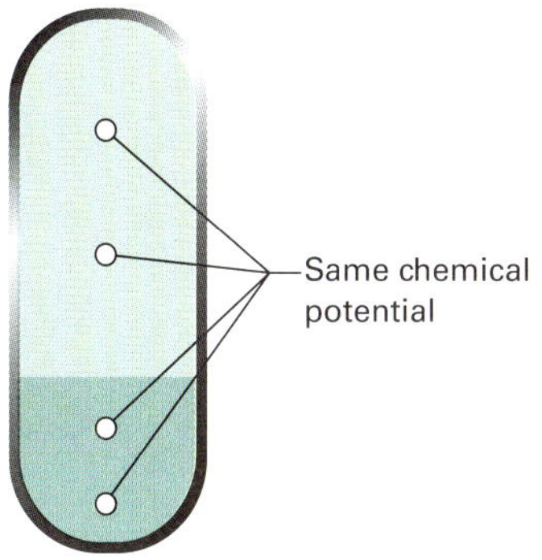

Figure 4A.3 When two or more phases are in equilibrium, the chemical potential of a substance (and, in a mixture, a component) is the same in each phase and is the same at all points in each phase.

fact, this quantity plays such an important role in this chapter and the rest of the text that we give it a special name and symbol, the **chemical potential**, μ (mu). For a one-component system, 'molar Gibbs energy' and 'chemical potential' are synonyms, so $\mu = G_m$, but in Topic 5A we see that chemical potential has a broader significance and a more general definition. The name 'chemical potential' is also instructive: as we develop the concept, we shall see that μ is a measure of the potential that a substance has for undergoing change in a system. In this chapter and Chapter 5, it reflects the potential of a substance to undergo physical change. In Chapter 6, we see that μ is the potential of a substance to undergo chemical change.

We base the entire discussion on the following consequence of the Second Law (Fig. 4A.3):

> At equilibrium, the chemical potential of a substance is the same throughout a sample, regardless of how many phases are present. *Criterion of phase equilibrium*

To see the validity of this remark, consider a system in which the chemical potential of a substance is μ_1 at one location and μ_2 at another location. The locations may be in the same or in different phases. When an infinitesimal amount dn of the substance is transferred from one location to the other, the Gibbs energy of the system changes by $-\mu_1 dn$ when material is removed from location 1, and it changes by $+\mu_2 dn$ when that material is added to location 2. The overall change is therefore $dG = (\mu_2 - \mu_1)dn$. If the chemical potential at location 1 is higher than that at location 2, the transfer is accompanied by a decrease in G, and so has a spontaneous tendency to occur. Only if $\mu_1 = \mu_2$ is there no change in G, and only then is the system at equilibrium.

Brief illustration 4A.3 Gibbs energy and phase transition

The standard molar Gibbs energy of formation of water vapour at 298 K (25 °C) is –229 kJ mol^{-1} and that of liquid water at the same temperature is –237 kJ mol^{-1}. It follows that there is a decrease in Gibbs energy when water vapour condenses to the liquid at 298 K, so condensation is spontaneous at that temperature (and 1 bar).

Self-test 4A.3 The standard Gibbs energies of formation of HN_3 at 298 K are +327 kJ mol^{-1} and +328 kJ mol^{-1} for the liquid and gas phases, respectively. Which phase of hydrogen azide is the more stable at that temperature and 1 bar?

Answer: Liquid

4A.2 Phase boundaries

The **phase diagram** of a pure substance shows the regions of pressure and temperature at which its various phases are thermodynamically stable (Fig. 4A.4). In fact, any two intensive variables may be used (such as temperature and magnetic field; in Topic 5A mole fraction is another variable), but in this Topic we concentrate on pressure and temperature. The lines separating the regions, which are called **phase boundaries** (or *coexistence curves*), show the values of p and T at which two phases coexist in equilibrium and their chemical potentials are equal.

(a) Characteristic properties related to phase transitions

Consider a liquid sample of a pure substance in a closed vessel. The pressure of a vapour in equilibrium with the liquid is called the **vapour pressure** of the substance (Fig. 4A.5). Therefore, the liquid–vapour phase boundary in a phase diagram shows how the vapour pressure of the liquid varies with temperature. Similarly, the solid–vapour phase boundary shows the temperature variation of the **sublimation vapour pressure**, the vapour pressure of the solid phase. The vapour pressure of a substance increases with temperature because at higher temperatures

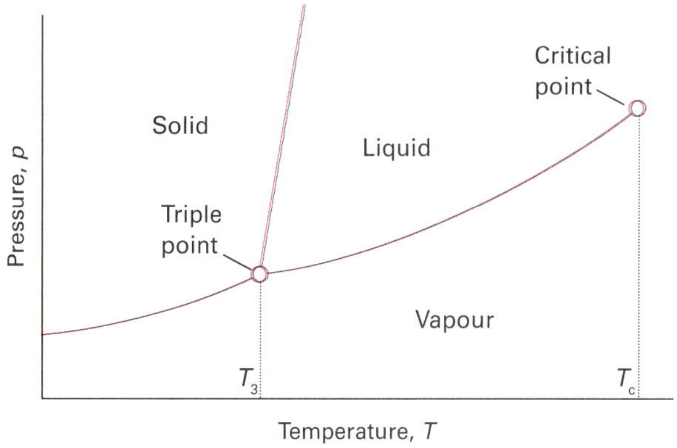

Figure 4A.4 The general regions of pressure and temperature where solid, liquid, or gas is stable (that is, has minimum molar Gibbs energy) are shown on this phase diagram. For example, the solid phase is the most stable phase at low temperatures and high pressures. In the following paragraphs we locate the precise boundaries between the regions.

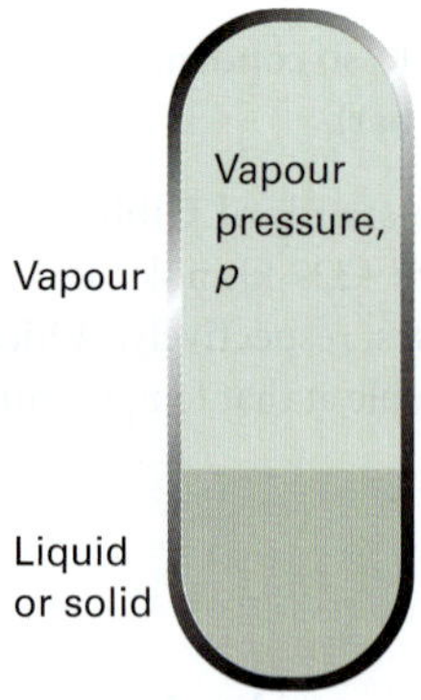

Figure 4A.5 The vapour pressure of a liquid or solid is the pressure exerted by the vapour in equilibrium with the condensed phase.

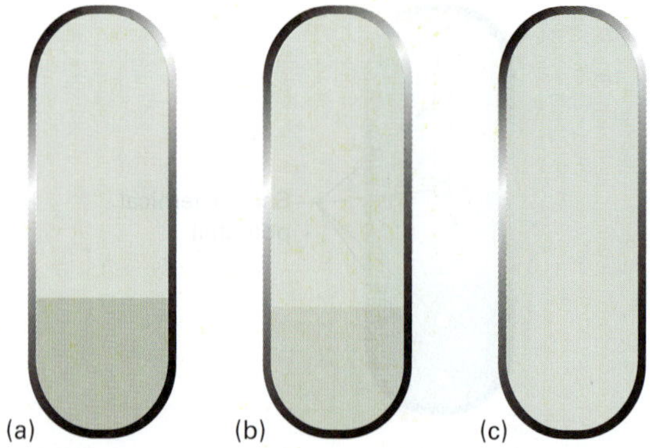

Figure 4A.6 (a) A liquid in equilibrium with its vapour. (b) When a liquid is heated in a sealed container, the density of the vapour phase increases and that of the liquid decreases slightly. There comes a stage (c) at which the two densities are equal and the interface between the fluids disappears. This disappearance occurs at the critical temperature. The container needs to be strong: the critical temperature of water is 374 °C and the vapour pressure is then 218 atm.

more molecules have sufficient energy to escape from their neighbours.

When a liquid is heated in an open vessel, the liquid vaporizes from its surface. When the vapour pressure is equal to the external pressure, vaporization can occur throughout the bulk of the liquid and the vapour can expand freely into the surroundings. The condition of free vaporization throughout the liquid is called **boiling**. The temperature at which the vapour pressure of a liquid is equal to the external pressure is called the **boiling temperature** at that pressure. For the special case of an external pressure of 1 atm, the boiling temperature is called the **normal boiling point**, T_b. With the replacement of 1 atm by 1 bar as standard pressure, there is some advantage in using the **standard boiling point** instead: this is the temperature at which the vapour pressure reaches 1 bar. Because 1 bar is slightly less than 1 atm (1.00 bar = 0.987 atm), the standard boiling point of a liquid is slightly lower than its normal boiling point. The normal boiling point of water is 100.0 °C; its standard boiling point is 99.6 °C. We need to distinguish normal and standard properties only for precise work in thermodynamics because any thermodynamic properties that we intend to add together must refer to the same conditions.

Boiling does not occur when a liquid is heated in a rigid, closed vessel. Instead, the vapour pressure, and hence the density of the vapour, rise as the temperature is raised (Fig. 4A.6). At the same time, the density of the liquid decreases slightly as a result of its expansion. There comes a stage when the density of the vapour is equal to that of the remaining liquid and the surface between the two phases disappears. The temperature at which the surface disappears is the **critical temperature**, T_c, of the substance. The vapour pressure at the critical temperature is called the **critical pressure**, p_c. At and above the critical temperature, a single uniform phase called a **supercritical fluid** fills the container and an interface no longer exists. That is, above the critical temperature, the liquid phase of the substance does not exist.

The temperature at which, under a specified pressure, the liquid and solid phases of a substance coexist in equilibrium is called the **melting temperature**. Because a substance melts at exactly the same temperature as it freezes, the melting temperature of a substance is the same as its **freezing temperature**. The freezing temperature when the pressure is 1 atm is called the **normal freezing point**, T_f, and its freezing point when the pressure is 1 bar is called the **standard freezing point**. The normal and standard freezing points are negligibly different for most purposes. The normal freezing point is also called the **normal melting point**.

There is a set of conditions under which three different phases of a substance (typically solid, liquid, and vapour) all simultaneously coexist in equilibrium. These conditions are represented by the **triple point**, a point at which the three phase boundaries meet. The temperature at the triple point is denoted T_3. The triple point of a pure substance is outside our control: it occurs at a single definite pressure and temperature characteristic of the substance.

As we can see from Fig. 4A.4, the triple point marks the lowest pressure at which a liquid phase of a substance can exist. If (as is common) the slope of the solid–liquid phase boundary is as shown in the diagram, then the triple point also marks the lowest temperature at which the liquid can exist; the critical temperature is the upper limit.

Brief illustration 4A.4 **The triple point**

The triple point of water lies at 273.16 K and 611 Pa (6.11 mbar, 4.58 Torr), and the three phases of water (ice, liquid water, and water vapour) coexist in equilibrium at no other combination of pressure and temperature. This invariance of the triple point was the basis of its use in the about-to-be superseded definition of the Kelvin scale of temperature (Topic 3A).

Self-test 4A.4 How many triple points are present (as far as it is known) in the full phase diagram for water shown later in this Topic in Fig. 4A.9?

Answer: 6

(b) The phase rule

In one of the most elegant arguments of the whole of chemical thermodynamics, which is presented in the following *Justification*, J.W. Gibbs deduced the **phase rule**, which gives the number of parameters that can be varied independently (at least to a small extent) while the number of phases in equilibrium is preserved. The phase rule is a general relation between the variance, F, the number of components, C, and the number of phases at equilibrium, P, for a system of any composition:

$$F = C - P + 2 \quad \text{The phase rule} \quad (4A.1)$$

A **component** is a *chemically independent* constituent of a system. The number of components, C, in a system is the minimum number of types of independent species (ions or molecules) necessary to define the composition of all the phases present in the system. In this chapter we deal only with one-component systems ($C=1$), so for this chapter

$$F = 3 - P \quad \textit{A one-component system} \quad \text{The phase rule} \quad (4A.2)$$

By a **constituent** of a system we mean a chemical species that is present. The **variance** (or *number of degrees of freedom*), F, of a system is the number of intensive variables that can be changed independently without disturbing the number of phases in equilibrium.

Brief illustration 4A.5 The number of components

A mixture of ethanol and water has two constituents. A solution of sodium chloride has three constituents: water, Na^+ ions, and Cl^- ions but only two components because the numbers of Na^+ and Cl^- ions are constrained to be equal by the requirement of charge neutrality.

Self-test 4A.5 How many components are present in an aqueous solution of acetic acid, allowing for its partial deprotonation and the autoprotolysis of water?

Answer: 2

In a single-component, single-phase system ($C=1$, $P=1$), the pressure and temperature may be changed independently without changing the number of phases, so $F=2$. We say that such a system is **bivariant**, or that it has two **degrees of freedom**. On the other hand, if two phases are in equilibrium (a liquid and its vapour, for instance) in a single-component system ($C=1$, $P=2$), the temperature (or the pressure) can be changed at will, but the change in temperature (or pressure) demands an accompanying change in pressure (or temperature) to preserve the number of phases in equilibrium. That is, the variance of the system has fallen to 1.

Justification 4A.1 The phase rule

Consider first the special case of a one-component system for which the phase rule is $F=3-P$. For two phases α and β in equilibrium ($P=2$, $F=1$) at a given pressure and temperature, we can write

$$\mu(\alpha; p,T) = \mu(\beta; p,T)$$

(For instance, when ice and water are in equilibrium, we have $\mu(\mathrm{s}; p,T)=\mu(\mathrm{l}; p,T)$ for H_2O.) This is an equation relating p and T, so only one of these variables is independent (just as the equation $x+y=xy$ is a relation for y in terms of x: $y=x/(x-1)$). That conclusion is consistent with $F=1$. For three phases of a one-component system in mutual equilibrium ($P=3$, $F=0$),

$$\mu(\alpha; p,T) = \mu(\beta; p,T) = \mu(\gamma; p,T)$$

This relation is actually two equations for two unknowns, $\mu(\alpha; p,T)=\mu(\beta; p,T)$ and $\mu(\beta; p,T)=\mu(\gamma; p,T)$, and therefore has a solution only for a single value of p and T (just as the pair of equations $x+y=xy$ and $3x-y=xy$ has the single solution $x=2$ and $y=2$). That conclusion is consistent with $F=0$. Four phases cannot be in mutual equilibrium in a one-component system because the three equalities

$$\mu(\alpha; p,T) = \mu(\beta; p,T)$$
$$\mu(\beta; p,T) = \mu(\gamma; p,T)$$
$$\mu(\gamma; p,T) = \mu(\delta; p,T)$$

are three equations for two unknowns (p and T) and are not consistent (just as $x+y=xy$, $3x-y=xy$, and $x+y=2xy^2$ have no solution).

Now consider the general case. We begin by counting the total number of intensive variables. The pressure, p, and temperature, T, count as 2. We can specify the composition of a phase by giving the mole fractions of $C-1$ components. We need specify only $C-1$ and not all C mole fractions because $x_1+x_2+\cdots+x_C=1$, and all mole fractions are known if all except one are specified. Because there are P phases, the total number of composition variables is $P(C-1)$. At this stage, the total number of intensive variables is $P(C-1)+2$.

At equilibrium, the chemical potential of a component J must be the same in every phase:

$$\mu(\alpha; p,T) = \mu(\beta; p,T) = \cdots \text{ for } P \text{ phases}$$

That is, there are $P-1$ equations of this kind to be satisfied for each component J. As there are C components, the total number of equations is $C(P-1)$. Each equation reduces our freedom to vary one of the $P(C-1)+2$ intensive variables. It follows that the total number of degrees of freedom is

$$F=P(C-1)+2-C(P-1)=C-P+2$$

which is eqn 4A.1.

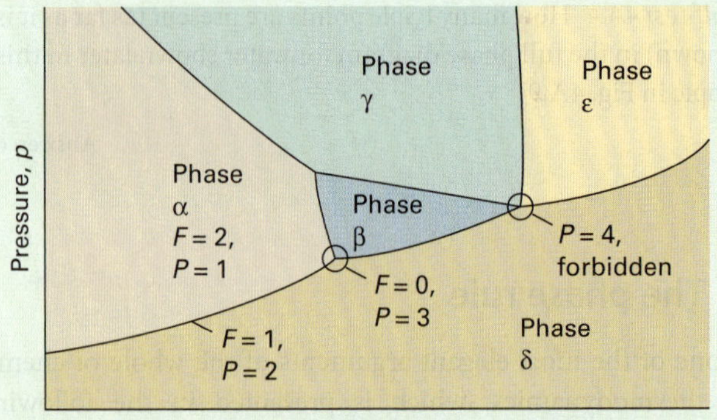

Figure 4A.7 The typical regions of a one-component phase diagram. The lines represent conditions under which the two adjoining phases are in equilibrium. A point represents the unique set of conditions under which three phases coexist in equilibrium. Four phases cannot mutually coexist in equilibrium.

4A.3 Three representative phase diagrams

For a one-component system, such as pure water, $F=3-P$. When only one phase is present, $F=2$ and both p and T can be varied independently (at least over a small range) without changing the number of phases. In other words, a single phase is represented by an *area* on a phase diagram. When two phases are in equilibrium $F=1$, which implies that pressure is not freely variable if the temperature is set; indeed, at a given temperature, a liquid has a characteristic vapour pressure. It follows that the equilibrium of two phases is represented by a *line* in the phase diagram. Instead of selecting the temperature, we could select the pressure, but having done so the two phases would be in equilibrium at a single definite temperature. Therefore, freezing (or any other phase transition) occurs at a definite temperature at a given pressure.

When three phases are in equilibrium, $F=0$ and the system is invariant. This special condition can be established only at a definite temperature and pressure that is characteristic of the substance and outside our control. The equilibrium of three phases is therefore represented by a *point*, the triple point, on a phase diagram. Four phases cannot be in equilibrium in a one-component system because F cannot be negative.

Brief illustration 4A.6 Characteristics of phase diagrams

Figure 4A.7 shows a reasonably typical phase diagram of a single pure substance, with one forbidden feature, the 'quadruple point' where phases β, γ, δ, and ε are said to be in equilibrium. Two triple points are shown (for the equilibria $\alpha \rightleftharpoons \beta \rightleftharpoons \gamma$ and $\alpha \rightleftharpoons \beta \rightleftharpoons \delta$, respectively), corresponding to $P=3$ and $F=0$. The lines represent various equilibria, including $\alpha \rightleftharpoons \beta$, $\alpha \rightleftharpoons \delta$, and $\gamma \rightleftharpoons \varepsilon$.

Self-test 4A.6 What is the minimum number of components necessary before five phases can be in mutual equilibrium in a system?

Answer: 3

(a) Carbon dioxide

The phase diagram for carbon dioxide is shown in Fig. 4A.8. The features to notice include the positive slope (up from left to right) of the solid–liquid boundary; the direction of this line is characteristic of most substances. This slope indicates that the melting temperature of solid carbon dioxide rises as the pressure is increased. Notice also that, as the triple point lies above 1 atm, the liquid cannot exist at normal atmospheric pressures whatever the temperature. As a result, the solid sublimes when left in the open (hence the name 'dry ice'). To obtain the liquid, it is necessary to exert a pressure of at least 5.11 atm. Cylinders of carbon dioxide generally contain the liquid or compressed gas; at 25 °C that implies a vapour pressure of 67 atm if both

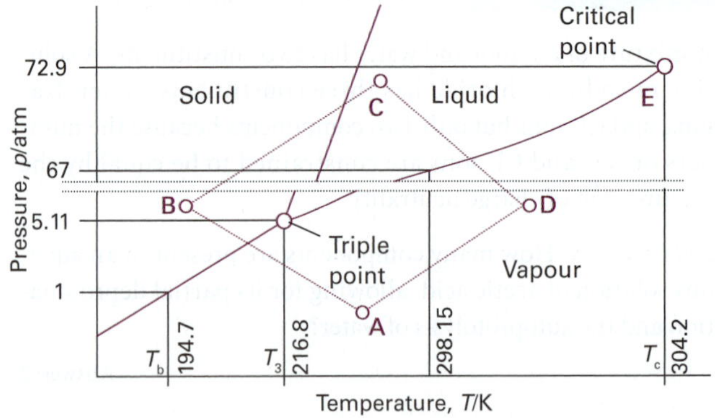

Figure 4A.8 The experimental phase diagram for carbon dioxide. Note that, as the triple point lies at pressures well above atmospheric, liquid carbon dioxide does not exist under normal conditions (a pressure of at least 5.11 atm must be applied). The path ABCD is discussed in *Brief illustration* 4A.7

gas and liquid are present in equilibrium. When the gas squirts through the throttle it cools by the Joule–Thomson effect, so when it emerges into a region where the pressure is only 1 atm, it condenses into a finely divided snow-like solid. That carbon dioxide gas cannot be liquefied except by applying high pressure reflects the weakness of the intermolecular forces between the nonpolar carbon dioxide molecules (Topic 16B).

Brief illustration 4A.7 A phase diagram 1

Consider the path ABCD in Fig. 4A.8. At A the carbon dioxide is a gas. When the temperature and pressure are adjusted to B, the vapour condenses directly to a solid. Increasing the pressure and temperature to C results in the formation of the liquid phase, which evaporates to the vapour when the conditions are changed to D.

Self-test 4A.7 Describe what happens on circulating around the critical point, Path E.

Answer: Liquid → $scCO_2$ → vapour → liquid

(b) Water

Figure 4A.9 shows the phase diagram for water. The liquid–vapour boundary in the phase diagram summarizes how the vapour pressure of liquid water varies with temperature. It also summarizes how the boiling temperature varies with pressure: we simply read off the temperature at which the vapour pressure is equal to the prevailing atmospheric pressure. The solid–liquid boundary shows how the melting temperature varies with the pressure; its very steep slope indicates that enormous pressures are needed to bring about significant changes. Notice that the line has a negative slope (down from left to right) up to 2 kbar, which means that the melting temperature falls as the pressure is raised. The reason for this almost unique behaviour can be traced to the decrease in volume that occurs on melting: it is more favourable for the solid to transform into the liquid as the pressure is raised. The decrease in volume is a result of the very open structure of ice: as shown in Fig. 4A.10, the water molecules are held apart, as well as together, by the hydrogen bonds between them but the hydrogen-bonded structure partially collapses on melting and the liquid is denser than the solid. Other consequences of its extensive hydrogen bonding are the anomalously high boiling point of water for a molecule of its molar mass and its high critical temperature and pressure.

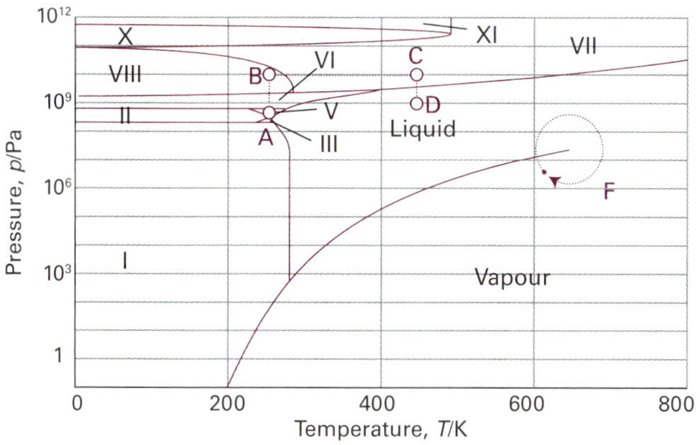

Figure 4A.9 The experimental phase diagram for water showing the different solid phases. The path ABCD is discussed in *Brief illustration* 4A.8.

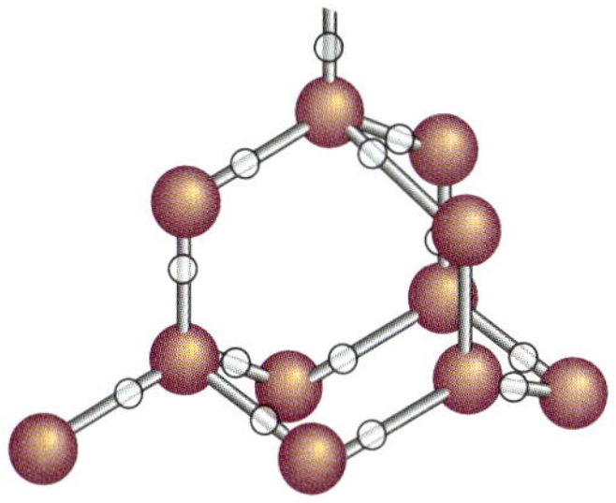

Figure 4A.10 A fragment of the structure of ice (ice-I). Each O atom is linked by two covalent bonds to H atoms and by two hydrogen bonds to a neighbouring O atom, in a tetrahedral array.

Figure 4A.9 shows that water has one liquid phase but many different solid phases other than ordinary ice ('ice I'). Some of these phases melt at high temperatures. Ice VII, for instance, melts at 100 °C but exists only above 25 kbar. Two further phases, Ice XIII and XIV, were identified in 2006 at −160 °C but have not yet been allocated regions in the phase diagram. Note that five more triple points occur in the diagram other than the one where vapour, liquid, and ice I coexist. Each one occurs at a definite pressure and temperature that cannot be changed. The solid phases of ice differ in the arrangement of the water molecules: under the influence of very high pressures, hydrogen bonds buckle and the H_2O molecules adopt different arrangements. These polymorphs of ice may contribute to the advance of glaciers, for ice at the bottom of glaciers experiences very high pressures where it rests on jagged rocks.

Brief illustration 4A.8 A phase diagram 2

Consider the path ABCD in Fig. 4A.9. At A, water is present as ice V. Increasing the pressure to B at the same temperature results in the formation of a polymorph, ice VIII. Heating to C leads to the formation of ice VII, and reduction in pressure to D results in the solid melting to liquid.

Self-test 4A.8 Describe what happens on circulating around the critical point, Path F.

Answer: Vapour → liquid → scH_2O → vapour

(c) Helium

When considering helium at low temperatures it is necessary to distinguish between the isotopes ^{3}He and ^{4}He. Figure 4A.11 shows the phase diagram of helium-4. Helium behaves unusually at low temperatures because the mass of its atoms is so low and their small number of electrons results in them interacting only very weakly with their neighbours. For instance, the solid and gas phases of helium are never in equilibrium however low the temperature: the atoms are so light that they vibrate with a large-amplitude motion even at very low temperatures and the solid simply shakes itself apart. Solid helium can be obtained, but only by holding the atoms together by applying pressure. The isotopes of helium behave differently for quantum mechanical reasons that are explained in Part 2. (The difference stems from the different nuclear spins of the isotopes and the role of the Pauli exclusion principle: helium-4 has $I=0$ and is a boson; helium-3 has $I=\frac{1}{2}$ and is a fermion.)

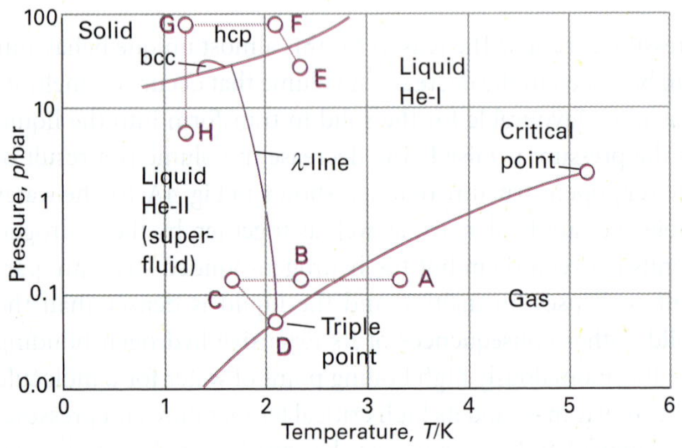

Figure 4A.11 The phase diagram for helium (^{4}He). The λ-line marks the conditions under which the two liquid phases are in equilibrium. Helium-II is the superfluid phase. Note that a pressure of over 20 bar must be exerted before solid helium can be obtained. The labels hcp and bcc denote different solid phases in which the atoms pack together differently: hcp denotes hexagonal closed packing and bcc denotes body-centred cubic (see Topic 18B for a description of these structures). The path ABCD is discussed in *Brief illustration* 4A.9.

Pure helium-4 has two liquid phases. The phase marked He-I in the diagram behaves like a normal liquid; the other phase, He-II, is a **superfluid**; it is so called because it flows without viscosity.[1] Provided we discount the liquid crystalline substances discussed in *Impact* I5.1 on line, helium is the only known substance with a liquid–liquid boundary, shown as the **λ-line** (lambda line) in Fig. 4A.11.

The phase diagram of helium-3 differs from the phase diagram of helium-4, but it also possesses a superfluid phase. Helium-3 is unusual in that melting is exothermic ($\Delta_{fus}H<0$) and therefore (from $\Delta_{fus}S=\Delta_{fus}H/T_f$) at the melting point the entropy of the liquid is lower than that of the solid.

Brief illustration 4A.9 A phase diagram 3

Consider the path ABCD in Fig. 4A.11. At A, helium is present as a vapour. On cooling to B it condenses to helium-I, and further cooling to C results in the formation of helium-II. Adjustment of the pressure and temperature to D results in a system in which three phases, helium-I, helium-II, and vapour, are in mutual equilibrium.

Self-test 4A.9 Describe what happens on the path EFGH.

Answer: He-I → solid → solid → He-II

[1] Water might also have a superfluid liquid phase.

Checklist of concepts

- ☐ 1. A **phase** is a form of matter that is uniform throughout in chemical composition and physical state.
- ☐ 2. A **phase transition** is the spontaneous conversion of one phase into another and may be studied by techniques that include thermal analysis.
- ☐ 3. The thermodynamic analysis of phases is based on the fact that at equilibrium, the chemical potential of a substance is the same throughout a sample.
- ☐ 4. A substance is characterized by a variety of parameters that can be identified on its **phase diagram**.
- ☐ 5. The **phase rule** relates the number of variables that may be changed while the phases of a system remain in mutual equilibrium.
- ☐ 6. Carbon dioxide is a typical substance but shows features that can be traced to its weak intermolecular forces.
- ☐ 7. Water shows anomalies that can be traced to its extensive hydrogen bonding.
- ☐ 8. Helium shows anomalies that can be traced to its low mass and weak interactions.

Checklist of equations

Property	Equation	Comment	Equation number
Chemical potential	$\mu = G_m$	For a pure substance	
Phase rule	$F = C - P + 2$		4A.1

4B Thermodynamic aspects of phase transitions

Contents

➤ Why do you need to know this material?

This Topic illustrates how thermodynamics is used to discuss the equilibria of the phases of one-component systems and shows how to make predictions about the effect of pressure on freezing and boiling points.

➤ What is the key idea?

The effect of temperature and pressure on the chemical potentials of phases in equilibrium is determined by the molar entropy and molar volume, respectively, of the phases.

➤ What do you need to know already?

You need to be aware that phases are in equilibrium when their chemical potentials are equal (Topic 4A) and that the variation of the molar Gibbs energy of a substance depends on its molar volume and entropy (Topic 3D). We draw on expressions for the entropy of transition (Topic 3B) and the perfect gas law (Topic 1A).

As explained in Topic 4A, the thermodynamic criterion of phase equilibrium is the equality of the chemical potentials of each phase. For a one-component system, the chemical potential is the same as the molar Gibbs energy of the phase. As Topic 3D explains how the Gibbs energy varies with temperature and pressure, by combining these two aspects, we can expect to be able to deduce how phase equilibria vary as the conditions are changed.

4B.1 The dependence of stability on the conditions

At very low temperatures and provided the pressure is not too low, the solid phase of a substance has the lowest chemical potential and is therefore the most stable phase. However, the chemical potentials of different phases change with temperature in different ways, and above a certain temperature the chemical potential of another phase (perhaps another solid phase, a liquid, or a gas) may turn out to be the lowest. When that happens, a transition to the second phase is spontaneous and occurs if it is kinetically feasible to do so.

(a) The temperature dependence of phase stability

The temperature dependence of the Gibbs energy is expressed in terms of the entropy of the system by eqn 3D.8 ($(\partial G/\partial T)_p = -S$). Because the chemical potential of a pure substance is just another name for its molar Gibbs energy, it follows that

$$\left(\frac{\partial \mu}{\partial T}\right)_p = -S_m \quad \text{Variation of chemical potential with } T \quad (4B.1)$$

This relation shows that, as the temperature is raised, the chemical potential of a pure substance decreases: $S_m > 0$ for all substances, so the slope of a plot of μ against T is negative.

Equation 4B.1 implies that because $S_m(g) > S_m(l)$ the slope of a plot of μ against temperature is steeper for gases than for liquids. Because $S_m(l) > S_m(s)$ almost always, the slope is also steeper for a liquid than the corresponding solid. These features are illustrated in Fig. 4B.1. The steep negative slope of $\mu(l)$ results in it falling below $\mu(s)$ when the temperature is high enough, and then the liquid becomes the stable phase: the solid melts. The chemical potential of the gas phase plunges steeply downwards as the temperature is raised (because the molar entropy of the vapour is so high), and there comes a temperature at which it lies lowest. Then the gas is the stable phase and vaporization is spontaneous.

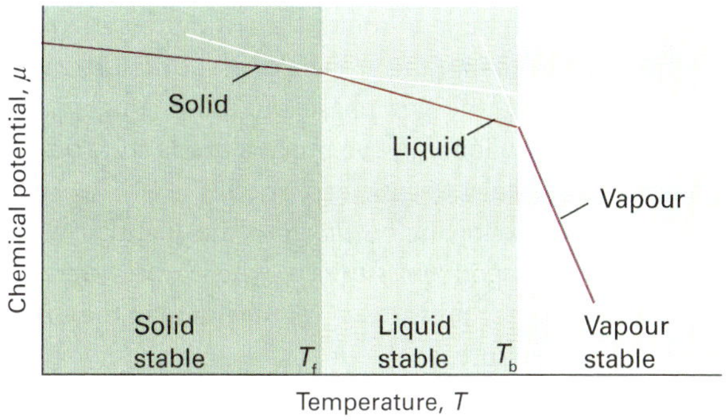

Figure 4B.1 The schematic temperature dependence of the chemical potential of the solid, liquid, and gas phases of a substance (in practice, the lines are curved). The phase with the lowest chemical potential at a specified temperature is the most stable one at that temperature. The transition temperatures, the melting and boiling temperatures (T_f and T_b, respectively), are the temperatures at which the chemical potentials of the two phases are equal.

Brief illustration 4B.1 The temperature variation of μ

The standard molar entropy of liquid water at 100 °C is 86.8 J K^{-1} mol^{-1} and that of water vapour at the same temperature is 195.98 J K^{-1} mol^{-1}. It follows that when the temperature is raised by 1.0 K the changes in chemical potential are

$$\delta\mu(l) \approx S_m(l)\delta T = 87\,\text{J mol}^{-1} \qquad \delta\mu(g) \approx S_m(g)\delta T = 196\,\text{J mol}^{-1}$$

At 100 °C the two phases are in equilibrium with equal chemical potentials, so at 1.0 K higher the chemical potential of the vapour is lower (by 109 J mol^{-1}) than that of the liquid and vaporization is spontaneous.

Self-test 4B.1 The standard molar entropy of liquid water at 0 °C is 65 J K^{-1} mol^{-1} and that of ice at the same temperature is 43 J K^{-1} mol^{-1}. What is the effect of increasing the temperature by 1.0 K?

Answer: $\delta\mu(l) \approx -65$ J mol^{-1}, $\delta\mu(s) \approx -43$ J mol^{-1}; ice melts

(b) The response of melting to applied pressure

Most substances melt at a higher temperature when subjected to pressure. It is as though the pressure is preventing the formation of the less dense liquid phase. Exceptions to this behaviour include water, for which the liquid is denser than the solid. Application of pressure to water encourages the formation of the liquid phase. That is, water freezes and ice melts at a lower temperature when it is under pressure.

We can rationalize the response of melting temperatures to pressure as follows. The variation of the chemical potential with pressure is expressed (from the second of eqns 3D.8, $(\partial G/\partial p)_T = V$) by

$$\left(\frac{\partial \mu}{\partial p}\right)_T = V_m \quad \text{Variation of chemical potential with } p \quad (4B.2)$$

This equation shows that the slope of a plot of chemical potential against pressure is equal to the molar volume of the substance. An increase in pressure raises the chemical potential of any pure substance (because $V_m > 0$). In most cases, $V_m(l) > V_m(s)$ and the equation predicts that an increase in pressure increases the chemical potential of the liquid more than that of the solid. As shown in Fig. 4B.2a, the effect of pressure in such a case is to raise the melting temperature slightly. For water, however, $V_m(l) < V_m(s)$, and an increase in pressure increases the chemical potential of the solid more than that of the liquid. In this case, the melting temperature is lowered slightly (Fig. 4B.2b).

Example 4B.1 Assessing the effect of pressure on the chemical potential

Calculate the effect on the chemical potentials of ice and water of increasing the pressure from 1.00 bar to 2.00 bar at 0 °C. The density of ice is 0.917 g cm^{-3} and that of liquid water is 0.999 g cm^{-3} under these conditions.

Method From eqn 4B.2 in the form $d\mu = V_m dp$, we know that the change in chemical potential of an incompressible substance when the pressure is changed by Δp is $\Delta\mu = V_m\Delta p$. Therefore, to answer the question, we need to know the molar volumes of the two phases of water. These values are obtained from the mass density, ρ, and the molar mass, M, by using $V_m = M/\rho$. We therefore use the expression $\Delta\mu = M\Delta p/\rho$.

Answer The molar mass of water is 18.02 g mol⁻¹ (1.802×10^{-2} kg mol⁻¹); therefore,

$$\Delta\mu(\text{ice}) = \frac{(1.802\times10^{-2}\,\text{kg mol}^{-1})\times(1.00\times10^5\,\text{Pa})}{917\,\text{kg m}^{-3}} = +1.97\,\text{J mol}^{-1}$$

$$\Delta\mu(\text{water}) = \frac{(1.802\times10^{-2}\,\text{kg mol}^{-1})\times(1.00\times10^5\,\text{Pa})}{999\,\text{kg m}^{-3}}$$

$$= +1.80\,\text{J mol}^{-1}$$

We interpret the numerical results as follows: the chemical potential of ice rises more sharply than that of water, so if they are initially in equilibrium at 1 bar, then there will be a tendency for the ice to melt at 2 bar.

Self-test 4B.2 Calculate the effect of an increase in pressure of 1.00 bar on the liquid and solid phases of carbon dioxide (molar mass 44.0 g mol⁻¹) in equilibrium with densities 2.35 g cm⁻³ and 2.50 g cm⁻³, respectively.

Answer: $\Delta\mu(\text{l}) = +1.87$ J mol⁻¹, $\Delta\mu(\text{s}) = +1.76$ J mol⁻¹; solid forms

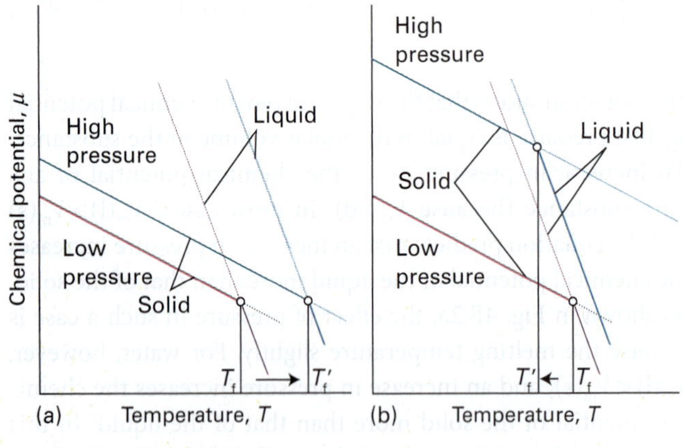

Figure 4B.2 The pressure dependence of the chemical potential of a substance depends on the molar volume of the phase. The lines show schematically the effect of increasing pressure on the chemical potential of the solid and liquid phases (in practice, the lines are curved), and the corresponding effects on the freezing temperatures. (a) In this case the molar volume of the solid is smaller than that of the liquid and μ(s) increases less than μ(l). As a result, the freezing temperature rises. (b) Here the molar volume is greater for the solid than the liquid (as for water), μ(s) increases more strongly than μ(l), and the freezing temperature is lowered.

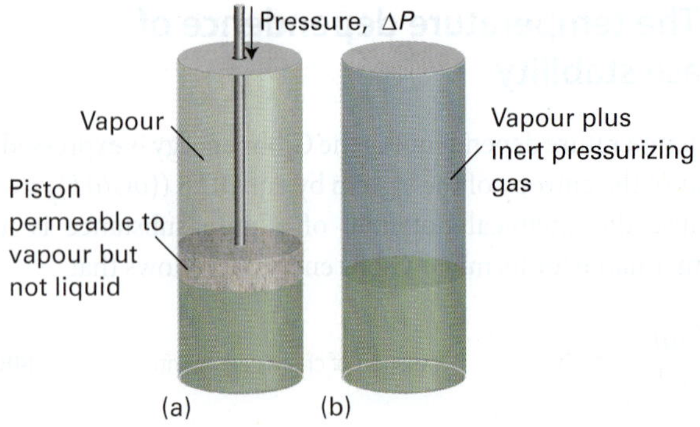

Figure 4B.3 Pressure may be applied to a condensed phase either (a) by compressing the condensed phase or (b) by subjecting it to an inert pressurizing gas. When pressure is applied, the vapour pressure of the condensed phase increases.

(c) The vapour pressure of a liquid subjected to pressure

When pressure is applied to a condensed phase, its vapour pressure rises: in effect, molecules are squeezed out of the phase and escape as a gas. Pressure can be exerted on the condensed phase mechanically or by subjecting it to the applied pressure of an inert gas (Fig. 4B.3). In the latter case, the vapour pressure is the partial pressure of the vapour in equilibrium with the condensed phase. We then speak of the **partial vapour pressure** of the substance. One complication (which we ignore here) is that, if the condensed phase is a liquid, then the pressurizing gas might dissolve and change the properties of the liquid. Another complication is that the gas phase molecules might attract molecules out of the liquid by the process of **gas solvation**, the attachment of molecules to gas-phase species.

As shown in the following *Justification*, the quantitative relation between the vapour pressure, p, when a pressure ΔP is applied and the vapour pressure, p^*, of the liquid in the absence of an additional pressure is

$$p = p^* e^{V_m(\text{l})\Delta P/RT} \quad \text{Effect of applied pressure } \Delta P \text{ on vapour pressure } p \quad (4B.3)$$

This equation shows how the vapour pressure increases when the pressure acting on the condensed phase is increased.

Justification 4B.1 The vapour pressure of a pressurized liquid

We calculate the vapour pressure of a pressurized liquid by using the fact that at equilibrium the chemical potentials of the liquid and its vapour are equal: $\mu(\text{l}) = \mu(\text{g})$. It follows that, for any change that preserves equilibrium, the resulting change in $\mu(\text{l})$ must be equal to the change in $\mu(\text{g})$; therefore, we can write $d\mu(\text{g}) = d\mu(\text{l})$. When the pressure P on the liquid is increased by dP, the chemical potential of the liquid changes by $d\mu(\text{l}) = V_m(\text{l})dP$. The chemical potential of the vapour changes by $d\mu(g) = V_m(g)dp$ where dp is the change in the vapour pressure we are trying to find. If we treat the vapour as a perfect gas, the molar volume can be replaced by $V_m(g) = RT/p$, and we obtain $d\mu(g) = RTdp/p$. Next, we equate the changes in chemical potentials of the vapour and the liquid:

$$\frac{RTdp}{p} = V_m(l)dP$$

We can integrate this expression once we know the limits of integration.

When there is no additional pressure acting on the liquid, P (the pressure experienced by the liquid) is equal to the normal vapour pressure p^*, so when $P = p^*$, $p = p^*$ too. When there is an additional pressure ΔP on the liquid, with the result that $P = p + \Delta P$, the vapour pressure is p (the value we want to find). Provided the effect of pressure on the vapour pressure is small (as will turn out to be the case) a good approximation is to replace the p in $p + \Delta P$ by p^* itself, and to set the upper limit of the integral to $p^* + \Delta P$. The integrations required are therefore as follows:

$$RT\int_{p^*}^{p}\frac{dp}{p} = \int_{p^*}^{p^*+\Delta P} V_m(l)\,dP$$

We now divide both sides by RT and assume that the molar volume of the liquid is the same throughout the small range of pressures involved:

$$\int_{p^*}^{p}\frac{dp}{p} = \frac{1}{RT}\int_{p^*}^{p^*+\Delta P} V_m(l)\,dP = \frac{V_m(l)}{RT}\int_{p^*}^{p^*+\Delta P} dP$$

Then both integrations are straightforward, and lead to

$$\ln\frac{p}{p^*} = \frac{V_m(l)}{RT}\Delta P$$

which rearranges to eqn 4B.3 because $e^{\ln x} = x$.

Brief illustration 4B.2 The effect of pressurization

For water, which has density 0.997 g cm^{-3} at 25 °C and therefore molar volume 18.1 cm^3 mol^{-1}, when the pressure is increased by 10 bar (that is, $\Delta P = 1.0\times10^6$ Pa)

$$\frac{V_m(l)\Delta P}{RT} = \frac{(1.81\times10^{-5}\,\mathrm{m^3\,mol^{-1}})\times(1.0\times10^6\,\mathrm{Pa})}{(8.3145\,\mathrm{J\,K^{-1}\,mol^{-1}})\times(298\,\mathrm{K})} = \frac{1.81\times1.0\times10^1}{8.3145\times298} = 0.0073\ldots$$

where we have used 1 J = 1 Pa m^3. It follows that $p = 1.0073p^*$, an increase of 0.73 per cent.

Self-test 4B.3 Calculate the effect of an increase in pressure of 100 bar on the vapour pressure of benzene at 25 °C, which has density 0.879 g cm^{-3}.

Answer: 43 per cent increase

4B.2 The location of phase boundaries

The precise locations of the phase boundaries—the pressures and temperatures at which two phases can coexist—can be found by making use of the fact that, when two phases are in equilibrium, their chemical potentials must be equal. Therefore, where the phases α and β are in equilibrium,

$$\mu(\alpha;p,T) = \mu(\beta;p,T) \qquad (4B.4)$$

By solving this equation for p in terms of T, we get an equation for the phase boundary.

(a) The slopes of the phase boundaries

It turns out to be simplest to discuss the phase boundaries in terms of their slopes, dp/dT. Let p and T be changed infinitesimally, but in such a way that the two phases α and β remain in equilibrium. The chemical potentials of the phases are initially equal (the two phases are in equilibrium). They remain equal when the conditions are changed to another point on the phase boundary, where the two phases continue to be in equilibrium (Fig. 4B.4). Therefore, the changes in the chemical potentials of the two phases must be equal and we can write $d\mu(\alpha) = d\mu(\beta)$. Because, from eqn 3D.7 ($dG = Vdp - SdT$), we know that $d\mu = -S_m dT + V_m dp$ for each phase, it follows that

$$-S_m(\alpha)dT + V_m(\alpha)dp = -S_m(\beta)dT + V_m(\beta)dp$$

where $S_m(\alpha)$ and $S_m(\beta)$ are the molar entropies of the phases and $V_m(\alpha)$ and $V_m(\beta)$ are their molar volumes. Hence

$$\{S_m(\beta) - S_m(\alpha)\}dT = \{V_m(\beta) - V_m(\alpha)\}dp$$

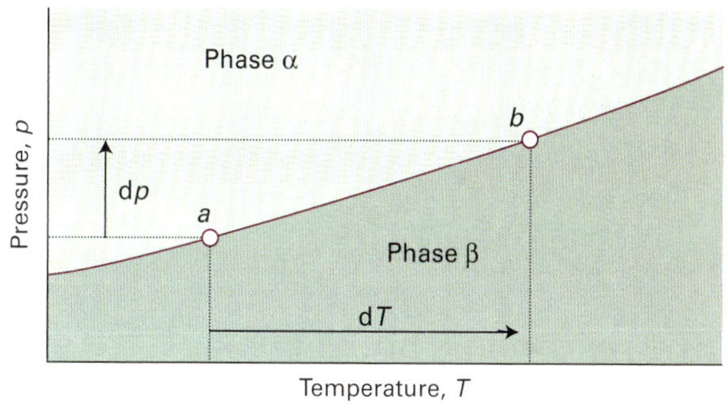

Figure 4B.4 When pressure is applied to a system in which two phases are in equilibrium (at *a*), the equilibrium is disturbed. It can be restored by changing the temperature, so moving the state of the system to *b*. It follows that there is a relation between d*p* and d*T* that ensures that the system remains in equilibrium as either variable is changed.

Then, with $\Delta_{trs}S = S_m(\beta) - S_m(\alpha)$ and $\Delta_{trs}V = V_m(\beta) - V_m(\alpha)$, which are the (molar) entropy and volume of transition, respectively,

$$\Delta_{trs}S\mathrm{d}T = \Delta_{trs}V\mathrm{d}p$$

This relation rearranges into the **Clapeyron equation**:

$$\frac{\mathrm{d}p}{\mathrm{d}T} = \frac{\Delta_{trs}S}{\Delta_{trs}V} \quad \text{Clapeyron equation} \quad (4B.5a)$$

The Clapeyron equation is an exact expression for the slope of the tangent to the boundary at any point and applies to any phase equilibrium of any pure substance. It implies that we can use thermodynamic data to predict the appearance of phase diagrams and to understand their form. A more practical application is to the prediction of the response of freezing and boiling points to the application of pressure, when it can be used in the form obtained by inverting both sides:

$$\frac{\mathrm{d}T}{\mathrm{d}p} = \frac{\Delta_{trs}V}{\Delta_{trs}S} \quad (4B.5b)$$

Brief illustration 4B.3 The Clapeyron equation

The standard volume and entropy of transition of water from ice to liquid are $-1.6\,\mathrm{cm^3\,mol^{-1}}$ and $+22\,\mathrm{J\,K^{-1}\,mol^{-1}}$, respectively, at 0 °C. The slope of the solid–liquid phase boundary at that temperature is therefore

$$\frac{\mathrm{d}T}{\mathrm{d}p} = \frac{-1.6\times10^{-6}\,\mathrm{m^3\,mol^{-1}}}{22\,\mathrm{J^{-1}\,mol^{-1}}} = -7.3\times10^{-8}\frac{\mathrm{K}}{\mathrm{J\,m^{-3}}} = -7.3\times10^{-8}\,\mathrm{K\,Pa^{-1}}$$

which corresponds to $-7.3\,\mathrm{mK\,bar^{-1}}$. An increase of 100 bar therefore results in a lowering of the freezing point of water by 0.73 K.

Self-test 4B.4 The standard volume and entropy of transition of water from liquid to vapour are $+30\,\mathrm{dm^3\,mol^{-1}}$ and $+109\,\mathrm{J\,K^{-1}\,mol^{-1}}$, respectively, at 100 °C. By how much does the boiling temperature change when the pressure is reduced from 1.0 bar to 0.80 bar?

Answer: −5.5 K

(b) The solid–liquid boundary

Melting (fusion) is accompanied by a molar enthalpy change $\Delta_{fus}H$ and occurs at a temperature T. The molar entropy of melting at T is therefore $\Delta_{fus}H/T$ (Topic 3B), and the Clapeyron equation becomes

$$\frac{\mathrm{d}p}{\mathrm{d}T} = \frac{\Delta_{fus}H}{T\Delta_{fus}V} \quad \text{Slope of solid–liquid boundary} \quad (4B.6)$$

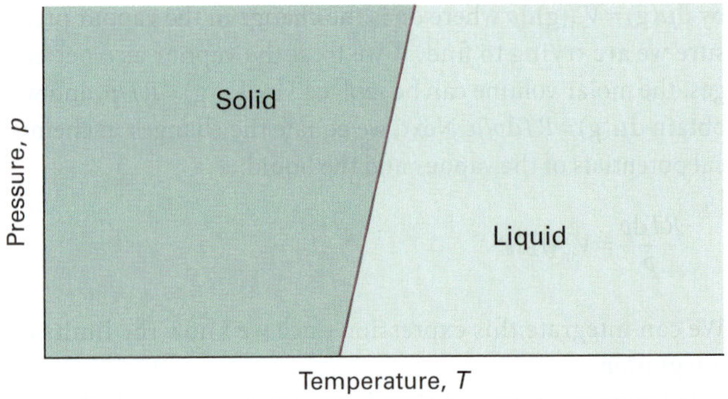

Figure 4B.5 A typical solid–liquid phase boundary slopes steeply upwards. This slope implies that, as the pressure is raised, the melting temperature rises. Most substances behave in this way.

where $\Delta_{fus}V$ is the change in molar volume that occurs on melting. The enthalpy of melting is positive (the only exception is helium-3) and the volume change is usually positive and always small. Consequently, the slope $\mathrm{d}p/\mathrm{d}T$ is steep and usually positive (Fig. 4B.5).

We can obtain the formula for the phase boundary by integrating $\mathrm{d}p/\mathrm{d}T$, assuming that $\Delta_{fus}H$ and $\Delta_{fus}V$ change so little with temperature and pressure that they can be treated as constant. If the melting temperature is T^* when the pressure is p^*, and T when the pressure is p, the integration required is

$$\int_{p^*}^{p}\mathrm{d}p = \frac{\Delta_{fus}H}{\Delta_{fus}V}\int_{T^*}^{T}\frac{\mathrm{d}T}{T}$$

Therefore, the approximate equation of the solid–liquid boundary is

$$p = p^* + \frac{\Delta_{fus}H}{\Delta_{fus}V}\ln\frac{T}{T^*} \quad (4B.7)$$

This equation was originally obtained by yet another Thomson—James, the brother of William, Lord Kelvin. When T is close to T^*, the logarithm can be approximated by using

$$\ln\frac{T}{T^*} = \ln\left(1 + \frac{T-T^*}{T^*}\right) \approx \frac{T-T^*}{T^*}$$

where we have used the expansion $\ln(1+x) = x - \tfrac{1}{2}x^2 + \cdots$ (*Mathematical background* 1) and neglected all but the leading term; therefore

$$p \approx p^* + \frac{\Delta_{fus}H}{T^*\Delta_{fus}V}(T-T^*) \quad (4B.8)$$

This expression is the equation of a steep straight line when p is plotted against T (as in Fig. 4B.5).

Brief illustration 4B.4 The solid–liquid boundary

The enthalpy of fusion of ice at 0 °C and 1 bar (273 K) is 6.008 kJ mol^{-1} and the volume of fusion is −1.6 cm^3 mol^{-1}. It follows that the solid–liquid phase boundary is given by the equation

$$p \approx 1\,\text{bar} + \frac{6.008\times10^3\,\text{J mol}^{-1}}{(273\,\text{K})\times(-1.6\times10^{-6}\,\text{m}^3\,\text{mol}^{-1})}(T-T^*)$$
$$\approx 1\,\text{bar} - 1.4\times10^7\,\text{Pa K}^{-1}(T-T^*)$$

That is,

$$p/\text{bar} = 1 - 140(T-T^*)/\text{K}$$

with $T^* = 273$ K. This expression is plotted in Fig. 4B.6.

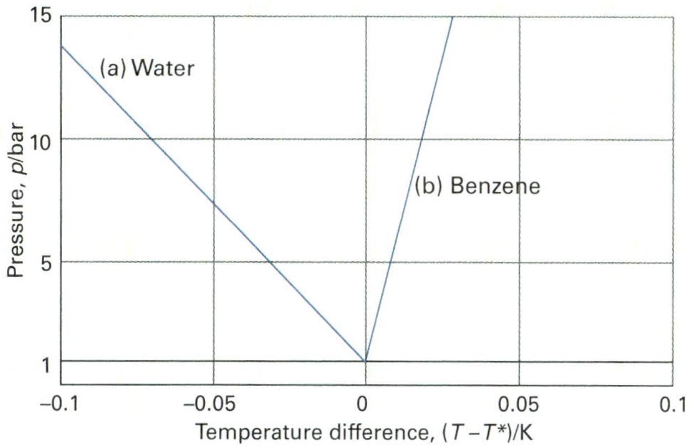

Figure 4B.6 The solid–liquid phase boundaries (the melting point curves) for water and benzene, as calculated in *Brief illustration* 4B.4.

Self-test 4B.5 The enthalpy of fusion of benzene is 10.59 kJ mol^{-1} at its melting point of 279 K and its volume of fusion is close to +0.50 cm^3 mol^{-1} (an estimated value). What is the equation of its solid–liquid phase boundary?

Answer: $p/\text{bar} = 1 + 760(T - T^*)$, as in Fig. 4B.6

(c) The liquid–vapour boundary

The entropy of vaporization at a temperature T is equal to $\Delta_{\text{vap}}H/T$; the Clapeyron equation for the liquid–vapour boundary is therefore

$$\frac{dp}{dT} = \frac{\Delta_{\text{vap}}H}{T\Delta_{\text{vap}}V} \quad \text{Slope of liquid–vapour boundary} \quad (4B.9)$$

The enthalpy of vaporization is positive; $\Delta_{\text{vap}}V$ is large and positive. Therefore, dp/dT is positive, but it is much smaller than for the solid–liquid boundary. It follows that dT/dp is large, and hence that the boiling temperature is more responsive to pressure than the freezing temperature.

Example 4B.2 Estimating the effect of pressure on the boiling temperature

Estimate the typical size of the effect of increasing pressure on the boiling point of a liquid.

Method To use eqn 4B.9 we need to estimate the right-hand side. At the boiling point, the term $\Delta_{\text{vap}}H/T$ is Trouton's constant (Topic 3B). Because the molar volume of a gas is so much greater than the molar volume of a liquid, we can write $\Delta_{\text{vap}}V = V_{\text{m}}(\text{g}) - V_{\text{m}}(\text{l}) \approx V_{\text{m}}(\text{g})$ and take for $V_{\text{m}}(\text{g})$ the molar volume of a perfect gas (at low pressures, at least).

Answer Trouton's constant has the value 85 J K^{-1} mol^{-1}. The molar volume of a perfect gas is about 25 dm^3 mol^{-1} at 1 atm and near but above room temperature. Therefore,

$$\frac{dp}{dT} \approx \frac{85\,\text{J K}^{-1}\,\text{mol}^{-1}}{2.5\times10^{-2}\,\text{m}^3\,\text{mol}^{-1}} = 3.4\times10^3\,\text{Pa K}^{-1}$$

We have used 1 J = 1 Pa m^3. This value corresponds to 0.034 atm K^{-1} and hence to $dT/dp = 29$ K atm^{-1}. Therefore, a change of pressure of +0.1 atm can be expected to change a boiling temperature by about +3 K.

Self-test 4B.6 Estimate dT/dp for water at its normal boiling point using the information in Table 3A.2 and $V_{\text{m}}(\text{g}) = RT/p$.

Answer: 28 K atm^{-1}

Because the molar volume of a gas is so much greater than the molar volume of a liquid, we can write $\Delta_{\text{vap}}V \approx V_{\text{m}}(\text{g})$ (as in *Example* 4B.2). Moreover, if the gas behaves perfectly, $V_{\text{m}}(\text{g}) = RT/p$. These two approximations turn the exact Clapeyron equation into

$$\frac{dp}{dT} = \frac{\Delta_{\text{vap}}H}{T(RT/p)} = \frac{p\Delta_{\text{vap}}H}{RT^2}$$

which, by using $dx/x = d\ln x$, rearranges into the **Clausius–Clapeyron equation** for the variation of vapour pressure with temperature:

$$\frac{d\ln p}{dT} = \frac{\Delta_{\text{vap}}H}{RT^2} \quad \textit{Vapour is a perfect gas} \quad \text{Clausius–Clapeyron equation} \quad (4B.10)$$

Like the Clapeyron equation (which is exact), the Clausius–Clapeyron equation (which is an approximation) is important for understanding the appearance of phase diagrams, particularly the location and shape of the liquid–vapour and solid–vapour phase boundaries. It lets us predict how the vapour pressure varies with temperature and how the boiling temperature varies with pressure. For instance, if we also assume that the enthalpy of vaporization is independent of temperature, eqn 4B.10 can be integrated as follows:

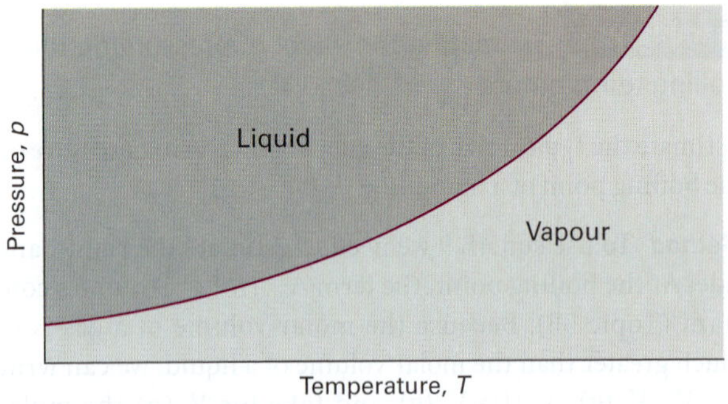

Figure 4B.7 A typical liquid–vapour phase boundary. The boundary can be regarded as a plot of the vapour pressure against the temperature. Note that, in some depictions of phase diagrams in which a logarithmic pressure scale is used, the phase boundary has the opposite curvature (see Fig. 4B.8). This phase boundary terminates at the critical point (not shown).

$$\int_{\ln p^*}^{\ln p} \mathrm{d}\ln p = \frac{\Delta_{\mathrm{vap}}H}{R}\int_{T^*}^{T}\frac{\mathrm{d}T}{T^2} = -\frac{\Delta_{\mathrm{vap}}H}{R}\left(\frac{1}{T}-\frac{1}{T^*}\right)$$

where p^* is the vapour pressure when the temperature is T^* and p the vapour pressure when the temperature is T. Therefore, because the integral on the left evaluates to $\ln(p/p^*)$, the two vapour pressures are related by

$$p = p^*\mathrm{e}^{-\chi} \qquad \chi = \frac{\Delta_{\mathrm{vap}}H}{R}\left(\frac{1}{T}-\frac{1}{T^*}\right) \qquad (4B.11)$$

Equation 4B.11 is plotted as the liquid–vapour boundary in Fig. 4B.7. The line does not extend beyond the critical temperature T_c, because above this temperature the liquid does not exist.

Brief illustration 4B.5 The Clausius–Clapeyron equation

Equation 4B.11 can be used to estimate the vapour pressure of a liquid at any temperature from its normal boiling point, the temperature at which the vapour pressure is 1.00 atm (101 kPa). The normal boiling point of benzene is 80 °C (353 K) and (from Table 3A.2), $\Delta_{\mathrm{vap}}H^{\ominus} = 30.8\,\mathrm{kJ\,mol^{-1}}$. Therefore, to calculate the vapour pressure at 20 °C (293 K), we write

$$\chi = \frac{3.08\times10^4\,\mathrm{J\,mol^{-1}}}{8.3145\,\mathrm{J\,K^{-1}\,mol^{-1}}}\left(\frac{1}{293\,\mathrm{K}}-\frac{1}{353\,\mathrm{K}}\right) = 2.14\ldots$$

and substitute this value into eqn 4B.11 with $p^* = 101\,\mathrm{kPa}$. The result is 12 kPa. The experimental value is 10 kPa.

A note on good practice Because exponential functions are so sensitive, it is good practice to carry out numerical calculations like this without evaluating the intermediate steps and using rounded values.

(d) The solid–vapour boundary

The only difference between this case and the last is the replacement of the enthalpy of vaporization by the enthalpy of sublimation, $\Delta_{\mathrm{sub}}H$. Because the enthalpy of sublimation is greater than the enthalpy of vaporization (recall that $\Delta_{\mathrm{sub}}H = \Delta_{\mathrm{fus}}H + \Delta_{\mathrm{vap}}H$), the equation predicts a steeper slope for the sublimation curve than for the vaporization curve at similar temperatures, which is near where they meet at the triple point (Fig. 4B.8).

Brief illustration 4B.6 The solid–vapour boundary

The enthalpy of fusion of ice at the triple point of water (6.1 mbar, 273 K) is negligibly different from its standard enthalpy of fusion at its freezing point, which is 6.008 kJ mol^{-1}. The enthalpy of vaporization at that temperature is 45.0 kJ mol^{-1} (once again, ignoring differences due to the pressure not being 1 bar). The enthalpy of sublimation is therefore 51.0 kJ mol^{-1}. Therefore, the equations for the slopes of (a) the liquid–vapour and (b) the solid–vapour phase boundaries at the triple point are

$$\text{(a)}\ \frac{\mathrm{d}\ln p}{\mathrm{d}T} = \frac{45.0\times10^3\,\mathrm{J\,mol^{-1}}}{(8.3145\,\mathrm{J\,K^{-1}\,mol^{-1}})\times(273\,\mathrm{K})^2} = 0.0726\,\mathrm{K^{-1}}$$

$$\text{(b)}\ \frac{\mathrm{d}\ln p}{\mathrm{d}T} = \frac{51.0\times10^3\,\mathrm{J\,mol^{-1}}}{(8.3145\,\mathrm{J\,K^{-1}\,mol^{-1}})\times(273\,\mathrm{K})^2} = 0.0823\,\mathrm{K^{-1}}$$

We see that the slope of $\ln p$ plotted against T is greater for the solid–vapour boundary than for the liquid–vapour boundary at the triple point.

Self-test 4B.7 Confirm that the same may be said for the plot of p against T at the triple point.

Answer: $\mathrm{d}p/\mathrm{d}T = p\,\mathrm{d}\ln p/\mathrm{d}T$, $p = p_3 = 6.1$ mbar

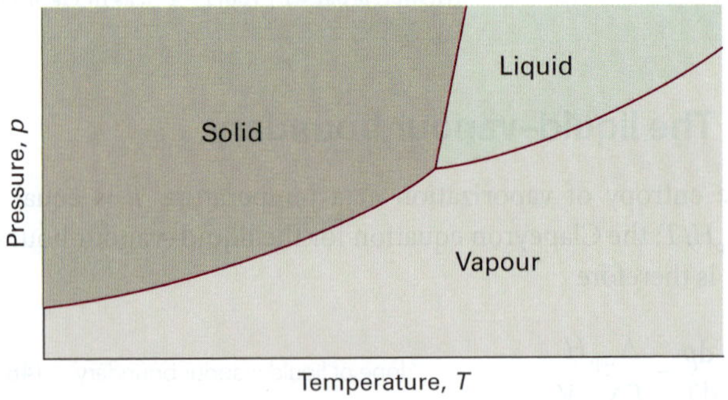

Figure 4B.8 Near the point where they coincide (at the triple point), the solid–vapour boundary has a steeper slope than the liquid–vapour boundary because the enthalpy of sublimation is greater than the enthalpy of vaporization and the temperatures that occur in the Clausius–Clapeyron equation for the slope have similar values.

4B.3 The Ehrenfest classification of phase transitions

There are many different types of phase transition, including the familiar examples of fusion and vaporization and the less familiar examples of solid–solid, conducting–superconducting, and fluid–superfluid transitions. We shall now see that it is possible to use thermodynamic properties of substances, and in particular the behaviour of the chemical potential, to classify phase transitions into different types. Classification is commonly a first step towards a molecular interpretation and the identification of common features. The classification scheme was originally proposed by Paul Ehrenfest, and is known as the **Ehrenfest classification**.

(a) The thermodynamic basis

Many familiar phase transitions, like fusion and vaporization, are accompanied by changes of enthalpy and volume. These changes have implications for the slopes of the chemical potentials of the phases at either side of the phase transition. Thus, at the transition from a phase α to another phase β,

$$\left(\frac{\partial\mu(\beta)}{\partial p}\right)_T-\left(\frac{\partial\mu(\alpha)}{\partial p}\right)_T=V_m(\beta)-V_m(\alpha)=\Delta_{trs}V$$
$$\left(\frac{\partial\mu(\beta)}{\partial T}\right)_p-\left(\frac{\partial\mu(\alpha)}{\partial T}\right)_p=-S_m(\beta)+S_m(\alpha)=-\Delta_{trs}S=-\frac{\Delta_{trs}H}{T_{trs}} \qquad (4B.12)$$

Because $\Delta_{trs}V$ and $\Delta_{trs}H$ are non-zero for melting and vaporization, it follows that for such transitions the slopes of the chemical potential plotted against either pressure or temperature are different on either side of the transition (Fig. 4B.9a). In other words, the first derivatives of the chemical potentials with respect to pressure and temperature are discontinuous at the transition.

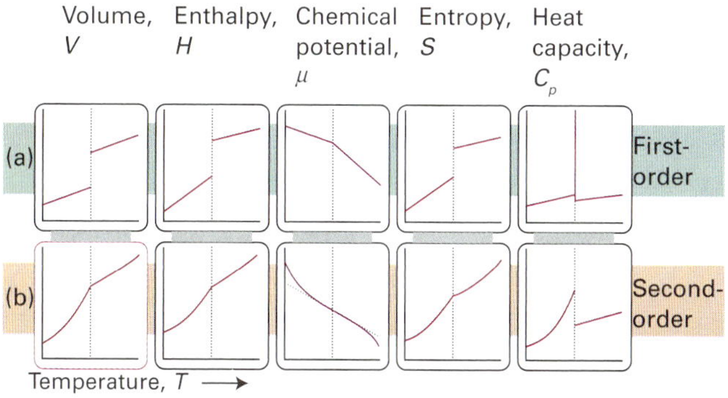

Figure 4B.9 The changes in thermodynamic properties accompanying (a) first-order and (b) second-order phase transitions.

Brief illustration 4B.7 Discontinuities of the transitions

The melting of water at its normal melting point of 0 °C has $\Delta_{trs}V=-1.6\,\mathrm{cm^3\,mol^{-1}}$ and $\Delta_{trs}H=6.008\,\mathrm{kJ\,mol^{-1}}$, so

$$\left(\frac{\partial\mu(\mathrm{l})}{\partial p}\right)_T-\left(\frac{\partial\mu(\mathrm{s})}{\partial p}\right)_T=\Delta_{fus}V=-1.6\,\mathrm{cm^3\,mol^{-1}}$$
$$\left(\frac{\partial\mu(\mathrm{l})}{\partial T}\right)_p-\left(\frac{\partial\mu(\mathrm{s})}{\partial T}\right)_p=-\frac{\Delta_{fus}H}{T_{fus}}=-\frac{6.008\times10^3\,\mathrm{J\,mol^{-1}}}{273\,\mathrm{K}}$$
$$=-22.0\,\mathrm{J\,mol^{-1}}$$

and both slopes are discontinuous.

Self-test 4B.8 Evaluate the difference in slopes at the normal boiling point.

Answer: $+31\,\mathrm{dm^3\,mol^{-1}}$, $-109\,\mathrm{J\,mol^{-1}}$

A transition for which the first derivative of the chemical potential with respect to temperature is discontinuous is classified as a **first-order phase transition**. The constant-pressure heat capacity, C_p, of a substance is the slope of a plot of the enthalpy with respect to temperature. At a first-order phase transition, H changes by a finite amount for an infinitesimal change of temperature. Therefore, at the transition the heat capacity is infinite. The physical reason is that heating drives the transition rather than raising the temperature. For example, boiling water stays at the same temperature even though heat is being supplied.

A **second-order phase transition** in the Ehrenfest sense is one in which the first derivative of μ with respect to temperature is continuous but its second derivative is discontinuous. A continuous slope of μ (a graph with the same slope on either side of the transition) implies that the volume and entropy (and hence the enthalpy) do not change at the transition (Fig. 4B.9b). The heat capacity is discontinuous at the transition but does not become infinite there. An example of a second-order transition is the conducting–superconducting transition in metals at low temperatures.[1]

The term **λ-transition** is applied to a phase transition that is not first-order yet the heat capacity becomes infinite at the transition temperature. Typically, the heat capacity of a system that shows such a transition begins to increase well before the transition (Fig. 4B.10), and the shape of the heat capacity curve resembles the Greek letter lambda. Examples of λ-transitions include order–disorder transitions in alloys, the onset of ferromagnetism, and the fluid–superfluid transition of liquid helium.

[1] A metallic conductor is a substance with an electrical conductivity that decreases as the temperature increases. A superconductor is a solid that conducts electricity without resistance. See Topic 18C for more details.

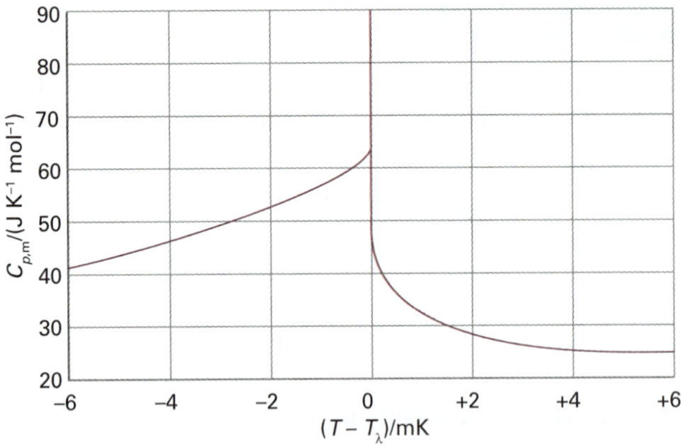

Figure 4B.10 The λ-curve for helium, where the heat capacity rises to infinity. The shape of this curve is the origin of the name λ-transition.

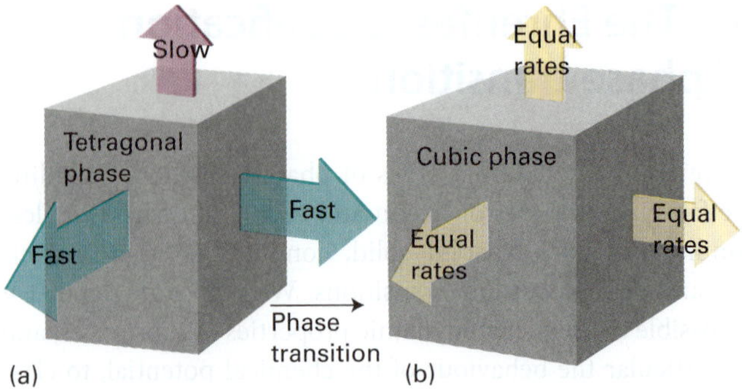

Figure 4B.11 One version of a second-order phase transition in which (a) a tetragonal phase expands more rapidly in two directions than a third, and hence becomes a cubic phase, which (b) expands uniformly in three directions as the temperature is raised. There is no rearrangement of atoms at the transition temperature, and hence no enthalpy of transition.

(b) Molecular interpretation

First-order transitions typically involve the relocation of atoms, molecules, or ions with a consequent change in the energies of their interactions. Thus, vaporization eliminates the attractions between molecules and a first-order phase transition from one ionic polymorph to another (as in the conversion of calcite to aragonite) involves the adjustment of the relative positions of ions.

One type of second-order transition is associated with a change in symmetry of the crystal structure of a solid. Thus, suppose the arrangement of atoms in a solid is like that represented in Fig. 4B.11a, with one dimension (technically, of the unit cell) longer than the other two, which are equal. Such a crystal structure is classified as tetragonal (see Topic 18A). Moreover, suppose the two shorter dimensions increase more than the long dimension when the temperature is raised. There may come a stage when the three dimensions become equal. At that point the crystal has cubic symmetry (Fig. 4B.11b), and at higher temperatures it will expand equally in all three directions (because there is no longer any distinction between them). The tetragonal→cubic phase transition has occurred, but as it has not involved a discontinuity in the interaction energy between the atoms or the volume they occupy, the transition is not first-order.

The order–disorder transition in β-brass (CuZn) is an example of a λ-transition. The low-temperature phase is an orderly array of alternating Cu and Zn atoms. The high-temperature phase is a random array of the atoms (Fig. 4B.12). At $T=0$ the order is perfect, but islands of disorder appear as the temperature is raised. The islands form because the transition is cooperative in the sense that, once two atoms have exchanged locations, it is easier for their neighbours to exchange their locations. The islands grow in extent and merge throughout the crystal at the transition temperature (742 K). The heat capacity increases as the transition temperature is approached because the cooperative nature of the transition means that it is increasingly easy for the heat supplied to drive the phase transition rather than to be stored as thermal motion.

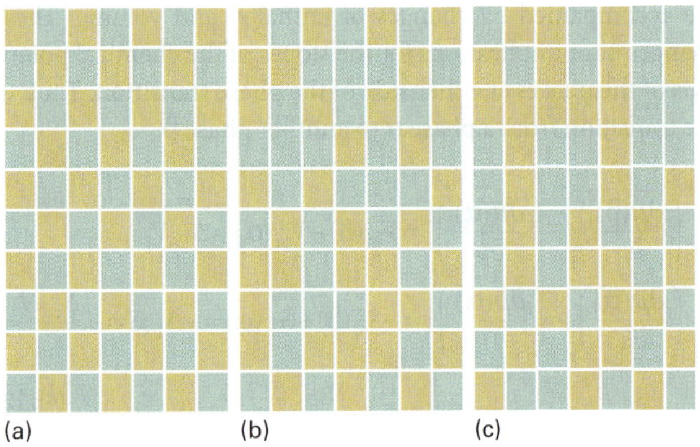

Figure 4B.12 An order–disorder transition. (a) At $T=0$, there is perfect order, with different kinds of atoms occupying alternate sites. (b) As the temperature is increased, atoms exchange locations and islands of each kind of atom form in regions of the solid. Some of the original order survives. (c) At and above the transition temperature, the islands occur at random throughout the sample.

Checklist of concepts

- ☐ 1. The chemical potential of a substance decreases with increasing temperature at a rate determined by its molar entropy.
- ☐ 2. The chemical potential of a substance increases with increasing pressure at a rate determined by its molar volume.
- ☐ 3. When pressure is applied to a condensed phase, its vapour pressure rises.
- ☐ 4. The **Clapeyron equation** is an expression for the slope of a phase boundary.
- ☐ 5. The **Clausius–Clapeyron equation** is an approximation that relates the slope of the liquid–vapour boundary to the enthalpy of vaporization.
- ☐ 6. According to the **Ehrenfest classification**, different types of phase transition are identified by the behaviour of thermodynamic properties at the transition temperature.
- ☐ 7. The classification reveals the type of molecular process occurring at the phase transition.

Checklist of equations

Property	Equation	Comment	Equation number
Variation of μ with temperature	$(\partial\mu/\partial T)_p = -S_m$		4B.1
Variation of μ with pressure	$(\partial\mu/\partial p)_T = V_m$		4B.2
Vapour pressure in the presence of applied pressure	$p = p^* e^{V_m(l)\Delta P/RT}$	$\Delta P = P_{applied} - p^*$	4B.3
Clapeyron equation	$dp/dT = \Delta_{trs}S/\Delta_{trs}V$		4B.5a
Clausius–Clapeyron equation	$d\ln p/dT = \Delta_{vap}H/RT^2$	Assumes $V_m(g) \gg V_m(l)$ and vapour is a perfect gas	4B.10

CHAPTER 4 Physical transformations of pure substances

TOPIC 4A Phase diagrams of pure substances

Discussion questions

4A.1 Describe how the concept of chemical potential unifies the discussion of phase equilibria.

4A.2 Why does the chemical potential change with pressure even if the system is incompressible (that is, remains at the same volume when pressure is applied)?

4A.3 Explain why four phases cannot be in equilibrium in a one-component system.

4A.4 Discuss what would be observed as a sample of water is taken along a path that encircles and is close to its critical point.

Exercises

4A.1(a) How many phases are present at each of the points marked in Fig. 4.1a?
4A.1(b) How many phases are present at each of the points marked in Fig. 4.1b?

4A.2(a) The difference in chemical potential between two regions of a system is +7.1 kJ mol^{-1}. By how much does the Gibbs energy change when 0.10 mmol of a substance is transferred from one region to the other?
4A.2(b) The difference in chemical potential between two regions of a system is −8.3 kJ mol^{-1}. By how much does the Gibbs energy change when 0.15 mmol of a substance is transferred from one region to the other?

4A.3(a) What is the maximum number of phases that can be in mutual equilibrium in a two-component system?
4A.3(b) What is the maximum number of phases that can be in mutual equilibrium in a four-component system?

For problems relating to one-component phase diagrams, see the Integrated activities section of this chapter.

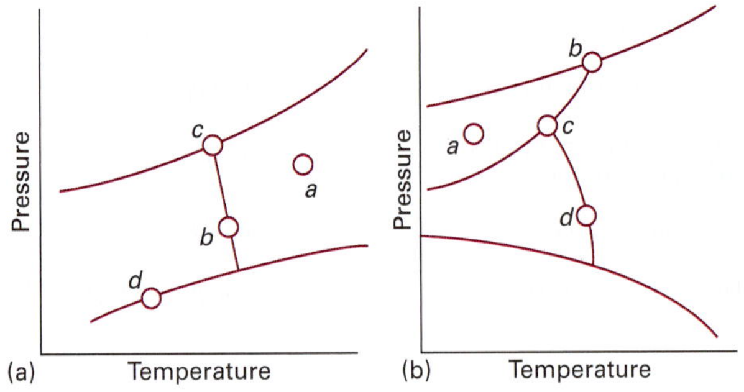

Figure 4.1 The phase diagrams referred to in (a) Exercise 4A.1(a) and (b) Exercise 4A.1(b).

TOPIC 4B Thermodynamic aspects of phase transitions

Discussion questions

4B.1 What is the physical reason for the fact that the chemical potential of a pure substance decreases as the temperatures is raised?

4B.2 What is the physical reason for the fact that the chemical potential of a pure substance increases as the pressure is raised?

4B.3 How may differential scanning calorimetry (DSC) be used to identify phase transitions?

4B.4 Distinguish between a first-order phase transition, a second-order phase transition, and a λ-transition at both molecular and macroscopic levels.

Exercises

4B.1(a) Estimate the difference between the normal and standard melting points of ice.
4B.1(b) Estimate the difference between the normal and standard boiling points of water.

4B.2(a) Water is heated from 25 °C to 100 °C. By how much does its chemical potential change?
4B.2(b) Iron is heated from 100 °C to 1000 °C. By how much does its chemical potential change? Take $S_m^{\ominus} = 53\,\mathrm{J\,K^{-1}\,mol^{-1}}$ for the entire range.

4B.3(a) By how much does the chemical potential of copper change when the pressure exerted on a sample is increased from 100 kPa to 10 MPa?
4B.3(b) By how much does the chemical potential of benzene change when the pressure exerted on a sample is increased from 100 kPa to 10 MPa?

4B.4(a) Pressure was exerted with a piston on water at 20 °C. The vapour pressure of water under 1.0 bar is 2.34 kPa. What is its vapour pressure when the pressure on the liquid is 20 MPa?
4B.4(b) Pressure was exerted with a piston on molten naphthalene at 95 °C. The vapour pressure of naphthalene under 1.0 bar is 2.0 kPa. What is its vapour pressure when the pressure on the liquid is 15 MPa?

4B.5(a) The molar volume of a certain solid is 161.0 $cm^3\,mol^{-1}$ at 1.00 atm and 350.75 K, its melting temperature. The molar volume of the liquid at this temperature and pressure is 163.3 $cm^3\,mol^{-1}$. At 100 atm the melting temperature changes to 351.26 K. Calculate the enthalpy and entropy of fusion of the solid.
4B.5(b) The molar volume of a certain solid is 142.0 $cm^3\,mol^{-1}$ at 1.00 atm and 427.15 K, its melting temperature. The molar volume of the liquid at this temperature and pressure is 152.6 $cm^3\,mol^{-1}$. At 1.2 MPa the melting temperature changes to 429.26 K. Calculate the enthalpy and entropy of fusion of the solid.

4B.6(a) The vapour pressure of dichloromethane at 24.1 °C is 53.3 kPa and its enthalpy of vaporization is 28.7 $kJ\,mol^{-1}$. Estimate the temperature at which its vapour pressure is 70.0 kPa.
4B.6(b) The vapour pressure of a substance at 20.0 °C is 58.0 kPa and its enthalpy of vaporization is 32.7 $kJ\,mol^{-1}$. Estimate the temperature at which its vapour pressure is 66.0 kPa.

4B.7(a) The vapour pressure of a liquid in the temperature range 200 K to 260 K was found to fit the expression $\ln(p/\text{Torr})=16.255-2501.8/(T/\text{K})$. What is the enthalpy of vaporization of the liquid?
4B.7(b) The vapour pressure of a liquid in the temperature range 200 K to 260 K was found to fit the expression $\ln(p/\text{Torr})=18.361-3036.8/(T/\text{K})$. What is the enthalpy of vaporization of the liquid?

4B.8(a) The vapour pressure of benzene between 10 °C and 30 °C fits the expression $\log(p/\text{Torr})=7.960-1780/(T/\text{K})$. Calculate (i) the enthalpy of vaporization and (ii) the normal boiling point of benzene.
4B.8(b) The vapour pressure of a liquid between 15 °C and 35 °C fits the expression $\log(p/\text{Torr})=8.750-1625/(T/\text{K})$. Calculate (i) the enthalpy of vaporization and (ii) the normal boiling point of the liquid.

4B.9(a) When benzene freezes at 5.5 °C its density changes from 0.879 $g\,cm^{-3}$ to 0.891 $g\,cm^{-3}$. Its enthalpy of fusion is 10.59 $kJ\,mol^{-1}$. Estimate the freezing point of benzene at 1000 atm.
4B.9(b) When a certain liquid freezes at −3.65 °C its density changes from 0.789 $g\,cm^{-3}$ to 0.801 $g\,cm^{-3}$. Its enthalpy of fusion is 8.68 $kJ\,mol^{-1}$. Estimate the freezing point of the liquid at 100 MPa.

4B.10(a) In July in Los Angeles, the incident sunlight at ground level has a power density of 1.2 $kW\,m^{-2}$ at noon. A swimming pool of area 50 m^2 is directly exposed to the sun. What is the maximum rate of loss of water? Assume that all the radiant energy is absorbed.
4B.10(b) Suppose the incident sunlight at ground level has a power density of 0.87 $kW\,m^{-2}$ at noon. What is the maximum rate of loss of water from a lake of area 1.0 ha? (1 ha $=10^4\,m^2$.) Assume that all the radiant energy is absorbed.

4B.11(a) An open vessel containing (i) water, (ii) benzene, (iii) mercury stands in a laboratory measuring 5.0 m × 5.0 m × 3.0 m at 25 °C. What mass of each substance will be found in the air if there is no ventilation? (The vapour pressures are (i) 3.2 kPa, (ii) 13.1 kPa, (iii) 0.23 Pa.)
4B.11(b) On a cold, dry morning after a frost, the temperature was −5 °C and the partial pressure of water in the atmosphere fell to 0.30 kPa. Will the frost sublime? What partial pressure of water would ensure that the frost remained?

4B.12(a) Naphthalene, $C_{10}H_8$, melts at 80.2 °C. If the vapour pressure of the liquid is 1.3 kPa at 85.8 °C and 5.3 kPa at 119.3 °C, use the Clausius–Clapeyron equation to calculate (i) the enthalpy of vaporization, (ii) the normal boiling point, and (iii) the enthalpy of vaporization at the boiling point.
4B.12(b) The normal boiling point of hexane is 69.0 °C. Estimate (i) its enthalpy of vaporization and (ii) its vapour pressure at 25 °C and 60 °C.

4B.13(a) Calculate the melting point of ice under a pressure of 50 bar. Assume that the density of ice under these conditions is approximately 0.92 $g\,cm^{-3}$ and that of liquid water is 1.00 $g\,cm^{-3}$.
4B.13(b) Calculate the melting point of ice under a pressure of 10 MPa. Assume that the density of ice under these conditions is approximately 0.915 $g\,cm^{-3}$ and that of liquid water is 0.998 $g\,cm^{-3}$.

4B.14(a) What fraction of the enthalpy of vaporization of water is spent on expanding the water vapour?
4B.14(b) What fraction of the enthalpy of vaporization of ethanol is spent on expanding its vapour?

Problems

4B.1 The temperature dependence of the vapour pressure of solid sulfur dioxide can be approximately represented by the relation $\log(p/\text{Torr})=10.5916-1871.2/(T/\text{K})$ and that of liquid sulfur dioxide by $\log(p/\text{Torr})=8.3186-1425.7/(T/\text{K})$. Estimate the temperature and pressure of the triple point of sulfur dioxide.

4B.2 Prior to the discovery that freon-12 (CF_2Cl_2) was harmful to the Earth's ozone layer, it was frequently used as the dispersing agent in spray cans for hair spray, etc. Its enthalpy of vaporization at its normal boiling point of −29.2 °C is 20.25 $kJ\,mol^{-1}$. Estimate the pressure that a can of hair spray using freon-12 had to withstand at 40 °C, the temperature of a can that has been standing in sunlight. Assume that $\Delta_{vap}H$ is a constant over the temperature range involved and equal to its value at −29.2 °C.

4B.3 The enthalpy of vaporization of a certain liquid is found to be 14.4 $kJ\,mol^{-1}$ at 180 K, its normal boiling point. The molar volumes of the liquid and the vapour at the boiling point are 115 $cm^3\,mol^{-1}$ and 14.5 $dm^3\,mol^{-1}$, respectively. (a) Estimate dp/dT from the Clapeyron equation and (b) the percentage error in its value if the Clausius–Clapeyron equation is used instead.

4B.4 Calculate the difference in slope of the chemical potential against temperature on either side of (a) the normal freezing point of water and (b) the normal boiling point of water. (c) By how much does the chemical potential of water supercooled to −5.0 °C exceed that of ice at that temperature?

4B.5 Calculate the difference in slope of the chemical potential against pressure on either side of (a) the normal freezing point of water and (b) the normal boiling point of water. The densities of ice and water at 0 °C are 0.917 $g\,cm^{-3}$ and 1.000 $g\,cm^{-3}$, and those of water and water vapour at 100 °C are 0.958 $g\,cm^{-3}$ and 0.598 $g\,dm^{-3}$, respectively. By how much does the chemical potential of water vapour exceed that of liquid water at 1.2 atm and 100 °C?

4B.6 The enthalpy of fusion of mercury is 2.292 $kJ\,mol^{-1}$, and its normal freezing point is 234.3 K with a change in molar volume of +0.517 $cm^{-3}\,mol^{-1}$ on melting. At what temperature will the bottom of a column of mercury (density 13.6 $g\,cm^{-3}$) of height 10.0 m be expected to freeze?

4B.7 50.0 dm^3 of dry air was slowly bubbled through a thermally insulated beaker containing 250 g of water initially at 25 °C. Calculate the final temperature. (The vapour pressure of water is approximately constant at 3.17 kPa throughout, and its heat capacity is 75.5 $J\,K^{-1}\,mol^{-1}$. Assume that the air is not heated or cooled and that water vapour is a perfect gas.)

4B.8 The vapour pressure, p, of nitric acid varies with temperature as follows:

$\theta/°C$	0	20	40	50	70	80	90	100
p/kPa	1.92	6.38	17.7	27.7	62.3	89.3	124.9	170.9

What are (a) the normal boiling point and (b) the enthalpy of vaporization of nitric acid?

4B.9 The vapour pressure of the ketone carvone ($M = 150.2\,g\,mol^{-1}$), a component of oil of spearmint, is as follows:

$\theta/°C$	57.4	100.4	133.0	157.3	203.5	227.5
$p/Torr$	1.00	10.0	40.0	100	400	760

What are (a) the normal boiling point and (b) the enthalpy of vaporization of carvone?

4B.10‡ In a study of the vapour pressure of chloromethane, A. Bah and N. Dupont-Pavlovsky (*J. Chem. Eng. Data* **40**, 869 (1995)) presented data for the vapour pressure over solid chloromethane at low temperatures. Some of that data is as follows:

T/K	145.94	147.96	149.93	151.94	153.97	154.94
p/Pa	13.07	18.49	25.99	36.76	50.86	59.56

Estimate the standard enthalpy of sublimation of chloromethane at 150 K. (Take the molar volume of the vapour to be that of a perfect gas, and that of the solid to be negligible.)

4B.11 Show that, for a transition between two incompressible solid phases, ΔG is independent of the pressure.

4B.12 The change in enthalpy is given by $dH = C_p dT + V dp$. The Clapeyron equation relates dp and dT at equilibrium, and so in combination the two equations can be used to find how the enthalpy changes along a phase boundary as the temperature changes and the two phases remain in equilibrium. Show that $d(\Delta H/T) = \Delta C_p \, d\ln T$.

4B.13 In the 'gas saturation method' for the measurement of vapour pressure, a volume V of gas (as measured at a temperature T and a pressure p) is bubbled slowly through the liquid that is maintained at the temperature T, and a mass loss m is measured. Show that the vapour pressure, p, of the liquid is related to its molar mass, M, by $p = AmP/(1+Am)$, where $A = RT/MPV$. The vapour pressure of geraniol ($M = 154.2\,g\,mol^{-1}$), which is a component of oil of roses, was measured at 110 °C. It was found that, when 5.00 dm³ of nitrogen at 760 Torr was passed slowly through the heated liquid, the loss of mass was 0.32 g. Calculate the vapour pressure of geraniol.

4B.14 The vapour pressure of a liquid in a gravitational field varies with the depth below the surface on account of the hydrostatic pressure exerted by the overlying liquid. Adapt eqn. 4B.3 to predict how the vapour pressure of a liquid of molar mass M varies with depth. Estimate the effect on the vapour pressure of water at 25 °C in a column 10 m high.

4B.15 Combine the 'barometric formula', $p = p_0 e^{-a/H}$, where $H = 8\,km$, for the dependence of the pressure on altitude, a, with the Clausius–Clapeyron equation, and predict how the boiling temperature of a liquid depends on the altitude and the ambient temperature. Take the mean ambient temperature as 20 °C and predict the boiling temperature of water at 3000 m.

4B.16 Figure 4B.1 gives a schematic representation of how the chemical potentials of the solid, liquid, and gaseous phases of a substance vary with temperature. All have a negative slope, but it is unlikely that they are truly straight lines as indicated in the illustration. Derive an expression for the curvatures (specifically, the second derivatives with respect to temperature) of these lines. Is there a restriction on the curvature of these lines? Which state of matter shows the greatest curvature?

4B.17 The Clapeyron equation does not apply to second-order phase transitions, but there are two analogous equations, the *Ehrenfest equations*, that do. They are:

$$\text{(a)}\ \frac{dp}{dT} = \frac{\alpha_2 - \alpha_1}{\kappa_{T;2} - \kappa_{T;1}} \qquad \text{(b)}\ \frac{dp}{dT} = \frac{C_{p,m;2} - C_{p,m;1}}{TV_m(\alpha_2 - \alpha_1)}$$

where α is the expansion coefficient, κ_T the isothermal compressibility, and the subscripts 1 and 2 refer to two different phases. Derive these two equations. Why does the Clapeyron equation not apply to second-order transitions?

4B.18 For a first-order phase transition, to which the Clapeyron equation does apply, prove the relation

$$C_S = C_p - \frac{\alpha V \Delta_{trs} H}{\Delta_{trs} V}$$

where $C_S = (\partial q/\partial T)_S$ is the heat capacity along the coexistence curve of two phases.

Integrated activities

4.1 Construct the phase diagram for benzene near its triple point at 36 Torr and 5.50 °C using the following data: $\Delta_{fus}H = 10.6\,kJ\,mol^{-1}$, $\Delta_{vap}H = 30.8\,kJ\,mol^{-1}$, $\rho(s) = 0.891\,g\,cm^{-3}$, $\rho(l) = 0.879\,g\,cm^{-3}$.

4.2‡ In an investigation of thermophysical properties of toluene, R.D. Goodwin (*J. Phys. Chem. Ref. Data* **18**, 1565 (1989)) presented expressions for two phase boundaries. The solid–liquid boundary is given by

$$p/\text{bar} = p_3/\text{bar} + 1000(5.60 + 11.727x)x$$

where $x = T/T_3 - 1$ and the triple point pressure and temperature are $p_3 = 0.4362\,\mu bar$ and $T_3 = 178.15\,K$. The liquid–vapour curve is given by:

$$\ln(p/\text{bar}) = -10.418/y + 21.157 - 15.996y + 14.015y^2 - 5.0120y^3 + 4.7334(1-y)^{1.70}$$

where $y = T/T_c = T/(593.95\,K)$. (a) Plot the solid–liquid and liquid–vapour phase boundaries. (b) Estimate the standard melting point of toluene. (c) Estimate the standard boiling point of toluene. (d) Compute the standard enthalpy of vaporization of toluene, given that the molar volumes of the liquid and vapour at the normal boiling point are 0.12 dm³ mol⁻¹ and 30.3 dm³ mol⁻¹, respectively.

4.3 Proteins are polymers of amino acids that can exist in ordered structures stabilized by a variety of molecular interactions. However, when certain conditions are changed, the compact structure of a polypeptide chain may collapse into a random coil. This structural change may be regarded as a phase transition occurring at a characteristic transition temperature, the *melting temperature*, T_m, which increases with the strength and number

‡ These problems were supplied by Charles Trapp and Carmen Giunta.

of intermolecular interactions in the chain. A thermodynamic treatment allows predictions to be made of the temperature T_m for the unfolding of a helical polypeptide held together by hydrogen bonds into a random coil. If a polypeptide has N amino acids, $N-4$ hydrogen bonds are formed to form an α-helix, the most common type of helix in naturally occurring proteins (see Topic 17A). Because the first and last residues in the chain are free to move, $N-2$ residues form the compact helix and have restricted motion. Based on these ideas, the molar Gibbs energy of unfolding of a polypeptide with $N \geq 5$ may be written as

$$\Delta_{\text{unfold}}G = (N-4)\Delta_{\text{hb}}H - (N-2)T\Delta_{\text{hb}}S$$

where $\Delta_{\text{hb}}H$ and $\Delta_{\text{hb}}S$ are, respectively, the molar enthalpy and entropy of dissociation of hydrogen bonds in the polypeptide. (a) Justify the form of the equation for the Gibbs energy of unfolding. That is, why are the enthalpy and entropy terms written as $(N-4)\Delta_{\text{hb}}H$ and $(N-2)\Delta_{\text{hb}}S$, respectively? (b) Show that T_m may be written as

$$T_m = \frac{(N-4)\Delta_{\text{hb}}H}{(N-2)\Delta_{\text{hb}}S}$$

(c) Plot $T_m/(\Delta_{\text{hb}}H_m/\Delta_{\text{hb}}S_m)$ for $5 \leq N \leq 20$. At what value of N does T_m change by less than 1 per cent when N increases by 1?

4.4‡ A substance as well-known as methane still receives research attention because it is an important component of natural gas, a commonly used fossil fuel. Friend et al. have published a review of thermophysical properties of methane (D.G. Friend, J.F. Ely, and H. Ingham, *J. Phys. Chem. Ref. Data* **18**, 583 (1989)), which included the following data describing the liquid–vapour phase boundary.

T/K	100	108	110	112	114	120	130	140	150	160	170	190
p/MPa	0.034	0.074	0.088	0.104	0.122	0.192	0.368	0.642	1.041	1.593	2.329	4.521

(a) Plot the liquid–vapour phase boundary. (b) Estimate the standard boiling point of methane. (c) Compute the standard enthalpy of vaporization of methane, given that the molar volumes of the liquid and vapour at the standard boiling point are 3.80×10^{-2} and $8.89\,\text{dm}^3\,\text{mol}^{-1}$, respectively.

4.5‡ Diamond is the hardest substance and the best conductor of heat yet characterized. For these reasons, it is used widely in industrial applications that require a strong abrasive. Unfortunately, it is difficult to synthesize diamond from the more readily available allotropes of carbon, such as graphite. To illustrate this point, calculate the pressure required to convert graphite into diamond at 25 °C. The following data apply to 25 °C and 100 kPa. Assume the specific volume, V_s, and κ_T are constant with respect to pressure changes.

	Graphite	**Diamond**
$\Delta_r G^\ominus/(\text{kJ mol}^{-1})$	0	+2.8678
$V_s/(\text{cm}^3\,\text{g}^{-1})$	0.444	0.284
κ_T/kPa	3.04×10^{-8}	0.187×10^{-8}

CHAPTER 5

Simple mixtures

Mixtures are an essential part of chemistry, either in their own right or as starting materials for chemical reactions. This group of Topics deals with the rich physical properties of mixtures and shows how to express them in terms of thermodynamic quantities.

5A The thermodynamic description of mixtures

The first Topic in this chapter develops the concept of chemical potential as an example of a partial molar quantity and explores how to use the chemical potential of a substance to describe the physical properties of mixtures. The underlying principle to keep in mind is that at equilibrium the chemical potential of a species is the same in every phase. We see, by making use of the experimental observations known as Raoult's and Henry's laws, how to express the chemical potential of a substance in terms of its mole fraction in a mixture.

5B The properties of solutions

In this Topic, the concept of chemical potential is applied to the discussion of the effect of a solute on certain thermodynamic properties of a solution. These properties include the lowering of vapour pressure of the solvent, the elevation of its boiling point, the depression of its freezing point, and the origin of osmotic pressure. We see that it is possible to construct a model of a certain class of real solutions called 'regular solutions', and see how they have properties that diverge from those of ideal solutions.

5C Phase diagrams of binary systems

One widely used device used to summarize the equilibrium properties of mixtures is the phase diagram. We see how to construct and interpret these diagrams. The Topic introduces systems of gradually increasing complexity. In each case we shall see how the phase diagram for the system summarizes empirical observations on the conditions under which the various phases of the system are stable.

5D Phase diagrams of ternary systems

Many modern materials (and ancient ones too) have more than two components. In this Topic we show how phase diagrams are extended to the description of systems of three components and how to interpret triangular phase diagrams.

5E Activities

The extension of the concept of chemical potential to real solutions involves introducing an effective concentration called an 'activity'. We see how the activity may be defined and measured. We shall also see how, in certain cases, the activity may be interpreted in terms of intermolecular interactions.

5F The activities of ions

One of the most important types of mixtures encountered in chemistry is an electrolyte solution. Such solutions often deviate considerably from ideal behaviour on account of the strong, long-range interactions between ions. In this Topic we show how a model can be used to estimate the deviations from ideal behaviour when the solution is very dilute, and how to extend the resulting expressions to more concentrated solutions.

What is the impact of this material?

We consider just two applications of this material, one from biology and the other from materials science, from among the huge number that could be chosen for this centrally important field. In *Impact* I5.1, we see how the phenomenon of osmosis contributes to the ability of biological cells to maintain their shapes. In *Impact* I5.2, we see how phase diagrams are used to describe the properties of the technologically important liquid crystals.

To read more about the impact of this material, scan the QR code, or go to bcs.whfreeman.com/webpub/chemistry/pchem10e/impact/pchem-5-1.html

5A The thermodynamic description of mixtures

Contents

➤ Why do you need to know this material?

Chemistry deals with a wide variety of mixtures, including mixtures of substances that can react together. Therefore, it is important to generalize the concepts introduced in Chapter 4 to deal with substances that are mingled together. This Topic also introduces the fundamental equation of chemical thermodynamics on which many of the applications of thermodynamics to chemistry are based.

➤ What is the key idea?

The chemical potential of a substance in a mixture is a logarithmic function of its concentration.

➤ What do you need to know already?

This Topic extends the concept of chemical potential to substances in mixtures by building on the concept introduced in the context of pure substances (Topic 4A). It makes use of the relation between entropy and the temperature dependence of the Gibbs energy (Topic 3D) and the concept of partial pressure (Topic 1A). It uses the notation of partial derivatives (*Mathematical background* 2) but does not draw on their advanced properties.

As a first step towards dealing with chemical reactions (which are treated in Topic 6A), here we consider mixtures of substances that do not react together. At this stage we deal mainly with **binary mixtures**, which are mixtures of two components, A and B. We shall therefore often be able to simplify equations by making use of the relation $x_A + x_B = 1$. In Topic 1A it is established that the partial pressure, which is the contribution of one component to the total pressure, is used to discuss the properties of mixtures of gases. For a more general description of the thermodynamics of mixtures we need to introduce other analogous 'partial' properties.

One preliminary remark is in order. Throughout this and related Topics we need to refer to various measures of concentration of a solute in a solution. The **molar concentration** (colloquially, the 'molarity', [J] or c_J) is the amount of solute divided by the volume of the solution and is usually expressed in moles per cubic decimetre (mol dm^{-3}; more informally, mol L^{-1}). We write $c^{\ominus} = 1\,\text{mol dm}^{-3}$. The term **molality**, b, is the amount of solute divided by the mass of solvent and is usually expressed in moles per kilogram of solvent (mol kg^{-1}). We write $b^{\ominus} = 1\,\text{mol kg}^{-1}$.

5A.1 Partial molar quantities

The easiest partial molar property to visualize is the 'partial molar volume', the contribution that a component of a mixture makes to the total volume of a sample.

(a) Partial molar volume

Imagine a huge volume of pure water at 25 °C. When a further 1 mol H_2O is added, the volume increases by 18 cm^3 and we can report that 18 cm^3 mol^{-1} is the molar volume of pure water. However, when we add 1 mol H_2O to a huge volume of pure ethanol, the volume increases by only 14 cm^3. The reason for the different increase in volume is that the volume occupied by a given number of water molecules depends on the identity of the molecules that surround them. In the latter case there is so much ethanol present that each H_2O molecule is surrounded by ethanol molecules. The network of hydrogen bonds that normally hold H_2O molecules at certain distances from each other in pure water does not form. The packing of the molecules in the mixture results in the H_2O molecules increasing the volume by only 14 cm^3. The quantity 14 cm^3 mol^{-1} is the partial molar volume of water in pure ethanol. In general, the **partial molar volume** of a substance A in a mixture is the change in volume per mole of A added to a large volume of the mixture.

The partial molar volumes of the components of a mixture vary with composition because the environment of each type of molecule changes as the composition changes from pure A to pure B. It is this changing molecular environment, and the consequential modification of the forces acting between molecules, that results in the variation of the thermodynamic properties of a mixture as its composition is changed. The partial molar volumes of water and ethanol across the full composition range at 25 °C are shown in Fig. 5A.1.

The partial molar volume, V_J, of a substance J at some general composition is defined formally as follows:

$$V_J = \left(\frac{\partial V}{\partial n_J}\right)_{p,T,n'} \quad \textit{Definition} \quad \text{Partial molar volume} \quad (5A.1)$$

where the subscript n' signifies that the amounts of all other substances present are constant. The partial molar volume is

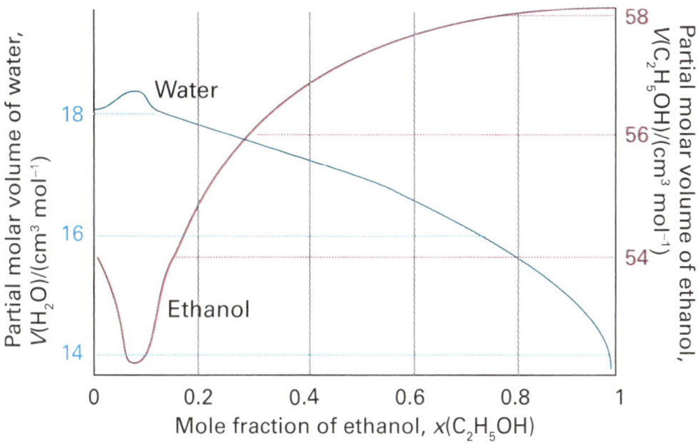

Figure 5A.1 The partial molar volumes of water and ethanol at 25 °C. Note the different scales (water on the left, ethanol on the right).

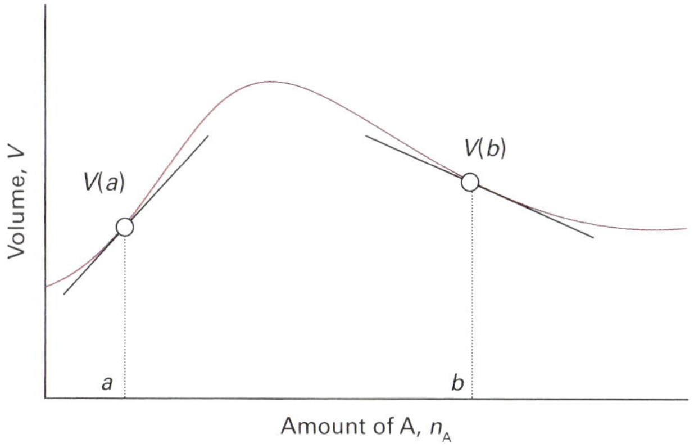

Figure 5A.2 The partial molar volume of a substance is the slope of the variation of the total volume of the sample plotted against the composition. In general, partial molar quantities vary with the composition, as shown by the different slopes at the compositions *a* and *b*. Note that the partial molar volume at *b* is negative: the overall volume of the sample decreases as A is added.

the slope of the plot of the total volume as the amount of J is changed, the pressure, temperature, and amount of the other components being constant (Fig. 5A.2). Its value depends on the composition, as we saw for water and ethanol.

A note on good practice The IUPAC recommendation is to denote a partial molar quantity by $\overline{X}$, but only when there is the possibility of confusion with the quantity X. For instance, to avoid confusion, the partial molar volume of NaCl in water could be written $\overline{V}$(NaCl, aq) to distinguish it from the total volume of the solution, V.

The definition in eqn 5A.1 implies that when the composition of the mixture is changed by the addition of dn_A of A and dn_B of B, then the total volume of the mixture changes by

$$\begin{aligned} dV &= \left(\frac{\partial V}{\partial n_A}\right)_{p,T,n_B} dn_A + \left(\frac{\partial V}{\partial n_B}\right)_{p,T,n_A} dn_B \\ &= V_A dn_A + V_B dn_B \end{aligned} \quad (5A.2)$$

Provided the relative composition is held constant as the amounts of A and B are increased, the partial molar volumes are both constant. In that case we can obtain the final volume by integration, treating V_A and V_B as constants:

$$\begin{aligned} V &= \int_0^{n_A} V_A dn_A + \int_0^{n_B} V_B dn_B = V_A \int_0^{n_A} dn_A + V_B \int_0^{n_B} dn_B \\ &= V_A n_A + V_B n_B \end{aligned} \quad (5A.3)$$

Although we have envisaged the two integrations as being linked (in order to preserve constant relative composition), because V is a state function the final result in eqn 5A.3 is valid however the solution is in fact prepared.

Partial molar volumes can be measured in several ways. One method is to measure the dependence of the volume on the composition and to fit the observed volume to a function of the amount of the substance. Once the function has been found, its slope can be determined at any composition of interest by differentiation.

Example 5A.1 Determining a partial molar volume

A polynomial fit to measurements of the total volume of a water/ethanol mixture at 25 °C that contains 1.000 kg of water is

$$v = 1002.93 + 54.6664x - 0.363\,94x^2 + 0.028\,256x^3$$

where $v = V/\text{cm}^3$, $x = n_E/\text{mol}$, and n_E is the amount of CH_3CH_2OH present. Determine the partial molar volume of ethanol.

Method Apply the definition in eqn 5A.1 taking care to convert the derivative with respect to n to a derivative with respect to x and keeping the units intact.

Answer The partial molar volume of ethanol, V_E, is

$$V_E = \left(\frac{\partial V}{\partial n_E}\right)_{p,T,n_W} = \left(\frac{\partial(V/\text{cm}^3)}{\partial(n_E/\text{mol})}\right)_{p,T,n_W} \frac{\text{cm}^3}{\text{mol}}$$

$$= \left(\frac{\partial v}{\partial x}\right)_{p,T,n_W} \text{cm}^3\,\text{mol}^{-1}$$

Then, because

$$\frac{dv}{dx} = 54.6664 - 2(0.363\,94)x + 3(0.028\,256)x^2$$

we can conclude that

$$V_E/(\text{cm}^3\text{mol}^{-1}) = 54.6664 - 0.727\,88x + 0.084\,768x^2$$

Figure 5A.3 shows a graph of this function.

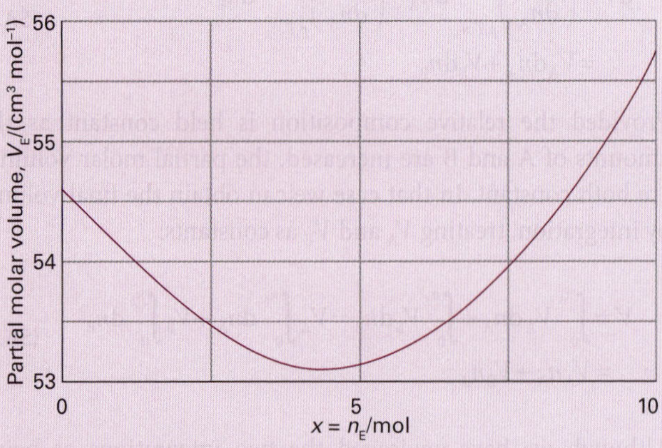

Figure 5A.3 The partial molar volume of ethanol, as expressed by the polynomial in *Example* 5A.1.

Self-test 5A.1 At 25 °C, the density of a 50 per cent by mass ethanol/water solution is 0.914 g cm^{-3}. Given that the partial molar volume of water in the solution is 17.4 cm^3 mol^{-1}, what is the partial molar volume of the ethanol?

Answer: 56.4 cm^3 mol^{-1}; 54.6 cm^3 mol^{-1} by the formula above

Molar volumes are always positive, but partial molar quantities need not be. For example, the limiting partial molar volume of $MgSO_4$ in water (its partial molar volume in the limit of zero concentration) is −1.4 cm^3 mol^{-1}, which means that the addition of 1 mol $MgSO_4$ to a large volume of water results in a decrease in volume of 1.4 cm^3. The mixture contracts because the salt breaks up the open structure of water as the Mg^{2+} and SO_4^{2-} ions become hydrated, and it collapses slightly.

(b) Partial molar Gibbs energies

The concept of a partial molar quantity can be extended to any extensive state function. For a substance in a mixture, the chemical potential is *defined* as the partial molar Gibbs energy:

$$\mu_J = \left(\frac{\partial G}{\partial n_J}\right)_{p,T,n'} \qquad \text{Definition} \quad \text{Chemical potential} \qquad (5A.4)$$

That is, the chemical potential is the slope of a plot of Gibbs energy against the amount of the component J, with the pressure and temperature (and the amounts of the other substances) held constant (Fig. 5A.4). For a pure substance we can write $G = n_J G_{J,m}$, and from eqn 5A.4 obtain $\mu_J = G_{J,m}$: in this case, the chemical potential is simply the molar Gibbs energy of the substance, as is used in Topic 4B.

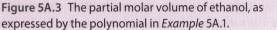

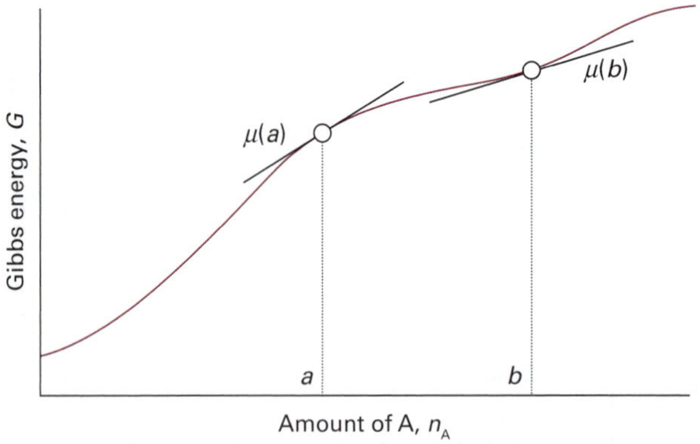

Figure 5A.4 The chemical potential of a substance is the slope of the total Gibbs energy of a mixture with respect to the amount of substance of interest. In general, the chemical potential varies with composition, as shown for the two values at *a* and *b*. In this case, both chemical potentials are positive.

By the same argument that led to eqn 5A.2, it follows that the total Gibbs energy of a binary mixture is

$$G=n_A\mu_A+n_B\mu_B \qquad (5A.5)$$

where μ_A and μ_B are the chemical potentials at the composition of the mixture. That is, the chemical potential of a substance in a mixture is the contribution of that substance to the total Gibbs energy of the mixture. Because the chemical potentials depend on composition (and the pressure and temperature), the Gibbs energy of a mixture may change when these variables change, and for a system of components A, B, etc., the equation $dG=Vdp-SdT$ becomes

$$dG=Vdp-SdT+\mu_A dn_A+\mu_B dn_B+\cdots$$

Fundamental equation of chemical thermodynamics (5A.6)

This expression is the **fundamental equation of chemical thermodynamics**. Its implications and consequences are explored and developed in this and the next two chapters.

At constant pressure and temperature, eqn 5A.6 simplifies to

$$dG=\mu_A dn_A+\mu_B dn_B+\cdots \qquad (5A.7)$$

We saw in Topic 3C that under the same conditions $dG=dw_{add,max}$. Therefore, at constant temperature and pressure,

$$dw_{add,max}=\mu_A dn_A+\mu_B dn_B+\cdots \qquad (5A.8)$$

That is, additional (non-expansion) work can arise from the changing composition of a system. For instance, in an electrochemical cell, the chemical reaction is arranged to take place in two distinct sites (at the two electrodes). The electrical work the cell performs can be traced to its changing composition as products are formed from reactants.

(c) The wider significance of the chemical potential

The chemical potential does more than show how G varies with composition. Because $G=U+pV-TS$, and therefore $U=-pV+TS+G$, we can write a general infinitesimal change in U for a system of variable composition as

$$\begin{aligned}dU&=-pdV-Vdp+SdT+TdS+dG\\&=-pdV-Vdp+SdT+TdS+\\&\quad(Vdp-SdT+\mu_A dn_A+\mu_B dn_B+\cdots)\\&=-pdV+TdS+\mu_A dn_A+\mu_B dn_B+\cdots\end{aligned}$$

This expression is the generalization of eqn 3D.1 (that $dU=TdS-pdV$) to systems in which the composition may change. It follows that at constant volume and entropy,

$$dU=\mu_A dn_A+\mu_B dn_B+\cdots \qquad (5A.9)$$

and hence that

$$\mu_J=\left(\frac{\partial U}{\partial n_J}\right)_{S,V,n'} \qquad (5A.10)$$

Therefore, not only does the chemical potential show how G changes when the composition changes, it also shows how the internal energy changes too (but under a different set of conditions). In the same way it is possible to deduce that

$$\text{(a) } \mu_J=\left(\frac{\partial H}{\partial n_J}\right)_{S,p,n'} \qquad \text{(b) } \mu_J=\left(\frac{\partial A}{\partial n_J}\right)_{T,V,n'} \qquad (5A.11)$$

Thus we see that the μ_J shows how all the extensive thermodynamic properties U, H, A, and G depend on the composition. This is why the chemical potential is so central to chemistry.

(d) The Gibbs–Duhem equation

Because the total Gibbs energy of a binary mixture is given by eqn 5A.5 and the chemical potentials depend on the composition, when the compositions are changed infinitesimally we might expect G of a binary system to change by

$$dG=\mu_A dn_A+\mu_B dn_B+n_A d\mu_A+n_B d\mu_B$$

However, we have seen that at constant pressure and temperature a change in Gibbs energy is given by eqn 5A.7. Because G is a state function, these two equations must be equal, which implies that at constant temperature and pressure

$$n_A d\mu_A+n_B d\mu_B=0 \qquad (5A.12a)$$

This equation is a special case of the **Gibbs–Duhem equation**:

$$\sum_J n_J d\mu_J=0$$

Gibbs–Duhem equation (5A.12b)

The significance of the Gibbs–Duhem equation is that the chemical potential of one component of a mixture cannot change independently of the chemical potentials of the other components. In a binary mixture, if one partial molar quantity increases, then the other must decrease, with the two changes related by

$$d\mu_B=-\frac{n_A}{n_B}d\mu_A \qquad (5A.13)$$

Brief illustration 5A.1 The Gibbs–Duhem equation

If the composition of a mixture is such that $n_A = 2n_B$, and a small change in composition results in μ_A changing by $\delta\mu_A = +1\,\text{J mol}^{-1}$, μ_B will change by

$$\delta\mu_B = -2\times(1\,\text{J mol}^{-1}) = -2\,\text{J mol}^{-1}$$

Self-test 5A.2 Suppose that $n_A = 0.3n_B$ and a small change in composition results in μ_A changing by $\delta\mu_A = -10\,\text{J mol}^{-1}$, by how much will μ_B change?

Answer: $+3\,\text{J mol}^{-1}$

The same line of reasoning applies to all partial molar quantities. We can see in Fig. 5A.1, for example, that where the partial molar volume of water increases, that of ethanol decreases. Moreover, as eqn 5A.13 shows, and as we can see from Fig 5A.1, a small change in the partial molar volume of A corresponds to a large change in the partial molar volume of B if n_A/n_B is large, but the opposite is true when this ratio is small. In practice, the Gibbs–Duhem equation is used to determine the partial molar volume of one component of a binary mixture from measurements of the partial molar volume of the second component.

Example 5A.2 Using the Gibbs–Duhem equation

The experimental values of the partial molar volume of K_2SO_4(aq) at 298 K are found to fit the expression

$$v_B = 32.280 + 18.216x^{1/2}$$

where $v_B = V_{K_2SO_4}/(\text{cm}^3\,\text{mol}^{-1})$ and x is the numerical value of the molality of K_2SO_4 ($x = b/b^\ominus$; see the remark in the introduction to this chapter). Use the Gibbs–Duhem equation to derive an equation for the molar volume of water in the solution. The molar volume of pure water at 298 K is 18.079 cm³ mol⁻¹.

Method Let A denote H_2O, the solvent, and B denote K_2SO_4, the solute. The Gibbs–Duhem equation for the partial molar volumes of two components is $n_A dV_A + n_B dV_B = 0$. This relation implies that $dv_A = -(n_B/n_A)dv_B$, and therefore that v_A can be found by integration:

$$v_A = v_A^* - \int_0^{v_B} \frac{n_B}{n_A} dv_B$$

where $v_A^* = V_A/(\text{cm}^3\,\text{mol}^{-1})$ is the numerical value of the molar volume of pure A. The first step is to change the variable v_B to $x = b/b^\ominus$ and then to integrate the right-hand side between $x = 0$ (pure B) and the molality of interest.

Answer It follows from the information in the question that, with B = K_2SO_4, $dv_B/dx = 9.108x^{-1/2}$. Therefore, the integration required is

$$v_A = v_A^* - 9.108\int_0^{b/b^\ominus} \frac{n_B}{n_A} x^{-1/2}dx$$

However, the ratio of amounts of A (H_2O) and B (K_2SO_4) is related to the molality of B, $b = n_B/(1\,\text{kg water})$ and $n_A = (1\,\text{kg water})/M_A$ where M_A is the molar mass of water, by

$$\frac{n_B}{n_A} = \frac{n_B}{(1\,\text{kg})/M_A} = \frac{n_B M_A}{1\,\text{kg}} = bM_A = xb^\ominus M_A$$

and hence

$$v_A = v_A^* - 9.108M_A b^\ominus \int_0^{b/b^\ominus} x^{1/2}dx$$

$$= v_A^* - \frac{2}{3}(9.108M_A b^\ominus)(b/b^\ominus)^{3/2}$$

It then follows, by substituting the data (including $M_A = 1.802\times10^{-2}\,\text{kg mol}^{-1}$, the molar mass of water), that

$$V_A/(\text{cm}^3\text{mol}^{-1}) = 18.079 - 0.1094(b/b^\ominus)^{3/2}$$

The partial molar volumes are plotted in Fig. 5A.5.

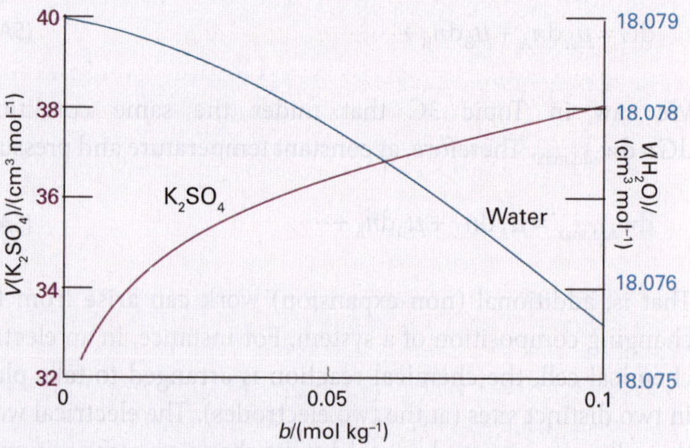

Figure 5A.5 The partial molar volumes of the components of an aqueous solution of potassium sulfate.

Self-test 5A.3 Repeat the calculation for a salt B for which $V_B/(\text{cm}^3\,\text{mol}^{-1}) = 6.218 + 5.146b - 7.147b^2$.

Answer: $V_A/(\text{cm}^3\,\text{mol}^{-1}) = 18.079 - 0.0464b^2 + 0.0859b^3$

5A.2 The thermodynamics of mixing

The dependence of the Gibbs energy of a mixture on its composition is given by eqn 5A.5, and we know that at constant temperature and pressure systems tend towards lower Gibbs energy. This is the link we need in order to apply thermodynamics to the discussion of spontaneous changes of composition, as in the mixing of two substances. One simple example of a spontaneous mixing process is that of two gases introduced into the same container. The mixing is spontaneous, so it must correspond to a decrease in G. We shall now see how to express this idea quantitatively.

(a) The Gibbs energy of mixing of perfect gases

Let the amounts of two perfect gases in the two containers be n_A and n_B; both are at a temperature T and a pressure p (Fig. 5A.6). At this stage, the chemical potentials of the two gases have their 'pure' values, which are obtained by applying the definition $\mu=G_m$ to eqn 3D.15 ($G_m(p)=G_m^\ominus+RT\ln(p/p^\ominus)$):

$$\mu=\mu^\ominus+RT\ln\frac{p}{p^\ominus} \qquad (5A.14a)$$

Perfect gas Variation of chemical potential with pressure

where $\mu^\ominus$ is the **standard chemical potential**, the chemical potential of the pure gas at 1 bar. It will be much simpler notationally if we agree to let p denote the pressure relative to $p^\ominus$; that is, to replace $p/p^\ominus$ by p, for then we can write

$$\mu=\mu^\ominus+RT\ln p \qquad (5A.14b)$$

To use the equations, we have to remember to replace p by $p/p^\ominus$ again. In practice, that simply means using the numerical value of p in bars. The Gibbs energy of the total system is then given by eqn 5A.5 as

$$G_i=n_A\mu_A+n_B\mu_B=n_A(\mu_A^\ominus+RT\ln p)+n_B(\mu_B^\ominus+RT\ln p) \qquad (5A.15a)$$

After mixing, the partial pressures of the gases are p_A and p_B, with $p_A+p_B=p$. The total Gibbs energy changes to

$$G_f=n_A(\mu_A^\ominus+RT\ln p_A)+n_B(\mu_B^\ominus+RT\ln p_B) \qquad (5A.15b)$$

The difference G_f-G_i, the **Gibbs energy of mixing**, $\Delta_{mix}G$, is therefore

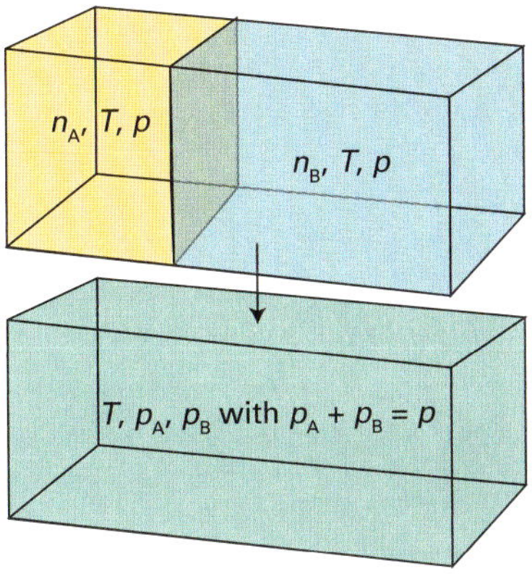

Figure 5A.6 The arrangement for calculating the thermodynamic functions of mixing of two perfect gases.

$$\Delta_{mix}G=n_ART\ln\frac{p_A}{p}+n_BRT\ln\frac{p_B}{p} \qquad (5A.15c)$$

At this point we may replace n_J by x_Jn, where n is the total amount of A and B, and use the relation between partial pressure and mole fraction (Topic 1A, $p_J=x_Jp$) to write $p_J/p=x_J$ for each component, which gives

$$\Delta_{mix}G=nRT(x_A\ln x_A+x_B\ln x_B) \qquad (5A.16)$$

Perfect gases Gibbs energy of mixing

Because mole fractions are never greater than 1, the logarithms in this equation are negative, and $\Delta_{mix}G<0$ (Fig. 5A.7). The conclusion that $\Delta_{mix}G$ is negative for all compositions confirms that perfect gases mix spontaneously in all proportions. However, the equation extends common sense by allowing us to discuss the process quantitatively.

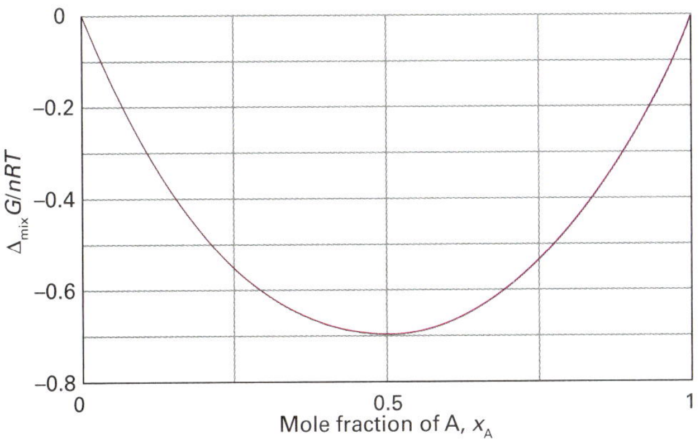

Figure 5A.7 The Gibbs energy of mixing of two perfect gases and (as discussed later) of two liquids that form an ideal solution. The Gibbs energy of mixing is negative for all compositions and temperatures, so perfect gases mix spontaneously in all proportions.

Example 5A.3 Calculating a Gibbs energy of mixing

A container is divided into two equal compartments (Fig. 5A.8). One contains 3.0 mol H_2(g) at 25 °C; the other contains 1.0 mol N_2(g) at 25 °C. Calculate the Gibbs energy of mixing when the partition is removed. Assume perfect behaviour.

Method Equation 5A.16 cannot be used directly because the two gases are initially at different pressures. We proceed by calculating the initial Gibbs energy from the chemical potentials. To do so, we need the pressure of each gas. Write the pressure of nitrogen as p; then the pressure of hydrogen as a multiple of p can be found from the gas laws. Next, calculate the Gibbs energy for the system when the partition is removed. The volume occupied by each gas doubles, so its initial partial pressure is halved.

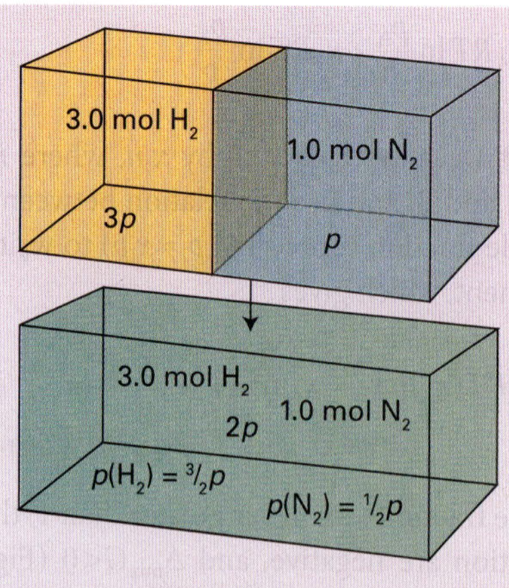

Figure 5A.8 The initial and final states considered in the calculation of the Gibbs energy of mixing of gases at different initial pressures.

Answer Given that the pressure of nitrogen is p, the pressure of hydrogen is $3p$; therefore, the initial Gibbs energy is

$$G_i = (3.0\ \text{mol})\{\mu^\ominus(H_2) + RT\ln 3p\} + (1.0\ \text{mol})\{\mu^\ominus(N_2) + RT\ln p\}$$

When the partition is removed and each gas occupies twice the original volume, the partial pressure of nitrogen falls to $\frac{1}{2}p$ and that of hydrogen falls to $\frac{3}{2}p$. Therefore, the Gibbs energy changes to

$$G_f = (3.0\ \text{mol})\{\mu^\ominus(H_2) + RT\ln \tfrac{3}{2}p\} + (1.0\ \text{mol})\{\mu^\ominus(N_2) + RT\ln \tfrac{1}{2}p\}$$

The Gibbs energy of mixing is the difference of these two quantities:

$$\begin{aligned}\Delta_{mix}G &= (3.0\,\text{mol})RT\ln\frac{\frac{3}{2}p}{3p} + (1.0\,\text{mol})RT\ln\frac{\frac{1}{2}p}{p}\\ &= -(3.0\,\text{mol})RT\ln 2 - (1.0\,\text{mol})RT\ln 2\\ &= -(4.0\,\text{mol})RT\ln 2 = -6.9\,\text{kJ}\end{aligned}$$

In this example, the value of $\Delta_{mix}G$ is the sum of two contributions: the mixing itself, and the changes in pressure of the two gases to their final total pressure, $2p$. When 3.0 mol H_2 mixes with 1.0 mol N_2 at the same pressure, with the volumes of the vessels adjusted accordingly, the change of Gibbs energy is −5.6 kJ. However, do not be misled into interpreting this negative change in Gibbs energy as a sign of spontaneity: in this case, the pressure changes, and $\Delta G < 0$ is a signpost of spontaneous change only at constant temperature and pressure.

Self-test 5A.4 Suppose that 2.0 mol H_2 at 2.0 atm and 25 °C and 4.0 mol N_2 at 3.0 atm and 25 °C are mixed by removing the partition between them. Calculate $\Delta_{mix}G$.

Answer: −9.7 kJ

(b) Other thermodynamic mixing functions

In Topic 3D it is shown that $(\partial G/\partial T)_{p,n} = -S$. It follows immediately from eqn 5A.16 that, for a mixture of perfect gases initially at the same pressure, the entropy of mixing, $\Delta_{mix}S$, is

$$\Delta_{mix}S = -\left(\frac{\partial \Delta_{mix}G}{\partial T}\right)_{p,n_A,n_B} = -nR(x_A\ln x_A + x_B\ln x_B)$$

Perfect gases Entropy of mixing (5A.17)

Because $\ln x < 0$, it follows that $\Delta_{mix}S > 0$ for all compositions (Fig. 5A.9).

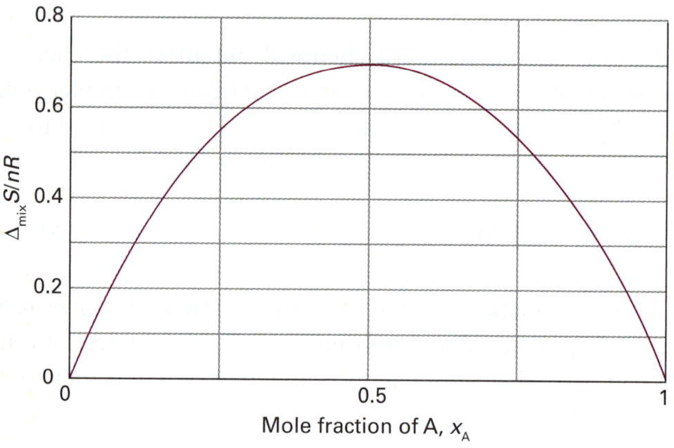

Figure 5A.9 The entropy of mixing of two perfect gases and (as discussed later) of two liquids that form an ideal solution. The entropy increases for all compositions and temperatures, so perfect gases mix spontaneously in all proportions. Because there is no transfer of heat to the surroundings when perfect gases mix, the entropy of the surroundings is unchanged. Hence, the graph also shows the total entropy of the system plus the surroundings when perfect gases mix.

Brief illustration 5A.2 The entropy of mixing

For equal amounts of perfect gas molecules that are mixed at the same pressure we set $x_A = x_B = \frac{1}{2}$ and obtain

$$\Delta_{mix}S = -nR\{\tfrac{1}{2}\ln\tfrac{1}{2} + \tfrac{1}{2}\ln\tfrac{1}{2}\} = nR\ln 2$$

with n the total amount of gas molecules. For 1 mol of each species, so $n = 2$ mol,

$$\Delta_{mix}S = (2\ \text{mol}) \times R\ln 2 = +11.5\,\text{J mol}^{-1}$$

An increase in entropy is what we expect when one gas disperses into the other and the disorder increases.

Self-test 5A.5 Calculate the change in entropy for the arrangement in *Example* 5A.3.

Answer: +23 J mol^{-1}

We can calculate the isothermal, isobaric (constant pressure) **enthalpy of mixing**, $\Delta_{\text{mix}}H$, the enthalpy change accompanying mixing, of two perfect gases from $\Delta G = \Delta H - T\Delta S$. It follows from eqns 5A.16 and 5A.17 that

$$\Delta_{\text{mix}}H = 0 \quad \textit{Perfect gases} \quad \text{Enthalpy of mixing} \quad (5A.18)$$

The enthalpy of mixing is zero, as we should expect for a system in which there are no interactions between the molecules forming the gaseous mixture. It follows that the whole of the driving force for mixing comes from the increase in entropy of the system because the entropy of the surroundings is unchanged.

5A.3 The chemical potentials of liquids

To discuss the equilibrium properties of liquid mixtures we need to know how the Gibbs energy of a liquid varies with composition. To calculate its value, we use the fact that, as established in Topic 4A, at equilibrium the chemical potential of a substance present as a vapour must be equal to its chemical potential in the liquid.

(a) Ideal solutions

We shall denote quantities relating to pure substances by a superscript *, so the chemical potential of pure A is written μ_A^* and as $\mu_A^*(\text{l})$ when we need to emphasize that A is a liquid. Because the vapour pressure of the pure liquid is p_A^* it follows from eqn 5A.14 that the chemical potential of A in the vapour (treated as a perfect gas) is $\mu_A^\ominus = +RT\ln p_A$ (with p_A to be interpreted as the relative pressure, $p_A/p^\ominus$). These two chemical potentials are equal at equilibrium (Fig. 5A.10), so we can write

$$\mu_A^* = \mu_A^\ominus + RT\ln p_A^* \quad (5A.19a)$$

If another substance, a solute, is also present in the liquid, the chemical potential of A in the liquid is changed to μ_A and its vapour pressure is changed to p_A. The vapour and solvent are still in equilibrium, so we can write

$$\mu_A = \mu_A^\ominus + RT\ln p_A \quad (5A.19b)$$

Next, we combine these two equations to eliminate the standard chemical potential of the gas. To do so, we write eqn 5A.19a as $\mu_A^\ominus = \mu_A^* - RT\ln p_A^*$ and substitute this expression into eqn 5A.19b to obtain

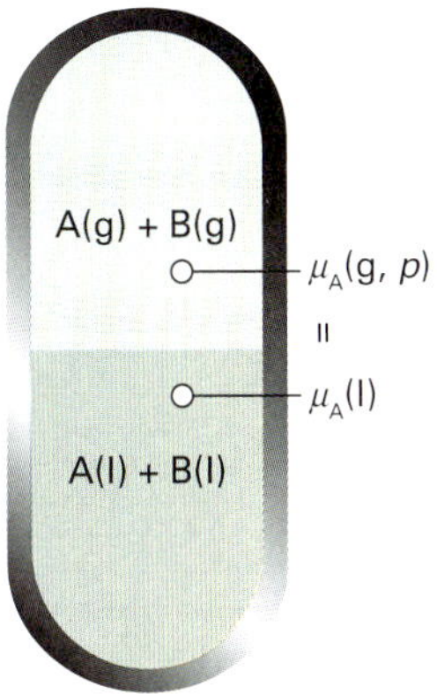

Figure 5A.10 At equilibrium, the chemical potential of the gaseous form of a substance A is equal to the chemical potential of its condensed phase. The equality is preserved if a solute is also present. Because the chemical potential of A in the vapour depends on its partial vapour pressure, it follows that the chemical potential of liquid A can be related to its partial vapour pressure.

$$\mu_A = \mu_A^* - RT\ln p_A^* + RT\ln p_A = \mu_A^* + RT\ln\frac{p_A}{p_A^*} \quad (5A.20)$$

In the final step we draw on additional experimental information about the relation between the ratio of vapour pressures and the composition of the liquid. In a series of experiments on mixtures of closely related liquids (such as benzene and methylbenzene), the French chemist François Raoult found that the ratio of the partial vapour pressure of each component to its vapour pressure as a pure liquid, p_A/p_A^*, is approximately equal to the mole fraction of A in the liquid mixture. That is, he established what we now call **Raoult's law**:

$$p_A = x_A p_A^* \quad \textit{Ideal solution} \quad \text{Raoult's law} \quad (5A.21)$$

This law is illustrated in Fig. 5A.11. Some mixtures obey Raoult's law very well, especially when the components are

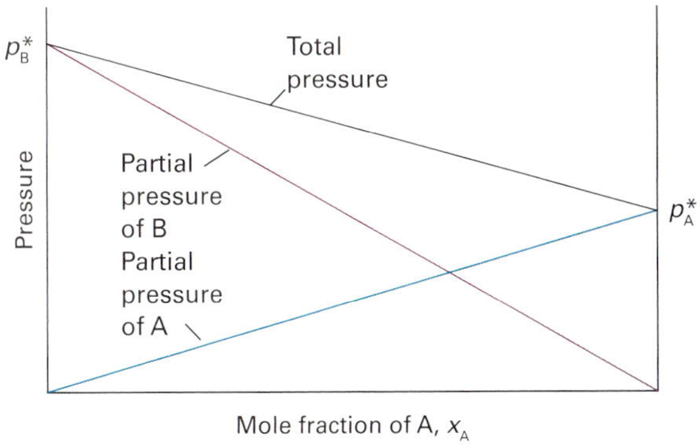

Figure 5A.11 The total vapour pressure and the two partial vapour pressures of an ideal binary mixture are proportional to the mole fractions of the components.

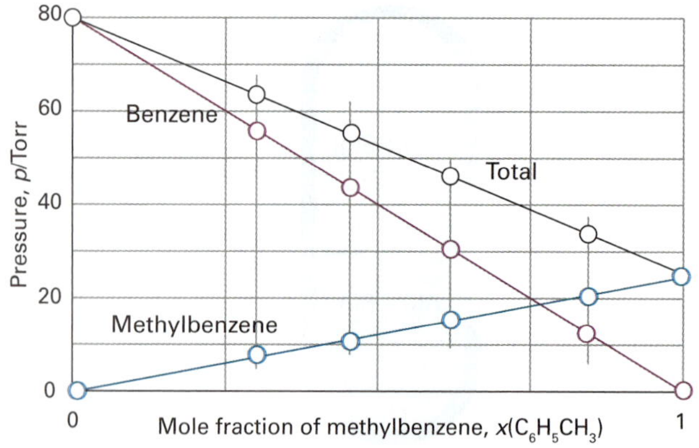

Figure 5A.12 Two similar liquids, in this case benzene and methylbenzene (toluene), behave almost ideally, and the variation of their vapour pressures with composition resembles that for an ideal solution.

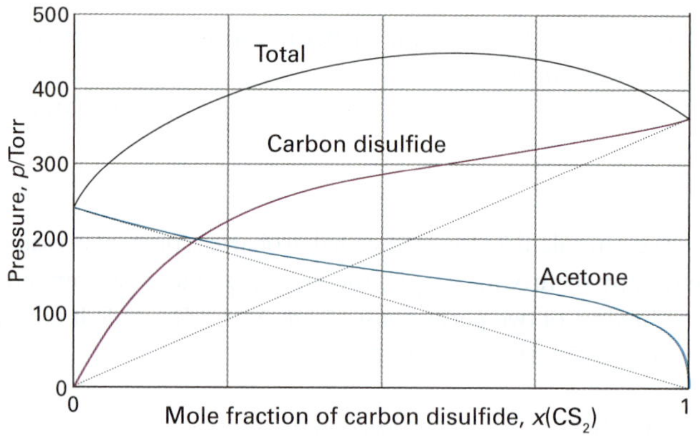

Figure 5A.13 Strong deviations from ideality are shown by dissimilar liquids, in this case carbon disulfide and acetone (propanone).

structurally similar (Fig. 5A.12). Mixtures that obey the law throughout the composition range from pure A to pure B are called **ideal solutions**.

Brief illustration 5A.3 Raoult's law

The vapour pressure of benzene at 20 °C is 75 Torr and that of methylbenzene is 21 Torr at the same temperature. In an equimolar mixture, $x_{\text{benzene}} = x_{\text{methylbenzene}} = \frac{1}{2}$ so the vapour pressure of each one in the mixture is

$$p_{\text{benzene}} = \tfrac{1}{2} \times 75\ \text{Torr} = 38\ \text{Torr}$$
$$p_{\text{methylbenzene}} = \tfrac{1}{2} \times 21\ \text{Torr} = 11\ \text{Torr}$$

The total vapour pressure of the mixture is 49 Torr. Given the two partial vapour pressures, it follows from the definition of partial pressure (Topic 1A) that the mole fractions in the vapour are $x_{\text{vap,benzene}} = (38\ \text{Torr})/(49\ \text{Torr}) = 0.78$ and $x_{\text{vap,methylbenzene}} = (11\ \text{Torr})/(49\ \text{Torr}) = 0.22$. The vapour is richer in the more volatile component (benzene).

Self-test 5A.6 At 90 °C the vapour pressure of 1,2-dimethylbenzene is 20 kPa and that of 1,3-dimethylbenzene is 18 kPa. What is the composition of the vapour when the liquid mixture has the composition $x_{12} = 0.33$ and $x_{13} = 0.67$?

Answer: $x_{\text{vap},12} = 0.35$, $x_{\text{vap},13} = 0.65$

For an ideal solution, it follows from eqns 5A.19a and 5A.21 that

$$\mu_A = \mu_A^* + RT \ln x_A \quad \textit{Ideal solution} \quad \text{Chemical potential} \quad (5A.22)$$

This important equation can be used as the *definition* of an ideal solution (so that it implies Raoult's law rather than stemming from it). It is in fact a better definition than eqn 5A.21 because it does not assume that the vapour is a perfect gas.

The molecular origin of Raoult's law is the effect of the solute on the entropy of the solution. In the pure solvent, the molecules have a certain disorder and a corresponding entropy; the vapour pressure then represents the tendency of the system and its surroundings to reach a higher entropy. When a solute is present, the solution has a greater disorder than the pure solvent because we cannot be sure that a molecule chosen at random will be a solvent molecule. Because the entropy of the solution is higher than that of the pure solvent, the solution has a lower tendency to acquire an even higher entropy by the solvent vaporizing. In other words, the vapour pressure of the solvent in the solution is lower than that of the pure solvent.

Some solutions depart significantly from Raoult's law (Fig. 5A.13). Nevertheless, even in these cases the law is obeyed increasingly closely for the component in excess (the solvent) as it approaches purity. The law is another example of a limiting law (in this case, achieving reliability as $x_A \rightarrow 1$) and is a good approximation for the properties of the solvent if the solution is dilute.

(b) Ideal–dilute solutions

In ideal solutions the solute, as well as the solvent, obeys Raoult's law. However, the English chemist William Henry found experimentally that, for real solutions at low concentrations, although the vapour pressure of the solute is proportional to its mole fraction, the constant of proportionality is not the vapour pressure of the pure substance (Fig. 5A.14). **Henry's law** is:

$$p_B = x_B K_B \quad \textit{Ideal–dilute solution} \quad \text{Henry's law} \quad (5A.23)$$

In this expression, x_B is the mole fraction of the solute and K_B is an empirical constant (with the dimensions of pressure)

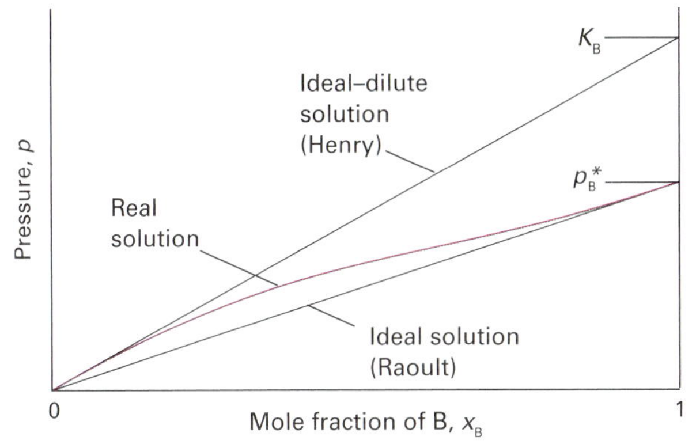

Figure 5A.14 When a component (the solvent) is nearly pure, it has a vapour pressure that is proportional to mole fraction with a slope p_B^* (Raoult's law). When it is the minor component (the solute) its vapour pressure is still proportional to the mole fraction, but the constant of proportionality is now K_B (Henry's law).

chosen so that the plot of the vapour pressure of B against its mole fraction is tangent to the experimental curve at $x_B=0$. Henry's law is therefore also a limiting law, achieving reliability as $x_B \rightarrow 0$.

Mixtures for which the solute B obeys Henry's law and the solvent A obeys Raoult's law are called **ideal–dilute solutions**. The difference in behaviour of the solute and solvent at low concentrations (as expressed by Henry's and Raoult's laws, respectively) arises from the fact that in a dilute solution the solvent molecules are in an environment very much like the one they have in the pure liquid (Fig. 5A.15). In contrast, the solute molecules are surrounded by solvent molecules, which is entirely different from their environment when pure. Thus, the solvent behaves like a slightly modified pure liquid, but the solute behaves entirely differently from its pure state unless the solvent and solute molecules happen to be very similar. In the latter case, the solute also obeys Raoult's law.

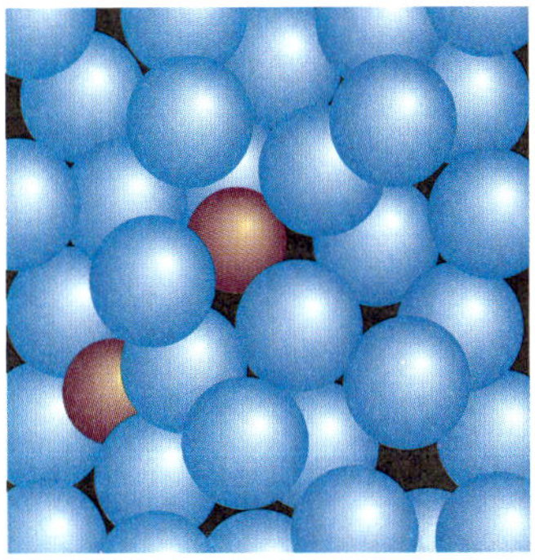

Figure 5A.15 In a dilute solution, the solvent molecules (the blue spheres) are in an environment that differs only slightly from that of the pure solvent. The solute particles (the purple spheres), however, are in an environment totally unlike that of the pure solute.

Example 5A.4 Investigating the validity of Raoult's and Henry's laws

The vapour pressures of each component in a mixture of propanone (acetone, A) and trichloromethane (chloroform, C) were measured at 35 °C with the following results:

x_C	0	0.20	0.40	0.60	0.80	1
p_C/kPa	0	4.7	11	18.9	26.7	36.4
p_A/kPa	46.3	33.3	23.3	12.3	4.9	0

Confirm that the mixture conforms to Raoult's law for the component in large excess and to Henry's law for the minor component. Find the Henry's law constants.

Method Both Raoult's and Henry's laws are statements about the form of the graph of partial vapour pressure against mole fraction. Therefore, plot the partial vapour pressures against mole fraction. Raoult's law is tested by comparing the data with the straight line $p_J = x_J p_J^*$ for each component in the region in which it is in excess (and acting as the solvent). Henry's law is tested by finding a straight line $p_J = x_J K_J^*$ that is tangent to each partial vapour pressure at low x, where the component can be treated as the solute.

Answer The data are plotted in Fig. 5A.16 together with the Raoult's law lines. Henry's law requires $K_A = 16.9$ kPa for propanone and $K_C = 20.4$ kPa for trichloromethane. Notice how the system deviates from both Raoult's and Henry's laws even for quite small departures from $x=1$ and $x=0$, respectively. We deal with these deviations in Topic 5E.

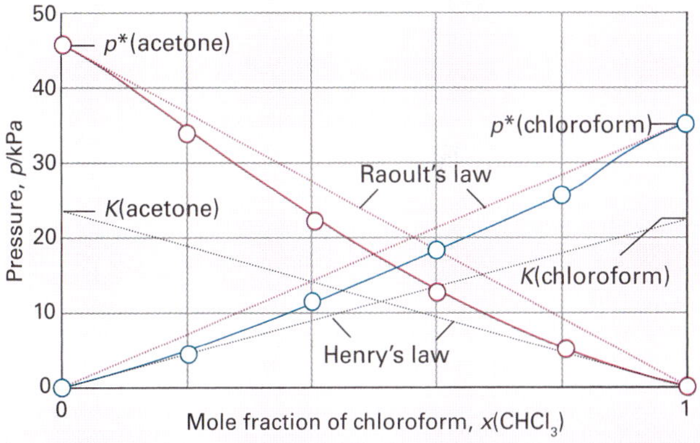

Figure 5A.16 The experimental partial vapour pressures of a mixture of chloroform (trichloromethane) and acetone (propanone) based on the data in *Example* 5A.3. The values of K are obtained by extrapolating the dilute solution vapour pressures, as explained in the *Example*.

Self-test 5A.7 The vapour pressure of chloromethane at various mole fractions in a mixture at 25 °C was found to be as follows:

x	0.005	0.009	0.019	0.024
p/kPa	27.3	48.4	101	126

Estimate Henry's law constant.

Answer: 5 MPa

For practical applications, Henry's law is expressed in terms of the molality, b, of the solute, $p_B = b_B K_B$. Some Henry's law data for this convention are listed in Table 5A.1. As well as providing a link between the mole fraction of solute and its partial pressure, the data in the table may also be used to calculate gas solubilities. A knowledge of Henry's law constants for gases in blood and fats is important for the discussion of respiration, especially when the partial pressure of oxygen is abnormal, as in diving and mountaineering, and for the discussion of the action of gaseous anaesthetics.

Table 5A.1* Henry's law constants for gases in water at 298 K, $K/(\text{kPa kg mol}^{-1})$

	$K/(\text{kPa kg mol}^{-1})$
CO_2	3.01×10^3
H_2	1.28×10^5
N_2	1.56×10^5
O_2	7.92×10^4

* More values are given in the *Resource section*.

Brief illustration 5A.4 Henry's law and gas solubility

To estimate the molar solubility of oxygen in water at 25 °C and a partial pressure of 21 kPa, its partial pressure in the atmosphere at sea level, we write

$$b_{O_2} = \frac{p_{O_2}}{K_{O_2}} = \frac{21\,\text{kPa}}{7.9\times10^4\,\text{kPa kg mol}^{-1}} = 2.9\times10^{-4}\,\text{mol kg}^{-1}$$

The molality of the saturated solution is therefore 0.29 mmol kg^{-1}. To convert this quantity to a molar concentration, we assume that the mass density of this dilute solution is essentially that of pure water at 25 °C, or $\rho = 0.997\,\text{kg dm}^{-3}$. It follows that the molar concentration of oxygen is

$$[O_2] = b_{O_2}\rho = (2.9\times10^{-4}\,\text{mol kg}^{-1})\times(0.997\,\text{kg dm}^{-3}) = 0.29\,\text{mmol dm}^{-3}$$

Self-test 5A.8 Calculate the molar solubility of nitrogen in water exposed to air at 25 °C; partial pressures were calculated in *Example* 1A.3 of Topic 1A.

Answer: 0.51 mmol dm^{-3}

Checklist of concepts

- ☐ 1. The **molar concentration** of a solute is the amount of solute divided by the volume of the solution.
- ☐ 2. The **molality** of a solute is the amount of solute divided by the mass of solvent.
- ☐ 3. The **partial molar volume** of a substance is the contribution to the volume that a substance makes when it is part of a mixture.
- ☐ 4. The **chemical potential** is the partial molar Gibbs energy and enables us to express the dependence of the Gibbs energy on the composition of a mixture.
- ☐ 5. The chemical potential also shows how, under a variety of different conditions, the thermodynamic functions vary with composition.
- ☐ 6. The **Gibbs–Duhem equation** shows how the changes in chemical potential of the components of a mixture are related.
- ☐ 7. The **Gibbs energy of mixing** is calculated by forming the difference of the Gibbs energies before and after mixing: the quantity is negative for perfect gases at the same pressure.
- ☐ 8. The **entropy of mixing** of perfect gases initially at the same pressure is positive and the enthalpy of mixing is zero.
- ☐ 9. **Raoult's law** provides a relation between the vapour pressure of a substance and its mole fraction in a mixture; it is the basis of the definition of an ideal solution.
- ☐ 10. **Henry's law** provides a relation between the vapour pressure of a solute and its mole fraction in a mixture; it is the basis of the definition of an ideal–dilute solution.

Checklist of equations

Property	Equation	Comment	Equation number
Partial molar volume	$V_J = (\partial V/\partial n_J)_{p,T,n'}$	Definition	5A.1
Chemical potential	$\mu_J = (\partial G/\partial n_J)_{p,T,n'}$	Definition	5A.4
Total Gibbs energy	$G = n_A\mu_A + n_B\mu_B$		5A.5

Property	Equation	Comment	Equation number
Fundamental equation of chemical thermodynamics	$dG = Vdp - SdT + \mu_A dn_A + \mu_B dn_B + \ldots$		5A.6
Gibbs–Duhem equation	$\sum_J n_J d\mu_J = 0$		5A.12b
Chemical potential of a gas	$\mu = \mu^{\ominus} + RT\ln(p/p^{\ominus})$	Perfect gas	5A.14a
Gibbs energy of mixing	$\Delta_{mix}G = nRT(x_A \ln x_A + x_B \ln x_B)$	Perfect gases and ideal solutions	5A.16
Entropy of mixing	$\Delta_{mix}S = -nR(x_A \ln x_A + x_B \ln x_B)$	Perfect gases and ideal solutions	5A.17
Enthalpy of mixing	$\Delta_{mix}H = 0$	Perfect gases and ideal solutions	5A.18
Raoult's law	$p_A = x_A p_A^*$	True for ideal solutions; limiting law as $x_A \to 1$	5A.21
Chemical potential of component	$\mu_A = \mu_A^* + RT\ln x_A$	Ideal solution	5A.22
Henry's law	$p_B = x_B K_B$	True for ideal-dilute solutions; limiting law as $x_B \to 0$	5A.23

Brief illustration 5B.2 Excess functions

Figure 5B.3 shows two examples of the composition dependence of molar excess functions. In Fig 5B.3a, the positive values of H^E, which implies that $\Delta_{mix}H>0$, indicate that the A–B interactions in the mixture are less attractive than the A–A and B–B interactions in the pure liquids (which are benzene and pure cyclohexane). The symmetrical shape of the curve reflects the similar strengths of the A–A and B–B interactions. Figure 5B.3b shows the composition dependence of the excess volume, V^E, of a mixture of tetrachloroethene and cyclopentane. At high mole fractions of cyclopentane, the solution contracts as tetrachloroethene is added because the ring structure of cyclopentane results in inefficient packing of the molecules but as tetrachloroethene is added, the molecules in the mixture pack together more tightly. Similarly, at high mole fractions of tetrachloroethene, the solution expands as cyclopentane is added because tetrachloroethene molecules are nearly flat and pack efficiently in the pure liquid but become disrupted as bulky ring cyclopentane is added.

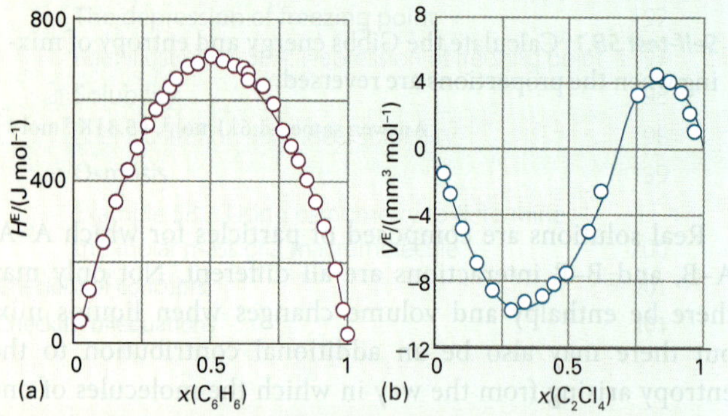

Figure 5B.3 Experimental excess functions at 25 °C. (a) H^E for benzene/cyclohexane; this graph shows that the mixing is endothermic (because $\Delta_{mix}H=0$ for an ideal solution). (b) The excess volume, V^E, for tetrachloroethene/cyclopentane; this graph shows that there is a contraction at low tetrachloroethene mole fractions, but an expansion at high mole fractions (because $\Delta_{mix}V=0$ for an ideal mixture).

Self-test 5B.2 Would you expect the excess volume of mixing of oranges and melons to be positive or negative?

Answer: Positive; close-packing disrupted

Deviations of the excess energies from zero indicate the extent to which the solutions are non-ideal. In this connection a useful model system is the **regular solution**, a solution for which $H^E \neq 0$ but $S^E=0$. We can think of a regular solution as one in which the two kinds of molecules are distributed randomly (as in an ideal solution) but have different energies of interactions with each other. To express this concept more quantitatively we can suppose that the excess enthalpy depends on composition as

$$H^E = n\xi RT x_A x_B \quad (5B.6)$$

where ξ (xi) is a dimensionless parameter that is a measure of the energy of AB interactions relative to that of the AA and BB interactions. (For H^E expressed as a molar quantity, discard the n.) The function given by eqn 5B.6 is plotted in Fig. 5B.4, and we see it resembles the experimental curve in Fig. 5B.3a. If $\xi<0$, mixing is exothermic and the solute–solvent interactions are more favourable than the solvent–solvent and solute–solute interactions. If $\xi>0$, then the mixing is endothermic. Because the entropy of mixing has its ideal value for a regular solution, the excess Gibbs energy is equal to the excess enthalpy, and the Gibbs energy of mixing is

$$\Delta_{mix}G = nRT(x_A \ln x_A + x_B \ln x_B + \xi x_A x_B) \quad (5B.7)$$

Figure 5B.5 shows how $\Delta_{mix}G$ varies with composition for different values of ξ. The important feature is that for $\xi>2$ the graph shows two minima separated by a maximum. The implication of this observation is that, provided $\xi>2$, then the

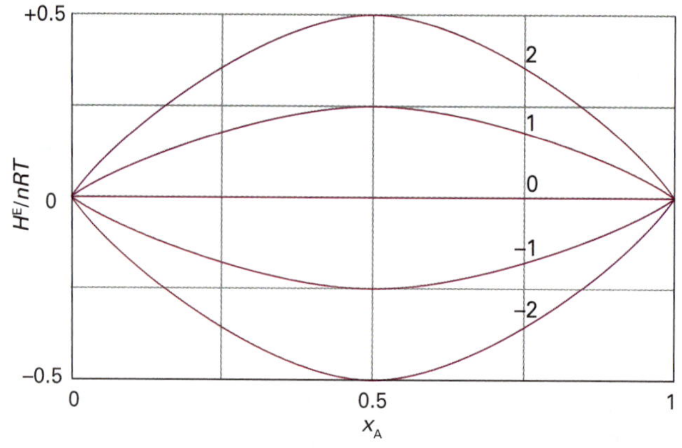

Figure 5B.4 The excess enthalpy according to a model in which it is proportional to $\xi x_A x_B$, for different values of the parameter ξ.

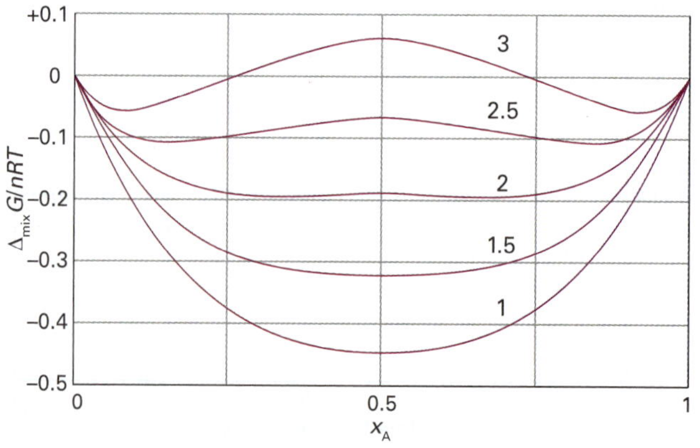

Figure 5B.5 The Gibbs energy of mixing for different values of the parameter ξ.

system will separate spontaneously into two phases with compositions corresponding to the two minima, for that separation corresponds to a reduction in Gibbs energy. We develop this point in Topic 5C.

Example 5B.1 Identifying the parameter for a regular solution

Identify the value of the parameter ξ that would be appropriate to model a mixture of benzene and cyclohexane at 25 °C and estimate the Gibbs energy of mixing to produce an equimolar mixture.

Method Refer to Fig. 5B.3a and identify the value at the curve maximum, and then relate it to eqn 5B.6 written as a molar quantity ($H^E = \xi RTx_Ax_B$). For the second part, assume that the solution is regular and that the Gibbs energy of mixing is given by eqn 5B.7.

Answer The experimental value occurs close to $x_A = x_B = \frac{1}{2}$ and its value is close to 710 J mol⁻¹. It follows that

$$\xi = \frac{H^E}{RTx_Ax_B} = \frac{701\,\text{J mol}^{-1}}{(8.3145\,\text{J K}^{-1}\,\text{mol}^{-1})\times(298\,\text{K})\times\frac{1}{2}\times\frac{1}{2}} = 1.13$$

The total Gibbs energy of mixing to achieve the stated composition (provided the solution is regular) is therefore

$$\begin{aligned}\Delta_{\text{mix}}G/n &= -RT\ln 2 + 701\,\text{J mol}^{-1}\\ &= -1.72\,\text{kJ mol}^{-1} + 0.701\,\text{kJ mol}^{-1} = -1.02\,\text{kJ mol}^{-1}\end{aligned}$$

Self-test 5B.3 Fit the entire data set, as best as can be inferred from the graph in Fig. 5B.3a, to an expression of the form in eqn 5B.6 by a curve-fitting procedure.

Answer: The best fit of the form $Ax(1-x)$ to the data pairs

X	0.1	0.2	0.3	0.4	0.5	0.6	0.7	0.8	0.9
H^E/(J mol⁻¹)	150	350	550	680	700	690	600	500	280

is $A = 690\,\text{J mol}^{-1}$

5B.2 Colligative properties

The properties we consider are the lowering of vapour pressure, the elevation of boiling point, the depression of freezing point, and the osmotic pressure arising from the presence of a solute. In dilute solutions these properties depend only on the number of solute particles present, not their identity. For this reason, they are called **colligative properties** (denoting 'depending on the collection'). In this development, we denote the solvent by A and the solute by B.

We assume throughout the following that the solute is not volatile, so it does not contribute to the vapour. We also assume that the solute does not dissolve in the solid solvent: that is, the pure solid solvent separates when the solution is frozen. The latter assumption is quite drastic, although it is true of many mixtures; it can be avoided at the expense of more algebra, but that introduces no new principles.

(a) The common features of colligative properties

All the colligative properties stem from the reduction of the chemical potential of the liquid solvent as a result of the presence of solute. For an ideal solution (one that obeys Raoult's law, Topic 5A; $p_A = x_A p_A^*$), the reduction is from μ_A^* for the pure solvent to $\mu_A = \mu_A^* + RT\ln x_A$ when a solute is present ($\ln x_A$ is negative because $x_A < 1$). There is no direct influence of the solute on the chemical potential of the solvent vapour and the solid solvent because the solute appears in neither the vapour nor the solid. As can be seen from Fig. 5B.6, the reduction in chemical potential of the solvent implies that the liquid–vapour equilibrium occurs at a higher temperature (the boiling point is raised) and the solid–liquid equilibrium occurs at a lower temperature (the freezing point is lowered).

The molecular origin of the lowering of the chemical potential is not the energy of interaction of the solute and solvent particles, because the lowering occurs even in an ideal solution (for which the enthalpy of mixing is zero). If it is not an enthalpy effect, it must be an entropy effect. The vapour pressure of the pure liquid reflects the tendency of the solution towards greater entropy, which can be achieved if the liquid vaporizes to form a gas. When a solute is present, there is an additional contribution to the entropy of the liquid, even in an ideal solution. Because the entropy of the liquid is already higher than that of the pure liquid, there is a weaker tendency to form the gas (Fig. 5B.7). The effect of the solute appears as a lowered vapour pressure, and hence a higher boiling point. Similarly, the enhanced molecular randomness of the solution opposes the tendency to freeze. Consequently, a lower temperature must be reached

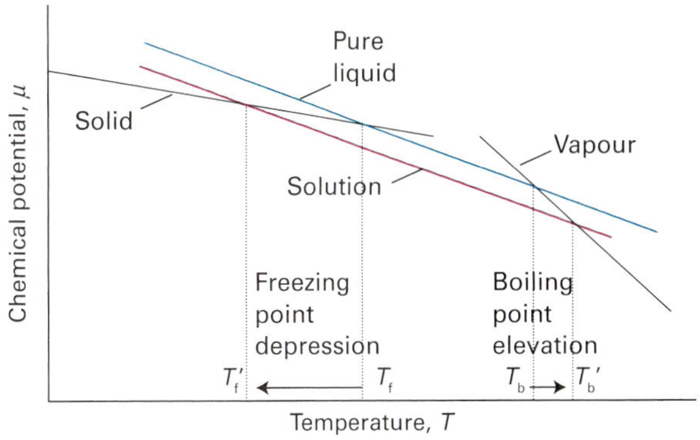

Figure 5B.6 The chemical potential of a solvent in the presence of a solute. The lowering of the liquid's chemical potential has a greater effect on the freezing point than on the boiling point because of the angles at which the lines intersect.

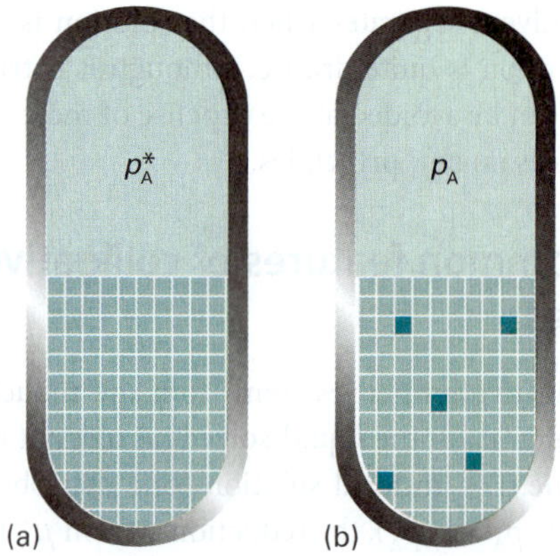

Figure 5B.7 The vapour pressure of a pure liquid represents a balance between the increase in disorder arising from vaporization and the decrease in disorder of the surroundings. (a) Here the structure of the liquid is represented highly schematically by the grid of squares. (b) When solute (the dark squares) is present, the disorder of the condensed phase is higher than that of the pure liquid, and there is a decreased tendency to acquire the disorder characteristic of the vapour.

before equilibrium between solid and solution is achieved. Hence, the freezing point is lowered.

The strategy for the quantitative discussion of the elevation of boiling point and the depression of freezing point is to look for the temperature at which, at 1 atm, one phase (the pure solvent vapour or the pure solid solvent) has the same chemical potential as the solvent in the solution. This is the new equilibrium temperature for the phase transition at 1 atm, and hence corresponds to the new boiling point or the new freezing point of the solvent.

(b) The elevation of boiling point

The heterogeneous equilibrium of interest when considering boiling is between the solvent vapour and the solvent in solution at 1 atm (Fig. 5B.8). The equilibrium is established at a temperature for which

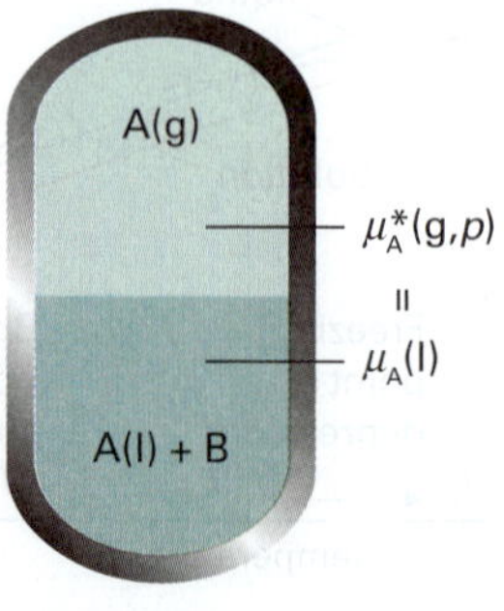

Figure 5B.8 The heterogeneous equilibrium involved in the calculation of the elevation of boiling point is between A in the pure vapour and A in the mixture, A being the solvent and B a non-volatile solute.

$$\mu_A^*(g) = \mu_A^*(l) + RT \ln x_A \quad (5B.8)$$

(The pressure of 1 atm is the same throughout, and will not be written explicitly.) We show in the following *Justification* that this equation implies that the presence of a solute at a mole fraction x_B causes an increase in normal boiling point from T^* to $T^* + \Delta T_b$, where

$$\Delta T_b = K x_B \quad K = \frac{RT^{*2}}{\Delta_{vap}H} \quad \textit{Ideal solution} \quad \text{Elevation of boiling point} \quad (5B.9)$$

Justification 5B.1 The elevation of the boiling point of a solvent

Equation 5B.8 can be rearranged into

$$\ln x_A = \frac{\mu_A^*(g) - \mu_A^*(l)}{RT} = \frac{\Delta_{vap}G}{RT}$$

where $\Delta_{vap}G$ is the Gibbs energy of vaporization of the pure solvent (A). First, to find the relation between a change in composition and the resulting change in boiling temperature, we differentiate both sides with respect to temperature and use the Gibbs–Helmholtz equation (Topic 3D, $(\partial(G/T)/\partial T)_p = -H/T^2$) to express the term on the right:

$$\frac{d \ln x_A}{dT} = \frac{1}{R}\frac{d(\Delta_{vap}G/T)}{dT} = -\frac{\Delta_{vap}H}{RT^2}$$

Now multiply both sides by dT and integrate from $x_A = 1$, corresponding to $\ln x_A = 0$ (and when $T = T^*$, the boiling point of pure A) to x_A (when the boiling point is T):

$$\int_0^{\ln x_A} d \ln x_A = -\frac{1}{R}\int_{T^*}^{T} \frac{\Delta_{vap}H}{T^2} dT$$

The left-hand side integrates to $\ln x_A$, which is equal to $\ln(1 - x_B)$. The right-hand side can be integrated if we assume that the enthalpy of vaporization is a constant over the small range of temperatures involved and can be taken outside the integral. Thus, we obtain

$$\ln(1 - x_B) = -\frac{\Delta_{vap}H}{R}\int_{T^*}^{T} \frac{1}{T^2} dT$$

and therefore

$$\ln(1 - x_B) = \frac{\Delta_{vap}H}{R}\left(\frac{1}{T} - \frac{1}{T^*}\right)$$

We now suppose that the amount of solute present is so small that $x_B \ll 1$, and use the expansion $\ln(1-x) = -x - \frac{1}{2}x^2 + \cdots \approx -x$ (*Mathematical background* 1) and hence obtain

$$x_B = \frac{\Delta_{vap}H}{R}\left(\frac{1}{T^*} - \frac{1}{T}\right)$$

Finally, because $T \approx T^*$, it also follows that

$$\frac{1}{T^*} - \frac{1}{T} = \frac{T - T^*}{TT^*} \approx \frac{T - T^*}{T^{*2}} = \frac{\Delta T_b}{T^{*2}}$$

with $\Delta T_b = T - T^*$. The previous equation then rearranges into eqn 5B.9.

Because eqn 5B.9 makes no reference to the identity of the solute, only to its mole fraction, we conclude that the elevation of boiling point is a colligative property. The value of ΔT does depend on the properties of the solvent, and the biggest changes occur for solvents with high boiling points. By Trouton's rule (Topic 3B), $\Delta_{vap}H/T^*$ is a constant; therefore eqn 5B.9 has the form $\Delta T \propto T^*$ and is independent of $\Delta_{vap}H$ itself. For practical applications of eqn 5B.9, we note that the mole fraction of B is proportional to its molality, b, in the solution, and write

$$\Delta T_b = K_b b \quad \textit{Empirical relation} \quad \text{Boiling point elevation} \quad (5B.10)$$

where K_b is the empirical **boiling-point constant** of the solvent (Table 5B.1).

Brief illustration 5B.3 Elevation of boiling point

The boiling-point constant of water is 0.51 K kg mol^{-1}, so a solute present at a molality of 0.10 mol kg^{-1} would result in an elevation of boiling point of only 0.051 K. The boiling-point constant of benzene is significantly larger, at 2.53 K kg mol^{-1}, so the elevation would be 0.25 K.

Self-test 5B.4 Identify the feature that accounts for the difference in boiling-point constants of water and benzene.

Answer: High enthalpy of vaporization of water; given molality corresponds to a smaller mole fraction

(c) The depression of freezing point

The heterogeneous equilibrium now of interest is between pure solid solvent A and the solution with solute present at a mole fraction x_B (Fig. 5B.9). At the freezing point, the chemical potentials of A in the two phases are equal:

Table 5B.1* Freezing-point (K_f) and boiling-point (K_b) constants

	K_f/(K kg mol^{-1})	K_b/(K kg mol^{-1})
Benzene	5.12	2.53
Camphor	40	
Phenol	7.27	3.04
Water	1.86	0.51

* More values are given in the *Resource section*.

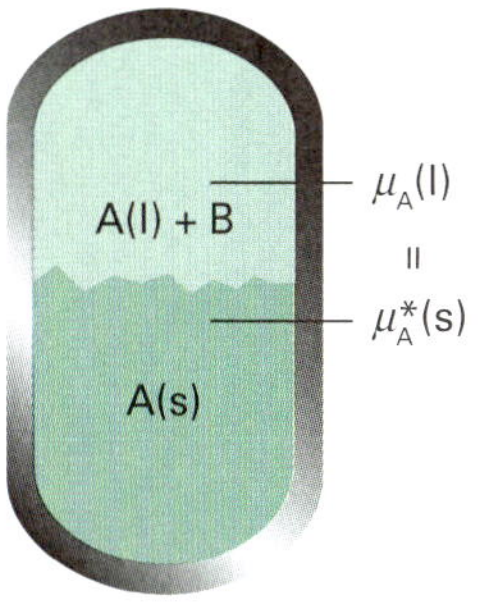

Figure 5B.9 The heterogeneous equilibrium involved in the calculation of the lowering of freezing point is between A in the pure solid and A in the mixture, A being the solvent and B a solute that is insoluble in solid A.

$$\mu_A^*(s) = \mu_A^*(l) + RT \ln x_A \quad (5B.11)$$

The only difference between this calculation and the last is the appearance of the solid's chemical potential in place of the vapour's. Therefore we can write the result directly from eqn 5B.9:

$$\Delta T_f = K' x_B \quad K' = \frac{RT^{*2}}{\Delta_{fus}H} \quad \text{Freezing point depression} \quad (5B.12)$$

where ΔT_f is the freezing point depression, $T^* - T$, and $\Delta_{fus}H$ is the enthalpy of fusion of the solvent. Larger depressions are observed in solvents with low enthalpies of fusion and high melting points. When the solution is dilute, the mole fraction is proportional to the molality of the solute, b, and it is common to write the last equation as

$$\Delta T_f = K_f b \quad \textit{Empirical relation} \quad \text{Freezing point depression} \quad (5B.13)$$

where K_f is the empirical **freezing-point constant** (Table 5B.1). Once the freezing-point constant of a solvent is known, the depression of freezing point may be used to measure the molar mass of a solute in the method known as **cryoscopy**; however, the technique is of little more than historical interest.

Brief illustration 5B.4 Depression of freezing point

The freezing-point constant of water is 1.86 K kg mol^{-1}, so a solute present at a molality of 0.10 mol kg^{-1} would result in a depression of freezing point of only 0.19 K. The freezing-point constant of camphor is significantly larger, at 40 K kg mol^{-1}, so the depression would be 4.0 K. Camphor was once widely used for estimates of molar mass by cryoscopy.

Self-test 5B.5 Why are freezing-point constants typically larger than the corresponding boiling-point constants of a solvent?

Answer: Enthalpy of fusion is smaller than the enthalpy of vaporization of a substance

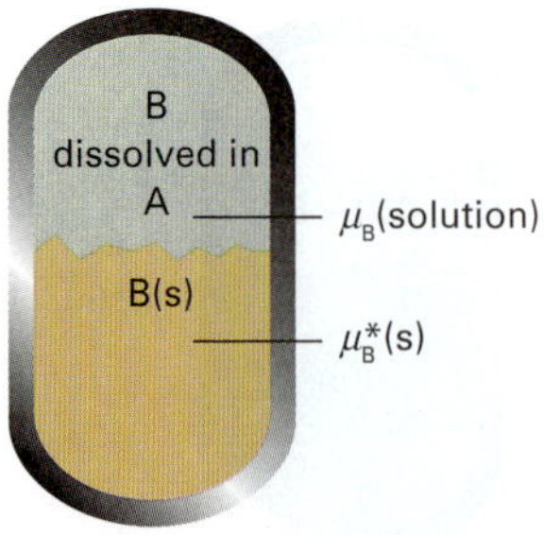

Figure 5B.10 The heterogeneous equilibrium involved in the calculation of the solubility is between pure solid B and B in the mixture.

(d) Solubility

Although solubility is not a colligative property (because solubility varies with the identity of the solute), it may be estimated by the same techniques as we have been using. When a solid solute is left in contact with a solvent, it dissolves until the solution is saturated. Saturation is a state of equilibrium, with the undissolved solute in equilibrium with the dissolved solute. Therefore, in a saturated solution the chemical potential of the pure solid solute, $\mu_B^*(s)$, and the chemical potential of B in solution, μ_B, are equal (Fig. 5B.10). Because the latter is related to the mole fraction in the solution by $\mu_B = \mu_B^*(l) + RT \ln x_B$, we can write

$$\mu_B^*(s) = \mu_B^*(l) + RT \ln x_B \qquad (5B.14)$$

This expression is the same as the starting equation of the last section, except that the quantities refer to the solute B, not the solvent A. We now show in the following *Justification* that

$$\ln x_B = \frac{\Delta_{fus}H}{R}\left(\frac{1}{T_f} - \frac{1}{T}\right) \qquad \text{Ideal solubility} \quad (5B.15)$$

where $\Delta_{fus}H$ is the enthalpy of fusion of the solute and T_f is its melting point.

Justification 5B.2 The solubility of an ideal solute

The starting point is the same as in *Justification* 5B.1 but the aim is different. In the present case, we want to find the mole fraction of B in solution at equilibrium when the temperature is T. Therefore, we start by rearranging eqn 5B.14 into

$$\ln x_B = \frac{\mu_B^*(s) - \mu_B^*(l)}{RT} = -\frac{\Delta_{fus}G}{RT}$$

As in *Justification* 5B.1, we relate the change in composition $d \ln x_B$ to the change in temperature by differentiation and use of the Gibbs–Helmholtz equation. Then we integrate from the melting temperature of B (when $x_B = 1$ and $\ln x_B = 0$) to the *lower* temperature of interest (when x_B has a value between 0 and 1):

$$\int_0^{\ln x_B} d \ln x_B = \frac{1}{R}\int_{T_f}^{T} \frac{\Delta_{fus}H}{T^2} dT$$

If we suppose that the enthalpy of fusion of B is constant over the range of temperatures of interest, it can be taken outside the integral, and we obtain eqn 5B.15.

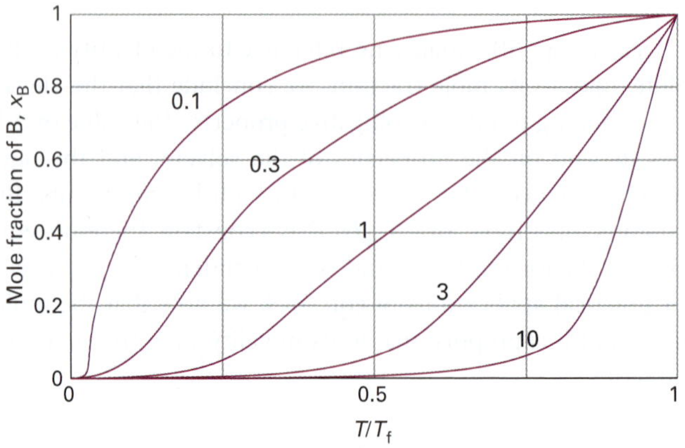

Figure 5B.11 The variation of solubility (the mole fraction of solute in a saturated solution) with temperature (T_f is the freezing temperature of the solute). Individual curves are labelled with the value of $\Delta_{fus}H/RT_f$.

Equation 5B.15 is plotted in Fig. 5B.11. It shows that the solubility of B decreases exponentially as the temperature is lowered from its melting point. The illustration also shows that solutes with high melting points and large enthalpies of melting have low solubilities at normal temperatures. However, the detailed content of eqn 5B.15 should not be treated too seriously because it is based on highly questionable approximations, such as the ideality of the solution. One aspect of its approximate character is that it fails to predict that solutes will have different solubilities in different solvents, for no solvent properties appear in the expression.

Brief illustration 5B.5 Ideal solubility

The ideal solubility of naphthalene in benzene is calculated from eqn 5B.15 by noting that the enthalpy of fusion of naphthalene is 18.80 kJ mol^{-1} and its melting point is 354 K. Then, at 20 °C,

$$\ln x_{naphthalene} = \frac{1.880 \times 10^4\ \mathrm{J\,mol^{-1}}}{8.3145\ \mathrm{J\,K^{-1}\,mol^{-1}}}\left(\frac{1}{354\ \mathrm{K}} - \frac{1}{293\ \mathrm{K}}\right) = -1.32\ldots$$

and therefore $x_{naphthalene} = 0.26$. This mole fraction corresponds to a molality of 4.5 mol kg^{-1} (580 g of naphthalene in 1 kg of benzene).

Self-test 5B.6 Plot the solubility of naphthalene as a function of temperature against mole fraction: in Topic 5C we see that such diagrams are 'temperature–composition phase diagrams'.

Answer: See Fig. 5B.12.

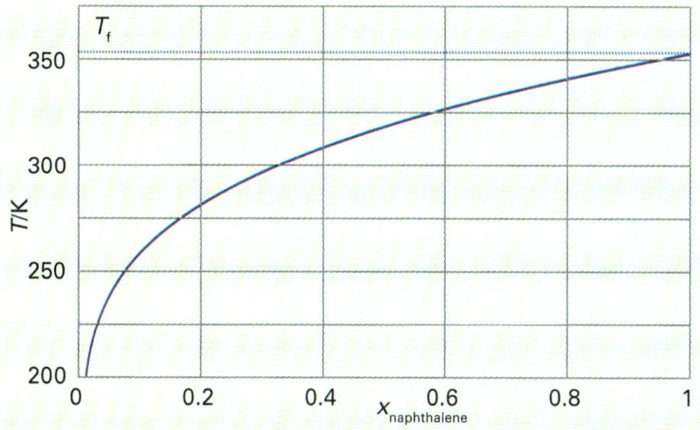

Figure 5B.12 The theoretical solubility of naphthalene in benzene, as calculated in *Self-test* 5B.6.

(e) Osmosis

The phenomenon of **osmosis** (from the Greek word for 'push') is the spontaneous passage of a pure solvent into a solution separated from it by a **semipermeable membrane**, a membrane permeable to the solvent but not to the solute (Fig. 5B.13). The **osmotic pressure**, Π, is the pressure that must be applied to the solution to stop the influx of solvent. Important examples of osmosis include transport of fluids through cell membranes, dialysis and **osmometry**, the determination of molar mass by the measurement of osmotic pressure. Osmometry is widely used to determine the molar masses of macromolecules.

In the simple arrangement shown in Fig. 5B.14, the opposing pressure arises from the head of solution that the osmosis itself produces. Equilibrium is reached when the hydrostatic pressure of the column of solution matches the osmotic pressure.

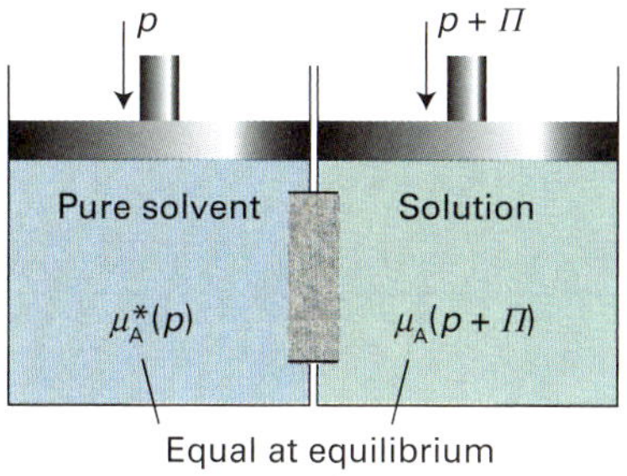

Figure 5B.13 The equilibrium involved in the calculation of osmotic pressure, Π, is between pure solvent A at a pressure p on one side of the semipermeable membrane and A as a component of the mixture on the other side of the membrane, where the pressure is $p+\Pi$.

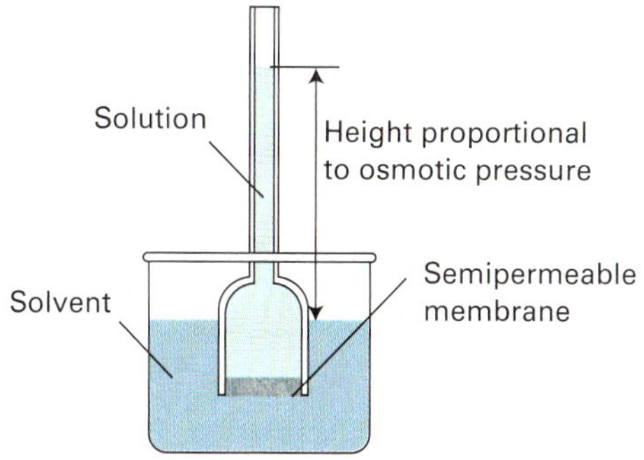

Figure 5B.14 In a simple version of the osmotic pressure experiment, A is at equilibrium on each side of the membrane when enough has passed into the solution to cause a hydrostatic pressure difference.

The complicating feature of this arrangement is that the entry of solvent into the solution results in its dilution, and so it is more difficult to treat than the arrangement in Fig. 5B.13, in which there is no flow and the concentrations remain unchanged.

The thermodynamic treatment of osmosis depends on noting that, at equilibrium, the chemical potential of the solvent must be the same on each side of the membrane. The chemical potential of the solvent is lowered by the solute, but is restored to its 'pure' value by the application of pressure. As shown in the following *Justification*, this equality implies that for dilute solutions the osmotic pressure is given by the **van 't Hoff equation**:

$$\Pi = [\mathrm{B}]RT \qquad \text{van 't Hoff equation} \qquad (5B.16)$$

where $[\mathrm{B}] = n_{\mathrm{B}}/V$ is the molar concentration of the solute.

Justification 5B.3 The van 't Hoff equation

On the pure solvent side the chemical potential of the solvent, which is at a pressure p, is $\mu_A^*(p)$. On the solution side, the chemical potential is lowered by the presence of the solute, which reduces the mole fraction of the solvent from 1 to x_A. However, the chemical potential of A is raised on account of the greater pressure, $p+\Pi$, that the solution experiences. At equilibrium the chemical potential of A is the same in both compartments, and we can write

$$\mu_A^*(p) = \mu_A(x_A, p+\Pi)$$

The presence of solute is taken into account in the normal way by using eqn 5B.1:

$$\mu_A(x_A, p+\Pi) = \mu_A^*(p+\Pi) + RT\ln x_A$$

Equation 3D.12b,

$$G_m(p_f) = G_m(p_i) + \int_{p_i}^{p_f} V_m \mathrm{d}p$$

written as

$$\mu_A^*(p+\Pi)=\mu_A^*(p)+\int_p^{p+\Pi} V_m\mathrm{d}p$$

where V_m is the molar volume of the pure solvent A, shows how to take the effect of pressure into account:. When these three equations are combined and the $\mu_A^*(p)$ are cancelled we are left with

$$-RT\ln x_A=\int_p^{p+\Pi} V_m\mathrm{d}p \qquad (5B.17)$$

This expression enables us to calculate the additional pressure Π that must be applied to the solution to restore the chemical potential of the solvent to its 'pure' value and thus to restore equilibrium across the semipermeable membrane. For dilute solutions, ln x_A may be replaced by $\ln(1-x_B)\approx -x_B$. We may also assume that the pressure range in the integration is so small that the molar volume of the solvent is a constant. That being so, V_m may be taken outside the integral, giving

$$RTx_B=\Pi V_m$$

When the solution is dilute, $x_B\approx n_B/n_A$. Moreover, because $n_AV_m=V$, the total volume of the solvent, the equation simplifies to eqn 5B.16.

Because the effect of osmotic pressure is so readily measurable and large, one of the most common applications of osmometry is to the measurement of molar masses of macromolecules, such as proteins and synthetic polymers. As these huge molecules dissolve to produce solutions that are far from ideal, it is assumed that the van 't Hoff equation is only the first term of a virial-like expansion, much like the extension of the perfect gas equation to real gases (in Topic 1C) to take into account molecular interactions:

$$\Pi=[\mathrm{J}]RT\{1+B[\mathrm{J}]+\ldots\} \qquad \text{Osmotic virial expansion} \qquad (5B.18)$$

(We have denoted the solute J to avoid too many different Bs in this expression.) The additional terms take the non-ideality into account; the empirical constant B is called the **osmotic virial coefficient**.

Example 5B.2 Using osmometry to determine the molar mass of a macromolecule

The osmotic pressures of solutions of poly(vinyl chloride), PVC, in cyclohexanone at 298 K are given below. The pressures are expressed in terms of the heights of solution (of mass density $\rho=0.980\,\mathrm{g\,cm^{-3}}$) in balance with the osmotic pressure. Determine the molar mass of the polymer.

$c/(\mathrm{g\,dm^{-3}})$	1.00	2.00	4.00	7.00	9.00
h/cm	0.28	0.71	2.01	5.10	8.00

Method The osmotic pressure is measured at a series of mass concentrations, c, and a plot of Π/c against c is used to determine the molar mass of the polymer. We use eqn 5B.18 with $[\mathrm{J}]=c/M$ where c is the mass concentration of the polymer and M is its molar mass. The osmotic pressure is related to the hydrostatic pressure by $\Pi=\rho gh$ (*Example* 1A.1) with $g=9.81\,\mathrm{m\,s^{-2}}$. With these substitutions, eqn 5B.18 becomes

$$\frac{h}{c}=\frac{RT}{\rho gM}\left\{1+\frac{Bc}{M}+\cdots\right\}=\frac{RT}{\rho gM}+\left(\frac{RTB}{\rho gM^2}\right)c+\cdots$$

Therefore, to find M, plot h/c against c, and expect a straight line with intercept $RT/\rho gM$ at $c=0$.

Answer The data give the following values for the quantities to plot:

$c/(\mathrm{g\,dm^{-3}})$	1.00	2.00	4.00	7.00	9.00
$(h/c)/(\mathrm{cm\,g^{-1}\,dm^3})$	0.28	0.36	0.503	0.729	0.889

The points are plotted in Fig. 5B.15. The intercept is at 0.20. Therefore,

$$\begin{aligned}M&=\frac{RT}{\rho g}\times\frac{1}{0.20\,\mathrm{cm\,g^{-1}\,dm^3}}\\&=\frac{(8.3145\,\mathrm{J\,K^{-1}\,mol^{-1}})\times(298\,\mathrm{K})}{(980\,\mathrm{kg\,m^{-3}})\times(9.81\,\mathrm{m\,s^{-2}})}\times\frac{1}{2.0\times10^{-3}\,\mathrm{m^4\,kg^{-1}}}\\&=1.3\times10^2\,\mathrm{kg\,mol^{-1}}\end{aligned}$$

where we have used $1\,\mathrm{kg\,m^2\,s^{-2}}=1\,\mathrm{J}$. Modern osmometers give readings of osmotic pressure in pascals, so the analysis of the data is more straightforward and eqn 5B.18 can be used directly. As explained in Topic 17D, the value obtained from osmometry is the 'number average molar mass'.

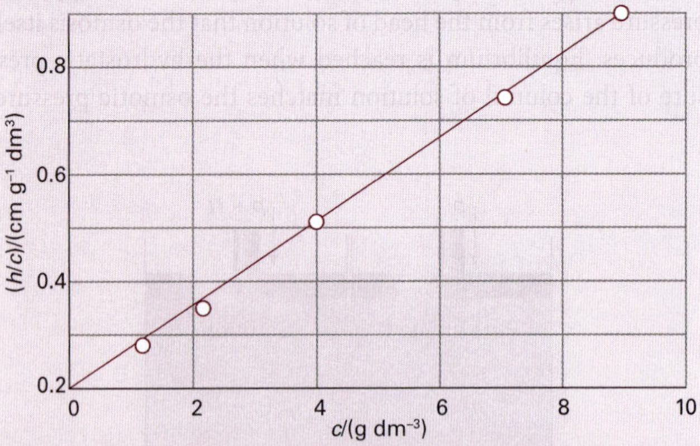

Figure 5B.15 The plot involved in the determination of molar mass by osmometry. The molar mass is calculated from the intercept at $c=0$.

Self-test 5B.7 Estimate the depression of freezing point of the most concentrated of these solutions, taking K_f as about $10\,\mathrm{K/(mol\,kg^{-1})}$.

Answer: 0.8 mK

Checklist of concepts

- ☐ 1. The **Gibbs energy of mixing** of two liquids to form an ideal solution is calculated in the same way as for two perfect gases.
- ☐ 2. The **enthalpy of mixing** is zero and the Gibbs energy is due entirely to the entropy of mixing.
- ☐ 3. A **regular solution** is one in which the entropy of mixing is the same as for an ideal solution but the enthalpy of mixing is non-zero.
- ☐ 4. A **colligative property** depends only on the number of solute particles present, not their identity.
- ☐ 5. All the colligative properties stem from the reduction of the chemical potential of the liquid solvent as a result of the presence of solute.
- ☐ 6. The **elevation of boiling point** is proportional to the molality of the solute.
- ☐ 7. The **depression of freezing point** is also proportional to the molality of the solute.
- ☐ 8. Solutes with high melting points and large enthalpies of melting have low solubilities at normal temperatures.
- ☐ 9. The **osmotic pressure** is the pressure that when applied to a solution prevents the influx of solvent through a semipermeable membrane.
- ☐ 10. The relation of the osmotic pressure to the molar concentration of the solute is given by the **van 't Hoff equation** and is a sensitive way of determining molar mass.

Checklist of equations

Property	Equation	Comment	Equation number
Gibbs energy of mixing	$\Delta_{\text{mix}}G = nRT(x_A \ln x_A + x_B \ln x_B)$	Ideal solutions	5B.3
Entropy of mixing	$\Delta_{\text{mix}}S = -nR(x_A \ln x_A + x_B \ln x_B)$	Ideal solutions	5B.4
Enthalpy of mixing	$\Delta_{\text{mix}}H = 0$	Ideal solutions	
Excess function	$X^E = \Delta_{\text{mix}}X - \Delta_{\text{mix}}X^{\text{ideal}}$	Definition	5B.5
Regular solution ($S^E = 0$)	$H^E = n\xi RTx_A x_B$	Model	5B.6
Elevation of boiling point	$\Delta T_b = K_b b$	Empirical, non-volatile solute	5B.10
Depression of freezing point	$\Delta T_f = K_f b$	Empirical, solute insoluble in solid solvent	5B.13
Ideal solubility	$\ln x_B = (\Delta_{\text{fus}}H/R)(1/T_f - 1/T)$	Ideal solution	58.15
van 't Hoff equation	$\Pi = [B]RT$	Valid as $[B] \to 0$	5B.16
Osmotic virial expansion	$\Pi = [J]RT\{1 + B[J] + \ldots\}$	Empirical	5B.18

5C Phase diagrams of binary systems

Contents

➤ Why do you need to know this material?

Phase diagrams are used widely in materials science, metallurgy, geology, and the chemical industry to summarize the composition of mixtures and it is important to be able to interpret them.

➤ What is the key idea?

A phase diagram is a map showing the conditions under which each phase of a system is the most stable.

➤ What do you need to know already?

It would be helpful to review the interpretation of one-component phase diagrams and the phase rule (Topic 4A). The early part of this Topic draws on Raoult's law (Topic 4B) and the concept of partial pressure (Topic 1A).

One-component phase diagrams are described in Topic 4A. The phase equilibria of binary systems are more complex because composition is an additional variable. However, they provide very useful summaries of phase equilibria for both ideal and empirically established real systems.

5C.1 Vapour pressure diagrams

The partial vapour pressures of the components of an ideal solution of two volatile liquids are related to the composition of the liquid mixture by Raoult's law (Topic 5A)

$$p_A = x_A p_A^* \qquad p_B = x_B p_B^* \tag{5C.1}$$

where p_A^* is the vapour pressure of pure A and p_B^* that of pure B. The total vapour pressure p of the mixture is therefore

$$p = p_A + p_B = x_A p_A^* + x_B p_B^* = p_B^* + (p_A^* - p_B^*)x_A$$

Total vapour pressure (5C.2)

This expression shows that the total vapour pressure (at some fixed temperature) changes linearly with the composition from p_B^* to p_A^* as x_A changes from 0 to 1 (Fig. 5C.1).

(a) The composition of the vapour

The compositions of the liquid and vapour that are in mutual equilibrium are not necessarily the same. Common sense suggests that the vapour should be richer in the more volatile component. This expectation can be confirmed as follows. The partial pressures of the components are given by eqn 1A.8 of Topic 1A ($p_J = x_J p$). It follows from that definition that the mole fractions in the gas, y_A and y_B, are

$$y_A = \frac{p_A}{p} \qquad y_B = \frac{p_B}{p} \tag{5C.3}$$

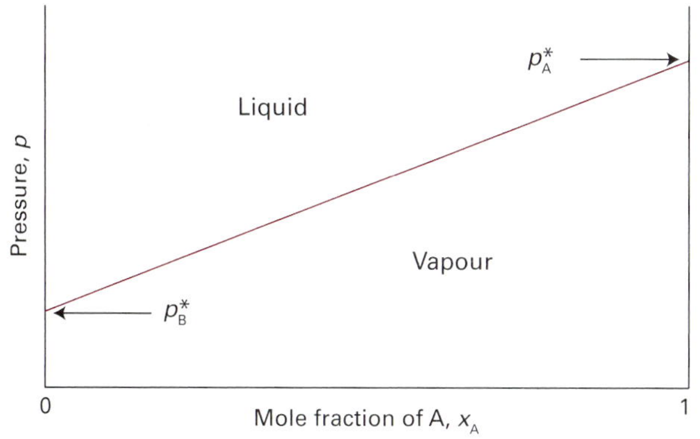

Figure 5C.1 The variation of the total vapour pressure of a binary mixture with the mole fraction of A in the liquid when Raoult's law is obeyed.

Provided the mixture is ideal, the partial pressures and the total pressure may be expressed in terms of the mole fractions in the liquid by using eqn 5C.1 for p_J and eqn 5C.2 for the total vapour pressure p, which gives

$$y_A = \frac{x_A p_A^*}{p_B^* + (p_A^* - p_B^*)x_A} \quad y_B = 1 - y_A$$ Composition of vapour (5C.4)

Figure 5C.2 shows the composition of the vapour plotted against the composition of the liquid for various values of $p_A^*/p_B^* > 1$. We see that in all cases $y_A > x_A$, that is, the vapour is richer than the liquid in the more volatile component. Note that if B is non-volatile, so that $p_B^* = 0$ at the temperature of interest, then it makes no contribution to the vapour ($y_B = 0$).

Equation 5C.3 shows how the total vapour pressure of the mixture varies with the composition of the liquid. Because we can relate the composition of the liquid to the composition of the vapour through eqn 5C.3, we can now also relate the total vapour pressure to the composition of the vapour:

$$p = \frac{p_A^* p_B^*}{p_A^* + (p_B^* - p_A^*)y_A}$$ Total vapour pressure (5C.5)

This expression is plotted in Fig. 5C.3.

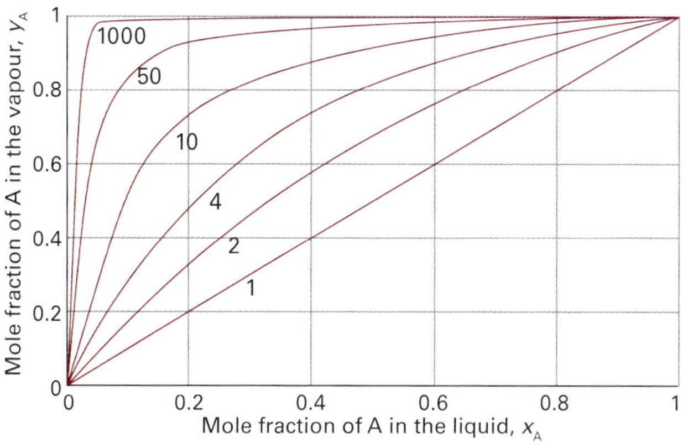

Figure 5C.2 The mole fraction of A in the vapour of a binary ideal solution expressed in terms of its mole fraction in the liquid, calculated using eqn 5C.4 for various values of p_A^*/p_B^* (the label on each curve) with A more volatile than B. In all cases the vapour is richer than the liquid in A.

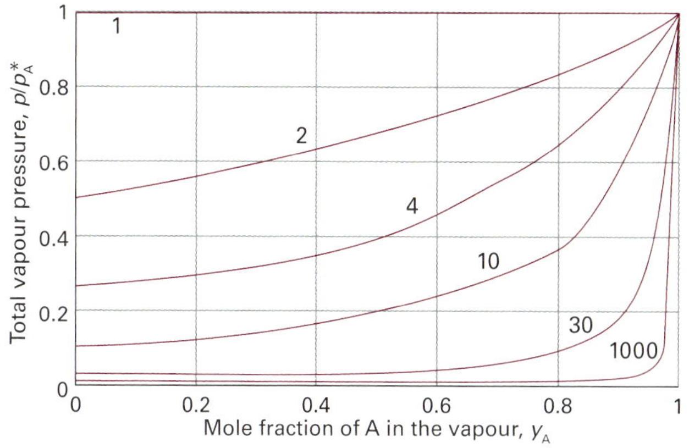

Figure 5C.3 The dependence of the vapour pressure of the same system as in Fig. 5C.2, but expressed in terms of the mole fraction of A in the vapour by using eqn 5C.5. Individual curves are labelled with the value of p_A^*/p_B^*.

Brief illustration 5C.1 The composition of the vapour

The vapour pressures of benzene and methylbenzene at 20 °C are 75 Torr and 21 Torr, respectively. The composition of the vapour in equilibrium with an equimolar liquid mixture ($x_{\text{benzene}} = x_{\text{methylbenzene}} = \frac{1}{2}$) is

$$y_{\text{benzene}} = \frac{\frac{1}{2}\times(75\,\text{Torr})}{21\,\text{Torr} + (75 - 21\,\text{Torr})\times\frac{1}{2}} = 0.78$$

$$y_{\text{methylbenzene}} = 1 - 0.78 = 0.22$$

The vapour pressure of each component is

$$p_{\text{benzene}} = \tfrac{1}{2}(75\,\text{Torr}) = 38\,\text{Torr}$$

$$p_{\text{methylbenzene}} = \tfrac{1}{2}(21\,\text{Torr}) = 10\,\text{Torr}$$

for a total vapour pressure of 48 Torr.

Self-test 5C.1 What is the composition of the vapour in equilibrium with a mixture in which the mole fraction of benzene is 0.75?

Answer: 0.91, 0.09

(b) The interpretation of the diagrams

If we are interested in distillation, both the vapour and the liquid compositions are of equal interest. It is therefore sensible

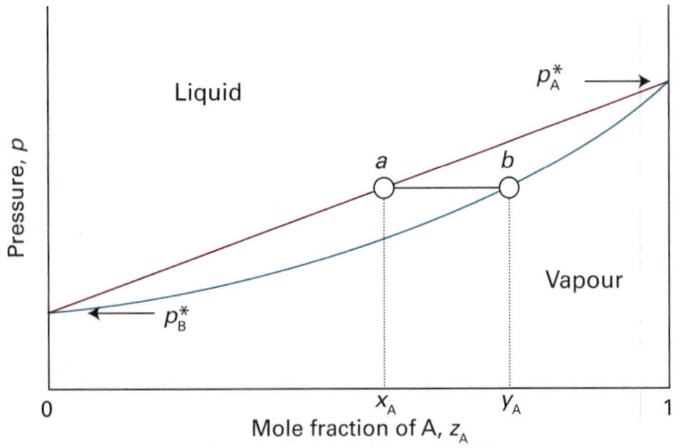

Figure 5C.4 The dependence of the total vapour pressure of an ideal solution on the mole fraction of A in the entire system. A point between the two lines corresponds to both liquid and vapour being present; outside that region there is only one phase present. The mole fraction of A is denoted z_A, as explained in the text.

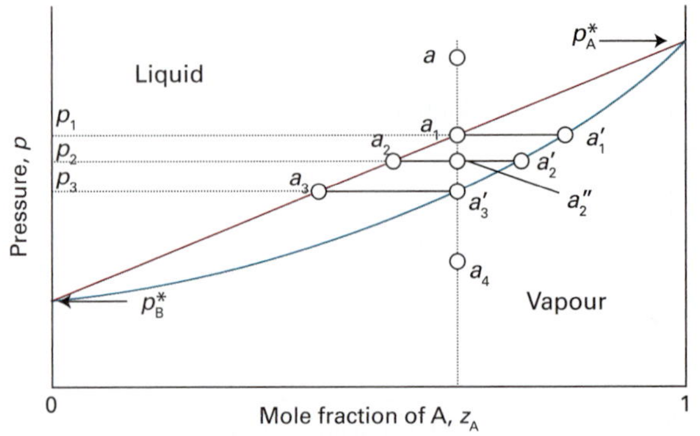

Figure 5C.5 The points of the pressure–composition diagram discussed in the text. The vertical line through a is an isopleth, a line of constant composition of the entire system.

to combine Figs. 5C.2 and 5C.3 into one (Fig. 5C.4). The point a indicates the vapour pressure of a mixture of composition x_A, and the point b indicates the composition of the vapour that is in equilibrium with the liquid at that pressure. A richer interpretation of the phase diagram is obtained, however, if we interpret the horizontal axis as showing the *overall* composition, z_A, of the system (essentially, the mole fraction showing how the mixture was prepared). If the horizontal axis of the vapour pressure diagram is labelled with z_A, then all the points down to the solid diagonal line in the graph correspond to a system that is under such high pressure that it contains only a liquid phase (the applied pressure is higher than the vapour pressure), so $z_A = x_A$, the composition of the liquid. On the other hand, all points below the lower curve correspond to a system that is under such low pressure that it contains only a vapour phase (the applied pressure is lower than the vapour pressure), so $z_A = y_A$.

Points that lie between the two lines correspond to a system in which there are two phases present, one a liquid and the other a vapour. To see this interpretation, consider the effect of lowering the pressure on a liquid mixture of overall composition a in Fig. 5C.5. The lowering of pressure can be achieved by drawing out a piston (Fig. 5C.6). The changes to the system do not affect the overall composition, so the state of the system moves down the vertical line that passes through a. This vertical line is called an **isopleth**, from the Greek words for 'equal abundance'. Until the point a_1 is reached (when the pressure has been reduced to p_1), the sample consists of a single liquid phase. At a_1 the liquid can exist in equilibrium with its vapour. As we have seen, the composition of the vapour phase is given by point a_1'. A line joining two points representing phases in equilibrium is called a **tie line**. The composition of the liquid is the same as initially (a_1 lies on the isopleth through a), so we have to conclude that at this pressure there is virtually no vapour present; however, the tiny amount of vapour that is present has the composition a_1'

Now consider the effect of lowering the pressure to p_2, so taking the system to a pressure and overall composition represented by the point a_2''. This new pressure is below the vapour pressure of the original liquid, so it vaporizes until the vapour pressure of the remaining liquid falls to p_2. Now we know that the composition of such a liquid must be a_2. Moreover, the composition of the vapour in equilibrium with that liquid must be given by the point a_2' at the other end of the tie line. If the pressure is reduced to p_3, a similar readjustment in composition takes place, and now the compositions of the liquid and vapour are represented by the points a_3 and a_3', respectively. The latter point corresponds to a system in which the composition of the vapour is the same as the overall composition, so we have

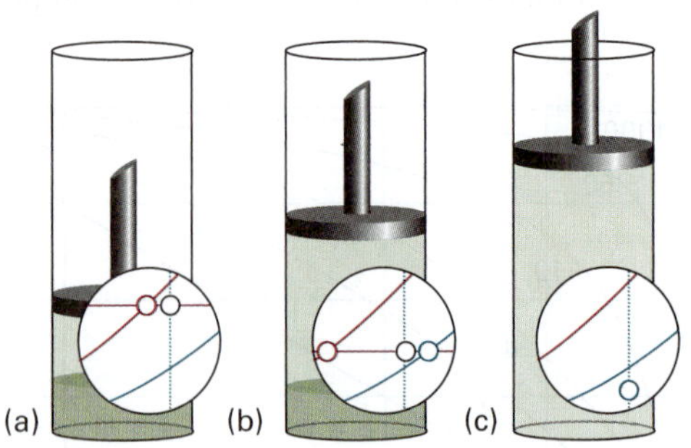

Figure 5C.6(a) A liquid in a container exists in equilibrium with its vapour. The superimposed fragment of the phase diagram shows the compositions of the two phases and their abundances (by the lever rule; see section 5C.1(c)). (b) When the pressure is changed by drawing out a piston, the compositions of the phases adjust as shown by the tie line in the phase diagram. (c) When the piston is pulled so far out that all the liquid has vaporized and only the vapour is present, the pressure falls as the piston is withdrawn and the point on the phase diagram moves into the one-phase region.

to conclude that the amount of liquid present is now virtually zero, but the tiny amount of liquid present has the composition a_3. A further decrease in pressure takes the system to the point a_4; at this stage, only vapour is present and its composition is the same as the initial overall composition of the system (the composition of the original liquid).

Example 5C.1 Constructing a vapour pressure diagram

The following temperature/composition data were obtained for a mixture of octane (O) and methylbenzene (M) at 1.00 atm, where x is the mole fraction in the liquid and y the mole fraction in the vapour at equilibrium.

θ/°C	110.9	112.0	114.0	115.8	117.3	119.0	121.1	123.0
x_M	0.908	0.795	0.615	0.527	0.408	0.300	0.203	0.097
y_M	0.923	0.836	0.698	0.624	0.527	0.410	0.297	0.164

The boiling points are 110.6 °C and 125.6 °C for M and O, respectively. Plot the temperature–composition diagram for the mixture. What is the composition of the vapour in equilibrium with the liquid of composition (a) $x_M=0.250$ and (b) $x_O=0.250$?

Method Plot the composition of each phase (on the horizontal axis) against the temperature (on the vertical axis). The two boiling points give two further points corresponding to $x_M=1$ and $x_M=0$, respectively. Use a curve-fitting program to draw the phase boundaries. For the interpretation, draw the appropriate tie-lines.

Answer The points are plotted in Fig. 5C.7. The two sets of points are fitted to the polynomials $a+bx+cx^2+dx^3$ with

$$\text{For the liquid line: } \theta/°\text{C}=125.422-22.9494x+6.64602x^2 +1.32623x^3+\ldots$$

$$\text{For the vapour line: } \theta/°\text{C}=125.485-11.9387x-12.5626x^2 +9.36542x^3+\ldots$$

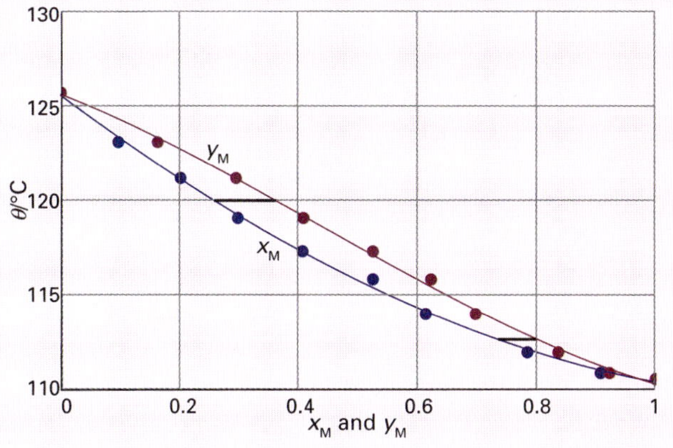

Figure 5C.7 The plot of data and the fitted curves for a mixture of octane and methylbenzene (M) in *Example* 5C.1.

The tie lines at $x_M=0.250$ and $x_O=0.250$ (corresponding to $x_M=0.750$) have been drawn on the graph starting at the lower (liquid curve). They intersect the upper (vapour curve) at $y_M=0.36$ and 0.80, respectively.

Self-test 5C.2 Repeat the analysis for the following data on hexane and heptane at 70 °C:

θ/°C	65	66	70	77	85	100
x_{hexane}	0	0.20	0.40	0.60	0.80	1
y_{hexane}	0	0.02	0.08	0.20	0.48	1

Answer: See Fig. 5C.8.

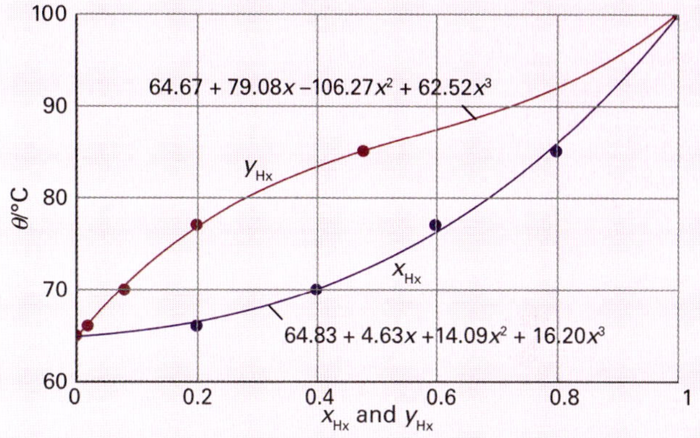

Figure 5C.8 The plot of data and the fitted curves for a mixture of hexane (Hx) and heptane in *Self-test* 5C.2.

(c) The lever rule

A point in the two-phase region of a phase diagram indicates not only qualitatively that both liquid and vapour are present, but represents quantitatively the relative amounts of each. To find the relative amounts of two phases α and β that are in equilibrium, we measure the distances l_α and l_β along the horizontal tie line, and then use the **lever rule** (Fig. 5C.9):

$$n_\alpha l_\alpha = n_\beta l_\beta \qquad \text{Lever rule} \qquad (5C.6)$$

Here n_α is the amount of phase α and n_β the amount of phase β. In the case illustrated in Fig. 5C.9, because $l_\beta \approx 2l_\alpha$, the amount of phase α is about twice the amount of phase β.

Justification 5C.1 The lever rule

To prove the lever rule we write the total amount of A and B molecules as $n=n_\alpha+n_\beta$, where n_α is the amount of molecules in phase α and n_β the amount in phase β. The mole fraction of A in phase α is $x_{A,\alpha}$, so the amount of A in that phase is

$n_\alpha x_{A,\alpha}$. Similarly, the amount of A in phase β is $n_\beta x_{A,\beta}$. The total amount of A is therefore

$$n_A = n_\alpha x_{A,\alpha} + n_\beta x_{A,\beta}$$

Let the composition of the entire mixture be expressed by the mole fraction z_A (this is the label on the horizontal axis, and reflects how the sample is prepared). The total amount of A molecules is therefore

$$n_A = nz_A = n_\alpha z_A + n_\beta z_A$$

By equating these two expressions it follows that

$$n_\alpha(x_{A,\alpha} - z_A) = n_\beta(z_A - x_{A,\beta})$$

which corresponds to eqn 5C.6.

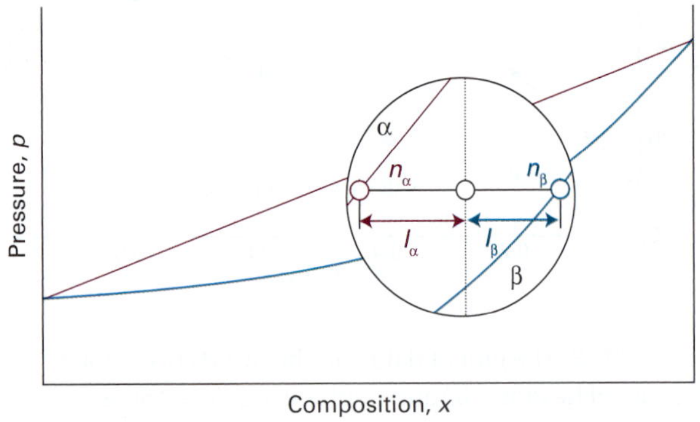

Figure 5C.9 The lever rule. The distances l_α and l_β are used to find the proportions of the amounts of phases α (such as vapour) and β (for example, liquid) present at equilibrium. The lever rule is so called because a similar rule relates the masses at two ends of a lever to their distances from a pivot ($m_\alpha l_\alpha = m_\beta l_\beta$ for balance).

Brief illustration 5C.2 The lever rule

At p_1 in Fig. 5C.5, the ratio l_{vap}/l_{liq} is almost infinite for this tie line, so n_{liq}/n_{vap} is also almost infinite, and there is only a trace of vapour present. When the pressure is reduced to p_2, the value of l_{vap}/l_{liq} is about 0.3, so $n_{liq}/n_{vap} \approx 0.3$ and the amount of liquid is about 0.3 times the amount of vapour. When the pressure has been reduced to p_3, the sample is almost completely gaseous and because $l_{vap}/l_{liq} \approx 0$ we conclude that there is only a trace of liquid present.

Self-test 5C.3 Suppose that in a phase diagram, when the sample was prepared with the mole fraction of component A equal to 0.40 it was found that the compositions of the two phases in equilibrium corresponded to the mole fractions $x_{A,\alpha} = 0.60$ and $x_{A,\beta} = 0.20$. What is the ratio of amounts of the two phases?

Answer: $n_\alpha/n_\beta = 1.0$

5C.2 Temperature–composition diagrams

To discuss distillation we need a **temperature–composition diagram**, a phase diagram in which the boundaries show the composition of the phases that are in equilibrium at various temperatures (and a given pressure, typically 1 atm). An example is shown in Fig. 5C.10. Note that the liquid phase now lies in the lower part of the diagram.

(a) The distillation of mixtures

Consider what happens when a liquid of composition a_1 in Fig. 5C.10 is heated. It boils when the temperature reaches T_2. Then the liquid has composition a_2 (the same as a_1) and the vapour (which is present only as a trace) has composition a_2'. The vapour is richer in the more volatile component A (the component with the lower boiling point). From the location of a_2, we can state the vapour's composition at the boiling point, and from the location of the tie line joining a_2 and a_2' we can read off the boiling temperature (T_2) of the original liquid mixture.

In a **simple distillation**, the vapour is withdrawn and condensed. This technique is used to separate a volatile liquid from a non-volatile solute or solid. In **fractional distillation**, the boiling and condensation cycle is repeated successively. This technique is used to separate volatile liquids. We can follow the changes that occur by seeing what happens when the first condensate of composition a_3 is reheated. The phase diagram shows that this mixture boils at T_3 and yields a vapour of composition a_3', which is even richer in the more volatile component. That vapour is drawn off, and the first drop condenses to a liquid of composition a_4. The cycle can then be repeated until

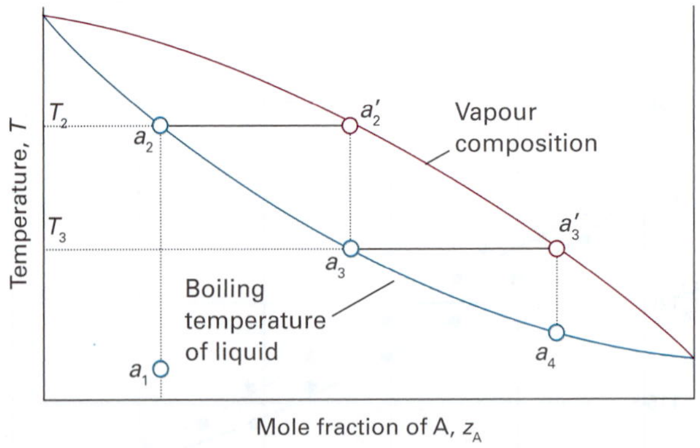

Figure 5C.10 The temperature–composition diagram corresponding to an ideal mixture with the component A more volatile than component B. Successive boilings and condensations of a liquid originally of composition a_1 lead to a condensate that is pure A. The separation technique is called fractional distillation.

in due course almost pure A is obtained in the vapour and pure B remains in the liquid.

The efficiency of a fractionating column is expressed in terms of the number of **theoretical plates**, the number of effective vaporization and condensation steps that are required to achieve a condensate of given composition from a given distillate.

Brief illustration 5C.3 Theoretical plates

To achieve the degree of separation shown in Fig. 5C.11a, the fractionating column must correspond to three theoretical plates. To achieve the same separation for the system shown in Fig. 5C.11b, in which the components have more similar partial pressures, the fractionating column must be designed to correspond to five theoretical plates.

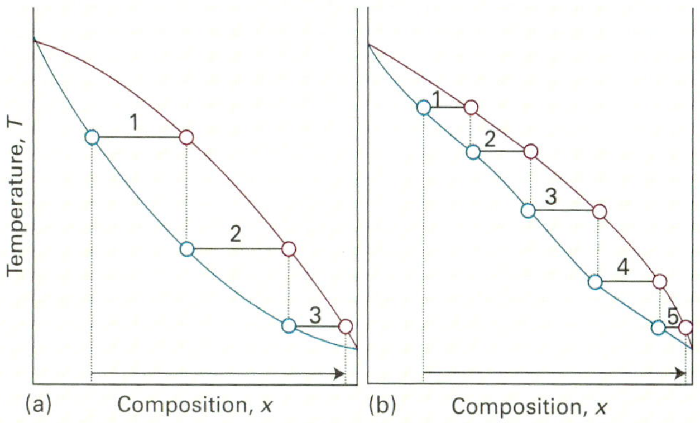

Figure 5C.11 The number of theoretical plates is the number of steps needed to bring about a specified degree of separation of two components in a mixture. The two systems shown correspond to (a) 3, (b) 5 theoretical plates.

Self-test 5C.4 Refer to Fig. 5C.11b: suppose the composition of the mixture corresponds to $z_A = 0.1$; how many theoretical plates would be required to achieve a composition $z_A = 0.9$?

Answer: 5

(b) Azeotropes

Although many liquids have temperature–composition phase diagrams resembling the ideal version in Fig. 5C.10, in a number of important cases there are marked deviations. A maximum in the phase diagram (Fig. 5C.12) may occur when the favourable interactions between A and B molecules reduce the vapour pressure of the mixture below the ideal value and so raise its boiling temperature: in effect, the A–B interactions stabilize the liquid. In such cases the excess Gibbs energy, G^E (Topic 5B), is negative (more favourable to mixing than ideal).

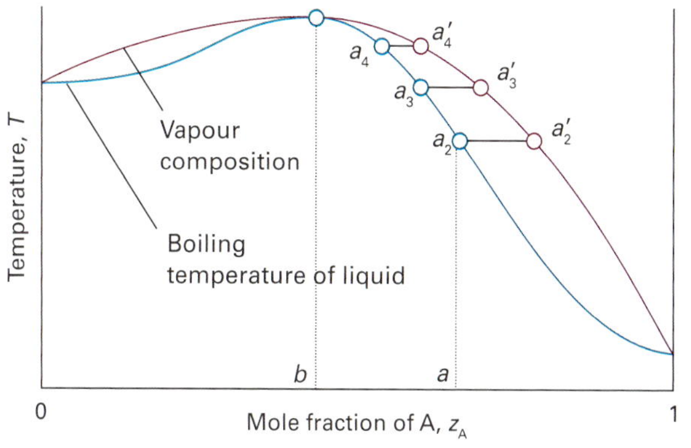

Figure 5C.12 A high-boiling azeotrope. When the liquid of composition *a* is distilled, the composition of the remaining liquid changes towards *b* but no further.

Phase diagrams showing a minimum (Fig. 5C.13) indicate that the mixture is destabilized relative to the ideal solution, the A–B interactions then being unfavourable; in this case, the boiling temperature is lowered. For such mixtures G^E is positive (less favourable to mixing than ideal), and there may be contributions from both enthalpy and entropy effects.

Deviations from ideality are not always so strong as to lead to a maximum or minimum in the phase diagram, but when they do there are important consequences for distillation. Consider a liquid of composition *a* on the right of the maximum in Fig. 5C.12. The vapour (at a_2') of the boiling mixture (at a_2) is richer in A. If that vapour is removed (and condensed elsewhere), then the remaining liquid will move to a composition that is richer in B, such as that represented by a_3, and the vapour in equilibrium with this mixture will have composition a_3'. As that vapour is removed, the composition of the boiling liquid shifts to a point such as a_4, and the composition of the vapour shifts to a_4'. Hence, as evaporation proceeds, the composition of the remaining liquid shifts towards B as A is drawn off. The boiling point of the liquid rises, and the vapour becomes richer in

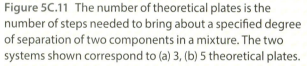

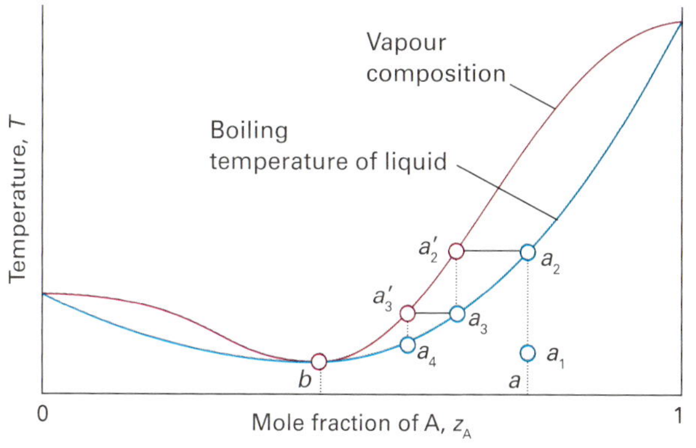

Figure 5C.13 A low-boiling azeotrope. When the mixture at *a* is fractionally distilled, the vapour in equilibrium in the fractionating column moves towards *b* and then remains unchanged.

B. When so much A has been evaporated that the liquid has reached the composition b, the vapour has the same composition as the liquid. Evaporation then occurs without change of composition. The mixture is said to form an **azeotrope**.[1] When the azeotropic composition has been reached, distillation cannot separate the two liquids because the condensate has the same composition as the azeotropic liquid.

The system shown in Fig. 5C.13 is also azeotropic, but shows its azeotropy in a different way. Suppose we start with a mixture of composition a_1, and follow the changes in the composition of the vapour that rises through a fractionating column (essentially a vertical glass tube packed with glass rings to give a large surface area). The mixture boils at a_2 to give a vapour of composition a_2'. This vapour condenses in the column to a liquid of the same composition (now marked a_3). That liquid reaches equilibrium with its vapour at a_3', which condenses higher up the tube to give a liquid of the same composition, which we now call a_4. The fractionation therefore shifts the vapour towards the azeotropic composition at b, but not beyond, and the azeotropic vapour emerges from the top of the column.

Brief illustration 5C.4 Azeotropes

Examples of the behaviour of the type shown in Fig. 5C.12 include (a) trichloromethane/propanone and (b) nitric acid/water mixtures. Hydrochloric acid/water is azeotropic at 80 per cent by mass of water and boils unchanged at 108.6 °C. Examples of the behaviour of the type shown in Fig. 5C.13 include (c) dioxane/water and (d) ethanol/water mixtures. Ethanol/water boils unchanged when the water content is 4 per cent by mass and the temperature is 78 °C.

Self-test 5C.5 Suggest a molecular interpretation of the two types of behaviour.

Answer: (a,b) favourable A–B interactions; (c,d) unfavourable A–B interactions

(c) Immiscible liquids

Finally we consider the distillation of two immiscible liquids, such as octane and water. At equilibrium, there is a tiny amount of A dissolved in B, and similarly a tiny amount of B dissolved in A: both liquids are saturated with the other component (Fig. 5C.14a). As a result, the total vapour pressure of the mixture is close to $p = p_A^* + p_B^*$. If the temperature is raised to the value at which this total vapour pressure is equal to the atmospheric pressure, boiling commences and the dissolved substances are purged from their solution. However, this boiling results in a vigorous agitation of the mixture, so each component is kept saturated in the other component, and the purging continues as the very dilute solutions are replenished. This intimate contact is essential: two immiscible liquids heated in a container like that shown in Fig. 5C.14b would not boil at the same temperature. The presence of the saturated solutions means that the 'mixture' boils at a lower temperature than either component would alone because boiling begins when the total vapour pressure reaches 1 atm, not when either vapour pressure reaches 1 atm. This distinction is the basis of **steam distillation**, which enables some heat-sensitive, water-insoluble organic compounds to be distilled at a lower temperature than their normal boiling point. The only snag is that the composition of the condensate is in proportion to the vapour pressures of the components, so oils of low volatility distil in low abundance.

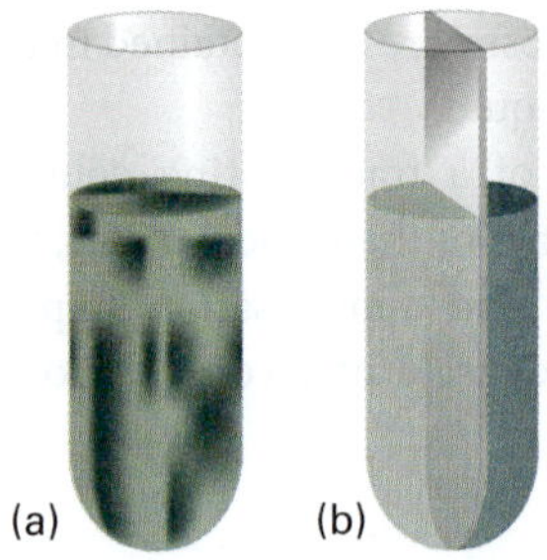

Figure 5C.14 The distillation of (a) two immiscible liquids can be regarded as (b) the joint distillation of the separated components, and boiling occurs when the sum of the partial pressures equals the external pressure.

5C.3 Liquid–liquid phase diagrams

Now we consider temperature–composition diagrams for systems that consist of pairs of **partially miscible** liquids, which are liquids that do not mix in all proportions at all temperatures. An example is hexane and nitrobenzene. The same principles of interpretation apply as to liquid–vapour diagrams.

(a) Phase separation

Suppose a small amount of a liquid B is added to a sample of another liquid A at a temperature T'. Liquid B dissolves completely, and the binary system remains a single phase. As more B is added, a stage comes at which no more dissolves. The sample now consists of two phases in equilibrium with each other, the most abundant one consisting of A saturated with B, the minor one a trace of B saturated with A. In the temperature–composition diagram drawn in Fig. 5C.15, the composition of the former is represented by the point a' and that of the latter by the point a''. The relative abundances of the two phases are given by the lever rule. When more B is added, A dissolves in it slightly. The compositions of the two phases in equilibrium remain a' and a''. A stage is reached when so much B is present

[1] The name comes from the Greek words for 'boiling without changing'.

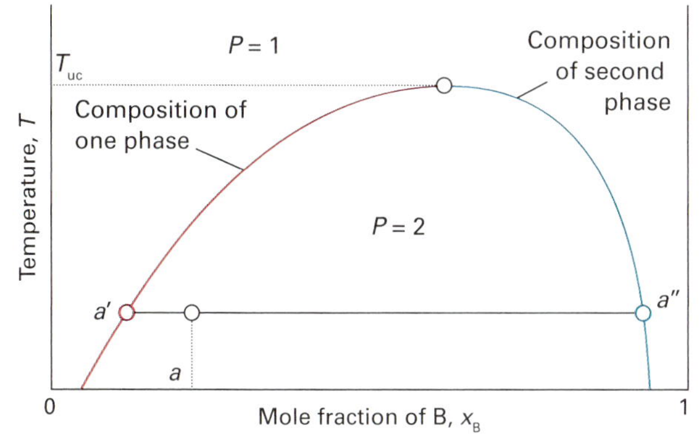

Figure 5C.15 The temperature–composition diagram for a mixture of A and B. The region below the curve corresponds to the compositions and temperatures at which the liquids are partially miscible. The upper critical temperature, T_{uc}, is the temperature above which the two liquids are miscible in all proportions.

that it can dissolve all the A, and the system reverts to a single phase. The addition of more B now simply dilutes the solution, and from then on a single phase remains.

The composition of the two phases at equilibrium varies with the temperature. For the system shown in Fig. 5C.15, raising the temperature increases the miscibility of A and B. The two-phase region therefore becomes narrower because each phase in equilibrium is richer in its minor component: the A-rich phase is richer in B and the B-rich phase is richer in A. We can construct the entire phase diagram by repeating the observations at different temperatures and drawing the envelope of the two-phase region.

Example 5C.2 Interpreting a liquid–liquid phase diagram

The phase diagram for the system nitrobenzene/hexane at 1 atm is shown in Fig. 5C.16. A mixture of 50 g of hexane (0.59 mol C_6H_{14}) and 50 g of nitrobenzene (0.41 mol $C_6H_5NO_2$) was prepared at 290 K. What are the compositions of the phases, and in what proportions do they occur? To what temperature must the sample be heated in order to obtain a single phase?

Method The compositions of phases in equilibrium are given by the points where the tie-line representing the temperature intersects the phase boundary. Their proportions are given by the lever rule (eqn 5C.6). The temperature at which the components are completely miscible is found by following the isopleth upwards and noting the temperature at which it enters the one-phase region of the phase diagram.

Answer We denote hexane by H and nitrobenzene by N; refer to Fig. 5C.16. The point $x_N=0.41$, $T=290$ K occurs in the two-phase region of the phase diagram. The horizontal tie line cuts the phase boundary at $x_N=0.35$ and $x_N=0.83$, so those are the compositions of the two phases. According to the lever rule, the ratio of amounts of each phase is equal to the ratio of the distances l_α and l_β:

$$\frac{n_\alpha}{n_\beta}=\frac{l_\beta}{l_\alpha}=\frac{0.83-0.41}{0.41-0.35}=\frac{0.42}{0.06}=7$$

That is, there is about 7 times more hexane-rich phase than nitrobenzene-rich phase. Heating the sample to 292 K takes it into the single-phase region. Because the phase diagram has been constructed experimentally, these conclusions are not based on any assumptions about ideality. They would be modified if the system were subjected to a different pressure.

Self-test 5C.6 Repeat the problem for 50 g of hexane and 100 g of nitrobenzene at 273 K.

Answer: $x_N=0.09$ and 0.95 in ratio 1:1.3; 294 K

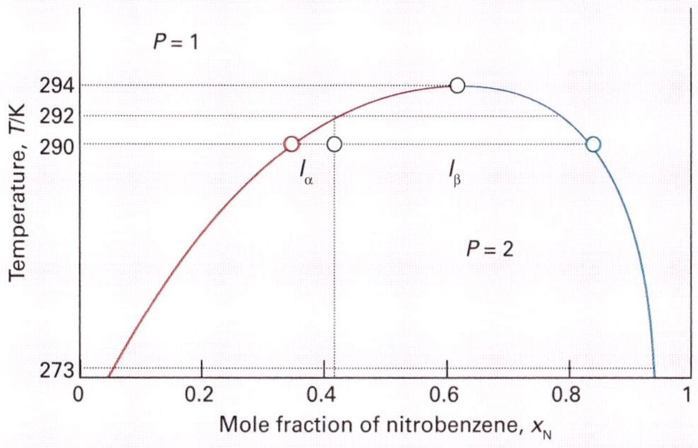

Figure 5C.16 The temperature–composition diagram for hexane and nitrobenzene at 1 atm, with the points and lengths discussed in the text.

(b) Critical solution temperatures

The **upper critical solution temperature**, T_{uc} (or *upper consolute temperature*), is the highest temperature at which phase separation occurs. Above the upper critical temperature the two components are fully miscible. This temperature exists because the greater thermal motion overcomes any potential energy advantage in molecules of one type being close together. One example is the nitrobenzene/hexane system shown in Fig. 5C.16. An example of a solid solution is the palladium/hydrogen system, which shows two phases, one a solid solution of hydrogen in palladium and the other a palladium hydride, up to 300 °C but forms a single phase at higher temperatures (Fig. 5C.17).

The thermodynamic interpretation of the upper critical solution temperature focuses on the Gibbs energy of mixing and its variation with temperature. The simple model of a real solution (specifically, of a regular solution) discussed in Topic 5B results in a Gibbs energy of mixing that behaves as shown in

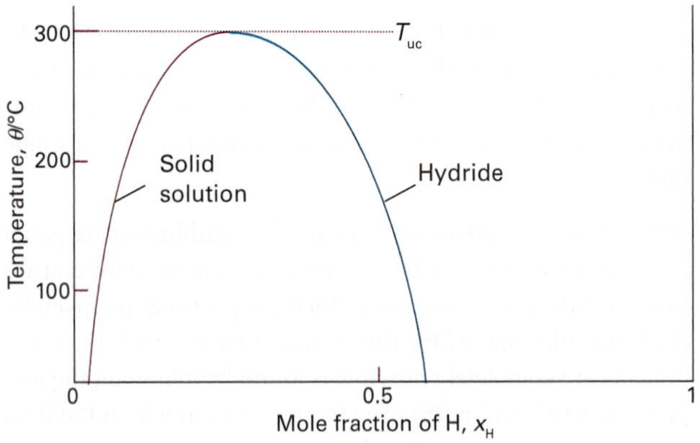

Figure 5C.17 The phase diagram for palladium and palladium hydride, which has an upper critical temperature at 300 °C.

Fig. 5C.18. Provided the parameter ξ introduced in eqn 5B.6 ($H^E = \xi RTx_Ax_B$) is greater than 2, the Gibbs energy of mixing has a double minimum. As a result, for $\xi > 2$ we can expect phase separation to occur. The same model shows that the compositions corresponding to the minima are obtained by looking for the conditions at which $\partial\Delta_{mix}G/\partial x = 0$, and a simple manipulation of eqn 5B.7 ($\Delta_{mix}G = nRT(x_A \ln x_A + x_B \ln x_B + \xi x_A x_B)$, with $x_B = 1 - x_A$) shows that we have to solve

$$\left(\frac{\partial\Delta_{mix}G}{\partial x_A}\right)_{T,p} = nRT\left(\frac{\partial\{x_A \ln x_A + (1-x_A)\ln(1-x_A) + \xi x_A(1-x_A)\}}{\partial x_A}\right)_{T,p}$$

$$= nRT\{\ln x_A + 1 - \ln(1-x_A) - 1 + \xi(1-2x_A)\}$$

$$= nRT\left\{\ln\frac{x_A}{1-x_A} + \xi(1-2x_A)\right\}$$

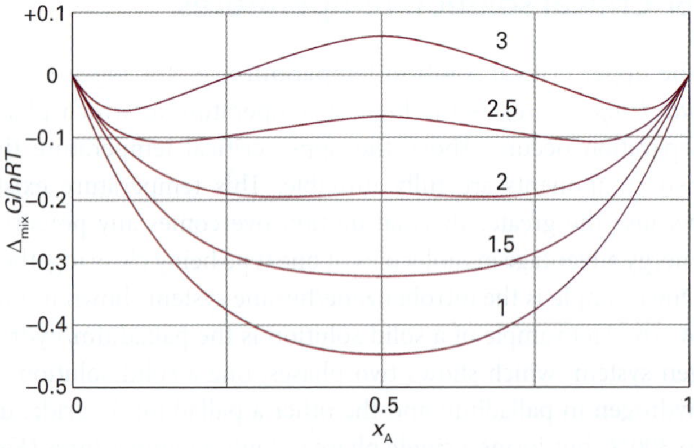

Figure 5C.18 The temperature variation of the Gibbs energy of mixing of a system that is partially miscible at low temperatures. A system of composition in the region $P = 2$ forms two phases with compositions corresponding to the two local minima of the curve. This illustration is a duplicate of Fig. 5B.5.

The Gibbs-energy minima therefore occurs where

$$\ln\frac{x_A}{1-x_A} = -\xi(1-2x_A) \qquad (5C.7)$$

This equation is an example of a 'transcendental equation', an equation that does not have a solution that can be expressed in a closed form. The solutions (the values of x_A that satisfy the equation) can be found numerically by using mathematical software or by plotting the terms on the left and right against x_A for a choice of values of ξ and identifying the values of x_A where the plots intersect (which is where the two expressions are equal) (Fig. 5C.19). The solutions found in this way are plotted in Fig. 5C.20. We see that, as ξ decreases, which can be interpreted as an increase in temperature provided the intermolecular forces remain constant (so that H^E remains constant), then the two minima move together and merge when $\xi = 2$.

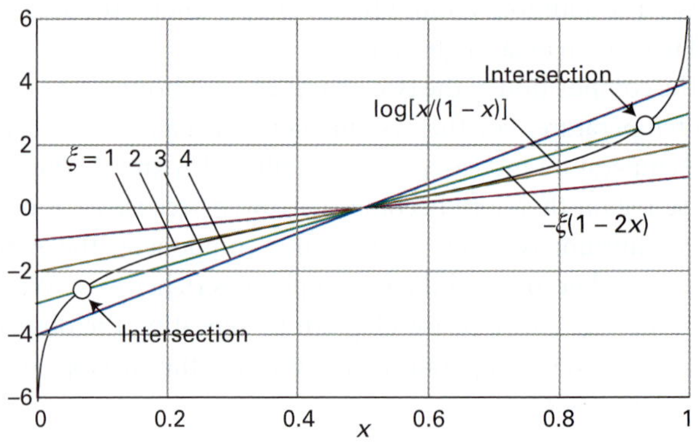

Figure 5C.19 The graphical procedure for solving eqn 5C.7. When $\xi < 2$, the only intersection occurs at $x = 0$. When $\xi \geq 2$, there are two solutions (those for $\xi = 3$ are marked).

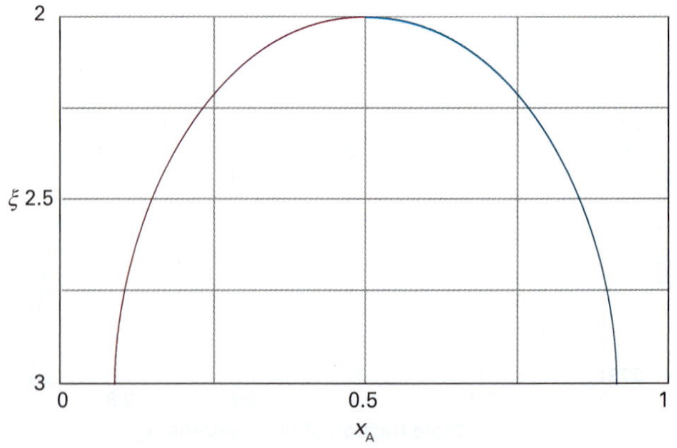

Figure 5C.20 The location of the phase boundary as computed on the basis of the ξ-parameter model introduced in Topic 5B.

Brief illustration 5C.5 Phase separation

In the system composed of benzene and cyclohexane treated in *Example* 5B.1 it is established that $\xi = 1.13$, so we do not expect a two-phase system; that is, the two components are completely miscible at the temperature of the experiment. The single solution of the equation

$$\ln \frac{x_A}{1-x_A} + 1.13(1-2x_A) = 0$$

is $x_A = \frac{1}{2}$, corresponding to a single minimum of the Gibbs energy of mixing, and there is no phase separation.

Self-test 5C.7 Would phase separation be expected if the excess enthalpy were modelled by the expression $H^E = \xi RT x_A^2 x_B^2$ (Fig. 5C.21a)?

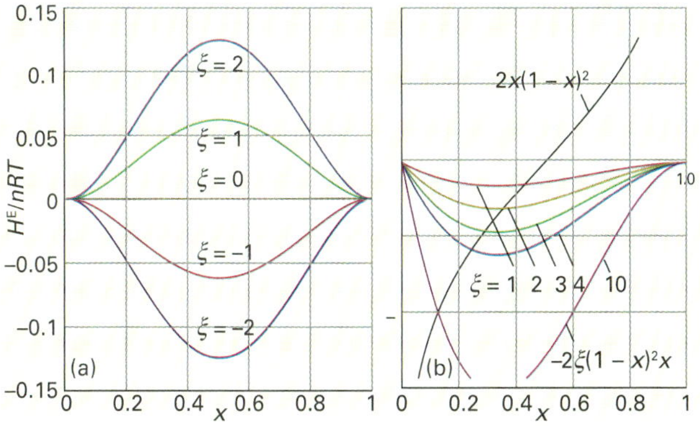

Figure 5C.21(a) The excess enthalpy and (b) the graphical solution of the resulting equation for the minima of the Gibbs energy of mixing in *Self-test* 5C.7.

Answer: No, see Fig. 5C.21b

Some systems show a **lower critical solution temperature**, T_{lc} (or *lower consolute temperature*), below which they mix in all proportions and above which they form two phases. An example is water and triethylamine (Fig. 5C.22). In this case, at low temperatures the two components are more miscible because they form a weak complex; at higher temperatures the complexes break up and the two components are less miscible.

Some systems have both upper and lower critical solution temperatures. They occur because, after the weak complexes have been disrupted, leading to partial miscibility, the thermal motion at higher temperatures homogenizes the mixture again, just as in the case of ordinary partially miscible liquids. The most famous example is nicotine and water, which are partially miscible between 61 °C and 210 °C (Fig. 5C.23).

(c) The distillation of partially miscible liquids

Consider a pair of liquids that are partially miscible and form a low-boiling azeotrope. This combination is quite common because both properties reflect the tendency of the two kinds of molecule to avoid each other. There are two possibilities: one in which the liquids become fully miscible before they boil; the other in which boiling occurs before mixing is complete.

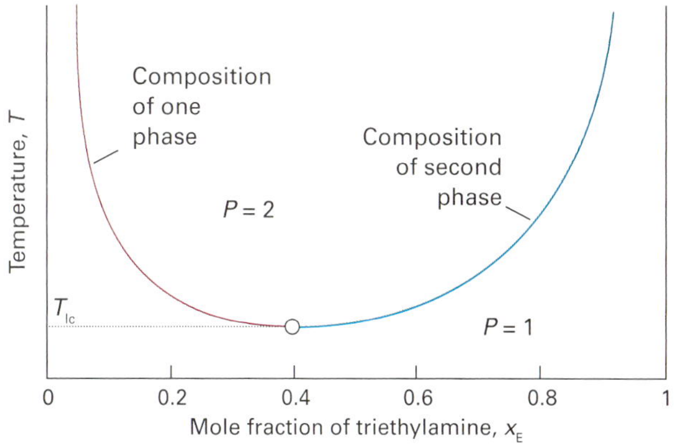

Figure 5C.22 The temperature–composition diagram for water and triethylamine. This system shows a lower critical temperature at 292 K. The labels indicate the interpretation of the boundaries.

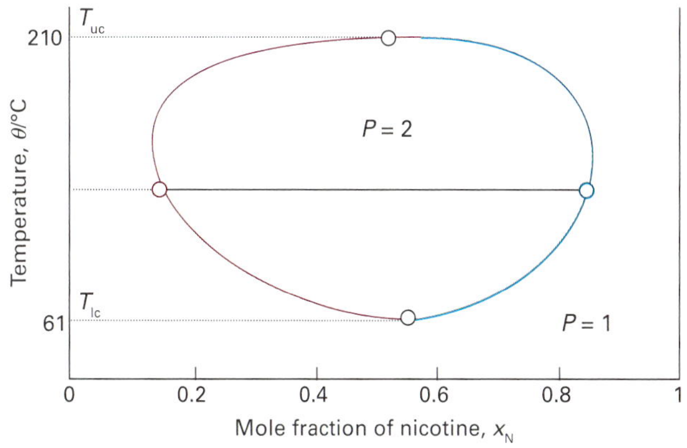

Figure 5C.23 The temperature–composition diagram for water and nicotine, which has both upper and lower critical temperatures. Note the high temperatures for the liquid (especially the water): the diagram corresponds to a sample under pressure.

Figure 5C.24 shows the phase diagram for two components that become fully miscible before they boil. Distillation of a mixture of composition a_1 leads to a vapour of composition b_1, which condenses to the completely miscible single-phase solution at b_2. Phase separation occurs only when this distillate is cooled to a point in the two-phase liquid region, such as b_3. This description applies only to the first drop of distillate. If distillation continues, the composition of the remaining liquid changes. In the end, when the whole sample has evaporated and condensed, the composition is back to a_1.

Figure 5C.25 shows the second possibility, in which there is no upper critical solution temperature. The distillate obtained from a liquid initially of composition a_1 has composition b_3 and is a two-phase mixture. One phase has composition b_3' and the other has composition b_3''.

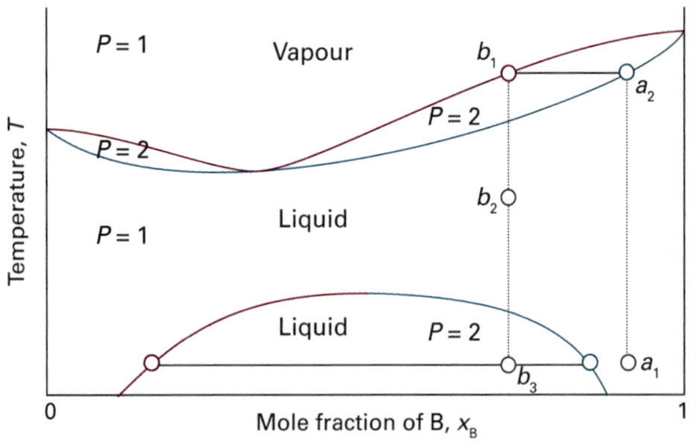

Figure 5C.24 The temperature–composition diagram for a binary system in which the upper critical temperature is less than the boiling point at all compositions. The mixture forms a low-boiling azeotrope.

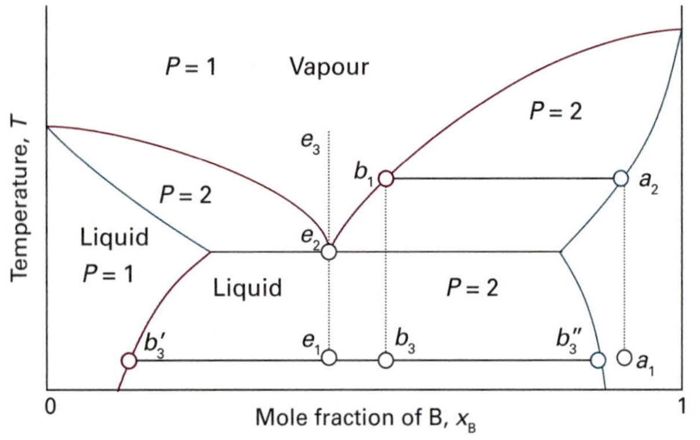

Figure 5C.25 The temperature–composition diagram for a binary system in which boiling occurs before the two liquids are fully miscible.

The behaviour of a system of composition represented by the isopleth e in Fig. 5C.25 is interesting. A system at e_1 forms two phases, which persist (but with changing proportions) up to the boiling point at e_2. The vapour of this mixture has the same composition as the liquid (the liquid is an azeotrope). Similarly,

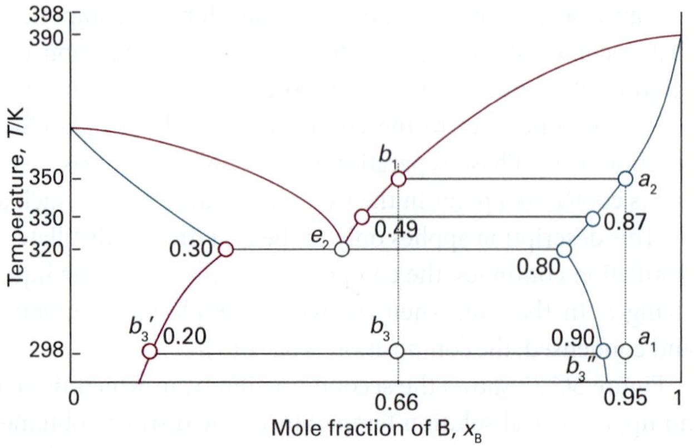

Figure 5C.26 The points of the phase diagram in Fig. 5C.25 that are discussed in *Example* 5C.3.

condensing a vapour of composition e_3 gives a two-phase liquid of the same overall composition. At a fixed temperature, the mixture vaporizes and condenses like a single substance.

Example 5C.3 Interpreting a phase diagram

State the changes that occur when a mixture of composition $x_B = 0.95$ (a_1) in Fig. 5C.26 is boiled and the vapour condensed.

Method The area in which the point lies gives the number of phases; the compositions of the phases are given by the points at the intersections of the horizontal tie line with the phase boundaries; the relative abundances are given by the lever rule .

Answer The initial point is in the one-phase region. When heated it boils at 350 K (a_2) giving a vapour of composition $x_B = 0.66$ (b_1). The liquid gets richer in B, and the last drop (of pure B) evaporates at 390 K. The boiling range of the liquid is therefore 350 to 390 K. If the initial vapour is drawn off, it has a composition $x_B = 0.66$. This composition would be maintained if the sample were very large, but for a finite sample it shifts to higher values and ultimately to $x_B = 0.95$. Cooling the distillate corresponds to moving down the $x_B = 0.66$ isopleth. At 330 K, for instance, the liquid phase has composition $x_B = 0.87$, the vapour $x_B = 0.49$; their relative proportions are 1:3. At 320 K the sample consists of three phases: the vapour and two liquids. One liquid phase has composition $x_B = 0.30$; the other has composition $x_B = 0.80$ in the ratio 0.62:1. Further cooling moves the system into the two-phase region, and at 298 K the compositions are 0.20 and 0.90 in the ratio 0.82:1. As further distillate boils over, the overall composition of the distillate becomes richer in B. When the last drop has been condensed the phase composition is the same as at the beginning.

Self-test 5C.8 Repeat the discussion, beginning at the point $x_B = 0.4$, $T = 298$ K.

5C.4 Liquid–solid phase diagrams

Knowledge of the temperature–composition diagrams for solid mixtures guides the design of important industrial processes, such as the manufacture of liquid crystal displays and semiconductors. In this section, we shall consider systems where solid and liquid phases may both be present at temperatures below the boiling point.

(a) Eutectics

Consider the two-component liquid of composition a_1 in Fig. 5C.27. The changes that occur as the system is cooled may be expressed as follows:

- $a_1 \rightarrow a_2$. The system enters the two-phase region labelled 'Liquid+B'. Pure solid B begins to come out of solution and the remaining liquid becomes richer in A

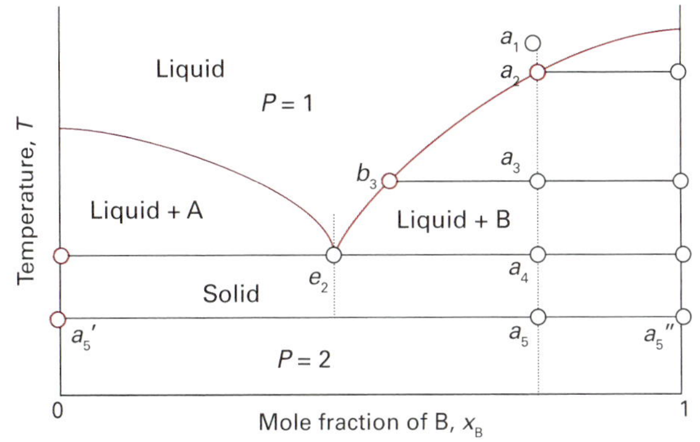

Figure 5C.27 The temperature–composition phase diagram for two almost immiscible solids and their completely miscible liquids. Note the similarity to Fig. 5C.25. The isopleth through e_2 corresponds to the eutectic composition, the mixture with lowest melting point.

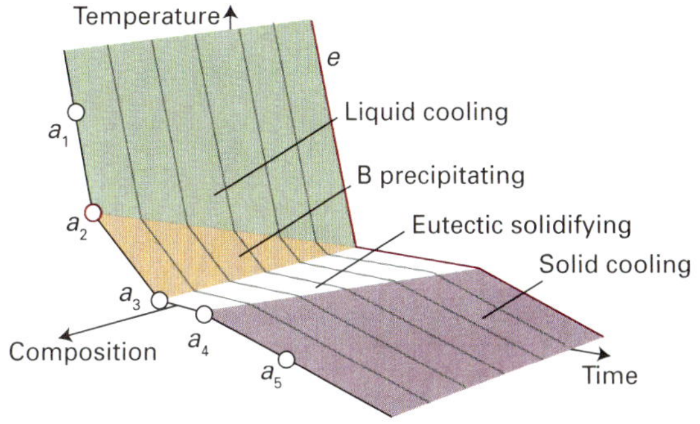

Figure 5C.28 The cooling curves for the system shown in Fig. 5C.27. For isopleth a, the rate of cooling slows at a_2 because solid B deposits from solution. There is a complete halt at a_4 while the eutectic solidifies. This halt is longest for the eutectic isopleth, e. The eutectic halt shortens again for compositions beyond e (richer in A). Cooling curves are used to construct the phase diagram.

Physical interpretation

- $a_2 \rightarrow a_3$. More of the solid B forms, and the relative amounts of the solid and liquid (which are in equilibrium) are given by the lever rule. At this stage there are roughly equal amounts of each. The liquid phase is richer in A than before (its composition is given by b_3) because some B has been deposited.
- $a_3 \rightarrow a_4$. At the end of this step, there is less liquid than at a_3, and its composition is given by e_2. This liquid now freezes to give a two-phase system of pure B and pure A.

The isopleth at e_2 in Fig. 5C.27 corresponds to the **eutectic** composition, the mixture with the lowest melting point.[2] A liquid with the eutectic composition freezes at a single temperature, without previously depositing solid A or B. A solid with the eutectic composition melts, without change of composition, at the lowest temperature of any mixture. Solutions of composition to the right of e_2 deposit B as they cool, and solutions to the left deposit A: only the eutectic mixture (apart from pure A or pure B) solidifies at a single definite temperature without gradually unloading one or other of the components from the liquid.

One technologically important eutectic is solder, which in one form has mass composition of about 67 per cent tin and 33 per cent lead and melts at 183 °C. The eutectic formed by 23 per cent NaCl and 77 per cent H_2O by mass melts at −21.1 °C. When salt is added to ice under isothermal conditions (for example, when spread on an icy road) the mixture melts if the temperature is above −21.1 °C (and the eutectic composition has been achieved). When salt is added to ice under adiabatic conditions (for example, when added to ice in a vacuum flask) the ice melts, but in doing so it absorbs heat from the rest of the mixture. The temperature of the system falls and, if enough salt is added, cooling continues down to the eutectic temperature. Eutectic formation occurs in the great majority of binary alloy systems, and is of great importance for the microstructure of solid materials. Although a eutectic solid is a two-phase system, it crystallizes out in a nearly homogeneous mixture of microcrystals. The two microcrystalline phases can be distinguished by microscopy and structural techniques such as X-ray diffraction (Topic 18A).

Thermal analysis is a very useful practical way of detecting eutectics. We can see how it is used by considering the rate of cooling down the isopleth through a_1 in Fig. 5C.27. The liquid cools steadily until it reaches a_2, when B begins to be deposited (Fig. 5C.28). Cooling is now slower because the solidification of B is exothermic and retards the cooling. When the remaining liquid reaches the eutectic composition, the temperature remains constant until the whole sample has solidified: this region of constant temperature is the eutectic halt. If the liquid has the eutectic composition e initially, the liquid cools steadily down to the freezing temperature of the eutectic, when there is a long **eutectic halt** as the entire sample solidifies (like the freezing of a pure liquid).

Brief illustration 5C.6 Interpreting a binary phase diagram

Figure 5C.29 is the phase diagram for the binary system silver/tin. The regions have been labelled to show which each one represents. When a liquid of composition a is cooled, solid silver with dissolved tin begins to precipitate at a_1 and the sample solidifies completely at a_2.

Self-test 5C.9 Describe what happens when the sample of composition b is cooled.

[2] The name comes from the Greek words for 'easily melted'.

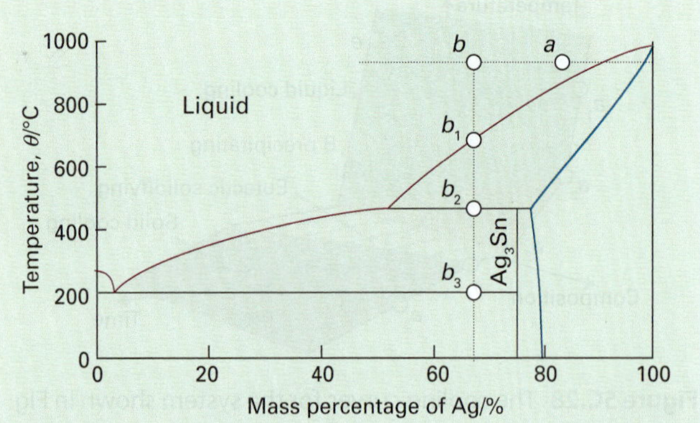

Figure 5C.29 The phase diagram for silver/tin discussed in *Brief illustration* 5C.6.

Answer: Solid Ag with dissolved Sn begins to precipitate at b_1, and the liquid becomes richer in Sn as the temperature falls further. At b_2 solid Ag_3Sn begins to precipitate, and the liquid becomes richer in Sn. At b_3 the system has its eutectic composition (a solid solution of Sn and Ag_3Sn) and it freezes without further change in composition.

Monitoring the cooling curves at different overall compositions gives a clear indication of the structure of the phase diagram. The solid–liquid boundary is given by the points at which the rate of cooling changes. The longest eutectic halt gives the location of the eutectic composition and its melting temperature.

(b) Reacting systems

Many binary mixtures react to produce compounds, and technologically important examples of this behaviour include the Group 13/15 (III/V) semiconductors, such as the gallium arsenide system, which forms the compound GaAs. Although three constituents are present, there are only two components because GaAs is formed from the reaction Ga + As → GaAs. We shall illustrate some of the principles involved with a system that forms a compound C that also forms eutectic mixtures with the species A and B (Fig. 5C.30).

A system prepared by mixing an excess of B with A consists of C and unreacted B. This is a binary B, C system, which we suppose forms a eutectic. The principal change from the eutectic phase diagram in Fig. 5C.27 is that the whole of the phase diagram is squeezed into the range of compositions lying between equal amounts of A and B ($x_B = 0.5$, marked C in Fig. 5C.30) and pure B. The interpretation of the information in the diagram is obtained in the same way as for Fig. 5C.27. The solid deposited on cooling along the isopleth *a* is the compound C. At temperatures below a_4 there are two solid phases, one consisting of C and the other of B. The pure compound C melts **congruently**, that is, the composition of the liquid it forms is the same as that of the solid compound.

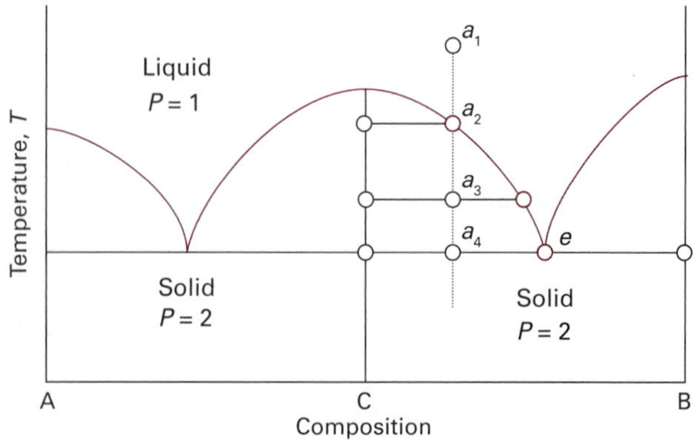

Figure 5C.30 The phase diagram for a system in which A and B react to form a compound C = AB. This resembles two versions of Fig. 5C.27 in each half of the diagram. The constituent C is a true compound, not just an equimolar mixture.

(c) Incongruent melting

In some cases the compound C is not stable as a liquid. An example is the alloy Na_2K, which survives only as a solid (Fig. 5C.31). Consider what happens as a liquid at a_1 is cooled:

- $a_1 \rightarrow a_2$. A solid solution rich in Na is deposited, and the remaining liquid is richer in K.
- $a_2 \rightarrow$ just below a_3. The sample is now entirely solid and consists of a solid solution rich in Na and solid Na_2K.

Now consider the isopleth through b_1:

- $b_1 \rightarrow b_2$. No obvious change occurs until the phase boundary is reached at b_2 when a solid solution rich in Na begins to deposit.

Physical interpretation

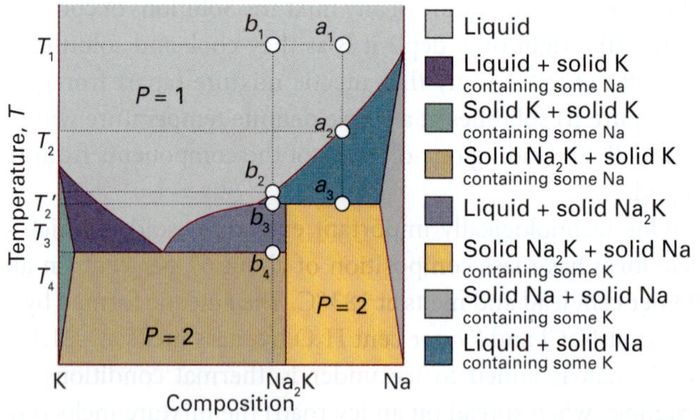

Figure 5C.31 The phase diagram for an actual system (sodium and potassium) like that shown in Fig. 5C.30, but with two differences. One is that the compound is Na_2K, corresponding to A_2B and not AB as in that illustration. The second is that the compound exists only as the solid, not as the liquid. The transformation of the compound at its melting point is an example of incongruent melting.

- $b_2 \rightarrow b_3$. A solid solution rich in Na deposits, but at b_3 a reaction occurs to form Na_2K: this compound is formed by the K atoms diffusing into the solid Na.
- b_3. At b_3, three phases are in mutual equilibrium: the liquid, the compound Na_2K, and a solid solution rich in Na. The horizontal line representing this three-phase equilibrium is called a **peritectic line**. At this stage the liquid Na/K mixture is in equilibrium with a little solid Na_2K, but there is still no liquid compound
- $b_3 \rightarrow b_4$. As cooling continues, the amount of solid compound increases until at b_4 the liquid reaches its eutectic composition. It then solidifies to give a two-phase solid consisting of a solid solution rich in K and solid Na_2K.

Physical interpretation

If the solid is reheated, the sequence of events is reversed. No liquid Na_2K forms at any stage because it is too unstable to exist as a liquid. This behaviour is an example of **incongruent melting**, in which a compound melts into its components and does not itself form a liquid phase.

Checklist of concepts

- ☐ 1. Raoult's law is used to calculate the total vapour pressure of a binary system of two volatile liquids.
- ☐ 2. The composition of the vapour in equilibrium with a binary mixture is calculated by using Dalton's law.
- ☐ 3. The composition of the vapour and the liquid phase in equilibrium are located at each end of a tie line.
- ☐ 4. The **lever rule** is used to deduce the relative abundances of each phase in equilibrium.
- ☐ 5. A phase diagram can be used to discuss the process of **fractional distillation**.
- ☐ 6. Depending on the relative strengths of the intermolecular forces, high- or low-boiling **azeotropes** may be formed.
- ☐ 7. The vapour pressure of a system composed of immiscible liquids is the sum of the vapour pressures of the pure liquids.
- ☐ 8. A phase diagram may be used to discuss the distillation of partially miscible liquids.
- ☐ 9. Phase separation of partially miscible liquids may occur when the temperature is below the upper critical solution temperature or above the lower critical solution temperature; the process may be discussed in terms of the model of a regular solution.
- ☐ 10. A phase diagram summarizes the temperature–composition properties of a binary system with solid and liquid phases; at the **eutectic composition** the liquid phase solidifies without change of composition.
- ☐ 11. The phase equilibria of binary systems in which the components react may also be summarized by a phase diagram.
- ☐ 12. In some cases, a solid compound does not survive melting.

Checklist of equations

Property	Equation	Comment	Equation number
Composition of vapour	$y_A = x_A p_A^*/(p_B^* + (p_A^* - p_B^*)x_A)$ $y_B = 1 - y_A$	Ideal solution	5C.4
Total vapour pressure	$p = p_A^* p_B^*/(p_A^* + (p_B^* - p_A^*)y_A)$	Ideal solution	5C.5
Lever rule	$n_\alpha l_\alpha = n_\beta l_\beta$		5C.6

5D Phase diagrams of ternary systems

Contents

➤ Why do you need to know this material?

Ternary phase diagrams have become important in materials science as more complex materials are investigated, such as the ceramics found to have superconducting properties.

➤ What is the key idea?

A phase diagram is a map showing the conditions under which each phase of a system is the most stable.

➤ What do you need to know already?

It would be helpful to review the interpretation of two-component phase diagrams (Topic 5C) and the phase rule (Topic 5A). The interpretation of the phase diagrams presented here uses the lever rule (Topic 5C).

This short Topic is a brief introduction to the depiction of phases of systems of three components. In terms of the phase rule (Topic 5A), $C=3$, so $F=5-P$. If we restrict systems to constant temperature and pressure, two degrees of freedom are discarded and we are left with $F''=3-P$. An area on a ternary phase diagram therefore represents a region where a single phase is present, a line represents the equilibrium between two phases of varying composition, and a point corresponds to a composition at which three phases are present in equilibrium.

5D.1 Triangular phase diagrams

The mole fractions of the three components of a ternary system satisfy $x_A+x_B+x_C=1$. A phase diagram drawn as an equilateral triangle ensures that this property is satisfied automatically because the sum of the distances to a point inside an equilateral triangle of side 1 and measured parallel to the edges is equal to 1 (Fig. 5D.1).

Figure 5D.1 shows how this approach works in practice. The edge AB corresponds to $x_C=0$, and likewise for the other two edges. Hence, each of the three edges corresponds to one of the three binary systems (A,B), (B,C), and (C,A). An interior point corresponds to a system in which all three components are present. The point P, for instance, represents $x_A=0.50$, $x_B=0.10$, $x_C=0.40$.

Any point on a straight line joining an apex to a point on the opposite edge (the broken line in Fig. 5D.1) represents a composition that is progressively richer in A the closer the point is to the A apex but for which the concentration ratio B:C remains constant. Therefore, if we wish to represent the changing composition of a system as A is added, we draw a line from the A apex to the point on BC representing the initial binary system. Any ternary system formed by adding A then lies at some point on this line.

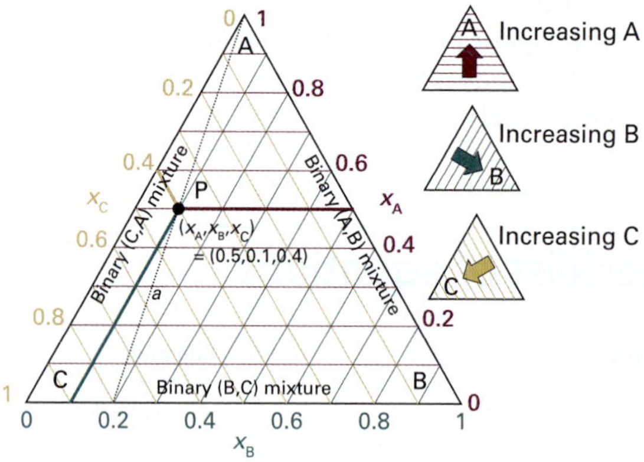

Figure 5D.1 The triangular coordinates used for the discussion of three-component systems. Each edge corresponds to a binary system. All points along the dotted line *a* correspond to mole fractions of C and B in the same ratio.

Brief illustration 5D.1 The representation of composition

The following points are represented on Fig. 5D.2:

Point	x_A	x_B	x_C
a	0.20	0.80	0
b	0.42	0.26	0.32
c	0.80	0.10	0.10
d	0.10	0.20	0.70
e	0.20	0.40	0.40
f	0.30	0.60	0.10

Note that the points d, e, f have $x_A/x_B = 0.50$ and lie on a straight line.

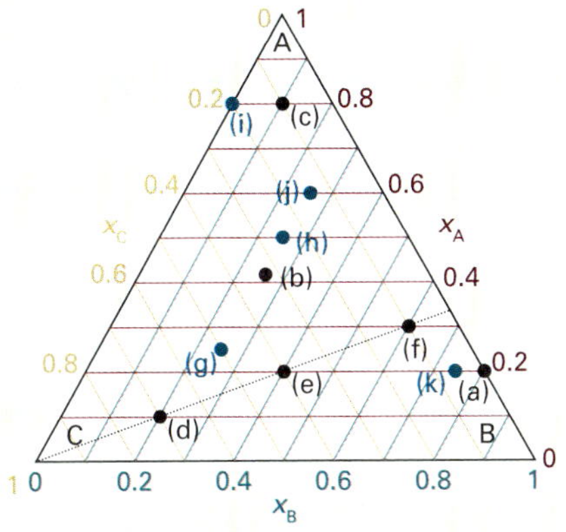

Figure 5D.2 The points referred to in *Brief illustration* 5D.1 (black) and *Self-test* 5D.1 (blue).

Self-test 5D.1 Mark the following points on the triangle.

Point	x_A	x_B	x_C
g	0.25	0.25	0.50
h	0.50	0.25	0.25
i	0.80	0	0.20
j	0.60	0.25	0.15
k	0.20	0.75	0.0.05

Answer: See Fig. 5D.2.

A single triangle represents the equilibria when one of the discarded degrees of freedom (the temperature, for instance) has a certain value. Different temperatures give rise to different equilibrium behaviour and therefore different triangular phase diagrams. Each one may therefore be regarded as a horizontal slice through a three-dimensional triangular prism, such as that shown in Fig. 5D.3.

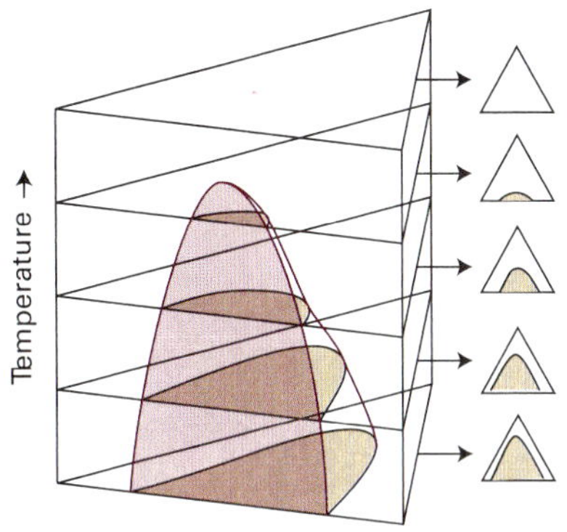

Figure 5D.3 When temperature is included as a variable, the phase diagram becomes a triangular prism. Horizontal sections through the prism correspond to the triangular phase diagrams being discussed.

5D.2 Ternary systems

Ternary phase diagrams are widely used in metallurgy and materials science in general. Although they can become quite complex, they can be interpreted in much the same way as binary diagrams. Here we give two examples.

(a) Partially miscible liquids

The phase diagram for a ternary system in which W (in due course: water) and A (in due course: acetic acid) are fully miscible, A and C (in due course: chloroform) are fully miscible, but W and C are only partially miscible is shown in Fig. 5D.4. This illustration is for the system water/acetic acid/chloroform at room temperature, which behaves in this way. It shows that the two fully miscible pairs, (A,W) and (A,C), form single-phase

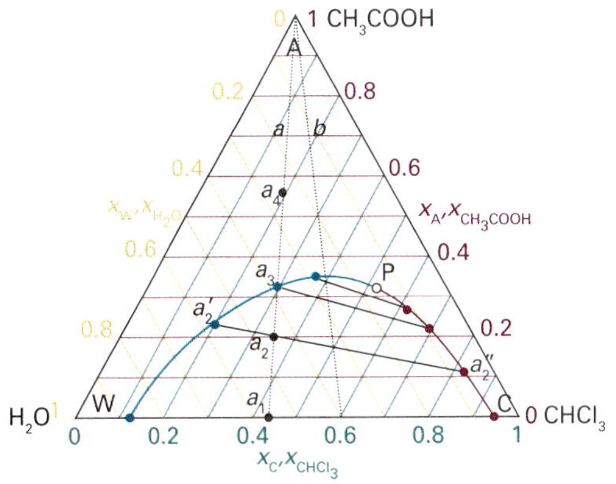

Figure 5D.4 The phase diagram, at fixed temperature and pressure, of the three-component system acetic acid, chloroform, and water. Only some of the tie lines have been drawn in the two-phase region. All points along the line *a* correspond to chloroform and water present in the same ratio.

regions and that (W,C) system (along the base of the triangle) has a two-phase region. The base of the triangle corresponds to one of the horizontal lines in a two-component phase diagram. The tie lines in the two-phase regions are constructed experimentally by determining the compositions of the two phases that are in equilibrium, marking them on the diagram, and then joining them with a straight line.

A single-phase system is formed when enough A is added to the binary (W,C) mixture. This effect is shown by following the line a in Fig. 5D.4:

Physical interpretation

- a_1. The system consists of two phases and the relative amounts of the two phases can be read off by using the lever rule.
- $a_1 \rightarrow a_2$. The addition of A takes the system along the line joining a_1 to the A apex. At a_2 the solution still has two phases, but there is slightly more W in the largely C phase (represented by the point a_2'') and more C in the largely W phase a_2' because the presence of A helps both to dissolve. The phase diagram shows that there is more A in the W-rich phase than in the C-rich phase (a_2' is closer than (a_2'') to the A apex).
- $a_2 \rightarrow a_3$. At a_3 two phases are present, but the C-rich layer is present only as a trace (lever rule).
- $a_3 \rightarrow a_4$. Further addition of A takes the system towards and beyond a_4, and only a single phase is present.

Brief illustration 5D.2 Partially miscible liquids

Suppose we have a mixture of water (W in Fig. 5D.4) and chloroform (C) with $x_W = 0.40$ and $x_C = 0.60$, and acetic acid (A) is added to it. The relative proportions of A and C remain constant, so the point representing the overall composition moves along the straight line b from $x_C = 0.60$ on the base to the acetic acid apex. The initial composition is in a two-phase region: one phase has the composition $(x_W, x_C, x_A) = (0.95, 0.05, 0)$ and the other has composition $(x_W, x_C, x_A) = (0.12, 0.88, 0)$. When sufficient acetic acid has been added to raise its mole fraction to 0.18 the system consists of two phases of composition (0.07, 0.82, 0.11) and (0.57, 0.20, 0.23) in almost equal abundance.

Self-test 5D.2 Specify the system when enough acid has been added to raise its mole fraction to 0.34.

Answer: A trace of a phase of composition (0.12, 0.61, 0.27) and a dominating phase of composition (0.28, 0.37, 0.35)

The point marked P in Fig. 5D.4 is called the **plait point**. It is yet another example of a critical point. At the plait point, the compositions of the two phases in equilibrium become identical. For convenience, the general interpretation of a triangular phase diagram is summarized in Fig. 5D.5.

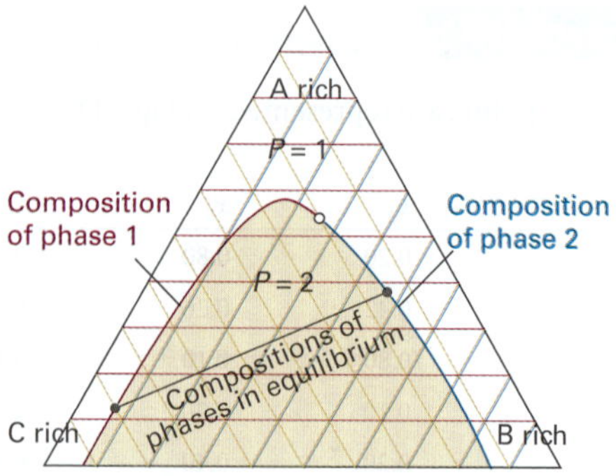

Figure 5D.5 The interpretation of a triangular phase diagram. The region inside the curved line consists of two phases, and the compositions of the two phases in equilibrium are given by the points at the ends of the tie lines (the tie lines are determined experimentally).

(b) Ternary solids

The triangular phase diagram in Fig. 5D.6 is typical of that for a solid alloy with varying compositions of three metals, A, B, and C.

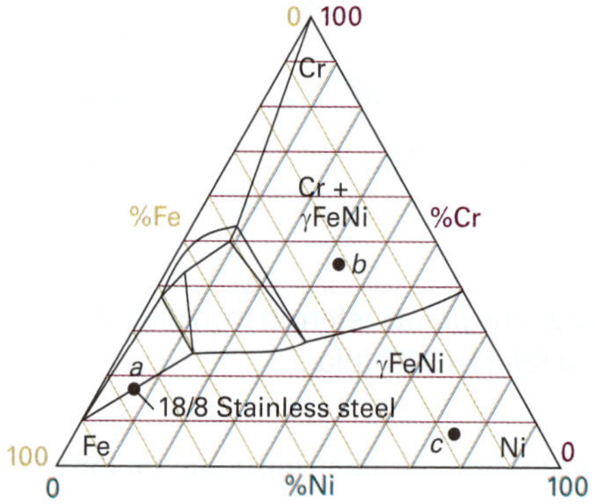

Figure 5D.6 A simplified triangular phase diagram of the ternary system represented by a stainless steel composed of iron, chromium, and nickel.

Brief illustration 5D.3 Stainless steel

Figure 5D.6 is a simplified version of the phase diagram for a stainless steel consisting of iron, chromium, and nickel. The axes are labelled with the mass percentage compositions instead of the mole fractions, but as the three percentages add up to 100 per cent, the interpretation of points in the triangle is essentially the same as for mole fractions. The point a corresponds to the composition 74 per cent Fe, 18 per cent Cr, and 8 per cent Ni. It corresponds to the most common form of stainless steel, '18-8 stainless steel'. The composition corresponding to point b lies in the two-phase region, one phase consisting of Cr and the other of the alloy γ-FeNi.

Self-test 5D.3 Identify the composition represented by point c.

Answer: Three phases, Fe, Ni, and γ-FeNi

Checklist of concepts

- ☐ 1. A phase diagram drawn as an equilateral triangle ensures that the property $x_A + x_B + x_C = 1$ is satisfied automatically.
- ☐ 2. At the **plait point**, the compositions of the two phases in equilibrium are identical.

at all temperatures and pressures. Deviations of the solute from ideality disappear as zero concentration is approached.

Example 5E.1 Measuring activity

Use the following information to calculate the activity and activity coefficient of trichloromethane (chloroform, C) in propanone (acetone, A) at 25 °C, treating it first as a solvent and then as a solute.

x_C	0	0.20	0.40	0.60	0.80	1
p_C/kPa	0	4.7	11	18.9	26.7	36.4
p_A/kPa	46.3	33.3	23.3	12.3	4.9	0

Method For the activity of chloroform as a solvent (the Raoult's law activity), form $a_C = p_C/p_C^*$ and $\gamma_C = a_C/x_C$. For its activity as a solute (the Henry's law activity), form $a_C = p_C/K_C$ and $\gamma_C = a_C/x_C$ with the new activity.

Answer Because $p_C^* = 36.4$ kPa and $K_C = 22.0$ kPa, we can construct the following tables. For instance, at $x_C = 0.20$, in the Raoult's law case we find $a_C = (4.7\,\text{kPa})/(36.4\,\text{kPa}) = 0.13$ and $\gamma_C = 0.13/0.20 = 0.65$; likewise, in the Henry's law case, $a_C = (4.7\,\text{kPa})/(22.0\,\text{kPa}) = 0.21$ and $\gamma_C = 0.21/0.20 = 1.05$.

From Raoult's law (chloroform regarded as the solvent):

a_C	0	0.13	0.30	0.52	0.73	1.00
γ_C		0.65	0.75	0.87	0.91	1.00

From Henry's law (chloroform regarded as the solute):

a_C	0	0.21	0.50	0.86	1.21	1.65
γ_C	1	1.05	1.25	1.43	1.51	1.65

These values are plotted in Fig. 5E.1. Notice that $\gamma_C \to 1$ as $x_C \to 1$ in the Raoult's law case, but that $\gamma_C \to 1$ as $x_C \to 0$ in the Henry's law case.

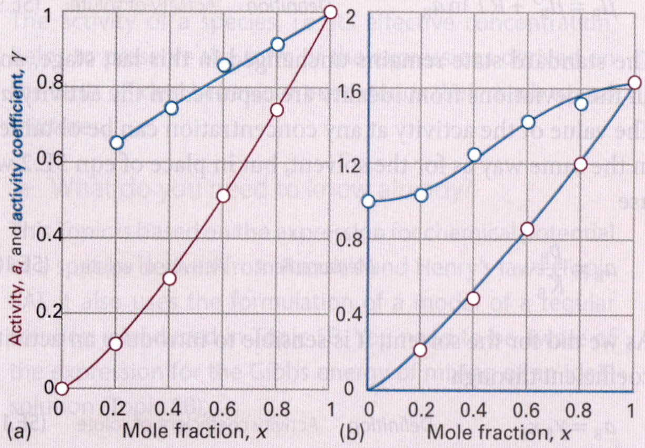

Figure 5E.1 The variation of activity and activity coefficient for a chloroform/acetone (trichloromethane/propanone) solution with composition according to (a) Raoult's law, (b) Henry's law.

Self-test 5E.3 Calculate the activities and activity coefficients for acetone according to the two conventions.

Answer: At $x_A = 0.60$, for instance $a_R = 0.50$; $\gamma_R = 0.83$; $a_H = 1.00$, $\gamma_H = 1.67$

(c) Activities in terms of molalities

The selection of a standard state is entirely arbitrary, so we are free to choose one that best suits our purpose and the description of the composition of the system. Because compositions are often expressed as molalities, b, in place of mole fractions it is convenient to write

$$\mu_B = \mu_B^\ominus + RT\ln b_B \tag{5E.13}$$

where $\mu_B^\ominus$ has a different value from the standard values introduced earlier. According to this definition, the chemical potential of the solute has its standard value $\mu_B^\ominus$ when the molality of B is equal to $b^\ominus$ (that is, at 1 mol kg^{-1}). Note that as $b_B \to 0$, $\mu_B \to \infty$; that is, as the solution becomes diluted, so the solute becomes increasingly stabilized. The practical consequence of this result is that it is very difficult to remove the last traces of a solute from a solution.

Now, as before, we incorporate deviations from ideality by introducing a dimensionless activity a_B, a dimensionless activity coefficient γ_B, and writing

$$a_B = \gamma_B \frac{b_B}{b^\ominus}, \text{ where } \gamma_B \to 1 \text{ as } b_B \to 0 \tag{5E.14}$$

at all temperatures and pressures. The standard state remains unchanged in this last stage and, as before, all the deviations from ideality are captured in the activity coefficient γ_B. We then arrive at the following succinct expression for the chemical potential of a real solute at any molality:

$$\mu_B = \mu_B^\ominus + RT\ln a_B \tag{5E.15}$$

(d) The biological standard state

One important illustration of the ability to choose a standard state to suit the circumstances arises in biological applications. The conventional standard state of hydrogen ions (unit activity, corresponding to pH = 0)[1] is not appropriate to normal biological conditions. Therefore, in biochemistry it is common to adopt the **biological standard state**, in which pH = 7 (an activity of 10^{-7}, neutral solution) and to label the corresponding standard thermodynamic functions as $G^\oplus$, $H^\oplus$, $\mu^\oplus$, and $S^\oplus$ (some texts use $X^{\circ\prime}$).

[1] Recall from introductory chemistry courses that pH = −log $a(H_3O^+)$.

To find the relation between the thermodynamic and biological standard values of the chemical potential of hydrogen ions we need to note from eqn 5E.15 that

$$\mu_{H^+} = \mu_{H^+}^{\ominus} + RT\ln a_{H^+} = \mu_{H^+}^{\ominus} - (RT\ln 10)\text{pH}$$

It follows that

$$\mu_{H^+}^{\oplus} = \mu_{H^+}^{\ominus} - 7RT\ln 10 \qquad \text{Relation between standard state and biological standard state} \qquad (5E.16)$$

Brief illustration 5E.3 The biological standard state

At 298 K, $7RT\ln 10 = 39.96\ \text{kJ mol}^{-1}$, so the two standard values differ by about $40\ \text{kJ mol}^{-1}$ and specifically $\mu_{H^+}^{\oplus} = \mu_{H^+}^{\ominus} - 39.96\ \text{kJ mol}^{-1}$. Thus, in a reaction of the form $A + 2\,H^+(aq) \rightarrow B$, the standard and biological standard Gibbs energies are related as follows:

$$\begin{aligned}\Delta_r G^{\oplus} &= \mu_B^{\ominus} - \{\mu_A^{\ominus} + 2\mu_{H^+}^{\oplus}\} = \mu_B^{\ominus} - \{\mu_A^{\ominus} + 2\mu_{H^+}^{\ominus} - 14RT\ln 10\} \\ &= \mu_B^{\ominus} - \{\mu_A^{\ominus} + 2\mu_{H^+}^{\ominus}\} + 14RT\ln 10 = \Delta_r G^{\ominus} + 14RT\ln 10 \\ &= \Delta_r G^{\ominus} + 79.92\ \text{kJ mol}^{-1}\end{aligned}$$

Self-test 5E.4 Find the relation between the standard and biological standard Gibbs energies of a reaction of the form $A \rightarrow B + 3\,H^+(aq)$.

Answer: $\Delta_r G^{\oplus} = \Delta_r G^{\ominus} - 119.88\ \text{kJ mol}^{-1}$

5E.3 The activities of regular solutions

The material on regular solutions in Topic 5B gives further insight into the origin of deviations from Raoult's law and its relation to activity coefficients. The starting point is the model expression for the excess (molar) enthalpy (eqn 5B.6, $H^E = \xi RTx_Ax_B$) and its implication for the Gibbs energy of mixing for a regular solution. We show in the following *Justification* that for this model the activity coefficients are given by

$$\ln\gamma_A = \xi x_B^2 \qquad \ln\gamma_B = \xi x_A^2 \qquad \text{Margules equations} \qquad (5E.17)$$

These relations are called the **Margules equations**.

Justification 5E.1 The Margules equations

The Gibbs energy of mixing to form an ideal solution is

$$\Delta_{mix}G = nRT\{x_A\ln x_A + x_B\ln x_B\}$$

(This is eqn 5B.16 of Topic 5B.) The corresponding expression for a non-ideal solution is

$$\Delta_{mix}G = nRT\{x_A\ln a_A + x_B\ln a_B\}$$

This relation follows in the same way as for an ideal mixture but with activities in place of mole fractions. If each activity is replaced by γx, this expression becomes

$$\begin{aligned}\Delta_{mix}G &= nRT\{x_A\ln x_A\gamma_A + x_B\ln x_B\gamma_B\} \\ &= nRT\{x_A\ln x_A + x_B\ln x_B + x_A\ln\gamma_A + x_B\ln\gamma_B\}\end{aligned}$$

Now we introduce the two expressions in eqn 5E.17, and use $x_A + x_B = 1$, which gives

$$\begin{aligned}\Delta_{mix}G &= nRT\{x_A\ln x_A + x_B\ln x_B + \xi x_Ax_B^2 + \xi x_Bx_A^2\} \\ &= nRT\{x_A\ln x_A + x_B\ln x_B + \xi x_Ax_B(x_A + x_B)\} \\ &= nRT\{x_A\ln x_A + x_B\ln x_B + \xi x_Ax_B\}\end{aligned}$$

Note that the activity coefficients behave correctly for dilute solutions: $\gamma_A \rightarrow 1$ as $x_B \rightarrow 0$ and $\gamma_B \rightarrow 1$ as $x_A \rightarrow 0$.

At this point we can use the Margules equations to write the activity of A as

$$a_A = \gamma_A x_A = x_A e^{\xi x_B^2} = x_A e^{\xi(1-x_A)^2} \qquad (5E.18)$$

with a similar expression for a_B. The activity of A, though, is just the ratio of the vapour pressure of A in the solution to the vapour pressure of pure A (eqn 5E.2, $a_A = p_A/p_A^*$), so we can write

$$p_A = p_A^* x_A e^{\xi(1-x_A)^2} \qquad (5E.19)$$

This function is plotted in Fig. 5E.2. We see that

Physical interpretation

- When $\xi = 0$, corresponding to an ideal solution, $p_A = p_A^* x_A$, in accord with Raoult's law.
- Positive values of ξ (endothermic mixing, unfavourable solute–solvent interactions) give vapour pressures higher than ideal.
- Negative values of ξ (exothermic mixing, favourable solute–solvent interactions) give a lower vapour pressure.

All the plots of eqn 5E.19 approach linearity and coincide with the Raoult's law line as $x_A \rightarrow 1$ and the exponential function in eqn 5E.19 approaches 1. When $x_A \ll 1$, eqn 5E.19 approaches

$$p_A = p_A^* x_A e^{\xi} \qquad (5E.20)$$

This expression has the form of Henry's law once we identify K with $e^{\xi}p_A^*$, which is different for each solute–solvent system.

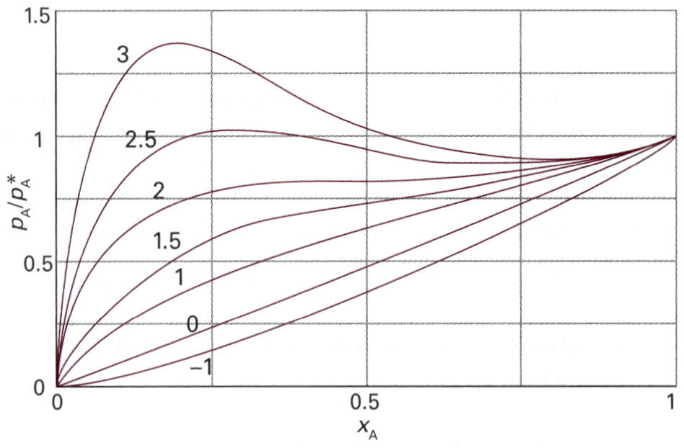

Figure 5E.2 The vapour pressure of a mixture based on a model in which the excess enthalpy is proportional to ξRTx_Ax_B. An ideal solution corresponds to $\xi=0$ and gives a straight line, in accord with Raoult's law. Positive values of ξ give vapour pressures higher than ideal. Negative values of ξ give a lower vapour pressure.

Brief illustration 5E.4 The Margules equations

In *Example* 5B.1 of Topic 5B it is established that $\xi=1.13$ for a mixture of benzene and cyclohexane at 25 °C. Because $\xi>0$ we can expect the vapour pressure of the mixture to be greater than its ideal value. The total vapour pressure of the mixture is therefore

$$p=p^*_{\text{benzene}}x_{\text{benzene}}e^{1.13(1-x_{\text{benzene}})^2}+p^*_{\text{cyclohexane}}x_{\text{cyclohexane}}e^{1.13(1-x_{\text{cyclohexane}})^2}$$

This expression is plotted in Fig. 5E.3a using $p_{\text{benzene}}=10.0\,\text{kPa}$ and $p^*_{\text{cyclohexane}}=10.4\,\text{kPa}$.

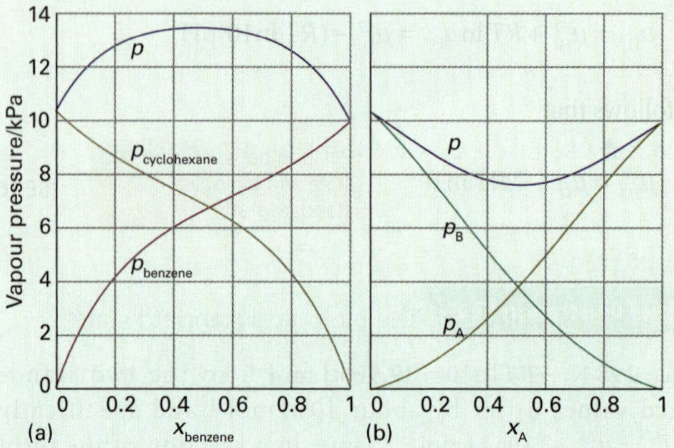

Figure 5E.3 The computed vapour pressure curves for a mixture of benzene and cyclohexane at 25 °C (a) as derived in *Brief illustration* 5E.4 and (b) *Self-test* 5E.5.

Self-test 5E.5 Suppose it is found that for a hypothetical mixture $\xi=-1.13$, but with other properties the same. Draw the vapour pressure diagram.

Answer: See Fig. 5E.3b

Checklist of concepts

- ☐ 1. The **activity** is an effective concentration that preserves the form of the expression for the chemical potential. See Table 5E.1.
- ☐ 2. The chemical potential of a solute in an ideal–dilute solution is defined on the basis of Henry's law.
- ☐ 3. The activity of a solute takes into account departures from Henry's law behaviour.
- ☐ 4. An alternative approach to the definition of the solute activity is based on the molality of the solute.
- ☐ 5. The **biological standard state** of a species in solution is defined as pH=7 (and 1 bar).
- ☐ 6. The **Margules equations** relate the activities of the components of a model regular solution to its composition. They lead to expressions for the vapour pressures of the components of a regular solution.

Table 5E.1 Activities and standard states: a summary*

Component	Basis	Standard state	Activity	Limits
Solid or liquid		Pure, 1 bar	$a=1$	
Solvent	Raoult	Pure solvent, 1 bar	$a=p/p^*$, $a=\gamma x$	$\gamma\to1$ as $x\to1$ (pure solvent)
Solute	Henry	(1) A hypothetical state of the pure solute	$a=p/K$, $a=\gamma x$	$\gamma\to1$ as $x\to0$
		(2) A hypothetical state of the solute at molality $b^\ominus$	$a=\gamma b/b^\ominus$	$\gamma\to1$ as $b\to0$
Gas	Fugacity†	Pure, a hypothetical state of 1 bar and behaving as a perfect gas	$f=\gamma p$	$\gamma\to1$ as $p\to0$

* In each case, $\mu=\mu^\ominus+RT\ln a$.

† Fugacity is discussed in Topic 3D.

Checklist of equations

Property	Equation	Comment	Equation number
Chemical potential of solvent	$\mu_A = \mu_A^{\star} + RT \ln a_A$	Definition	5E.1
Activity of solvent	$a_A = p_A/p_A^{\star}$	$a_A \to x_A$ as $x_A \to 1$	5E.2
Activity coefficient of solvent	$a_A = \gamma_A x_A$	$\gamma_A \to 1$ as $x_A \to 1$	5E.4
Chemical potential of solute	$\mu_B = \mu_B^{\ominus} + RT \ln a_B$	Definition	5E.9
Activity of solute	$a_B = p_B/K_B$	$a_B \to x_B$ as $x_B \to 0$	5E.10
Activity coefficient of solute	$a_B = \gamma_B x_B$	$\gamma_B \to 1$ as $x_B \to 0$	5E.11
Conversion to biological standard state	$\mu_{H^+}^{\oplus} = \mu_{H^+}^{\ominus} - 7RT \ln 10$		5E.16
Margules equations	$\ln \gamma_A = \xi x_B^2$, $\ln \gamma_B = \xi x_A^2$	Regular solution	5E.17
Vapour pressure	$p_A = p_A^{\star} x_A e^{\xi(1-x_A)^2}$	Regular solution	5E.19

5F The activities of ions

Contents

➤ Why do you need to know this material?

Interactions between ions are so strong that the approximation of replacing activities by molalities is valid only in very dilute solutions (less than 1 mmol kg^{-1} in total ion concentration) and in precise work activities themselves must be used. We need, therefore, to pay special attention to the activities of ions in solution, especially in preparation for the discussion of electrochemical phenomena.

➤ What is the key idea?

The chemical potential of an ion is lowered as a result of its electrostatic interaction with its ionic atmosphere.

➤ What do you need to know already?

This Topic builds on the relation between chemical potential and mole fraction (Topic 5A) and on the relation between Gibbs free energy and non-expansion work (Topic 3D). If you intend to work through the derivation of the Debye–Hückel theory, you need to be familiar with some concepts from electrostatics, including the Coulomb potential and its relation to charge density through Poisson's equation (which is explained in the Topic); this Topic also draws on the Boltzmann distribution (*Foundations* B and, in much more detail, Topic 15A).

If the chemical potential of the cation M^+ is denoted μ_+ and that of the anion X^- is denoted μ_-, the total molar Gibbs energy of the ions in the electrically neutral solution is the sum of these partial molar quantities. The molar Gibbs energy of an *ideal* solution of such ions is

$$G_m^{ideal} = \mu_+^{ideal} + \mu_-^{ideal} \tag{5F.1}$$

with $\mu_J^{ideal} = \mu_J^{\ominus} + RT \ln x_J$. However, for a *real* solution of M^+ and X^- of the same molality we write $\mu_J = \mu_J^{\ominus} + RT \ln a_J$ with $a_J = \gamma_J x_J$, which implies that $\mu_J = \mu_J^{ideal} + RT \ln \gamma_J$. It then follows that

$$\begin{aligned} G_m &= \mu_+ + \mu_- = \mu_+^{ideal} + \mu_-^{ideal} + RT\ln\gamma_+ + RT\ln\gamma_- \\ &= G_m^{ideal} + RT\ln\gamma_+\gamma_- \end{aligned} \tag{5F.2}$$

All the deviations from ideality are contained in the last term.

5F.1 Mean activity coefficients

There is no experimental way of separating the product $\gamma_+\gamma_-$ into contributions from the cations and the anions. The best we can do experimentally is to assign responsibility for the non-ideality equally to both kinds of ion. Therefore, for a 1,1-electrolyte, we introduce the 'mean activity coefficient' as the geometric mean of the individual coefficients, where the geometric mean of x^p and y^q is $(x^p y^q)^{1/(p+q)}$. Thus:

$$\gamma_\pm = (\gamma_+\gamma_-)^{1/2} \tag{5F.3}$$

and express the individual chemical potentials of the ions as

$$\mu_+ = \mu_+^{ideal} + RT\ln\gamma_\pm \quad \mu_- = \mu_-^{ideal} + RT\ln\gamma_\pm \tag{5F.4}$$

The sum of these two chemical potentials is the same as before, eqn 5F.2, but now the non-ideality is shared equally.

We can generalize this approach to the case of a compound M_pX_q that dissolves to give a solution of p cations and q anions from each formula unit. The molar Gibbs energy of the ions is the sum of their partial molar Gibbs energies:

$$G_m = p\mu_+ + q\mu_- = G_m^{ideal} + pRT\ln\gamma_+ + qRT\ln\gamma_- \tag{5F.5}$$

If we introduce the **mean activity coefficient** now defined in a more general way as

$$\gamma_\pm=(\gamma_+^p\gamma_-^q)^{1/s} \quad s=p+q \qquad \textit{Definition} \quad \text{Mean activity coefficient} \qquad (5F.6)$$

and write the chemical potential of each ion as

$$\mu_i=\mu_i^{\text{ideal}}+RT\ln\gamma_\pm \qquad (5F.7)$$

we get the same expression as in eqn 5F.2 for G_m when we write $G_m=p\mu_++q\mu_-$. However, both types of ion now share equal responsibility for the non-ideality.

(a) The Debye–Hückel limiting law

The long range and strength of the Coulombic interaction between ions means that it is likely to be primarily responsible for the departures from ideality in ionic solutions and to dominate all the other contributions to non-ideality. This domination is the basis of the **Debye–Hückel theory** of ionic solutions, which was devised by Peter Debye and Erich Hückel in 1923. We give here a qualitative account of the theory and its principal conclusions. The calculation itself, which is a profound example of how a seemingly intractable problem can be formulated and then resolved by drawing on physical insight, is described the following section.

Oppositely charged ions attract one another. As a result, anions are more likely to be found near cations in solution, and vice versa (Fig. 5F.1). Overall, the solution is electrically neutral, but near any given ion there is an excess of counter ions (ions of opposite charge). Averaged over time, counter ions are more likely to be found near any given ion. This time-averaged, spherical haze around the central ion, in which counter ions outnumber ions of the same charge as the central ion, has a net charge equal in magnitude but opposite in sign to that on the central ion, and is called its **ionic atmosphere**. The energy, and therefore the chemical potential, of any given central ion is lowered as a result of its electrostatic interaction with its ionic atmosphere. This lowering of energy appears as the difference between the molar Gibbs energy G_m and the ideal value G_m^{ideal} of the solute, and hence can be identified with $RT\ln\gamma_\pm$. The stabilization of ions by their interaction with their ionic atmospheres is part of the explanation why chemists commonly use dilute solutions, in which the stabilization is less important, to achieve precipitation of ions from electrolyte solutions.

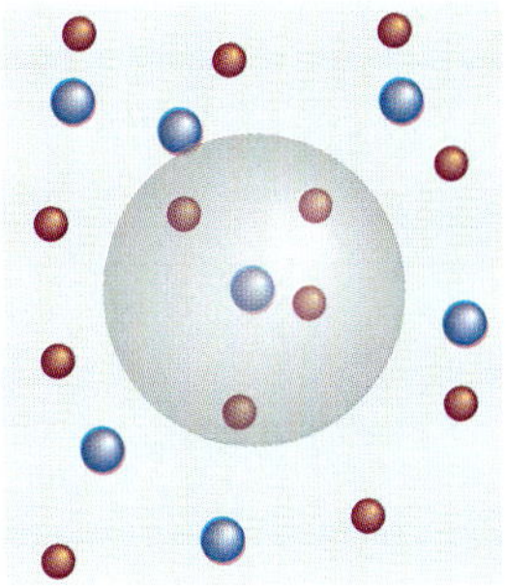

Figure 5F.1 The picture underlying the Debye–Hückel theory is of a tendency for anions to be found around cations, and of cations to be found around anions (one such local clustering region is shown by the grey sphere). The ions are in ceaseless motion, and the diagram represents a snapshot of their motion. The solutions to which the theory applies are far less concentrated than shown here.

The model leads to the result that at very low concentrations the activity coefficient can be calculated from the **Debye–Hückel limiting law**

$$\log\gamma_\pm=-A|z_+z_-|I^{1/2} \qquad \text{Debye–Hückel limiting law} \qquad (5F.8)$$

where $A=0.509$ for an aqueous solution at 25 °C and I is the dimensionless **ionic strength** of the solution:

$$I=\frac{1}{2}\sum_i z_i^2(b_i/b^\ominus) \qquad \textit{Definition} \quad \text{Ionic strength} \qquad (5F.9)$$

In this expression, z_i is the charge number of an ion i (positive for cations and negative for anions) and b_i is its molality. The ionic strength occurs widely wherever ionic solutions are discussed, as we shall see. The sum extends over all the ions present in the solution. For solutions consisting of two types of ion at molalities b_+ and b_-,

$$I=\tfrac{1}{2}(b_+z_+^2+b_-z_-^2)/b^\ominus \qquad (5F.10)$$

The ionic strength emphasizes the charges of the ions because the charge numbers occur as their squares. Table 5F.1 summarizes the relation of ionic strength and molality in an easily usable form.

Brief illustration 5F.1 The limiting law

The mean activity coefficient of 5.0 mmol kg^{-1} KCl(aq) at 25 °C is calculated by writing $I=\frac{1}{2}(b_++b_-)/b^\ominus=b/b^\ominus$, where b is the molality of the solution (and $b_+=b_-=b$). Then, from eqn 5F.8,

$$\log\gamma_\pm=-0.509\times(5.0\times10^{-3})^{1/2}=-0.03\ldots$$

Hence, $\gamma_\pm=0.92$. The experimental value is 0.927.

Self-test 5F.1 Calculate the ionic strength and the mean activity coefficient of 1.00 mmol kg^{-1} $CaCl_2$(aq) at 25 °C.

Answer: 3.00 mmol kg^{-1}, 0.880

The name 'limiting law' is applied to eqn 5F.8 because ionic solutions of moderate molalities may have activity

Table 5F.1 Ionic strength and molality, $I=kb/b^{\ominus}$

k	X^-	X^{2-}	X^{3-}	X^{4-}
M^+	1	3	6	10
M^{2+}	3	4	15	12
M^{3+}	6	15	9	42
M^{4+}	10	12	42	16

For example, the ionic strength of an M_2X_3 solution of molality b, which is understood to give M^{3+} and X^{2-} ions in solution is $15b/b^{\ominus}$.

Table 5F.2* Mean activity coefficients in water at 298 K

$b/b^{\ominus}$	KCl	$CaCl_2$
0.001	0.966	0.888
0.01	0.902	0.732
0.1	0.770	0.524
1.0	0.607	0.725

* More values are given in the *Resource section*.

coefficients that differ from the values given by this expression, yet all solutions are expected to conform as $b \to 0$. Table 5F.2 lists some experimental values of activity coefficients for salts of various valence types. Figure 5F.2 shows some of these values plotted against $I^{1/2}$, and compares them with the theoretical straight lines calculated from eqn 5F.8. The agreement at very low molalities (less than about 1 mmol kg^{-1}, depending on charge type) is impressive and convincing evidence in support of the model. Nevertheless, the departures from the theoretical curves above these molalities are large, and show that the approximations are valid only at very low concentrations.

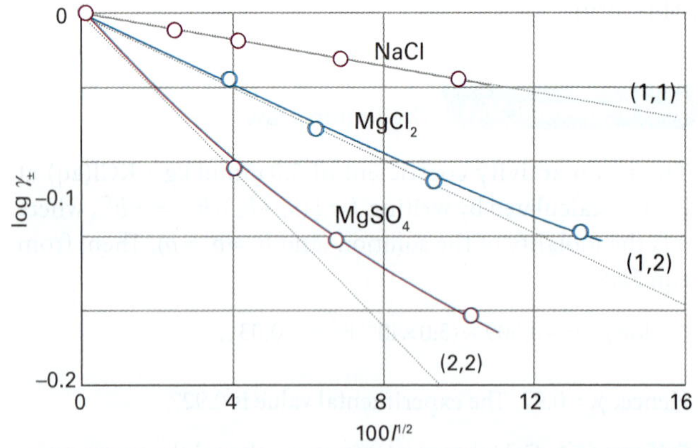

Figure 5F.2 An experimental test of the Debye–Hückel limiting law. Although there are marked deviations for moderate ionic strengths, the limiting slopes as $I \to 0$ are in good agreement with the theory, so the law can be used for extrapolating data to very low molalities.

(b) Extensions of the limiting law

When the ionic strength of the solution is too high for the limiting law to be valid, the activity coefficient may be estimated from the **extended Debye–Hückel law** (sometimes called the *Truesdell–Jones equation*):

$$\log \gamma_{\pm} = -\frac{A|z_+z_-|I^{1/2}}{1+BI^{1/2}} \quad \text{Extended Debye–Hückel law} \quad (5F.11a)$$

where B is a dimensionless constant. A more flexible extension is the **Davies equation** proposed by C.W. Davies in 1938:

$$\log \gamma_{\pm} = -\frac{A|z_+z_-|I^{1/2}}{1+BI^{1/2}} + CI \quad \text{Davies equation} \quad (5F.11b)$$

where C is another dimensionless constant. Although B can be interpreted as a measure of the closest approach of the ions, it (like C) is best regarded as an adjustable empirical parameter. A curve drawn on the basis of the Davies equation is shown in Fig. 5F.3. It is clear that eqn 5F.11 accounts for some activity coefficients over a moderate range of dilute solutions (up to about 0.1 mol kg^{-1}); nevertheless it remains very poor near 1 mol kg^{-1}.

Current theories of activity coefficients for ionic solutes take an indirect route. They set up a theory for the dependence of the activity coefficient of the solvent on the concentration of the solute, and then use the Gibbs–Duhem equation (eqn 5A.12a, $n_A d\mu_A + n_B d\mu_B = 0$) to estimate the activity coefficient of the solute. The results are reasonably reliable for solutions with molalities greater than about 0.1 mol kg^{-1} and are valuable for the discussion of mixed salt solutions, such as sea water.

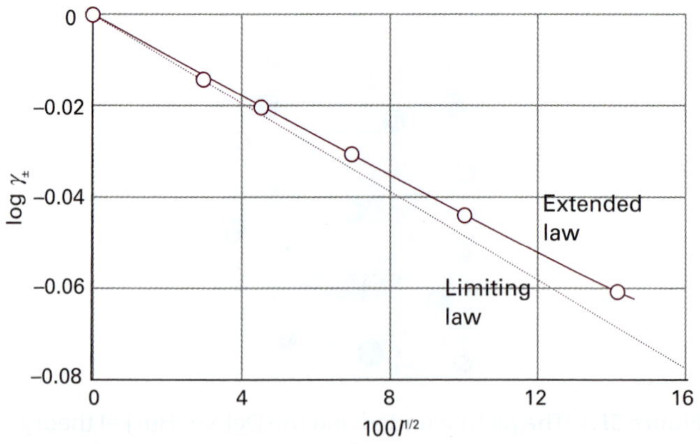

Figure 5F.3 The Davies equation gives agreement with experiment over a wider range of molalities than the limiting law (as shown here for a 1,1-electrolyte), but it fails at higher molalities.

5F.2 The Debye–Hückel theory

The strategy for the calculation is to establish the relation between the work needed to charge an ion and its chemical potential, and then to relate that work to the ion's interaction with the atmosphere of counter ions that has assembled around it as a result of the competition of the attraction between oppositely charged ions, the repulsion of like-charged ions, and the distributing effect of thermal motion.

(a) The work of charging

Imagine a solution in which all the ions have their actual positions, but in which their Coulombic interactions have been turned off and so are behaving 'ideally'. The difference in molar Gibbs energy between the ideal and real solutions is equal to w_e, the electrical work of charging the system in this arrangement. For a salt M_pX_q, we write

$$w_e = \underbrace{(p\mu_+ + q\mu_-)}_{G_m,\ \text{charged}} - \underbrace{(p\mu_+^{\text{ideal}} + q\mu_-^{\text{ideal}})}_{G_m^{\text{ideal}},\ \text{uncharged}} = p(\mu_+ - \mu_+^{\text{ideal}}) + q(\mu_- - \mu_-^{\text{ideal}}) \qquad (5F.12)$$

From eqn 5F.7 we write

$$\mu_+ - \mu_+^{\text{ideal}} = \mu_- - \mu_-^{\text{ideal}} = RT \ln \gamma_\pm$$

So it follows that

$$\ln \gamma_\pm = \frac{w_e}{sRT} \qquad s = p + q \qquad (5F.13)$$

This equation tells us that we must first find the final distribution of the ions and then the work of charging them in that distribution.

Brief illustration 5F.2 The work of charging

The measured mean activity coefficient of 5.0 mmol kg^{-1} KCl(aq) at 25 °C is 0.927. It follows that the average work involved in charging the ions in their environment in the solution is given by eqn 5F.13 in the form

$$w_e = sRT \ln \gamma_\pm = 2 \times (8.3145\ \text{J K}^{-1}\,\text{mol}^{-1}) \times (298\ \text{K}) \times \ln 0.927 = -0.38\ \text{kJ mol}^{-1}$$

Self-test 5F.2 The measured mean activity coefficient of 0.1 mol kg^{-1} Na_2SO_4(aq) at 25 °C is 0.445. What is the work of charging the ions in the solution?

Answer: −6.02 kJ mol^{-1}

(b) The potential due to the charge distribution

As explained in *Foundations* B, the Coulomb potential at a distance r from an isolated ion of charge z_ie in a medium of permittivity ε is

$$\phi_i = \frac{Z_i}{r} \qquad Z_i = \frac{z_i e}{4\pi\varepsilon} \qquad (5F.14)$$

The ionic atmosphere causes the potential to decay with distance more sharply than this expression implies. Such shielding is a familiar problem in electrostatics, and its effect is taken into account by replacing the Coulomb potential by the **shielded Coulomb potential**, an expression of the form

$$\phi_i = \frac{Z_i}{r} e^{-r/r_D} \qquad \text{Shielded Coulomb potential} \qquad (5F.15)$$

where r_D is called the **Debye length**. When r_D is large, the shielded potential is virtually the same as the unshielded potential. When it is small, the shielded potential is much smaller than the unshielded potential, even for short distances (Fig. 5F.4). We establish in the following *Justification* that

$$r_D = \left(\frac{\varepsilon RT}{2\rho F^2 I b^\ominus}\right)^{1/2} \qquad \text{Debye length} \qquad (5F.16)$$

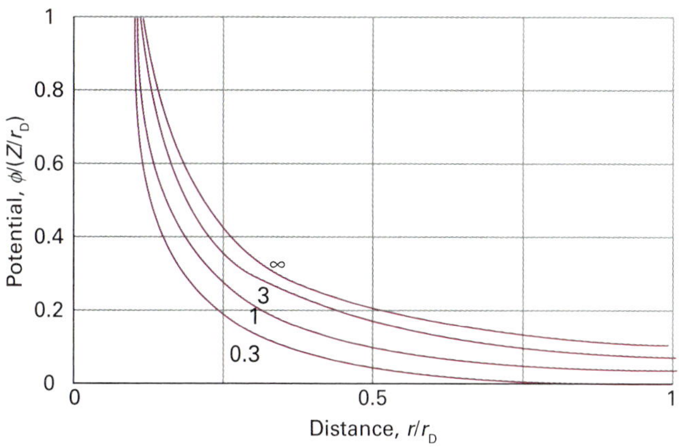

Figure 5F.4 The variation of the shielded Coulomb potential with distance for different values of the Debye length, r_D/a. The smaller the Debye length, the more sharply the potential decays to zero. In each case, a is an arbitrary unit of length.

Brief illustration 5F.3 The Debye length

To estimate the Debye length in an aqueous solution of ionic strength 0.100 and density 1.000 g cm^{-3} at 25 °C we write

$$r_D = \left(\frac{\overbrace{80.10 \times (8.854\times10^{-12}\,\text{J}^{-1}\,\text{C}^2\,\text{m}^{-1})}^{\varepsilon} \times (8.3145\,\text{J K}^{-1}\,\text{mol}^{-1}) \times (298\,\text{K})}{2 \times \underbrace{(1000\,\text{kg m}^{-3})}_{1.000\,\text{g cm}^{-3}} \times (9.649\times10^4\,\text{C mol}^{-1})^2 \times (0.100) \times \underbrace{(1\,\text{mol kg}^{-1})}_{b^{\ominus}}} \right)^{1/2}$$

$$= 9.72\times10^{-10}\,\text{m, or } 0.972\,\text{nm}$$

Self-test 5F.3 Estimate the Debye length in an ethanol solution of ionic strength 0.100 and density 0.789 g cm^{-3} at 25 °C. Use $\varepsilon_r = 25.3$.

Answer: 0.615 nm

Justification 5F.1 The Debye length

To calculate r_D, we need to know how the **charge density**, ρ_i, of the ionic atmosphere, the charge in a small region divided by the volume of the region, varies with distance from the ion. This step draws on another standard result of electrostatics, in which charge density and potential are related by **Poisson's equation**:

$$\nabla^2\phi = -\frac{\rho}{\varepsilon} \qquad \text{Poisson's equation}$$

where $\nabla^2 = \partial^2/\partial x^2 + \partial^2/\partial y^2 + \partial^2/\partial z^2$. Because we are considering only a spherical ionic atmosphere, we can use a simplified form of this equation in which the charge density varies only with distance from the central ion:

$$\frac{1}{r^2}\frac{\mathrm{d}}{\mathrm{d}r}\left(r^2\frac{\mathrm{d}\phi_i}{\mathrm{d}r}\right) = -\frac{\rho_i}{\varepsilon}$$

Substitution of the expression for the shielded potential, eqn 5F.15, results in

$$r_D^2 = -\frac{\varepsilon\phi_i}{\rho_i}$$

To solve this equation we need to relate ρ_i and ϕ_i.

For this next step we draw on the fact that the energy of an ion depends on its closeness to the central ion, and then use the Boltzmann distribution (*Foundations* B) to work out the probability that an ion will be found at each distance. The energy of an ion j of charge z_je at a distance where it experiences the potential ϕ_i of the central ion i relative to its energy when it is far away in the bulk solution is its charge times the potential z_je,ϕ_i. Therefore, according to the Boltzmann distribution, the ratio of the molar concentration, c_j, of ions at a distance r and the molar concentration in the bulk, c_j°, where the energy is zero, is

$$\frac{c_j}{c_j^{\circ}} = \mathrm{e}^{-z_je\phi_i/kT}$$

The charge density, ρ_i, at a distance r from the ion i is the molar concentration of each type of ion multiplied by the charge per mole of ions, z_ieN_A. The quantity $eN_A = F$, the magnitude of the charge per mole of electrons, is Faraday's constant. It follows that

$$\rho_i = c_+z_+F + c_-z_-F = c_+^{\circ}z_+F\mathrm{e}^{-z_+e\phi_i/kT} + c_-^{\circ}z_-F\mathrm{e}^{-z_-e\phi_i/kT}$$

At this stage we need to simplify the expression to avoid the awkward exponential terms. Because the average electrostatic interaction energy is small compared with kT we may use the expansion $\mathrm{e}^x = 1 + x + \cdots$ and write the charge density as

$$\rho_i = c_+^{\circ}z_+F\left(1 - \frac{z_+e\phi_i}{kT} + \cdots\right) + c_-^{\circ}z_-F\left(1 - \frac{z_-e\phi_i}{kT} + \cdots\right)$$

$$= \overbrace{(c_+^{\circ}z_+ + c_-^{\circ}z_-)}^{0}F - (c_+^{\circ}z_+^2 + c_-^{\circ}z_-^2)\frac{Fe\phi_i}{kT} + \cdots$$

The first term in the expansion is zero because it is the charge density in the bulk, uniform solution, and the solution is electrically neutral. Replacing e by F/N_A and N_Ak by R results in the following expression:

$$\rho_i = -(c_+^{\circ}z_+^2 + c_-^{\circ}z_-^2)\frac{F^2\phi_i}{RT}$$

The unwritten terms are assumed to be too small to be significant. This one remaining term, in blue, can be expressed in terms of the ionic strength, eqn 5F.9, by noting that in the dilute aqueous solutions we are considering there is little difference between molality and molar concentration, and $c \approx b\rho$, where ρ is the mass density of the solvent

$$c_+^{\circ}z_+^2 + c_-^{\circ}z_-^2 \approx \overbrace{(b_+^{\circ}z_+^2 + b_-^{\circ}z_-^2)}^{2Ib^{\ominus}}\rho = 2Ib^{\ominus}\rho$$

With these approximations, the last equation becomes

$$\rho_i = -\frac{2Ib^{\ominus}\rho F^2\phi_i}{RT}$$

We can now substitute this expression into $r_D^2 = -\varepsilon\phi_i/\rho_i$, when the ϕ_i cancel and we obtain eqn 5F.16.

(c) The activity coefficient

To calculate the activity coefficient we need to find the electrical work of charging the central ion when it is surrounded by its atmosphere. This calculation is carried through in the following *Justification*, which leads to the conclusion that the work of charging an ion i when it is surrounded by the atmosphere it has assembled is

$$w_{e,i} = -\frac{z_i^2F^2}{8\pi N_A\varepsilon r_D} \qquad \text{Work of charging} \qquad (5F.17)$$

Justification 5F.2 The work of charging

To calculate the work of charging the central ion we need to know the potential at the ion due to its atmosphere, ϕ_{atmos}. This potential is the difference between the total potential, given by eqn 5F.15, and the potential due to the central ion itself:

$$\phi_{\text{atmosphere}} = \phi - \phi_{\text{central ion}} = Z_i\left(\frac{e^{-r/r_D}}{r} - \frac{1}{r}\right),\ Z_i = \frac{z_i e}{4\pi\varepsilon}$$

The potential at the central ion (at $r=0$) is obtained by taking the limit of this expression as $r \to 0$ and is

$$\phi_{\text{atmosphere}}(0) = Z_i \lim_{r\to 0}\left(\frac{1-r/r_D+\cdots}{r} - \frac{1}{r}\right) = -\frac{Z_i}{r_D}$$

This expression shows us that the potential due to the ionic atmosphere is equivalent to the potential arising from a single charge of equal magnitude but opposite sign to that of the central ion and located at a distance r_D from the ion. If the charge of the central ion were Q and not $z_i e$, then the potential due to its atmosphere would be

$$\phi_{\text{atmosphere}}(0) = -\frac{Q}{4\pi\varepsilon r_D}$$

The work of adding a charge dQ to a region where the electrical potential is $\phi_{\text{atmosphere}}(0)$, Table 2A.1 (from d$w = \phi$d$Q$), is

$$\mathrm{d}w_e = \phi_{\text{atmosphere}}(0)\mathrm{d}Q$$

Therefore, the total molar work of fully charging the ion i in the presence of its atmosphere is

$$w_{e,i} = N_A \int_0^{z_i e} \phi_{\text{atmosphere}}(0)\mathrm{d}Q = -\frac{N_A}{4\pi\varepsilon r_D}\int_0^{z_i e} Q\mathrm{d}Q$$
$$= -\frac{N_A z_i^2 e^2}{8\pi\varepsilon r_D} = -\frac{z_i^2 F^2}{8\pi N_A \varepsilon r_D}$$

where in the last step we have used $F = N_A e$. This expression is eqn. 5F.17.

We can now collect the various pieces of this calculation and arrive at an expression for the mean activity coefficient. It follows from eqn 5F.13 with $w_e = pw_{e,+} + qw_{e,-}$, the total work of charging p cations and q anions in the presence of their atmospheres, that the mean activity coefficient of the ions is

$$\ln\gamma_\pm = \frac{pw_{e,+} + qw_{e,-}}{sRT} = -\frac{(pz_+^2 + qz_-^2)F^2}{8\pi N_A sRT\varepsilon r_D}$$

However, for neutrality $pz_+ + qz_- = 0$; therefore

$$pz_+^2 + qz_-^2 = \underbrace{pz_+}_{-qz_-} z_+ + \underbrace{qz_-}_{-pz_+} z_- = -\underbrace{(p+q)}_{s}\underbrace{z_+z_-}_{-|z_+z_-|} = s|z_+z_-|$$

It then follows that

$$\ln\gamma_\pm = -\frac{|z_+z_-|F^2}{8\pi N_A RT\varepsilon r_D} \qquad (5F.18)$$

The replacement of r_D with the expression in eqn 5F.16 gives

$$\ln\gamma_\pm = -\frac{|z_+z_-|F^2}{8\pi N_A RT\varepsilon}\left(\frac{2\rho F^2 I b^{\ominus}}{\varepsilon RT}\right)^{1/2} \qquad (5F.19a)$$
$$= -|z_+z_-|\left\{\frac{F^3}{4\pi N_A}\left(\frac{\rho b^{\ominus}}{2\varepsilon^3 R^3 T^3}\right)^{1/2}\right\} I^{1/2}$$

where we have grouped terms in such a way as to show that this expression is beginning to take the form of eqn 5F.8 ($\log\gamma_\pm = -|z_+z_-|AI^{1/2}$). Indeed, conversion to common logarithms (by using $\ln x = \ln 10 \times \log x$) gives

$$\log\gamma_\pm = -|z_+z_-|\overbrace{\left\{\frac{F^3}{4\pi N_A \ln 10}\left(\frac{\rho b^{\ominus}}{2\varepsilon^3 R^3 T^3}\right)^{1/2}\right\}}^{A} I^{1/2} \qquad (5F.19b)$$

which is eqn 5F.8 with

$$A = \frac{F^3}{4\pi N_A \ln 10}\left(\frac{\rho b^{\ominus}}{2\varepsilon^3 R^3 T^3}\right)^{1/2} \qquad (5F.20)$$

Brief illustration 5F.4 The Debye–Hückel constant

To evaluate the constant A for water at 25.00 °C, we use $\rho = 0.9971\ \mathrm{g\,cm^{-3}}$ and $\varepsilon = 78.54\varepsilon_0$ to find

$$A = \frac{(9.649\times10^4\ \mathrm{C\,mol^{-1}})^3}{4\pi\times(6.022\times10^{23}\ \mathrm{mol^{-1}})\ln 10}$$
$$\times\left(\frac{(997.1\ \mathrm{kg\,m^{-3}})\times(1\ \mathrm{mol\,kg^{-1}})}{2\times(78.54\times8.854\times10^{-12}\ \mathrm{J^{-1}\,C^2\,m^{-1}})^3\times(8.3145\ \mathrm{J\,K^{-1}\,mol^{-1}})^3\times(298.15\ \mathrm{K})^3}\right)^{1/2}$$
$$= 0.5086$$

Self-test 5F.4 Evaluate the constant A for ethanol at 25 °C, when $\varepsilon_r = 25.3$ and $\rho = 0.789\ \mathrm{g\,cm^{-3}}$.

Answer: 2.47

Checklist of concepts

☐ 1. **Mean activity coefficients** apportion deviations from ideality equally to the cations and anions in an ionic solution.

☐ 2. The **Debye–Hückel theory** ascribes deviations from ideality to the Coulombic interaction of an ion with the ionic atmosphere that assembles around it.

☐ 3. The **Debye–Hückel limiting law** is extended by including two further empirical constants.

Checklist of equations

Property	Equation	Comment	Equation number
Mean activity coefficient	$\gamma_\pm = (\gamma_+^p \gamma_-^q)^{1/(p+q)}$	Definition	5F.6
Debye–Hückel limiting law	$\log \gamma_\pm = -A\lvert z_+ z_-\rvert I^{1/2}$	Valid as $I \to 0$	5F.8
Ionic strength	$I = \frac{1}{2}\sum_i z_i^2 (b_i / b^\ominus)$	Definition	5F.9
Davies equation	$\log \gamma_\pm = -A\lvert z_+ z_-\rvert I^{1/2}/(1+BI^{1/2}) + CI$	A, B, C empirical constants	5F.11

CHAPTER 5 Simple mixtures

TOPIC 5A The thermodynamic description of mixtures

Discussion questions

5A.1 Explain the concept of partial molar quantity, and justify the remark that the partial molar property of a solute depends on the properties of the solvent too.

5A.2 Explain how thermodynamics relates non-expansion work to a change in composition of a system.

5A.3 Are there any circumstances under which two (real) gases will not mix spontaneously?

5A.4 Explain how Raoult's law and Henry's law are used to specify the chemical potential of a component of a mixture.

5A.5 Explain the molecular origin of Raoult's law and Henry's law.

Exercises

5A.1(a) A polynomial fit to measurements of the total volume of a binary mixture of A and B is

$$v = 987.93 + 35.6774x - 0.45923x^2 + 0.017325x$$

where $v = V/\text{cm}^3$, $x = n_B/\text{mol}$, and n_B is the amount of B present. Determine the partial molar volumes of A and B.

5A.1(b) A polynomial fit to measurements of the total volume of a binary mixture of A and B is

$$v = 778.55 - 22.5749x + 0.56892x^2 + 0.01023x^3 + 0.00234x$$

where $v = V/\text{cm}^3$, $x = n_B/\text{mol}$, and n_B is the amount of B present. Determine the partial molar volumes of A and B.

5A.2(a) The volume of an aqueous solution of NaCl at 25 °C was measured at a series of molalities b, and it was found that the volume fitted the expression $v = 1003 + 16.62x + 1.77x^{3/2} + 0.12x^2$ where $v = V/\text{cm}^3$, V is the volume of a solution formed from 1.000 kg of water and $x = b/b^{\ominus}$. Calculate the partial molar volume of the components in a solution of molality 0.100 mol kg^{-1}.

5A.2(b) At 18 °C the total volume V of a solution formed from $MgSO_4$ and 1.000 kg of water fits the expression $v = 1001.21 + 34.69(x - 0.070)^2$, where $v = V/\text{cm}^3$ and $x = b/bx = b/b^{\ominus}$. Calculate the partial molar volumes of the salt and the solvent when in a solution of molality 0.050 mol kg^{-1}.

5A.3(a) Suppose that $n_A = 0.10n_B$ and a small change in composition results in μ_A changing by $\delta\mu_A = +12\ \text{J mol}^{-1}$, by how much will μ_B change?

5A.3(b) Suppose that $n_A = 0.22n_B$ and a small change in composition results in μ_A changing by $\delta\mu_A = -15\ \text{J mol}^{-1}$, by how much will μ_B change?

5A.4(a) Consider a container of volume 5.0 dm^3 that is divided into two compartments of equal size. In the left compartment there is nitrogen at 1.0 atm and 25 °C; in the right compartment there is hydrogen at the same temperature and pressure. Calculate the entropy and Gibbs energy of mixing when the partition is removed. Assume that the gases are perfect.

5A.4(b) Consider a container of volume 250 cm^3 that is divided into two compartments of equal size. In the left compartment there is argon at 100 kPa and 0 °C; in the right compartment there is neon at the same temperature and pressure. Calculate the entropy and Gibbs energy of mixing when the partition is removed. Assume that the gases are perfect.

5A.5(a) Air is a mixture with mass percentage composition 75.5 (N_2), 23.2 (O_2), 1.3 (Ar). Calculate the entropy of mixing when it is prepared from the pure (and perfect) gases.

5A.5(b) When carbon dioxide is taken into account, the mass percentage composition of air is 75.52 (N_2), 23.15 (O_2), 1.28 (Ar), and 0.046 (CO_2). What is the change in entropy from the value in the preceding exercise?

5A.6(a) The vapour pressure of benzene at 20 °C is 10 kPa and that of methylbenzene is 2.8 kPa at the same temperature. What is the vapour pressure of a mixture of equal masses of each component?

5A.6(b) At 90 °C the vapour pressure of 1,2-dimethylbenzene is 20 kPa and that of 1,3-dimethylbenzene is 18 kPa. What is the composition of the vapour of an equimolar mixture of the two components?

5A.7(a) The partial molar volumes of acetone (propanone) and chloroform (trichloromethane) in a mixture in which the mole fraction of $CHCl_3$ is 0.4693 are 74.166 cm^3 mol^{-1} and 80.235 cm^3 mol^{-1}, respectively. What is the volume of a solution of mass 1.000 kg?

5A.7(b) The partial molar volumes of two liquids A and B in a mixture in which the mole fraction of A is 0.3713 are 188.2 cm^3 mol^{-1} and 176.14 cm^3 mol^{-1}, respectively. The molar masses of the A and B are 241.1 g mol^{-1} and 198.2 g mol^{-1}. What is the volume of a solution of mass 1.000 kg?

5A.8(a) At 25 °C, the density of a 50 per cent by mass ethanol–water solution is 0.914 g cm^{-3}. Given that the partial molar volume of water in the solution is 17.4 cm^3 mol^{-1}, calculate the partial molar volume of the ethanol.

5A.8(b) At 20 °C, the density of a 20 per cent by mass ethanol/water solution is 968.7 kg m^{-3}. Given that the partial molar volume of ethanol in the solution is 52.2 cm^3 mol^{-1}, calculate the partial molar volume of the water.

5A.9(a) At 300 K, the partial vapour pressure of HCl (that is, the partial pressure of the HCl vapour) in liquid $GeCl_4$ is as follows:

x_{HCl}	0.005	0.012	0.019
p_{HCl}/kPa	32.0	76.9	121.8

Show that the solution obeys Henry's law in this range of mole fractions, and calculate Henry's law constant at 300 K.

5A.9(b) At 310 K, the partial vapour pressure of a substance B dissolved in a liquid A is as follows:

x_B	0.010	0.015	0.020
p_B/kPa	82.0	122.0	166.1

Show that the solution obeys Henry's law in this range of mole fractions, and calculate Henry's law constant at 310 K.

5A.10(a) Calculate the molar solubility of nitrogen in benzene exposed to air at 25 °C; partial pressures were calculated in *Example* 1A.3 of Topic 1A.
5A.10(b) Calculate the molar solubility of methane at 1.0 bar in benzene at 25 °C.

5A.11(a) Use Henry's law and the data in Table 5A.1 to calculate the solubility (as a molality) of CO_2 in water at 25 °C when its partial pressure is (i) 0.10 atm, (ii) 1.00 atm.
5A.11(b) The mole fractions of N_2 and O_2 in air at sea level are approximately 0.78 and 0.21. Calculate the molalities of the solution formed in an open flask of water at 25 °C.

5A.12(a) A water carbonating plant is available for use in the home and operates by providing carbon dioxide at 5.0 atm. Estimate the molar concentration of the soda water it produces.
5A.12(b) After some weeks of use, the pressure in the water carbonating plant mentioned in the previous exercise has fallen to 2.0 atm. Estimate the molar concentration of the soda water it produces at this stage.

Problems

5A.1 The experimental values of the partial molar volume of a salt in water are found to fit the expression $v_B = 5.117 + 19.121x^{1/2}$, where $v_B = V_B/(\text{cm}^3\,\text{mol}^{-1})$ and x is the numerical value of the molality of B ($x = b/b^{\ominus}$). Use the Gibbs–Duhem equation to derive an equation for the molar volume of water in the solution. The molar volume of pure water at the same temperature is 18.079 cm³ mol⁻¹.

5A.2 The compound *p*-azoxyanisole forms a liquid crystal. 5.0 g of the solid was placed in a tube, which was then evacuated and sealed. Use the phase rule to prove that the solid will melt at a definite temperature and that the liquid crystal phase will make a transition to a normal liquid phase at a definite temperature.

5A.3 The following table gives the mole fraction of methylbenzene (A) in liquid and gaseous mixtures (x_A and y_A, respectively) with butanone at equilibrium at 303.15 K and the total pressure p. Take the vapour to be perfect and calculate the partial pressures of the two components. Plot them against their respective mole fractions in the liquid mixture and find the Henry's law constants for the two components.

x_A	0	0.0898	0.2476	0.3577	0.5194	0.6036	0.7188	0.8019	0.9105	1
y_A	0	0.0410	0.1154	0.1762	0.2772	0.3393	0.4450	0.5435	0.7284	1
p/kPa	36.066	34.121	30.900	28.626	25.239	23.402	20.6984	18.592	15.496	12.295

5A.4 The densities of aqueous solutions of copper(II) sulfate at 20 °C were measured as set out below. Determine and plot the partial molar volume of $CuSO_4$ in the range of the measurements.

$m(CuSO_4)$/g	5	10	15	20
$\rho/(\text{g cm}^{-3})$	1.051	1.107	1.167	1.230

where $m(CuSO_4)$ is the mass of $CuSO_4$ dissolved in 100 g of solution.

5A.5 Haemoglobin, the red blood protein responsible for oxygen transport, binds about 1.34 cm³ of oxygen per gram. Normal blood has a haemoglobin concentration of 150 g dm⁻³. Haemoglobin in the lungs is about 97 per cent saturated with oxygen, but in the capillary is only about 75 per cent saturated. What volume of oxygen is given up by 100 cm³ of blood flowing from the lungs in the capillary?

5A.6 Use the data from *Example* 5A.1 to determine the value of b at which V_E has a minimum value.

TOPIC 5B The properties of solutions

Discussion questions

5B.1 Explain what is meant by a regular solution; what additional features distinguish a real solution from a regular solution?

5B.2 Explain the physical origin of colligative properties.

5B.3 Colligative properties are independent of the identity of the solute. Why, then, can osmometry be used to determine the molar mass of a solute?

Exercises

5B.1(a) Predict the partial vapour pressure of HCl above its solution in liquid germanium tetrachloride of molality 0.10 mol kg⁻¹. For data, see Exercise 5A.10(a).
5B.1(b) Predict the partial vapour pressure of the component B above its solution in A in Exercise 5A.10(b) when the molality of B is 0.25 mol kg⁻¹. The molar mass of A is 74.1 g mol⁻¹.

5B.2(a) The vapour pressure of benzene is 53.3 kPa at 60.6 °C, but it fell to 51.5 kPa when 19.0 g of a non-volatile organic compound was dissolved in 500 g of benzene. Calculate the molar mass of the compound.
5B.2(b) The vapour pressure of 2-propanol is 50.00 kPa at 338.8 K, but it fell to 49.62 kPa when 8.69 g of a non-volatile organic compound was dissolved in 250 g of 2-propanol. Calculate the molar mass of the compound.

5B.3(a) The addition of 100 g of a compound to 750 g of CCl_4 lowered the freezing point of the solvent by 10.5 K. Calculate the molar mass of the compound.
5B.3(b) The addition of 5.00 g of a compound to 250 g of naphthalene lowered the freezing point of the solvent by 0.780 K. Calculate the molar mass of the compound.

5B.4(a) The osmotic pressure of an aqueous solution at 300 K is 120 kPa. Calculate the freezing point of the solution.
5B.4(b) The osmotic pressure of an aqueous solution at 288 K is 99.0 kPa. Calculate the freezing point of the solution.

5B.5(a) Calculate the Gibbs energy, entropy, and enthalpy of mixing when 0.50 mol C_6H_{14} (hexane) is mixed with 2.00 mol C_7H_{16} (heptane) at 298 K; treat the solution as ideal.
5B.5(b) Calculate the Gibbs energy, entropy, and enthalpy of mixing when 1.00 mol C_6H_{14} (hexane) is mixed with 1.00 mol C_7H_{16} (heptane) at 298 K; treat the solution as ideal.

5B.6(a) What proportions of hexane and heptane should be mixed (i) by mole fraction, (ii) by mass in order to achieve the greatest entropy of mixing?
5B.6(b) What proportions of benzene and ethylbenzene should be mixed (i) by mole fraction, (ii) by mass in order to achieve the greatest entropy of mixing?

5B.7(a) The enthalpy of fusion of anthracene is 28.8 kJ mol^{-1} and its melting point is 217 °C. Calculate its ideal solubility in benzene at 25 °C.
5B.7(b) Predict the ideal solubility of lead in bismuth at 280 °C given that its melting point is 327 °C and its enthalpy of fusion is 5.2 kJ mol^{-1}.

5B.8(a) The osmotic pressure of solutions of polystyrene in toluene were measured at 25 °C and the pressure was expressed in terms of the height of the solvent of density 1.004 g cm^{-3}:

$c/(\mathrm{g\,dm^{-3}})$	2.042	6.613	9.521	12.602
h/cm	0.592	1.910	2.750	3.600

Calculate the molar mass of the polymer.

5B.8(b) The molar mass of an enzyme was determined by dissolving it in water, measuring the osmotic pressure at 20 °C, and extrapolating the data to zero concentration. The following data were obtained:

$c/(\mathrm{mg\,cm^{-3}})$	3.221	4.618	5.112	6.722
h/cm	5.746	8.238	9.119	11.990

Calculate the molar mass of the enzyme.

5B.9(a) A dilute solution of bromine in carbon tetrachloride behaves as an ideal dilute solution. The vapour pressure of pure CCl_4 is 33.85 Torr at 298 K. The Henry's law constant when the concentration of Br_2 is expressed as a mole fraction is 122.36 Torr. Calculate the vapour pressure of each component, the total pressure, and the composition of the vapour phase when the mole fraction of Br_2 is 0.050, on the assumption that the conditions of the ideal dilute solution are satisfied at this concentration.

5B.9(b) The vapour pressure of a pure liquid A is 23 kPa at 20 °C and its Henry's law constant in liquid B is 73 kPa. Calculate the vapour pressure of each component, the total pressure, and the composition of the vapour phase when the mole fraction of A is 0.066 on the assumption that the conditions of the ideal dilute solution are satisfied at this concentration.

5B.10(a) At 90 °C, the vapour pressure of methylbenzene is 53.3 kPa and that of 1,2-dimethylbenzene is 20.0 kPa. What is the composition of a liquid mixture that boils at 90 °C when the pressure is 0.50 atm? What is the composition of the vapour produced?
5B.10(b) At 90 °C, the vapour pressure of 1,2-dimethylbenzene is 20 kPa and that of 1,3-dimethylbenzene is 18 kPa What is the composition of a liquid mixture that boils at 90 °C when the pressure is 19 kPa? What is the composition of the vapour produced?

5B.11(a) The vapour pressure of pure liquid A at 300 K is 76.7 kPa and that of pure liquid B is 52.0 kPa. These two compounds form ideal liquid and gaseous mixtures. Consider the equilibrium composition of a mixture in which the mole fraction of A in the vapour is 0.350. Calculate the total pressure of the vapour and the composition of the liquid mixture.
5B.11(b) The vapour pressure of pure liquid A at 293 K is 68.8 kPa and that of pure liquid B is 82.1 kPa. These two compounds form ideal liquid and gaseous mixtures. Consider the equilibrium composition of a mixture in which the mole fraction of A in the vapour is 0.612. Calculate the total pressure of the vapour and the composition of the liquid mixture.

5B.12(a) It is found that the boiling point of a binary solution of A and B with $x_A = 0.6589$ is 88 °C. At this temperature the vapour pressures of pure A and B are 127.6 kPa and 50.60 kPa, respectively. (i) Is this solution ideal? (ii) What is the initial composition of the vapour above the solution?
5B.12(b) It is found that the boiling point of a binary solution of A and B with $x_A = 0.4217$ is 96 °C. At this temperature the vapour pressures of pure A and B are 110.1 kPa and 76.5 kPa, respectively. (i) Is this solution ideal? (ii) What is the initial composition of the vapour above the solution?

5B.13(a) Dibromoethene (DE, $p^*_{DE} = 22.9$ kPa at 358 K) and dibromopropene (DP, $p^*_{DP} = 17.1$ kPa at 358 K) form a nearly ideal solution. If $x_{DE} = 0.60$, what is (i) p_{total} when the system is all liquid, (ii) the composition of the vapour when the system is still almost all liquid.
5B.13(b) Benzene and toluene form nearly ideal solutions. Consider an equimolar solution of benzene and toluene. At 20 °C the vapour pressures of pure benzene and toluene are 9.9 kPa and 2.9 kPa, respectively. The solution is boiled by reducing the external pressure below the vapour pressure. Calculate (i) the pressure when boiling begins, (ii) the composition of each component in the vapour, and (iii) the vapour pressure when only a few drops of liquid remain. Assume that the rate of vaporization is low enough for the temperature to remain constant at 20 °C.

Problems

5B.1 Potassium fluoride is very soluble in glacial acetic acid and the solutions have a number of unusual properties. In an attempt to understand them, freezing point depression data were obtained by taking a solution of known molality and then diluting it several times (J. Emsley, *J. Chem. Soc. A*, 2702 (1971)). The following data were obtained:

$b/(\mathrm{mol\,kg^{-1}})$	0.015	0.037	0.077	0.295	0.602
ΔT/K	0.115	0.295	0.470	1.381	2.67

Calculate the apparent molar mass of the solute and suggest an interpretation. Use $\Delta_{fus}H = 11.4$ kJ mol^{-1} and $T_f^* = 290$ K.

5B.2 In a study of the properties of an aqueous solution of $Th(NO_3)_4$ (by A. Apelblat, D. Azoulay, and A. Sahar, *J. Chem. Soc. Faraday Trans., I*, 1618 (1973)), a freezing point depression of 0.0703 K was observed for an aqueous solution of molality 9.6 mmol kg^{-1}. What is the apparent number of ions per formula unit?

5B.3‡ Aminabhavi et al. examined mixtures of cyclohexane with various long-chain alkanes (T.M. Aminabhavi et al., *J. Chem. Eng. Data* **41**, 526 (1996)). Among their data are the following measurements of the density of a mixture of cyclohexane and pentadecane as a function of mole fraction of cyclohexane (x_c) at 298.15 K:

x_c	0.6965	0.7988	0.9004
$\rho/(\mathrm{g\,cm^{-3}})$	0.7661	0.7674	0.7697

Compute the partial molar volume for each component in a mixture which has a mole fraction of cyclohexane of 0.7988.

5B.4‡ Comelli and Francesconi examined mixtures of propionic acid with various other organic liquids at 313.15 K (F. Comelli and R. Francesconi, *J. Chem. Eng. Data* **41**,101 (1996)). They report the excess volume of mixing

‡ These problems were provided by Charles Trapp and Carmen Giunta.

propionic acid with oxane as $V^{E} = x_1x_2\{a_0 + a_1(x_1 - x_2)\}$, where x_1 is the mole fraction of propionic acid, x_2 that of oxane, $a_0 = -2.4697\,\text{cm}^3\,\text{mol}^{-1}$ and $a_1 = 0.0608\,\text{cm}^3\,\text{mol}^{-1}$. The density of propionic acid at this temperature is $0.97174\,\text{g}\,\text{cm}^{-3}$; that of oxane is $0.86398\,\text{g}\,\text{cm}^{-3}$. (a) Derive an expression for the partial molar volume of each component at this temperature. (b) Compute the partial molar volume for each component in an equimolar mixture.

5B.5‡ Equation 5B.15 indicates, after it has been converted into an expression for x_B, that solubility is an exponential function of temperature. The data in the table below gives the solubility, S, of calcium acetate in water as a function of temperature.

$\theta/°\text{C}$	0	20	40	60	80
$S/(\text{g (100 g solvent)}^{-1})$	36.4	34.9	33.7	32.7	31.7

Determine the extent to which the data fit the exponential $S = S_0 e^{\tau/T}$ and obtain values for S_0 and τ. Express these constants in terms of properties of the solute.

5B.6 The excess Gibbs energy of solutions of methylcyclohexane (MCH) and tetrahydrofuran (THF) at 303.15 K were found to fit the expression

$$G^{E} = RTx(1-x)\left\{0.4857 - 0.1077(2x-1) + 0.0191(2x-1)^2\right\}$$

where x is the mole fraction of the methylcyclohexane. Calculate the Gibbs energy of mixing when a mixture of 1.00 mol of MCH and 3.00 mol of THF is prepared.

5B.7‡ Figure 5.1 shows $\Delta_{\text{mix}}G(x_{\text{Pb}},T)$ for a mixture of copper and lead. (a) What does the graph reveal about the miscibility of copper and lead and the spontaneity of solution formation? What is the variance (F) at (i) 1500 K, (ii) 1100 K? (b) Suppose that at 1500 K a mixture of composition (i) $x_{\text{Pb}} = 0.1$, (ii) $x_{\text{Pb}} = 0.7$, is slowly cooled to 1100 K. What is the equilibrium composition of the final mixture? Include an estimate of the relative amounts of each phase. (c) What is the solubility of (i) lead in copper, (ii) copper in lead at 1100 K?

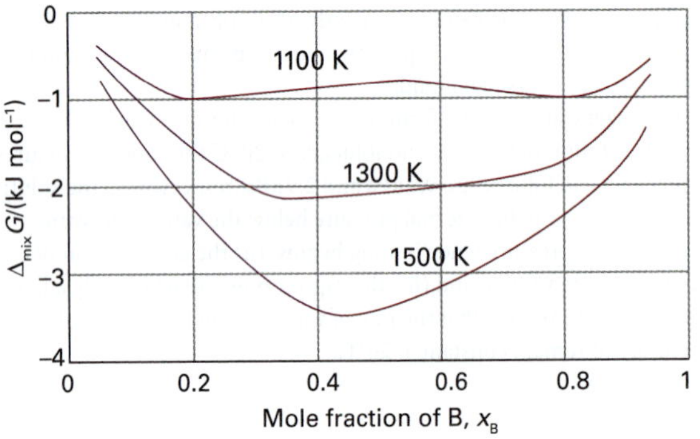

Figure 5.1 The Gibbs energy of mixing of copper and lead.

5B.8 The excess Gibbs energy of a certain binary mixture is equal to $gRTx(1-x)$ where g is a constant and x is the mole fraction of a solute A. Find an expression for the chemical potential of A in the mixture and sketch its dependence on the composition.

5B.9 Use the Gibbs–Helmholtz equation to find an expression for d ln x_A in terms of dT. Integrate d ln x_A from $x_A = 0$ to the value of interest, and integrate the right-hand side from the transition temperature for the pure liquid A to the value in the solution. Show that, if the enthalpy of transition is constant, then eqns 5B.9 and 5B.12 are obtained.

5B.10‡ Polymer scientists often report their data in a variety of units. For example, in the determination of molar masses of polymers in solution by osmometry, osmotic pressures are often reported in grams per square centimetre ($\text{g}\,\text{cm}^{-2}$) and concentrations in grams per cubic centimetre ($\text{g}\,\text{cm}^{-3}$). (a) With these choices of units, what would be the units of R in the van 't Hoff equation? (b) The data in the table below on the concentration dependence of the osmotic pressure of polyisobutene in chlorobenzene at 25 °C have been adapted from J. Leonard and H. Daoust (*J. Polymer Sci.* **57**, 53 (1962)). From these data, determine the molar mass of polyisobutene by plotting Π/c against c. (c) 'Theta solvents' are solvents for which the second osmotic coefficient is zero; for 'poor' solvents the plot is linear and for good solvents the plot is nonlinear. From your plot, how would you classify chlorobenzene as a solvent for polyisobutene? Rationalize the result in terms of the molecular structure of the polymer and solvent. (d) Determine the second and third osmotic virial coefficients by fitting the curve to the virial form of the osmotic pressure equation. (e) Experimentally, it is often found that the virial expansion can be represented as

$$\Pi/c = RT/M(1 + B_c' + gB'^2c^2 + \ldots)$$

and in good solvents, the parameter g is often about 0.25. With terms beyond the second power ignored, obtain an equation for $(\Pi/c)^{1/2}$ and plot this quantity against c. Determine the second and third virial coefficients from the plot and compare to the values from the first plot. Does this plot confirm the assumed value of g?

$10^{-2}(\Pi/c)/(\text{g}\,\text{cm}^{-2}/\text{g}\,\text{cm}^{-3})$	2.6	2.9	3.6	4.3	6.0	12.0
$c/(\text{g}\,\text{cm}^{-3})$	0.0050	0.010	0.020	0.033	0.057	0.10

$10^{-2}(\Pi/c/(\text{g}\,\text{cm}^{-2}//\text{g}\,\text{cm}^{-3})$	19.0	31.0	38.0	52	63
$c/(\text{g}\,\text{cm}^{-3})$	0.145	0.195	0.245	0.27	0.29

5B.11‡ K. Sato, F.R. Eirich, and J.E. Mark (*J. Polymer Sci., Polym. Phys.* **14**, 619 (1976)) have reported the data in the table below for the osmotic pressures of polychloroprene ($\rho = 1.25\,\text{g}\,\text{cm}^{-3}$) in toluene ($\rho = 0.858\,\text{g}\,\text{cm}^{-3}$) at 30 °C. Determine the molar mass of polychloroprene and its second osmotic virial coefficient.

$c/(\text{mg}\,\text{cm}^{-3})$	1.33	2.10	4.52	7.18	9.87
$\Pi/(\text{N}\,\text{m}^{-2})$	30	51	132	246	390

5B.12 Use mathematical software, a spreadsheet, or the *Living graphs* on the web site for this book to draw graphs of $\Delta_{\text{mix}}G$ against x_A at different temperatures in the range 298 K to 500 K. For what value of x_A does $\Delta_{\text{mix}}G$ depend on temperature most strongly?

5B.13 Using the graph in Fig. 5B.4, fix ξ and vary the temperature. For what value of x_A does the excess enthalpy depend on temperature most strongly?

5B.14 Derive an expression for the temperature coefficient of the solubility, dx_B/dT, and plot it as a function of temperature for several values of the enthalpy of fusion.

5B.15 Calculate the osmotic virial coefficient B from the data in *Example* 5B.2.

TOPIC 5C Phase diagrams of binary systems

Discussion questions

5C.1 Draw phase diagrams for the following types of systems. Label the regions of the diagrams, stating what materials (possibly compounds or azeotropes) are present and whether they are solid liquid or gas:

(a) two-component, temperature–composition, solid–liquid diagram, one compound AB formed that melts congruently, negligible solid–solid solubility;

(b) two-component, temperature–composition, solid–liquid diagram, one compound of formula AB_2 that melts incongruently, negligible solid–solid solubility;
(c) two-component, constant temperature–composition, liquid–vapour diagram, formation of an azeotrope at $x_B = 0.333$, complete miscibility.

Exercises

5C.1(a) The following temperature–composition data were obtained for a mixture of octane (O) and methylbenzene (M) at 1.00 atm, where x is the mole fraction in the liquid and y the mole fraction in the vapour at equilibrium.

θ/°C	110.9	112.0	114.0	115.8	117.3	119.0	121.1	123.0
x_M	0.908	0.795	0.615	0.527	0.408	0.300	0.203	0.097
y_M	0.923	0.836	0.698	0.624	0.527	0.410	0.297	0.164

The boiling points are 110.6 °C and 125.6 °C for M and O, respectively. Plot the temperature–composition diagram for the mixture. What is the composition of the vapour in equilibrium with the liquid of composition (i) $x_M = 0.250$ and (ii) $x_O = 0.250$?

5C.1(b) The following temperature–composition data were obtained for a mixture of two liquids A and B at 1.00 atm, where x is the mole fraction in the liquid and y the mole fraction in the vapour at equilibrium.

θ/°C	125	130	135	140	145	150
x_A	0.91	0.65	0.45	0.30	0.18	0.098
y_A	0.99	0.91	0.77	0.61	0.45	0.25

The boiling points are 124 °C for A and 155 °C for B. Plot the temperature/composition diagram for the mixture. What is the composition of the vapour in equilibrium with the liquid of composition (i) $x_A = 0.50$ and (ii) $x_B = 0.33$?

5C.2(a) Methylethyl ether (A) and diborane, B_2H_6 (B), form a compound which melts congruently at 133 K. The system exhibits two eutectics, one at 25 mol per cent B and 123 K and a second at 90 mol per cent B and 104 K. The melting points of pure A and B are 131 K and 110 K, respectively. Sketch the phase diagram for this system. Assume negligible solid–solid solubility.

5C.2(b) Sketch the phase diagram of the system NH_3/N_2H_4 given that the two substances do not form a compound with each other, that NH_3 freezes at −78 °C and N_2H_4 freezes at +2 °C, and that a eutectic is formed when the mole fraction of N_2H_4 is 0.07 and that the eutectic melts at −80 °C.

5C.3(a) Figure 5.2 shows the phase diagram for two partially miscible liquids, which can be taken to be that for water (A) and 2-methyl-1-propanol (B). Describe what will be observed when a mixture of composition $x_B = 0.8$ is heated, at each stage giving the number, composition, and relative amounts of the phases present.

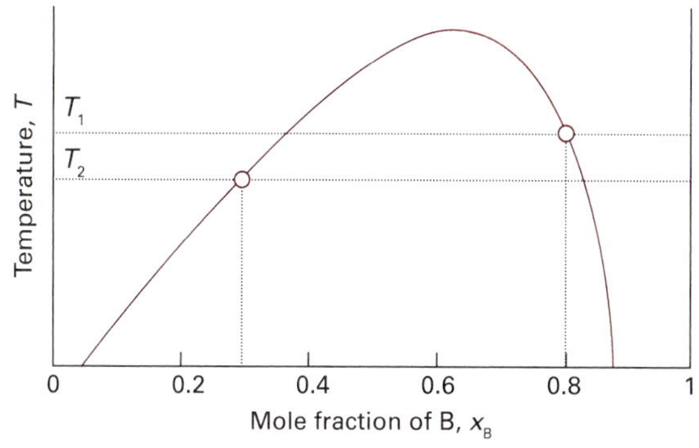

Figure 5.2 The phase diagram for two partially miscible liquids.

5C.2 What molecular features determine whether a mixture of two liquids will show high- and low-boiling azeotropic behaviour?

5C.3 What factors determine the number of theoretical plates required to achieve a desired degree of separation in fractional distillation?

5C.3(b) Refer to Fig. 5.2 again. Describe what will be observed when a mixture of composition $x_B = 0.3$ is heated, at each stage giving the number, composition, and relative amounts of the phases present.

5C.4(a) Indicate on the phase diagram in Fig. 5.3 the feature that denotes incongruent melting. What is the composition of the eutectic mixture and at what temperature does it melt?

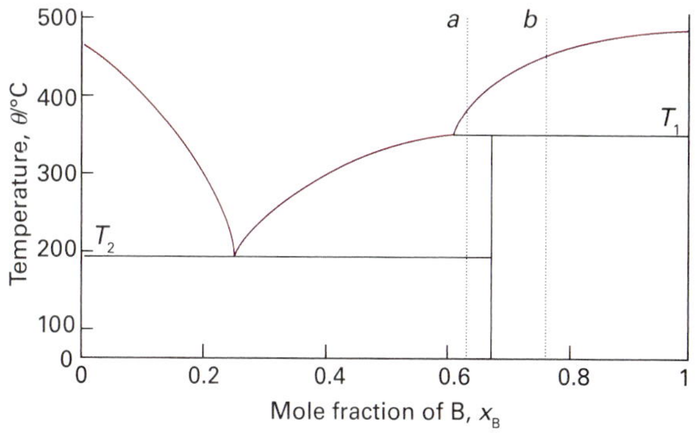

Figure 5.3 The phase diagram referred to in Exercise 5C.4(a).

5C.4(b) Indicate on the phase diagram in Fig. 5.4 the feature that denotes incongruent melting. What are the compositions of any eutectic mixtures and at what temperatures do they melt?

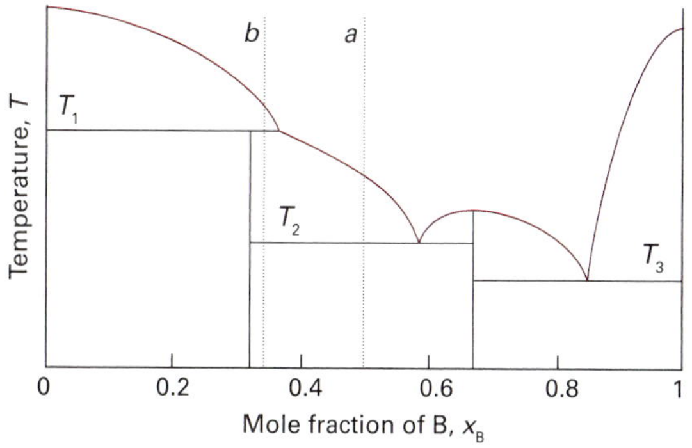

Figure 5.4 The phase diagram referred to in Exercise 5C.4(b).

5C.5(a) Sketch the cooling curves for the isopleths a and b in Fig. 5.3.

5C.5(b) Sketch the cooling curves for the isopleths a and b in Fig. 5.4.

5C.6(a) Use the phase diagram in Fig. 5.3 to state (i) the solubility of Ag in Sn at 800 °C and (ii) the solubility of Ag_3Sn in Ag at 460 °C, (iii) the solubility of Ag_3Sn in Ag at 300 °C.

5C.6(b) Use the phase diagram in Fig. 5.3 to state (i) the solubility of B in A at 500 °C and (ii) the solubility of B in A at 390 °C, (iii) the solubility of AB_2 in B at 300 °C.

5C.7(a) Figure 5.5 shows the experimentally determined phase diagrams for the nearly ideal solution of hexane and heptane. (i) Label the regions of the diagrams to which phases are present. (ii) For a solution containing 1 mol each of hexane and heptane molecules, estimate the vapour pressure at 70 °C when vaporization on reduction of the external pressure just begins. (iii) What is the vapour pressure of the solution at 70 °C when just one drop of liquid remains. (iv) Estimate from the figures the mole fraction of hexane in the liquid and vapour phases for the conditions of part b. (v) What are the mole fractions for the conditions of part c? (vi) At 85 °C and 760 Torr, what are the amounts substance in the liquid and vapour phases when $z_{\text{heptane}} = 0.40$?

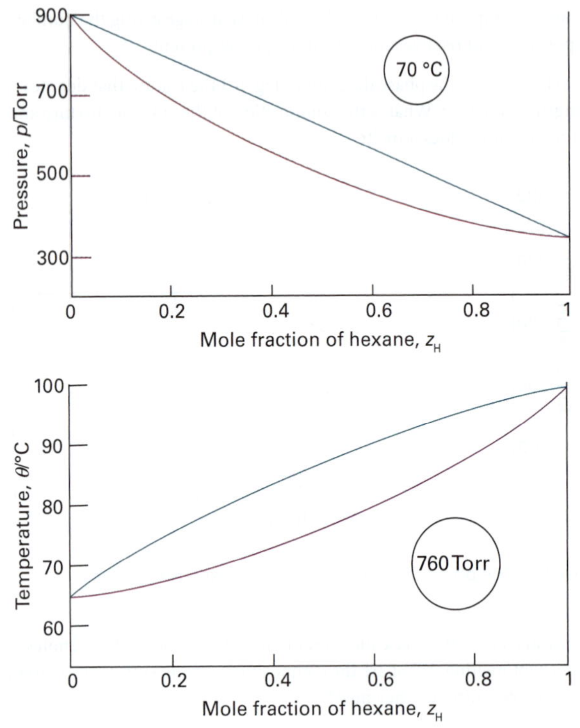

Figure 5.5 Phase diagrams for the nearly ideal solution of hexane and heptane.

5C.7(b) Uranium tetrafluoride and zirconium tetrafluoride melt at 1035 °C and 912 °C respectively. They form a continuous series of solid solutions with a minimum melting temperature of 765 °C and composition $x(ZrF_4) = 0.77$. At 900 °C, the liquid solution of composition $x(ZrF_4) = 0.28$ is in equilibrium with a solid solution of composition $x(ZrF_4) = 0.14$. At 850 °C the two compositions are 0.87 and 0.90, respectively. Sketch the phase diagram for this system and state what is observed when a liquid of composition $x(ZrF_4) = 0.40$ is cooled slowly from 900 °C to 500 °C.

5C.8(a) Methane (melting point 91 K) and tetrafluoromethane (melting point 89 K) do not form solid solutions with each other, and as liquids they are only partially miscible. The upper critical temperature of the liquid mixture is 94 K at $x(CF_4) = 0.43$ and the eutectic temperature is 84 K at $x(CF_4) = 0.88$. At 86 K, the phase in equilibrium with the tetrafluoromethane-rich solution changes from solid methane to a methane-rich liquid. At that temperature, the two liquid solutions that are in mutual equilibrium have the compositions $x(CF_4) = 0.10$ and $x(CF_4) = 0.80$. Sketch the phase diagram.

5C.8(b) Describe the phase changes that take place when a liquid mixture of 4.0 mol B_2H_6 (melting point 131 K) and 1.0 mol CH_3OCH_3 (melting point 135 K) is cooled from 140 K to 90 K. These substances form a compound $(CH_3)_2OB_2H_6$ that melts congruently at 133 K. The system exhibits one eutectic at $x(B_2H_6) = 0.25$ and 123 K and another at $x(B_2H_6) = 0.90$ and 104 K.

5C.9(a) Refer to the information in Exercise 5C.8(a) and sketch the cooling curves for liquid mixtures in which $x(CF_4)$ is (i) 0.10, (ii) 0.30, (iii) 0.50, (iv) 0.80, and (v) 0.95.

5C.9(b) Refer to the information in Exercise 5C.8(b) and sketch the cooling curves for liquid mixtures in which $x(B_2H_6)$ is (i) 0.10, (ii) 0.30, (iii) 0.50, (iv) 0.80, and (v) 0.95.

5C.10(a) Hexane and perfluorohexane show partial miscibility below 22.70 °C. The critical concentration at the upper critical temperature is $x = 0.355$, where x is the mole fraction of C_6F_{14}. At 22.0 °C the two solutions in equilibrium have $x = 0.24$ and $x = 0.48$, respectively, and at 21.5 °C the mole fractions are 0.22 and 0.51. Sketch the phase diagram. Describe the phase changes that occur when perfluorohexane is added to a fixed amount of hexane at (i) 23 °C, (ii) 22 °C.

5C.10(b) Two liquids, A and B, show partial miscibility below 52.4 °C. The critical concentration at the upper critical temperature is $x = 0.459$, where x is the mole fraction of A. At 40.0 °C the two solutions in equilibrium have $x = 0.22$ and $x = 0.60$, respectively, and at 42.5 °C the mole fractions are 0.24 and 0.48. Sketch the phase diagram. Describe the phase changes that occur when B is added to a fixed amount of A at (i) 48 °C, (ii) 52.4 °C.

Problems

5C.1‡ 1-Butanol and chlorobenzene form a minimum-boiling azeotropic system. The mole fraction of 1-butanol in the liquid (x) and vapour (y) phases at 1.000 atm is given below for a variety of boiling temperatures (H. Artigas et al., *J. Chem. Eng. Data* **42**, 132 (1997)).

T/K	396.57	393.94	391.60	390.15	389.03	388.66	388.57
x	0.1065	0.1700	0.2646	0.3687	0.5017	0.6091	0.7171
y	0.2859	0.3691	0.4505	0.5138	0.5840	0.6409	0.7070

Pure chlorobenzene boils at 404.86 K. (a) Construct the chlorobenzene-rich portion of the phase diagram from the data. (b) Estimate the temperature at which a solution for which the mole fraction of 1-butanol is 0.300 begins to boil. (c) State the compositions and relative proportions of the two phases present after a solution initially 0.300 1-butanol is heated to 393.94 K.

5C.2‡ An, Zhao, Jiang, and Shen investigated the liquid–liquid coexistence curve of *N,N*-dimethylacetamide and heptane (X. An et al., *J. Chem. Thermodynamics* **28**, 1221 (1996)). Mole fractions of *N,N*-dimethylacetamide in the upper (x_1) and lower (x_2) phases of a two-phase region are given opposite as a function of temperature:

T/K	309.820	309.422	309.031	308.006	306.686
x_1	0.473	0.400	0.371	0.326	0.239
x_2	0.529	0.601	0.625	0.657	0.690
T/K	304.553	301.803	299.097	296.000	294.534
x_1	0.255	0.218	0.193	0.168	0.157
x_2	0.724	0.758	0.783	0.804	0.814

(a) Plot the phase diagram. (b) State the proportions and compositions of the two phases that form from mixing 0.750 mol of *N,N*-dimethylacetamide with 0.250 mol of heptane at 296.0 K. To what temperature must the mixture be heated to form a single-phase mixture?

5C.3 Phosphorus and sulfur form a series of binary compounds. The best characterized are P_4S_3, P_4S_7, and P_4S_{10}, all of which melt congruently. Assuming that only these three binary compounds of the two elements exist, (a) draw schematically only the P/S phase diagram. Label each region of the diagram with the substance that exists in that region and indicate its phase. Label the horizontal axis as x_S and give the numerical values of x_S that correspond to the compounds. The melting point of pure phosphorus is 44 °C and that of pure sulfur is 119 °C. (b) Draw, schematically, the cooling curve for a mixture of composition $x_S = 0.28$. Assume that a eutectic occurs at $x_S = 0.2$ and negligible solid–solid solubility.

5C.4 The following table gives the break and halt temperatures found in the cooling curves of two metals A and B. Construct a phase diagram consistent with the data of these curves. Label the regions of the diagram, stating what phases and substances are present. Give the probable formulas of any compounds that form.

$100x_B$	$\theta_{break}/°C$	$\theta_{halt,1}/°C$	$\theta_{halt,2}/°C$
0		1100	
10.0	1060	700	
20.0	1000	700	
30.0	940	700	400
40.0	850	700	400
50.0	750	700	400
60.0	670	400	
70.0	550	400	
80.0		400	
90.0	450	400	
100.0		500	

5C.5 Consider the phase diagram in Fig. 5.6, which represents a solid–liquid equilibrium. Label all regions of the diagram according to the chemical species that exist in that region and their phases. Indicate the number of species and phases present at the points labelled *b*, *d*, *e*, *f*, *g*, and *k*. Sketch cooling curves for compositions $x_B = 0.16, 0.23, 0.57, 0.67$, and 0.84.

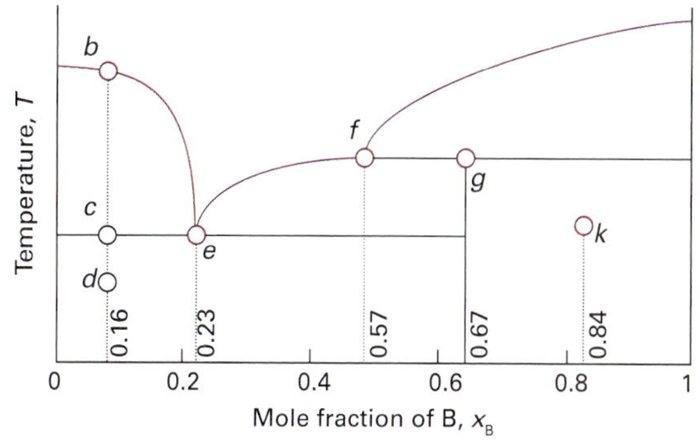

Figure 5.6 The phase diagram referred to in Problem 5C.5.

5C.6 Sketch the phase diagram for the Mg/Cu system using the following information: θ_f(Mg) = 648 °C, θ_f(Cu) = 1085 °C; two intermetallic compounds are formed with $\theta_f(MgCu_2)$ = 800 °C and $\theta_f(Mg_2Cu)$ = 580 °C; eutectics of mass percentage Mg composition and melting points 10 per cent (690 °C), 33 per cent (560 °C), and 65 per cent (380 °C). A sample of Mg/Cu alloy containing 25 per cent Mg by mass was prepared in a crucible heated to 800 °C in an inert atmosphere. Describe what will be observed if the melt is cooled slowly to room temperature. Specify the composition and relative abundances of the phases and sketch the cooling curve.

5C.7‡ The temperature–composition diagram for the Ca/Si binary system is shown in Fig. 5.7. (a) Identify eutectics, congruent melting compounds, and incongruent melting compounds. (b) If a 20 per cent by atom composition melt of silicon at 1500 °C is cooled to 1000 °C, what phases (and phase composition) would be at equilibrium? Estimate the relative amounts of each phase. (c) Describe the equilibrium phases observed when an 80 per cent by atom composition Si melt is cooled to 1030 °C. What phases, and relative amounts, would be at equilibrium at a temperature (i) slightly higher than 1030 °C, (ii) slightly lower than 1030 °C? Draw a graph of the mole percentages of both Si(s) and $CaSi_2$(s) as a function of mole percentage of melt that is freezing at 1030 °C.

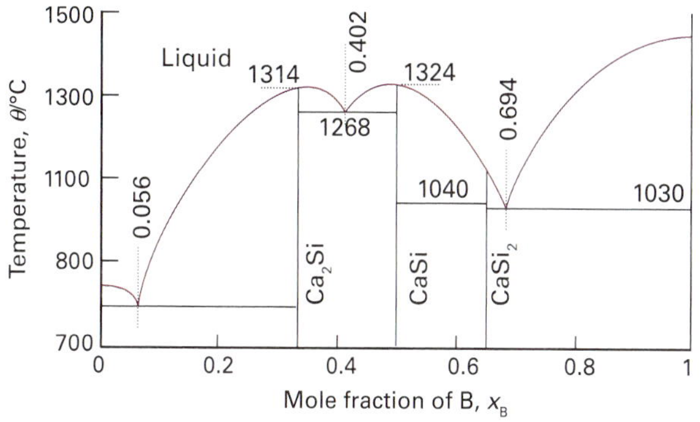

Figure 5.7 The temperature–composition diagram for the Ca/Si binary system.

5C.8 Iron(II) chloride (melting point 677 °C) and potassium chloride (melting point 776 °C) form the compounds $KFeCl_3$ and K_2FeCl_4 at elevated temperatures. $KFeCl_3$ melts congruently at 380 °C and K_2FeCl_4 melts incongruently at 399 °C. Eutectics are formed with compositions $x = 0.38$ (melting point 351 °C) and $x = 0.54$ (melting point 393 °C), where x is the mole fraction of $FeCl_2$. The KCl solubility curve intersects the K_2FeCl_4 curve at $x = 0.34$. Sketch the phase diagram. State the phases that are in equilibrium when a mixture of composition $x = 0.36$ is cooled from 400 °C to 300 °C.

5C.9 To reproduce the results of Fig. 5C.2, first rearrange eqn 5C.4 so that y_A is expressed as a function of x_A and the ratio p_A^*/p_B^*. Then plot y_A against x_A for several values of ratio $p_A^*/p_B^* > 1$.

5C.10 To reproduce the results of Fig. 5C.3, first rearrange eqn 5C.5 so that the ratio p_A^*/p_B^* is expressed as a function of y_A and the ratio p_A^*/p_B^*. Then plot p_A/p_A^* against y_A for several values of $p_A^*/p_B^* > 1$.

5C.11 Working from eqn 5B.7, write an expression for T_{min}, the temperature at which $\Delta_{mix}G$ has a minimum, as a function of ξ and x_A. Then, plot T_{min} against x_A for several values of ξ. Provide a physical interpretation for any maxima or minima that you observe in these plots.

5C.12 Use eqn 5C.7 to generate plots of ξ against x_A by one of two methods: (a) solve the transcendental equation $\ln\{(x/(1-x)\} + \xi(1-2x) = 0$ numerically, or (b) plot the first term of the transcendental equation against the second and identify the points of intersection as ξ is changed.

TOPIC 5D Phase diagrams of ternary systems

Discussion questions

5D.1 What is the maximum number of phases that can be in equilibrium in a ternary system?

5D.2 Does the lever rule apply to a ternary system?

5D.3 Could a regular tetrahedron be used to depict the properties of a four-component system?

Exercises

5D.1(a) Mark the following features on triangular coordinates: (i) the point (0.2, 0.2, 0.6), (ii) the point (0, 0.2, 0.8), (iii) the point at which all three mole fractions are the same.
5D.1(b) Mark the following features on triangular coordinates: (i) the point (0.6, 0.2, 0.2), (ii) the point (0.8, 0.2, 0), (iii) the point (0.25, 0.25, 0.50).

5D.2(a) Mark the following points on a ternary phase diagram for the system $NaCl/Na_2SO_4\cdot 10H_2O/H_2O$: (i) 25 per cent by mass NaCl, 25 per cent $Na_2SO_4\cdot 10H_2O$, and the rest H_2O; (ii) the line denoting the same relative composition of the two salts but with changing amounts of water.
5D.2(b) Mark the following points on a ternary phase diagram for the system $NaCl/Na_2SO_4\cdot 10H_2O/H_2O$: (i) 33 per cent by mass NaCl, 33 per cent $Na_2SO_4\cdot 10H_2O$, and the rest H_2O; (ii) the line denoting the same relative composition of the two salts but with changing amounts of water.

5D.3(a) Refer to the ternary phase diagram in Fig. 5D.4. How many phases are present, and what are their compositions and relative abundances, in a mixture that contains 2.3 g of water, 9.2 g of chloroform, and 3.1 g of acetic acid? Describe what happens when (i) water, (iii) acetic acid is added to the mixture.
5D.3(b) Refer to the ternary phase diagram in Fig. 5D.4. How many phases are present, and what are their compositions and relative abundances, in a mixture that contains 55.0 g of water, 8.8 g of chloroform, and 3.7 g of acetic acid? Describe what happens when (i) water, (ii) acetic acid is added to the mixture.

5D.4(a) Figure 5.8 shows the phase diagram for the ternary system $NH_4Cl/(NH_4)_2SO_4/H_2O$ at 25 °C. Identify the number of phases present for mixtures of compositions (i) (0.2, 0.4, 0.4), (ii) (0.4, 0.4, 0.2), (iii) (0.2, 0.1, 0.7), (iv) (0.4, 0.16, 0.44). The numbers are mole fractions of the three components in the order $(NH_4Cl,(NH_4)_2SO_4,H_2O)$.

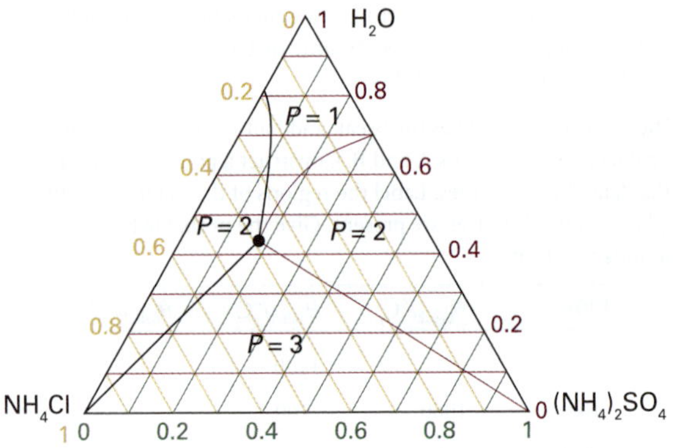

Figure 5.8 The phase diagram for the ternary system $NH_4Cl/(NH_4)_2SO_4/H_2O$ at 25 °C.

5D.4(b) Refer to Fig. 5.8 and identify the number of phases present for mixtures of compositions (i) (0.4, 0.1, 0.5), (ii) (0.8, 0.1, 0.1), (iii) (0, 0.3,0.7), (iv) (0.33, 0.33, 0.34). The numbers are mole fractions of the three components in the order $(NH_4Cl,(NH_4)_2SO_4,H_2O)$.

5D.5(a) Referring to Fig. 5.8, deduce the molar solubility of (i) NH_4Cl, (ii) $(NH_4)_2SO_4$ in water at 25 °C.
5D.5(b) Describe what happens when (i) $(NH_4)_2SO_4$ is added to a saturated solution of NH_4Cl in water in the presence of excess NH_4Cl, (ii) water is added to a mixture of 25 g of NH_4Cl and 75 g of $(NH_4)_2SO_4$.

Problems

5D.1 At a certain temperature, the solubility of I_2 in liquid CO_2 is $x(I_2) = 0.03$. At the same temperature its solubility in nitrobenzene is 0.04. Liquid carbon dioxide and nitrobenzene are miscible in all proportions, and the solubility of I_2 in the mixture varies linearly with the proportion of nitrobenzene. Sketch a phase diagram for the ternary system.

5D.2 The binary system nitroethane/decahydronaphthalene (DEC) shows partial miscibility, with the two-phase region lying between $x = 0.08$ and $x = 0.84$, where x is the mole fraction of nitroethane. The binary system liquid carbon dioxide/DEC is also partially miscible, with its two-phase region lying between $y = 0.36$ and $y = 0.80$, where y is the mole fraction of DEC. Nitroethane and liquid carbon dioxide are miscible in all proportions. The addition of liquid carbon dioxide to mixtures of nitroethane and DEC increases the range of miscibility, and the plait point is reached when z, the mole fraction of CO_2, is 0.18 and $x = 0.53$. The addition of nitroethane to mixtures of carbon dioxide and DEC also results in another plait point at $x = 0.08$ and $y = 0.52$. (a) Sketch the phase diagram for the ternary system, (b) For some binary mixtures of nitroethane and liquid carbon dioxide the addition of arbitrary amounts of DEC will not cause phase separation. Find the range of concentration for such binary mixtures.

5D.3 Prove that a straight line from the apex A of a ternary phase diagram to the opposite edge BC represents mixtures of constant ratio of B and C, however much A is present.

TOPIC 5E Activities

Discussion questions

5E.1 What are the contributions that account for the difference between activity and concentration?

5E.2 How is Raoult's law modified so as to describe the vapour pressure of real solutions?

5E.3 Summarize the ways in which activities may be measured.

Exercises

5E.1(a) Substances A and B are both volatile liquids with $p_A^* = 300$ Torr, $p_B^* = 250$ Torr, and $K_B = 200$ Torr (concentration expressed in mole fraction). When $x_A = 0.9$, $b_B = 2.22\ \mathrm{mol\,kg^{-1}}$, $p_A = 250$ Torr, and $p_B = 25$ Torr. Calculate the activities and activity coefficients of A and B. Use the mole fraction, Raoult's law basis system for A and the Henry's law basis system (both mole fractions and molalities) for B.

5E.1(b) Given that $p^*(H_2O)=0.02308$ atm and $p(H_2O)=0.02239$ atm in a solution in which 0.122 kg of a involatile solute ($M=241$ g mol^{-1}) is dissolved in 0.920 kg water at 293 K, calculate the activity and activity coefficient of water in the solution.

5E.2(a) By measuring the equilibrium between liquid and vapour phases of an acetone(A)/methanol(M) solution at 57.2 °C at 1.00 atm, it was found that $x_A=0.400$ when $y_A=0.516$. Calculate the activities and activity coefficients of both components in this solution on the Raoult's law basis. The vapour pressures of the pure components at this temperature are: $p_A^*=105$ kPa and $p_M^*=73.5$ kPa. (x_A is the mole fraction in the liquid and y_A the mole fraction in the vapour.)

5E.2(b) By measuring the equilibrium between liquid and vapour phases of a solution at 30 °C at 1.00 atm, it was found that $x_A=0.220$ when $y_A=0.314$. Calculate the activities and activity coefficients of both components in this solution on the Raoult's law basis. The vapour pressures of the pure components at this temperature are: $p_A^*=73.0$ kPa and $p_B^*=92.1$ kPa. (x_A is the mole fraction in the liquid and y_A the mole fraction in the vapour.)

5E.3(a) Find the relation between the standard and biological standard Gibbs energies of a reaction of the form $A \rightarrow 2B + 2\,H^+(aq)$.

5E.3(b) Find the relation between the standard and biological standard Gibbs energies of a reaction of the form $2\,A \rightarrow B + 4\,H^+(aq)$.

5E.4(a) Suppose it is found that for a hypothetical regular solution that $\xi=1.40$, $p_A^*=15.0$ kPa and $p_B^*=11.6$ kPa. Draw the vapour-pressure diagram.

5E.4(b) Suppose it is found that for a hypothetical regular solution that $\xi=-1.40$, $p_A^*=15.0$ kPa and $p_B^*=11.6$ kPa. Draw the vapour-pressure diagram.

Problems

5E.1‡ Francesconi, Lunelli, and Comelli studied the liquid–vapour equilibria of trichloromethane and 1,2-epoxybutane at several temperatures (Francesconi et al., *J. Chem. Eng. Data* **41**, 310 (1996)). Among their data are the following measurements of the mole fractions of trichloromethane in the liquid phase (x_T) and the vapour phase (y_T) at 298.15 K as a function of pressure.

p/kPa	23.40	21.75	20.25	18.75	18.15	20.25	22.50	26.30
x	0	0.129	0.228	0.353	0.511	0.700	0.810	1
y	0	0.065	0.145	0.285	0.535	0.805	0.915	1

Compute the activity coefficients of both components on the basis of Raoult's law.

5E.2 The *osmotic coefficient* ϕ is defined as $\phi=-(x_A/x_B)\ln a_A$. By writing $r=x_B/x_A$, and using the Gibbs–Duhem equation, show that we can calculate the activity of B from the activities of A over a composition range by using the formula

$$\ln\frac{a_B}{r}=\phi-\phi(0)+\int_0^r \frac{\phi-1}{r}\,dr$$

5E.3 Show that the osmotic pressure of a real solution is given by $\Pi V=-RT\ln a_A$. Go on to show that, provided the concentration of the solution is low, this expression takes the form $\Pi V=\phi RT[B]$ and hence that the osmotic coefficient ϕ (which is defined in Problem 5E.2) may be determined from osmometry.

5E.4 Use mathematical software, a spreadsheet, or the *Living graphs* on the web site for this book to plot p_A/p_A^* against x_A with $\xi=2.5$ by using eqn 5E.19 and then eqn 5E.20. Above what value of x_A do the values of p_A/p_A^* given by these equations differ by more than 10 per cent?

TOPIC 5F The activities of ions

Discussion questions

5F.1 Why do the activity coefficients of ions in solution differ from 1? Why are they less than 1 in dilute solutions?

5F.2 Describe the general features of the Debye–Hückel theory of electrolyte solutions.

5F.3 Suggest an interpretation of the additional terms in extended versions of the Debye–Hückel limiting law.

Exercises

5F.1(a) Calculate the ionic strength of a solution that is 0.10 mol kg^{-1} in KCl(aq) and 0.20 mol kg^{-1} in $CuSO_4$(aq).

5F.1(b) Calculate the ionic strength of a solution that is 0.040 mol kg^{-1} in $K_3[Fe(CN)_6]$(aq), 0.030 mol kg^{-1} in KCl(aq), and 0.050 mol kg^{-1} in NaBr(aq).

5F.2(a) Calculate the masses of (i) $Ca(NO_3)_2$ and, separately, (ii) NaCl to add to a 0.150 mol kg^{-1} solution of KNO_3(aq) containing 500 g of solvent to raise its ionic strength to 0.250.

5F.2(b) Calculate the masses of (i) KNO_3 and, separately, (ii) $Ba(NO_3)_2$ to add to a 0.110 mol kg^{-1} solution of KNO_3(aq) containing 500 g of solvent to raise its ionic strength to 1.00.

5F.3(a) Estimate the mean ionic activity coefficient and activity of a solution at 25 °C that is 0.010 mol kg^{-1} $CaCl_2$(aq) and 0.030 mol kg^{-1} NaF(aq).

5F.3(b) Estimate the mean ionic activity coefficient and activity of a solution at 25 °C that is 0.020 mol kg^{-1} NaCl(aq) and 0.035 mol kg^{-1} $Ca(NO_3)_2$(aq).

5F.4(a) The mean activity coefficients of HBr in three dilute aqueous solutions at 25 °C are 0.930 (at 5.0 mmol kg^{-1}), 0.907 (at 10.0 mmol kg^{-1}), and 0.879 (at 20.0 mmol kg^{-1}). Estimate the value of B in eqn 5F.11a.

5F.4(b) The mean activity coefficients of KCl in three dilute aqueous solutions at 25 °C are 0.927 (at 5.0 mmol kg^{-1}), 0.902 (at 10.0 mmol kg^{-1}), and 0.816 (at 50.0 mmol kg^{-1}). Estimate the value of B in eqn 5F.11a.

Problems

5F.1 The mean activity coefficients for aqueous solutions of NaCl at 25 °C are given opposite. Confirm that they support the Debye–Hückel limiting law and that an improved fit is obtained with the Davies equation.

b/(mmol kg^{-1})	1.0	2.0	5.0	10.0	20.0
$\gamma_\pm$	0.9649	0.9519	0.9275	0.9024	0.8712

5F.2 Consider the plot of log $\gamma_\pm$ against $I^{1/2}$ with $B=1.50$ and $C=0$ in the Davies equation as a representation of experimental data for a certain MX electrolyte. Over what range of ionic strengths does the application of the limiting law lead to an error in the value of the activity coefficient of less than 10 per cent of the value predicted by the extended law?

Integrated activities

5.1 The table below lists the vapour pressures of mixtures of iodoethane (I) and ethyl acetate (A) at 50 °C. Find the activity coefficients of both components on (a) the Raoult's law basis, (b) the Henry's law basis with iodoethane as solute.

x_I	0	0.0579	0.1095	0.1918	0.2353	0.3718
p_I/kPa	0	3.73	7.03	11.7	14.05	20.72
p_A/kPa	37.38	35.48	33.64	30.85	29.44	25.05

x_I	0.5478	0.6349	0.8253	0.9093	1.0000
p_I/kPa	28.44	31.88	39.58	43.00	47.12
p_A/kPa	19.23	16.39	8.88	5.09	0

5.2 Plot the vapour pressure data for a mixture of benzene (B) and acetic acid (A) given below and plot the vapour pressure/composition curve for the mixture at 50 °C. Then confirm that Raoult's and Henry's laws are obeyed in the appropriate regions. Deduce the activities and activity coefficients of the components on the Raoult's law basis and then, taking B as the solute, its activity and activity coefficient on a Henry's law basis. Finally, evaluate the excess Gibbs energy of the mixture over the composition range spanned by the data.

x_A	0.0160	0.0439	0.0835	0.1138	0.1714
p_A/kPa	0.484	0.967	1.535	1.89	2.45
p_B/kPa	35.05	34.29	33.28	32.64	30.90

x_A	0.2973	0.3696	0.5834	0.6604	0.8437	0.9931
p_A/kPa	3.31	3.83	4.84	5.36	6.76	7.29
p_B/kPa	28.16	26.08	20.42	18.01	10.0	0.47

5.3‡ Chen and Lee studied the liquid–vapour equilibria of cyclohexanol with several gases at elevated pressures (J.-T. Chen and M.-J. Lee, *J. Chem. Eng. Data* **41**, 339 (1996)). Among their data are the following measurements of the mole fractions of cyclohexanol in the vapour phase (y) and the liquid phase (x) at 393.15 K as a function of pressure.

p/bar	10.0	20.0	30.0	40.0	60.0	80.0
y_{cyc}	0.0267	0.0149	0.0112	0.009 47	0.008 35	0.009 21
x_{cyc}	0.9741	0.9464	0.9204	0.892	0.836	0.773

Determine the Henry's law constant of CO_2 in cyclohexanol, and compute the activity coefficient of CO_2.

5.4‡ The following data have been obtained for the liquid–vapour equilibrium compositions of mixtures of nitrogen and oxygen at 100 kPa.

T/K	77.3	78	80	82	84	86	88	90.2
$x(O_2)$	0	10	34	54	70	82	92	100
$y(O_2)$	0	2	11	22	35	52	73	100
$p^*(O_2)$/Torr	154	171	225	294	377	479	601	760

Plot the data on a temperature–composition diagram and determine the extent to which it fits the predictions for an ideal solution by calculating the activity coefficients of O_2 at each composition.

5.5 Use the Gibbs–Duhem equation to derive the *Gibbs–Duhem–Margules equation*

$$\left(\frac{\partial \ln f_A}{\partial \ln x_A}\right)_{p,T}=\left(\frac{\partial \ln f_B}{\partial \ln x_B}\right)_{p,T}$$

where f is the fugacity. Use the relation to show that when the fugacities are replaced by pressures, that if Raoult's law applies to one component in a mixture it must also apply to the other.

5.6 Use the Gibbs–Duhem equation to show that the partial molar volume (or any partial molar property) of a component B can be obtained if the partial molar volume (or other property) of A is known for all compositions up to the one of interest. Do this by proving that

$$V_B=V_B^*-\int_{V_A^*}^{V_A}\frac{x_A}{1-x_A}\mathrm{d}V_A$$

where the x_A are functions of the V_A. Use the following data (which are for 298 K) to evaluate the integral graphically to find the partial molar volume of acetone at $x=0.500$.

$x(CHCl_3)$	0	0.194	0.385	0.559	0.788	0.889	1.000
$V_m/(cm^3\,mol^{-1})$	73.99	75.29	76.50	77.55	79.08	79.82	80.67

5.7 Show that the freezing-point depression of a real solution in which the solvent of molar mass M has activity a_A obeys

$$\frac{\mathrm{d}\ln a_A}{\mathrm{d}(\Delta T)}=-\frac{M}{K_f}$$

and use the Gibbs–Duhem equation to show that

$$\frac{\mathrm{d}\ln a_B}{\mathrm{d}(\Delta T)}=-\frac{1}{b_B K_f}$$

where a_B is the solute activity and b_B is its molality. Use the Debye–Hückel limiting law to show that the osmotic coefficient (ϕ, Problem 5E.2) is given by $\phi=1-\frac{1}{3}A'I$ with $A'=2.303A$ and $I=b/b^{\ominus}$.

5.8 For the calculation of the solubility c of a gas in a solvent, it is often convenient to use the expression $c=Kp$, where K is the Henry's law constant. Breathing air at high pressures, such as in scuba diving, results in an increased concentration of dissolved nitrogen. The Henry's law constant for the solubility of nitrogen is 0.18 μg/(g H_2O atm). What mass of nitrogen is dissolved in 100 g of water saturated with air at 4.0 atm and 20 °C? Compare your answer to that for 100 g of water saturated with air at 1.0 atm. (Air is 78.08 mole per cent N_2.) If nitrogen is four times as soluble in fatty tissues as in water, what is the increase in nitrogen concentration in fatty tissue in going from 1 atm to 4 atm?

5.9 Dialysis may be used to study the binding of small molecules to macromolecules, such as an inhibitor to an enzyme, an antibiotic to DNA, and any other instance of cooperation or inhibition by small molecules attaching to large ones. To see how this is possible, suppose inside the dialysis bag the molar concentration of the macromolecule M is [M] and the total concentration of small molecule A is $[A]_{in}$. This total concentration is the sum of the concentrations of free A and bound A, which we write $[A]_{free}$ and $[A]_{bound}$, respectively. At equilibrium, $\mu_{A,free} = \mu_{A,out}$, which implies that $[A]_{free} = [A]_{out}$, provided the activity coefficient of A is the same in both solutions. Therefore, by measuring the concentration of A in the solution outside the bag, we can find the concentration of unbound A in the macromolecule solution and, from the difference $[A]_{in} - [A]_{free} = [A]_{in} - [A]_{out}$, the concentration of bound A. Now we explore the quantitative consequences of the experimental arrangement just described.

(a) The average number of A molecules bound to M molecules, ν, is

$$\nu = \frac{[A]_{bound}}{[M]} = \frac{[A]_{in} - [A]_{out}}{[M]}$$

The bound and unbound A molecules are in equilibrium, $M + A \rightleftharpoons MA$. Recall from introductory chemistry that we may write the equilibrium constant for binding, K, as

$$K = \frac{[MA]}{[M]_{free}[A]_{free}}$$

Now show that

$$K = \frac{\nu}{(1-\nu)[A]_{out}}$$

(b) If there are N *identical* and *independent* binding sites on each macromolecule, each macromolecule behaves like N separate smaller macromolecules, with the same value of K for each site. It follows that the average number of A molecules per site is ν/N. Show that, in this case, we may write the *Scatchard equation*:

$$\frac{\nu}{[A]_{out}} = KN - K\nu$$

(c) To apply the Scatchard equation, consider the binding of ethidium bromide (E^-) to a short piece of DNA by a process called *intercalation*, in which the aromatic ethidium cation fits between two adjacent DNA base pairs. An equilibrium dialysis experiment was used to study the binding of ethidium bromide (EB) to a short piece of DNA. A 1.00 μmol dm^{-3} aqueous solution of the DNA sample was dialysed against an excess of EB. The following data were obtained for the total concentration of EB:

[EB]/(μmol dm^{-3})					
Side without DNA	0.042	0.092	0.204	0.526	1.150
Side with DNA	0.292	0.590	1.204	2.531	4.150

From these data, make a Scatchard plot and evaluate the intrinsic equilibrium constant, K, and total number of sites per DNA molecule. Is the identical and independent sites model for binding applicable?

5.10 The form of the Scatchard equation given Problem 5.9 applies only when the macromolecule has identical and independent binding sites. For non-identical independent binding sites, the Scatchard equation is

$$\frac{\nu}{[A]_{out}} = \sum_i \frac{N_i K_i}{1 + K_i[A]_{out}}$$

Plot $\nu/[A]$ for the following cases. (a) There are four independent sites on an enzyme molecule and the intrinsic binding constant is $K = 1.0 \times 10^7$. (b) There are a total of six sites per polymer. Four of the sites are identical and have an intrinsic binding constant of 1×10^5. The binding constants for the other two sites are 2×10^6.

5.11 The addition of a small amount of a salt, such as $(NH_4)_2SO_4$, to a solution containing a charged protein increases the solubility of the protein in water. This observation is called the *salting-in effect*. However, the addition of large amounts of salt can decrease the solubility of the protein to such an extent that the protein precipitates from solution. This observation is called the *salting-out effect* and is used widely by biochemists to isolate and purify proteins. Consider the equilibrium $PX_\nu(s) \rightleftharpoons P^{\nu+}(aq) + \nu\, X^-(aq)$, where $P^{\nu+}$ is a polycationic protein of charge $\nu+$ and X^- is its counter ion. Use Le Chatelier's principle and the physical principles behind the Debye–Hückel theory to provide a molecular interpretation for the salting-in and salting-out effects.

5.12 Some polymers can form liquid crystal mesophases with unusual physical properties. For example, liquid crystalline Kevlar (**1**) is strong enough to be the material of choice for bulletproof vests and is stable at temperatures up to 600 K. What molecular interactions contribute to the formation, thermal stability, and mechanical strength of liquid crystal mesophases in Kevlar?

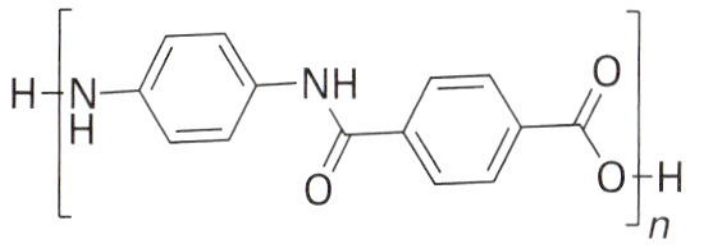

1 Kevlar

CHAPTER 6

Chemical equilibrium

Chemical reactions tend to move towards a dynamic equilibrium in which both reactants and products are present but have no further tendency to undergo net change. In some cases, the concentration of products in the equilibrium mixture is so much greater than that of the unchanged reactants that for all practical purposes the reaction is 'complete'. However, in many important cases the equilibrium mixture has significant concentrations of both reactants and products.

6A The equilibrium constant

This Topic develops the concept of chemical potential and shows how it is used to account for the equilibrium composition of chemical reactions. The equilibrium composition corresponds to a minimum in the Gibbs energy plotted against the extent of reaction. By locating this minimum we establish the relation between the equilibrium constant and the standard Gibbs energy of reaction.

6B The response of equilibria to the conditions

The thermodynamic formulation of equilibrium enables us to establish the quantitative effects of changes in the conditions. One very important aspect of equilibrium is the control that can be exercised by varying the conditions, such as the pressure or temperature.

6C Electrochemical cells

Because many reactions involve the transfer of electrons, they can be studied (and utilized) by allowing them to take place in a cell equipped with electrodes, with the spontaneous reaction forcing electrons through an external circuit. We shall see that the electric potential of the cell is related to the reaction Gibbs energy, so providing an electrical procedure for the determination of thermodynamic quantities.

6D Electrode potentials

Electrochemistry is in part a major application of thermodynamic concepts to chemical equilibria as well as being of great technological importance. As elsewhere in thermodynamics, we see how to report electrochemical data in a compact form and apply it to problems of real chemical significance, especially to the prediction of the spontaneous direction of reactions and the calculation of equilibrium constants.

What is the impact of this material?

The thermodynamic description of spontaneous reactions has numerous practical and theoretical applications. We highlight two applications. One is to the discussion of biochemical processes, where one reaction drives another (*Impact* I6.1). That, ultimately, is why we have to eat, for we see that the reaction that takes place when one substance is oxidized can drive nonspontaneous reactions, such as protein synthesis, forward. Another makes use of the great sensitivity of electrochemical processes to the concentration of electroactive materials, and we see how specially designed electrodes are used in analysis (*Impact* I6.2).

To read more about the impact of this material, scan the QR code, or go to bcs.whfreeman.com/webpub/chemistry/pchem10e/impact/pchem-6-1.html

6A The equilibrium constant

Contents

➤ Why do you need to know this material?

Equilibrium constants lie at the heart of chemistry and are a key point of contact between thermodynamics and laboratory chemistry. The material in this Topic shows how they arise and explains the thermodynamic properties that determine their values.

➤ What is the key idea?

The composition of a reaction mixture tends to change until the Gibbs energy is a minimum.

➤ What do you need to know already?

Underlying the whole discussion is the expression of the direction of spontaneous change in terms of the Gibbs energy of a system (Topic 3C).This material draws on the concept of chemical potential and its dependence on the concentration or pressure of the substance (Topic 5A). You need to know how to express the total Gibbs energy of a mixture in terms of the chemical potentials of its components (Topic 5A).

As explained in Topic 3C, the direction of spontaneous change at constant temperature and pressure is towards lower values of the Gibbs energy, *G*. The idea is entirely general, and in this Topic we apply it to the discussion of chemical reactions. There is a tendency of a mixture of reactants to undergo reaction until the Gibbs energy of the mixture has reached a minimum: that state corresponds to a state of chemical equilibrium. The equilibrium is dynamic in the sense that the forward and reverse reactions continue, but at matching rates. As always in the application of thermodynamics, spontaneity is a *tendency*: there might be kinetic reasons why that tendency is not realized.

6A.1 The Gibbs energy minimum

We locate the equilibrium composition of a reaction mixture by calculating the Gibbs energy of the reaction mixture and identifying the composition that corresponds to minimum *G*. Here we proceed in two steps: first, we consider a very simple equilibrium, and then we generalize it.

(a) The reaction Gibbs energy

Consider the equilibrium $A \rightleftharpoons B$. Even though this reaction looks trivial, there are many examples of it, such as the isomerization of pentane to 2-methylbutane and the conversion of L-alanine to D-alanine.

Suppose an infinitesimal amount $d\xi$ of A turns into B, then the change in the amount of A present is $dn_A = -d\xi$ and the change in the amount of B present is $dn_B = +d\xi$. The quantity ξ (xi) is called the **extent of reaction;** it has the dimensions of amount of substance and is reported in moles. When the extent of reaction changes by a measurable amount $\Delta\xi$, the amount of A present changes from $n_{A,0}$ to $n_{A,0} - \Delta\xi$ and the amount of B changes from $n_{B,0}$ to $n_{B,0} + \Delta\xi$. In general, the amount of a component J changes by $\nu_J \Delta\xi$, where ν_J is the stoichiometric number of the species J (positive for products, negative for reactants).

Brief illustration 6A.1 The extent of reaction

If initially 2.0 mol A is present and we wait until $\Delta\xi = +1.5$ mol, then the amount of A remaining will be 0.5 mol. The amount of B formed will be 1.5 mol.

Self-test 6A.1 Suppose the reaction is 3 A→2 B and that initially 2.5 mol A is present. What is the composition when $\Delta\xi = +0.5$ mol?

Answer: 1.0 mol A, 1.0 mol B

The **reaction Gibbs energy**, $\Delta_r G$, is defined as the slope of the graph of the Gibbs energy plotted against the extent of reaction:

$$\Delta_r G = \left(\frac{\partial G}{\partial \xi}\right)_{p,T} \quad \textit{Definition} \quad \text{Reaction Gibbs energy} \quad (6A.1)$$

Although Δ normally signifies a *difference* in values, here it signifies a *derivative*, the slope of G with respect to ξ. However, to see that there is a close relationship with the normal usage, suppose the reaction advances by $d\xi$. The corresponding change in Gibbs energy is

$$dG = \mu_A dn_A + \mu_B dn_B = -\mu_A d\xi + \mu_B d\xi = (\mu_B - \mu_A)d\xi$$

This equation can be reorganized into

$$\left(\frac{\partial G}{\partial \xi}\right)_{p,T} = \mu_B - \mu_A$$

That is,

$$\Delta_r G = \mu_B - \mu_A \quad (6A.2)$$

We see that $\Delta_r G$ can also be interpreted as the difference between the chemical potentials (the partial molar Gibbs energies) of the reactants and products *at the current composition of the reaction mixture.*

Because chemical potentials vary with composition, the slope of the plot of Gibbs energy against extent of reaction, and therefore the reaction Gibbs energy, changes as the reaction proceeds. The spontaneous direction of reaction lies in the direction of decreasing G (that is, down the slope of G plotted against ξ). Thus we see from eqn 6A.2 that the reaction A→B is spontaneous when $\mu_A > \mu_B$, whereas the reverse reaction is spontaneous when $\mu_B > \mu_A$. The slope is zero, and the reaction is at equilibrium and spontaneous in neither direction, when

$$\Delta_r G = 0 \quad \text{Condition of equilibrium} \quad (6A.3)$$

This condition occurs when $\mu_B = \mu_A$ (Fig. 6A.1). It follows that, if we can find the composition of the reaction mixture that ensures $\mu_B = \mu_A$, then we can identify the composition of the reaction mixture at equilibrium. Note that the chemical potential is now fulfilling the role its name suggests: it represents the potential for chemical change, and equilibrium is attained when these potentials are in balance.

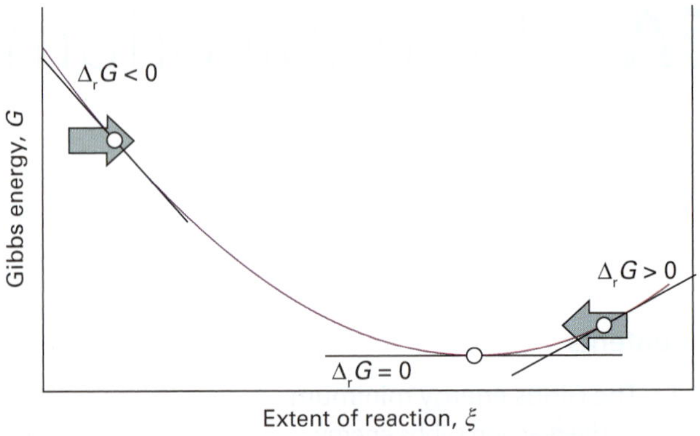

Figure 6A.1 As the reaction advances (represented by motion from left to right along the horizontal axis) the slope of the Gibbs energy changes. Equilibrium corresponds to zero slope at the foot of the valley.

(b) Exergonic and endergonic reactions

The spontaneity of a reaction at constant temperature and pressure can be expressed in terms of the reaction Gibbs energy:

- If $\Delta_r G < 0$, the forward reaction is spontaneous.
- If $\Delta_r G > 0$, the reverse reaction is spontaneous.
- If $\Delta_r G = 0$, the reaction is at equilibrium.

A reaction for which $\Delta_r G < 0$ is called **exergonic** (from the Greek words for work producing). The name signifies that, because the process is spontaneous, it can be used to drive another process, such as another reaction, or used to do non-expansion work. A simple mechanical analogy is a pair of weights joined by a string (Fig. 6A.2): the lighter of the pair of weights will be pulled up as the heavier weight falls down. Although the lighter weight has a natural tendency to move downward, its coupling to the heavier weight results in it being raised. In biological cells, the oxidation of carbohydrates act as

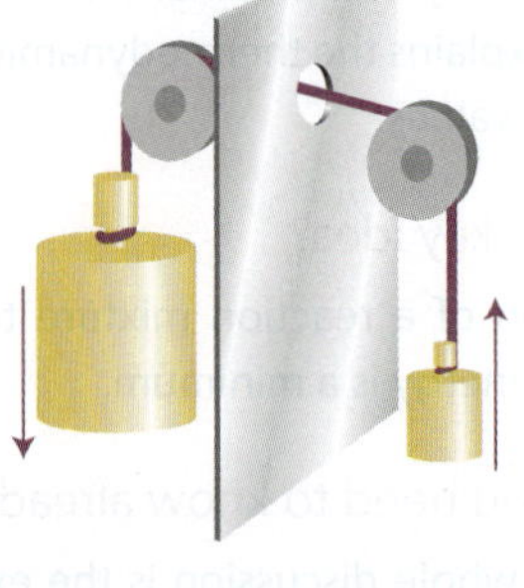

Figure 6A.2 If two weights are coupled as shown here, then the heavier weight will move the lighter weight in its non-spontaneous direction: overall, the process is still spontaneous. The weights are the analogues of two chemical reactions: a reaction with a large negative ΔG can force another reaction with a smaller ΔG to run in its non-spontaneous direction.

the heavy weight that drives other reactions forward and results in the formation of proteins from amino acids, muscle contraction, and brain activity. A reaction for which $\Delta_r G>0$ is called **endergonic** (signifying work consuming). The reaction can be made to occur only by doing work on it, such as electrolysing water to reverse its spontaneous formation reaction.

Brief illustration 6A.2 Exergonic and endergonic reactions

The standard Gibbs energy of the reaction $H_2(g)+\frac{1}{2}O_2(g)\rightarrow H_2O(l)$ at 298 K is −237 kJ mol⁻¹, so the reaction is exergonic and in a suitable device (a fuel cell, for instance) operating at constant temperature and pressure could produce 237 kJ of electrical work for each mole of H_2 molecules that react. The reverse reaction, for which $\Delta_r G^{\ominus}=+237\ \text{kJ mol}^{-1}$ is endergonic and at least 237 kJ of work must be done to achieve it.

Self-test 6A.2 Classify the formation of methane from its elements as exergonic or endergonic under standard conditions at 298 K.

Answer: Endergonic

6A.2 The description of equilibrium

With the background established, we are now ready to see how to apply thermodynamics to the description of chemical equilibrium.

(a) Perfect gas equilibria

When A and B are perfect gases we can use eqn 5A.14b ($\mu=\mu^{\ominus}+RT\ln p$, with p interpreted as $p/p^{\ominus}$) to write

$$\begin{aligned}\Delta_r G&=\mu_B-\mu_A=(\mu_B^{\ominus}+RT\ln p_B)-(\mu_A^{\ominus}+RT\ln p_A)\\&=\Delta_r G^{\ominus}+RT\ln\frac{p_B}{p_A}\end{aligned}\qquad(6A.4)$$

If we denote the ratio of partial pressures by Q, we obtain

$$\Delta_r G=\Delta_r G^{\ominus}+RT\ln Q\qquad Q=\frac{p_B}{p_A}\qquad(6A.5)$$

The ratio Q is an example of a 'reaction quotient', a quantity we define more formally shortly. It ranges from 0 when $p_B=0$ (corresponding to pure A) to infinity when $p_A=0$ (corresponding to pure B). The standard reaction Gibbs energy, $\Delta_r G^{\ominus}$ (Topic 3C), is the difference in the standard molar Gibbs energies of the reactants and products, so for our reaction

$$\Delta_r G^{\ominus}=G_m^{\ominus}(B)-G_m^{\ominus}(A)=\mu_B^{\ominus}-\mu_A^{\ominus}\qquad(6A.6)$$

Note that in the definition of $\Delta_r G^{\ominus}$, the Δ_r has its normal meaning as the difference 'products − reactants'. In Topic 3C we saw that the difference in standard molar Gibbs energies of the products and reactants is equal to the difference in their standard Gibbs energies of formation, so in practice we calculate $\Delta_r G^{\ominus}$ from

$$\Delta_r G^{\ominus}=\Delta_f G^{\ominus}(B)-\Delta_f G^{\ominus}(A)\qquad(6A.7)$$

At equilibrium, $\Delta_r G=0$. The ratio of partial pressures at equilibrium is denoted K, and eqn 6A.5 becomes

$$0=\Delta_r G^{\ominus}+RT\ln K$$

which rearranges to

$$RT\ln K=-\Delta_r G^{\ominus}\qquad K=\left(\frac{p_B}{p_A}\right)_{\text{equilibrium}}\qquad(6A.8)$$

This relation is a special case of one of the most important equations in chemical thermodynamics: it is the link between tables of thermodynamic data, such as those in the *Resource section*, and the chemically important 'equilibrium constant', K (again, a quantity we define formally shortly).

Brief illustration 6A.3 The equilibrium constant

The standard Gibbs energy of the isomerization of pentane to 2-methylbutane at 298 K, the reaction $CH_3(CH_2)_3CH_3(g)\rightarrow(CH_3)_2CHCH_2CH_3(g)$, is close to −6.7 kJ mol⁻¹ (this is an estimate based on enthalpies of formation; its actual value is not listed). Therefore, the equilibrium constant for the reaction is

$$K=e^{-(-6.7\times10^3\,\text{J mol}^{-1})/(8.3145\,\text{J K}^{-1}\,\text{mol}^{-1})\times(298\,\text{K})}=e^{2.7\ldots}=15$$

Self-test 6A.3 Suppose it is found that at equilibrium the partial pressures of A and B in the gas-phase reaction $A\rightleftharpoons B$ are equal. What is the value of $\Delta_r G^{\ominus}$?

Answer: 0

In molecular terms, the minimum in the Gibbs energy, which corresponds to $\Delta_r G=0$, stems from the Gibbs energy of mixing of the two gases. To see the role of mixing, consider the reaction $A\rightarrow B$. If only the enthalpy were important, then H and therefore G would change linearly from its value for pure reactants to its value for pure products. The slope of this straight line is a constant and equal to $\Delta_r G^{\ominus}$ at all stages of the reaction and there is no intermediate minimum in the graph (Fig. 6A.3). However, when the entropy is taken into account, there is an additional contribution to the Gibbs energy that is given by eqn 5A.16 ($\Delta_{mix}G=nRT(x_A\ln x_A+x_B\ln x_B)$). This expression makes a U-shaped contribution to the total change in Gibbs energy.

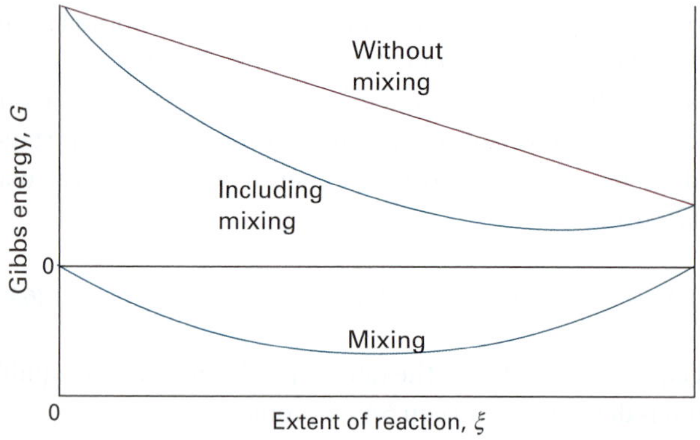

Figure 6A.3 If the mixing of reactants and products is ignored, then the Gibbs energy changes linearly from its initial value (pure reactants) to its final value (pure products) and the slope of the line is $\Delta_r G^\ominus$. However, as products are produced, there is a further contribution to the Gibbs energy arising from their mixing (lowest curve). The sum of the two contributions has a minimum. That minimum corresponds to the equilibrium composition of the system.

As can be seen from Fig. 6A.3, when it is included there is an intermediate minimum in the total Gibbs energy, and its position corresponds to the equilibrium composition of the reaction mixture.

We see from eqn 6A.8 that, when $\Delta_r G^\ominus > 0$, $K<1$. Therefore, at equilibrium the partial pressure of A exceeds that of B, which means that the reactant A is favoured in the equilibrium. When $\Delta_r G^\ominus < 0$, $K>1$, so at equilibrium the partial pressure of B exceeds that of A. Now the product B is favoured in the equilibrium.

A note on good practice A common remark is that 'a reaction is spontaneous if $\Delta_r G^\ominus < 0$'. However, whether or not a reaction is spontaneous at a particular composition depends on the value of $\Delta_r G$ at that composition, not $\Delta_r G^\ominus$. It is far better to interpret the sign of $\Delta_r G^\ominus$ as indicating whether K is greater or smaller than 1. The forward reaction is spontaneous ($\Delta_r G<0$) when $Q<K$ and the reverse reaction is spontaneous when $Q>K$.

(b) The general case of a reaction

We can now extend the argument that led to eqn 6A.8 to a general reaction. First, we note that a chemical reaction may be expressed symbolically in terms of (signed) stoichiometric numbers as

$$0=\sum_J \nu_J J$$ *Symbolic form* Chemical equation (6A.9)

where J denotes the substances and the ν_J are the corresponding stoichiometric numbers in the chemical equation. In the reaction 2 A+B→3 C+D, for instance, these numbers have the values $\nu_A=-2$, $\nu_B=-1$, $\nu_C=+3$, and $\nu_D=+1$. A stoichiometric number is positive for products and negative for reactants. Then we define the extent of reaction ξ so that, if it changes by $\Delta\xi$, then the change in the amount of any species J is $\nu_J\Delta\xi$.

With these points in mind and with the reaction Gibbs energy, $\Delta_r G$, defined in the same way as before (eqn 6A.1) we show in the following *Justification* that the Gibbs energy of reaction can always be written

$$\Delta_r G=\Delta_r G^\ominus + RT\ln Q$$ Reaction Gibbs energy at an arbitrary stage (6A.10)

with the standard reaction Gibbs energy calculated from

$$\Delta_r G^\ominus = \sum_{\text{Products}} \nu\Delta_f G^\ominus - \sum_{\text{Reactants}} \nu\Delta_f G^\ominus$$ *Practical implementation* Reaction Gibbs energy (6A.11a)

where the ν are the (positive) stoichiometric coefficients. More formally,

$$\Delta_r G^\ominus = \sum_J \nu_J \Delta_f G^\ominus(J)$$ *Formal expression* Reaction Gibbs energy (6A.11b)

where the ν_J are the (signed) stoichiometric numbers. The reaction quotient, Q, has the form

$$Q=\frac{\text{activities of products}}{\text{activities of reactants}}$$ *General form* Reaction quotient (6A.12a)

with each species raised to the power given by its stoichiometric coefficient. More formally, to write the general expression for Q we introduce the symbol Π to denote the product of what follows it (just as Σ denotes the sum), and define Q as

$$Q=\prod_J a_J^{\nu_J}$$ *Definition* Reaction quotient (6A.12b)

Because reactants have negative stoichiometric numbers, they automatically appear as the denominator when the product is written out explicitly. Recall from Table 5E.1 that, for pure solids and liquids, the activity is 1, so such substances make no contribution to Q even though they may appear in the chemical equation.

Brief illustration 6A.4 The reaction quotient

Consider the reaction 2 A+3 B→C+2 D, in which case $\nu_A=-2$, $\nu_B=-3$, $\nu_C=+1$, and $\nu_D=+2$. The reaction quotient is then

$$Q=a_A^{-2}a_B^{-3}a_C a_D^2=\frac{a_C a_D^2}{a_A^2 a_B^3}$$

Self-test 6A.4 Write the reaction quotient for A+2 B→3 C.

Answer: $Q=a_C^3/a_A a_B^2$

Justification 6A.1 The dependence of the reaction Gibbs energy on the reaction quotient

Consider a reaction with stoichiometric numbers ν_J. When the reaction advances by $d\xi$, the amounts of reactants and products change by $dn_J = \nu_J d\xi$. The resulting infinitesimal change in the Gibbs energy at constant temperature and pressure is

$$dG = \sum_J \mu_J dn_J = \sum_J \mu_J \nu_J d\xi = \left(\sum_J \mu_J \nu_J\right) d\xi$$

It follows that

$$\Delta_r G = \left(\frac{\partial G}{\partial \xi}\right)_{p,T} = \sum_J \nu_J \mu_J$$

To make progress, we note that the chemical potential of a species J is related to its activity by eqn 5E.9 ($\mu_J = \mu_J^{\ominus} + RT \ln a_J$). When this expression is substituted into eqn 6A.11 we obtain

$$\Delta_r G = \overbrace{\sum_J \nu_J \mu_J^{\ominus}}^{\Delta_r G^{\ominus}} + RT \sum_J \nu_J \ln a_J$$
$$= \Delta_r G^{\ominus} + RT \sum_J \nu_J \ln a_J = \Delta_r G^{\ominus} + RT \ln \overbrace{\prod_J a_J^{\nu_J}}^{Q}$$
$$= \Delta_r G^{\ominus} + RT \ln Q$$

In the second line we use first $a \ln x = \ln x^a$ and then $\ln x + \ln y + \ldots = \ln xy\ldots$, so

$$\sum_i \ln x_i = \ln\left(\prod_i x_i\right)$$

Now we conclude the argument, starting from eqn 6A.10. At equilibrium, the slope of G is zero: $\Delta_r G = 0$. The activities then have their equilibrium values and we can write

$$K = \left(\prod_J a_J^{\nu_J}\right)_{\text{equilibrium}} \quad \textit{Definition} \quad \text{Equilibrium constant} \quad (6A.13)$$

This expression has the same form as Q but is evaluated using equilibrium activities. From now on, we shall not write the 'equilibrium' subscript explicitly, and will rely on the context to make it clear that for K we use equilibrium values and for Q we use the values at the specified stage of the reaction. An equilibrium constant K expressed in terms of activities (or fugacities) is called a **thermodynamic equilibrium constant**. Note that, because activities are dimensionless numbers, the thermodynamic equilibrium constant is also dimensionless. In elementary applications, the activities that occur in eqn 6A.13 are often replaced as follows:

State	Measure	Approximation for a_J	Definition
Solute	molality	$b_J/b_J^{\ominus}$	$b^{\ominus} = 1\ \text{mol kg}^{-1}$
	molar concentration	$[J]/c^{\ominus}$	$c^{\ominus} = 1\ \text{mol dm}^{-3}$
Gas phase	partial pressure	$p_J/p^{\ominus}$	$p^{\ominus} = 1\ \text{bar}$

In such cases, the resulting expressions are only approximations. The approximation is particularly severe for electrolyte solutions, for in them activity coefficients differ from 1 even in very dilute solutions (Topic 5F).

Brief illustration 6A.5 The equilibrium constant

The equilibrium constant for the heterogeneous equilibrium $CaCO_3(s) \rightleftharpoons CaO(s) + CO_2(g)$ is

$$K = a_{CaCO_3(s)}^{-1} a_{CaO(s)} a_{CO_2(g)} = \frac{\overbrace{a_{CaO(s)}}^{1} a_{CO_2(g)}}{\underbrace{a_{CaCO_3(s)}}_{1}} = a_{CO_2(g)}$$

Provided the carbon dioxide can be treated as a perfect gas, we can go on to write

$$K = p_{CO_2}/p^{\ominus}$$

and conclude that in this case the equilibrium constant is the numerical value of the decomposition vapour pressure of calcium carbonate.

Self-test 6A.5 Write the equilibrium constant for the reaction $N_2(g) + 3\,H_2(g) \rightleftharpoons 2\,NH_3(g)$, with the gases treated as perfect.

Answer: $K = a_{NH_3}^2/a_{N_2}a_{H_2}^3 = p_{NH_3}^2 p^{\ominus 2}/p_{N_2}p_{H_2}^3$

At this point we set $\Delta_r G = 0$ in eqn 6A.10 and replace Q by K. We immediately obtain

$$\Delta_r G^{\ominus} = -RT \ln K \quad \text{Thermodynamic equilibrium constant} \quad (6A.14)$$

This is an exact and highly important thermodynamic relation, for it enables us to calculate the equilibrium constant of any reaction from tables of thermodynamic data, and hence to predict the equilibrium composition of the reaction mixture. In Topic 15F we see that the right-hand side of eqn 6A.14 may be expressed in terms of spectroscopic data for gas-phase species; so this expression also provides a link between spectroscopy and equilibrium composition.

Example 6A.1 Calculating an equilibrium constant

Calculate the equilibrium constant for the ammonia synthesis reaction, $N_2(g) + 3\,H_2(g) \rightleftharpoons 2\,NH_3(g)$, at 298 K and show how K is related to the partial pressures of the species at equilibrium when the overall pressure is low enough for the gases to be treated as perfect.

Method Calculate the standard reaction Gibbs energy from eqn 6A.10 and convert it to the value of the equilibrium constant by using eqn 6A.14. The expression for the equilibrium constant is obtained from eqn 6A.13, and because the gases are taken to be perfect, we replace each activity by the ratio $p_J/p^{\ominus}$, where p_J is the partial pressure of species J.

Answer The standard Gibbs energy of the reaction is

$$\begin{aligned}\Delta_r G^{\ominus} &= 2\Delta_f G^{\ominus}(NH_3,g) - \{\Delta_f G^{\ominus}(N_2,g) + 3\Delta_f G^{\ominus}(H_2,g)\}\\ &= 2\Delta_f G^{\ominus}(NH_3,g) = 2\times(-16.45\,\text{kJ mol}^{-1})\end{aligned}$$

Then,

$$\ln K = -\frac{2\times(-1.645\times10^4\,\text{J mol}^{-1})}{(8.3145\,\text{J K}^{-1}\,\text{mol}^{-1})\times(298\,\text{K})} = \frac{2\times1.645\times10^4}{8.3145\times298} = 13.2\ldots$$

Hence, $K = 5.8\times10^5$. This result is thermodynamically exact. The thermodynamic equilibrium constant for the reaction is

$$K = \frac{a_{NH_3}^2}{a_{N_2}a_{H_2}^3}$$

and this ratio has the value we have just calculated. At low overall pressures, the activities can be replaced by the ratios $p_J/p^{\ominus}$ and an approximate form of the equilibrium constant is

$$K = \frac{(p_{NH_3}/p^{\ominus})^2}{(p_{N_2}/p^{\ominus})(p_{H_2}/p^{\ominus})^3} = \frac{p_{NH_3}^2 p^{\ominus 2}}{p_{N_2}p_{H_2}^3}$$

Self-test 6A.6 Evaluate the equilibrium constant for $N_2O_4(g) \rightleftharpoons 2\,NO_2(g)$ at 298 K.

Answer: $K = 0.15$

Example 6A.2 Estimating the degree of dissociation at equilibrium

The *degree of dissociation* (or *extent of dissociation*, α) is defined as the fraction of reactant that has decomposed; if the initial amount of reactant is n and the amount at equilibrium is n_{eq}, then $\alpha = (n - n_{eq})/n$. The standard reaction Gibbs energy for the decomposition $H_2O(g) \rightarrow H_2(g) + \frac{1}{2}O_2(g)$ is +118.08 kJ mol^{-1} at 2300 K. What is the degree of dissociation of H_2O at 2300 K and 1.00 bar?

Method The equilibrium constant is obtained from the standard Gibbs energy of reaction by using eqn 6A.11, so the task is to relate the degree of dissociation, α, to K and then to find its numerical value. Proceed by expressing the equilibrium compositions in terms of α, and solve for α in terms of K. Because the standard reaction Gibbs energy is large and positive, we can anticipate that K will be small, and hence that $\alpha \ll 1$, which opens the way to making approximations to obtain its numerical value.

Answer The equilibrium constant is obtained from eqn 6A.14 in the form

$$\ln K = -\frac{\Delta_r G^{\ominus}}{RT} = -\frac{1.1808\times10^5\,\text{J mol}^{-1}}{(8.3145\,\text{J K}^{-1}\,\text{mol}^{-1})\times(2300\,\text{K})}$$

$$= -\frac{1.1808\times10^5}{8.3145\times2300} = -6.17\ldots$$

It follows that $K = 2.08\times10^{-3}$. The equilibrium composition can be expressed in terms of α by drawing up the following table:

	H_2O	H_2	$+\frac{1}{2}O_2$	
Initial amount	n	0	0	
Change to reach equilibrium	$-\alpha n$	$+\alpha n$	$+\frac{1}{2}\alpha n$	
Amount at equilibrium	$(1-\alpha)n$	αn	$\frac{1}{2}\alpha n$	Total: $(1+\frac{1}{2}\alpha)n$
Mole fraction, x_J	$\frac{1-\alpha}{1+\frac{1}{2}\alpha}$	$\frac{\alpha}{1+\frac{1}{2}}$	$\frac{\frac{1}{2}\alpha}{1+\frac{1}{2}\alpha}$	
Partial pressure, p_J	$\frac{(1-\alpha)p}{1+\frac{1}{2}\alpha}$	$\frac{\alpha p}{1+\frac{1}{2}}$	$\frac{\frac{1}{2}\alpha p}{1+\frac{1}{2}\alpha}$	

where, for the entries in the last row, we have used $p_J = x_J p$ (eqn 1A.8). The equilibrium constant is therefore

$$K = \frac{p_{H_2}p_{O_2}^{1/2}}{p_{H_2O}} = \frac{\alpha^{3/2}p^{1/2}}{(1-\alpha)(2+\alpha)^{1/2}}$$

In this expression, we have written p in place of $p/p^{\ominus}$, to simplify its appearance. Now make the approximation that $\alpha \ll 1$, and hence obtain

$$K \approx \frac{\alpha^{3/2}p^{1/2}}{2^{1/2}}$$

Under the stated condition, $p = 1.00$ bar (that is, $p/p^{\ominus} = 1.00$), so $\alpha \approx (2^{1/2}K)^{2/3} = 0.0205$. That is, about 2 per cent of the water has decomposed.

A note on good practice Always check that the approximation is consistent with the final answer. In this case $\alpha \ll 1$, in accord with the original assumption.

Self-test 6A.7 Given that the standard Gibbs energy of reaction at 2000 K is +135.2 kJ mol^{-1} for the same reaction, suppose that steam at 200 kPa is passed through a furnace tube at that temperature. Calculate the mole fraction of O_2 present in the output gas stream.

Answer: 0.00221

(c) The relation between equilibrium constants

Equilibrium constants in terms of activities are exact, but it is often necessary to relate them to concentrations. Formally, we need to know the activity coefficients, and then to use $a_J = \gamma_J x_J$, $a_J = \gamma_J b_J / b^{\ominus}$, or $a_J = [J]/c^{\ominus}$, where x_J is a mole fraction, b_J is a molality, and [J] is a molar concentration. For example, if we were interested in the composition in terms of molality for an equilibrium of the form $A + B \rightleftharpoons C + D$, where all four species are solutes, we would write

$$K = \frac{a_C a_D}{a_A a_B} = \frac{\gamma_C \gamma_D}{\gamma_A \gamma_B} \times \frac{b_C b_D}{b_A b_B} = K_\gamma K_b \qquad (6A.15)$$

The activity coefficients must be evaluated at the equilibrium composition of the mixture (for instance, by using one of the Debye–Hückel expressions, Topic 5F), which may involve a complicated calculation, because the activity coefficients are known only if the equilibrium composition is already known. In elementary applications, and to begin the iterative calculation of the concentrations in a real example, the assumption is often made that the activity coefficients are all so close to unity that $K_\gamma = 1$. Then we obtain the result widely used in elementary chemistry that $K \approx K_b$, and equilibria are discussed in terms of molalities (or molar concentrations) themselves.

A special case arises when we need to express the equilibrium constant of a gas-phase reaction in terms of molar concentrations instead of the partial pressures that appear in the thermodynamic equilibrium constant. Provided we can treat the gases as perfect, the p_J that appear in K can be replaced by $[J]RT$, and

$$K = \prod_J a_J^{\nu_J} = \prod_J \left(\frac{p_J}{p^{\ominus}}\right)^{\nu_J} = \prod_J [J]^{\nu_J} \left(\frac{RT}{p^{\ominus}}\right)^{\nu_J} = \prod_J [J]^{\nu_J} \times \prod_J \left(\frac{RT}{p^{\ominus}}\right)^{\nu_J}$$

(Products can always be factorized like that: *abcdef* is the same as *abc*×*def*.) The (dimensionless) equilibrium constant K_c is defined as

$$K_c = \prod_J \left(\frac{[J]}{c^{\ominus}}\right)^{\nu_J} \qquad \textit{Definition} \quad K_c \text{ for gas-phase reactions} \qquad (6A.16)$$

It follows that

$$K = K_c \times \prod_J \left(\frac{c^{\ominus} RT}{p^{\ominus}}\right)^{\nu_J} \qquad (6A.17a)$$

If now we write $\Delta\nu = \sum_J \nu_J$, which is easier to think of as ν(products) − ν(reactants), then the relation between K and K_c for a gas-phase reaction is

$$K = K_c \times \left(\frac{c^{\ominus} RT}{p^{\ominus}}\right)^{\Delta\nu} \qquad \text{Relation between } K \text{ and } K_c \text{ for gas-phase reactions} \qquad (6A.17b)$$

The term in parentheses works out as $T/(12.03\,\text{K})$.

Brief illustration 6A.6 The relation between equilibrium constants

For the reaction $N_2(g) + 3\,H_2(g) \rightarrow 2\,NH_3(g)$, $\Delta\nu = 2 - 4 = -2$, so

$$K = K_c \times \left(\frac{T}{12.03\,\text{K}}\right)^{-2} = K_c \times \left(\frac{12.03\,\text{K}}{T}\right)^{2}$$

At 298.15 K the relation is

$$K = K_c \times \left(\frac{12.03\,\text{K}}{298.15\,\text{K}}\right)^{2} = \frac{K_c}{614.2}$$

so $K_c = 614.2K$. Note that both K and K_c are dimensionless.

Self-test 6A.8 Find the relation between K and K_c for the equilibrium $H_2(g) + \frac{1}{2}O_2(g) \rightarrow H_2O(l)$ at 298 K.

Answer: $K_c = 123K$

(d) Molecular interpretation of the equilibrium constant

Deeper insight into the origin and significance of the equilibrium constant can be obtained by considering the Boltzmann distribution of molecules over the available states of a system composed of reactants and products (*Foundations* B). When atoms can exchange partners, as in a reaction, the available states of the system include arrangements in which the atoms are present in the form of reactants and in the form of products: these arrangements have their characteristic sets of energy levels, but the Boltzmann distribution does not distinguish between their identities, only their energies. The atoms distribute themselves over both sets of energy levels in accord with the Boltzmann distribution (Fig. 6A.4). At a given temperature, there will be a specific distribution of populations, and hence a specific composition of the reaction mixture.

It can be appreciated from the illustration that, if the reactants and products both have similar arrays of molecular energy levels, then the dominant species in a reaction mixture at equilibrium will be the species with the lower set of energy levels. However, the fact that the Gibbs energy occurs in the expression is a signal that entropy plays a role as well as energy. Its role can be appreciated by referring to Fig. 6A.4. In Fig. 6A.4b we see that, although the B energy levels lie higher than the A energy levels, in this instance they are much more closely spaced. As a result, their total population may be considerable and B could even dominate in the reaction mixture at equilibrium. Closely spaced energy levels correlate with a high

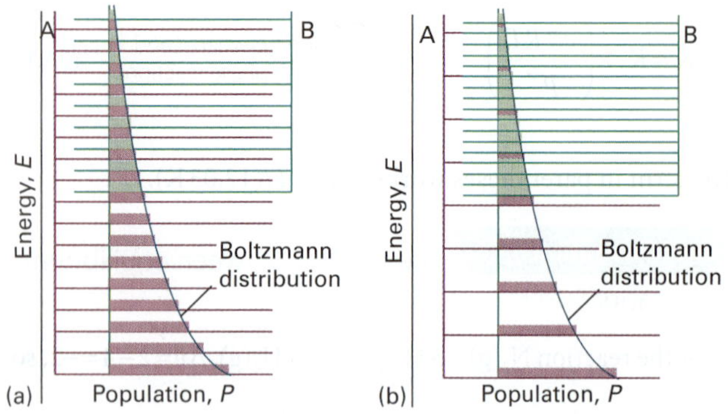

Figure 6A.4 The Boltzmann distribution of populations over the energy levels of two species A and B with similar densities of energy levels. The reaction A→B is endothermic in this example. (a) The bulk of the population is associated with the species A, so that species is dominant at equilibrium. (b) Even though the reaction A→B is endothermic, the density of energy levels in B is so much greater than that in A that the population associated with B is greater than that associated with A, so B is dominant at equilibrium.

entropy (Topic 15E), so in this case we see that entropy effects dominate adverse energy effects. This competition is mirrored in eqn 6A.14, as can be seen most clearly by using $\Delta_r G^\ominus = \Delta_r H^\ominus - T\Delta_r S^\ominus$ and writing it in the form

$$K = e^{-\Delta_r H^\ominus/RT} e^{\Delta_r S^\ominus/R} \qquad (6A.18)$$

Note that a positive reaction enthalpy results in a lowering of the equilibrium constant (that is, an endothermic reaction can be expected to have an equilibrium composition that favours the reactants). However, if there is positive reaction entropy, then the equilibrium composition may favour products, despite the endothermic character of the reaction.

Brief illustration 6A.7 Contributions to K

In *Example* 6A.1 it is established that $\Delta_r G^\ominus = -33.0\,\mathrm{kJ\,mol^{-1}}$ for the reaction $N_2(g) + 3\,H_2(g) \rightleftharpoons 2\,NH_3(g)$ at 298 K. From the tables of data in the *Resource section*, we can find that $\Delta_r H^\ominus = -92.2\,\mathrm{kJ\,mol^{-1}}$ and $\Delta_r S^\ominus = -198.8\,\mathrm{J\,K^{-1}\,mol^{-1}}$. The contributions to K are therefore

$$\begin{aligned} K &= e^{-(-9.22\times 10^4\,\mathrm{J\,mol^{-1}})/(8.3145\,\mathrm{J\,K^{-1}\,mol^{-1}})\times(298\,\mathrm{K})} \\ &\quad \times e^{(-198.8\,\mathrm{J\,K^{-1}\,mol^{-1}})/(8.3145\,\mathrm{J\,K^{-1}\,mol^{-1}})} \\ &= e^{37.2\ldots} \times e^{-23.9\ldots} \end{aligned}$$

We see that the exothermic character of the reaction encourages the formation of products (it results in a large increase in entropy of the surroundings) but the decrease in entropy of the system as H atoms are pinned to N atoms opposes their formation.

Self-test 6A.9 Analyse the equilibrium $N_2O_4(g) \rightleftharpoons 2\,NO_2(g)$ similarly.

Answer: $K = e^{-26.7\ldots} \times e^{21.1\ldots}$; enthalpy opposes, entropy encourages

Checklist of concepts

☐ 1. The **reaction Gibbs energy** is the slope of the plot of Gibbs energy against extent of reaction.

☐ 2. Reactions are either **exergonic** or **endergonic**.

☐ 3. The reaction Gibbs energy depends logarithmically on the **reaction quotient**.

☐ 4. When the reaction Gibbs energy is zero the reaction quotient has a value called the **equilibrium constant**.

☐ 5. Under ideal conditions, the thermodynamic equilibrium constant may be approximated by expressing it in terms of concentrations and partial pressures.

Checklist of equations

Property	Equation	Comment	Equation number
Reaction Gibbs energy	$\Delta_r G = (\partial G/\partial \xi)_{p,T}$	Definition	6A.1
Reaction Gibbs energy	$\Delta_r G = \Delta_r G^\ominus + RT \ln Q$		6A.10
Standard reaction Gibbs energy	$\Delta_r G^\ominus = \sum_{\text{Products}} \nu\Delta_f G^\ominus - \sum_{\text{Reactants}} \nu\Delta_f G^\ominus = \sum_J \nu_J \Delta_f G^\ominus(J)$	ν are positive; ν_J are signed	6A.11

Property	Equation	Comment	Equation number
Reaction quotient	$Q=\prod_{J} a_J^{\nu_J}$	Definition; evaluated at arbitrary stage of reaction	6A.12
Thermodynamic equilibrium constant	$K=\left(\prod_{J} a_J^{\nu_J}\right)_{\text{equilibrium}}$	Definition	6A.13
Equilibrium constant	$\Delta_r G^{\ominus}=-RT\ln K$		6A.14
Relation between K and K_c	$K=K_c(c^{\ominus}RT/p^{\ominus})^{\Delta\nu}$	Gas-phase reactions; perfect gases	6A.17b

6B The response of equilibria to the conditions

Contents

➤ Why do you need to know this material?

Chemists, and chemical engineers designing a chemical plant, need to know how an equilibrium will respond to changes in the conditions, such as a change in pressure or temperature. The variation with temperature also provides a way to determine the enthalpy and entropy of a reaction.

➤ What is the key idea?

A system at equilibrium, when subjected to a disturbance, responds in a way that tends to minimize the effect of the disturbance.

➤ What do you need to know already?

This Topic builds on the relation between the equilibrium constant and the standard Gibbs energy of reaction (Topic 6A). To express the temperature dependence of K it draws on the Gibbs–Helmholtz equation (Topic 3D).

The equilibrium constant for a reaction is not affected by the presence of a catalyst or an enzyme (a biological catalyst). As explained in detail in Topics 20H and 22C, catalysts increase the rate at which equilibrium is attained but do not affect its position. However, it is important to note that in industry reactions rarely reach equilibrium, partly on account of the rates at which reactants mix. The equilibrium constant is also independent of pressure, but as we shall see, that does not necessarily mean that the composition at equilibrium is independent of pressure. The equilibrium constant does depend on the temperature in a manner that can be predicted from the standard reaction enthalpy.

6B.1 The response to pressure

The equilibrium constant depends on the value of $\Delta_r G^\ominus$, which is defined at a single, standard pressure. The value of $\Delta_r G^\ominus$, and hence of K, is therefore independent of the pressure at which the equilibrium is actually established. In other words, at a given temperature K is a constant.

The conclusion that K is independent of pressure does not necessarily mean that the equilibrium composition is independent of the pressure, and the effect depends on how the pressure is applied.

The pressure within a reaction vessel can be increased by injecting an inert gas into it. However, so long as the gases are perfect, this addition of gas leaves all the partial pressures of the reacting gases unchanged: the partial pressures of a perfect gas is the pressure it would exert if it were alone in the container, so the presence of another gas has no effect. It follows that pressurization by the addition of an inert gas has no effect on the equilibrium composition of the system (provided the gases are perfect).

Alternatively, the pressure of the system may be increased by confining the gases to a smaller volume (that is, by compression). Now the individual partial pressures are changed but their ratio (as it appears in the equilibrium constant) remains the same. Consider, for instance, the perfect gas equilibrium $A \rightleftharpoons 2B$, for which the equilibrium constant is

$$K = \frac{p_B^2}{p_A p^\ominus}$$

The right-hand side of this expression remains constant only if an increase in p_A cancels an increase in the *square* of p_B. This relatively steep increase of p_A compared to p_B will occur if the equilibrium composition shifts in favour of A at the expense of B. Then the number of A molecules will increase as the volume of the container is decreased and its partial pressure will rise

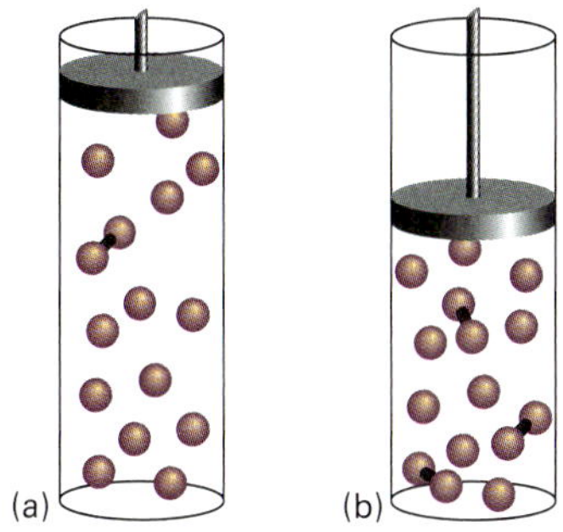

Figure 6B.1 When a reaction at equilibrium is compressed (from *a* to *b*), the reaction responds by reducing the number of molecules in the gas phase (in this case by producing the dimers represented by the linked spheres).

more rapidly than can be ascribed to a simple change in volume alone (Fig. 6B.1).

The increase in the number of A molecules and the corresponding decrease in the number of B molecules in the equilibrium $A \rightleftharpoons 2\,B$ is a special case of a principle proposed by the French chemist Henri Le Chatelier, which states that:

> A system at equilibrium, when subjected to a disturbance, responds in a way that tends to minimize the effect of the disturbance.

Le Chatelier's principle

The principle implies that, if a system at equilibrium is compressed, then the reaction will adjust so as to minimize the increase in pressure. This it can do by reducing the number of particles in the gas phase, which implies a shift $A \leftarrow 2\,B$.

To treat the effect of compression quantitatively, we suppose that there is an amount n of A present initially (and no B). At equilibrium the amount of A is $(1-\alpha)n$ and the amount of B is $2\alpha n$, where α is the degree of dissociation of A into 2B. It follows that the mole fractions present at equilibrium are

$$x_A = \frac{(1-\alpha)n}{(1-\alpha)n + 2\alpha n} = \frac{1-\alpha}{1+\alpha} \qquad x_B = \frac{2\alpha}{1+\alpha}$$

The equilibrium constant for the reaction is

$$K = \frac{p_B^2}{p_A p^{\ominus}} = \frac{x_B^2 p^2}{x_A p p^{\ominus}} = \frac{4\alpha^2 (p/p^{\ominus})}{1-\alpha^2}$$

which rearranges to

$$\alpha = \left(\frac{1}{1+4p/Kp^{\ominus}}\right)^{1/2} \qquad (6B.1)$$

This formula shows that, even though K is independent of pressure, the amounts of A and B do depend on pressure (Fig. 6B.2). It also shows that as p is increased, α decreases, in accord with Le Chatelier's principle.

Brief illustration 6B.1 Le Chatelier's principle

To predict the effect of an increase in pressure on the composition of the ammonia synthesis at equilibrium, *Example* 6A.1, we note that the number of gas molecules decreases (from 4 to 2). So, Le Chatelier's principle predicts that an increase in pressure will favour the product. The equilibrium constant is

$$K = \frac{p_{NH_3}^2 p^{\ominus}}{p_{N_2} p_{H_2}^3} = \frac{x_{NH_3}^2 p^2 p^{\ominus 2}}{x_{N_2} x_{H_2}^3 p^4} = \frac{x_{NH_3}^2 p^{\ominus 2}}{x_{N_2} x_{H_2}^3 p^2} = K_x \times \frac{p^{\ominus 2}}{p^2}$$

where K_x is the part of the equilibrium constant expression that contains the equilibrium mole fractions of reactants and products (note that, unlike K itself, K_x is not an equilibrium constant). Therefore, doubling the pressure must increase K_x by a factor of 4 to preserve the value of K.

Self-test 6B.1 Predict the effect of a tenfold pressure increase on the equilibrium composition of the reaction $3\,N_2(g) + H_2(g) \rightleftharpoons 2\,N_3H(g)$.

Answer: 100-fold increase in K_x

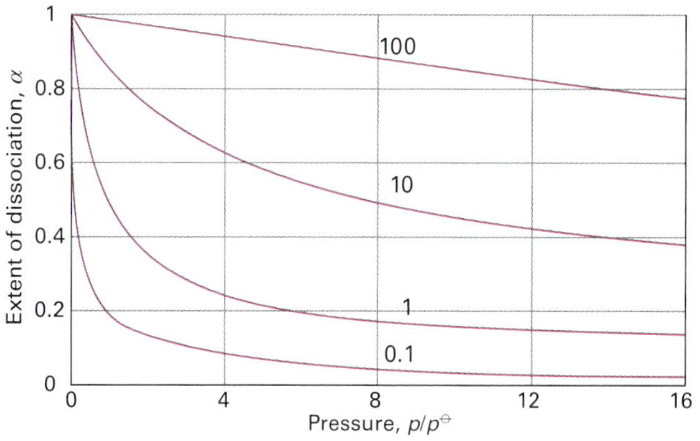

Figure 6B.2 The pressure dependence of the degree of dissociation, α, at equilibrium for an $A(g) \rightleftharpoons 2\,B(g)$ reaction for different values of the equilibrium constant K. The value $\alpha=0$ corresponds to pure A; $\alpha=1$ corresponds to pure B

6B.2 The response to temperature

Le Chatelier's principle predicts that a system at equilibrium will tend to shift in the endothermic direction if the temperature is raised, for then energy is absorbed as heat and the rise in temperature is opposed. Conversely, an equilibrium can be expected to shift in the exothermic direction if the temperature is lowered, for then energy is released and the reduction in temperature is opposed. These conclusions can be summarized as follows:

> Exothermic reactions: increased temperature favours the reactants.

Endothermic reactions: increased temperature favours the products.

We shall now justify these remarks thermodynamically and see how to express the changes quantitatively.

(a) The van 't Hoff equation

The **van 't Hoff equation**, which is derived in the following *Justification*, is an expression for the slope of a plot of the equilibrium constant (specifically, ln *K*) as a function of temperature. It may be expressed in either of two ways:

$$\frac{\mathrm{d}\ln K}{\mathrm{d}T}=\frac{\Delta_r H^\ominus}{RT^2} \qquad \text{van 't Hoff equation} \qquad (6B.2a)$$

$$\frac{\mathrm{d}\ln K}{\mathrm{d}(1/T)}=-\frac{\Delta_r H^\ominus}{R} \qquad \textit{Alternative version} \quad \text{van 't Hoff equation} \qquad (6B.2b)$$

Justification 6B.1 The van 't Hoff equation

From eqn 6A.14, we know that

$$\ln K=-\frac{\Delta_r G^\ominus}{RT}$$

Differentiation of ln *K* with respect to temperature then gives

$$\frac{\mathrm{d}\ln K}{\mathrm{d}T}=-\frac{1}{R}\frac{\mathrm{d}(\Delta_r G^\ominus/T)}{\mathrm{d}T}$$

The differentials are complete (that is, they are not partial derivatives) because *K* and $\Delta_r G^\ominus$ depend only on temperature, not on pressure. To develop this equation we use the Gibbs–Helmholtz equation (eqn 3D.10, $\mathrm{d}(\Delta G/T)=-\Delta H/T^2$) in the form

$$\frac{\mathrm{d}(\Delta_r G^\ominus/T)}{\mathrm{d}T}=-\frac{\Delta_r H^\ominus}{R}$$

where $\Delta_r H^\ominus$ is the standard reaction enthalpy at the temperature *T*. Combining the two equations gives the van 't Hoff equation, eqn 6B.2a. The second form of the equation is obtained by noting that

$$\frac{\mathrm{d}(1/T)}{\mathrm{d}T}=-\frac{1}{T^2}, \quad \text{so } \mathrm{d}T=-T^2\mathrm{d}(1/T)$$

It follows that eqn 6B.2a can be rewritten as

$$-\frac{\mathrm{d}\ln K}{T^2\mathrm{d}(1/T)}=\frac{\Delta_r H^\ominus}{RT^2}$$

which simplifies into eqn 6B.2b.

Equation 6B.2a shows that $\mathrm{d}\ln K/\mathrm{d}T<0$ (and therefore that $\mathrm{d}K/\mathrm{d}T<0$) for a reaction that is exothermic under standard conditions ($\Delta_r H^\ominus<0$). A negative slope means that ln *K*, and therefore *K* itself, decreases as the temperature rises. Therefore, as asserted above, in the case of an exothermic reaction the equilibrium shifts away from products. The opposite occurs in the case of endothermic reactions.

Insight into the thermodynamic basis of this behaviour comes from the expression $\Delta_r G^\ominus=\Delta_r H^\ominus - T\Delta_r S^\ominus$ written in the form $-\Delta_r G^\ominus/T=-\Delta_r H^\ominus/T+\Delta_r S^\ominus$. When the reaction is exothermic, $-\Delta_r H^\ominus/T$ corresponds to a positive change of entropy of the surroundings and favours the formation of products. When the temperature is raised, $-\Delta_r H^\ominus/T$ decreases and the increasing entropy of the surroundings has a less important role. As a result, the equilibrium lies less to the right. When the reaction is endothermic, the principal factor is the increasing entropy of the reaction system. The importance of the unfavourable change of entropy of the surroundings is reduced if the temperature is raised (because then $\Delta_r H^\ominus/T$ is smaller), and the reaction is able to shift towards products.

These remarks have a molecular basis that stems from the Boltzmann distribution of molecules over the available energy levels (*Foundations* B, and in more detail in Topic 15F). The typical arrangement of energy levels for an endothermic reaction is shown in Fig. 6B.3a. When the temperature is increased, the Boltzmann distribution adjusts and the populations change as shown. The change corresponds to an increased population of the higher energy states at the expense of the population of the lower energy states. We see that the states that arise from the B molecules become more populated at the expense of the A molecules. Therefore, the total population of B states increases, and B becomes more abundant in the equilibrium mixture. Conversely, if the reaction is exothermic (Fig. 6B.3b),

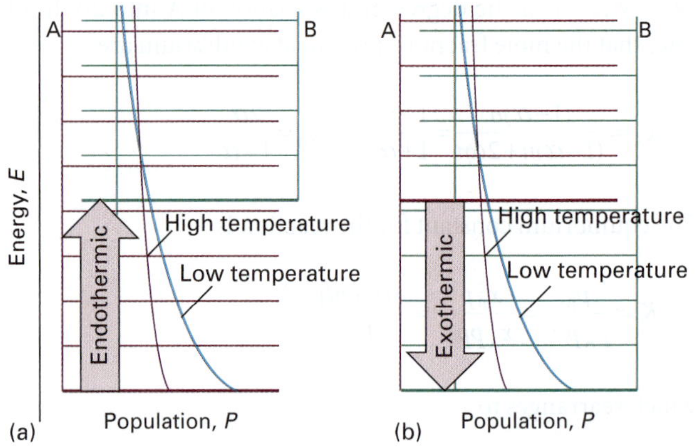

Figure 6B.3 The effect of temperature on a chemical equilibrium can be interpreted in terms of the change in the Boltzmann distribution with temperature and the effect of that change in the population of the species. (a) In an endothermic reaction, the population of B increases at the expense of A as the temperature is raised. (b) In an exothermic reaction, the opposite happens.

then an increase in temperature increases the population of the A states (which start at higher energy) at the expense of the B states, so the reactants become more abundant.

The temperature dependence of the equilibrium constant provides a non-calorimetric method of determining $\Delta_r H^{\ominus}$. A drawback is that the reaction enthalpy is actually temperature-dependent, so the plot is not expected to be perfectly linear. However, the temperature dependence is weak in many cases, so the plot is reasonably straight. In practice, the method is not very accurate, but it is often the only method available.

Example 6B.1 Measuring a reaction enthalpy

The data below show the temperature variation of the equilibrium constant of the reaction $Ag_2CO_3(s) \rightleftharpoons Ag_2O(s) + CO_2(g)$. Calculate the standard reaction enthalpy of the decomposition.

T/K	350	400	450	500
K	3.98×10^{-4}	1.41×10^{-2}	1.86×10^{-1}	1.48

Method It follows from eqn 6B.2b that, provided the reaction enthalpy can be assumed to be independent of temperature, a plot of $-\ln K$ against $1/T$ should be a straight line of slope $\Delta_r H^{\ominus}/R$.

Answer We draw up the following table:

T/K	350	400	450	500
$(10^3\,\text{K})/T$	2.86	2.50	2.22	2.00
$-\ln K$	6.83	4.26	1.68	−0.39

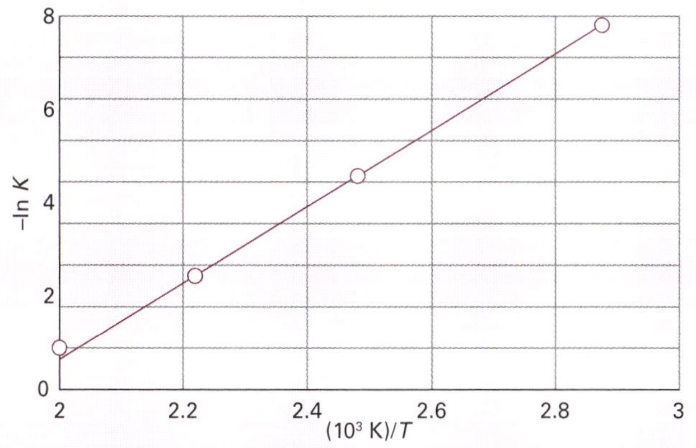

Figure 6B.4 When $-\ln K$ is plotted against $1/T$, a straight line is expected with slope equal to $\Delta_r H^{\ominus}/R$ if the standard reaction enthalpy does not vary appreciably with temperature. This is a non-calorimetric method for the measurement of reaction enthalpies.

These points are plotted in Fig. 6B.4. The slope of the graph is $+9.6\times10^3$, so

$$\Delta_r H^{\ominus} = (+9.6\times10^3\,\text{K})\times R = +80\,\text{kJ mol}^{-1}$$

Self-test 6B.2 The equilibrium constant of the reaction $2\,SO_2(g) + O_2(g) \rightleftharpoons 2\,SO_3(g)$ is 4.0×10^{24} at 300 K, 2.5×10^{10} at 500 K, and 3.0×10^4 at 700 K. Estimate the reaction enthalpy at 500 K.

Answer: −200 kJ mol⁻¹

(b) The value of *K* at different temperatures

To find the value of the equilibrium constant at a temperature T_2 in terms of its value K_1 at another temperature T_1, we integrate eqn 6B.2b between these two temperatures:

$$\ln K_2 - \ln K_1 = -\frac{1}{R}\int_{1/T_1}^{1/T_2} \Delta_r H^{\ominus}\,\mathrm{d}(1/T) \qquad (6B.4)$$

If we suppose that $\Delta_r H^{\ominus}$ varies only slightly with temperature over the temperature range of interest, then we may take it outside the integral. It follows that

$$\ln K_2 - \ln K_1 = -\frac{\Delta_r H^{\ominus}}{R}\left(\frac{1}{T_2} - \frac{1}{T_1}\right) \qquad \text{Temperature dependence of } K \qquad (6B.5)$$

Brief illustration 6B.2 The temperature dependence of *K*

To estimate the equilibrium constant or the synthesis of ammonia at 500 K from its value at 298 K (6.1×10^5 for the reaction written as $N_2(g) + 3\,H_2(g) \rightleftharpoons 2\,NH_3(g)$) we use the standard reaction enthalpy, which can be obtained from Table 2C.2 in the *Resource section* by using $\Delta_r H^{\ominus} = 2\Delta_f H^{\ominus}(NH_3,g)$ and assume that its value is constant over the range of temperatures. Then, with $\Delta_r H^{\ominus} = -92.2\,\text{kJ mol}^{-1}$, from eqn 6B.3 we find

$$\ln K_2 = \ln(6.1\times10^5) - \left(\frac{-9.22\times10^4\,\text{J mol}^{-1}}{8.3145\,\text{J K}^{-1}\,\text{mol}^{-1}}\right)\times\left(\frac{1}{500\,\text{K}} - \frac{1}{298\,\text{K}}\right)$$
$$= -1.7\ldots$$

It follows that $K_2 = 0.18$, a lower value than at 298 K, as expected for this exothermic reaction.

Self-test 6B.3 The equilibrium constant for $N_2O_4(g) \rightleftharpoons 2\,NO_2(g)$ was calculated in *Self-test* 6A.6. Estimate its value at 100 °C.

Answer: 15

Checklist of concepts

- ☐ 1. The thermodynamic equilibrium constant is independent of pressure.
- ☐ 2. The response of composition to changes in the conditions is summarized by **Le Chatelier's principle**.
- ☐ 3. The dependence of the equilibrium constant on the temperature is expressed by the **van 't Hoff equation** and can be explained in terms of the distribution of molecules over the available states.

Checklist of equations

Property	Equation	Comment	Equation number
van 't Hoff equation	$\mathrm{d}\ln K/\mathrm{d}T = \Delta_r H^\ominus/RT^2$		6B.2a
	$\mathrm{d}\ln K/\mathrm{d}(1/T) = -\Delta_r H^\ominus/R$	Alternative version	6B.2b
Temperature dependence of equilibrium constant	$\ln K_2 - \ln K_1 = -(\Delta_r H^\ominus/R)(1/T_2 - 1/T_1)$	$\Delta_r H^\ominus$ assumed constant	6B.5

6C Electrochemical cells

Contents

➤ Why do you need to know this material?

One very special case of the material treated in Topic 6B that has enormous fundamental, technological, and economic significance concerns reactions that take place in electrochemical cells. Moreover, the ability to make very precise measurements of potential differences ('voltages') means that electrochemical methods can be used to determine thermodynamic properties of reactions that may be inaccessible by other methods.

➤ What is the key idea?

The electrical work that a reaction can perform at constant pressure and temperature is equal to the reaction Gibbs energy.

➤ What do you need to know already?

This Topic develops the relation between the Gibbs energy and non-expansion work (Topic 3C). You need to be aware of how to calculate the work of moving a charge through an electrical potential difference (Topic 2A). The equations make use of the definition of the reaction quotient *Q* and the equilibrium constant *K* (Topic 6A).

An **electrochemical cell** consists of two **electrodes**, or metallic conductors, in contact with an **electrolyte**, an ionic conductor (which may be a solution, a liquid, or a solid). An electrode and its electrolyte comprise an **electrode compartment**. The two electrodes may share the same compartment. The various kinds of electrode are summarized in Table 6C.1. Any 'inert metal' shown as part of the specification is present to act as a source or sink of electrons, but takes no other part in the reaction other than acting as a catalyst for it. If the electrolytes are different, the two compartments may be joined by a **salt bridge**, which is a tube containing a concentrated electrolyte solution (for instance, potassium chloride in agar jelly) that completes the electrical circuit and enables the cell to function. A **galvanic cell** is an electrochemical cell that produces electricity as a result of the spontaneous reaction occurring inside it. An **electrolytic cell** is an electrochemical cell in which a non-spontaneous reaction is driven by an external source of current.

6C.1 Half-reactions and electrodes

It will be familiar from introductory chemistry courses that **oxidation** is the removal of electrons from a species, a **reduction** is the addition of electrons to a species, and a **redox reaction** is a

Table 6C.1 Varieties of electrode

Electrode type	Designation	Redox couple	Half-reaction
Metal/metal ion	$M(s)\vert M^+(aq)$	M^+/M	$M^+(aq)+e^- \rightarrow M(s)$
Gas	$Pt(s)\vert X_2(g)\vert X^+(aq)$	X^+/X_2	$X^+(aq)+e^- \rightarrow \frac{1}{2}X_2(g)$
	$Pt(s)\vert X_2(g)\vert X^-(aq)$	X_2/X^-	$\frac{1}{2}X_2(g)+e^- \rightarrow X^-(aq)$
Metal/insoluble salt	$M(s)\vert MX(s)\vert X^-(aq)$	$MX/M,X^-$	$MX(s)+e^- \rightarrow M(s)+X^-(aq)$
Redox	$Pt(s)\vert M^+(aq),M^{2+}(aq)$	M^{2+}/M^+	$M^{2+}(aq)+e^- \rightarrow M^+(aq)$

reaction in which there is a transfer of electrons from one species to another. The electron transfer may be accompanied by other events, such as atom or ion transfer, but the net effect is electron transfer and hence a change in oxidation number of an element. The **reducing agent** (or *reductant*) is the electron donor; the **oxidizing agent** (or *oxidant*) is the electron acceptor. It should also be familiar that any redox reaction may be expressed as the difference of two reduction **half-reactions**, which are conceptual reactions showing the gain of electrons. Even reactions that are not redox reactions may often be expressed as the difference of two reduction half-reactions. The reduced and oxidized species in a half-reaction form a **redox couple**. In general we write a couple as Ox/Red and the corresponding reduction half-reaction as

$$\mathrm{Ox} + \nu\mathrm{e}^- \rightarrow \mathrm{Red} \quad (6C.1)$$

Brief illustration 6C.1 Redox couples

The dissolution of silver chloride in water $AgCl(s) \rightarrow Ag^+(aq) + Cl^-(aq)$, which is not a redox reaction, can be expressed as the difference of the following two reduction half-reactions:

$$AgCl(s) + e^- \rightarrow Ag(s) + Cl^-(aq)$$
$$Ag^+(aq) + e^- \rightarrow Ag(s)$$

The redox couples are $AgCl/Ag,Cl^-$ and Ag^+/Ag, respectively.

Self-test 6C.1 Express the formation of H_2O from H_2 and O_2 in acidic solution (a redox reaction) as the difference of two reduction half-reactions.

Answer: $4\,H^+(aq) + 4\,e^- \rightarrow 2\,H_2(g)$, $O_2(g) + 4\,H^+(aq) + 4\,e^- \rightarrow 2\,H_2O(l)$

We shall often find it useful to express the composition of an electrode compartment in terms of the reaction quotient, Q, for the half-reaction. This quotient is defined like the reaction quotient for the overall reaction (Topic 6A, $Q = \prod_J a_J^{\nu_J}$), but the electrons are ignored because they are stateless.

Brief illustration 6C.2 The reaction quotient

The reaction quotient for the reduction of O_2 to H_2O in acid solution, $O_2(g) + 4\,H^+(aq) + 4\,e^- \rightarrow 2\,H_2O(l)$, is

$$Q = \frac{a_{H_2O}^2}{a_{H^+}^4 a_{O_2}} \approx \frac{p^{\ominus}}{a_{H^+}^4 p_{O_2}}$$

The approximations used in the second step are that the activity of water is 1 (because the solution is dilute) and the oxygen behaves as a perfect gas, so $a_{O_2} \approx p_{O_2}/p^{\ominus}$.

Self-test 6C.2 Write the half-reaction and the reaction quotient for a chlorine gas electrode.

Answer: $Cl_2(g) + 2\,e^- \rightarrow 2\,Cl^-(aq)$, $Q \approx a_{Cl^-}^2 p^{\ominus}/p_{Cl_2}$

The reduction and oxidation processes responsible for the overall reaction in a cell are separated in space: oxidation takes place at one electrode and reduction takes place at the other. As the reaction proceeds, the electrons released in the oxidation $\mathrm{Red}_1 \rightarrow \mathrm{Ox}_1 + \nu\mathrm{e}^-$ at one electrode travel through the external circuit and re-enter the cell through the other electrode. There they bring about reduction $\mathrm{Ox}_2 + \nu\mathrm{e}^- \rightarrow \mathrm{Red}_2$. The electrode at which oxidation occurs is called the **anode**; the electrode at which reduction occurs is called the **cathode**. In a galvanic cell, the cathode has a higher potential than the anode: the species undergoing reduction, Ox_2, withdraws electrons from its electrode (the cathode, Fig. 6C.1), so leaving a relative positive charge on it (corresponding to a high potential). At the anode, oxidation results in the transfer of electrons to the electrode, so giving it a relative negative charge (corresponding to a low potential).

6C.2 Varieties of cells

The simplest type of cell has a single electrolyte common to both electrodes (as in Fig. 6C.1). In some cases it is necessary to immerse the electrodes in different electrolytes, as in the 'Daniell cell' in which the redox couple at one electrode is Cu^{2+}/Cu and at the other is Zn^{2+}/Zn (Fig. 6C.2). In an **electrolyte concentration cell**, the electrode compartments are identical except for the concentrations of the electrolytes. In an **electrode concentration cell** the electrodes themselves have different concentrations, either because they are gas electrodes operating at different pressures or because they are amalgams (solutions in mercury) with different concentrations.

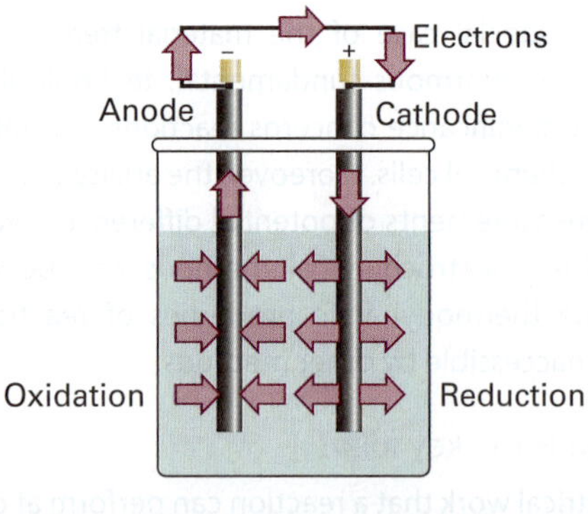

Figure 6C.1 When a spontaneous reaction takes place in a galvanic cell, electrons are deposited in one electrode (the site of oxidation, the anode) and collected from another (the site of reduction, the cathode), and so there is a net flow of current which can be used to do work. Note that the + sign of the cathode can be interpreted as indicating the electrode at which electrons enter the cell, and the – sign of the anode is where the electrons leave the cell.

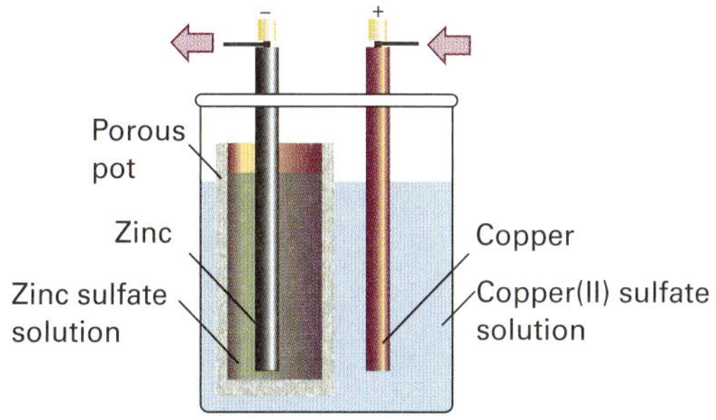

Figure 6C.2 One version of the Daniell cell. The copper electrode is the cathode and the zinc electrode is the anode. Electrons leave the cell from the zinc electrode and enter it again through the copper electrode.

(a) Liquid junction potentials

In a cell with two different electrolyte solutions in contact, as in the Daniell cell, there is an additional source of potential difference across the interface of the two electrolytes. This potential is called the **liquid junction potential**, E_{lj}. Another example of a junction potential is that between different concentrations of hydrochloric acid. At the junction, the mobile H^+ ions diffuse into the more dilute solution. The bulkier Cl^- ions follow, but initially do so more slowly, which results in a potential difference at the junction. The potential then settles down to a value such that, after that brief initial period, the ions diffuse at the same rates. Electrolyte concentration cells always have a liquid junction; electrode concentration cells do not.

The contribution of the liquid junction to the potential can be reduced (to about 1 to 2 mV) by joining the electrolyte compartments through a salt bridge (Fig. 6C.3). The reason for the success of the salt bridge is that provided the ions dissolved in the jelly have similar mobilities, then the liquid junction potentials at either end are largely independent of the concentrations of the two dilute solutions, and so nearly cancel.

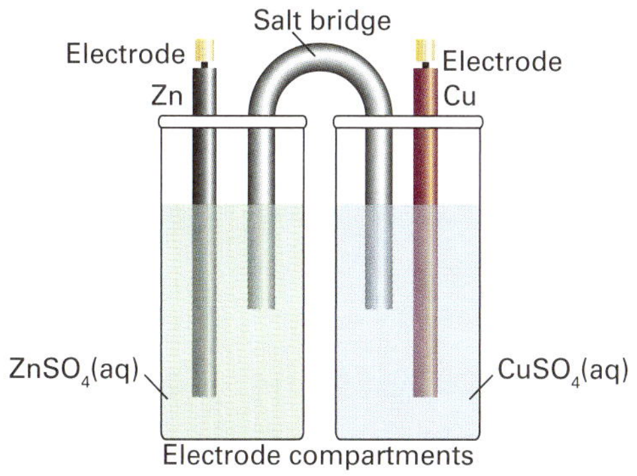

Figure 6C.3 The salt bridge, essentially an inverted U-tube full of concentrated salt solution in a jelly, has two opposing liquid junction potentials that almost cancel.

(b) Notation

We use the following notation for cells:

$\vert$	A phase boundary
$\vdots$	A liquid junction
$\Vert$	An interface for which it is assumed that the junction potential has been eliminated

Brief illustration 6C.3 Cell notation

A cell in which two electrodes share the same electrolyte is

$$Pt(s)|H_2(g)|HCl(aq)|AgCl|Ag(s)$$

The cell in Fig. 6C.2 is denoted

$$Zn(s)|ZnSO_4(aq)\vdots CuSO_4(aq)|Cu(s)$$

The cell in Fig. 6C.3 is denoted

$$Zn(s)|ZnSO_4(aq)||CuSO_4(aq)|Cu(s)$$

An example of an electrolyte concentration cell in which the liquid junction potential is assumed to be eliminated is

$$Pt(s)|H_2(g)|HCl(aq,b_1)||HCl(aq,b_2)|H_2(g)|Pt(s)$$

Self-test 6C.3 Write the symbolism for a cell in which the half-reactions are $4\,H^+(aq)+4\,e^- \rightarrow 2\,H_2(g)$ and $O_2(g)+4\,H^+(aq)+4\,e^- \rightarrow 2\,H_2O(l)$, (a) with a common electrolyte, (b) with separate compartments joined by a salt bridge.

Answer: (a) $Pt(s)|H_2(g)|HCl(aq,b)|O_2(g)|Pt(s)$; (b) $Pt(s)|H_2(g)|HCl(aq,b_1)||HCl(aq,b_2)|O_2(g)|Pt(s)$

6C.3 The cell potential

The current produced by a galvanic cell arises from the spontaneous chemical reaction taking place inside it. The **cell reaction** is the reaction in the cell written on the assumption that the right-hand electrode is the cathode, and hence that the spontaneous reaction is one in which reduction is taking place in the right-hand compartment. Later we see how to predict if the right-hand electrode is in fact the cathode; if it is, then the cell reaction is spontaneous as written. If the left-hand electrode turns out to be the cathode, then the reverse of the corresponding cell reaction is spontaneous.

To write the cell reaction corresponding to a cell diagram, we first write the right-hand half-reaction as a reduction (because we have assumed that to be spontaneous). Then we subtract from it the left-hand reduction half-reaction (for, by implication, that electrode is the site of oxidation).

Brief illustration 6C.4 The cell reaction

For the cell $Zn(s)|ZnSO_4(aq)||CuSO_4(aq)|Cu(s)$ the two electrodes and their reduction half-reactions are

$$\text{Right-hand electrode}: Cu^{2+}(aq)+2e^- \rightarrow Cu(s)$$
$$\text{Left-hand electrode}: Zn^{2+}(aq)+2e^- \rightarrow Zn(s)$$

Hence, the overall cell reaction is the difference Right – Left:

$$Cu^{2+}(aq)+2e^- - Zn^{2+}(aq)-2e^- \rightarrow Cu(s)-Zn(s)$$

which, after cancellation of the $2e^-$, rearranges to

$$Cu^{2+}(aq)+Zn(s) \rightarrow Cu(s)+Zn^{2+}(aq)$$

Self-test 6C.4 Construct the overall cell reaction for the cells:

(a) $Pt(s)|H_2(g)|HCl(aq,b)|O_2(g)|Pt(s)$;

(b) $Pt(s)|H_2(g)|HCl(aq,b_L)||HCl(aq,b_R)|O_2(g)|Pt(s)$.

Answer: (a) $2\,H_2(g)+O_2(g)\rightarrow 2\,H_2O(l)$;
(b) $2\,H_2(g)+O_2(g)+4\,H^+(b_R)\rightarrow 2\,H_2O(l)+4\,H^+(b_L)$

(a) The Nernst equation

A cell in which the overall cell reaction has not reached chemical equilibrium can do electrical work as the reaction drives electrons through an external circuit. The work that a given transfer of electrons can accomplish depends on the potential difference between the two electrodes. When the potential difference is large, a given number of electrons travelling between the electrodes can do a large amount of electrical work. When the potential difference is small, the same number of electrons can do only a small amount of work. A cell in which the overall reaction is at equilibrium can do no work, and then the potential difference is zero.

According to the discussion in Topic 3C, we know that the maximum non-expansion work a system can do at constant temperature and pressure is given by eqn 3C.16b ($w_{e,max}=\Delta G$). In electrochemistry, the non-expansion work is identified with electrical work, the system is the cell, and ΔG is the Gibbs energy of the cell reaction, $\Delta_r G$. Maximum work is produced when a change occurs reversibly. It follows that, to draw thermodynamic conclusions from measurements of the work that a cell can do, we must ensure that the cell is operating reversibly. Moreover, it is established in Topic 6A that the reaction Gibbs energy is actually a property related, through $RT \ln Q$, to a specified composition of the reaction mixture. Therefore, to make use of $\Delta_r G$ we must ensure that the cell is operating reversibly at a specific, constant composition. Both these conditions are achieved by measuring the cell potential when it is balanced by an exactly opposing source of potential so that the cell reaction occurs reversibly, the composition is constant, and no current flows: in effect, the cell reaction is poised for change, but not actually changing. The resulting potential difference is called the **cell potential**, E_{cell}, of the cell.

A note on good practice The cell potential was formerly, and is still widely, called the *electromotive force* (emf) of the cell. IUPAC prefers the term 'cell potential' because a potential difference is not a force.

As we show in the following *Justification*, the relation between the reaction Gibbs energy and the cell potential is

$$-\nu F E_{cell} = \Delta_r G \qquad \text{The cell potential} \quad (6C.2)$$

where F is Faraday's constant, $F=eN_A$, and ν is the stoichiometric coefficient of the electrons in the half-reactions into which the cell reaction can be divided. This equation is the key connection between electrical measurements on the one hand and thermodynamic properties on the other. It will be the basis of all that follows.

Justification 6C.1 The relation between the cell potential and the reaction Gibbs energy

We consider the change in G when the cell reaction advances by an infinitesimal amount $d\xi$ at some composition. From *Justification* 6A.1, specifically the equation $\Delta_r G=(\partial G/\partial \xi)_{T,p}$, we can write (at constant temperature and pressure)

$$dG = \Delta_r G d\xi$$

The maximum non-expansion (electrical) work that the reaction can do as it advances by $d\xi$ at constant temperature and pressure is therefore

$$dw_e = \Delta_r G d\xi$$

This work is infinitesimal, and the composition of the system is virtually constant when it occurs.

Suppose that the reaction advances by $d\xi$, then $\nu d\xi$ electrons must travel from the anode to the cathode. The total charge transported between the electrodes when this change occurs is $-\nu e N_A d\xi$ (because $\nu d\xi$ is the amount of electrons in moles and the charge per mole of electrons is $-eN_A$). Hence, the total charge transported is $-\nu F d\xi$ because $eN_A = F$. The work done when an infinitesimal charge $-\nu F d\xi$ travels from the anode to the cathode is equal to the product of the charge and the potential difference E_{cell} (see Table 2A.1, the entry $dw = Q d\phi$):

$$dw_e = -\nu F E_{cell} d\xi$$

When this relation is equated to the one above ($dw_e = \Delta_r G d\xi$), the advancement $d\xi$ cancels, and we obtain eqn 6C.2.

It follows from eqn 6C.2 that, by knowing the reaction Gibbs energy at a specified composition, we can state the cell

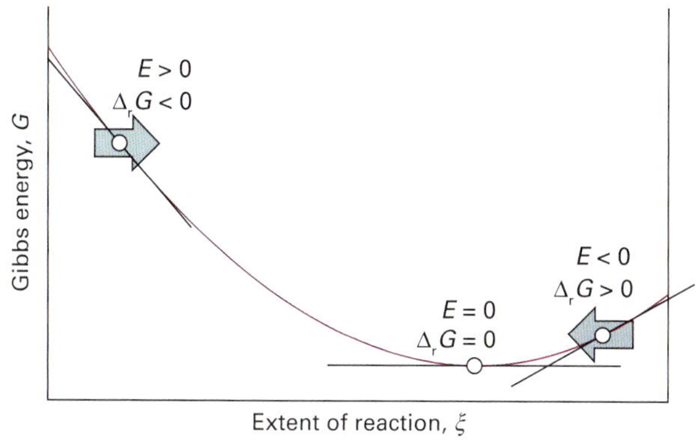

Figure 6C.4 A spontaneous reaction occurs in the direction of decreasing Gibbs energy. When expressed in terms of a cell potential, the spontaneous direction of change can be expressed in terms of the cell potential, E_{cell}. The reaction is spontaneous as written (from left to right on the illustration) when $E_{cell} > 0$. The reverse reaction is spontaneous when $E_{cell} < 0$. When the cell reaction is at equilibrium, the cell potential is zero.

potential at that composition. Note that a negative reaction Gibbs energy, corresponding to a spontaneous cell reaction, corresponds to a positive cell potential. Another way of looking at the content of eqn 6C.2 is that it shows that the driving power of a cell (that is, its potential) is proportional to the slope of the Gibbs energy with respect to the extent of reaction. It is plausible that a reaction that is far from equilibrium (when the slope is steep) has a strong tendency to drive electrons through an external circuit (Fig. 6C.4). When the slope is close to zero (when the cell reaction is close to equilibrium), the cell potential is small.

Brief illustration 6C.5 The reaction Gibbs energy

Equation 6C.2 provides an electrical method for measuring a reaction Gibbs energy at any composition of the reaction mixture: we simply measure the cell potential and convert it to $\Delta_r G$. Conversely, if we know the value of $\Delta_r G$ at a particular composition, then we can predict the cell potential. For example, if $\Delta_r G = -1\times10^2\,\text{kJ mol}^{-1}$ and $\nu = 1$, then

$$E_{cell} = -\frac{\Delta_r G}{\nu F} = -\frac{(-1\times10^5\,\text{J mol}^{-1})}{1\times(9.6485\times10^4\,\text{C mol}^{-1})} = 1\,\text{V}$$

where we have used $1\,\text{J} = 1\,\text{C V}$.

Self-test 6C.5 Estimate the potential of a fuel cell in which the reaction is $H_2(g) + \frac{1}{2}O_2(g) \rightarrow H_2O(l)$.

Answer: 1.2 V

We can go on to relate the cell potential to the activities of the participants in the cell reaction. We know that the reaction Gibbs energy is related to the composition of the reaction mixture by eqn 6A.10 ($\Delta_r G = \Delta_r G^\ominus + RT \ln Q$); it follows, on division of both sides by $-\nu F$ and recognizing that $\Delta_r G/(-\nu F) = E_{cell}$, that

$$E_{cell} = -\frac{\Delta_r G^\ominus}{\nu F} - \frac{RT}{\nu F}\ln Q$$

The first term on the right is written

$$E^\ominus_{cell} = -\frac{\Delta_r G^\ominus}{\nu F} \qquad \textit{Definition} \quad \text{Standard cell potential} \quad (6C.3)$$

and called the **standard cell potential**. That is, the standard cell potential is the standard reaction Gibbs energy expressed as a potential difference (in volts). It follows that

$$E_{cell} = E^\ominus_{cell} - \frac{RT}{\nu F}\ln Q \qquad \text{Nernst equation} \quad (6C.4)$$

This equation for the cell potential in terms of the composition is called the **Nernst equation**; the dependence that it predicts is summarized in Fig. 6C.5. One important application of the Nernst equation is to the determination of the pH of a solution and, with a suitable choice of electrodes, of the concentration of other ions (Topic 6D).

We see from eqn 6C.4 that the standard cell potential can be interpreted as the cell potential when all the reactants and products in the cell reaction are in their standard states, for then all activities are 1, so $Q = 1$ and $\ln Q = 0$. However, the fact that the standard cell potential is merely a disguised form of the standard reaction Gibbs energy (eqn 6C.3) should always be kept in mind and underlies all its applications.

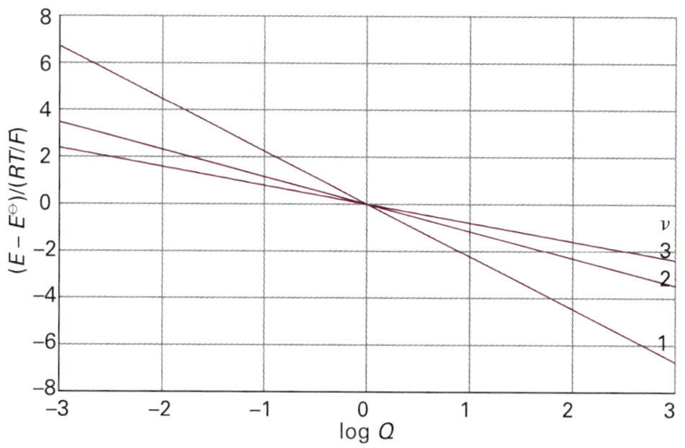

Figure 6C.5 The variation of cell potential with the value of the reaction quotient for the cell reaction for different values of ν (the number of electrons transferred). At 298 K, $RT/F = 25.69$ mV, so the vertical scale refers to multiples of this value.

Brief illustration 6C.6 The Nernst equation

Because $RT/F = 25.7\,\text{mV}$ at 25 °C, a practical form of the Nernst equation is

$$E_{\text{cell}} = E^{\ominus}_{\text{cell}} - \frac{25.7\,\text{mV}}{\nu}\ln Q$$

It then follows that, for a reaction in which $\nu = 1$, if Q is increased by a factor of 10, then the cell potential decreases by 59.2 mV.

Self-test 6C.6 By how much does the cell potential change when Q is decreased by a factor of 10 for a reaction in which $\nu = 2$?

Answer: −29.6 V

An important feature of a standard cell potential is that it is unchanged if the chemical equation for the cell reaction is multiplied by a numerical factor. A numerical factor increases the value of the standard Gibbs energy for the reaction. However, it also increases the number of electrons transferred by the same factor, and by eqn 6D.2 the value of $E^{\ominus}_{\text{cell}}$ remains unchanged. A practical consequence is that a cell potential is independent of the physical size of the cell. In other words, the cell potential is an intensive property.

(b) Cells at equilibrium

A special case of the Nernst equation has great importance in electrochemistry and provides a link to the discussion of equilibrium in Topic 6A. Suppose the reaction has reached equilibrium; then $Q = K$, where K is the equilibrium constant of the cell reaction. However, a chemical reaction at equilibrium cannot do work, and hence it generates zero potential difference between the electrodes of a galvanic cell. Therefore, setting $E_{\text{cell}} = 0$ and $Q = K$ in the Nernst equation gives

$$E^{\ominus}_{\text{cell}} = \frac{RT}{\nu F}\ln K \qquad \text{Equilibrium constant and standard cell potential} \qquad (6C.5)$$

This very important equation (which could also have been obtained more directly by substituting eqn 6A.14, $\Delta_r G^{\ominus} = -RT\ln K$, into eqn 6C.3) lets us predict equilibrium constants from measured standard cell potentials. However, before we use it extensively, we need to establish a further result.

Brief illustration 6C.7 Equilibrium constants

Because the standard potential of the Daniell cell is +1.10 V, the equilibrium constant for the cell reaction $Cu^{2+}(aq) + Zn(s) \rightarrow Cu(s) + Zn^{2+}(aq)$, for which $\nu = 2$, is $K = 1.5 \times 10^{37}$ at 298 K. We conclude that the displacement of copper by zinc goes virtually to completion. Note that a cell potential of about 1 V is easily measurable but corresponds to an equilibrium constant that would be impossible to measure by direct chemical analysis.

Self-test 6C.7 What would be the standard cell potential for a reaction with $K = 1$?

Answer: 0

6C.4 The determination of thermodynamic functions

The standard potential of a cell is related to the standard reaction Gibbs energy through eqn 6C.3 (written as $-\nu F E^{\ominus}_{\text{cell}} = \Delta_r G^{\ominus}$). Therefore, by measuring $E^{\ominus}_{\text{cell}}$ we can obtain this important thermodynamic quantity. Its value can then be used to calculate the Gibbs energy of formation of ions by using the convention explained in Topic 3C, that $\Delta_f G^{\ominus}(H^+,aq) = 0$.

Brief illustration 6C.8 The reaction Gibbs energy

The reaction taking place in the cell

$$Pt(s)|H_2(g)|H^+(aq)||Ag^+(aq)|Ag(s) \quad E^{\ominus}_{\text{cell}} = +0.7996\,\text{V}$$

is

$$Ag^+(aq) + \tfrac{1}{2}H_2(g) \rightarrow H^+(aq) + Ag(s) \quad \Delta_r G^{\ominus} = -\Delta_f G^{\ominus}(Ag^+,aq)$$

Therefore, with $\nu = 1$, we find

$$\begin{aligned}\Delta_f G^{\ominus}(Ag^+,aq) &= -(-FE^{\ominus}) \\ &= (9.6485\times10^4\,\text{C mol}^{-1})\times(0.7996\,\text{V}) \\ &= +77.15\,\text{kJ mol}^{-1}\end{aligned}$$

which is in close agreement with the value in Table 2C.2 of the *Resource section*.

Self-test 6C.8 Derive the value of $\Delta_f G^{\ominus}(H_2O,l)$ at 298 K from the standard potential of the cell $Pt(s)|H_2(g)|HCl(aq)|O_2(g)|Pt$, $E^{\ominus}_{\text{cell}} = +1.23\,\text{V}$.

Answer: −237 kJ mol⁻¹

The temperature coefficient of the standard cell potential, $dE^{\ominus}_{\text{cell}}/dT$, gives the standard entropy of the cell reaction. This conclusion follows from the thermodynamic relation $(\partial G/\partial T)_p = -S$ derived in Topic 3D and eqn 6C.3, which combine to give

$$\frac{dE^{\ominus}_{\text{cell}}}{dT} = \frac{\Delta_r S^{\ominus}}{\nu F} \qquad \text{Temperature coefficient of standard cell potential} \qquad (6C.6)$$

The derivative is complete (not partial) because $E^{\ominus}_{\text{cell}}$ like $\Delta_r G^{\ominus}$, is independent of the pressure. Hence we have an electrochemical technique for obtaining standard reaction entropies and through them the entropies of ions in solution.

Finally, we can combine the results obtained so far and use them to obtain the standard reaction enthalpy:

$$\Delta_r H^{\ominus} = \Delta_r G^{\ominus} + T\Delta_r S^{\ominus} = -\nu F\left(E^{\ominus}_{\text{cell}} - T\frac{dE^{\ominus}_{\text{cell}}}{dT}\right) \qquad (6C.7)$$

This expression provides a non-calorimetric method for measuring $\Delta_r H^{\ominus}$ and, through the convention $\Delta_f H^{\ominus}(H^+, aq) = 0$ the standard enthalpies of formation of ions in solution (Topic 2C).

Example 6C.1 Using the temperature coefficient of the cell potential

The standard potential of the cell Pt(s)|H_2(g)|HBr(aq)|AgBr(s)|Ag(s) was measured over a range of temperatures, and the data were found to fit the following polynomial:

$$E^{\ominus}_{\text{cell}}/\text{V} = 0.07131 - 4.99\times10^{-4}(T/\text{K} - 298) - 3.45\times10^{-6}(T/\text{K} - 298)^2$$

The cell reaction is $\text{AgBr(s)} + \frac{1}{2}\text{H}_2\text{(g)} \rightarrow \text{Ag(s)} + \text{HBr(aq)}$. Evaluate the standard reaction Gibbs energy, enthalpy, and entropy at 298 K.

Method The standard Gibbs energy of reaction is obtained by using eqn 6C.2 after evaluating $E^{\ominus}_{\text{cell}}$ at 298 K and by using 1 V C = 1 J. The standard entropy of reaction is obtained by using eqn 6C.6, which involves differentiating the polynomial with respect to T and then setting $T = 298$ K. The reaction enthalpy is obtained by combining the values of the standard Gibbs energy and entropy.

Answer At $T = 298$ K, $E^{\ominus}_{\text{cell}}/\text{V} = 0.07131$ V, so

$$\begin{aligned}\Delta_r G^{\ominus} &= -\nu F E^{\ominus}_{\text{cell}} = -(1)\times(9.6485\times10^4\,\text{C mol}^{-1})\times(0.07131\,\text{V})\\ &= -6.880\times10^3\,\text{C V mol}^{-1} = -6.880\,\text{kJ mol}^{-1}\end{aligned}$$

The temperature coefficient of the cell potential is

$$\frac{dE^{\ominus}_{\text{cell}}}{dT} = -4.99\times10^{-4}\,\text{V K}^{-1} - 2(3.45\times10^{-6})(T/\text{K} - 298)\,\text{V K}^{-1}$$

At $T = 298$ K this expression evaluates to

$$\frac{dE^{\ominus}_{\text{cell}}}{dT} = -4.99\times10^{-4}\,\text{V K}^{-1}$$

So, from eqn 6C.6 the reaction entropy is

$$\begin{aligned}\Delta_r S^{\ominus} &= \nu F\frac{dE^{\ominus}_{\text{cell}}}{dT} = (1)\times(9.6485\times10^4\,\text{C mol}^{-1})\times(-4.99\times10^{-4}\,\text{V})\\ &= -48.2\,\text{J K}^{-1}\,\text{mol}^{-1}\end{aligned}$$

The negative value stems in part from the elimination of gas in the cell reaction. It then follows that

$$\begin{aligned}\Delta_r H^{\ominus} &= \Delta_r G^{\ominus} + T\Delta_r S^{\ominus} = -6.880\,\text{kJ mol}^{-1}\\ &\quad + (298\,\text{K})\times(-0.0482\,\text{kJ K}^{-1}\,\text{mol}^{-1})\\ &= -21.2\,\text{kJ mol}^{-1}\end{aligned}$$

One difficulty with this procedure lies in the accurate measurement of small temperature coefficients of cell potential. Nevertheless, it is another example of the striking ability of thermodynamics to relate the apparently unrelated, in this case to relate electrical measurements to thermal properties.

Self-test 6C.9 Predict the standard potential of the Harned cell at 303 K from tables of thermodynamic data.

Answer: +0.2222 V

Checklist of concepts

- ☐ 1. A **redox reaction** is expressed as the difference of two reduction **half-reactions**; each one defines a redox couple.
- ☐ 2. **Galvanic cells** are classified as **electrolyte concentration** and **electrode concentration cells**.
- ☐ 3. A **liquid junction potential** arises at the junction of two electrolyte solutions.
- ☐ 4. The cell notation specifies the structure of a cell.
- ☐ 5. The **Nernst equation** relates the cell potential to the composition of the reaction mixture.
- ☐ 6. The **standard cell potential** may be used to calculate the equilibrium constant of the cell reaction.
- ☐ 7. The temperature coefficient of the cell potential is used to measure thermodynamic properties of electroactive species.

Checklist of equations

Property	Equation	Comment	Equation number
Cell potential and reaction Gibbs energy	$-\nu FE_{cell}=\Delta_r G$	Constant temperature and pressure	6C.2
Standard cell potential	$E^{\ominus}_{cell}=-\Delta_r G^{\ominus}/\nu F$	Definition	6C.3
Nernst equation	$E_{cell}=E^{\ominus}_{cell}-(RT/\nu F)\ln Q$		6C.4
Equilibrium constant of cell reaction	$E^{\ominus}_{cell}=(RT/\nu F)\ln K$		6C.5
Temperature coefficient of cell potential	$dE^{\ominus}_{cell}/dT=\Delta_r S^{\ominus}/\nu F$		6C.6

6D Electrode potentials

Contents

➤ Why do you need to know this material?

A very powerful, compact, and widely used way to report standard cell potentials is to ascribe a potential to each electrode. Electrode potentials are widely used in chemistry to assess the oxidizing and reducing power of redox couples and to infer thermodynamic properties, including equilibrium constants.

➤ What is the key idea?

Each electrode of a cell can be supposed to make a characteristic contribution to the cell potential; redox couples with low electrode potentials tend to reduce those with higher potentials.

➤ What do you need to know already?

This Topic develops the concepts in Topic 6D, so you need to understand the concept of cell potential and standard cell potential (Topic 6D); it makes use of the Nernst equation (Topic 6D). The measurement of standard potentials makes use of the Debye–Hückel limiting law (Topic 5F).

As explained in Topic 6C, a galvanic cell is a combination of two electrodes each of which can be considered to make a characteristic contribution to the overall cell potential. Although it is not possible to measure the contribution of a single electrode, we can define the potential of one of the electrodes as zero and then assign values to others on that basis.

6D.1 Standard potentials

The specially selected electrode is the **standard hydrogen electrode** (SHE):

$$\mathrm{Pt(s)}|\mathrm{H_2(g)}|\mathrm{H^+(aq)} \quad E^{\ominus}=0 \qquad \textit{Convention} \quad \text{Standard potentials} \qquad (6D.1)$$

at all temperatures. To achieve the standard conditions, the activity of the hydrogen ions must be 1 (that is, pH=0) and the pressure (more precisely, the fugacity) of the hydrogen gas must be 1 bar. The **standard potential**, $E^{\ominus}(X)$, of another couple X is then assigned by constructing a cell in which it is the right-hand electrode and the standard hydrogen electrode is the left-hand electrode:

$$\mathrm{Pt(s)}|\mathrm{H_2(g)}|\mathrm{H^+(aq)}||\mathrm{X} \quad E^{\ominus}(\mathrm{X})=E^{\ominus}_{\mathrm{cell}} \qquad \textit{Convention} \quad \text{Standard potentials} \qquad (6D.2)$$

The standard potential of a cell of the form L||R, where L is the left-hand electrode of the cell as written (not as arranged on the bench) and R is the right-hand electrode, is then given by the difference of the two standard potentials:

$$\mathrm{L}||\mathrm{R} \quad E^{\ominus}_{\mathrm{cell}}=E^{\ominus}(\mathrm{R})-E^{\ominus}(\mathrm{L}) \qquad \text{Standard cell potential} \qquad (6D.3)$$

A list of standard potentials at 298 K is given in Table 6D.1, and longer lists in numerical and alphabetical order are in the *Resource section*.

Table 6D.1* Standard potentials at 298 K, $E^{\ominus}$/V

Couple	$E^{\ominus}$/V
$Ce^{4+}(aq)+e^- \rightarrow Ce^{3+}(aq)$	+1.61
$Cu^{2+}(aq)+2e^- \rightarrow Cu(s)$	+0.34
$H^+(aq)+e^- \rightarrow \frac{1}{2}H_2(g)$	0
$AgCl(s)+e^- \rightarrow Ag(s)+Cl^-(aq)$	+0.22
$Zn^{2+}(aq)+2e^- \rightarrow Zn(s)$	−0.76
$Na^+(aq)+e^- \rightarrow Na(s)$	−2.71

* More values are given in the *Resource section*.

Brief illustration 6D.1 Standard electrode potentials

The cell Ag(s)|AgCl(s)|HCl(aq)|O_2(g)|Pt(s) can be regarded as formed from the following two electrodes, with their standard potentials taken from the *Resource section*:

Electrode	Half-reaction	Standard potential
R: Pt(s)|O_2(g)|H^+(aq)	$O_2(g)+4\,H^+(aq)+4\,e^- \to 2\,H_2O(l)$	+1.23 V
L: Ag(s)|AgCl(s)|Cl^-(aq)	$AgCl(s)+e^- \to Ag(s)+Cl^-(aq)$	+0.22 V
	$E^\ominus_{cell}=$	+1.01 V

Self-test 6D.1 What is the standard potential of the cell Pt(s)|Fe^{2+}(aq),Fe^{3+}(aq)||Ce^{4+}(aq),Ce^{3+}(aq)|Pt(s)?

Answer: +0.84 V

(a) The measurement procedure

The procedure for measuring a standard potential can be illustrated by considering a specific case, the silver chloride electrode. The measurement is made on the 'Harned cell':

$$\mathrm{Pt(s)|H_2(g)|HCl(aq,}b\mathrm{)|AgCl(s)|Ag(s)}$$
$$\tfrac{1}{2}\mathrm{H_2(g)+AgCl(s)\to HCl(aq)+Ag(s)}$$
$$E^\ominus_{cell}=E^\ominus(\mathrm{AgCl/Ag,Cl^-})-E^\ominus(\mathrm{SHE})=E^\ominus(\mathrm{AgCl/Ag,Cl^-}),\nu=1$$

for which the Nernst equation is

$$E_{cell}=E^\ominus(\mathrm{AgCl/Ag,Cl^-})-\frac{RT}{F}\ln\frac{a_{H^+}a_{Cl^-}}{a_{H_2}^{1/2}}$$

We shall set $a_{H_2}=1$ from now on, and for simplicity write the standard potential of the AgCl/Ag,Cl^- electrode as $E^\ominus$; then

$$E_{cell}=E^\ominus-\frac{RT}{F}\ln a_{H^+}a_{Cl^-}$$

We show in the following *Justification* that as the molality $b\to 0$,

$$\overbrace{E_{cell}+\frac{2RT}{F}\ln b}^{y}=\overbrace{E^\ominus}^{\text{intercept}}+\overbrace{C\times b^{1/2}}^{\text{slope}\times x} \quad (6D.4)$$

where C is a constant. To use this equation, which has the form $y=\text{intercept}+\text{slope}\times x$ with $x=b^{1/2}$, the expression on the left is evaluated at a range of molalities, plotted against $b^{1/2}$, and extrapolated to $b=0$. The intercept at $b^{1/2}=0$ is the value of $E^\ominus$ for the silver/silver chloride electrode. In precise work, the $b^{1/2}$ term is brought to the left, and a higher-order correction term from extended versions of the Debye–Hückel law is used on the right.

Justification 6D.1 The Harned cell potential

The activities in the expression for E_{cell} can be expressed in terms of the molality b of HCl(aq) through $a_{H^+}=\gamma_\pm b/b^\ominus$ and $a_{Cl^-}=\gamma_\pm b/b^\ominus$ as established in Topic 5E, so, with $b/b^\ominus$ replaced by b

$$E_{cell}=E^\ominus-\frac{RT}{F}\ln b^2-\frac{RT}{F}\ln\gamma_\pm^2$$
$$=E^\ominus-\frac{2RT}{F}\ln b-\frac{2RT}{F}\ln\gamma_\pm$$

and therefore

$$E_{cell}+\frac{2RT}{F}\ln b=E^\ominus-\frac{2RT}{F}\ln\gamma_\pm$$

From the Debye–Hückel limiting law for a 1,1-electrolyte (eqn 5F.8, $\log\gamma_\pm=-A|z_+z_-|I^{1/2}$), as $b\to 0$

$$\log\gamma_\pm=-A|z_+z_-|I^{1/2}=-A(b/b^\ominus)^{1/2}$$

Therefore, because $\ln x=\ln 10\log x$,

$$\ln\gamma_\pm=\ln 10\log\gamma_\pm=-A\ln 10(b/b^\ominus)^{1/2}$$

and the equation for E_{cell} becomes

$$E_{cell}+\frac{2RT}{F}\ln b=E^\ominus+\frac{2ART\ln 10}{F(b^\ominus)^{1/2}}b^{1/2}\quad \text{as } b\to 0$$

With the term in blue denoted C, this equation becomes eqn 6D.4.

Example 6D.1 Evaluating a standard electrode potential

The potential of the Harned cell at 25 °C has the following values:

$b/(10^{-3}b^\ominus)$	3.215	5.619	9.138	25.63
E_{cell}/V	0.520 53	0.492 57	0.468 60	0.418 24

Determine the standard potential of the silver–silver chloride electrode.

Method As explained in the text, evaluate $y=E_{cell}+(2RT/F)\ln b$ and plot it against $b^{1/2}$; then extrapolate to $b=0$. Use $2RT/F=0.051\,39$ V.

Answer To determine the standard potential of the cell we draw up the following table:

$b/(10^{-3}b^\ominus)$	3.215	5.619	9.138	25.63
$\{b/(10^{-3}b^\ominus)\}^{1/2}$	1.793	2.370	3.023	5.063
E_{cell}/V	0.520 53	0.492 57	0.468 60	0.418 24
y/V	0.2256	0.2263	0.2273	0.2299

The data are plotted in Fig. 6D.1; as can be seen, they extrapolate to $E^\ominus=+0.2232$ V (the value obtained, to preserve the precision of the data, by linear regression).

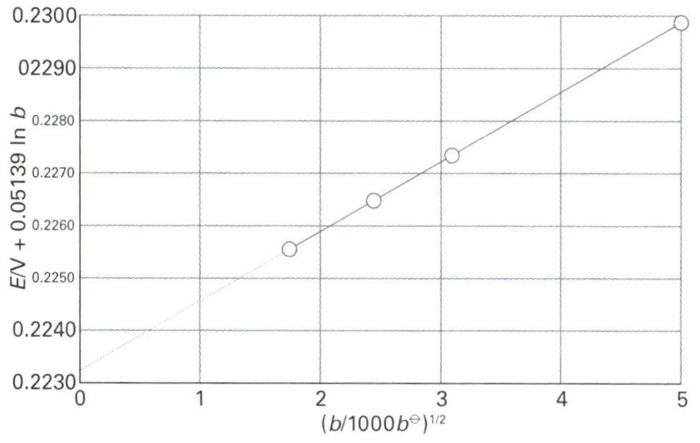

Figure 6D.1 The plot and the extrapolation used for the experimental measurement of a standard cell potential. The intercept at $b^{1/2}=0$ is $E^{\ominus}_{\text{cell}}$.

Self-test 6D.2 The data below are for the cell $\text{Pt(s)}|\text{H}_2\text{(g)}|\text{HBr(aq,}b\text{)}|\text{AgBr(s)}|\text{Ag(s)}$ at 25 °C. Determine the standard cell potential.

$b/(10^{-4}b^{\ominus})$	4.042	8.444	37.19
E_{cell}/V	0.47381	0.43636	0.36173

Answer: +0.071 V

(b) Combining measured values

The standard potentials in Table 6D.1 may be combined to give values for couples that are not listed there. However, to do so, we must take into account the fact that different couples may correspond to the transfer of different numbers of electrons. The procedure is illustrated in the following *Example*.

Example 6D.2 Evaluating a standard potential from two others

Given that the standard potentials of the Cu^{2+}/Cu and Cu^{+}/Cu couples are +0.340 V and +0.522 V, respectively, evaluate $E^{\ominus}(\text{Cu}^{2+},\text{Cu}^{+})$.

Method First, we note that reaction Gibbs energies may be added (as in a Hess's law analysis of reaction enthalpies). Therefore, we should convert the $E^{\ominus}$ values to $\Delta_r G^{\ominus}$ values by using eqn 6C.2 ($-\nu FE^{\ominus}=\Delta_r G^{\ominus}$), add them appropriately, and then convert the overall $\Delta_r G^{\ominus}$ to the required $E^{\ominus}$ by using eqn 6D.2 again. This roundabout procedure is necessary because, as we shall see, although the factor F cancels, the factor ν in general does not.

Answer The electrode half-reactions are as follows:

(a) $\text{Cu}^{2+}\text{(aq)}+2\,\text{e}^- \rightarrow \text{Cu(s)} \quad E^{\ominus}(\text{a})=+0.340\,\text{V},$
so $\Delta_r G^{\ominus}(\text{a})=-2(0.340\,\text{V})F$

(b) $\text{Cu}^{+}\text{(aq)}+\text{e}^- \rightarrow \text{Cu(s)} \quad E^{\ominus}(\text{b})=+0.522\,\text{V},$
so $\Delta_r G^{\ominus}(\text{b})=-(0.522\,\text{V})F$

The required reaction is

(c) $\text{Cu}^{2+}\text{(aq)}+\text{e}^- \rightarrow \text{Cu}^{+}\text{(aq)} \quad E^{\ominus}(\text{c})=-\Delta_r G^{\ominus}(\text{c})/F$

Because (c) = (a) − (b), the standard Gibbs energy of reaction (c) is

$$-\Delta_r G^{\ominus}(\text{c})=\Delta_r G^{\ominus}(\text{a})-\Delta_r G^{\ominus}(\text{b})=-(0.680\,\text{V})F-(-0.522\,\text{V})F$$
$$=(+0.158\,\text{V})F$$

Therefore, $E^{\ominus}(\text{c})=+0.158$ V.

Self-test 6D.3 Evaluate $E^{\ominus}(\text{Fe}^{3+},\text{Fe}^{2+})$ from $E^{\ominus}(\text{Fe}^{3+},\text{Fe})$ and $E^{\ominus}(\text{Fe}^{2+},\text{Fe})$.

Answer: +0.76 V

The generalization of the calculation in *Example* 6D.2 is

$$\nu_c E^{\ominus}(\text{c})=\nu_a E^{\ominus}(\text{a})-\nu_b E^{\ominus}(\text{b}) \quad \text{Combination of standard potentials} \quad (6D.5)$$

with the ν_r the stoichiometric coefficients of the electrons in each half-reaction

6D.2 Applications of standard potentials

Cell potentials are a convenient source of data on equilibrium constants and the Gibbs energies, enthalpies, and entropies of reactions. In practice the standard values of these quantities are the ones normally determined.

(a) The electrochemical series

We have seen that for two redox couples, Ox_L/Red_L and Ox_R/Red_R, and the cell

$$\text{L}\|\text{R}=\text{Ox}_L/\text{Red}\|\text{Ox}_R/\text{Red}_R$$
$$\text{Ox}_R+\nu\text{e}^-\rightarrow\text{Red}_R \quad \text{Ox}_L+\nu\text{e}^-\rightarrow\text{Red}_L \quad \text{Cell convention} \quad (6D.6a)$$
$$E^{\ominus}_{\text{cell}}=E^{\ominus}(\text{R})-E^{\ominus}(\text{L})$$

that the cell reaction

$$\text{R}-\text{L}: \text{Red}_L+\text{Ox}_R\rightarrow\text{Ox}_L+\text{Red}_R \quad (6D.6b)$$

has $K>1$ as written if $E^{\ominus}_{\text{cell}}>0$, and therefore if $E^{\ominus}(\text{L})<E^{\ominus}(\text{R})$. Because in the cell reaction Red_L reduces Ox_R, we can conclude that:

> Red_L has a thermodynamic tendency (in the sense $K>1$) to reduce Ox_R if $E^{\ominus}(\text{L})<E^{\ominus}(\text{R})$

More briefly: low reduces high.

Table 6D.2* The electrochemical series of the metals

Least strongly reducing
Gold
Platinum
Silver
Mercury
Copper
(Hydrogen)
Lead
Tin
Nickel
Iron
Zinc
Chromium
Aluminium
Magnesium
Sodium
Calcium
Potassium
Most strongly reducing

* The complete series can be inferred from Table 6D.1 in the *Resource section*.

Table 6D.2 shows a part of the **electrochemical series**, the metallic elements (and hydrogen) arranged in the order of their reducing power as measured by their standard potentials in aqueous solution. A metal low in the series (with a lower standard potential) can reduce the ions of metals with higher standard potentials. This conclusion is qualitative. The quantitative value of K is obtained by doing the calculations we have described previously and review below.

Brief illustration 6D.2 The electrochemical series

Zinc lies above magnesium in the electrochemical series, so zinc cannot reduce magnesium ions in aqueous solution. Zinc can reduce hydrogen ions, because hydrogen lies higher in the series. However, even for reactions that are thermodynamically favourable, there may be kinetic factors that result in very slow rates of reaction.

Self-test 6D.4 Can nickel reduce hydrogen ions to hydrogen gas?

Answer: Yes

(b) The determination of activity coefficients

Once the standard potential of an electrode in a cell is known, we can use it to determine mean activity coefficients by measuring the cell potential with the ions at the concentration of interest. For example, the mean activity coefficient of the ions in hydrochloric acid of molality b is obtained from the relation

$$E_{\text{cell}} + \frac{2RT}{F}\ln b = E^{\ominus} - \frac{2RT}{F}\ln \gamma_{\pm}$$

in *Justification* 6D.1 in the form

$$\ln \gamma_{\pm} = \frac{E^{\ominus}_{\text{cell}} - E_{\text{cell}}}{2RT/F} - \ln b \qquad (6D.7)$$

Brief illustration 6D.3 Activity coefficients

The data in *Example* 6D.1 include the fact that $E_{\text{cell}} = 0.468\,60$ V when $b = 9.138\times10^{-3}b^{\ominus}$. Because $2RT/F = 0.051\,39$ V, and in the *Example* it is established that $E^{\ominus}_{\text{cell}} = 0.2232$ V, the mean activity coefficient at this molality is

$$\ln \gamma_{\pm} = \frac{0.2232\,\text{V} - 0.46860\,\text{V}}{0.05139\,\text{V}} - \ln(9.138\times10^{-3}) = -0.0788\ldots$$

Therefore, $\gamma_{\pm} = 0.9242$.

Self-test 6D.5 Evaluate the mean activity coefficient when $b = 5.619\times10^{-3}b^{\ominus}$.

Answer: 0.9417

(c) The determination of equilibrium constants

The principal use for standard potentials is to calculate the standard potential of a cell formed from any two electrodes. To do so, we construct $E^{\ominus}_{\text{cell}} = E^{\ominus}(\text{R}) - E^{\ominus}(\text{L})$ and use eqn 6C.5 of Topic 6C ($E^{\ominus}_{\text{cell}} = (RT/\nu F)\ln K$, arranged into $\ln K = \nu F E^{\ominus}_{\text{cell}}/RT$).

Brief illustration 6D.4 Equilibrium constants

A disproportionation reaction is a reaction in which a species is both oxidized and reduced. To study the disproportionation $2\,Cu^{+}(aq) \rightarrow Cu(s) + Cu^{2+}(aq)$ at 298 K we combine the following electrodes:

R: $Cu(s)\vert Cu^{+}(aq)$	$Cu^{+}(aq) + e^{-} \rightarrow Cu(s)$	$E^{\ominus}(\text{R}) = +0.52$ V
L: $Pt(s)\vert Cu^{2+}(aq),Cu^{+}(aq)$	$Cu^{2+}(aq) + e^{-} \rightarrow Cu^{+}(s)$	$E^{\ominus}(\text{R}) = +0.16$ V

The standard potential of the cell is therefore

$$E^{\ominus}_{\text{cell}} = 0.52\,\text{V} - 0.16\,\text{V} = +0.36\,\text{V}$$

We can now calculate the equilibrium constant of the cell reaction. Because $\nu = 1$, from eqn 6C.5 with $RT/F = 0.025\,693$ V,

$$\ln K = \frac{0.36\,\text{V}}{0.025693\,\text{V}} = 14.0\ldots$$

Hence, $K = 1.2\times10^{6}$.

Self-test 6D.6 Evaluate the equilibrium constant for the reaction $Sn(s) + Sn^{4+}(aq) \rightleftharpoons 2\,Sn^{2+}(aq)$.

Answer: 6.5×10^{9}

Checklist of concepts

- ☐ 1. The **standard potential** of a couple is the cell potential in which it forms the right-hand electrode and the left-hand electrode is a **standard hydrogen electrode**.
- ☐ 2. The **electrochemical series** lists the metallic elements in the order of their reducing power as measured by their standard potentials in aqueous solution: low reduces high.
- ☐ 3. The cell potential is used to measure the activity coefficient of electroactive ions.
- ☐ 4. The standard cell potential is used to infer the equilibrium constant of the cell reaction.

Checklist of equations

Property	Equation	Comment	Equation number
Standard cell potential	$E^{\ominus}_{\text{cell}} = E^{\ominus}(\text{R}) - E^{\ominus}(\text{L})$	Cell: L‖R	6D.3
Combined potentials	$\nu_{\text{c}}E^{\ominus}(\text{c}) = \nu_{\text{a}}E^{\ominus}(\text{a}) - \nu_{\text{b}}E^{\ominus}(\text{b})$		6D.5

CHAPTER 6 Chemical equilibrium

TOPIC 6A The equilibrium constant

Discussion questions

6A.1 Explain how the mixing of reactants and products affects the position of chemical equilibrium.

6A.2 What is the justification for not including a pure liquid or solid in the expression for an equilibrium constant?

Exercises

6A.1(a) Consider the reaction A → 2 B. Initially, 1.50 mol A is present and no B. What are the amounts of A and B when the extent of reaction is 0.60 mol?
6A.1(b) Consider the reaction 2 A → B. Initially, 1.75 mol A and 0.12 mol B are present. What are the amounts of A and B when the extent of reaction is 0.30 mol?

6A.2(a) When the reaction A → 2 B advances by 0.10 mol (that is, $\Delta\xi = +0.10$ mol) the Gibbs energy of the system changes by −6.4 kJ mol^{-1}. What is the Gibbs energy of reaction at this stage of the reaction?
6A.2(b) When the reaction 2 A → B advances by 0.051 mol (that is, $\Delta\xi = +0.051$ mol) the Gibbs energy of the system changes by −2.41 kJ mol^{-1}. What is the Gibbs energy of reaction at this stage of the reaction?

6A.3(a) The standard Gibbs energy of the reaction $N_2(g) + 3\,H_2(g) \rightarrow 2\,NH_3(g)$ is −32.9 kJ mol^{-1} at 298 K. What is the value of $\Delta_r G$ when Q = (i) 0.010, (ii) 1.0, (iii) 10.0, (iv) 100 000, (v) 1 000 000? Estimate (by interpolation) the value of K from the values you calculate. What is the actual value of K?
6A.3(b) The standard Gibbs energy of the reaction $2\,NO_2(g) \rightarrow N_2O_4(g)$ is −4.73 kJ mol^{-1} at 298 K. What is the value of $\Delta_r G$ when Q = (i) 0.10, (ii) 1.0, (iii) 10, (iv) 100? Estimate (by interpolation) the value of K from the values you calculate. What is the actual value of K?

6A.4(a) At 2257 K and 1.00 bar total pressure, water is 1.77 per cent dissociated at equilibrium by way of the reaction $2\,H_2O(g) \rightleftharpoons 2\,H_2(g) + O_2(g)$. Calculate K.
6A.4(b) For the equilibrium, $N_2O_4(g) \rightleftharpoons 2\,NO_2(g)$, the degree of dissociation, α, at 298 K is 0.201 at 1.00 bar total pressure. Calculate K.

6A.5(a) Dinitrogen tetroxide is 18.46 per cent dissociated at 25 °C and 1.00 bar in the equilibrium $N_2O_4(g) \rightleftharpoons 2\,NO_2(g)$. Calculate K at (i) 25 °C, (ii) 100 °C given that $\Delta_r H^\ominus = +56.2$ kJ mol^{-1} over the temperature range.
6A.5(b) Molecular bromine is 24 per cent dissociated at 1600 K and 1.00 bar in the equilibrium $Br_2(g) \rightleftharpoons 2\,Br(g)$. Calculate K at (i) 1600 °C, (ii) 2000 °C given that $\Delta_r H^\ominus = +112$ kJ mol^{-1} over the temperature range.

6A.6(a) From information in the *Resource section*, calculate the standard Gibbs energy and the equilibrium constant at (i) 298 K and (ii) 400 K for the reaction $PbO(s) + CO(g) \rightleftharpoons Pb(s) + CO_2(g)$. Assume that the reaction enthalpy is independent of temperature.
6A.6(b) From information in the *Resource section*, calculate the standard Gibbs energy and the equilibrium constant at (i) 25 °C and (ii) 50 °C for the reaction $CH_4(g) + 3\,Cl_2(g) \rightleftharpoons CHCl_3(l) + 3\,HCl(g)$. Assume that the reaction enthalpy is independent of temperature.

6A.7(a) Establish the relation between K and K_c for the reaction $H_2CO(g) \rightleftharpoons CO(g) + H_2(g)$.
6A.7(b) Establish the relation between K and K_c for the reaction $3\,N_2(g) + H_2(g) \rightleftharpoons 2\,HN_3(g)$.

6A.8(a) In the gas-phase reaction $2\,A + B \rightleftharpoons 3\,C + 2\,D$, it was found that, when 1.00 mol A, 2.00 mol B, and 1.00 mol D were mixed and allowed to come to equilibrium at 25 °C, the resulting mixture contained 0.90 mol C at a total pressure of 1.00 bar. Calculate (i) the mole fractions of each species at equilibrium, (ii) K_x, (iii) K, and (iv) $\Delta_r G^\ominus$.
6A.8(b) In the gas-phase reaction $A + B \rightleftharpoons C + 2\,D$, it was found that, when 2.00 mol A, 1.00 mol B, and 3.00 mol D were mixed and allowed to come to equilibrium at 25 °C, the resulting mixture contained 0.79 mol C at a total pressure of 1.00 bar. Calculate (i) the mole fractions of each species at equilibrium, (ii) K_x, (iii) K, and (iv) $\Delta_r G^\ominus$.

6A.9(a) The standard reaction Gibbs energy of the isomerization of borneol ($C_{10}H_{17}OH$) to isoborneol in the gas phase at 503 K is +9.4 kJ mol^{-1}. Calculate the reaction Gibbs energy in a mixture consisting of 0.15 mol of borneol and 0.30 mol of isoborneol when the total pressure is 600 Torr.
6A.9(b) The equilibrium pressure of H_2 over solid uranium and uranium hydride, UH_3, at 500 K is 139 Pa. Calculate the standard Gibbs energy of formation of $UH_3(s)$ at 500 K.

6A.10(a) The standard Gibbs energy of formation of $NH_3(g)$ is −16.5 kJ mol^{-1} at 298 K. What is the reaction Gibbs energy when the partial pressures of the N_2, H_2, and NH_3 (treated as perfect gases) are 3.0 bar, 1.0 bar, and 4.0 bar, respectively? What is the spontaneous direction of the reaction in this case?
6A.10(b) The standard Gibbs energy of formation of $PH_3(g)$ is +13.4 kJ mol^{-1} at 298 K. What is the reaction Gibbs energy when the partial pressures of the H_2 and PH_3 (treated as perfect gases) are 1.0 bar and 0.60 bar, respectively? What is the spontaneous direction of the reaction in this case?

6A.11(a) For $CaF_2(s) \rightleftharpoons Ca^{2+}(aq) + 2\,F^-(aq)$, $K = 3.9 \times 10^{-11}$ at 25 °C and the standard Gibbs energy of formation of $CaF_2(s)$ is −1167 kJ mol^{-1}. Calculate the standard Gibbs energy of formation of $CaF_2(aq)$.
6A.11(b) For $PbI_2(s) \rightleftharpoons Pb^{2+}(aq) + 2\,I^-(aq)$, $K = 1.4 \times 10^{-8}$ at 25 °C and the standard Gibbs energy of formation of $PbI_2(s)$ is −173.64 kJ mol^{-1}. Calculate the standard Gibbs energy of formation of $PbI_2(aq)$.

Problems

6A.1 The equilibrium constant for the reaction $I_2(s) + Br_2(g) \rightleftharpoons 2\,IBr(g)$ is 0.164 at 25 °C. (a) Calculate $\Delta_r G^\ominus$ for this reaction. (b) Bromine gas is introduced into a container with excess solid iodine. The pressure and temperature are held at 0.164 atm and 25 °C, respectively. Find the partial pressure of IBr(g) at equilibrium. Assume that all the bromine is in the liquid form and that the vapour pressure of iodine is negligible. (c) In fact, solid iodine has a measurable vapour pressure at 25 °C. In this case, how would the calculation have to be modified?

6A.2 Calculate the equilibrium constant of the reaction $CO(g) + H_2(g) \rightleftharpoons H_2CO(g)$ given that, for the production of liquid formaldehyde, $\Delta_r G^\ominus = +28.95$ kJ mol^{-1} at 298 K and that the vapour pressure of formaldehyde is 1500 Torr at that temperature.

6A.3 A sealed container was filled with 0.300 mol H_2(g), 0.400 mol I_2(g), and 0.200 mol HI(g) at 870 K and total pressure 1.00 bar. Calculate the amounts of the components in the mixture at equilibrium given that $K=870$ for the reaction $H_2(g)+I_2(g) \rightleftharpoons 2\,HI(g)$.

6A.4‡ Nitric acid hydrates have received much attention as possible catalysts for heterogeneous reactions that bring about the Antarctic ozone hole. Standard reaction Gibbs energies are as follows:

(i) $H_2O(g) \rightarrow H_2O(s)$ $\quad \Delta_r G^\ominus$ −23.6 kJ mol^{-1}

(ii) $H_2O(g)+HNO_3(g) \rightarrow HNO_3{\cdot}H_2O(s)$ $\quad \Delta_r G^\ominus$ −57.2 kJ mol^{-1}

(iii) $2\,H_2O(g)+HNO_3(g) \rightarrow HNO_3{\cdot}2H_2O(s)$ $\quad \Delta_r G^\ominus$ −85.6 kJ mol^{-1}

(iv) $3\,H_2O(g)+HNO_3(g) \rightarrow HNO_3{\cdot}3H_2O(s)$ $\quad \Delta_r G^\ominus$ −112.8 kJ mol^{-1}

Which solid is thermodynamically most stable at 190 K if $p_{H_2}=0.13\,\mu$bar and $p_{HNO_3}=0.41$ nbar *Hint*: Try computing $\Delta_r G$ for each reaction under the prevailing conditions; if more than one solid form spontaneously, examine $\Delta_r G$ for the conversion of one solid to another.

6A.5 Express the equilibrium constant of a gas-phase reaction $A+3\,B \rightleftharpoons 2\,C$ in terms of the equilibrium value of the extent of reaction, ξ, given that initially A and B were present in stoichiometric proportions. Find an expression for ξ as a function of the total pressure, p, of the reaction mixture and sketch a graph of the expression obtained.

TOPIC 6B The response to equilibria to the conditions

Discussion questions

6B.1 Suggest how the thermodynamic equilibrium constant may respond differently to changes in pressure and temperature from the equilibrium constant expressed in terms of partial pressures.

6B.2 Account for Le Chatelier's principle in terms of thermodynamic quantities.

6B.3 Explain the molecular basis of the van 't Hoff equation for the temperature dependence of K.

Exercises

6B.1(a) The standard reaction enthalpy of $Zn(s)+H_2O(g) \rightarrow ZnO(s)+H_2(g)$ is approximately constant at +224 kJ mol^{-1} from 920 K up to 1280 K. The standard reaction Gibbs energy is +33 kJ mol^{-1} at 1280 K. Estimate the temperature at which the equilibrium constant becomes greater than 1.

6B.1(b) The standard enthalpy of a certain reaction is approximately constant at +125 kJ mol^{-1} from 800 K up to 1500 K. The standard reaction Gibbs energy is +22 kJ mol^{-1}at 1120 K. Estimate the temperature at which the equilibrium constant becomes greater than 1.

6B.2(a) The equilibrium constant of the reaction $2\,C_3H_6(g) \rightleftharpoons C_2H_4(g)+C_4H_8(g)$ is found to fit the expression $\ln K=A+B/T+C/T^2$ between 300 K and 600 K, with $A=-1.04$, $B=-1088$ K, and $C=1.51\times10^5$ K^2. Calculate the standard reaction enthalpy and standard reaction entropy at 400 K.

6B.2(b) The equilibrium constant of a reaction is found to fit the expression $\ln K=A+B/T+C/T^3$ between 400 K and 500 K with $A=-2.04$, $B=-1176$ K, and $C=2.1\times10^7$ K^3. Calculate the standard reaction enthalpy and standard reaction entropy at 450 K.

6B.3(a) Calculate the percentage change in K_x for the reaction $H_2CO(g) \rightleftharpoons CO(g)+H_2(g)$ when the total pressure is increased from 1.0 bar to 2.0 bar at constant temperature.

6B.3(b) Calculate the percentage change in K_x for the reaction $CH_3OH(g)+NOCl(g) \rightleftharpoons HCl(g)+CH_3NO_2(g)$ when the total pressure is increased from 1.0 bar to 2.0 bar at constant temperature.

6B.4(a) The equilibrium constant for the gas-phase isomerization of borneol ($C_{10}H_{17}OH$) to isoborneol at 503 K is 0.106. A mixture consisting of 7.50 g of borneol and 14.0 g of isoborneol in a container of volume 5.0 dm^3 is heated to 503 K and allowed to come to equilibrium. Calculate the mole fractions of the two substances at equilibrium.

6B.4(b) The equilibrium constant for the reaction $N_2(g)+O_2(g) \rightleftharpoons 2\,NO(g)$ is 1.69×10^{-3} at 2300 K. A mixture consisting of 5.0 g of nitrogen and 2.0 g of oxygen in a container of volume 1.0 dm^3 is heated to 2300 K and allowed to come to equilibrium. Calculate the mole fraction of NO at equilibrium.

6B.5(a) What is the standard enthalpy of a reaction for which the equilibrium constant is (i) doubled, (ii) halved when the temperature is increased by 10 K at 298 K?

6B.5(b) What is the standard enthalpy of a reaction for which the equilibrium constant is (i) doubled, (ii) halved when the temperature is increased by 15 K at 310 K?

6B.6(a) Estimate the temperature at which $CaCO_3$(calcite) decomposes.

6B.6(b) Estimate the temperature at which $CuSO_4{\cdot}5H_2O$ undergoes dehydration.

6B.7(a) The dissociation vapour pressure of a salt $A_2B(s) \rightleftharpoons A_2(g)+B(g)$ at 367 °C is 208 kPa but at 477 °C it has risen to 547 kPa. Calculate (i) the equilibrium constant, (ii) the standard reaction Gibbs energy, (iii) the standard enthalpy, (iv) the standard entropy of dissociation, all at 422 °C. Assume that the vapour behaves as a perfect gas and that $\Delta H^\ominus$ and $\Delta S^\ominus$ are independent of temperature in the range given.

6B.7(b) The dissociation vapour pressure of NH_4Cl at 427 °C is 608 kPa but at 459 °C it has risen to 1115 kPa. Calculate (i) the equilibrium constant, (ii) the standard reaction Gibbs energy, (iii) the standard enthalpy, (iv) the standard entropy of dissociation, all at 427 °C. Assume that the vapour behaves as a perfect gas and that $\Delta H^\ominus$ and $\Delta S^\ominus$ are independent of temperature in the range given.

Problems

6B.1 Consider the dissociation of methane, CH_4(g), into the elements H_2(g) and C(s, graphite). (a) Given that $\Delta_f H^\ominus(CH_4,g)=-74.85$ kJ mol^{-1} and that $\Delta_r S^\ominus=-80.67$ J K^{-1} mol^{-1} at 298 K, calculate the value of the equilibrium constant at 298 K. (b) Assuming that $\Delta_r H^\ominus$ is independent of temperature, calculate K at 50 °C. (c) Calculate the degree of dissociation, α, of methane

‡ These problems were supplied by Charles Trapp and Carmen Giunta.

at 25 °C and a total pressure of 0.010 bar. (d) Without doing any numerical calculations, explain how the degree of dissociation for this reaction will change as the pressure and temperature are varied.

6B.2 The equilibrium pressure of H_2 over U(s) and UH_3(s) between 450 K and 715 K fits the expression $\ln(p/\text{Pa}) = A + B/T + C\ln(T/\text{K})$, with $A = 69.32$, $B = -1.464 \times 10^4$ K, and $C = -5.65$. Find an expression for the standard enthalpy of formation of UH_3(s) and from it calculate $\Delta_f C_p^{\ominus}$.

6B.3 The degree of dissociation, α, of CO_2(g) into CO(g) and O_2(g) at high temperatures was found to vary with temperature as follows:

T/K	1395	1443	1498
$\alpha/10^{-4}$	1.44	2.50	4.71

Assuming $\Delta_r H^{\ominus}$ to be constant over this temperature range, calculate K, $\Delta_r G^{\ominus}$, $\Delta_r H^{\ominus}$, and $\Delta_r S^{\ominus}$. Make any justifiable approximations.

6B.4 The standard reaction enthalpy for the decomposition of $CaCl_2{\cdot}NH_3$(s) into $CaCl_2$(s) and NH_3(g) is nearly constant at +78 kJ mol^{-1} between 350 K and 470 K. The equilibrium pressure of NH_3 in the presence of $CaCl_2{\cdot}NH_3$ is 1.71 kPa at 400 K. Find an expression for the temperature dependence of $\Delta_r G^{\ominus}$ in the same range.

6B.5 Acetic acid was evaporated in a container of volume 21.45 cm^3 at 437 K and at an external pressure of 101.9 kPa, and the container was then sealed. The mass of acid present in the sealed container was 0.0519 g. The experiment was repeated with the same container but at 471 K, and it was found that 0.0380 g of acetic acid was present. Calculate the equilibrium constant for the dimerization of the acid in the vapour and the enthalpy of vaporization.

6B.6 The dissociation of I_2 can be monitored by measuring the total pressure, and three sets of results are as follows:

T/K	973	1073	1173
$100p$/atm	6.244	6.500	9.181
$10^4 n_I$	2.4709	2.4555	2.4366

where n_I is the amount of I atoms per mole of I_2 molecules in the mixture, which occupied 342.68 cm^3. Calculate the equilibrium constants of the dissociation and the standard enthalpy of dissociation at the mean temperature.

6B.7‡ The 1980s saw reports of $\Delta_f G^{\ominus}(SiH_2)$ ranging from 243 to 289 kJ mol^{-1}. If the standard enthalpy of formation is uncertain by this amount, by what factor is the equilibrium constant for the formation of SiH_2 from its elements uncertain at (a) 298 K, (b) 700 K?

6B.8 Find an expression for the standard reaction Gibbs energy at a temperature T' in terms of its value at another temperature T and the coefficients a, b, and c in the expression for the molar heat capacity listed in Table 2B.1. Evaluate the standard Gibbs energy of formation of H_2O(l) at 372 K from its value at 298 K.

6B.9 Derive an expression for the temperature dependence of K_c for a gas-phase reaction.

TOPIC 6C Electrochemical cells

Discussion questions

6C.1 Explain why reactions that are not redox reactions may be used to generate an electric current.

6C.2 Explain the role of a salt bridge.

6C.3 Why is it necessary to measure the cell potential under zero-current conditions?

6C.4 Can you identify other contributions to the cell potential when a current is being drawn from the cell?

Exercises

6C.1(a) Write the cell reaction and electrode half-reactions and calculate the standard potential of each of the following cells:

(i) Zn|$ZnSO_4$(aq)||$AgNO_3$(aq)|Ag
(ii) Cd|$CdCl_2$(aq)||HNO_3(aq)|H_2(g)|Pt
(iii) Pt|$K_3[Fe(CN)_6]$(aq),$K_4[Fe(CN)_6]$(aq)||$CrCl_3$(aq)|Cr

6C.1(b) Write the cell reaction and electrode half-reactions and calculate the standard potential of each the following cells:

(i) Pt|Cl_2(g)|HCl(aq)||K_2CrO_4(aq)|Ag_2CrO_4(s)|Ag
(ii) Pt|Fe^{3+}(aq),Fe^{2+}(aq)||Sn^{4+}(aq),Sn^{2+}(aq)|Pt
(iii) Cu|Cu^{2+}(aq)||Mn^{2+}(aq),H^+(aq)|MnO_2(s)|Pt

6C.2(a) Devise cells in which the following are the reactions and calculate the standard cell potential in each case:

(i) Zn(s) + $CuSO_4$(aq) $\rightarrow$ $ZnSO_4$(aq) + Cu(s)
(ii) 2 AgCl(s) + H_2(g) $\rightarrow$ 2 HCl(aq) + 2 Ag(s)
(iii) 2 H_2(g) + O_2(g) $\rightarrow$ 2 H_2O(l)

6C.2(b) Devise cells in which the following are the reactions and calculate the standard cell potential in each case:

(i) 2 Na(s) + 2 H_2O(l) $\rightarrow$ 2 NaOH(aq) + H_2(g)
(ii) H_2(g) + I_2(g) $\rightarrow$ 2 HI(aq)
(iii) H_3O^+(aq) + OH^-(aq) $\rightarrow$ 2 H_2O(l)

6C.3(a) Use the Debye–Hückel limiting law and the Nernst equation to estimate the potential of the cell Ag|AgBr(s)|KBr(aq, 0.050 mol kg^{-1})||$Cd(NO_3)_2$(aq, 0.010 mol kg^{-1})|Cd at 25 °C.

6C.3(b) Consider the cell Pt|H_2(g,$p^{\ominus}$)| HCl(aq)|AgCl(s)|Ag, for which the cell reaction is 2 AgCl(s) + H_2(g) $\rightarrow$ 2 Ag(s) + 2 HCl(aq). At 25 °C and a molality of HCl of 0.010 mol kg^{-1}, $E_{\text{cell}} = +0.4658$ V. (i) Write the Nernst equation for the cell reaction. (ii) Calculate $\Delta_r G$ for the cell reaction. (iii) Assuming that the Debye–Hückel limiting law holds at this concentration, calculate $E^{\ominus}(Cl^-, AgCl, Ag)$.

Problems

6C.1 A fuel cell develops an electric potential from the chemical reaction between reagents supplied from an outside source. What is the cell potential of a cell fuelled by (a) hydrogen and oxygen, (b) the combustion of butane at 1.0 bar and 298 K?

6C.2 Although the hydrogen electrode may be conceptually the simplest electrode and is the basis for our reference state of electrical potential in electrochemical systems, it is cumbersome to use. Therefore, several substitutes for it have been devised. One of these alternatives is the quinhydrone electrode (quinhydrone, Q·QH_2, is a complex of quinone, $C_6H_4O_2$=Q, and hydroquinone, $C_6H_4O_2H_2=QH_2$). The electrode half-reaction is $Q(aq)+2\,H^+(aq)+2\,e^- \rightarrow QH_2(aq)$, $E^\ominus=+0.6994$ V. If the cell $Hg|Hg_2Cl_2(s)|HCl(aq)|Q{\cdot}QH_2|Au$ is prepared, and the measured cell potential is +0.190 V, what is the pH of the HCl solution? Assume that the Debye–Hückel limiting law is applicable.

6C.3 Fuel cells provide electrical power for spacecraft (as in the NASA space shuttles) and also show promise as power sources for automobiles. Hydrogen and carbon monoxide have been investigated for use in fuel cells, so their solubilities in molten salts are of interest. Their solubilities in a molten $NaNO_3/KNO_3$ mixture were found to fit the following expressions:

$$\log s_{H_2} = -5.39 - \frac{980}{T/K} \qquad \log s_{CO} = -5.98 - \frac{980}{T/K}$$

where s is the solubility in mol cm^{-3} bar^{-1}. Calculate the standard molar enthalpies of solution of the two gases at 570 K.

TOPIC 6D Electrode potentials

Discussion questions

6D.1 Describe a method for the determination of the standard potential of a redox couple.

6D.2 Devise a method for the determination of the pH of an aqueous solution.

Exercises

6D.1(a) Calculate the equilibrium constants of the following reactions at 25 °C from standard potential data:

(i) $Sn(s)+Sn^{4+}(aq) \rightleftharpoons 2\,Sn^{2+}(aq)$
(ii) $Sn(s)+2\,AgCl(s) \rightleftharpoons SnCl_2(aq)+2\,Ag(s)$

6D.1(b) Calculate the equilibrium constants of the following reactions at 25 °C from standard potential data:

(i) $Sn(s)+CuSO_4(aq) \rightleftharpoons Cu(s)+SnSO_4(aq)$
(ii) $Cu^{2+}(aq)+Cu(s) \rightleftharpoons 2\,Cu^+(aq)$

6D.2(a) The potential of the cell Ag|AgI(s)|AgI(aq)|Ag is +0.9509 V at 25 °C. Calculate (i) the solubility product of AgI and (ii) its solubility.

6D.2(b) The potential of the cell $Bi|Bi_2S_3(s)|Bi_2S_3(aq)|Bi$ is −0.96 V at 25 °C. Calculate (i) the solubility product of Bi_2S_3 and (ii) its solubility.

Problems

6D.1 The potential of the cell $Pt|H_2(g,p^\ominus)|HCl(aq,b)|Hg_2Cl_2(s)|Hg(l)$ has been measured with high precision with the following results at 25 °C:

b/(mmol kg^{-1})	1.6077	3.0769	5.0403	7.6938	10.9474
E/V	0.60080	0.56825	0.54366	0.52267	0.50532

Determine the standard cell potential and the mean activity coefficient of HCl at these molalities. (Make a least-squares fit of the data to the best straight line.)

6D.2 The standard potential of the AgCl/Ag,Cl⁻ couple fits the expression

$$E^\ominus/V = 0.23659 - 4.8564\times10^{-4}(\theta/°C) - 3.4205\times10^{-6}(\theta/°C)^2 + 5.869\times10^{-9}(\theta/°C)^3$$

Calculate the standard Gibbs energy and enthalpy of formation of $Cl^-(aq)$ and its entropy at 298 K.

Integrated activities

6.1‡ Thorn et al. (*J. Phys. Chem.* **100**, 14178 (1996)) carried out a study of $Cl_2O(g)$ by photoelectron ionization. From their measurements, they report $\Delta_f H^\ominus(Cl_2O)=+77.2$ kJ mol^{-1}. They combined this measurement with literature data on the reaction $Cl_2O\,(g)+H_2O(g) \rightarrow 2\,HOCl(g)$, for which $K=8.2\times10^{-2}$ and $\Delta_r S^\ominus=+16.38$ J K^{-1} mol^{-1}, and with readily available thermodynamic data on water vapour to report a value for $\Delta_f H^\ominus$(HOCl). Calculate that value. All quantities refer to 298 K.

6.2 Given that $\Delta_r G^\ominus=-212.7$ kJ mol^{-1} for the reaction in the Daniell cell at 25 °C, and $b(CuSO_4)=1.0\times10^{-3}$ mol kg^{-1} and $b(ZnSO_4)=3.0\times10^{-3}$ mol kg^{-1}, calculate (a) the ionic strengths of the solutions, (b) the mean ionic activity coefficients in the compartments, (c) the reaction quotient, (d) the standard cell potential, and (e) the cell potential. (Take $\gamma_+=\gamma_-=\gamma_\pm$ in the respective compartments.)

6.3 Consider the cell, $Zn(s)|ZnCl_2(0.0050\,mol\,kg^{-1})|Hg_2Cl_2(s)|Hg(l)$, for which the cell reaction is $Hg_2Cl_2(s)+Zn(s)\rightarrow 2\,Hg(l)+2\,Cl^-(aq)+Zn^{2+}(aq)$. Given that $E^\ominus(Zn^{2+},Zn)=-0.7628\,V$, $E^\ominus(Hg_2Cl_2,Hg)=+0.2676\,V$, and that the cell potential is +1.2272 V, (a) write the Nernst equation for the cell. Determine (b) the standard cell potential, (c) Δ_rG, $\Delta_rG^\ominus$, and K for the cell reaction, (d) the mean ionic activity and activity coefficient of $ZnCl_2$ from the measured cell potential, and (e) the mean ionic activity coefficient of $ZnCl_2$ from the Debye–Hückel limiting law. (f) Given that $(\partial E_{cell}/\partial T)_p=-4.52\times10^{-4}\,V\,K^{-1}$, calculate Δ_rS and Δ_rH.

6.4 Careful measurements of the potential of the cell $Pt|H_2(g,p^\ominus)|NaOH(aq, 0.0100\,mol\,kg^{-1}),NaCl(aq, 0.011\,25\,mol\,kg^{-1})|AgCl(s)|Ag(s)$ have been reported. Among the data is the following information:

θ/°C	20.0	25.0	30.0
E_{cell}/V	1.04774	1.04864	1.04942

Calculate pK_w at these temperatures and the standard enthalpy and entropy of the autoprotolysis of water at 25.0 °C.

6.5 Measurements of the potential of cells of the type $Ag|AgX(s)|MX(b_1)|M_xHg|MX(b_2)|AgX(s)|Ag$, where M_xHg denotes an amalgam and the electrolyte is an LiCl in ethylene glycol, are given below. Estimate the activity coefficient at the concentration marked * and then use this value to calculate activity coefficients from the measured cell potential at the other concentrations. Base your answer on the Davies equation (eqn 5F.11) with $A=1.461$, $B=1.70$, $C=0.20$, and $I=b/b^\ominus$. For $b_2=0.09141\,mol\,kg^{-1}$:

b_1/(mol kg^{-1})	0.0555	0.09141	0.1652	0.2171	1.040	1.350*
E/V	−0.0220	0.0000	0.0263	0.0379	0.1156	0.1336

6.6‡ The table below summarizes the potential of the cell $Pd|H_2(g, 1\,bar)|BH(aq, b), B(aq, b)|AgCl(s)|Ag$. Each measurement is made at equimolar concentrations of 2-aminopyridinium chloride (BH) and 2-aminopyridine (B). The data are for 25 °C and it is found that $E^\ominus=0.222\,51\,V$. Use the data to determine pK_a for the acid at 25 °C and the mean activity coefficient ($\gamma_\pm$) of BH as a function of molality (b) and ionic strength (I). Use the Davies equation (eqn 5F.11) with $A=0.5091$ and B and C are parameters that depend upon the ions. Draw a graph of the mean activity coefficient with $b=0.04\,mol\,kg^{-1}$ and $0\le I\le 0.1$.

b/(mol kg^{-1})	0.01	0.02	0.03	0.04	0.05
E_{cell}(25 °C)/V	0.74452	0.72853	0.71928	0.71314	0.70809
b/(mol kg^{-1})	0.06	0.07	0.08	0.09	0.10
E_{cell}(25 °C)/V	0.70380	0.70059	0.69790	0.69571	0.69338

Hint: Use mathematical software or a spreadsheet.

6.7 Here we investigate the molecular basis for the observation that the hydrolysis of ATP is exergonic at pH = 7.0 and 310 K. (a) It is thought that the exergonicity of ATP hydrolysis is due in part to the fact that the standard entropies of hydrolysis of polyphosphates are positive. Why would an increase in entropy accompany the hydrolysis of a triphosphate group into a diphosphate and a phosphate group? (b) Under identical conditions, the Gibbs energies of hydrolysis of H_4ATP and $MgATP^{2-}$, a complex between the Mg^{2+} ion and ATP^{4-}, are less negative than the Gibbs energy of hydrolysis of ATP^{4-}. This observation has been used to support the hypothesis that electrostatic repulsion between adjacent phosphate groups is a factor that controls the exergonicity of ATP hydrolysis. Provide a rationale for the hypothesis and discuss how the experimental evidence supports it. Do these electrostatic effects contribute to the Δ_rH or Δ_rS terms that determine the exergonicity of the reaction? *Hint*. In the $MgATP^{2-}$complex, the Mg^{2+} ion and ATP^{4-} anion form two bonds: one that involves a negatively charged oxygen belonging to the terminal phosphate group of ATP^{4-} and another that involves a negatively charged oxygen belonging to the phosphate group adjacent to the terminal phosphate group of ATP^{4-}.

6.8 To get a sense of the effect of cellular conditions on the ability of ATP to drive biochemical processes, compare the standard Gibbs energy of hydrolysis of ATP to ADP with the reaction Gibbs energy in an environment at 37 °C in which pH = 7.0 and the ATP, ADP, and P_i^- concentrations are all 1.0 mmol dm^{-3}.

6.9 Under biochemical standard conditions, aerobic respiration produces approximately 38 molecules of ATP per molecule of glucose that is completely oxidized. (a) What is the percentage efficiency of aerobic respiration under biochemical standard conditions? (b) The following conditions are more likely to be observed in a living cell: $p_{CO_2}=5.3\times10^{-2}$ atm, $p_{O_2}=0.132$ atm, [glucose] = 5.6 pmol dm^{-3}, [ATP] = [ADP] = [P_i] = 0.10 mmol dm^{-3}, pH = 7.4, $T=310$ K. Assuming that activities can be replaced by the numerical values of molar concentrations, calculate the efficiency of aerobic respiration under these physiological conditions. (c) A typical diesel engine operates between $T_c=873$ K and $T_h=1923$ K with an efficiency that is approximately 75 per cent of the theoretical limit of $1-T_c/T_h$ (see Topic 3A). Compare the efficiency of a typical diesel engine with that of aerobic respiration under typical physiological conditions (see part b). Why is biological energy conversion more or less efficient than energy conversion in a diesel engine?

6.10 In anaerobic bacteria, the source of carbon may be a molecule other than glucose and the final electron acceptor is some molecule other than O_2. Could a bacterium evolve to use the ethanol/nitrate pair instead of the glucose/O_2 pair as a source of metabolic energy?

6.11 The standard potentials of proteins are not commonly measured by the methods described in this chapter because proteins often lose their native structure and function when they react on the surfaces of electrodes. In an alternative method, the oxidized protein is allowed to react with an appropriate electron donor in solution. The standard potential of the protein is then determined from the Nernst equation, the equilibrium concentrations of all species in solution, and the known standard potential of the electron donor. We illustrate this method with the protein cytochrome *c*. The one-electron reaction between cytochrome *c*, cyt, and 2,6-dichloroindophenol, D, can be followed spectrophotometrically because each of the four species in solution has a distinct absorption spectrum. We write the reaction as $cyt_{ox}+D_{red}\rightleftharpoons cyt_{red}+D_{ox}$, where the subscripts 'ox' and 'red' refer to oxidized and reduced states, respectively. (a) Consider $E^\ominus_{cyt}$ and $E^\ominus_D$ to be the standard potentials of cytochrome *c* and D, respectively. Show that, at equilibrium, a plot of $\ln([D_{ox}]_{eq}/[D_{red}]_{eq})$ versus $\ln([cyt_{ox}]_{eq}/[cyt_{red}]_{eq})$ is linear with slope of 1 and y-intercept $F(E^\ominus_{cyt}-E^\ominus_D)/RT$, where equilibrium activities are replaced by the numerical values of equilibrium molar concentrations. (b) The following data were obtained for the reaction between oxidized cytochrome *c* and reduced D in a pH 6.5 buffer at 298 K. The ratios $[D_{ox}]_{eq}/[D_{red}]_{eq}$ and $[cyt_{ox}]_{eq}/[cyt_{red}]_{eq}$ were adjusted by titrating a solution containing oxidized cytochrome *c* and reduced D with a solution of sodium ascorbate, which is a strong reductant. From the data and the standard potential of D of 0.237 V, determine the standard potential cytochrome *c* at pH 6.5 and 298 K.

$[D_{ox}]_{eq}/[D_{red}]_{eq}$	0.00279	0.00843	0.0257	0.0497	0.0748	0.238	0.534
$[cyt_{ox}]_{eq}/[cyt_{red}]_{eq}$	0.0106	0.0230	0.0894	0.197	0.335	0.809	1.39

6.12‡ The dimerization of ClO in the Antarctic winter stratosphere is believed to play an important part in that region's severe seasonal depletion of ozone. The following equilibrium constants are based on measurements on the reaction $2\,ClO(g)\rightarrow(ClO)_2(g)$.

T/K	233	248	258	268	273	280
K	4.13×10^8	5.00×10^7	1.45×10^7	5.37×10^6	3.20×10^6	9.62×10^5
T/K	288	295	303			
K	4.28×10^5	1.67×10^5	6.02×10^4			

(a) Derive the values of $\Delta_rH^\ominus$ and $\Delta_rS^\ominus$ for this reaction. (b) Compute the standard enthalpy of formation and the standard molar entropy of $(ClO)_2$ given $\Delta_fH^\ominus(ClO,g)=+101.8\,kJ\,mol^{-1}$ and $S^\ominus_m(ClO,g)=266.6\,J\,K^{-1}\,mol^{-1}$.

6.13[‡] Suppose that an iron catalyst at a particular manufacturing plant produces ammonia in the most cost-effective manner at 450 °C when the pressure is such that $\Delta_r G$ for the reaction $\frac{1}{2}N_2(g)+\frac{3}{2}H_2(g)\rightarrow NH_3(g)$ is equal to $-500\,J\,mol^{-1}$. (a) What pressure is needed? (b) Now suppose that a new catalyst is developed that is most cost-effective at 400 °C when the pressure gives the same value of $\Delta_r G$. What pressure is needed when the new catalyst is used? What are the advantages of the new catalyst? Assume that (i) all gases are perfect gases or that (ii) all gases are van der Waals gases. Isotherms of $\Delta_r G(T, p)$ in the pressure range $100\,atm \leq p \leq 400\,atm$ are needed to derive the answer. (c) Do the isotherms you plotted confirm Le Chatelier's principle concerning the response of equilibrium changes in temperature and pressure?

PART THREE

Change

In Part 3 we consider the processes by which chemical change occurs and one form of matter is converted into another. We prepare the ground for a discussion of the rates of reactions by considering the motion of molecules in gases and in liquids. Then we establish the precise meaning of reaction rate and see how the overall rate, and the complex behaviour of some reactions, may be expressed in terms of elementary steps and the atomic events that take place when molecules meet. Of enormous importance in industry are reactions on solid surfaces, such as redox reactions at electrodes and various chemical transformations accelerated by solid catalysts. We discuss these processes in the final chapter of the text.

CHAPTER 19

Molecules in motion

This chapter provides techniques for discussing the motion of all kinds of particles in all kinds of fluids. It makes extensive use of the kinetic theory of gases treated in Topic 1B.

19A Transport in gases

We set the scene by showing that molecular motion in fluids (both gases and liquids) shows a number of similarities. We concentrate on the 'transport properties' of a substance, its ability to transfer matter, energy, or some other property from one place to another. These properties include diffusion, thermal conduction, viscosity, and effusion and we see that their rates are expressed in terms of the kinetic theory of gases.

19B Motion in liquids

Molecular motion in liquids is different from that in gases on account of the intermolecular forces, which now play an important role and govern, for instance the viscosity. One way to probe motion in liquids is to drag ions through a solvent by applying an electric field, and we see how the conductivities and mobilities of ions give some insight into motion in liquids.

19C Diffusion

One very important type of motion in fluids is diffusion, and it turns out that it can be discussed in a uniform way by introducing the concept of a 'thermodynamic force'. The spread of molecules can be explored by setting up and solving the 'diffusion equation', and that equation can be interpreted in terms of the molecules undergoing a random walk.

What is the impact of this material?

A great deal of chemistry, chemical engineering, and biology depends on the ability of molecules and ions to migrate through media of various kinds. In *Impact* I19.1 we see how conductivity measurements are used to analyse the motion of nutrients and other matter through biological membranes.

To read more about the impact of this material, scan the QR code, or go to bcs.whfreeman.com/webpub/chemistry/pchem10e/impact/pchem-19-1.html

19A Transport in gases

Contents

➤ Why do you need to know this material?

The transport of properties by gas molecules plays an important role in the atmosphere. The Topic also extends the approach of kinetic theory, showing how to extract quantitative expressions from simple models.

➤ What is the key idea?

A molecule carries properties through space for about the distance of its mean free path.

➤ What do you need to know already?

This Topic builds on and extends the kinetic theory of gases (Topic 1B) and you need to be familiar with the expressions from that Topic for the mean speed of molecules and with the significance of the mean free path and its pressure-dependence.

Transport properties are commonly expressed in terms of a number of equations that are empirical summaries of experimental observations. These equations apply to all kinds of properties and media. In the following sections, we introduce the equations for the general case and then show how to calculate the parameters that appear in them on the basis of a model of molecular behaviour in gases. A more general approach is taken in Topic 19C.

19A.1 The phenomenological equations

By a 'phenomenological equation', a term encountered commonly in the study of fluids, we mean an equation that summarizes empirical observations on phenomena without, initially at least, being based on an understanding of the molecular processes responsible for the property.

The rate of migration of a property is measured by its **flux**, J, the quantity of that property passing through a given area in a given time interval divided by the area and the duration of the interval. If matter is flowing (as in diffusion), we speak of a **matter flux** of so many molecules per square metre per second (number or amount $\mathrm{m^{-2}\,s^{-1}}$); if the property is energy (as in thermal conduction), then we speak of the **energy flux** and express it in joules per square metre per second ($\mathrm{J\,m^{-2}\,s^{-1}}$), and so on. To calculate the total quantity of each property transferred through a given area A in a given time interval Δt, we multiply the flux by the area and the time interval and form $JA\Delta t$.

Experimental observations on transport properties show that the flux of a property is usually proportional to the first derivative of some other related property. For example, the flux of matter diffusing parallel to the z-axis of a container is found to be proportional to the first derivative of the concentration:

$$J(\text{matter}) \propto \frac{d\mathcal{N}}{dz} \qquad \text{Fick's first law of diffusion} \qquad (19A.1)$$

where $\mathcal{N}$ is the number density of particles with units number per metre cubed ($\mathrm{m^{-3}}$). The proportionality of the flux of matter to the concentration gradient is sometimes called **Fick's first law of diffusion**: the law implies that diffusion will be faster when the concentration varies steeply with position than when the concentration is nearly uniform. There is no net flux if the concentration is uniform ($d\mathcal{N}/dz=0$). Similarly, the rate of thermal conduction (the flux of the energy associated with

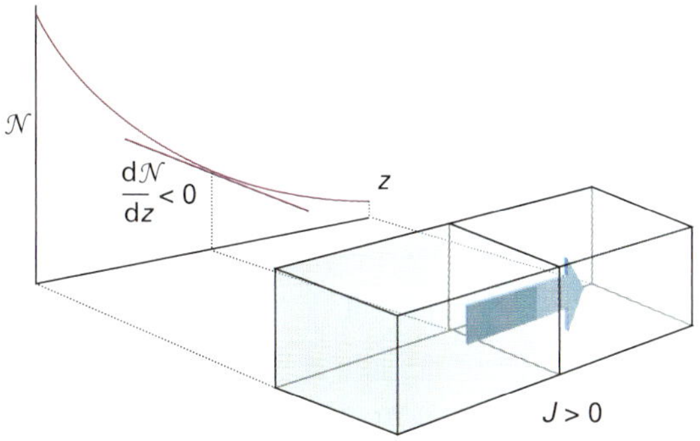

Figure 19A.1 The flux of particles down a concentration gradient. Fick's first law states that the flux of matter (the number of particles passing through an imaginary window in a given interval divided by the area of the window and the length of the interval) is proportional to the density gradient at that point.

thermal motion) is found to be proportional to the temperature gradient:

$$J(\text{energy of thermal motion}) \propto \frac{dT}{dz} \quad \text{Flux of energy} \quad (19A.2)$$

A positive value of J signifies a flux towards positive z; a negative value of J signifies a flux towards negative z. Because matter flows down a concentration gradient, from high concentration to low concentration, J is positive if $d\mathcal{N}/dz$ is negative (Fig. 19A.1). Therefore, the coefficient of proportionality in eqn 19A.1 must be negative, and we write it $-D$:

$$J(\text{matter}) = -D\frac{d\mathcal{N}}{dz} \quad \text{Fick's first law in terms of the diffusion coefficient} \quad (19A.3)$$

The constant D is the called the **diffusion coefficient**; its SI units are metre squared per second ($m^2\ s^{-1}$). Energy migrates down a temperature gradient, and the same reasoning leads to

$$J(\text{energy of thermal motion}) = -\kappa\frac{dT}{dz} \quad \text{Flux of energy in terms of the coefficient of thermal conductivity} \quad (19A.4)$$

where κ is the **coefficient of thermal conductivity**. The SI units of κ are joules per kelvin per metre per second ($J\ K^{-1}\ m^{-1}\ s^{-1}$) or, because $1\ J\ s^{-1} = 1\ W$, watts per kelvin per metre ($W\ K^{-1}\ m^{-1}$). Some experimental values are given in Table 19A.1.

Brief illustration 19A.1 Energy flux

Suppose that there is a temperature difference of 10 K between two metal plates that are separated by 1.0 cm in air (for which $\kappa = 24.1\ mW\ K^{-1}\ m^{-1}$). The temperature gradient is

$$\frac{dT}{dz} = -\frac{-10\ K}{1.0\times10^{-2}\ m} = -1.0\times10^{3}\ K\ m^{-1}$$

Therefore, because for air the energy flux is

$$J(\text{energy of thermal motion}) = -(24.1\ mW\ K^{-1}\ m^{-1}) \times(-1.0\times10^{3}\ K\ m^{-1}) = +24\ W\ m^{-2}$$

As a result, in 1.0 h (3600 s) the transfer of energy through an area of the opposite walls of 1.0 cm^2 is

$$\text{Transfer} = (24\ W\ m^{-2})\times(1.0\times10^{-4}\ m^2)\times(3600\ s) = 8.6\ J$$

Self-test 19A.1 The thermal conductivity of glass is 0.92 $W\ K^{-1}\ m^{-1}$. What is the rate of energy transfer through a window pane of thickness 0.50 cm and area 1.0 m^2 when the room is at 22 °C and the exterior is at 0 °C?

Answer: 2.8 Kw

Table 19A.1* Transport properties of gases at 1 atm

	$\kappa/(mW\ K^{-1}\ m^{-1})$	$\eta/\mu P^{\ddagger}$	
	273 K	273 K	293 K
Ar	16.3	210	223
CO_2	14.5	136	147
He	144.2	187	196
N_2	24.0	166	176

* More values are given in the *Resource section*.
‡ $1\ \mu P = 10^{-7}\ kg\ m^{-1}\ s^{-1}$.

To see the connection between the flux of momentum and the viscosity, consider a fluid in a state of **Newtonian flow**, which can be imagined as occurring by a series of layers moving past one another (Fig. 19A.2). The layer next to the wall

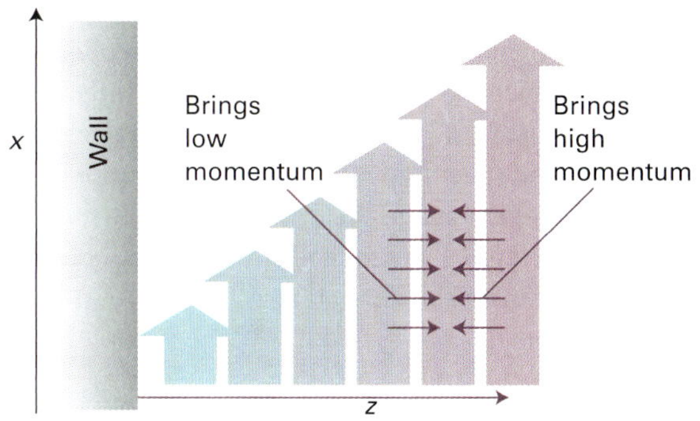

Figure 19A.2 The viscosity of a fluid arises from the transport of linear momentum. In this illustration the fluid is undergoing Newtonian (laminar) flow, and particles bring their initial momentum when they enter a new layer. If they arrive with high x-component of momentum they accelerate the layer; if with low x-component of momentum they retard the layer.

of the vessel is stationary, and the velocity of successive layers varies linearly with distance, z, from the wall. Molecules ceaselessly move between the layers and bring with them the x-component of linear momentum they possessed in their original layer. A layer is retarded by molecules arriving from a more slowly moving layer because they have a low momentum in the x-direction. A layer is accelerated by molecules arriving from a more rapidly moving layer. We interpret the net retarding effect as the fluid's viscosity.

Because the retarding effect depends on the transfer of the x-component of linear momentum into the layer of interest, the viscosity depends on the flux of this x-component in the z-direction. The flux of the x-component of momentum is proportional to dv_x/dz because there is no net flux when all the layers move at the same velocity. We can therefore write

$$J(x\text{-component of momentum}) = -\eta\frac{dv_x}{dz}$$

Momentum flux in terms of the coefficient of viscosity (19A.5)

The constant of proportionality, η, is the **coefficient of viscosity** (or simply 'the viscosity'). Its units are kilograms per metre per second (kg m^{-1} s^{-1}). Viscosities are often reported in poise (P), where 1 P = 10^{-1} kg m^{-1} s^{-1}. Some experimental values are given in Table 19A.1.

Although it is not strictly a transport property, closely related to diffusion is **effusion**, the escape of matter through a small hole. The essential empirical observations on effusion are summarized by **Graham's law of effusion**, which states that the rate of effusion is inversely proportional to the square root of the molar mass, M.

19A.2 The transport parameters

Here we derive expressions for the diffusion characteristics of a perfect gas on the basis of a model, the kinetic-molecular theory. All the expressions depend on knowing the **collision flux**, Z_W, the rate at which molecules strike a region in the gas (which may be an imaginary window in the gas, a part of a wall, or a hole in a wall) and specifically the number of collisions divided by the area of the region and the time interval. We show in the following *Justification* that the collision flux of molecules of mass m at a pressure p and temperature T is

$$Z_W = \frac{p}{(2\pi mkT)^{1/2}}$$

Perfect gas Collision flux (19A.6)

Brief illustration 19A.2 The collision flux

The collision flux of O_2 molecules, with $m = M/N_A$ and $M = 32.00$ g mol^{-1}, at 25 °C and at 1.00 bar is

$$Z_W = \frac{1.00\times10^5\ \overbrace{\text{Pa}}^{\text{kg m}^{-1}\text{s}^{-2}}}{\left\{2\pi\times\dfrac{32.00\times10^{-3}\ \text{kg mol}^{-1}}{6.022\times10^{23}\ \text{mol}^{-1}}\times(1.381\times10^{-23}\ \text{J K}^{-1})\times(298\ \text{K})\right\}^{1/2}}$$
$$= 2.70\times10^{27}\ \text{m}^{-2}\ \text{s}^{-1}$$

This flux corresponds to 2.70×10^{23} cm^{-2} s^{-1}.

Self-test 19A.2 Evaluate the collision flux of H_2 molecules under the same conditions.

Answer: 1.07×10^{28} m^{-2} s^{-1}

Justification 19A.1 The collision flux

Consider a wall of area A perpendicular to the x-axis (Fig. 19A.3). If a molecule has $v_x > 0$ (that is, it is travelling in the direction of positive x), then it will strike the wall within an interval Δt if it lies within a distance $v_x\Delta t$ of the wall. Therefore, all molecules in the volume $Av_x\Delta t$, and with positive x-component of velocities, will strike the wall in the interval Δt. The total number of collisions in this interval is therefore the volume $Av_x\Delta t$ multiplied by the number density, $\mathcal{N}$, of molecules. However, to take account of the presence of a range of velocities in the sample, we must sum the result over all the positive values of v_x weighted by the probability distribution of velocities given in *Justification* 1B.2:

$$f(v_x) = \left(\frac{m}{2\pi kT}\right)^{1/2} e^{-mv_x^2/2kT}$$

That is,

$$\text{Number of collisions} = \mathcal{N}A\Delta t\int_0^\infty v_x f(v_x)\,dv_x$$

The collision flux is the number of collisions divided by A and Δt, so

$$Z_W = \mathcal{N}\int_0^\infty v_x f(v_x)\,dv_x$$

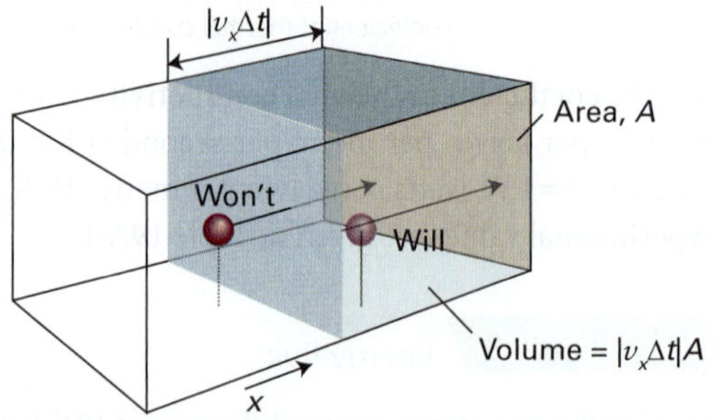

Figure 19A.3 A molecule will reach the wall on the right within an interval Δt if it is within a distance $v_x\Delta t$ of the wall and travelling to the right.

Then, because

$$\int_0^{\infty} v_x f(v_x)\mathrm{d}v_x = \left(\frac{m}{2\pi kT}\right)^{1/2} \int_0^{\infty} v_x \mathrm{e}^{-mv_x^2/2kT}\mathrm{d}v_x \overset{\text{Integral G.2}}{=} \left(\frac{kT}{2\pi m}\right)^{1/2}$$

it follows that

$$Z_\mathrm{W} = \mathcal{N}\left(\frac{kT}{2\pi m}\right)^{1/2}$$

Substitution of $\mathcal{N}=p/kT$ then gives eqn 19A.6.

According to eqn 19A.6, the collision flux increases with pressure, because the rate of collisions on the region of interest increases with pressure. The flux decreases with increasing mass of the molecules because heavy molecules move more slowly that light molecules. Caution, however, is needed with the interpretation of the role of temperature, for it is wrong to conclude that because $T^{1/2}$ appears in the denominator that the collision flux decreases with increasing temperature. If the system has constant volume, the pressure increases with temperature ($p \propto T$), so the collision flux is in fact proportional to $T/T^{1/2} = T^{1/2}$, and increases with temperature (because the molecules are moving faster).

(a) The diffusion coefficient

Consider the arrangement depicted in Fig. 19A.4. The molecules passing through the area A at $z=0$ have travelled an average of about one mean free path λ since their last collision. Therefore, the number density where they originated is $\mathcal{N}(z)$ evaluated at $z=-\lambda$. This number density is approximately

$$\mathcal{N}(-\lambda) = \mathcal{N}(0) - \lambda\left(\frac{\mathrm{d}\mathcal{N}}{\mathrm{d}z}\right)_0 \qquad (19A.7a)$$

where we have used a Taylor expansion of the form $f(x)=f(0)+(\mathrm{d}f/\mathrm{d}x)_0 x + \cdots$ truncated after the second term (see

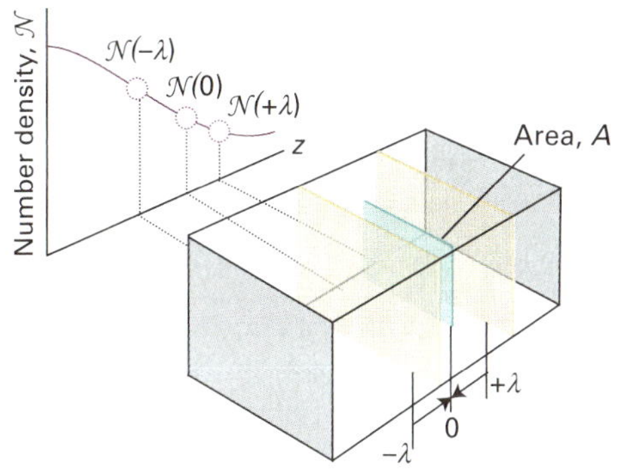

Figure 19A.4 The calculation of the rate of diffusion of a gas considers the net flux of molecules through a plane of area A as a result of arrivals from an average distance λ away in each direction, where λ is the mean free path.

Mathematical Background 1). Similarly, the number density at an equal distance on the other side of the area is

$$\mathcal{N}(\lambda) = \mathcal{N}(0) + \lambda\left(\frac{\mathrm{d}\mathcal{N}}{\mathrm{d}z}\right)_0 \qquad (19A.7b)$$

The average number of impacts on the imaginary region of area A_0 during an interval Δt is $Z_\mathrm{W}A_0\Delta t$, where Z_W is the collision flux. Therefore, the flux from left to right, $J(\mathrm{L}\rightarrow\mathrm{R})$, arising from the supply of molecules on the left, is

$$J(\mathrm{L}\rightarrow\mathrm{R}) = \frac{\overbrace{\tfrac{1}{4}\mathcal{N}(-\lambda)v_\mathrm{mean}}^{Z_\mathrm{W}} A_0\Delta t}{A_0\Delta t} = \tfrac{1}{4}\mathcal{N}(-\lambda)v_\mathrm{mean} \qquad (19A.8a)$$

There is also a flux of molecules from right to left. On average, the molecules making the journey have originated from $z=+\lambda$ where the number density is $\mathcal{N}(\lambda)$. Therefore,

$$J(\mathrm{L}\leftarrow\mathrm{R}) = \tfrac{1}{4}\mathcal{N}(\lambda)v_\mathrm{mean} \qquad (19A.8b)$$

The net flux from left to right is

$$\begin{aligned} J_z &= J(\mathrm{L}\rightarrow\mathrm{R}) - J(\mathrm{L}\leftarrow\mathrm{R}) \\ &= \tfrac{1}{4}v_\mathrm{mean}\{\mathcal{N}(-\lambda)-\mathcal{N}(\lambda)\} \\ &= \tfrac{1}{4}v_\mathrm{mean}\left\{\left[\mathcal{N}(0)-\lambda\left(\frac{\mathrm{d}\mathcal{N}}{\mathrm{d}z}\right)_0\right]-\left[\mathcal{N}(0)+\lambda\left(\frac{\mathrm{d}\mathcal{N}}{\mathrm{d}z}\right)_0\right]\right\} \end{aligned}$$

That is,

$$J_z = -\tfrac{1}{2}v_\mathrm{mean}\lambda\left(\frac{\mathrm{d}\mathcal{N}}{\mathrm{d}z}\right)_0 \qquad (19A.9)$$

This equation shows that the flux is proportional to the first derivative of the concentration, in agreement with Fick's law, eqn 19A.1.

At this stage it looks as though we can pick out a value of the diffusion coefficient by comparing eqns 19A.9 and 19A.3, so obtaining $D=\tfrac{1}{2}\lambda v_\mathrm{mean}$. It must be remembered, however, that the calculation is quite crude, and is little more than an assessment of the order of magnitude of D. One aspect that has not been taken into account is illustrated in Fig. 19A.5, which shows that although a molecule may have begun its journey very close to the window, it could have a long flight before it gets there. Because the path is long, the molecule is likely to collide before reaching the window, so it ought to be added to the graveyard of other molecules that have collided. To take this effect into account involves a lot of work, but the end result is the appearance of a factor of $\tfrac{2}{3}$ representing the lower flux. The modification results in

$$D = \tfrac{1}{3}\lambda v_\mathrm{mean} \qquad \text{Diffusion coefficient} \qquad (19A.10)$$

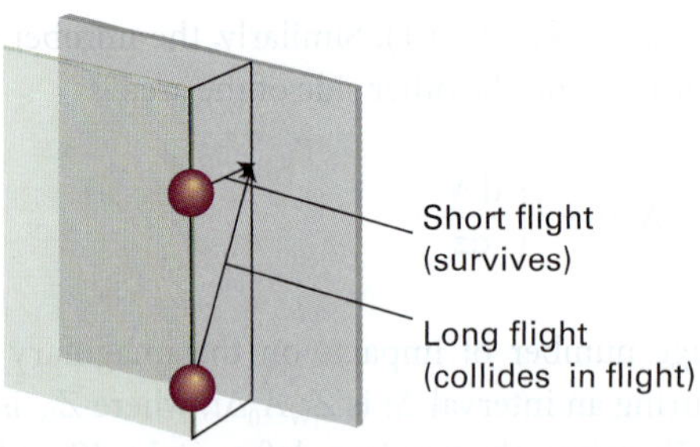

Figure 19A.5 One approximation ignored in the simple treatment is that some particles might make a long flight to the plane even though they are only a short perpendicular distance away, and therefore they have a higher chance of colliding during their journey.

Brief illustration 19A.3 The diffusion coefficient

In *Brief illustration* 1B.4 of Topic 1B it is established that the mean free path of N_2 molecules in a gas at 1.0 bar is 95 nm; in *Example* 1B.1 of the same Topic it is calculated that the mean speed of N_2 molecules at 25 °C is 475 m s^{-1}. Therefore, the diffusion coefficient for N_2 molecules under these conditions is

$$D=\tfrac{1}{3}\times(9.5\times10^{-8}\,\text{m})\times475\,\text{m s}^{-1}=1.5\times10^{-5}\,\text{m}^2\text{s}^{-1}$$

The experimental value (for N_2 in O_2) is 2.0×10^{-5} m^2 s^{-1}.

Self-test 19A.3 Evaluate D for H_2 under the same conditions.

Answer: 9.0×10^{-5} m^2 s^{-1}

There are three points to note about eqn 19A.10:

Physical interpretation

- The mean free path, λ, decreases as the pressure is increased (eqn 1B.13 of Topic 1B, $\lambda=kT/\sigma p$), so D decreases with increasing pressure and, as a result, the gas molecules diffuse more slowly.
- The mean speed, v_{mean}, increases with the temperature (eqn 1B.8 of Topic 1B, $v_{\text{mean}}=(8RT/\pi M)^{1/2}$), so D also increases with temperature. As a result, molecules in a hot sample diffuse more quickly than those in a cool sample (for a given concentration gradient).
- Because the mean free path increases when the collision cross-section σ of the molecules decreases (eqn 1B.13 again, $\lambda=kT/\sigma p$), the diffusion coefficient is greater for small molecules than for large molecules.

(b) Thermal conductivity

According to the equipartition theorem (*Foundations* C), each molecule carries an average energy $\varepsilon=\nu kT$, where ν is a number of the order of 1. For atoms, $\nu=\tfrac{3}{2}$. When one molecule passes through the imaginary window, it transports that average energy. We suppose that the number density is uniform but that the temperature is not. Molecules arrive from the left after travelling a mean free path from their last collision in a hotter region, and therefore with a higher energy. Molecules also arrive from the right after travelling a mean free path from a cooler region. The two opposing energy fluxes are therefore

$$J(\text{L}\leftarrow\text{R})=\overbrace{\tfrac{1}{4}\mathcal{N}v_{\text{mean}}}^{z_{\text{W}}}\,\varepsilon(-\lambda)\quad J(\text{L}\leftarrow\text{R})=\overbrace{\tfrac{1}{4}\mathcal{N}v_{\text{mean}}}^{z_{\text{W}}}\,\varepsilon(\lambda)\qquad(19\text{A}.11)$$

and the net flux is

$$\begin{aligned}J_z&=J(\text{L}\rightarrow\text{R})-J(\text{L}\leftarrow\text{R})\\&=\tfrac{1}{4}v_{\text{mean}}\mathcal{N}\{\varepsilon(-\lambda)-\varepsilon(\lambda)\}\\&=\tfrac{1}{4}v_{\text{mean}}\mathcal{N}\left\{\left[\varepsilon(0)-\lambda\left(\frac{\text{d}\varepsilon}{\text{d}z}\right)_0\right]-\left[\varepsilon(0)+\lambda\left(\frac{\text{d}\varepsilon}{\text{d}z}\right)_0\right]\right\}\end{aligned}$$

That is,

$$J_z=-\tfrac{1}{2}v_{\text{mean}}\lambda\mathcal{N}\left(\frac{\text{d}\varepsilon}{\text{d}z}\right)_0=-\tfrac{1}{2}\nu v_{\text{mean}}\lambda\mathcal{N}k\left(\frac{\text{d}T}{\text{d}z}\right)_0\qquad(19\text{A}.12)$$

The energy flux is proportional to the temperature gradient, as we wanted to show. As before, we multiply by $\tfrac{2}{3}$ to take long flight paths into account, and after comparison of this equation with eqn 19A.4 arrive at

$$\kappa=\tfrac{1}{3}\nu v_{\text{mean}}\lambda\mathcal{N}k\qquad\text{Thermal conductivity}\qquad(19\text{A}.13\text{a})$$

If we now identify $\mathcal{N}=nN_{\text{A}}/V=[\text{J}]N_{\text{A}}$, where [J] is the molar concentration of the carrier particles J and N_{A} is Avogadro's constant, and identify νkN_{A} as the molar constant-volume heat capacity of a perfect gas (which follows from $C_{V,\text{m}}=N_{\text{A}}(\partial\varepsilon/\partial T)_V$), this expression becomes

$$\kappa=\tfrac{1}{3}v_{\text{mean}}\lambda[\text{J}]C_{V,\text{m}}\qquad\text{Thermal conductivity}\qquad(19\text{A}.13\text{b})$$

Yet another form is found by recognizing that $\mathcal{N}=p/kT$ and the expression for D in eqn 19A.10, for then

$$\kappa=\frac{\nu pD}{T}\qquad\text{Thermal conductivity}\qquad(19\text{A}.13\text{c})$$

Brief illustration 19A.4 The thermal conductivity

In *Brief illustration* 19A.3 we calculated $D=1.5\times10^{-5}$ m^2 s^{-1} for N_2 molecules at 25 °C. To use eqn 19A.13c note that for N_2 molecules $\nu=\tfrac{5}{2}$ (there are three translational modes and two rotational modes). Therefore, at 1.0 bar

$$\kappa=\frac{\tfrac{5}{2}\times(1.0\times10^{5}\,\overbrace{\text{Pa}}^{\text{J m}^{-3}})\times(1.5\times10^{-5}\,\text{m}^2\,\text{s}^{-1})}{298\,\text{K}}=1.3\times10^{-2}\,\text{J K}^{-1}\,\text{m}^{-1}\,\text{s}^{-1}$$

or 13 mW K^{-1} m^{-1}. The experimental value is 26 mW K^{-1} m^{-1}.

Self-test 19A.4 Estimate the thermal conductivity of argon gas at 25 °C and 1.0 bar.

Answer: 8 mW K^{-1} m^{-1}

To interpret eqn 19A.13, we note that:

- Because λ is inversely proportional to the pressure (eqn 1B.13 of Topic 1B, $\lambda = kT/\sigma p$), and hence inversely proportional to the molar concentration of the gas, and $\mathcal{N}$ is proportional to the pressure ($\mathcal{N} = p/kT$), the thermal conductivity, which is proportional to the product λp, is independent of the pressure.
- The thermal conductivity is greater for gases with a high heat capacity (eqn 19A.13b) because a given temperature gradient then corresponds to a greater energy gradient.

Physical interpretation

The physical reason for the pressure independence of the thermal conductivity is that it can be expected to be large when many molecules are available to transport the energy, but the presence of so many molecules limits their mean free path and they cannot carry the energy over a great distance. These two effects balance. The thermal conductivity is indeed found experimentally to be independent of the pressure, except when the pressure is very low, when $\kappa \propto p$. At low pressures λ exceeds the dimensions of the apparatus, and the distance over which the energy is transported is determined by the size of the container and not by collisions with the other molecules present. The flux is still proportional to the number of carriers, but the length of the journey no longer depends on λ, so $\kappa \propto$ [A], which implies that $\kappa \propto p$.

(c) Viscosity

Molecules travelling from the right in Fig. 19A.6 (from a fast layer to a slower one) transport a momentum $mv_x(\lambda)$ to their new layer at $z = 0$; those travelling from the left transport $mv_x(-\lambda)$ to it. If it is assumed that the density is uniform, the collision flux is $\frac{1}{4}\mathcal{N}v_{\text{mean}}$. Those arriving from the right on average carry a momentum

$$mv_x(\lambda) = mv_x(0) + m\lambda\left(\frac{dv_x}{dz}\right)_0 \qquad (19A.14a)$$

Those arriving from the left bring a momentum

$$mv_x(-\lambda) = mv_x(0) - m\lambda\left(\frac{dv_x}{dz}\right)_0 \qquad (19A.14b)$$

The net flux of x-momentum in the z-direction is therefore

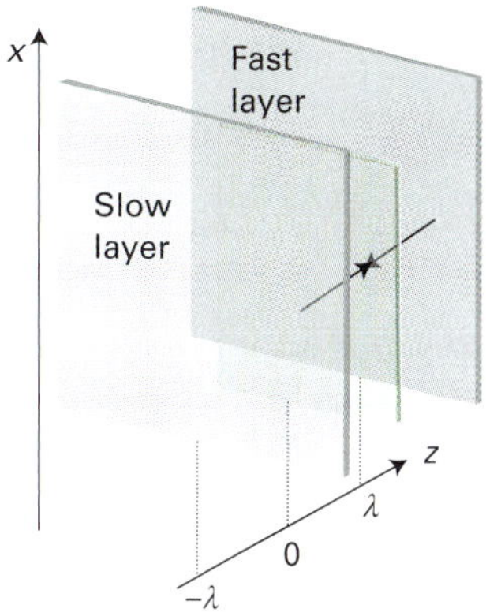

Figure 19A.6 The calculation of the viscosity of a gas examines the net x-component of momentum brought to a plane from faster and slower layers on average a mean free path away in each direction.

$$J_z = \tfrac{1}{4}v_{\text{mean}}\mathcal{N}\left\{\left[mv_x(0) - m\lambda\left(\frac{dv_x}{dz}\right)_0\right] - \left[mv_x(0) + m\lambda\left(\frac{dv_x}{dz}\right)_0\right]\right\} = -\tfrac{1}{2}v_{\text{mean}}\lambda m\mathcal{N}\left(\frac{dv_x}{dz}\right)_0 \qquad (19A.15)$$

The flux is proportional to the velocity gradient, as we wished to show. Comparison of this expression with eqn 19A.5, and multiplication by $\frac{2}{3}$ in the normal way, leads to

$$\eta = \tfrac{1}{3}v_{\text{mean}}\lambda m\mathcal{N} \qquad \text{Viscosity} \quad (19A.16a)$$

Two alternative forms of this expression (after using $mN_A = M$) are

$$\eta = MD[\text{J}] \qquad \text{Viscosity} \quad (19A.16b)$$

$$\eta = \frac{pMD}{RT} \qquad \text{Viscosity} \quad (19A.16c)$$

where [J] is the molar concentration of the gas molecules and M is their molar mass.

Brief illustration 19A.5 The viscosity

We have already calculated $D = 1.5\times10^{-5}$ m^2 s^{-1} for N_2 at 25 °C. Because $M = 28.02$ g mol^{-1}, for the gas at 1.0 bar, eqn 19A.17c gives

$$\eta = \frac{(1.0\times10^5\ \overbrace{\text{Pa}}^{\text{J m}^{-3}})\times(28.02\times10^{-3}\ \text{kg mol}^{-1})\times(1.5\times10^{-5}\ \text{m}^2\,\text{s}^{-1})}{(8.3145\ \text{J K}^{-1}\,\text{mol}^{-1})\times(298\ \text{K})}$$
$$= 1.7\times10^{-5}\ \text{kg m}^{-1}\,\text{s}^{-1}$$

or 171 µP. The experimental value is 176 µP.

Self-test 19A.5 Evaluate the viscosity of benzene vapour at 0.10 bar and 25 °C.

Answer: 140 μP

We can interpret eqn 19A.16a as follows:

- Because $\lambda \propto 1/p$ (eqn 1B.13, $\lambda = kT/\sigma p$)) and [A] $\propto p$, it follows that $\eta \propto \lambda\mathcal{N}$ is independent of p. That is, the viscosity is independent of the pressure.
- Because $v_{\text{mean}} \propto T^{1/2}$ (eqn 1B.8), $\eta \propto T^{1/2}$. That is, the viscosity of a gas *increases* with temperature.

Physical interpretation

The physical reason for the pressure independence of the viscosity is the same as for the thermal conductivity: more molecules are available to transport the momentum, but they carry it less far on account of the decrease in mean free path. The increase of viscosity with temperature is explained when we remember that at high temperatures the molecules travel more quickly, so the flux of momentum is greater. By contrast, as discussed in Topic 19B, the viscosity of a liquid *decreases* with increase in temperature because intermolecular interactions must be overcome.

(d) Effusion

Because the mean speed of molecules is inversely proportional to $M^{1/2}$, the rate at which they strike the area of the hole through which they are effusing is also inversely proportional to $M^{1/2}$, in accord with Graham's law. However, by using the expression for the collision flux, we can obtain a more detailed expression for the rate of effusion and hence use effusion data more effectively.

When a gas at a pressure p and temperature T is separated from a vacuum by a small hole, the rate of escape of its molecules is equal to the rate at which they strike the area of the hole, which is the product of the collision flux and the area of the hole, A_0:

$$\text{Rate of effusion} = Z_W A_0$$

$$= \frac{pA_0}{(2\pi mkT)^{1/2}} \overset{m=M/N_A,\ k=R/N_A}{=} \frac{pA_0N_A}{(2\pi MRT)^{1/2}}$$

Rate of effusion (19A.17)

This rate is inversely proportional to $M^{1/2}$, in accord with Graham's law. Do not conclude from eqn 19A.17, however, that effusion is slower at high temperatures than at low. Because $p \propto T$, the rate is in fact proportional to $T^{1/2}$ and increases with temperature.

Equation 19A.17 is the basis of the **Knudsen method** for the determination of the vapour pressures of liquids and solids, particularly of substances with very low vapour pressures. Thus, if the vapour pressure of a sample is p, and it is enclosed in a cavity with a small hole, then the rate of loss of mass from the container is proportional to p.

Example 19A.1 Calculating the vapour pressure from a mass loss

Caesium (m.p. 29 °C, b.p. 686 °C) was introduced into a container and heated to 500 °C. When a hole of diameter 0.50 mm was opened in the container for 100 s, a mass loss of 385 mg was measured. Calculate the vapour pressure of liquid caesium at 500 K.

Method The pressure of vapour is constant inside the container despite the effusion of atoms because the hot liquid metal replenishes the vapour. The rate of effusion is therefore constant, and given by eqn 19A.17. To express the rate in terms of mass, multiply the number of atoms that escape by the mass of each atom.

Answer The mass loss Δm in an interval Δt is related to the collision flux by

$$\Delta m = Z_W A_0 m \Delta t$$

where A_0 is the area of the hole and m is the mass of one atom. It follows that

$$Z_W = \frac{\Delta m}{A_0 m \Delta t}$$

Because Z_W is related to the pressure by eqn 19A.17, we can write

$$p = \left(\frac{2\pi RT}{M}\right)^{1/2} \frac{\Delta m}{A_0 \Delta t}$$

Because $M = 132.9\ \text{g mol}^{-1}$, substitution of the data gives $p = 8.7\ \text{kPa}$ (using $1\ \text{Pa} = 1\ \text{N m}^{-2} = 1\ \text{J m}^{-1}$), or 65 Torr.

Self-test 19A.6 How long would it take 1.0 g of Cs atoms to effuse out of the oven under the same conditions?

Answer: 260 s

Checklist of concepts

- ☐ 1. **Flux** is the quantity of a property passing through a given area in a given time interval divided by the area and the duration of the interval.
- ☐ 2. **Diffusion** is the migration of matter down a concentration gradient.

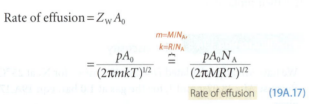

- ☐ 3. **Fick's first law of diffusion** states that the flux of matter is proportional to the concentration gradient.
- ☐ 4. **Thermal conduction** is the migration of energy down a temperature gradient and the flux of energy is proportional to the temperature gradient.
- ☐ 5. **Viscosity** is the migration of linear momentum down a velocity gradient and the flux of momentum is proportional to the velocity gradient.
- ☐ 6. **Effusion** is the emergence of a gas from a container through a small hole.
- ☐ 7. **Graham's law of effusion** states that the rate of effusion is inversely proportional to the square root of the molar mass.
- ☐ 8. The coefficients of diffusion, thermal conductivity, and viscosity of a perfect gas are proportional to the product of the mean free path and mean speed.

Checklist of equations

Property	Equation	Comment	Equation number
Fick's first law of diffusion	$J=-D\mathrm{d}\mathcal{N}/\mathrm{d}z$		19A.3
Flux of energy of thermal motion	$J=-\kappa\mathrm{d}T/\mathrm{d}z$		19A.4
Flux of momentum	$J=-\eta\mathrm{d}v_x/\mathrm{d}z$		19A.5
Diffusion coefficient of a perfect gas	$D=\frac{1}{3}\lambda v_{\mathrm{mean}}$	KMT*	19A.10
Coefficient of thermal conductivity of a perfect gas	$\kappa=\frac{1}{3}v_{\mathrm{mean}}\lambda[\mathrm{J}]C_{V,\mathrm{m}}$	KMT and equipartition	19A.13b
Coefficient of viscosity of a perfect gas	$\eta=\frac{1}{3}v_{\mathrm{mean}}\lambda m\mathcal{N}$	KMT	19A.16a
Rate of effusion	Rate $\propto 1/M^{1/2}$	Graham's law	19A.17

* KMT indicates that the equation is based on the kinetic theory of gases.

19B Motion in liquids

Contents

➤ Why do you need to know this material?

Liquids are central to chemical reactions, and it is important to know how the mobility of their molecules and solutes in them varies with the conditions. Ionic motion is a way of exploring this motion as forces to move them can be applied electrically. From electrical measurements the properties of diffusing neutral molecules may also be inferred.

➤ What is the key idea?

The viscosity of a liquid decreases with increasing temperature; ions reach a terminal velocity when the electrical force on them is balanced by the drag due to the viscosity of the solvent.

➤ What do you need to know already?

The discussion of viscosity starts with the definition of viscosity coefficient introduced in Topic 19A. One derivation uses the same argument about flux as was used in Topic 19A. The final section quotes the relation between the drift speed and a generalized force acting on a solute particle, which is derived in Topic 19C. You need to be aware of several concepts from electrostatics, which are introduced in *Foundations* B.

In this Topic we consider two aspects of motion in liquids. First, we deal with pure liquids, and examine how the mobilities of their molecules, as measured by their viscosity, varies with temperature. Then we consider the motion of solutes. A particularly simple and to some extent controllable type of motion through a liquid, is that of an ion, and we shall see that the information that motion provides can be used to infer the behaviour of uncharged species too.

19B.1 Experimental results

The motion of molecules in liquids can be studied experimentally by a variety of methods. Relaxation time measurements in NMR (Topic 14C) and EPR can be interpreted in terms of the mobilities of the molecules, and have been used to show that big molecules in viscous fluids typically rotate in a series of small (about 5°) steps, whereas small molecules in non-viscous fluids typically jump through about 1 radian (57°) in each step. Another important technique is **inelastic neutron scattering**, in which the energy neutrons collect or discard as they pass through a sample is interpreted in terms of the motion of its particles. The same technique is used to examine the internal dynamics of macromolecules.

(a) Liquid viscosity

The coefficient of viscosity, η (eta), is introduced in Topic 19A as a phenomenological coefficient, the constant of proportionality between the flux of linear momentum and the velocity gradient in a fluid:

$$J_z(x\text{-component of momentum}) = -\eta \frac{dv_x}{dz} \quad \text{Viscosity} \quad (19B.1)$$

(This is eqn 19A.5 of Topic 19A.) Some values for liquids are given in Table 19B.1. The SI units of viscosity are kilograms per metre per second (kg m^{-1} s^{-1}), but they may also be reported in the equivalent units of pascal seconds (Pa s). The non-SI unit poise (P) and centipoise (cP) are still widely encountered: 1 P = 10^{-1} Pa s and so 1 cP = 1 mPa s.

Unlike in a gas, for a molecule to move in a liquid it must acquire at least a minimum energy (an 'activation energy' in the language of Topic 20D) to escape from its neighbours. The probability that a molecule has at least an energy E_a is proportional to $e^{-E_a/RT}$, so the mobility of the molecules in the liquid

Table 19B.1* Viscosities of liquids at 298 K, $\eta/(10^{-3}\,\mathrm{kg\,m^{-1}\,s^{-1}})$

	$\eta/(10^{-3}\,\mathrm{kg\,m^{-1}\,s^{-1}})$
Benzene	0.601
Mercury	1.55
Pentane	0.224
Water‡	0.891

* More values are given in the *Resource section*.
‡ The viscosity of water corresponds to 0.891 cP.

should follow this type of temperature dependence. Because the coefficient of viscosity is inversely proportional to the mobility of the particles, we should expect that

$$\eta = \eta_0 e^{E_a/RT} \quad \text{Temperature dependence of viscosity (liquid)} \quad (19B.2)$$

(Note the positive sign of the exponent, because the viscosity is *inversely* proportion to the mobility.) This expression implies that the viscosity should decrease sharply with increasing temperature. Such a variation is found experimentally, at least over reasonably small temperature ranges (Fig. 19B.1). The activation energy typical of viscosity is comparable to the mean potential energy of intermolecular interactions.

One problem with the interpretation of viscosity measurements is that the change in density of the liquid as it is heated makes a pronounced contribution to the temperature variation of the viscosity. Thus, the temperature dependence of viscosity at constant volume, when the density is constant, is much less than that at constant pressure. The intermolecular interactions between the molecules of the liquid govern the magnitude of E_a, but the problem of calculating it is immensely difficult and still largely unsolved. At low temperatures, the viscosity of water decreases as the pressure is increased. This behaviour is consistent with the need to rupture hydrogen bonds for migration to occur.

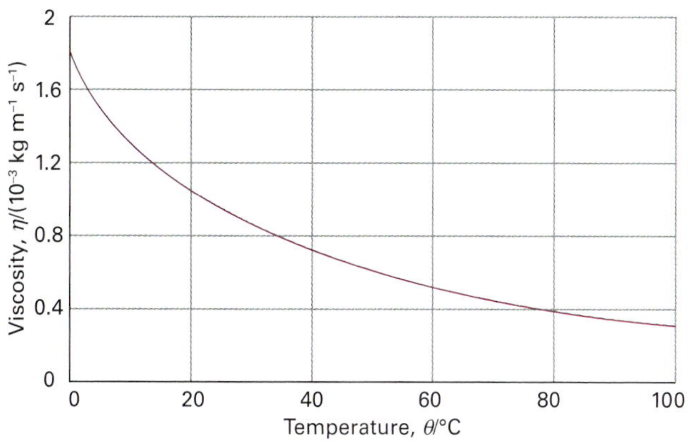

Figure 19B.1 The experimental temperature dependence of the viscosity of water. As the temperature is increased, more molecules are able to escape from the potential wells provided by their neighbours, and so the liquid becomes more fluid.

Brief illustration 19B.1 Liquid viscosity

The viscosity of water at 25 °C and 50 °C is 0.890 mPa s and 0.547 mPa s, respectively. It follows from eqn 19B.2 that the activation energy for molecular migration is the solution of

$$\frac{\eta(T_2)}{\eta(T_1)} = e^{(E_a/R)(1/T_2 - 1/T_1)}$$

which is

$$E_a = \frac{R\ln\{\eta(T_2)/\eta(T_1)\}}{1/T_2 - 1/T_1} = \frac{(8.3145\,\mathrm{J\,K^{-1}\,mol^{-1}})\ln(0.547/0.890)}{1/320\,\mathrm{K} - 1/298\,\mathrm{K}}$$

or 17.5 kJ mol^{-1}. That value is comparable to the strength of a hydrogen bond.

Self-test 19B.1 The corresponding values of the viscosity of benzene are 0.604 mPa s and 0.436 mPa s. Evaluate the activation energy for viscosity.

Answer: 11.7 kJ mol^{-1}

(b) Electrolyte solutions

Further insight into the nature of molecular motion can be obtained by studying the net transport of charged species through solution, for ions can be dragged through the solvent by the application of a potential difference between two electrodes immersed in the sample. By studying the transport of charge through electrolyte solutions it is possible to build up a picture of the events that occur in them and, in some cases, to extrapolate the conclusions to species that have zero charge; that is, to neutral molecules.

The fundamental measurement used to study the motion of ions is that of the electrical resistance, R, of the solution. The **conductance**, G, of a solution is the inverse of its resistance R: $G = 1/R$. As resistance is expressed in ohms, Ω, the conductance of a sample is expressed in Ω^{-1}. The reciprocal ohm used to be called the mho, but its SI designation is now the siemens, S, and $1\,\mathrm{S} = 1\,\Omega^{-1} = 1\,\mathrm{C\,V^{-1}\,s^{-1}}$. It is found that the conductance of a sample decreases with its length l and increases with its cross-sectional area A. We therefore write

$$G = \kappa\frac{A}{l} \quad \textit{Definition of } \kappa \quad \text{Conductivity} \quad (19B.3)$$

where κ (kappa) is the electrical **conductivity**. With the conductance in siemens and the dimensions in metres, it follows that the SI units of κ are siemens per metre (S m^{-1}).

The conductivity of a solution depends on the number of ions present, and it is normal to introduce the **molar conductivity**, Λ_m, which is defined as

$$\Lambda_m = \frac{\kappa}{c} \quad \textit{Definition} \quad \text{Molar conductivity} \quad (19B.4)$$

where c is the molar concentration of the added electrolyte. The SI unit of molar conductivity is siemens metre-squared per mole ($S\,m^2\,mol^{-1}$), and typical values are about $10\,mS\,m^2\,mol^{-1}$ (where $1\,mS = 10^{-3}\,S$).

The values of the molar conductivity as calculated by eqn 19B.4 are found to vary with the concentration. One reason for this variation is that the number of ions in the solution might not be proportional to the nominal concentration of the electrolyte. For instance, the concentration of ions in a solution of a weak electrolyte depends on the concentration of the solute in a complicated way, and doubling the concentration of the solute added does not double the number of ions. Secondly, because ions interact strongly with one another, the conductivity of a solution is not exactly proportional to the number of ions present.

In an extensive series of measurements during the nineteenth century, Friedrich Kohlrausch established the **Kohlrausch law**, that at low concentrations the molar conductivities of strong electrolytes vary linearly with the square root of the concentration:

$$\Lambda_m = \Lambda_m^\circ - \mathcal{K}c^{1/2} \quad \text{Kohlrausch law} \quad (19B.5)$$

He also established that Λ_m°, the **limiting molar conductivity**, the molar conductivity in the limit of zero concentration, is the sum of contributions from its individual ions. If the limiting molar conductivity of the cations is denoted λ_+ and that of the anions λ_-, then his **law of the independent migration of ions** states that

$$\Lambda_m^\circ = \nu_+\lambda_+ + \nu_-\lambda_- \quad \textit{Limiting law} \quad \text{Law of the independent migration of ions} \quad (19B.6)$$

where ν_+ and ν_- are the numbers of cations and anions per formula unit of electrolyte. For example, $\nu_+ = \nu_- = 1$ for HCl, NaCl, and $CuSO_4$, but $\nu_+ = 1$, $\nu_- = 2$ for $MgCl_2$.

Example 19B.1 Determining the limiting molar conductivity

The conductivity of KCl(aq) at 25 °C is $14.668\,mS\,m^{-1}$ when $c = 1.0000\,mmol\,dm^{-3}$ and $71.740\,mS\,m^{-1}$ when $c = 5.0000\,mmol\,dm^{-3}$. Determine the values of the limiting molar conductivity Λ_m° and the Kohlrausch constant $\mathcal{K}$.

Method Use eqn 19B.4 to determine the molar conductivities at the two concentrations, then use the Kohlrausch law, eqn 19B.5, in the form

$$\Lambda_m(c_2) - \Lambda_m(c_1) = \mathcal{K}(c_1^{1/2} - c_2^{1/2})$$

to determine $\mathcal{K}$. Then find Λ_m° from the law in the form

$$\Lambda_m^\circ = \Lambda_m + \mathcal{K}c^{1/2}$$

With more data available, a better procedure is to perform a linear regression.

Answer It follows that the molar conductivity of KCl when $c = 1.0000\,mmol\,dm^{-3}$ (which is the same as $1.0000\,mol\,m^{-3}$) is

$$\Lambda_m = \frac{14.688\,mS\,m^{-1}}{1.0000\,mol\,m^{-3}} = 14.688\,mS\,m^2\,mol^{-1}$$

Similarly, when $c = 5.0000\,mol\,dm^{-3}$ the molar conductivity is $14.348\,mS\,m^2\,mol^{-1}$. It then follows that

$$\begin{aligned}\mathcal{K} &= \frac{\Lambda_m(c_2) - \Lambda_m(c_1)}{c_1^{1/2} - c_2^{1/2}} \\ &= \frac{(14.348 - 14.688)\,mS\,m^2\,mol^{-1}}{(0.0010000^{1/2} - 0.0050000^{1/2})\,(mol\,dm^{-3})^{1/2}} \\ &= 8.698\,mS\,m^2\,mol^{-1}/(mol\,dm^{-3})^{1/2}\end{aligned}$$

(It is best to keep this awkward but convenient array of units as they are rather than converting them to the equivalent $10^{-3/2}\,S\,m^{7/2}\,mol^{-3/2}$.) Now we find the limiting value from the data for $c = 0.0100\,mol\,dm^{-3}$:

$$\begin{aligned}\Lambda_m^\circ &= 14.688\,mS\,m^2\,mol^{-1} + 8.698\frac{mS\,m^2\,mol^{-1}}{(mol\,dm^{-3})^{1/2}} \\ &\times (0.001\,0000\,mol\,dm^{-3})^{1/2} = 14.963\,mS\,m^2\,mol^{-1}\end{aligned}$$

Self-test 19B.2 The conductivity of $KClO_4$(aq) at 25 °C is $13.780\,mS\,m^{-1}$ when $c = 1.000\,mmol\,dm^{-3}$ and $67.045\,mS\,m^{-1}$ when $c = 5.000\,mmol\,dm^{-3}$. Determine the values of the limiting molar conductivity Λ_m° and the Kohlrausch constant $\mathcal{K}$ for this system.

Answer: $\mathcal{K} = 9.491\,mS\,m^2\,mol^{-1}/(mol\,dm^{-3})^{1/2}$, $\Lambda_m^\circ = 14.08\,mS\,m^2\,mol^{-1}$

19B.2 The mobilities of ions

To interpret conductivity measurements we need to know why ions move at different rates, why they have different molar conductivities, and why the molar conductivities of strong electrolytes decrease with the square root of the molar concentration. The central idea in this section is that although the motion of an ion remains largely random, the presence of an electric field biases its motion, and the ion undergoes net migration through the solution.

(a) The drift speed

When the potential difference between two planar electrodes a distance l apart is $\Delta\phi$, the ions in the solution between them experience a uniform electric field of magnitude

$$\mathcal{E} = \frac{\Delta\phi}{l} \quad (19B.7)$$

In such a field, an ion of charge ze experiences a force of magnitude

$$\mathcal{F} = ze\mathcal{E} = \frac{ze\Delta\phi}{l} \qquad \text{Electric force} \qquad (19B.8)$$

where here and throughout this section we disregard the sign of the charge number and so avoid notational complications.

A cation responds to the application of the field by accelerating towards the negative electrode and an anion responds by accelerating towards the positive electrode. However, this acceleration is short lived. As the ion moves through the solvent it experiences a frictional retarding force, $\mathcal{F}_{\text{fric}}$, proportional to its speed. For a spherical particle of radius a travelling at a speed s, this force is given by **Stokes' law**, which was derived by considering the hydrodynamics of the passage of a sphere through a continuous fluid:

$$\mathcal{F}_{\text{fric}} = fs \qquad f = 6\pi\eta a \qquad \text{Stokes' law} \qquad (19B.9)$$

where η is the viscosity. In writing eqn 19B.9, we have assumed that it applies on a molecular scale, and independent evidence from magnetic resonance suggests that it often gives at least the right order of magnitude.

The two forces act in opposite directions, and the ions quickly reach a terminal speed, the **drift speed**, when the accelerating force is balanced by the viscous drag. The net force is zero when $fs = ze\mathcal{E}$, or

$$s = \frac{ze\mathcal{E}}{f} \qquad \text{Drift speed} \qquad (19B.10)$$

It follows that the drift speed of an ion is proportional to the strength of the applied field. We write

$$s = u\mathcal{E} \qquad \textit{Definition of u} \quad \text{Mobility} \qquad (19B.11)$$

where u is called the **mobility** of the ion (Table 19B.2). Comparison of the last two equations shows that

$$u = \frac{ze}{f} = \frac{ze}{6\pi\eta a} \qquad \text{Mobility} \qquad (19B.12)$$

Table 19B.2* Ionic mobilities in water at 298 K, $u/(10^{-8}\,\text{m}^2\,\text{s}^{-1}\,\text{V}^{-1})$

	$u/(10^{-8}\,\text{m}^2\,\text{s}^{-1}\,\text{V}^{-1})$		$u/(10^{-8}\,\text{m}^2\,\text{s}^{-1}\,\text{V}^{-1})$
H^+	36.23	OH^-	20.64
Na^+	5.19	Cl^-	7.91
K^+	7.62	Br^-	8.09
Zn^{2+}	5.47	SO_4^{2-}	8.29

* More values are given in the *Resource section*.

Brief illustration 19B.2 Ion mobility

For an order of magnitude estimate we can take $z=1$ and a the radius of an ion such as Cs^+ (which might be typical of a smaller ion plus its hydration sphere), which is 170 pm. For the viscosity, we use $\eta = 1.0$ cP (1.0 mPa s, Table 19B.1). Then

$$u = \frac{1.6\times10^{-19}\,\overbrace{\text{C}}^{\text{J V}^{-1}}}{6\pi\times\left(1.0\times10^{-3}\,\underbrace{\text{Pa}}_{\text{J m}^{-3}}\,\text{s}\right)\times(170\times10^{-12}\,\text{m})} = 5.0\times10^{-8}\,\text{m}^2\,\text{V}^{-1}\,\text{s}^{-1}$$

This value means that when there is a potential difference of 1 V across a solution of length 1 cm (so $\mathcal{E} = 100\,\text{V m}^{-1}$), the drift speed is typically about $5\,\mu\text{m s}^{-1}$. That speed might seem slow, but not when expressed on a molecular scale, for it corresponds to an ion passing about 10^4 solvent molecules per second.

Self-test 19B.3 The mobility of an SO_4^{2-} ion in water at 25 °C is $8.29\times10^{-8}\,\text{m}^2\,\text{V}^{-1}\,\text{s}^{-1}$. What is its effective radius?

Answer: 205 pm

Because the drift speed governs the rate at which charged species are transported, we might expect the conductivity to decrease with increasing solution viscosity and ion size. Experiments confirm these predictions for bulky ions (such as R_4N^+ and RCO_2^-) but not for small ions. For example, the mobilities of the alkali metal ions in water increase from Li^+ to Cs^+ (Table 19B.2) even though the ionic radii increase. The paradox is resolved when we realize that the radius a in the Stokes formula is the **hydrodynamic radius** (or 'Stokes radius') of the ion, its effective radius in the solution taking into account all the H_2O molecules it carries in its hydration shell. Small ions give rise to stronger electric fields than large ones (the electric field at the surface of a sphere of radius r is proportional to ze/r^2 and it follows that the smaller the radius the stronger the field), so small ions are more extensively solvated than big ions. Thus, an ion of small ionic radius may have a large hydrodynamic radius because it drags many solvent molecules through the solution as it migrates. The hydrating H_2O molecules are often very labile, however, and NMR and isotope studies have shown that the exchange between the coordination sphere of the ion and the bulk solvent is very rapid for ions of low charge but may be slow for ions of high charge.

The proton, although it is very small, has a very high mobility (Table 19B.2)! Proton and ^{17}O-NMR show that the times characteristic of protons hopping from one molecule to the next are about 1.5 ps, which is comparable to the time that inelastic neutron scattering shows it takes a water molecule to

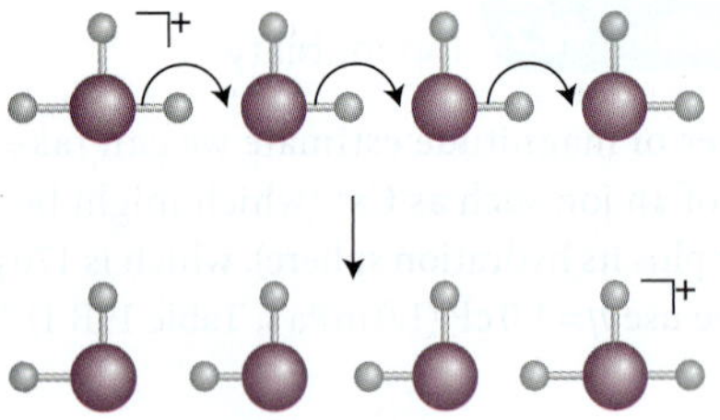

Figure 19B.2 A highly schematic diagram showing the effective motion of a proton in water.

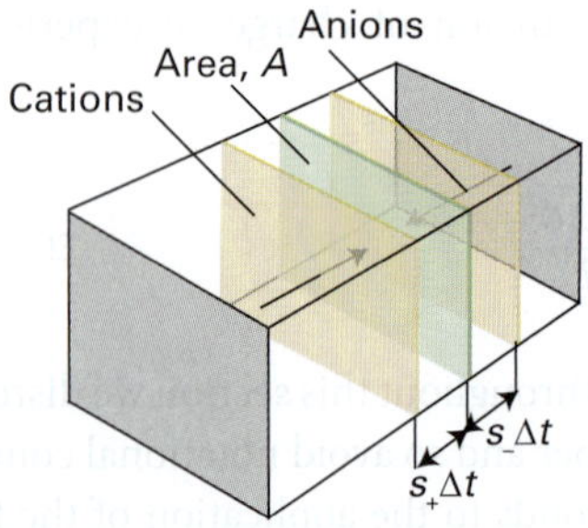

Figure 19B.3 In the calculation of the current, all the cations within a distance $s_+\Delta t$ (that is, those in the volume $s_+A\Delta t$) will pass through the area *A*. The anions in the corresponding volume the other side of the window will also contribute to the current similarly.

reorient through about 1 radian (1 to 2 ps). According to the **Grotthuss mechanism**, there is an effective motion of a proton that involves the rearrangement of bonds in a group of water molecules (Fig. 19B.2). However, the actual mechanism is still highly contentious. The mobility of NH_4^+ in liquid ammonia is also anomalous and presumably occurs by an analogous mechanism.

(b) Mobility and conductivity

Ionic mobilities provide a link between measurable and theoretical quantities. As a first step we establish in the following *Justification* the relation between an ion's mobility and its molar conductivity:

$$\lambda = zuF \quad \text{Ion conductivity} \quad (19B.13)$$

where *F* is Faraday's constant ($F = N_Ae$).

Justification 19B.1 The relation between ionic mobility and molar conductivity

To keep the calculation simple, we ignore signs in the following, and concentrate on the magnitudes of quantities.

Consider a solution of a fully dissociated strong electrolyte at a molar concentration *c*. Let each formula unit give rise to ν_+ cations of charge z_+e and ν_- anions of charge z_-e. The molar concentration of each type of ion is therefore νc (with $\nu = \nu_+$ or ν_-), and the number density of each type is $\nu c N_A$. The number of ions of one kind that pass through an imaginary window of area *A* during an interval Δt is equal to the number within the distance $s\Delta t$ (Fig. 19B.3), and therefore to the number in the volume $s\Delta tA$. (The same sort of argument is used in Topic 19A in the discussion of the transport properties of gases.) The number of ions of that kind in this volume is equal to $s\Delta tA\nu cN_A$. The flux through the window (the number of this type of ion passing through the window divided by the area of the window and the duration of the interval) is therefore

$$J(\text{ions}) = \frac{s\Delta tA\nu cN_A}{\Delta tA} = s\nu cN_A$$

Each ion carries a charge *ze*, so the flux of charge is

$$J(\text{charge}) = zs\nu ceN_A = zs\nu cF$$

Because $s = u\mathcal{E}$, the flux is

$$J(\text{charge}) = zu\nu cF\mathcal{E}$$

The current, *I*, through the window due to the ions we are considering is the charge flux times the area:

$$I = JA = zu\nu cF\mathcal{E}A$$

Because the electric field is the potential gradient (eqn 19B.7, $\mathcal{E} = \Delta\phi/l$), we can write

$$I = \frac{zu\nu cFA\Delta\phi}{l}$$

Current and potential difference are related by Ohm's law, $\Delta\phi = IR$, so it follows that

$$I = \frac{\Delta\phi}{R} = G\Delta\phi = \frac{\kappa A\Delta\phi}{l}$$

where we have used eqn 19B.3 in the form $\kappa = Gl/A$. Comparison of the last two expressions gives $\kappa = zu\nu cF$. Division by the molar concentration of ions, νc, then results in eqn 19B.13.

Equation 19B.13 applies to the cations and to the anions. Therefore, for the solution itself in the limit of zero concentration (when there are no ionic interactions),

$$\Lambda_m^\circ = (z_+u_+\nu_+ + z_-u_-\nu_-)F \quad (19B.14a)$$

For a symmetrical *z*:*z* electrolyte (for example, $CuSO_4$ with $z = 2$), this equation simplifies to

$$\Lambda_m^\circ = z(u_+ + u_-)F \quad (19B.14b)$$

Brief illustration 19B.3 Ionic conductivity

Earlier, we estimated the typical ionic mobility as $5.0\times10^{-8}\,\mathrm{m^2\,V^{-1}\,s^{-1}}$; so, with $z=1$ for both the cation and anion, we can estimate that a typical limiting molar conductivity should be about

$$\begin{aligned}\lambda &= (5.0\times10^{-8}\,\mathrm{m^2\,V^{-1}\,s^{-1}})\times(9.648\times10^{4}\,\mathrm{C\,mol^{-1}})\\ &= 4.8\times10^{-3}\,\mathrm{m^2\,V^{-1}\,s^{-1}\,C\,mol^{-1}}\end{aligned}$$

But $1\,\mathrm{V\,C^{-1}\,s}=1\,\mathrm{S}$ (see the remark preceding eqn 19B.3), so $\lambda\approx5\,\mathrm{mS\,m^2\,mol^{-1}}$, and about twice that value for Λ_m°, in accord with experiment. The experimental value for KCl, for instance, is $15\,\mathrm{mS\,m^2\,mol^{-1}}$.

Self-test 19B.4 Estimate the ionic conductivity of an SO_4^{2-} ion in water at 25 °C from its mobility (Table 19B.2).

Answer: $16\,\mathrm{mS\,m^2\,mol^{-1}}$

(c) The Einstein relations

An important relation between the drift speed s and a force $\mathcal{F}$ of any kind acting on a particle is derived in Topic 19C:

$$s=\frac{D\mathcal{F}}{RT} \qquad \text{Drift speed} \qquad (19\text{B}.15)$$

where D is the diffusion coefficient for the species (Table 19B.3). We have seen that an ion in solution has a drift speed $s=u\mathcal{E}$ when it experiences a force $N_\mathrm{A}ez\mathcal{E}$ from an electric field of strength $\mathcal{E}$. Therefore, substituting these known values into eqn 19B.15 and using $N_\mathrm{A}e=F$ gives $u\mathcal{E}=DFz\mathcal{E}/RT$ and hence, on cancelling the $\mathcal{E}$, we obtain the **Einstein relation**:

$$u=\frac{zDF}{RT} \qquad \text{Einstein relation} \qquad (19\text{B}.16)$$

The Einstein relation provides a link between the molar conductivity of an electrolyte and the diffusion coefficients of its ions. First, by using eqns 19B.13 and 19B.16 we write

$$\lambda=zuF=\frac{z^2DF^2}{RT} \qquad (19\text{B}.17)$$

for each type of ion. Then, from $\Lambda_\mathrm{m}^\circ=\nu_+\lambda_++\nu_-\lambda_-$, the limiting molar conductivity is

$$\Lambda_\mathrm{m}^\circ=(\nu_+z_+^2D_++\nu_-z_-^2D_-)\frac{F^2}{RT} \qquad \text{Nernst–Einstein equation} \qquad (19\text{B}.18)$$

which is the **Nernst–Einstein equation**. An application of this equation is to the determination of ionic diffusion coefficients from conductivity measurements; another is to the prediction of conductivities using models of ionic diffusion.

Table 19B.3* Diffusion coefficients at 298 K, $D/(10^{-9}\,\mathrm{m^2\,s^{-1}})$

Molecules in liquids		Ions in water			
I_2 in hexane	4.05	K^+	1.96	Br^-	2.08
in benzene	2.13	H^+	9.31	Cl^-	2.03
Glycine in water	1.055	Na^+	1.33	I^-	2.05
H_2O in water	2.26			OH^-	5.03
Sucrose in water	0.5216				

* More values are given in the *Resource section*.

Equations 19B.12 ($u=ze/f$) and 19B.16 ($u=zDF/RT$ in the form $u=zDe/kT$) relate the mobility of an ion to the frictional force and to the diffusion coefficient, respectively. We can combine the two expressions and cancel the ze and obtain the **Stokes–Einstein equation**:

$$D=\frac{kT}{f} \qquad \text{Stokes–Einstein equation} \qquad (19\text{B}.19\text{a})$$

If the frictional force is described by Stokes' law, then we also obtain a relation between the diffusion coefficient and the viscosity of the medium:

$$D=\frac{kT}{6\pi\eta a} \qquad \text{Stokes–Einstein equation} \qquad (19\text{B}.19\text{b})$$

An important feature of eqn 19B.19b is that it makes no reference to the charge of the diffusing species. Therefore, the equation also applies in the limit of vanishingly small charge; that is, it also applies to neutral molecules. This feature is taken further in Topic 19C. It must not be forgotten, however, that both equations depend on the assumption that the viscous drag is proportional to the speed.

Brief illustration 19B.4 Mobility and diffusion

From Table 19B.2, the mobility of SO_4^{2-} is $8.29\times10^{-8}\,\mathrm{m^2\,V^{-1}\,s^{-1}}$. It follows from eqn 19B.16 in the form $D=uRT/zF$ that the diffusion coefficient for the ion in water at 25 °C is

$$\begin{aligned}D&=\frac{(8.29\times10^{-8}\,\mathrm{m^2\,V^{-1}\,s^{-1}})\times(8.3145\,\mathrm{J\,K^{-1}\,mol^{-1}})\times(298\,\mathrm{K})}{2\times\left(9.649\times10^{4}\,\underbrace{\mathrm{C}}_{\mathrm{J\,V^{-1}}}\,\mathrm{mol^{-1}}\right)}\\ &=1.06\times10^{-9}\,\mathrm{m^2\,s^{-1}}\end{aligned}$$

Self-test 19B.5 Repeat the calculation for the NH_4^+ ion.

Answer: $1.96\times10^{-9}\,\mathrm{m^2\,s^{-1}}$

Checklist of concepts

- ☐ 1. The **viscosity of a liquid** decreases with increasing temperature.
- ☐ 2. The **conductance**, G, of a solution is the inverse of its resistance.
- ☐ 3. **Kohlrausch's law** states that at low concentrations the molar conductivities of strong electrolytes vary linearly with the square root of the concentration.
- ☐ 4. The **law of the independent migration of ions** states the molar conductivity in the limit of zero concentration, is the sum of contributions from its individual ions.
- ☐ 5. An ion reaches a **drift speed** when the acceleration due to the electrical force is balanced by the viscous drag.
- ☐ 6. The **hydrodynamic radius** of an ion may be greater than its geometrical radius due to solvation.
- ☐ 7. The high mobility of a proton in water is explained by the **Grotthuss mechanism**.

Checklist of equations

Property	Equation	Comment	Equation number
Viscosity of a liquid	$\eta=\eta_0 e^{E_a/RT}$	Narrow temperature range	19B.2
Conductivity	$\kappa=Gl/A,\ G=1/R$	Definition	19B.3
Molar conductivity	$\Lambda_m=\kappa/c$	Definition	19B.4
Kohlrausch's law	$\Lambda_m=\Lambda_m^\circ-\mathcal{K}c^{1/2}$	Empirical observation	19B.5
Law of independent migration of ions	$\Lambda_m^\circ=\nu_+\lambda_++\nu_-\lambda_-$	Limiting law	19B.6
Stokes' law	$\mathcal{F}_{fric}=fs \quad f=6\pi\eta a$	Hydrodynamic radius	19B.9
Drift speed	$s=u\mathcal{E}$	Defines u	19B.11
Ion mobility	$u=ze/6\pi\eta a$	Asumes Stokes' law	19B.12
Conductivity and mobility	$\lambda=zuF$		19B.13
Molar conductivity and mobility	$\Lambda_m^\circ=(z_+u_+\nu_++z_-u_-\nu_-)F$		19B.14a
Drift speed	$s=DF/RT$		19B.15
Einstein relation	$u=zDF/RT$		19B.16
Nernst–Einstein relation	$\Lambda_m^\circ=(\nu_+z_+^2D_++\nu_-z_-^2D_-)(F^2/RT)$		19B.18
Stokes–Einstein relation	$D=kT/f$		19B.19a

19C Diffusion

Contents

➤ Why do you need to know this material?

Diffusion is a hugely important process both in the atmosphere and in solution, and it is important to be able to predict the spread of one material through another when discussing reactions in solution and the spread of substances into the environment. The interpretation of diffusion in terms of a random walk also gives insight into the molecular basis of the process.

➤ What is the key idea?

Particles tend to spread and achieve a uniform distribution.

➤ What do you need to know already?

This Topic draws on arguments relating to flux that are treated in Topic 19A, particularly the way to calculate the flux of particles through a window of given area. This Topic goes into more detail about the diffusion coefficient, which was introduced in Topic 19A and used in Topic 19B. It uses the concept of chemical potential (Topic 5A) to discuss the direction of spontaneous change. One of the *Justifications* develops the random walk model introduced in Topic 17A.

That solutes in gases, liquids, and solids have a tendency to spread can be discussed from three points of view. One viewpoint is from the Second Law of thermodynamics and the tendency for entropy to increase or, if the temperature and pressure are constant, for the Gibbs energy to decrease. When this law is applied to solutes it appears that there is a force acting to disperse the solute. That force is illusory, but it provides an interesting and useful approach to discussing diffusion. The second approach is to set up a differential equation for the change in concentration in a region by considering the flux of material through its boundaries. The resulting 'diffusion equation' can then be solved (in principle, at least) for various configurations of the system, including taking into account the shape of a reaction vessel. The third approach is more mechanistic, and is to imagine diffusion as taking place in a series of random small steps.

19C.1 The thermodynamic view

It is established in Topic 3C that, at constant temperature and pressure, the maximum non-expansion work that can be done per mole when a substance moves from one location to another differing in molar Gibbs energy by $\mathrm{d}G_\mathrm{m}$ is $\mathrm{d}w_\mathrm{e}=\mathrm{d}G_\mathrm{m}$. In terms of the chemical potentials of the substance in the two locations (its partial molar Gibbs energy), $\mathrm{d}w_\mathrm{e}=\mathrm{d}\mu$. In a system in which the chemical potential depends on the position x,

$$\mathrm{d}w=\mathrm{d}\mu=\left(\frac{\partial\mu}{\partial x}\right)_{T,p}\mathrm{d}x$$

It is also shown in Topic 2A that in general work can always be expressed in terms of an opposing force (which here we write $\mathcal{F}$), and that $\mathrm{d}w=-\mathcal{F}\mathrm{d}x$. By comparing these two expressions for $\mathrm{d}w$ we see that the slope of the chemical potential with respect to position can be interpreted as an effective force per mole of molecules. We write this **thermodynamic force** as

$$\mathcal{F}=-\left(\frac{\partial\mu}{\partial x}\right)_{T,p} \qquad \text{Thermodynamic force} \qquad (19\text{C}.1)$$

There is not necessarily a real force pushing the particles down the slope of the chemical potential. As we shall see, the force may represent the spontaneous tendency of the molecules to disperse as a consequence of the Second Law and the hunt for maximum entropy.

In a solution in which the activity of the solute is a, the chemical potential is $\mu=\mu^{\ominus}+RT\ln a$. If the solution is not uniform the activity depends on the position and we can write

$$\mathcal{F}=-RT\left(\frac{\partial \ln a}{\partial x}\right)_{T,p} \qquad (19C.2a)$$

If the solution is ideal, a may be replaced by the molar concentration c, and then

$$\mathcal{F}=-RT\left(\frac{\partial \ln c}{\partial x}\right)_{T,p} \overset{\mathrm{d}\ln y/\mathrm{d}x=(1/y)(\mathrm{d}y/\mathrm{d}x)}{=} -\frac{RT}{c}\left(\frac{\partial c}{\partial x}\right)_{T,p} \qquad (19C.2b)$$

Brief illustration 19C.1 The thermodynamic force

Suppose a linear concentration gradient is set up across a container at 25 °C, with points separated by 1.0 cm differing in concentration by 0.10 mol dm^{-3} around a mean value of 1.0 mol dm^{-3}. According to eqn 19C.2b, the solute experiences a thermodynamic force of magnitude

$$|\mathcal{F}|=\frac{(8.3145\,\mathrm{J\,K^{-1}\,mol^{-1}})\times(298\,\mathrm{K})}{1.0\,\mathrm{mol\,dm^{-3}}}\times\frac{0.10\,\mathrm{mol\,dm^{-3}}}{1.0\times10^{-2}\,\mathrm{m}}$$

$$=2.5\times10^{4}\,\underbrace{\mathrm{J\,m^{-1}}}_{\mathrm{N}}\,\mathrm{mol^{-1}}$$

or 25 kN mol^{-1}. Note that the thermodynamic force is a molar quantity.

Self-test 19C.1 Suppose that the concentration of a solute decreases exponentially as $c(x)=c_0e^{-x/\lambda}$. Derive an expression for the thermodynamic force.

Answer: $\mathcal{F}=RT/\lambda$

In Topic 19A it is established that Fick's first law of diffusion, which we write here in the form

$$J(\text{number})=-D\frac{\mathrm{d}\mathcal{N}}{\mathrm{d}x} \qquad \text{Fick's first law} \qquad (19C.3)$$

can be deduced from the kinetic model of gases. Here we generalize that result and show that it applies to the diffusion of species in condensed phases too. To do so, we suppose that the flux of diffusing particles is a response to a thermodynamic force due to a concentration gradient. The diffusing particles reach a steady 'drift speed', s, when the thermodynamic force, $\mathcal{F}$, is matched by the drag due to the viscosity of the medium. This drift speed is proportional to the thermodynamic force, and we write $s\propto\mathcal{F}$. However, the particle flux, J, is proportional to the drift speed, and the thermodynamic force is proportional to the concentration gradient, $\mathrm{d}c/\mathrm{d}x$. The chain of proportionalities ($J\propto s$, $s\propto\mathcal{F}$, and $\mathcal{F}\propto\mathrm{d}c/\mathrm{d}x$) implies that $J\propto\mathrm{d}c/\mathrm{d}x$, which is the content of Fick's law.

If we divide both sides of eqn 19C.3 by Avogadro's constant, so converting numbers into amounts (numbers of moles), noting that $\mathcal{N}/N_A=(N/V)/N_A=(nN_A/V)/N_A=n/V=c$, the molar concentration, then Fick's law becomes

$$J(\text{amount})=-D\frac{\mathrm{d}c}{\mathrm{d}x} \qquad (19C.4)$$

In this expression, D is the diffusion coefficient and $\mathrm{d}c/\mathrm{d}x$ is the slope of the molar concentration. The flux is related to the drift speed by

$$J(\text{amount})=sc \qquad (19C.5)$$

This relation follows from the argument used in Topic 19A. Thus, all particles within a distance $s\Delta t$, and therefore in a volume $s\Delta tA$, can pass through a window of area A in an interval Δt. Hence, the amount of substance that can pass through the window in that interval is $s\Delta tAc$. The particle flux is this quantity divided by the area A and the time interval Δt, and is therefore simply sc.

By combining the last two equations for J(amount) and using eqn 19C.2b

$$sc=-D\frac{\mathrm{d}c}{\mathrm{d}x}=\frac{Dc\mathcal{F}}{RT} \quad \text{or} \quad s=\frac{D\mathcal{F}}{RT} \qquad (19C.6)$$

Therefore, once we know the effective force and the diffusion coefficient, D, we can calculate the drift speed of the particles (and vice versa) whatever the origin of the force. This equation is used in Topic 19B, where the force is applied electrically to an ion.

Brief illustration 19C.2 The thermodynamic force and the drift speed

Laser measurements showed that a molecule has a drift speed of 1.0 μm s^{-1} in water at 25 °C, with diffusion coefficient 5.0×10^{-9} m^2 s^{-1}. The corresponding thermodynamic force from eqn 19C.6 in the form $\mathcal{F}=sRT/D$ is

$$\mathcal{F}=\frac{(1.0\times10^{-6}\,\mathrm{m\,s^{-1}})\times(8.3145\,\mathrm{J\,K^{-1}\,mol^{-1}})\times(298\,\mathrm{K})}{(5.0\times10^{-9}\,\mathrm{m^2\,s^{-1}})}$$

$$=5.0\times10^{5}\,\underbrace{\mathrm{J\,m^{-1}}}_{\mathrm{N}}\,\mathrm{mol^{-1}}$$

or about 500 kN mol^{-1}.

Self-test 19C.2 What thermodynamic force would achieve a drift speed of 10 μm s^{-1} in water at 25 °C?

Answer: 5.0 MN mol^{-1}

19C.2 The diffusion equation

We now turn to the discussion of time-dependent diffusion processes, where we are interested in the spreading of inhomogeneities with time. One example is the temperature of a metal bar that has been heated at one end: if the source of heat is removed, then the bar gradually settles down into a state of uniform temperature. When the source of heat is maintained and the bar can radiate, it settles down into a steady state of non-uniform temperature. Another example (and one more relevant to chemistry) is the concentration distribution in a solvent to which a solute is added. We shall focus on the description of the diffusion of particles, but similar arguments apply to the diffusion of physical properties, such as temperature. Our aim is to obtain an equation for the rate of change of the concentration of particles in an inhomogeneous region.

(a) Simple diffusion

The central equation of this section is the **diffusion equation**, also called 'Fick's second law of diffusion', which relates the rate of change of concentration at a point to the spatial variation of the concentration at that point:

$$\frac{\partial c}{\partial t}=D\frac{\partial^2 c}{\partial x^2} \qquad \text{Diffusion equation} \qquad (19C.7)$$

We show in the following *Justification* that the diffusion equation follows from Fick's first law of diffusion.

Justification 19C.1 The diffusion equation

Consider a thin slab of cross-sectional area A that extends from x to $x+\lambda$ (Fig. 19C.1). Let the concentration at x be c at the time t. The rate at which the amount (in moles) of particles enter the slab is JA, so the rate of increase in molar concentration inside the slab (which has volume $A\lambda$) on account of the flux from the left is

$$\frac{\partial c}{\partial t}=\frac{JA}{A\lambda}=\frac{J}{\lambda}$$

There is also an outflow through the right-hand window. The flux through that window is J', and the rate of change of concentration that results is

$$\frac{\partial c}{\partial t}=-\frac{J'}{\lambda}$$

The net rate of change of concentration is therefore

$$\frac{\partial c}{\partial t}=\frac{J-J'}{\lambda}$$

Each flux is proportional to the concentration gradient at the respective window. So, by using Fick's first law, we can write

$$J-J'=-D\frac{\partial c}{\partial x}+D\frac{\partial c'}{\partial x}$$

The concentration at the right-hand window is related to that on the left by

$$c'=c+\left(\frac{\partial c}{\partial x}\right)\lambda$$

which implies that

$$J-J'=-D\frac{\partial c}{\partial x}+D\frac{\partial}{\partial x}\left\{c+\left(\frac{\partial c}{\partial x}\right)\lambda\right\}=D\lambda\frac{\partial^2 c}{\partial x^2}$$

When this relation is substituted into the expression for the rate of change of concentration in the slab, we get eqn 19C.7.

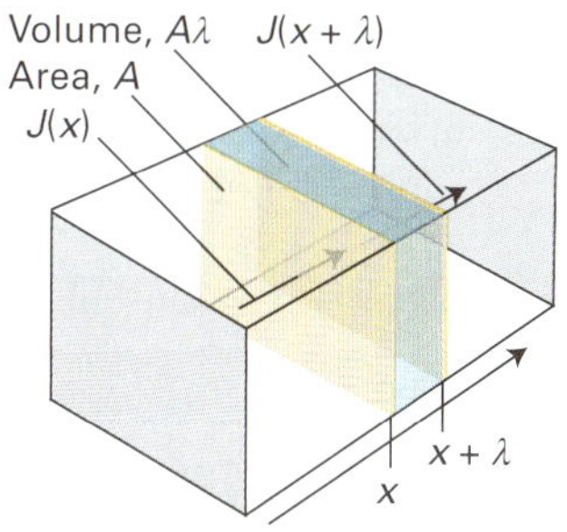

Figure 19C.1 The net flux in a region is the difference between the flux entering from the region of high concentration (on the left) and the flux leaving to the region of low concentration (on the right).

The diffusion equation shows that the rate of change of concentration is proportional to the curvature (more precisely, to the second derivative) of the concentration with respect to distance. If the concentration changes sharply from point to point (i.e. if the distribution is highly wrinkled), then the concentration changes rapidly with time. Where the curvature is positive (a dip, Fig. 19C.2), the change in concentration is positive; the dip tends to fill. Where the curvature is negative (a heap), the change in concentration is negative; the heap tends to spread. If the curvature is zero, then the concentration is constant in time. If the concentration decreases linearly with distance, then the concentration at any point is constant because the inflow of particles is exactly balanced by the outflow.

The diffusion equation can be regarded as a mathematical formulation of the intuitive notion that there is a natural tendency for the wrinkles in a distribution to disappear. More succinctly: Nature abhors a wrinkle.

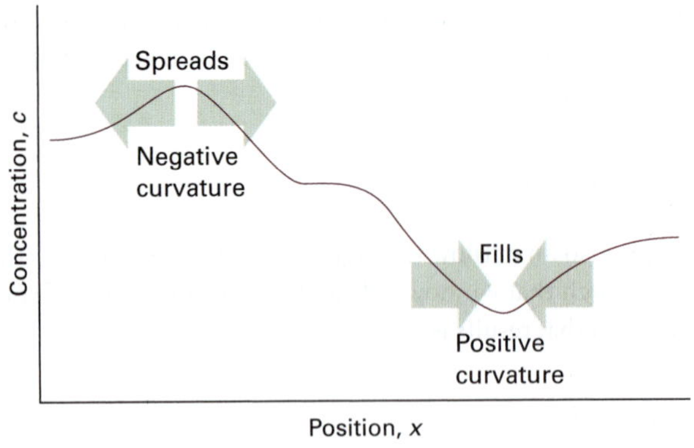

Figure 19C.2 Nature abhors a wrinkle. The diffusion equation tells us that peaks in a distribution (regions of negative curvature) spread and troughs (regions of positive curvature) fill in.

Brief illustration 19C.3 The diffusion equation

If a concentration falls linearly across a small region of space, in the sense that $c=c_0-ax$ then $\partial^2c/\partial x^2=0$ and consequently $\partial c/\partial t=0$. The concentration in the region is constant because the inward flow through one window is matched by the outward flow through the other window (Fig. 19C.3a). If the concentration varies as $c=c_0-\frac{1}{2}ax^2$ then $\partial^2c/\partial x^2=-a$ and consequently $\partial c/\partial t=-Da$. Now the concentration decreases, because there is a greater outward flow than inward flow (Fig. 19C.3b).

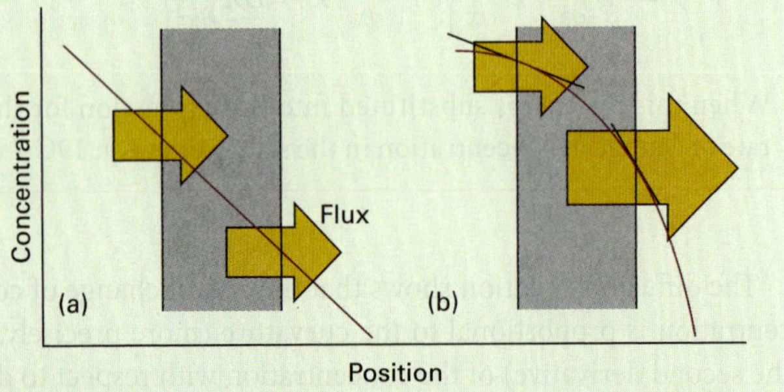

Figure 19C.3 The two instances treated in *Brief illustration* 19C.3: (a) linear concentration gradient, (b) parabolic concentration gradient.

Self-test 19C.3 What is the change in concentration when the concentration falls exponentially across a region? Take $c=c_0e^{-x/\lambda}$.

Answer: $\partial c/\partial t=-(D/\lambda^2)c$

(b) Diffusion with convection

The transport of particles arising from the motion of a streaming fluid is called **convection**. If for the moment we ignore diffusion, then the flux of particles through an area A in an interval Δt when the fluid is flowing at a velocity v can be calculated in the way we have used several times elsewhere (such as in Topic 19A, by counting the particles within a distance $v\Delta t$), and is

$$J_{\text{conv}}=\frac{cAv\Delta t}{A\Delta t}=cv \quad \text{Convective flux} \quad (19\text{C}.8)$$

This J is called the **convective flux**. The rate of change of concentration in a slab of thickness l and area A is, by the same argument as before and assuming that the velocity does not depend on the position,

$$\frac{\partial c}{\partial t}=\frac{J_{\text{conv}}-J'_{\text{conv}}}{\lambda}=\frac{cv}{\lambda}-\left\{c+\left(\frac{\partial c}{\partial x}\right)\lambda\right\}\frac{v}{\lambda}=-\left(\frac{\partial c}{\partial x}\right)v \quad \text{Convection} \quad (19\text{C}.9)$$

When both diffusion and convection occur, the total change of concentration in a region is the sum of the two effects, and the **generalized diffusion equation** is

$$\frac{\partial c}{\partial t}=D\frac{\partial^2 c}{\partial x^2}-v\frac{\partial c}{\partial x} \quad \text{Generalized diffusion equation} \quad (19\text{C}.10)$$

A further refinement, which is important in chemistry, is the possibility that the concentrations of particles may change as a result of reaction. When reactions are included in eqn 19C.10 (which we do in Topic 21B) we get a powerful differential equation for discussing the properties of reacting, diffusing, convecting systems, which is the basis of reactor design in chemical industry and of the utilization of resources in living cells.

Brief illustration 19C.4 Convection

Here we continue the discussion of the systems treated in *Brief illustration* 19C.3 and suppose that there is a convective flow v. If the concentration falls linearly across a small region of space, in the sense that $c=c_0-ax$ then $\partial c/\partial x=-a$ and the change in concentration in the region is $\partial c/\partial t=av$. There is now an increase in the region because the inward convective flow outweighs the outward flow, and there is no diffusion. If $a=0.010\,\text{mol dm}^{-3}\,\text{m}^{-1}$ and $v=+1.0\,\text{mm s}^{-1}$,

$$\frac{\partial c}{\partial t}=(0.010\,\text{mol dm}^{-3}\,\text{m}^{-1})\times(1.0\times10^{-3}\,\text{m s}^{-1})$$
$$=1.0\times10^{-5}\,\text{mol dm}^{-3}\,\text{s}^{-1}$$

and the concentration increases at the rate of $10\,\mu\text{mol dm}^{-3}\,\text{s}^{-1}$.

Self-test 19C.4 What rate of flow is needed to replenish the concentration when the concentration varies exponentially as $c=c_0e^{-x/\lambda}$ across the region?

Answer: $v=D/\lambda$

(c) Solutions of the diffusion equation

The diffusion equation is a second-order differential equation with respect to space and a first-order differential equation with respect to time. Therefore, we must specify two boundary conditions for the spatial dependence and a single initial condition for the time dependence (see *Mathematical background* 4 following Chapter 8).

As an illustration, consider a solvent in which the solute is initially coated on one surface of the container (for example, a layer of sugar on the bottom of a deep beaker of water). The single initial condition is that at $t=0$ all N_0 particles are concentrated on the yz-plane (of area A) at $x=0$. The two boundary conditions are derived from the requirements (1) that the concentration must everywhere be finite and (2) that the total amount (number of moles) of particles present is n_0 (with $n_0=N_0/N_A$) at all times. These requirements imply that the flux of particles is zero at the top and bottom surfaces of the system. Under these conditions it is found that

$$c(x,t)=\frac{n_0}{A(\pi Dt)^{1/2}}e^{-x^2/4Dt} \quad \text{One-dimensional diffusion} \quad (19C.11)$$

as may be verified by direct substitution. Figure 19C.4 shows the shape of the concentration distribution at various times, and it is clear that the concentration spreads and tends to uniformity.

Another useful result is for a localized concentration of solute in a three-dimensional solvent (a sugar lump suspended in a large flask of water). The concentration of diffused solute is spherically symmetrical, and at a radius r is

$$c(x,t)=\frac{n_0}{8(\pi Dt)^{3/2}}e^{-r^2/4Dt} \quad \text{Three-dimensional diffusion} \quad (19C.12)$$

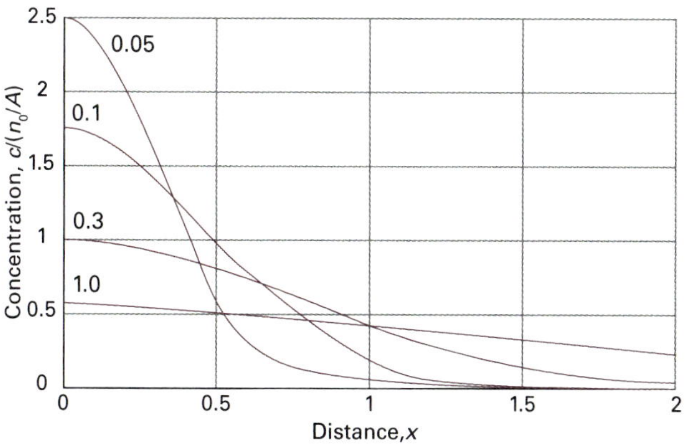

Figure 19C.4 The concentration profiles above a plane from which a solute is diffusing. The curves are plots of eqn 19C.11 and are labelled with different values of Dt. The units of Dt and x are arbitrary, but are related so that Dt/x^2 is dimensionless. For example, if x is in metres, Dt would be in metres^2; so, for $D=10^{-9}$ m^2 s^{-1}, $Dt=0.1$ m^2 corresponds to $t=10^8$ s.

Other chemically (and physically) interesting arrangements, such as transport of substances across biological membranes can be treated. In many cases the solutions are more cumbersome.

The solutions of the diffusion equation are useful for experimental determinations of diffusion coefficients. In the **capillary technique**, a capillary tube, open at one end and containing a solution, is immersed in a well-stirred larger quantity of solvent, and the change of concentration in the tube is monitored. The solute diffuses from the open end of the capillary at a rate that can be calculated by solving the diffusion equation with the appropriate boundary conditions, so D may be determined. In the **diaphragm technique**, the diffusion occurs through the capillary pores of a sintered glass diaphragm separating the well-stirred solution and solvent. The concentrations are monitored and then related to the solutions of the diffusion equation corresponding to this arrangement. Diffusion coefficients may also be measured by a number of techniques, including NMR spectroscopy.

The solutions of the diffusion equation can be used to predict the concentration of particles (or the value of some other physical quantity, such as the temperature in a non-uniform system) at any location. We can also use them to calculate the average displacement of the particles in a given time.

Example 19C.1 Calculating the average displacement

Calculate the average displacement of particles in a time t in a one-dimensional system if they have a diffusion constant D.

Method We need to calculate the probability that a particle will be found at a certain distance from the origin, and then calculate the average by weighting each distance by that probability.

Answer The number of particles in a slab of thickness dx and area A at x, where the molar concentration is c, is $cAN_A dx$. The probability that any of the $N_0=n_0N_A$ particles is in the slab is therefore $cAN_A dx/N_0$. If the particle is in the slab, it has travelled a distance x from the origin. Therefore, the average displacement of all the particles is the sum of each x weighted by the probability of its occurrence:

$$\langle x\rangle=\int_0^\infty x\frac{c(x,t)AN_A}{\underbrace{N_0}_{n_0N_A}}dx=\frac{1}{(\pi Dt)^{1/2}}\int_0^\infty xe^{-x^2/4Dt}dx \overset{\text{Integral G.2}}{=} 2\left(\frac{Dt}{\pi}\right)^{1/2}$$

The average displacement varies as the square root of the lapsed time.

Self-test 19C.5 Derive an expression for the root-mean-square distance travelled by diffusing particles in a time t in a one-dimensional system. You will need Integral G.3 from the *Resource section*.

Answer: $\langle x^2\rangle^{1/2}=(2Dt)^{1/2}$

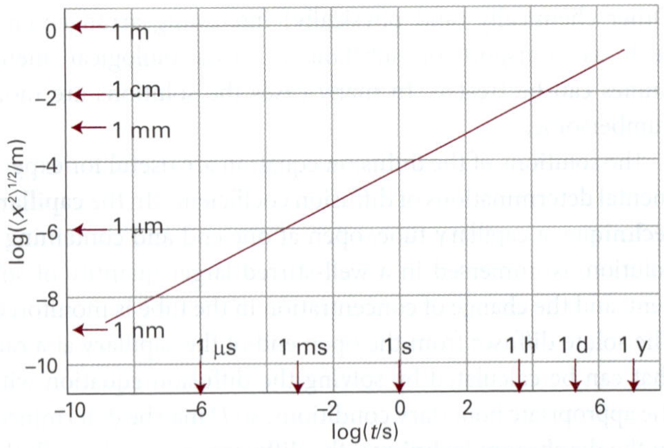

Figure 19C.5 The root-mean-square distance covered by particles with $D=5\times10^{-10}$ m² s⁻¹. Note the great slowness of diffusion.

As shown in *Example* 19C.1, the average displacement of a diffusing particle in a time t in a one-dimensional system is

$$x=2\left(\frac{Dt}{\pi}\right)^{1/2} \quad \textit{One dimension} \quad \text{Mean displacement} \quad (19C.13)$$

and the root-mean-square displacement in the same time is

$$\langle x^2\rangle^{1/2} = (2Dt)^{1/2} \quad \textit{One dimension} \quad \text{Root-mean-square displacement} \quad (19C.14)$$

The latter is a valuable measure of the spread of particles when they can diffuse in both directions from the origin (for then $\langle x\rangle=0$ at all times). The root-mean-square displacement of particles with a typical diffusion coefficient in a liquid ($D=5\times10^{-10}\,\text{m}^2\,\text{s}^{-1}$) is illustrated in Fig. 19C.5, which shows how long it takes for diffusion to increase the net distance travelled on average to about 1 cm in an unstirred solution. The graph shows that diffusion is a very slow process (which is why solutions are stirred, to encourage mixing by convection). The diffusion of pheromones in still air is also very slow, and greatly accelerated by convection.

19C.3 The statistical view

An intuitive picture of diffusion is of the particles moving in a series of small steps and gradually migrating from their original positions. We explore this idea by using a model in which the particles can jump through a distance λ in a time τ. The total distance travelled by a particle in a time t is therefore $t\lambda/\tau$. However, the particle will not necessarily be found at that distance from the origin. The direction of each step may be different, and the net distance travelled must take the changing directions into account.

If we simplify the discussion by allowing the particles to travel only along a straight line (the x-axis), and for each step (to the left or the right) to be through the same distance λ, then we obtain the **one-dimensional random walk**. The same model can be used to discuss the random coil structures of denatured polymers (Topic 17B).

We show in the following *Justification* that the probability of a particle being at a distance x from the origin after a time t is

$$P(x,t)=\left(\frac{2\tau}{\pi t}\right)^{1/2} e^{-x^2\tau/2t\lambda^2} \quad \textit{One dimension} \quad \text{Probability} \quad (19C.15)$$

Justification 19C.2 The one-dimensional random walk

The starting point for this calculation is the expression derived in *Justification* 17A.1, with steps in place of the bonds treated there, for the probability that the net distance $n\lambda$ reached from the origin with $n=N_R-N_L$ after N steps each of length λ, with N_R steps to the right and N_L to the left is

$$P(n\lambda)=\frac{N!}{(N-N_R)!N_R!2^N}$$

As in *Justification* 17A.1, this expression can be developed by making use of Stirling's approximation (Topic 15A) in the form

$$\ln x! \approx \ln(2\pi)^{1/2}+\left(x+\tfrac{1}{2}\right)\ln x - x$$

but here we use the parameter

$$\mu=\frac{N_R}{N}-\tfrac{1}{2}\ll 1$$

which is small because almost half the steps are to the right. The smallness of μ allows us to use the expansion

$$\ln\left(\tfrac{1}{2}\pm\mu\right)=-\ln 2\pm 2\mu-2\mu^2+\cdots$$

and retain terms through $O(\mu^2)$ in the overall expression for $\ln P(n\lambda)$. The final result, after quite a lot of algebra (see Problem 19C.11) is

$$P(n\lambda)=\frac{2^{N+1}e^{-2N\mu^2}}{2^N(2\pi N)^{1/2}}=\frac{2e^{-2N\mu^2}}{(2\pi N)^{1/2}}$$

At this point we recognize that

$$N\mu^2=\frac{(2N_R-N)^2}{4N}=\frac{(N_R-N_L)^2}{4N}=\frac{n^2}{4N}$$

The net distance from the origin is $x=n\lambda$ and the number of steps taken in a time t is $N=t/\tau$, so $N\mu^2=\tau x^2/4t\lambda^2$. Substitution of these quantities into the expression for P gives eqn 19C.15.

The differences of detail between eqns 19C.11 (for one-dimensional diffusion) and 19C.15 arises from the fact that in the present calculation the particles can migrate in either direction from the origin. Moreover, they can be found only at discrete points separated by λ instead of being anywhere on a continuous line. The fact that the two expressions are so similar suggests that diffusion can indeed be interpreted as the outcome of a large number of steps in random directions.

We can now relate the coefficient D to the step length λ and the rate at which the jumps occur. Thus, by comparing the two exponents in eqn 19C.11 and eqn 19C.15 we can immediately write down the **Einstein–Smoluchowski equation**:

$$D=\frac{\lambda^2}{2\tau} \qquad \text{Einstein–Smoluchowski equation} \qquad (19C.16)$$

Brief illustration 19C.5 Random walk

Suppose an SO_4^{2-} ion jumps through its own diameter each time it makes a move in an aqueous solution, then because $D=1.1\times10^{-9}\ m^2\ s^{-1}$ and $a=250$ pm (as deduced from mobility measurements, Topic 19B), it follows from $\lambda=2a$ that

$$\tau=\frac{(2a)^2}{2D}=\frac{2a^2}{D}=\frac{2\times(250\times10^{-12}\ \text{pm})^2}{1.1\times10^{-9}\ \text{m}^2\ \text{s}^{-1}}=1.1\times10^{-10}\ \text{s}$$

or $\tau=110$ ps. Because τ is the time for one jump, the ion makes about 1×10^{10} jumps per second.

Self-test 19C.6 What would be the diffusion constant for a similar ion that is 50 per cent larger than SO_4^{2-} and leaps through its own diameter at only 30 per cent of the rate?

Answer: $2.1\times10^{-9}\ m^2\ s^{-1}$

The Einstein–Smoluchowski equation is the central connection between the microscopic details of particle motion and the macroscopic parameters relating to diffusion. It also brings us back full circle to the properties of the perfect gas treated in Topic 19A. For if we interpret λ/τ as v_{mean}, the mean speed of the molecules, and interpret λ as a mean free path, then we can recognize in the Einstein–Smoluchowski equation essentially the same expression as we obtained from the kinetic model of gases (eqn 19A.10 of Topic 19A, $D=\frac{1}{3}\lambda v_{mean}$). That is, the diffusion of a perfect gas is a random walk with an average step size equal to the mean free path.

Checklist of concepts

- ☐ 1. A **thermodynamic force** represents the spontaneous tendency of the molecules to disperse as a consequence of the Second Law and the hunt for maximum entropy.
- ☐ 2. The **diffusion equation** (Fick's second law; see below) can be regarded as a mathematical formulation of the notion that there is a natural tendency for concentration to become uniform.
- ☐ 3. **Convection** is the transport of particles arising from the motion of a streaming fluid.
- ☐ 4. An intuitive picture of diffusion is of the particles moving in a series of small steps, a **random walk**, and gradually migrating from their original positions.

Checklist of equations

Property	Equation	Comment	Equation number
Thermodynamic force	$\mathcal{F}=-(\partial\mu/\partial x)_{T,p}$	Definition	19C.1
Fick's first law	$J(\text{amount})=-D\text{d}c/\text{d}x$		19C.4
Convective flux	$J=sc$		19C.5
Drift speed	$s=D\mathcal{F}/RT$		19C.6
Diffusion equation	$\partial c/\partial t=D\partial^2c/\partial x^2$	One dimension	19C.7
Generalized diffusion equation	$\partial c/\partial t=D\partial^2c/\partial x^2-v\partial c/\partial x$	One dimension	19C.10

Property	Equation	Comment	Equation number
Mean displacement	$\langle x\rangle = 2(Dt/\pi)^{1/2}$	One-dimensional diffusion	19C.13
Root-mean-square displacement	$\langle x^2\rangle^{1/2} = (2Dt)^{1/2}$	One-dimensional diffusion	19C.14
Probability of displacement	$P(x,t) = (2\tau/\pi t)^{1/2}\,e^{-x^2 t/2\tau\lambda^2}$	One-dimensional random walk	19C.15
Einstein–Smoluchowski equation	$D = \lambda^2/2\tau$	One-dimensional random walk	19C.16

CHAPTER 19 Molecules in motion

TOPIC 19A Transport properties of a perfect gas

Discussion questions

19A.1 Explain how Fick's first law arises from the concentration gradient of gas molecules.

19A.2 Provide molecular interpretations for the dependencies of the diffusion constant and the viscosity on the temperature, pressure, and size of gas molecules.

19A.3 What might be the effect of molecular interactions on the transport properties of a gas?

Exercises

19A.1(a) Calculate the thermal conductivity of argon ($C_{V,m}=12.5\,\mathrm{J\,K^{-1}\,mol^{-1}}$, $\sigma=0.36\,\mathrm{nm^2}$) at 298 K.
19A.1(b) Calculate the thermal conductivity of nitrogen ($C_{V,m}=20.8\,\mathrm{J\,K^{-1}\,mol^{-1}}$, $\sigma=0.43\,\mathrm{nm^2}$) at 298 K.

19A.2(a) Calculate the diffusion constant of argon at 20 °C and (i) 1.00 Pa, (ii) 100 kPa, (iii) 10.0 MPa. If a pressure gradient of $1.0\,\mathrm{bar\,m^{-1}}$ is established in a pipe, what is the flow of gas due to diffusion?
19A.2(b) Calculate the diffusion constant of nitrogen at 20 °C and (i) 100.0 Pa, (ii) 100 kPa, (iii) 20.0 MPa. If a pressure gradient of $1.20\,\mathrm{bar\,m^{-1}}$ is established in a pipe, what is the flow of gas due to diffusion?

19A.3(a) Calculate the flux of energy arising from a temperature gradient of $10.5\,\mathrm{K\,m^{-1}}$ in a sample of argon in which the mean temperature is 280 K.
19A.3(b) Calculate the flux of energy arising from a temperature gradient of $8.5\,\mathrm{K\,m^{-1}}$ in a sample of hydrogen in which the mean temperature is 290 K.

19A.4(a) Use the experimental value of the thermal conductivity of neon (Table 19A.2) to estimate the collision cross-section of Ne atoms at 273 K.
19A.4(b) Use the experimental value of the thermal conductivity of nitrogen (Table 19A.2) to estimate the collision cross-section of N_2 molecules at 298 K.

19A.5(a) In a double-glazed window, the panes of glass are separated by 1.0 cm. What is the rate of transfer of heat by conduction from the warm room (28 °C) to the cold exterior (−15 °C) through a window of area $1.0\,\mathrm{m^2}$? What power of heater is required to make good the loss of heat?
19A.5(b) Two sheets of copper of area $2.00\,\mathrm{m^2}$ are separated by 5.00 cm in air. What is the rate of transfer of heat by conduction from the warm sheet (70 °C) to the cold sheet (0 °C).

19A.6(a) Use the experimental value of the coefficient of viscosity for neon (Table 19A.1) to estimate the collision cross-section of Ne atoms at 273 K.
19A.6(b) Use the experimental value of the coefficient of viscosity for nitrogen (Table 19A.1) to estimate the collision cross-section of the molecules at 273 K.

19A.7(a) Calculate the viscosity of air at (i) 273 K, (ii) 298 K, (iii) 1000 K. Take $\sigma\approx 0.40\,\mathrm{nm^2}$. (The experimental values are 173 μP at 273 K, 182 μP at 20 °C, and 394 μP at 600 °C.)
19A.7(b) Calculate the viscosity of benzene vapour at (i) 273 K, (ii) 298 K, (iii) 1000 K. Take $\sigma\approx 0.88\,\mathrm{nm^2}$.

19A.8(a) A solid surface with dimensions 2.5 mm × 3.0 mm is exposed to argon gas at 90 Pa and 500 K. How many collisions do the Ar atoms make with this surface in 15 s?
19A.8(b) A solid surface with dimensions 3.5 mm × 4.0 cm is exposed to helium gas at 111 Pa and 1500 K. How many collisions do the He atoms make with this surface in 10 s?

19A.9(a) An effusion cell has a circular hole of diameter 2.50 mm. If the molar mass of the solid in the cell is $260\,\mathrm{g\,mol^{-1}}$ and its vapour pressure is 0.835 Pa at 400 K, by how much will the mass of the solid decrease in a period of 2.00 h?
19A.9(b) An effusion cell has a circular hole of diameter 3.00 mm. If the molar mass of the solid in the cell is $300\,\mathrm{g\,mol^{-1}}$ and its vapour pressure is 0.224 Pa at 450 K, by how much will the mass of the solid decrease in a period of 24.00 h?

19A.10(a) A solid compound of molar mass $100\,\mathrm{g\,mol^{-1}}$ was introduced into a container and heated to 400 °C. When a hole of diameter 0.50 mm was opened in the container for 400 s, a mass loss of 285 mg was measured. Calculate the vapour pressure of the compound at 400 °C.
19A.10(b) A solid compound of molar mass $200\,\mathrm{g\,mol^{-1}}$ was introduced into a container and heated to 300 °C. When a hole of diameter 0.50 mm was opened in the container for 500 s, a mass loss of 277 mg was measured. Calculate the vapour pressure of the compound at 300 °C.

19A.11(a) A manometer was connected to a bulb containing carbon dioxide under slight pressure. The gas was allowed to escape through a small pinhole, and the time for the manometer reading to drop from 75 cm to 50 cm was 52 s. When the experiment was repeated using nitrogen (for which $M=28.02\,\mathrm{g\,mol^{-1}}$) the same fall took place in 42 s. Calculate the molar mass of carbon dioxide.
19A.11(b) A manometer was connected to a bulb containing nitrogen under slight pressure. The gas was allowed to escape through a small pinhole, and the time for the manometer reading to drop from 65.1 cm to 42.1 cm was 18.5 s. When the experiment was repeated using a fluorocarbon gas, the same fall took place in 82.3 s. Calculate the molar mass of the fluorocarbon.

19A.12(a) A space vehicle of internal volume $3.0\,\mathrm{m^3}$ is struck by a meteor and a hole of radius 0.10 mm is formed. If the oxygen pressure within the vehicle is initially 80 kPa and its temperature 298 K, how long will the pressure take to fall to 70 kPa?
19A.12(b) A container of internal volume $22.0\,\mathrm{m^3}$ was punctured, and a hole of radius 0.050 mm was formed. If the nitrogen pressure within the container is initially 122 kPa and its temperature 293 K, how long will the pressure take to fall to 105 kPa?

Problems

19A.1‡ A. Fenghour et al. (*J. Phys. Chem. Ref. Data* **24**, 1649 (1995)) compiled an extensive table of viscosity coefficients for ammonia in the liquid and vapour phases. Deduce the effective molecular diameter of NH_3 based on each of the following vapour-phase viscosity coefficients: (a) $\eta=9.08\times10^{-6}\,\mathrm{kg\,m^{-1}\,s^{-1}}$ at 270 K and 1.00 bar; (b) $\eta=1.749\times10^{-5}\,\mathrm{kg\,m^{-1}\,s^{-1}}$ at 490 K and 10.0 bar.

‡ These problems were provided by Charles Trapp and Carmen Giunta.

19A.2 Calculate the ratio of the thermal conductivities of gaseous hydrogen at 300 K to gaseous hydrogen at 10 K. Be circumspect, and think about the modes of motion that are thermally active at the two temperatures.

19A.3 Interstellar space is quite different from the gaseous environments we commonly encounter on Earth. For instance, a typical density of the medium is about 1 atom cm^{-3} and that atom is typically H; the effective temperature due to stellar background radiation is about 10 kK. Estimate the diffusion coefficient and thermal conductivity of H under these conditions. *Comment*: Energy is in fact transferred much more effectively by radiation.

19A.4 A Knudsen cell was used to determine the vapour pressure of germanium at 1000 °C. During an interval of 7200 s the mass loss through a hole of radius 0.50 mm amounted to 43 μg. What is the vapour pressure of germanium at 1000 °C? Assume the gas to be monatomic.

19A.5 An atomic beam is designed to function with (a) cadmium, (b) mercury. The source is an oven maintained at 380 K, there being a small slit of dimensions $1.0\,cm \times 1.0 \times 10^{-3}\,cm$. The vapour pressure of cadmium is 0.13 Pa and that of mercury is 12 Pa at this temperature. What is the atomic current (the number of atoms per second) in the beams?

19A.6 Derive an expression that shows how the pressure of a gas inside an effusion oven (a heated chamber with a small hole in one wall) varies with time if the oven is not replenished as the gas escapes. Then show that $t_{1/2}$, the time required for the pressure to decrease to half its initial value, is independent of the initial pressure. *Hint.* Begin by setting up a differential equation relating dp/dt to $p = NkT/V$, and then integrating it.

TOPIC 19B Motion in liquids

Discussion questions

19B.1 Discuss the difference between the hydrodynamic radius of an ion and its ionic radius and explain why a small ion can have a large hydrodynamic radius.

19B.2 Discuss the mechanism of proton conduction in water. How could the model be tested?

19B.3 Why is a proton less mobile in liquid ammonia than in water?

Exercises

19B.1(a) The viscosity of water at 20 °C is 1.002 cP and 0.7975 cP at 30 °C. What is the energy of activation for the transport process?
19B.1(b) The viscosity of mercury at 20 °C is 1.554 cP and 1.450 cP at 40 °C. What is the energy of activation for the transport process?

19B.2(a) The mobility of a chloride ion in aqueous solution at 25 °C is $7.91 \times 10^{-8}\,m^2\,s^{-1}\,V^{-1}$. Calculate the molar ionic conductivity.
19B.2(b) The mobility of an acetate ion in aqueous solution at 25 °C is $4.24 \times 10^{-8}\,m^2\,s^{-1}\,V^{-1}$. Calculate the molar ionic conductivity.

19B.3(a) The mobility of a Rb^+ ion in aqueous solution is $7.92 \times 10^{-8}\,m^2\,s^{-1}\,V^{-1}$ at 25 °C. The potential difference between two electrodes placed in the solution is 25.0 V. If the electrodes are 7.00 mm apart, what is the drift speed of the Rb^+ ion?
19B.3(b) The mobility of a Li^+ ion in aqueous solution is $4.01 \times 10^{-8}\,m^2\,s^{-1}\,V^{-1}$ at 25 °C. The potential difference between two electrodes placed in the solution is 24.0 V. If the electrodes are 5.0 mm apart, what is the drift speed of the ion?

19B.4(a) The limiting molar conductivities of NaI, $NaNO_3$, and $AgNO_3$ are $12.69\,mS\,m^2\,mol^{-1}$, $12.16\,mS\,m^2\,mol^{-1}$ and $13.34\,mS\,m^2\,mol^{-1}$, respectively (all at 25 °C). What is the limiting molar conductivity of AgI at this temperature?
19B.4(b) The limiting molar conductivities of KF, KCH_3CO_2, and $Mg(CH_3CO_2)_2$ are $12.89\,mS\,m^2\,mol^{-1}$, $11.44\,mS\,m^2\,mol^{-1}$ and $18.78\,mS\,m^2\,mol^{-1}$, respectively (all at 25 °C). What is the limiting molar conductivity of MgF_2 at this temperature?

19B.5(a) At 25 °C the molar ionic conductivities of Li^+, Na^+, and K^+ are $3.87\,mS\,m^2\,mol^{-1}$, $5.01\,mS\,m^2\,mol^{-1}$, and $7.35\,mS\,m^2\,mol^{-1}$, respectively. What are their mobilities?
19B.5(b) At 25 °C the molar ionic conductivities of F^-, Cl^-, and Br^- are $5.54\,mS\,m^2\,mol^{-1}$, $7.635\,mS\,m^2\,mol^{-1}$, and $7.81\,mS\,m^2\,mol^{-1}$, respectively. What are their mobilities?

19B.6(a) Estimate the effective radius of a sucrose molecule in water at 25 °C given that its diffusion coefficient is $5.2 \times 10^{-10}\,m^2\,s^{-1}$ and that the viscosity of water is 1.00 cP.
19B.6(b) Estimate the effective radius of a glycine molecule in water at 25 °C given that its diffusion coefficient is $1.055 \times 10^{-9}\,m^2\,s^{-1}$ and that the viscosity of water is 1.00 cP.

19B.7(a) The mobility of a NO_3^- ion in aqueous solution at 25°C is $7.40 \times 10^{-8}\,m^2\,s^{-1}\,V^{-1}$. Calculate its diffusion coefficient in water at 25 °C.
19B.7(b) The mobility of a $CH_3CO_2^-$ ion in aqueous solution at 25°C is $4.24 \times 10^{-8}\,m^2\,s^{-1}\,V^{-1}$. Calculate its diffusion coefficient in water at 25 °C.

Problems

19B.1 The viscosity of benzene varies with temperature as shown in the following table. Use the data to infer the activation energy for viscosity.

θ/°C	10	20	30	40	50	60	70
η/cP	0.758	0.652	0.564	0.503	0.442	0.392	0.358

19B.2 An empirical expression that reproduces the viscosity of water in the range 20–100 °C is

$$\log\frac{\eta}{\eta_{20}} = \frac{1.3272(20-\theta/°C)-0.001053(20-\theta/°C)^2}{\theta/°C+105}$$

where η_{20} is the viscosity at 20 °C. Explore (by using mathematical software) the possibility of fitting an exponential curve to this expression and hence identifying an activation energy for the viscosity.

19B.3 The conductivity of aqueous ammonium chloride at a series of concentrations is listed in the following table. Deduce the molar conductivity and determine the parameters that occur in Kohlrausch's law.

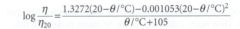

$c/(\text{mol dm}^{-3})$	1.334	1.432	1.529	1.672	1.725
$\kappa/(\text{mS cm}^{-1})$	131	139	147	156	164

19B.4 Conductivities are often measured by comparing the resistance of a cell filled with the sample to its resistance when filled with some standard solution, such as aqueous potassium chloride. The conductivity of water is 76 mS m^{-1} at 25 °C and the conductivity of 0.100 mol dm^{-3} KCl(aq) is 1.1639 S m^{-1}. A cell had a resistance of 33.21 Ω when filled with 0.100 mol dm^{-3} KCl(aq) and 300.0 Ω when filled with 0.100 mol dm^{-3} CH_3COOH(aq). What is the molar conductivity of acetic acid at that concentration and temperature?

19B.5 The resistances of a series of aqueous NaCl solutions, formed by successive dilution of a sample, were measured in a cell with cell constant (the constant C in the relation $\kappa = C/R$) equal to 0.2063 cm^{-1}. The following values were found:

$c/(\text{mol dm}^{-3})$	0.00050	0.0010	0.0050	0.010	0.020	0.050
R/Ω	3314	1669	342.1	174.1	89.08	37.14

Verify that the molar conductivity follows Kohlrausch's law and find the limiting molar conductivity. Determine the coefficient $\mathcal{K}$. Use the value of $\mathcal{K}$ (which should depend only on the nature, not the identity, of the ions) and the information that $\lambda(\text{Na}^+) = 5.01$ mS m^2 mol^{-1} and $\lambda(\text{I}^-) = 7.68$ mS m^2 mol^{-1} to predict (a) the molar conductivity, (b) the conductivity, (c) the resistance it would show in the cell of 0.010 mol dm^{-3} NaI(aq) at 25 °C.

19B.6 What are the drift speeds of Li^+, Na^+, and K^+ in water when a potential difference of 100 V is applied across a 5.00 cm conductivity cell? How long would it take an ion to move from one electrode to the other? In conductivity measurements it is normal to use alternating current: what are the displacements of the ions in (a) centimetres, (b) solvent diameters, about 300 pm, during a half cycle of 2.0 kHz applied potential difference?

19B.7‡ G. Bakale, et al. (*J. Phys. Chem.* **100**, 12477 (1996)) measured the mobility of singly charged C_{60}^- ions in a variety of nonpolar solvents. In cyclohexane at 22 °C, the mobility is 1.1 cm^2 V^{-1} s^{-1}. Estimate the effective radius of the C_{60}^- ion. The viscosity of the solvent is 0.93×10^{-3} kg m^{-1} s^{-1}. Suggest a reason why there is a substantial difference between this number and the van der Waals radius of neutral C_{60}.

19B.8 Estimate the diffusion coefficients and the effective hydrodynamic radii of the alkali metal cations in water from their mobilities at 25 °C. Estimate the approximate number of water molecules that are dragged along by the cations. Ionic radii are given Table 18B.2.

19B.9 Nuclear magnetic resonance can be used to determine the mobility of molecules in liquids. A set of measurements on methane in carbon tetrachloride showed that its diffusion coefficient is 2.05×10^{-9} m^2 s^{-1} at 0 °C and 2.89×10^{-9} m^2 s^{-1} at 25 °C. Deduce what information you can about the mobility of methane in carbon tetrachloride.

19B.10‡ A dilute solution of a weak (1,1)-electrolyte contains both neutral ion pairs and ions in equilibrium (AB ⇌ $A^+ + B^-$). Prove that molar conductivities are related to the degree of ionization by the equations:

$$\frac{1}{\Lambda_m} = \frac{1}{\Lambda_m(\alpha)} + \frac{(1-\alpha)\Lambda_m^\circ}{\alpha^2 \Lambda_m(\alpha)^2}, \quad \Lambda_m(\alpha) = \Lambda_m^\circ - \mathcal{K}(\alpha c)^{1/2}$$

where Λ_m° is the molar conductivity at infinite dilution and $\mathcal{K}$ is the constant in Kohlrausch's law.

TOPIC 19C Diffusion

Discussion questions

19C.1 Describe the origin of the thermodynamic force. To what extent can it be regarded as an actual force?

19C.2 Account physically for the form of the diffusion equation.

Exercises

19C.1(a) The diffusion coefficient of glucose in water at 25 °C is 6.73×10^{-10} m^2 s^{-1}. Estimate the time required for a glucose molecule to undergo a root-mean-square displacement of 5.0 mm.

19C.1(b) The diffusion coefficient of H_2O in water at 25 °C is 2.26×10^{-9} m^2 s^{-1}. Estimate the time required for an H_2O molecule to undergo a root-mean-square displacement of 1.0 cm.

19C.2(a) A layer of 20.0 g of sucrose is spread uniformly over a surface of area 5.0 cm^2 and covered in water to a depth of 20 cm. What will be the molar concentration of sucrose molecules at 10 cm above the original layer at (i) 10 s, (ii) 24 h? Assume diffusion is the only transport process and take $D = 5.216 \times 10^{-9}$ m^2 s^{-1}.

19C.2(b) A layer of 10.0 g of iodine is spread uniformly over a surface of area 10.0 cm^2 and covered in hexane to a depth of 10 cm. What will be the molar concentration of sucrose molecules at 5.0 cm above the original layer at (i) 10 s, (ii) 24 h? Assume diffusion is the only transport process and take $D = 4.05 \times 10^{-9}$ m^2 s^{-1}.

19C.3(a) Suppose the concentration of a solute decays linearly along the length of a container. Calculate the thermodynamic force on the solute at 25 °C and 10 cm and 20 cm given that the concentration falls to half its value in 10 cm.

19C.3(b) Suppose the concentration of a solute increases as x^2 along the length of a container. Calculate the thermodynamic force on the solute at 25 °C and 8 cm and 16 cm given that the concentration falls to half its value in 8 cm.

19C.4(a) Suppose the concentration of a solute follows a Gaussian distribution (proportional to e^{-x^2}) along the length of a container. Calculate the thermodynamic force on the solute at 20 °C and 5.0 cm given that the concentration falls to half its value in 5.0 cm.

19C.4(b) Suppose the concentration of a solute follows a Gaussian distribution (proportional to e^{-x^2}) along the length of a container. Calculate the thermodynamic force on the solute at 18 °C and 10.0 cm given that the concentration falls to half its value in 10.0 cm.

19C.5(a) The diffusion coefficient of CCl_4 in heptane at 25 °C is 3.17×10^{-9} m^2 s^{-1}. Estimate the time required for a CCl_4 molecule to have a root-mean-square displacement of 5.0 mm.

19C.5(b) The diffusion coefficient of I_2 in hexane at 25 °C is 4.05×10^{-9} m^2 s^{-1}. Estimate the time required for an iodine molecule to have a root-mean-square displacement of 1.0 cm.

19C.6(a) Estimate the effective radius of a sucrose molecule in water 25 °C given that its diffusion coefficient is 5.2×10^{-10} m^2 s^{-1} and that the viscosity of water is 1.00 cP.

19C.6(b) Estimate the effective radius of a glycine molecule in water at 25 °C given that its diffusion coefficient is $1.055 \times 10^{-9}\,m^2\,s^{-1}$ and that the viscosity of water is 1.00 cP.

19C.7(a) The diffusion coefficient for molecular iodine in benzene is $2.13 \times 10^{-9}\,m^2\,s^{-1}$. How long does a molecule take to jump through about one molecular diameter (approximately the fundamental jump length for translational motion)?

19C.7(b) The diffusion coefficient for CCl_4 in heptane is $3.17 \times 10^{-9}\,m^2\,s^{-1}$. How long does a molecule take to jump through about one molecular diameter (approximately the fundamental jump length for translational motion)?

19C.8(a) What are the root-mean-square distances travelled by an iodine molecule in benzene and by a sucrose molecule in water at 25 °C in 1.0 s?
19C.8(b) About how long, on average, does it take for the molecules in Exercise 19C.8(a) to drift to a point (i) 1.0 mm, (ii) 1.0 cm from their starting points?

Problems

19C.1 A dilute solution of potassium permanganate in water at 25 °C was prepared. The solution was in a horizontal tube of length 10 cm, and at first there was a linear gradation of intensity of the purple solution from the left (where the concentration was $0.100\,mol\,dm^{-3}$) to the right (where the concentration was $0.050\,mol\,dm^{-3}$). What is the magnitude and sign of the thermodynamic force acting on the solute (a) close to the left face of the container, (b) in the middle, (c) close to the right face? Give the force per mole and force per molecule in each case.

19C.2 A dilute solution of potassium permanganate in water at 25 °C was prepared. The solution was in a horizontal tube of length 10 cm, and at first there was a Gaussian distribution of concentration around the centre of the tube at $x=0$, $c(x)=c_0 e^{-ax^2}$, with $c_0=0.100\,mol\,dm^{-3}$ and $a=0.10\,cm^{-2}$. Determine the thermodynamic force acting on the solute as a function of location, x, and plot the result. Give the force per mole and force per molecule in each case. What do you expect to be the consequence of the thermodynamic force?

19C.3 Instead of a Gaussian 'heap' of solute, as in Problem 19C.2, suppose that there is a Gaussian dip, a distribution of the form $c(x)=c_0(1-e^{-ax^2})$. Repeat the calculation in Problem 19C.2 and its consequences.

19C.4 A lump of sucrose of mass 10.0 g is suspended in the middle of a spherical flask of water of radius 10 cm at 25 °C. What is the concentration of sucrose at the wall of the flask after (a) 1.0 h, (b) 1.0 week? Take $D=5.22 \times 10^{-10}\,m^2\,s^{-1}$.

19C.5 Confirm that eqn 19C.11 is a solution of the diffusion equation with the correct initial value.

19C.6 Confirm that

$$c(x)=\frac{c_0}{(4\pi Dt)^{1/2}}e^{-(x-x_0-vt)^2/4Dt}$$

is a solution of the diffusion equation with convection (eqn 19C.10) with all the solute concentrated at $x=x_0$ at $t=0$ and plot the concentration profile at a series of times to show how the distribution spreads and its centroid drifts.

19C.7 Calculate the relation between $\langle x^2\rangle^{1/2}$ and $\langle x^4\rangle^{1/4}$ for diffusing particles at a time t if they have a diffusion constant D.

19C.8 The thermodynamic force has a direction as well as a magnitude, and in a three-dimensional ideal system eqn 19C.7 becomes $\mathcal{F}=-RT\Delta(\ln c)$. What is the thermodynamic force acting to bring about the diffusion summarized by eqn 19C.12 (that of a solute initially suspended at the centre of a flask of solvent)? *Hint*: Use $\nabla = \boldsymbol{i}\partial/\partial x+\boldsymbol{j}\partial/\partial y+\boldsymbol{k}\partial/\partial z$.

19C.9 The diffusion equation is valid when many elementary steps are taken in the time interval of interest, but the random walk calculation lets us discuss distributions for short times as well as for long. Use the expression $P(n\lambda)=N!/(N-N_R)!N_R!2^N$ to calculate the probability of being six paces from the origin (that is, at $x=6\lambda$) after (a) four, (b) six, (c) twelve steps.

19C.10 Use mathematical software to calculate $P(n\lambda)$ in a one-dimensional random walk, and evaluate the probability of being at $x=n\lambda$ for $n=6, 10, 14, \ldots, 60$. Compare the numerical value with the analytical value in the limit of a large number of steps. At what value of n is the discrepancy no more than 0.1 per cent? Use $n=6$ and $N=6, 8, \ldots, 180$.

19C.11 Supply the intermediate mathematical steps in *Justification* 19C.2.

19C.12 The diffusion coefficient of a particular kind of t-RNA molecule is $D=1.0 \times 10^{-11}\,m^2\,s^{-1}$ in the medium of a cell interior. How long does it take molecules produced in the cell nucleus to reach the walls of the cell at a distance 1.0 μm, corresponding to the radius of the cell?

19C.13‡ In this problem, we examine a model for the transport of oxygen from air in the lungs to blood. First, show that, for the initial and boundary conditions $c(x,t)=c(x,0)=c_o$, $(0<x<\infty)$ and $c(0,t)=c_s$ $(0\le t\le\infty)$ where c_o and c_s are constants, the concentration, $c(x,t)$, of a species is given by

$$c(x,t)=c_o+(c_s-c_o)\{1-\mathrm{erf}(\xi)\}$$

where erf(ξ) is the error function (see the collection of integrals in the *Resource section*) and the concentration $c(x,t)$ evolves by diffusion from the yz-plane of constant concentration, such as might occur if a condensed phase is absorbing a species from a gas phase. Now draw graphs of concentration profiles at several different times of your choice for the diffusion of oxygen into water at 298 K (when $D=2.10 \times 10^{-9}\,m^2\,s^{-1}$) on a spatial scale comparable to passage of oxygen from lungs through alveoli into the blood. Use $c_o=0$ and set c_s equal to the solubility of oxygen in water. *Hint*: Use mathematical software.

Integrated activities

19.1 Use mathematical software, a spreadsheet, or the *Living graphs* on the web site of this book to generate a family of curves similar to that shown in Fig. 19C.4 but by using eqn 19C.14, which describes diffusion in three dimensions.

19.2 In Topic 20D it is shown that a general expression for the activation energy of a chemical reaction is $E_a = RT^2(\mathrm{d}\ln k/\mathrm{d}T)$. Confirm that the same expression may be used to extract the activation energy from eqn 19B.2 for the viscosity and then apply the expression to deduce the temperature-dependence of the activation energy when the viscosity of water is given by the empirical expression in Problem 19B.2. Plot this activation energy as a function of temperature. Suggest an explanation of the temperature dependence of E_a.

CHAPTER 20

Chemical kinetics

This chapter introduces the principles of 'chemical kinetics', the study of reaction rates. The rate of a chemical reaction might depend on variables under our control, such as the pressure, the temperature, and the presence of a catalyst, and we may be able to optimize the rate by the appropriate choice of conditions. Here we begin to see how such manipulations are possible. In the remaining chapters of the text we develop this material in more detail and apply it to more complicated or more specialized cases.

20A The rates of chemical reactions

This Topic discusses the definition of reaction rate and outlines the techniques for its measurement. The results of such measurements show that reaction rates depend on the concentration of reactants (and products) and 'rate constants' that are characteristic of the reaction. This dependence can be expressed in terms of differential equations known as 'rate laws'.

20B Integrated rate laws

'Integrated rate laws' are the solutions of the differential equations that describe rate laws. They are used to predict the concentrations of species at any time after the start of the reaction and to provide procedures for measuring rate constants. This Topic explores simple yet very useful integrated rate laws that appear throughout the chapter.

20C Reactions approaching equilibrium

In this Topic we see that in general the rate laws must take into account both the forward and reverse reactions and that they give rise to expressions that describe the approach to equilibrium, when the forward and reverse rates are equal. A result of the analysis is a useful relation, which can be explored experimentally, between the equilibrium constant of the overall process and the rate constants of the forward and reverse reactions in the proposed mechanism.

20D The Arrhenius equation

The rate constants of most reactions increase with increasing temperature. In this Topic we see that the 'Arrhenius equation' captures this temperature dependence by using only two parameters that can be determined experimentally.

20E Reaction mechanisms

The study of reaction rates also leads to an understanding of the 'mechanisms' of reactions, their analysis into a sequence of elementary steps. In this Topic we see how to construct rate laws from a proposed mechanism. The elementary steps themselves have simple rate laws which can be combined together by invoking the concept of the 'rate-determining step' of a reaction or making either the 'steady-state approximation' or the existence of a 'pre-equilibrium'.

20F Examples of mechanisms

This Topic develops two examples of reaction mechanisms. The first describes a special class of reactions in the gas phase that depend on the collisions between reactants. The second gives insight into the formation of polymers and shows how the kinetics of their formation affects their properties.

20G Photochemistry

'Photochemistry' is the study of reactions that are initiated by light. In this Topic we explore mechanisms of photochemical reactions, with special emphasis on electron and energy transfer processes.

20H Enzymes

In this Topic we discuss the general mechanism of action of 'enzymes', which are biological catalysts. We show how to assemble expressions for their influence on the rate of reactions and the effect of substances that inhibit their function.

What is the impact of this material?

Plants, algae, and some species of bacteria evolved apparatus that perform 'photosynthesis', the capture of visible and near-infrared radiation for the purpose of synthesizing complex molecules in the cell. In *Impact* I20.1 we explore plant photosynthesis in detail.

To read more about the impact of this material, scan the QR code, or go to bcs.whfreeman.com/webpub/chemistry/pchem10e/impact/pchem-20-1.html

20A The rates of chemical reactions

Contents

➤ Why do you need to know this material?

Studies of the rates of disappearance of reactants and appearance of products allow us to predict how quickly a reaction mixture approaches equilibrium. Furthermore, studies of reaction rates lead to detailed descriptions of the molecular events that transform reactants into products.

➤ What is the key idea?

Reaction rates can be expressed mathematically in terms of the concentrations of reactants and, in some cases, products.

➤ What do you need to know already?

This introductory Topic is the foundation of a sequence: all you need to be aware of initially is the significance of stoichiometric numbers (Topic 2C). For more background on the spectroscopic determination of concentration, refer to Topic 12A.

This Topic introduces the principles of **chemical kinetics**, the study of reaction rates, by showing how the rates of reactions are defined and measured. The results of such measurements show that reaction rates depend on the concentration of reactants (and products) in characteristic ways that can be expressed in terms of differential equations known as rate laws.

20A.1 Monitoring the progress of a reaction

The first steps in the kinetic analysis of reactions are to establish the stoichiometry of the reaction and identify any side reactions. The basic data of chemical kinetics are then the concentrations of the reactants and products at different times after a reaction has been initiated.

(a) General considerations

The rates of most chemical reactions are sensitive to the temperature (as described in Topic 20D), so in conventional experiments the temperature of the reaction mixture must be held constant throughout the course of the reaction. This requirement puts severe demands on the design of an experiment. Gas-phase reactions, for instance, are often carried out in a vessel held in contact with a substantial block of metal. Liquid-phase reactions, including flow reactions, must be carried out in an efficient thermostat. Special efforts have to be made to study reactions at low temperatures, as in the study of the kinds of reactions that take place in interstellar clouds. Thus, supersonic expansion of the reaction gas can be used to attain temperatures as low as 10 K. For work in the liquid-phase and the solid-phase, very low temperatures are often reached by flowing cold liquid or cold gas around the reaction vessel. Alternatively, the entire reaction vessel is immersed in a thermally insulated container filled with a cryogenic liquid, such as liquid helium (for work at around 4 K) or liquid nitrogen (for work at around 77 K). Non-isothermal conditions are sometimes employed. For instance, the shelf life of an expensive pharmaceutical may be explored by slowly raising the temperature of a single sample.

Spectroscopy is widely applicable to the study of reaction kinetics, and is especially useful when one substance in the reaction mixture has a strong characteristic absorption in a conveniently accessible region of the electromagnetic spectrum. For example, the progress of the reaction

$$H_2(g) + Br_2(g) \rightarrow 2HBr(g)$$

can be followed by measuring the absorption of visible light by bromine. A reaction that changes the number or type of ions present in a solution may be followed by monitoring the electrical conductivity of the solution. The replacement of neutral molecules by ionic products can result in dramatic changes in the conductivity, as in the reaction

$$(CH_3)_3CCl(aq) + H_2O(l) \rightarrow (CH_3)_3COH(aq) + H^+(aq) + Cl^-(aq)$$

If hydrogen ions are produced or consumed, the reaction may be followed by monitoring the pH of the solution.

Other methods of determining composition include emission spectroscopy (Topic 13B), mass spectrometry (Topic 17D), gas chromatography, nuclear magnetic resonance (Topics 14B and 14C), and electron paramagnetic resonance (for reactions involving radicals or paramagnetic d-metal ions; see Topic 14D). A reaction in which at least one component is a gas might result in an overall change in pressure in a system of constant volume, so its progress may be followed by recording the variation of pressure with time.

Example 20A.1 Monitoring the variation in pressure

Predict how the total pressure varies during the gas-phase decomposition $2\,N_2O_5(g) \rightarrow 4\,NO_2(g) + O_2(g)$ in a constant-volume container.

Method The total pressure (at constant volume and temperature and assuming perfect gas behaviour) is proportional to the number of gas-phase molecules. Therefore, because each mole of N_2O_5 gives rise to $\frac{5}{2}$ mol of gas molecules, we can expect the pressure to rise to $\frac{5}{2}$ times its initial value. To confirm this conclusion, express the progress of the reaction in terms of the fraction, α, of N_2O_5 molecules that have reacted.

Answer Let the initial pressure be p_0 and the initial amount of N_2O_5 molecules present be n. When a fraction α of the N_2O_5 molecules has decomposed, the amounts of the components in the reaction mixture are:

	N_2O_5	NO_2	O_2	Total
Amount:	$n(1-\alpha)$	$2\alpha n$	$\frac{1}{2}\alpha n$	$n(1+\frac{3}{2}\alpha)$

When $\alpha=0$ the pressure is p_0, so at any stage the total pressure is

$$p = (1+\tfrac{3}{2}\alpha)p_0$$

When the reaction is complete ($\alpha=1$), the pressure will have risen to $\frac{5}{2}$ times its initial value.

Self-test 20A.1 Repeat the calculation for $2\,NOBr(g) \rightarrow 2\,NO(g) + Br_2(g)$.

Answer: $p = (1+\frac{1}{2}\alpha)p_0$

(b) Special techniques

The method used to monitor concentrations depends on the species involved and the rapidity with which their concentrations change. Many reactions reach equilibrium over periods of minutes or hours, and several techniques may then be used to follow the changing concentrations. In a **real-time analysis** the composition of the system is analysed while the reaction is in progress. Either a small sample is withdrawn or the bulk solution is monitored. In the **flow method** the reactants are mixed as they flow together in a chamber (Fig. 20A.1). The reaction continues as the thoroughly mixed solutions flow through the outlet tube, and observation of the composition at different positions along the tube is equivalent to the observation of the reaction mixture at different times after mixing. The disadvantage of conventional flow techniques is that a large volume of reactant solution is necessary. This requirement makes the study of fast reactions particularly difficult because to spread the reaction over a length of tube the flow must be rapid. This disadvantage is avoided by the **stopped-flow technique**, in which the reagents are mixed very quickly in a small chamber fitted with a syringe instead of an outlet tube (Fig. 20A.2). The flow ceases when the plunger of the syringe reaches a stop and the reaction continues in the mixed solutions. Observations, commonly using spectroscopic techniques such as ultraviolet–visible absorption, circular dichroism, and fluorescence emission (all introduced in Topics 13A and 13B), are made on the sample as a function of time. The technique allows for the study

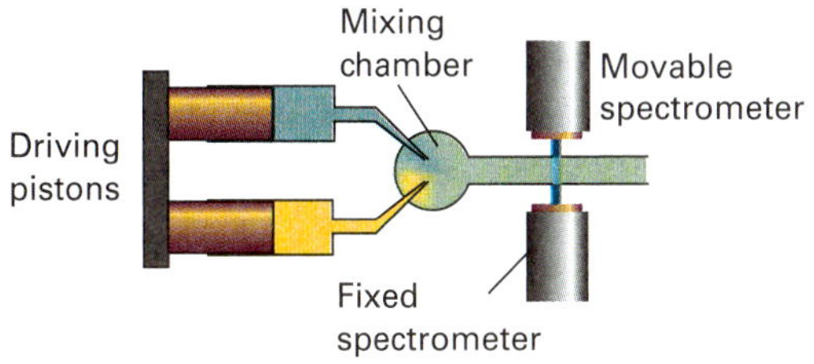

Figure 20A.1 The arrangement used in the flow technique for studying reaction rates. The reactants are injected into the mixing chamber at a steady rate. The location of the spectrometer corresponds to different times after initiation.

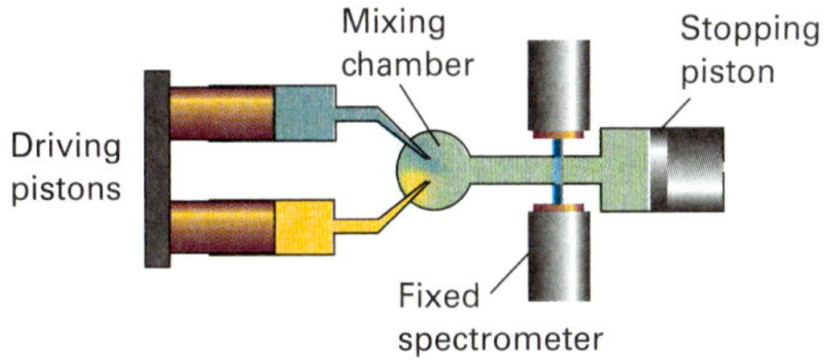

Figure 20A.2 In the stopped-flow technique the reagents are driven quickly into the mixing chamber by the driving pistons and then the time-dependence of the concentrations is monitored.

of reactions that occur on the millisecond to second timescale. The suitability of the stopped-flow method to the study of small samples means that it is appropriate for many biochemical reactions; it has been widely used to study the kinetics of protein folding and enzyme action.

Very fast reactions can be studied by **flash photolysis**, in which the sample is exposed to a brief flash of light that initiates the reaction and then the contents of the reaction chamber are monitored. The apparatus used for flash photolysis studies is based on the experimental design for time-resolved spectroscopy, in which reactions occurring on a picosecond or femtosecond timescale may be monitored by using electronic absorption or emission, infrared absorption, or Raman scattering (Topic 13C).

In contrast to real-time analysis, **quenching methods** are based on quenching, or stopping, the reaction after it has been allowed to proceed for a certain time. In this way the composition is analysed at leisure and reaction intermediates may be trapped. These methods are suitable only for reactions that are slow enough for there to be little reaction during the time it takes to quench the mixture. In the **chemical quench flow method**, the reactants are mixed in much the same way as in the flow method but the reaction is quenched by another reagent, such as solution of acid or base, after the mixture has travelled along a fixed length of the outlet tube. Different reaction times can be selected by varying the flow rate along the outlet tube. An advantage of the chemical quench flow method over the stopped-flow method is that spectroscopic fingerprints are not needed in order to measure the concentration of reactants and products. Once the reaction has been quenched, the solution may be examined by 'slow' techniques, such as gel electrophoresis, mass spectrometry, and chromatography. In the **freeze quench method**, the reaction is quenched by cooling the mixture within milliseconds and the concentrations of reactants, intermediates, and products are measured spectroscopically.

20A.2 The rates of reactions

Reaction rates depend on the composition and the temperature of the reaction mixture. The next few sections look at these observations in more detail.

(a) The definition of rate

Consider a reaction of the form A+2 B→3 C+D, in which at some instant the molar concentration of a participant J is [J] and the volume of the system is constant. The instantaneous **rate of consumption** of one of the reactants at a given time is −d[R]/dt, where R is A or B. This rate is a positive quantity (Fig. 20A.3). The **rate of formation** of one of the products (C or D,

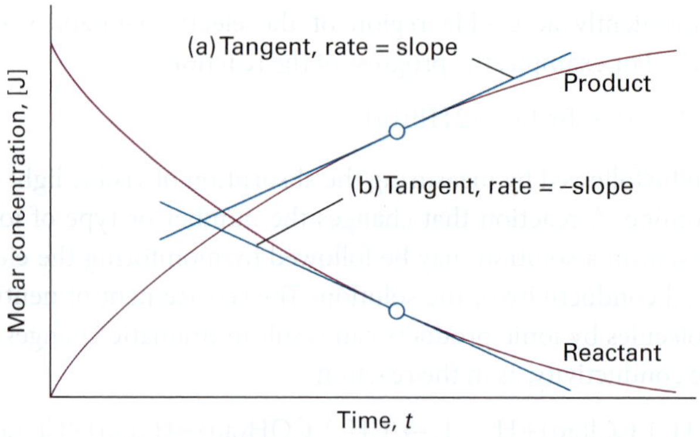

Figure 20A.3 The definition of (instantaneous) rate as the slope of the tangent drawn to the curve showing the variation of concentration of (a) products, (b) reactants with time. For negative slopes, the sign is changed when reporting the rate, so all reaction rates are positive.

which we denote P) is d[P]/dt (note the difference in sign). This rate is also positive.

It follows from the stoichiometry of the reaction A+2 B→3 C+D that

$$\frac{d[D]}{dt}=\frac{1}{3}\frac{d[C]}{dt}=-\frac{d[A]}{dt}=-\frac{1}{2}\frac{d[B]}{dt}$$

so there are several rates connected with the reaction. The undesirability of having different rates to describe the same reaction is avoided by using the extent of reaction, ξ (xi, the quantity introduced in Topic 6A):

$$\xi=\frac{n_J-n_{J,0}}{\nu_J} \qquad \textit{Definition} \quad \text{Extent of reaction} \qquad (20A.1)$$

where ν_J is the stoichiometric number of species J (Topic 2C; remember that ν_J is negative for reactants and positive for products), and defining the unique **rate of reaction**, v, as the rate of change of the extent of reaction:

$$v=\frac{1}{V}\frac{d\xi}{dt} \qquad \textit{Definition} \quad \text{Rate of reaction} \qquad (20A.2)$$

where V is the volume of the system. It follows that

$$v=\frac{1}{\nu_J}\times\frac{1}{V}\frac{dn_J}{dt} \qquad (20A.3a)$$

For a homogeneous reaction in a constant-volume system the volume V can be taken inside the differential and we use $[J]=n_J/V$ to write

$$v=\frac{1}{\nu_J}\frac{d[J]}{dt} \qquad (20A.3b)$$

For a heterogeneous reaction we use the (constant) surface area, A, occupied by the species in place of V and then use $\sigma_J = n_J/A$ to write

$$v = \frac{1}{\nu_J}\frac{d\sigma_J}{dt} \quad (20A.3c)$$

In each case there is now a single rate for the entire reaction (for the chemical equation as written). With molar concentrations in moles per cubic decimetre and time in seconds, reaction rates of homogeneous reactions are reported in moles per cubic decimetre per second ($mol\,dm^{-3}\,s^{-1}$) or related units. For gas-phase reactions, such as those taking place in the atmosphere, concentrations are often expressed in molecules per cubic centimetre ($molecules\,cm^{-3}$) and rates in molecules per cubic centimetre per second ($molecules\,cm^{-3}\,s^{-1}$). For heterogeneous reactions, rates are expressed in moles per square metre per second ($mol\,m^{-2}\,s^{-1}$) or related units.

Brief illustration 20A.1 Reaction rates from balanced chemical equations

If the rate of formation of NO in the reaction $2\,NOBr(g) \rightarrow 2\,NO(g) + Br_2(g)$ is reported as $0.16\,mmol\,dm^{-3}\,s^{-1}$, we use $\nu_{NO} = +2$ to report that $v = 0.080\,mmol\,dm^{-3}\,s^{-1}$. Because $\nu_{NOBr} = -2$ it follows that $d[NOBr]/dt = -0.16\,mmol\,dm^{-3}\,s^{-1}$. The rate of consumption of NOBr is therefore $0.16\,mmol\,dm^{-3}\,s^{-1}$, or $9.6\times10^{16}\,molecules\,cm^{-3}\,s^{-1}$.

Self-test 20A.2 The rate of change of molar concentration of CH_3 radicals in the reaction $2\,CH_3(g) \rightarrow CH_3CH_3(g)$ was reported as $d[CH_3]/dt = -1.2\,mol\,dm^{-3}\,s^{-1}$ under particular conditions. What is (a) the rate of reaction and (b) the rate of formation of CH_3CH_3?

Answer: (a) $0.60\,mol\,dm^{-3}\,s^{-1}$, (b) $0.60\,mol\,dm^{-3}\,s^{-1}$

(b) Rate laws and rate constants

The rate of reaction is often found to be proportional to the concentrations of the reactants raised to a power. For example, the rate of a reaction may be proportional to the molar concentrations of two reactants A and B, so we write

$$v = k_r[A][B] \quad (20A.4)$$

with each concentration raised to the first power. The coefficient k_r is called the **rate constant** for the reaction. The rate constant is independent of the concentrations but depends on the temperature. An experimentally determined equation of this kind is called the **rate law** of the reaction. More formally, a rate law is an equation that expresses the rate of reaction as a function of the concentrations of all the species present in the overall chemical equation for the reaction at the time of interest:

$$v = f([A],[B],\ldots) \quad (20A.5a)$$

General form — Rate law in terms of concentrations

For homogeneous gas-phase reactions, it is often more convenient to express the rate law in terms of partial pressures, which are related to molar concentrations by $p_J = RT[J]$. In this case, we write

$$v = f(p_A, p_B,\ldots) \quad (20A.5b)$$

General form — Rate law in terms of partial pressures

The rate law of a reaction is determined experimentally, and cannot in general be inferred from the chemical equation for the reaction. The reaction of hydrogen and bromine, for example, has a very simple stoichiometry, $H_2(g) + Br_2(g) \rightarrow 2\,HBr(g)$, but its rate law is complicated:

$$v = \frac{k_a[H_2][Br_2]^{3/2}}{[Br_2] + k_b[HBr]} \quad (20A.6)$$

In certain cases the rate law does reflect the stoichiometry of the reaction; but that is either a coincidence or reflects a feature of the underlying reaction mechanism (see Topic 20E).

A note on good (or, at least, our) practice We denote a general rate constant k_r to distinguish it from the Boltzmann constant k. In some texts k is used for the former and k_B for the latter. When expressing the rate constants in a more complicated rate law, such as that in eqn 20A.6, we use k_a, k_b, and so on.

The units of k_r are always such as to convert the product of concentrations into a rate expressed as a change in concentration divided by time. For example, if the rate law is the one shown in eqn 20A.4, with concentrations expressed in $mol\,dm^{-3}$, then the units of k_r will be $dm^3\,mol^{-1}\,s^{-1}$ because

$$dm^3\,mol^{-1}s^{-1} \times mol\,dm^{-3} \times mol\,dm^{-3} = mol\,dm^{-3}\,s^{-1}$$

In gas-phase studies, including studies of the processes taking place in the atmosphere, concentrations are commonly expressed in $molecules\,cm^{-3}$, so the rate constant for the reaction above would be expressed in $cm^3\,molecule^{-1}\,s^{-1}$. We can use the approach just developed to determine the units of the rate constant from rate laws of any form. For example, the rate constant for a reaction with rate law of the form $k_r[A]$ is commonly expressed in s^{-1}.

Brief illustration 20A.2 The units of rate constants

The rate constant for the reaction $O(g) + O_3(g) \rightarrow 2\,O_2(g)$ is $8.0\times10^{-15}\,cm^3\,molecule^{-1}\,s^{-1}$ at 298 K. To express this rate constant in $dm^3\,mol^{-1}\,s^{-1}$, we make use of the two relations

1 cm = 10^{-1} dm and 1 molecule = (1 mol)/(6.022 × 10^{23}). It follows that

$$k_r = 8.0\times10^{-15}\,\text{cm}^3\,\text{molecule}^{-1}\,\text{s}^{-1}$$
$$=8.0\times10^{-15}(10^{-1}\,\text{dm})^3\left(\frac{1\,\text{mol}}{6.022\times10^{23}}\right)^{-1}\text{s}^{-1}$$
$$=8.0\times10^{-15}\times10^{-3}\times6.022\times10^{23}\,\text{dm}^3\,\text{mol}^{-1}\,\text{s}^{-1}$$
$$=4.8\times10^{6}\,\text{dm}^3\,\text{mol}^{-1}\,\text{s}^{-1}$$

Self-test 20A.3 A reaction has a rate law of the form $k_r[A]^2[B]$. What are the units of the rate constant if the reaction rate is measured in mol dm^{-3} s^{-1}?

Answer: dm^6 mol^{-2} s^{-1}

A practical application of a rate law is that once we know the law and the value of the rate constant, we can predict the rate of reaction from the composition of the mixture. Moreover, as demonstrated in Topic 20B, by knowing the rate law, we can go on to predict the composition of the reaction mixture at a later stage of the reaction. Moreover, a rate law is a guide to the mechanism of the reaction, for any proposed mechanism must be consistent with the observed rate law. This application is developed in Topic 20E.

(c) Reaction order

Many reactions are found to have rate laws of the form

$$v=k_r[A]^a[B]^b\cdots \qquad (20A.7)$$

The power to which the concentration of a species (a product or a reactant) is raised in a rate law of this kind is the **order** of the reaction with respect to that species. A reaction with the rate law in eqn 20A.4 is first order in A and first order in B. The **overall order** of a reaction with a rate law like that in eqn 20A.7 is the sum of the individual orders, $a+b+\cdots$. The rate law in eqn 20A.4 is therefore second order overall.

A reaction need not have an integral order, and many gas-phase reactions do not. For example, a reaction having the rate law

$$v=k_r[A]^{1/2}[B] \qquad (20A.8)$$

is half order in A, first order in B, and three-halves order overall.

Brief illustration 20A.3 The interpretation of rate laws

The experimentally determined rate law for the gas-phase reaction $H_2(g)+Br_2(g)\rightarrow 2\,HBr(g)$ is given by eqn 20A.6. Although the reaction is first-order in H_2, it has an indefinite order with respect to both Br_2 and HBr and an indefinite order overall.

Self-test 20A.4 Repeat this analysis for a typical rate law for the action of an enzyme E on a substrate S: $v=k_r[E][S]/([S]+K_M)$, where K_M is a constant.

Answer: First order in E; no specific order with respect to S

Some reactions obey a zero-order rate law, and therefore have a rate that is independent of the concentration of the reactant (so long as some is present). Thus, the catalytic decomposition of phosphine (PH_3) on hot tungsten at high pressures has the rate law

$$v=k_r \qquad (20A.9)$$

PH_3 decomposes at a constant rate until it has almost entirely disappeared. Zero-order reactions typically occur when there is a bottle-neck of some kind in the mechanism, as in heterogeneous reactions when the surface is saturated and the subsequent reaction slow and in a number of enzyme reactions when there is a large excess of substrate relative to the enzyme.

As we saw in *Brief illustration* 20A.3, when a rate law is not of the form in eqn 20A.7, the reaction does not have an overall order and may not even have definite orders with respect to each participant.

These remarks point to three important tasks:

- To identify the rate law and obtain the rate constant from the experimental data. We concentrate on this aspect in this Topic.
- To construct reaction mechanisms that are consistent with the rate law. We introduce the techniques for doing so in Topic 20E.
- To account for the values of the rate constants and explain their temperature dependence. This task is undertaken in Topic 20D.

(d) The determination of the rate law

The determination of a rate law is simplified by the **isolation method** in which the concentrations of all the reactants except one are in large excess. If B is in large excess in a reaction between A and B, for example, then to a good approximation its concentration is constant throughout the reaction. Although the true rate law might be $v=k_r[A][B]$, we can approximate [B] by $[B]_0$, its initial value, and write

$$v=k_r'[A] \qquad k_r'=k_r[B]_0 \qquad (20A.10)$$

which has the form of a first-order rate law. Because the true rate law has been forced into first-order form by assuming that the concentration of B is constant, eqn 20A.10 is called a **pseudofirst-order rate law**. The dependence of the rate on the concentration of each of the reactants may be found by isolating them in turn (by having all the other substances present in large excess), and so constructing a picture of the overall rate law.

In the **method of initial rates**, which is often used in conjunction with the isolation method, the rate is measured at the beginning of the reaction for several different initial concentrations of reactants. We shall suppose that the rate law for a reaction with A isolated is $v = k_r'[\mathrm{A}]^a$, then its initial rate, v_0, is given by the initial values of the concentration of A, and we write $v = k_r'[\mathrm{A}]_0^a$. Taking logarithms gives:

$$\log v_0 = \log k_r' + a \log [\mathrm{A}]_0 \qquad (20\mathrm{A}.11)$$

A plot of the logarithms of the initial rates against the logarithms of a series of initial concentrations of A should be a straight line with slope a.

Example 20A.2 Using the method of initial rates

The recombination of iodine atoms in the gas phase in the presence of argon was investigated and the order of the reaction was determined by the method of initial rates. The initial rates of reaction of $2\,\mathrm{I(g)} + \mathrm{Ar(g)} \rightarrow \mathrm{I_2(g)} + \mathrm{Ar(g)}$ were as follows:

$[\mathrm{I}]_0/(10^{-5}\,\mathrm{mol\,dm^{-3}})$		1.0	2.0	4.0	6.0
$v_0/(\mathrm{mol\,dm^{-3}\,s^{-1}})$	(a)	8.70×10^{-4}	3.48×10^{-3}	1.39×10^{-2}	3.13×10^{-2}
	(b)	4.35×10^{-3}	1.74×10^{-2}	6.96×10^{-2}	1.57×10^{-1}
	(c)	8.69×10^{-3}	3.47×10^{-2}	1.38×10^{-1}	3.13×10^{-1}

The Ar concentrations are (a) $1.0\,\mathrm{mmol\,dm^{-3}}$, (b) $5.0\,\mathrm{mmol\,dm^{-3}}$, and (c) $10.0\,\mathrm{mmol\,dm^{-3}}$. Determine the orders of reaction with respect to the I and Ar atom concentrations and the rate constant.

Method Plot the logarithm of the initial rate, $\log v_0$, against $\log [\mathrm{I}]_0$ for a given concentration of Ar and, separately, against $\log [\mathrm{Ar}]_0$ for a given concentration of I. The slopes of the two lines are the orders of reaction with respect to I and Ar, respectively. The intercepts with the vertical axis give $\log k_r$.

Answer The plots are shown in Fig. 20A.4. The slopes are 2 and 1, respectively, so the (initial) rate law is $v_0 = k_r[\mathrm{I}]_0^2[\mathrm{Ar}]_0$. This rate law signifies that the reaction is second order in [I], first order in [Ar], and third order overall. The intercept corresponds to $k_r = 9\times10^9\,\mathrm{mol^{-2}\,dm^6\,s^{-1}}$.

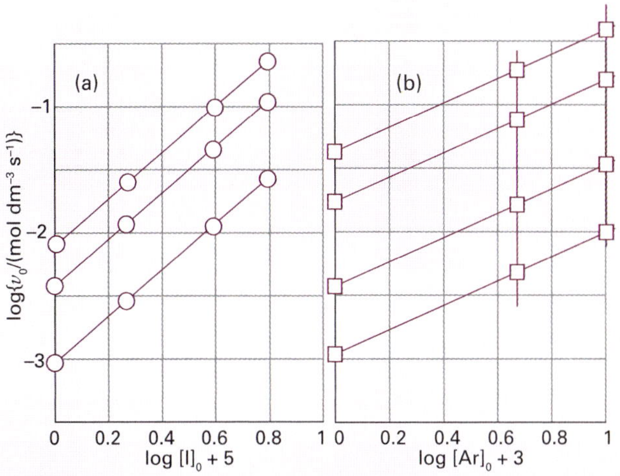

Figure 20A.4 The plot of $\log v_0$ against (a) $\log [\mathrm{I}]_0$ for a given $[\mathrm{Ar}]_0$, and (b) $\log [\mathrm{Ar}]_0$ for a given $[\mathrm{I}]_0$.

A note on good practice The units of k_r come automatically from the calculation, and are always such as to convert the product of concentrations to a rate in concentration/time (for example, $\mathrm{mol\,dm^{-3}\,s^{-1}}$).

Self-test 20A.5 The initial rate of a reaction depended on concentration of a substance J as follows:

$[\mathrm{J}]_0/(\mathrm{mmol\,dm^{-3}})$	5.0	8.2	17	30
$v_0/(10^{-7}\,\mathrm{mol\,dm^{-3}\,s^{-1}})$	3.6	9.6	41	130

Determine the order of the reaction with respect to J and calculate the rate constant.

Answer: 2, $1.4\times10^{-2}\,\mathrm{dm^3\,mol^{-1}\,s^{-1}}$

The method of initial rates might not reveal the full rate law, for once the products have been generated they might participate in the reaction and affect its rate. For example, products participate in the synthesis of HBr, because eqn 20A.6 shows that the full rate law depends on the concentration of HBr. To avoid this difficulty, the rate law should be fitted to the data throughout the reaction. The fitting may be done, in simple cases at least, by using a proposed rate law to predict the concentration of any component at any time, and comparing it with the data. A rate law should also be tested by observing whether the addition of products or, for gas phase reactions, a change in the surface-to-volume ratio in the reaction chamber affects the rate.

Checklist of concepts

☐ 1. The rates of chemical reactions are measured by using techniques that monitor the concentrations of species present in the reaction mixture. Examples include **real-time** and **quenching** procedures, **flow** and **stopped-flow** techniques, and **flash photolysis**.

☐ 2. The **instantaneous rate** of a reaction is the slope of the tangent to the graph of concentration against time (expressed as a positive quantity).

☐ 3. A **rate law** is an expression for the reaction rate in terms of the concentrations of the species that occur in the overall chemical reaction.

Checklist of equations

Property	Equation	Comment	Equation number
Extent of reaction	$\xi=(n_J-n_{J,0})/\nu_J$	Definition	20A.1
Rate of a reaction	$v=(1/V)(d\xi/dt),\ \xi=(n_J-n_{J,0})/\nu_J$	Definition	20A.2
Rate law (in some cases)	$v=k_r[A]^a[B]^b\cdots$	$a, b, \ldots$: orders; $a+b+\ldots$: overall order	20A.7
Method of initial rates	$\log v_0=\log k_r'+a\log[A]_0$		20A.11

20B Integrated rate laws

Contents

➤ Why do you need to know this material?

You need the integrated rate law if you want to predict the composition of a reaction mixture as it approaches equilibrium. It is also used to determine the rate law and rate constants of a reaction, which is a necessary step in the formulation of the mechanism of the reaction.

➤ What is the key idea?

A comparison between experimental data and the integrated form of the rate law is used to verify a proposed rate law and determine the order and rate constant of a reaction.

➤ What do you need to know already?

You need to be familiar with the concepts of rate law, reaction order, and rate constant (Topic 20A). The manipulation of simple rate laws requires only elementary techniques of integration (see the *Resource section* for standard integrals).

Rate laws (Topic 20A) are differential equations. We must integrate them if we want to find the concentrations as a function of time. Even the most complex rate laws may be integrated numerically. However, in a number of simple cases analytical solutions, known as **integrated rate laws**, are easily obtained and prove to be very useful. We examine a few of these simple cases here.

20B.1 First-order reactions

As shown in the following *Justification*, the integrated form of the first-order rate law

$$\frac{d[A]}{dt} = -k_r[A] \qquad (20B.1a)$$

is

$$\ln\frac{[A]}{[A]_0} = -k_r t \qquad [A] = [A]_0 e^{-k_r t} \qquad \text{Integrated first-order rate law} \qquad (20B.1b)$$

where $[A]_0$ is the initial concentration of A (at $t=0$).

Justification 20B.1 First-order integrated rate law

First, we rearrange eqn 20B.1a into

$$\frac{d[A]}{[A]^2} = -k_r dt$$

This expression can be integrated directly because k_r is a constant independent of t. Initially (at $t=0$) the concentration of A is $[A]_0$, and at a later time t it is $[A]$, so we make these values the limits of the integrals and write

$$\int_{[A]_0}^{[A]} \frac{d[A]}{[A]} = -k_r \int_0^t dt$$

Because the integral of $1/x$ is $\ln x + \text{constant}$, eqn 20B.1b is obtained immediately.

Equation 20B.1b shows that if $\ln([A]/[A]_0)$ is plotted against t, then a first-order reaction will give a straight line of slope $-k_r$. Some rate constants determined in this way are given in Table 20B.1. The second expression in eqn 20B.1b shows that in a first-order reaction the reactant concentration decreases exponentially with time with a rate determined by k_r (Fig. 20B.1).

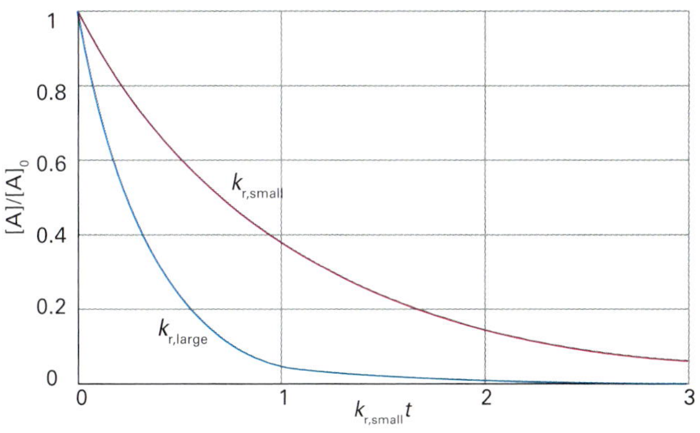

Figure 20B.1 The exponential decay of the reactant in a first-order reaction. The larger the rate constant, the more rapid is the decay: here $k_{r,large} = 3k_{r,small}$.

Table 20B.1* Kinetic data for first-order reactions

Reaction	Phase	θ/°C	k_r/s^{-1}	$t_{1/2}$
$2\,N_2O_5 \rightarrow 4\,NO_2 + O_2$	g	25	3.38×10^{-5}	5.70 h
	Br_2(l)	25	4.27×10^{-5}	4.51 h
$C_2H_6 \rightarrow 2\,CH_3$	g	700	5.36×10^{-4}	21.6 min

* More values are given in the *Resource section*.

A useful indication of the rate of a first-order chemical reaction is the **half-life**, $t_{1/2}$, of a substance, the time taken for the concentration of a reactant to fall to half its initial value. This quantity is readily obtained from the integrated rate law. Thus, the time for [A] to decrease from $[A]_0$ to $\frac{1}{2}[A]_0$ in a first-order reaction is given by eqn 20B.1b as

$$k_r t_{1/2} = -\ln\frac{\frac{1}{2}[A]_0}{[A]_0} = -\ln\tfrac{1}{2} = \ln 2$$

Hence

$$t_{1/2} = \frac{\ln 2}{k_r} \qquad \textit{First-order reaction} \quad \text{Half-life} \qquad (20B.2)$$

(Note that $\ln 2 = 0.693$.) The main point to note about this result is that for a first-order reaction, the half-life of a reactant is independent of its initial concentration. Therefore, if the concentration of A at some *arbitrary* stage of the reaction is [A], then it will have fallen to $\frac{1}{2}[A]$ after a further interval of $(\ln 2)/k_r$. Some half-lives are given in Table 20B.1.

Another indication of the rate of a first-order reaction is the **time constant**, τ (tau), the time required for the concentration of a reactant to fall to 1/e of its initial value. From eqn 20B.1b it follows that

$$k_r\tau = -\ln\frac{\frac{1}{e}[A]_0}{[A]_0} = -\ln\tfrac{1}{e} = 1$$

That is, the time constant of a first-order reaction is the reciprocal of the rate constant:

$$\tau = \frac{1}{k_r} \qquad \textit{First-order reaction} \quad \text{Time constant} \qquad (20B.3)$$

Example 20B.1 Analysing a first-order reaction

The variation in the partial pressure of azomethane with time was followed at 600 K, with the results given below. Confirm that the decomposition $CH_3N_2CH_3(g) \rightarrow CH_3CH_3(g) + N_2(g)$ is first order in azomethane, and find the rate constant, half-life, and time constant at 600 K.

t/s	0	1000	2000	3000	4000
p/Pa	10.9	7.63	5.32	3.71	2.59

Method As indicated in the text, to confirm that a reaction is first order, plot $\ln([A]/[A]_0)$ against time and expect a straight line. Because the partial pressure of a gas is proportional to its concentration, an equivalent procedure is to plot $\ln(p/p_0)$ against t. If a straight line is obtained, its slope can be identified with $-k_r$. The half-life and time constant are then calculated from k_r by using eqns 20B.2 and 20B.3, respectively.

Answer We draw up the following table:

t/s	0	1000	2000	3000	4000
$\ln(p/p_0)$	1	−0.357	−0.717	−1.078	−1.437

Figure 20B.2 shows the plot of $\ln(p/p_0)$ against t. The plot is straight, confirming a first-order reaction, and its slope is -3.6×10^{-4}. Therefore, $k_r = 3.6\times10^{-4}\,s^{-1}$.

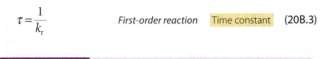

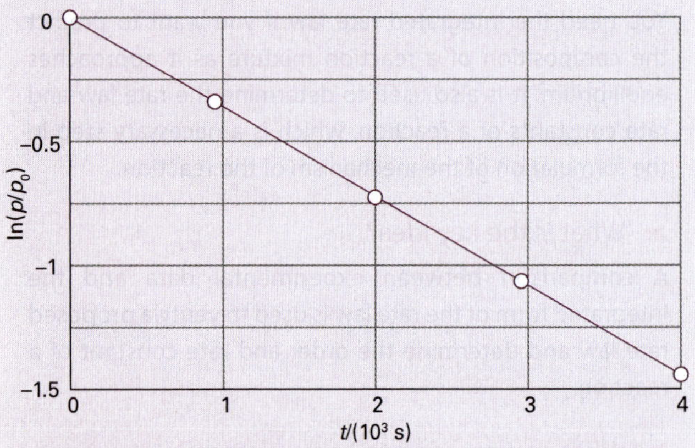

Figure 20B.2 The determination of the rate constant of a first-order reaction: a straight line is obtained when ln [A] (or, as here, ln p/p_0) is plotted against t; the slope gives k_r.

A note on good practice Because the horizontal and vertical axes of graphs are labelled with pure numbers, the slope of a graph is always dimensionless. For a graph of the form $y = b + mx$ we can write $y = b + (m \text{ units})(x/\text{units})$, where 'units' are the units of x, and identify the (dimensionless) slope with 'm units'. Then m = slope/units. In the present case, because the graph shown here is a plot of $\ln(p/p_0)$ against t/s (with 'units' = s) and k_r is the negative value of the slope of $\ln(p/p_0)$ against t itself, $k_r = -\text{slope/s}$.

It follows from eqns 5.2 and 5.3 that the half-life and time constant are, respectively,

$$t_{1/2} = \frac{\ln 2}{3.6\times10^{4}\,s^{-1}} = 1.9\times10^{-5}\,s \qquad \tau = \frac{1}{3.6\times10^{4}\,s^{-1}} = 2.8\times10^{-5}\,s$$

Self-test 20B.1 In a particular experiment, it was found that the concentration of N_2O_5 in liquid bromine varied with time as follows:

t/s	0	200	400	600	1000
$[N_2O_5]$/(mol dm^{-3})	0.110	0.073	0.048	0.032	0.014

Confirm that the reaction is first order in N_2O_5 and determine the rate constant.

Answer: $k_r = 2.1\times10^{-3}\,s^{-1}$

20B.2 Second-order reactions

We show in the following *Justification* that the integrated form of the second-order rate law

$$\frac{d[A]}{dt} = -k_r[A]^2 \quad (20B.4a)$$

is either of the following two forms:

$$\frac{1}{[A]} - \frac{1}{[A]_0} = k_r t \quad \textit{Second-order reaction} \quad \text{Integrated rate law} \quad (20B.4b)$$

$$[A] = \frac{[A]_0}{1+k_r t[A]_0} \quad \textit{Second-order reaction; alternative form} \quad \text{Integrated rate law} \quad (20B.4c)$$

where $[A]_0$ is the initial concentration of A (at $t=0$).

Justification 20B.2 Second-order integrated rate law

To integrate eqn 20B.4a we rearrange it into

$$\frac{d[A]}{[A]^2} = -k_r dt$$

The concentration is $[A]_0$ at $t=0$ and [A] at a general time t later. Therefore,

$$-\int_{[A]_0}^{[A]} \frac{d[A]}{[A]^2} = k_r \int_0^t dt$$

Because the integral of $1/x^2$ is $-1/x+$constant, we obtain eqn 20B.4b by substitution of the limits

$$\left.\frac{1}{[A]} + \text{constant}\right|_{[A]_0}^{[A]} = \frac{1}{[A]} - \frac{1}{[A]_0} = k_r t$$

We can then rearrange this expression into eqn 20B.4c.

Equation 20B.4b shows that to test for a second-order reaction we should plot $1/[A]$ against t and expect a straight line.

Table 20B.2* Kinetic data for second-order reactions

Reaction	Phase	θ/°C	k_r/(dm^3 mol^{-1} s^{-1})
$2\,NOBr \rightarrow 2\,NO + Br_2$	g	10	0.80
$2\,I \rightarrow I_2$	g	23	7×10^9
$CH_3Cl + CH_3O^-$	CH_3OH(l)	20	2.29×10^{-6}

* More values are given in the *Resource section*.

The slope of the graph is k_r. Some rate constants determined in this way are given in Table 20B.2. The rearranged form, eqn 20B.4c, lets us predict the concentration of A at any time after the start of the reaction. It shows that the concentration of A approaches zero more slowly than in a first-order reaction with the same initial rate (Fig. 20B.3).

It follows from eqn 20B.4b by substituting $t=t_{1/2}$ and $[A]=\frac{1}{2}[A]_0$ that the half-life of a species A that is consumed in a second-order reaction is

$$t_{1/2} = \frac{1}{k_r[A]_0} \quad \textit{Second-order reaction} \quad \text{Half-life} \quad (20B.5)$$

Therefore, unlike a first-order reaction, the half-life of a substance in a second-order reaction varies with the initial concentration. A practical consequence of this dependence is that species that decay by second-order reactions (which includes some environmentally harmful substances) may persist in low concentrations for long periods because their half-lives are long when their concentrations are low. In general, for an nth-order reaction (with n neither 0 nor 1) of the form A → products, the half-life is related to the rate constant and the initial concentration of A by (see Problem 20B.16)

$$t_{1/2} = \frac{2^{n-1}-1}{(n-1)k_r[A]_0^{n-1}} \quad \textit{nth-order reaction} \quad \text{Half-life} \quad (20B.6)$$

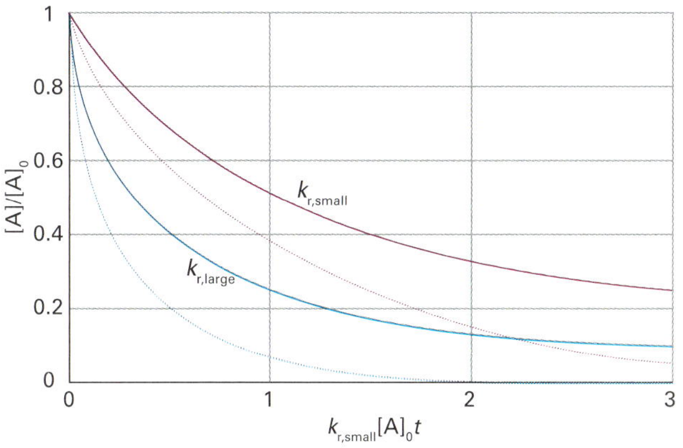

Figure 20B.3 The variation with time of the concentration of a reactant in a second-order reaction. The dotted lines are the corresponding decays in a first-order reaction with the same initial rate. For this illustration, $k_{r,large} = 3k_{r,small}$.

Another type of second-order reaction is one that is first order in each of two reactants A and B:

$$\frac{d[A]}{dt} = -k_r[A][B] \qquad (20B.7)$$

To integrate this rate law we need to know how the concentration of B is related to that of A. For example, if the reaction is $A+B \to P$, where P denotes products, and the initial concentrations are $[A]_0$ and $[B]_0$, then it is shown in the following *Justification* that at a time t after the start of the reaction, the concentrations satisfy the relation

$$\ln\frac{[B]/[B]_0}{[A]/[A]_0} = ([B]_0 - [A]_0)k_r t \qquad (20B.8)$$

Second-order reaction of the type $A+B \to P$ — Integrated rate law

Therefore, a plot of the expression on the left against t should be a straight line from which k_r can be obtained. As shown in the following *Brief illustration*, the rate constant may be estimated quickly by using data from only two measurements.

Brief illustration 20B.1 Second-order reactions

Consider a second-order reaction of the type $A+B \to P$ carried out in a solution. Initially, the concentrations of reactants were $[A]_0 = 0.075\,\text{mol dm}^{-3}$ and $[B]_0 = 0.050\,\text{mol dm}^{-3}$. After 1.0 h the concentration of B fell to $[B] = 0.020\,\text{mol dm}^{-3}$. Because $\Delta[B] = \Delta[A]$, it follows that during this time interval

$$\Delta[B] = (0.020 - 0.050)\,\text{mol dm}^{-3} = -0.030\,\text{mol dm}^{-3}$$
$$\Delta[A] = -0.030\,\text{mol dm}^{-3}$$

Therefore, the concentrations of A and B after 1.0 h are

$$[A] = \Delta[A] + [A]_0 = (-0.030 + 0.075)\,\text{mol dm}^{-3} = 0.045\,\text{mol dm}^{-3}$$
$$[B] = 0.020\,\text{mol dm}^{-3}$$

It follows from rearrangement of eqn 20B.8 that

$$k_r(3600\,\text{s}) = \frac{1}{(0.050 - 0.075)\,\text{mol dm}^{-3}}\ln\frac{0.020/0.050}{0.045/0.075}$$

where we have used 1 hr = 3600 s. Solving this expression for the rate constant gives

$$k_r = 4.5\times10^{-3}\,\text{dm}^3\,\text{mol}^{-1}\,\text{s}^{-1}$$

Self-test 20B.2 Calculate the half-life of the reactants for the reaction.

Answer: $t_{1/2}(A) = 5.1\times10^3\,\text{s}$, $t_{1/2}(B) = 2.1\times10^3\,\text{s}$

Justification 20B.3 Overall second-order rate law

It follows from the reaction stoichiometry that when the concentration of A has fallen to $[A]_0 - x$, the concentration of B will have fallen to $[B]_0 - x$ (because each A that disappears entails the disappearance of one B). It follows that

$$\frac{d[A]}{dt} = -k_r([A]_0 - x)([B]_0 - x)$$

Because $[A] = [A]_0 - x$, it follows that $d[A]/dt = -dx/dt$ and the rate law may be written as

$$\frac{dx}{dt} = k_r([A]_0 - x)([B]_0 - x)$$

The initial condition is that $x=0$ when $t=0$; so the integration required is

$$\int_0^x \frac{dx}{([A]_0 - x)([B]_0 - x)} = k_r\int_0^t dt$$

The integral on the right is simply $k_r t$. The integral on the left is evaluated by using the method of partial fractions (see *The chemist's toolkit* 20B.1):

$$\int_0^x \frac{dx}{([A]_0 - x)([B]_0 - x)} = \frac{1}{[B]_0 - [A]_0}\left\{\ln\frac{[A]_0}{[A]_0 - x} - \ln\frac{[B]_0}{[B]_0 - x}\right\}$$

The two logarithms can be combined as follows:

$$\ln\frac{[A]_0}{\underbrace{[A]_0 - x}_{[A]}} - \ln\frac{[B]_0}{\underbrace{[B]_0 - x}_{[B]}} = \ln\frac{[A]_0}{[A]} - \ln\frac{[B]_0}{[B]}$$
$$= \ln\frac{1}{[A]/[A]_0} - \ln\frac{1}{[B]/[B]_0}$$
$$= \ln\frac{[B]/[B]_0}{[A]/[A]_0}$$

Combining all the results so far gives eqn 20B.8. Similar calculations may be carried out to find the integrated rate laws for other orders, and some are listed in Table 20B.3.

The chemist's toolkit 20B.1 Integration by the method of partial fractions

To solve an integral of the form

$$I = \int \frac{1}{(a-x)(b-x)}dx$$

where a and b are constants, we use the method of partial fractions in which a fraction that is the product of terms (as in the denominator of this integrand) is written as a sum of fractions. To implement this procedure we write the integrand as

$$\frac{1}{(a-x)(b-x)}=\frac{1}{b-a}\left(\frac{1}{a-x}-\frac{1}{b-x}\right)$$

Then we integrate each term on the right. It follows that

$$I=\frac{1}{b-a}\left(\int\frac{dx}{a-x}-\int\frac{dx}{b-x}\right)\overset{\text{Integral A.2}}{=}\frac{1}{b-a}\left(\ln\frac{1}{a-x}-\ln\frac{1}{b-x}\right)a$$
$$+\text{constant}$$

Table 20B.3 Integrated rate laws

Order	Reaction	Rate law*	$t_{1/2}$
0	$A \to P$	$v=k_r$ $k_r t=x$ for $0\leq x\leq [A]_0$	$[A]_0/2k_r$
1	$A \to P$	$v=k_r[A]$ $k_r t=\ln\frac{[A]_0}{[A]_0-x}$	$(\ln 2)/k_r$
2	$A \to P$	$v=k_r[A]^2$ $k_r t=\frac{x}{[A]_0([A]_0-x)}$	$1/k_r[A]_0$
	$A+B \to P$	$v=k_r[A][B]$ $k_r t=\frac{1}{[B]_0-[A]_0}\ln\frac{[A]_0([B]_0-x)}{([A]_0-x)[B]_0}$	
	$A+2\,B \to P$	$v=k_r[A][B]$ $k_r t=\frac{1}{[B]_0-2[A]_0}\ln\frac{[A]_0([B]_0-2x)}{([A]_0-x)[B]_0}$	
	$A \to P$ with autocatalysis	$v=k_r[A][P]$ $k_r t=\frac{1}{[A]_0+[P]_0}\ln\frac{[A]_0([P]_0+x)}{([A]_0-x)[P]_0}$	
3	$A+2\,B \to P$	$v=k_r[A][B]^2$ $k_r t=\frac{2x}{(2[A]_0-[B]_0)([B]_0-2x)[B]_0}+\frac{1}{(2[A]_0-[B]_0)^2}\ln\frac{[A]_0([B]_0-2x)}{([A]_0-x)[B]_0}$	
$n\geq 2$	$A \to P$	$v=k_r[A]^n$ $k_r t=\frac{1}{n-1}\left\{\frac{1}{([A]_0-x)^{n-1}}-\frac{1}{[A]_0^{n-1}}\right\}$	$\frac{2^{n-1}-1}{(n-1)k_r[A]_0^{n-1}}$

* $x=[P]$ and $v=dx/dt$

Checklist of concepts

☐ 1. An **integrated rate law** is an expression for the concentration of a reactant or product as a function of time (Table 20B.3).

☐ 2. The **half-life** of a reactant is the time it takes for its concentration to fall to half its initial value.

☐ 3. Analysis of experimental data using integrated rate laws allow for the prediction of the composition of a reaction system at any stage, the verification of the rate law, and the determination of the rate constant.

Checklist of equations

Property	Equation	Comment	Equation number
Integrated rate law	$\ln([A]/[A]_0)=-k_rt$ or $[A]=[A]_0e^{-k_rt}$	First order, $A\rightarrow P$	20B.1b
Half-life	$t_{1/2}=(\ln 2)/k_r$	First order, $A\rightarrow P$	20B.2
Time constant	$\tau=1/k_r$	First order	20B.3
Integrated rate law	$1/[A]-1/[A]_0=k_rt$ or $[A]=[A]_0/(1+k_rt[A]_0)$	Second order, $A\rightarrow P$	20B.4b,c
Half-life	$t_{1/2}=1/k_r[A]_0$	Second order, $A\rightarrow P$	20B.5
	$t_{1/2}=(2^{n-1}-1)/(n-1)k_r[A]_0^{n-1}$	nth order, $n\neq 0,1$	20B.6
Integrated rate law	$\ln\{([B]/[B]_0)/([A]/[A]_0)\}=([B]_0-[A]_0)k_rt$	Second order, $A+B\rightarrow P$	20B.8

20C Reactions approaching equilibrium

Contents

➤ Why do you need to know this material?

All reactions approach equilibrium, so it is important to be able to describe the changing composition as they approach this composition. Analysis of the time dependence shows that there is an important relation between the rate constants and the equilibrium constant.

➤ What is the key idea?

Both forward and reverse reactions must be incorporated into a reaction scheme to account for the approach to equilibrium.

➤ What do you need to know already?

You need to be familiar with the concepts of rate law, reaction order, and rate constant (Topic 20A), integrated rate laws (Topic 20B), and equilibrium constants (Topic 6A). As in Topic 20B, the manipulation of simple rate laws requires only elementary techniques of integration.

In practice, most kinetic studies are made on reactions that are far from equilibrium and if products are in low concentration the reverse reactions are unimportant. Close to equilibrium the products may be so abundant that the reverse reaction must be taken into account.

20C.1 First-order reactions approaching equilibrium

We can explore the variation of the composition with time as a reaction approaches equilibrium by considering a reaction in which A forms B and both forward and reverse reactions are first order (as in some isomerizations):

$$\begin{aligned} &\mathrm{A}\rightarrow\mathrm{B} \quad v=k_\mathrm{r}[\mathrm{A}] \\ &\mathrm{B}\rightarrow\mathrm{A} \quad v=k_\mathrm{r}'[\mathrm{B}] \end{aligned} \tag{20C.1}$$

The concentration of A is reduced by the forward reaction (at a rate $k_\mathrm{r}[\mathrm{A}]$) but it is increased by the reverse reaction (at a rate $k_\mathrm{r}'[\mathrm{B}]$). The net rate of change at any stage is therefore

$$\frac{\mathrm{d}[\mathrm{A}]}{\mathrm{d}t}=-k_\mathrm{r}[\mathrm{A}]+k_\mathrm{r}'[\mathrm{B}] \tag{20C.2}$$

If the initial concentration of A is $[\mathrm{A}]_0$, and no B is present initially, then at all times $[\mathrm{A}]+[\mathrm{B}]=[\mathrm{A}]_0$. Therefore,

$$\begin{aligned} \frac{\mathrm{d}[\mathrm{A}]}{\mathrm{d}t}&=-k_\mathrm{r}[\mathrm{A}]+k_\mathrm{r}'([\mathrm{A}]_0-[\mathrm{A}]) \\ &=-(k_\mathrm{r}+k_\mathrm{r}')[\mathrm{A}]+k_\mathrm{r}'[\mathrm{A}]_0 \end{aligned} \tag{20C.3}$$

The solution of this first-order differential equation (as may be checked by differentiation, Problem 20C.1) is

$$[\mathrm{A}]=\frac{k_\mathrm{r}'+k_\mathrm{r}\mathrm{e}^{-(k_\mathrm{r}+k_\mathrm{r}')t}}{k_\mathrm{r}+k_\mathrm{r}'}[\mathrm{A}]_0 \tag{20C.4}$$

Figure 20C.1 shows the time dependence predicted by this equation, with $[\mathrm{B}]=[\mathrm{A}]_0-[\mathrm{A}]$.

As $t\rightarrow\infty$, the concentrations reach their equilibrium values, which are given by eqn 20C.4 as:

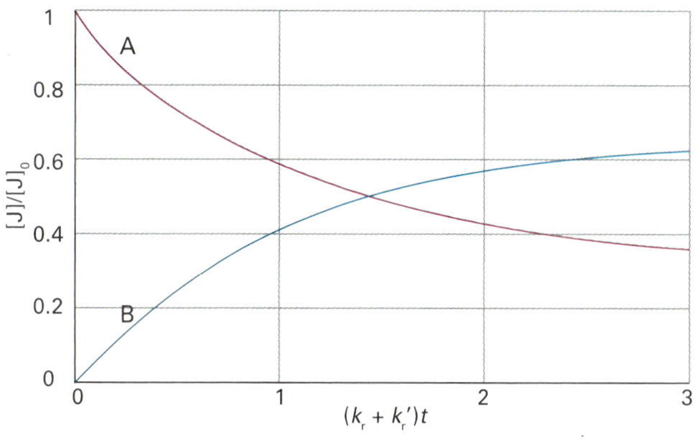

Figure 20C.1 The approach of concentrations to their equilibrium values as predicted by eqn 20C.4 for a reaction A ⇌ B that is first order in each direction, and for which $k_\mathrm{r}=2k_\mathrm{r}'$.

$$[A]_{eq} = \frac{k'_r[A]_0}{k_r + k'_r} \qquad [B]_{eq} = [A]_0 - [A]_{eq} = \frac{k_r[A]_0}{k_r + k'_r} \tag{20C.5}$$

It follows that the equilibrium constant of the reaction is

$$K = \frac{[B]_{eq}}{[A]_{eq}} = \frac{k_r}{k'_r} \tag{20C.6}$$

(As explained in Topic 5E, we are justified in replacing activities with the numerical values of molar concentrations if the system is treated as ideal.) Exactly the same conclusion can be reached—more simply, in fact—by noting that, at equilibrium, the forward and reverse rates must be the same, so

$$k_r[A] = k'_r[B] \tag{20C.7}$$

This relation rearranges into eqn 20C.6. The theoretical importance of eqn 20C.6 is that it relates a thermodynamic quantity, the equilibrium constant, to quantities relating to rates. Its practical importance is that if one of the rate constants can be measured, then the other may be obtained if the equilibrium constant is known.

Equation 20C.6 is valid even if the forward and reverse reactions have different orders, but in that case we need to be careful with units. For instance, if the reaction $A + B \rightarrow C$ is second order forward and first order in reverse, then the condition for equilibrium is $k_r[A]_{eq}[B]_{eq} = k'_r[C]_{eq}$ and the dimensionless equilibrium constant in full dress is

$$K = \frac{[C]_{eq}/c^{\ominus}}{([A]_{eq}/c^{\ominus})([B]_{eq}/c^{\ominus})} = \left(\frac{[C]}{[A][B]}\right)_{eq} c^{\ominus} = \frac{k_r}{k'_r} \times c^{\ominus}$$

The presence of $c^{\ominus} = 1\ \text{mol dm}^{-3}$ in the last term ensures that the ratio of second-order to first-order rate constants, with their different units, is turned into a dimensionless quantity.

Brief illustration 20C.1 The equilibrium constant from rate constants

The rates of the forward and reverse reactions for a dimerization reaction were found to be $8.0\times10^{8}\ \text{dm}^3\ \text{mol}^{-1}\ \text{s}^{-1}$ (second-order) and $2.0\times10^{6}\ \text{s}^{-1}$ (first-order). The equilibrium constant for the dimerization is therefore

$$K = \frac{8.0\times10^{8}\ \text{dm}^3\ \text{mol}^{-1}\ \text{s}^{-1}}{2.0\times10^{6}\ \text{s}^{-1}} \times 1\ \text{mol dm}^{-3} = 4.0\times10^{2}$$

Self-test 20C.1 The equilibrium constant for the attachment of a drug molecule to a protein was measured as 2.0×10^{2}. In a separate experiment, the rate constant for the second-order attachment was found to be $1.5\times10^{8}\ \text{dm}^3\ \text{mol}^{-1}\ \text{s}^{-1}$. What is the rate constant for the loss of the drug molecule from the protein?

Answer: $7.5\times10^{5}\ \text{s}^{-1}$

For a more general reaction, the overall equilibrium constant can be expressed in terms of the rate constants for all the intermediate stages of the reaction mechanism (see Problem 20C.4):

$$K = \frac{k_a}{k'_a} \times \frac{k_b}{k'_b} \times \cdots \qquad \text{Equilibrium constant in terms of the rate constants} \tag{20C.8}$$

where the k_r are the rate constants for the individual steps and the k'_r are those for the corresponding reverse steps.

20C.2 Relaxation methods

The term **relaxation** denotes the return of a system to equilibrium. It is used in chemical kinetics to indicate that an externally applied influence has shifted the equilibrium position of a reaction, normally suddenly, and that the reaction is adjusting to the equilibrium composition characteristic of the new conditions (Fig. 20C.2). We shall consider the response of reaction rates to a **temperature jump**, a sudden change in temperature. We know from Topic 6B that the equilibrium composition of a reaction depends on the temperature (provided $\Delta_r H^{\ominus}$ is nonzero), so a shift in temperature acts as a perturbation on the system. One way of achieving a temperature jump is to discharge a capacitor through a sample that has been made conducting by the addition of ions, but laser or microwave discharges can also be used. Temperature jumps of between 5 and 10 K can be achieved in about 1 μs with electrical discharges. The high energy output of pulsed lasers is sufficient to generate temperature jumps of between 10 and 30 K within nanoseconds in aqueous samples. Some equilibria are also sensitive to pressure, and **pressure-jump techniques** may then also be used.

We show in the following *Justification* that when a sudden temperature increase is applied to a simple $A \rightleftharpoons B$ equilibrium that is first order in each direction, the composition relaxes exponentially to the new equilibrium composition:

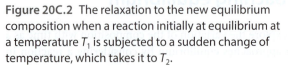

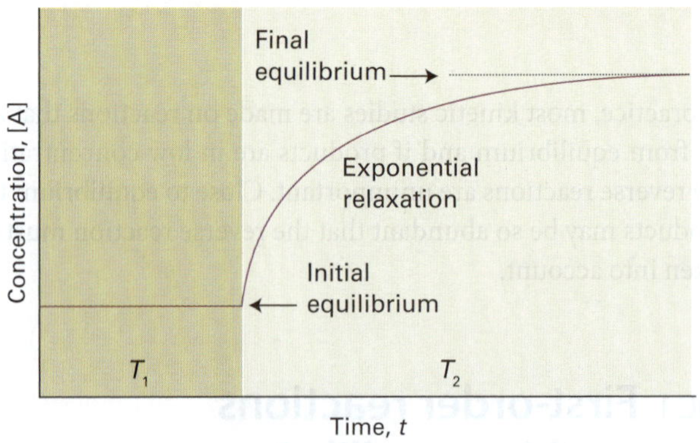

Figure 20C.2 The relaxation to the new equilibrium composition when a reaction initially at equilibrium at a temperature T_1 is subjected to a sudden change of temperature, which takes it to T_2.

$$x = x_0 e^{-t/\tau} \qquad \tau = \frac{1}{k_r + k_r'}$$ *First-order reaction* Relaxation after a temperature jump (20C.9)

where x_0 is the departure from equilibrium immediately after the temperature jump, x is the departure from equilibrium at the new temperature after a time t, and k_r and k_r' are the forward and reverse rate constants, respectively, at the new temperature.

Justification 20C.1 Relaxation to equilibrium

When the temperature of a system at equilibrium is increased suddenly, the rate constants change from their earlier values to the new values k_r and k_r' characteristic of that temperature, but the concentrations of A and B remain for an instant at their old equilibrium values. As the system is no longer at equilibrium, it readjusts to the new equilibrium concentrations, which are now given by

$$k_r[A]_{eq} = k_r'[B]_{eq}$$

and it does so at a rate that depends on the new rate constants. We write the deviation of [A] from its new equilibrium value as x, so $[A] = [A]_{eq} + x$ and $[B] = [B]_{eq} - x$. The concentration of A then changes as follows:

$$\frac{d[A]}{dt} = -k_r[A] + k_r'[B]$$
$$= -k_r([A]_{eq} + x) + k_r'([B]_{eq} - x)$$
$$= -(k_r + k_r')x$$

because the two terms involving the equilibrium concentrations cancel. Because $d[A]/dt = dx/dt$, this equation is a first-order differential equation with the solution that resembles eqn 20A.1b and is given in eqn 20C.9.

Equation 20C.9 shows that the concentrations of A and B relax into the new equilibrium at a rate determined by the sum of the two new rate constants. Because the equilibrium constant under the new conditions is $K \approx k_r/k_r'$, its value may be combined with the relaxation time measurement to find the individual k_r and k_r'.

Example 20C.1 Analysing a temperature-jump experiment

The equilibrium constant for the autoprotolysis of water, $H_2O(l) \rightleftharpoons H^+(aq) + OH^-(aq)$, is $K_w = a(H^+)a(OH^-) = 1.008 \times 10^{-14}$ at 298 K, where we have used the exact expression in terms of activities. After a temperature-jump, the reaction returns to equilibrium with a relaxation time of 37 μs at 298 K and pH ≈ 7. Given that the forward reaction is first order and the reverse is second order overall, calculate the rate constants for the forward and reverse reactions.

Method We need to derive an expression for the relaxation time, τ (the time constant for return to equilibrium), in terms of k_r (forward, first-order reaction) and k_r' (reverse, second-order reaction). We can proceed as above, but it will be necessary to make the assumption that the deviation from equilibrium (x) is so small that terms in x^2 can be neglected. Relate k_r and k_r' through the equilibrium constant, but be careful with units because K_w is dimensionless.

Answer The forward rate at the final temperature is $k_r[H_2O]$ and the reverse rate is $k_r'[H^+][OH^-]$. The net rate of deprotonation of H_2O is

$$\frac{d[H_2O]}{dt} = -k_r[H_2O] + k_r'[H^+][OH^-]$$

We write $[H_2O] = [H_2O]_{eq} + x$, $[H^+] = [H^+]_{eq} - x$, and $[OH^-] = [OH^-]_{eq} - x$, and obtain

$$\frac{dx}{dt} = -\{k_r + k_r'([H^+]_{eq} + [OH^-]_{eq})\}x$$
$$-k_r[H_2O]_{eq} + k_r'[H^+]_{eq}[OH^-]_{eq} + k_r'x^2$$
$$\approx -\{k_r + k_r'([H^+]_{eq} + [OH^-]_{eq})\}x$$

where we have neglected the term in x^2 because it is so small and have used the equilibrium condition $k_r[H_2O]_{eq} = k_r'[H^+]_{eq}[OH^-]_{eq}$ to eliminate the terms (in blue) that are independent of x. It follows that

$$\frac{1}{\tau} = k_r + k_r'([H^+]_{eq} + [OH^-]_{eq})$$

At this point we note that

$$K_w = a(H^+)a(OH^-) \approx ([H^+]_{eq}/c^{\ominus})([OH^-]_{eq}/c^{\ominus})$$
$$= [H^+]_{eq}[OH^-]_{eq}/c^{\ominus 2}$$

with $c^{\ominus} = 1\ \text{mol dm}^{-3}$. For this electrically neutral system, $[H^+] = [OH^-]$, so the concentration of each type of ion is $K_w^{1/2}c^{\ominus}$, and hence

$$\frac{1}{\tau} = k_r + k_r'(K_w^{1/2}c^{\ominus} + K_w^{1/2}c^{\ominus}) = k_r'\left\{\frac{k_r}{k_r'} + 2K_w^{1/2}c^{\ominus}\right\}$$

At this point we note that

$$\frac{k_r}{k_r'} = \frac{[H^+]_{eq}[OH^-]_{eq}}{[H_2O]_{eq}} = \frac{K_w c^{\ominus 2}}{[H_2O]_{eq}}$$

and therefore

$$\frac{1}{\tau} = 2k_r'\left(1 + \overbrace{\frac{K_w^{1/2}c^{\ominus}}{2[H_2O]_{eq}}}^{K}\right)K_w^{1/2}c^{\ominus} = 2k_r'(1+K)K_w^{1/2}c^{\ominus}$$

The molar concentration of pure water is 55.6 mol dm^{-3}, so $[H_2O]_{eq}/c^{\ominus} = 55.6$ and

$$K = \frac{(1.008 \times 10^{-14})^{1/2}}{2 \times 55.6} = 9.03 \times 10^{-10}$$

which implies that $1+K$ may be replaced by 1 and therefore that

$$k_r' \approx \frac{1}{2\tau K_w^{1/2} c^{\ominus}}$$

$$= \frac{1}{2(3.7\times10^{-5}\,\text{s})\times(1.008\times10^{-14})^{1/2}\times(1\,\text{mol}\,\text{dm}^{-3})}$$

$$= 1.4\times10^{11}\,\text{dm}^3\,\text{mol}^{-1}\,\text{s}^{-1}$$

It follows from the expression for k_r/k_r' that

$$k_r = \frac{K_w c^{\ominus 2} k_r'}{[\text{H}_2\text{O}]_{eq}}$$

$$= \frac{(1.008\times10^{-14})\times(1\,\text{mol}\,\text{dm}^{-3})^2\times(1.4\times10^{11}\,\text{dm}^3\,\text{mol}^{-1}\,\text{s}^{-1})}{55.6\,\text{mol}\,\text{dm}^{-3}}$$

$$= 2.5\times10^{-5}\,\text{s}^{-1}$$

The reaction is faster in ice, where $k_r' = 8.6\times10^{12}\,\text{dm}^3\,\text{mol}^{-1}\,\text{s}^{-1}$.

A note on good practice Notice how we keep track of units through the use of $c^{\ominus}$: K and K_w are dimensionless; k_r' is expressed in $\text{dm}^3\,\text{mol}^{-1}\,\text{s}^{-1}$ and k_r is expressed in s^{-1}.

Self-test 20C.2 Derive an expression for the relaxation time of a concentration when the reaction $\text{A}+\text{B}\rightleftharpoons\text{C}+\text{D}$ is second order in both directions.

Answer: $1/\tau = k_r([\text{A}]+[\text{B}])_{eq} + k_r'([\text{C}]+[\text{D}])_{eq}$

Checklist of concepts

☐ 1. There is a relation between the equilibrium constant, a thermodynamic quantity, and the rate constants of the forward and reverse reactions (see Checklist of equations).

☐ 2. In **relaxation methods** of kinetic analysis, the equilibrium position of a reaction is first shifted suddenly and then allowed to readjust to the equilibrium composition characteristic of the new conditions.

Checklist of equations

Property	Equation	Comment	Equation number
Equilibrium constant in terms of rate constants	$K = k_a/k_a' \times k_b/k_b' \times \cdots$	include $c^{\ominus}$ as appropriate	20C.8
Relaxation of an equilibrium $\text{A}\rightleftharpoons\text{B}$ after a temperature jump	$x = x_0 e^{-t/\tau}$ $\tau = 1/(k_r + k_r')$	First order in each direction	20C.9

20D The Arrhenius equation

Contents

➤ Why do you need to know this material?

The rates of reactions depend on the temperature. Exploration of this dependence leads to the formulation of theories that can help you understand the details of the processes that occur when reactant molecules meet and why a collection of reactants under specific conditions leads to certain products but not others.

➤ What is the key idea?

The temperature dependence of the rate of a reaction is summarized by the activation energy, the minimum energy needed for reaction to occur in an encounter between reactants.

➤ What do you need to know already?

You need to know that the rate of a chemical reaction is expressed by a rate constant (Topic 20A).

In this Topic we interpret the common experimental observation that chemical reactions usually go faster as the temperature is raised. We also begin to see how exploration of the temperature dependence of reaction rates can reveal some details of the energy requirements for molecular encounters that lead to the formation of products.

20D.1 The temperature dependence of reaction rates

It is found experimentally for many reactions that a plot of $\ln k_r$ against $1/T$ gives a straight line with a negative slope, indicating that an increase in $\ln k_r$ (and therefore an increase in k_r) results from a decrease in $1/T$ (that is, an increase in T). This behaviour is normally expressed mathematically by introducing two parameters, one representing the intercept and the other the slope of the straight line, and writing the **Arrhenius equation**

$$\ln k_r = \ln A - \frac{E_a}{RT} \qquad \text{Arrhenius equation} \qquad (20D.1)$$

The parameter A, which corresponds to the intercept of the line at $1/T=0$ (at infinite temperature, Fig. 20D.1), is called the **pre-exponential factor** or the 'frequency factor'. The parameter E_a, which is obtained from the slope of the line ($-E_a/R$), is called the **activation energy**. Collectively the two quantities are called the **Arrhenius parameters** (Table 20D.1).

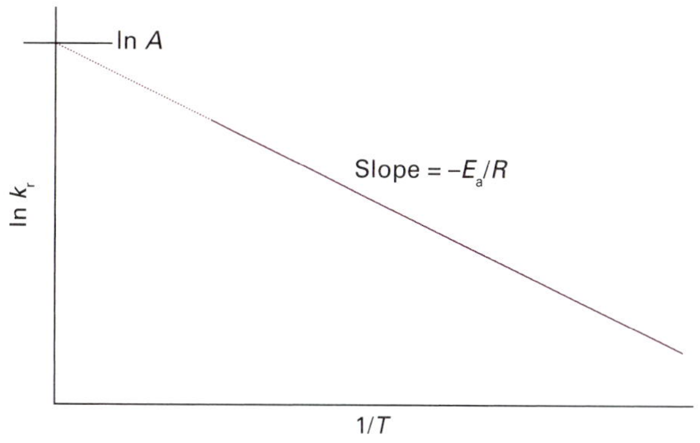

Figure 20D.1 A plot of $\ln k_r$ against $1/T$ is a straight line when the reaction follows the behaviour described by the Arrhenius equation (eqn 20D.1). The slope gives $-E_a/R$ and the intercept at $1/T=0$ gives $\ln A$.

Example 20D.1 Determining the Arrhenius parameters

The rate of the second-order decomposition of acetaldehyde (ethanal, CH_3CHO) was measured over the temperature range 700–1000 K, and the rate constants are reported below. Find E_a and A.

T/K	700	730	760	790	810	840	910	1000
k_r/(dm^3 mol^{-1} s^{-1})	0.011	0.035	0.105	0.343	0.789	2.17	20.0	145

Method According to eqn 20D.1, the data can be analysed by plotting $\ln(k_r/\text{dm}^3\,\text{mol}^{-1}\,\text{s}^{-1})$ against $1/(T/\text{K})$, or more conveniently $(10^3\,\text{K})/T$, and getting a straight line. Obtain the activation energy from the dimensionless slope by writing $-E_a/R$ = slope/units, where in this case 'units' $= 1/(10^3\,\text{K})$, so $E_a = -\text{slope}\times R\times 10^3\,\text{K}$. The intercept at $1/T=0$ is $\ln(A/\text{dm}^3\,\text{mol}^{-1}\,\text{s}^{-1})$. Use a least-squares procedure to determine the plot parameters.

Answer We draw up the following table:

$(10^3\,\text{K})/T$	1.43	1.37	1.32	1.27	1.23	1.19	1.10	1.00
$\ln(k_r/\text{dm}^3\,\text{mol}^{-1}\,\text{s}^{-1})$	−4.51	−3.35	−2.25	−1.07	−0.24	0.77	3.00	4.98

Now plot $\ln k_r$ against $1/T$ (Fig. 20D.2). The least-squares fit results in a line with slope −22.7 and intercept 27.7. Therefore,

$$E_a = 22.7\times(8.3145\,\text{J K}^{-1}\,\text{mol}^{-1})\times(10^3\,\text{K}) = 189\,\text{kJ mol}^{-1}$$
$$A = e^{27.7}\,\text{dm}^3\,\text{mol}^{-1}\,\text{s}^{-1} = 1.1\times10^{12}\,\text{dm}^3\,\text{mol}^{-1}\,\text{s}^{-1}$$

Note that A has the same units as k_r.

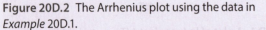

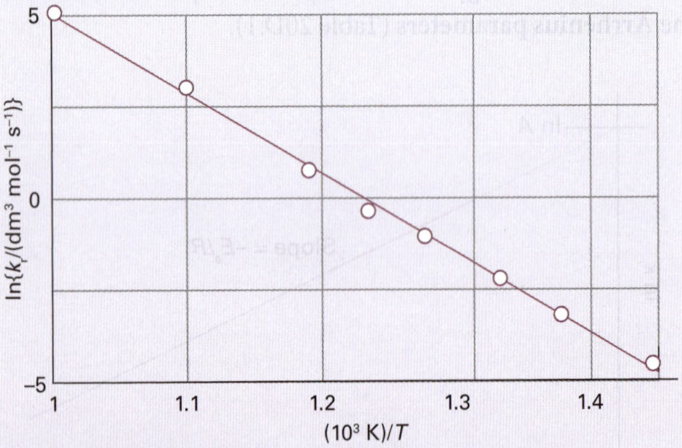

Figure 20D.2 The Arrhenius plot using the data in *Example* 20D.1.

Self-test 20D.1 Determine A and E_a from the following data:

T/K	300	350	400	450	500
k_r/(dm^3 mol^{-1} s^{-1})	7.9×10^6	3.0×10^7	7.9×10^7	1.7×10^8	3.2×10^8

Answer: $8\times10^{10}\,\text{dm}^3\,\text{mol}^{-1}\,\text{s}^{-1}$, $23\,\text{kJ mol}^{-1}$

Once the activation energy of a reaction is known, it is a simple matter to predict the value of a rate constant $k_{r,2}$ at a temperature T_2 from its value $k_{r,1}$ at another temperature T_1. To do so, we write

Table 20D.1* Arrhenius parameters

(1) First-order reactions	A/s^{-1}	E_a/(kJ mol^{-1})
$CH_3NC \rightarrow CH_3CN$	3.98×10^{13}	160
$2\,N_2O_5 \rightarrow 4\,NO_2 + O_2$	4.94×10^{13}	103.4
(2) Second-order reactions	**A/(dm^3 mol^{-1} s^{-1})**	**E_a/(kJ mol^{-1})**
$OH + H_2 \rightarrow H_2O + H$	8.0×10^{10}	42
$NaC_2H_5O + CH_3I$ in ethanol	2.42×10^{11}	81.6

* More values are given in the *Resource section*.

$$\ln k_{r,2} = \ln A - \frac{E_a}{RT_2}$$

and then subtract eqn 20D.1 (with T identified as T_1 and k_r as $k_{r,1}$), so obtaining

$$\ln k_{r,2} - \ln k_{r,1} = -\frac{E_a}{RT_2} + \frac{E_a}{RT_1}$$

We can rearrange this expression to

$$\ln\frac{k_{r,2}}{k_{r,1}} = \frac{E_a}{R}\left(\frac{1}{T_1} - \frac{1}{T_2}\right) \quad \text{Temperature dependence of the rate constant} \quad (20D.2)$$

Brief illustration 20D.1 The Arrhenius equation

For a reaction with an activation energy of 50 kJ mol^{-1}, an increase in the temperature from 25 °C to 37 °C (body temperature) corresponds to

$$\ln\frac{k_{r,2}}{k_{r,1}} = \frac{50\times10^3\,\text{J mol}^{-1}}{8.3145\,\text{J K}^{-1}\,\text{mol}^{-1}}\left(\frac{1}{298\,\text{K}} - \frac{1}{310\,\text{K}}\right) = \frac{50\times10^3}{8.3145}\left(\frac{1}{298} - \frac{1}{310}\right) = 0.781\ldots$$

By taking natural antilogarithms (that is, by forming e^x), $k_{r,2} = 2.18k_{r,1}$. This result corresponds to slightly more than a doubling of the rate constant as the temperature is increased from 298 K to 310 K.

Self-test 20D.2 The activation energy of one of the reactions in a biochemical process is 87 kJ mol^{-1}. What is the change in rate constant when the temperature falls from 37 °C to 15 °C?

Answer: $k_r(15\,°\text{C}) = 0.076k_r(37\,°\text{C})$

The fact that E_a is given by the slope of the plot of $\ln k_r$ against $1/T$ leads to the following conclusions:

- The stronger the temperature dependence of the rate constant (that is, the steeper the slope), the higher the activation energy.

- A high activation energy signifies that the rate constant depends strongly on temperature.
- If a reaction has zero activation energy, its rate is independent of temperature.
- A negative activation energy indicates that the rate decreases as the temperature is raised.

Physical interpretation

The temperature dependence of some reactions is 'non-Arrhenius' in the sense that a straight line is not obtained when $\ln k_r$ is plotted against $1/T$. However, it is still possible to define an activation energy at any temperature as

$$E_a = RT^2\left(\frac{d\ln k_r}{dT}\right) \quad \textit{Definition} \quad \text{Activation energy} \quad (20D.3)$$

This definition reduces to the earlier one (as the slope of a straight line) for a temperature-independent activation energy (see Problem 20D.1). However, the definition in eqn 20D.3 is more general than that in eqn 20D.1, because it allows E_a to be obtained from the slope (at the temperature of interest) of a plot of $\ln k_r$ against $1/T$ even if the Arrhenius plot is not a straight line. Non-Arrhenius behaviour is sometimes a sign that quantum mechanical tunnelling (Topic 8A) is playing a significant role in the reaction. In biological reactions it might signal that an enzyme has undergone a structural change and has become less efficient.

20D.2 The interpretation of the Arrhenius parameters

For the present Topic we shall regard the Arrhenius parameters as purely empirical quantities that enable us to summarize the variation of rate constants with temperature. However, it is useful to have an interpretation in mind. Topics 21A–21F provide a more elaborate interpretation.

(a) A first look at the energy requirements of reactions

To interpret E_a we consider how the molecular potential energy changes in the course of a chemical reaction that begins with a collision between molecules of A and molecules of B (Fig. 20D.3). In the gas phase that is an actual collision; in solution it is best regarded as a close encounter, possibly with excess energy, and might involve the solvent too. As the reaction event proceeds, A and B come into contact, distort, and begin to exchange or discard atoms. The **reaction coordinate** summarizes the collection of motions, such as changes in interatomic distances and bond angles, that are directly involved in the formation of products from reactants. (The reaction coordinate is essentially a geometrical concept and quite distinct from the extent of reaction.) The potential energy rises to a maximum and the cluster of atoms that corresponds to the region close to the maximum is called the **activated complex**.

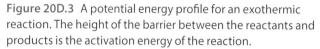

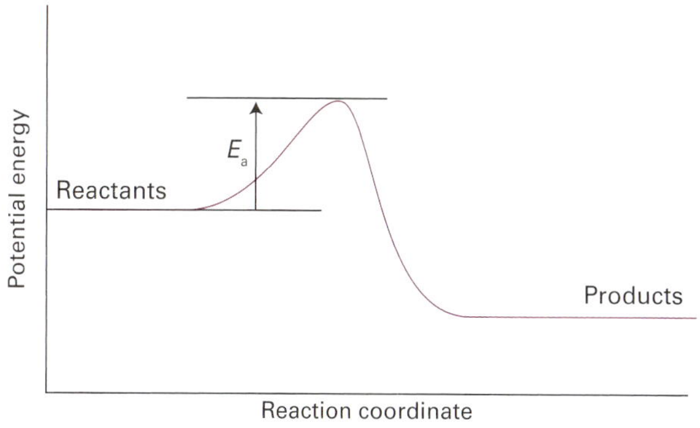

Figure 20D.3 A potential energy profile for an exothermic reaction. The height of the barrier between the reactants and products is the activation energy of the reaction.

After the maximum, the potential energy falls as the atoms rearrange in the cluster and reaches a value characteristic of the products. The climax of the reaction is at the peak of the potential energy, which corresponds to the activation energy E_a. Here two reactant molecules have come to such a degree of closeness and distortion that a small further distortion will send them in the direction of products. This crucial configuration is called the **transition state** of the reaction. Although some molecules entering the transition state might revert to reactants, if they pass through this configuration then it is inevitable that products will emerge from the encounter. (The terms 'activated complex' and 'transition state' are often used as synonyms; however, we shall preserve a distinction.)

We conclude from the preceding discussion that *the activation energy is the minimum energy reactants must have in order to form products*. For example, in a reaction mixture there are numerous molecular encounters each second, but only very few are sufficiently energetic to lead to reaction. The fraction of close encounters between reactants with energy in excess of E_a is given by the Boltzmann distribution (*Foundations* B and Topic 15A) as $e^{-E_a/RT}$. This interpretation is confirmed by comparing this expression with the Arrhenius equation written in the form

$$k_r = Ae^{-E_a/RT} \quad \textit{Alternative form} \quad \text{Arrhenius equation} \quad (20D.4)$$

which is obtained by taking antilogarithms of both sides of eqn 20D.1. We show in the following *Justification* that the exponential factor in eqn 20D.4 can be interpreted as the fraction of encounters that have enough energy to lead to reaction. This point is explored further for gas-phase reactions in Topic 21A and for reactions in solution in Topic 21C.

Justification 20D.1 Interpreting the activation energy

Suppose the energy levels available to the system form a uniform array of separation ε (Fig. 20D.4). The Boltzmann distribution is

$$\frac{N_i}{N}=\frac{e^{-i\varepsilon\beta}}{q}=(1-e^{-\varepsilon\beta})e^{-i\varepsilon\beta}$$

where $\beta=1/kT$ and we have used the result in eqn 15B.2a for the partition function q. The total number of molecules in states with energy of at least $i_{min}\varepsilon$ is

$$\sum_{i=i_{min}}^{\infty}N_i=\sum_{i=0}^{\infty}N_i-\sum_{i=0}^{i_{min}-1}N_i=N-\frac{N}{q}\sum_{i=0}^{i_{min}-1}e^{-i\varepsilon\beta}$$

The sum of the (blue) finite geometrical series is

$$\sum_{i=0}^{i_{min}-1}e^{-i\varepsilon\beta}=\frac{1-e^{-i_{min}\varepsilon\beta}}{1-e^{-\varepsilon\beta}}=q(1-e^{-i_{min}\varepsilon\beta})$$

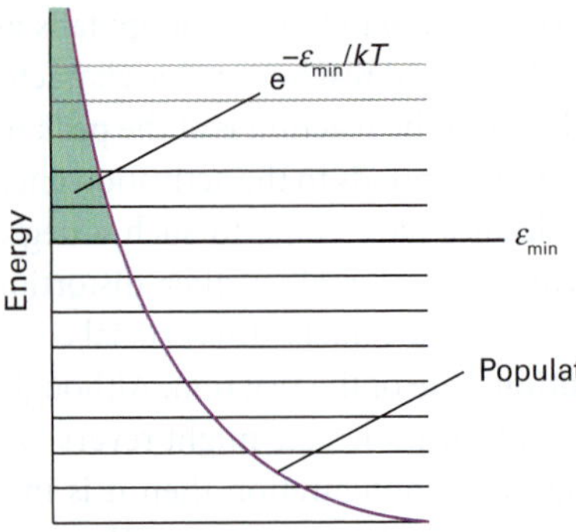

Figure 20D.4 Equally spaced energy levels of an idealized system. As shown in *Justification* 20D.1, the fraction of molecules with energy of at least $e^{-\varepsilon_{min}/kT}$.

Therefore, the fraction of molecules in states with energy of at least $\varepsilon_{min}=i_{min}\varepsilon$ is

$$\frac{1}{N}\sum_{i=i_{min}}^{\infty}N_i=1-(1-e^{-i_{min}\varepsilon\beta})=e^{-i_{min}\varepsilon\beta}=e^{-\varepsilon_{min}/kT}$$

which has the form of eqn 20D.4.

Brief illustration 20D.2 The fraction of reactive collisions

From *Justification* 20D.1 the fraction of molecules with energy at least ε_{min} is $e^{-\varepsilon_{min}/kT}$, By multiplying ε_{min} and k by N_A, Avogadro's constant, and identifying $N_A\varepsilon_{min}$ with E_a, then the fraction f of molecular collisions that occur with a kinetic energy E_a becomes $f=e^{-E_a/RT}$. With $E_a=50\,kJ\,mol^{-1}=5.0\times10^4\,J\,mol^{-1}$ and $T=298\,K$, we calculate

$$f=e^{-(5.0\times10^4\,J\,mol^{-1})/(8.3145\,J\,K^{-1}\,mol^{-1}\times298\,K)}=1.7\times10^{-9}$$

or about 1 in a billion.

Self-test 20D.3 At what temperature would $f=0.10$ if $E_a=50\,kJ\,mol^{-1}$?

Answer: $T=2612\,K$

The pre-exponential factor is a measure of the rate at which collisions occur irrespective of their energy. Hence, the product of A and the exponential factor, $e^{-E_a/RT}$ gives the rate of *successful* collisions. We develop these remarks in Topics 21A and 21C, and see that they have their analogues for reactions that take place in liquids.

(b) The effect of a catalyst on the activation energy

The Arrhenius equation tells us that the rate constant of a reaction can be increased by increasing the temperature or by decreasing the energy of activation. Changing the temperature of a reaction mixture is an easy strategy. Reducing the energy of activation is more challenging, but is possible if a reaction takes place in the presence of a suitable **catalyst,** a substance that accelerates a reaction but undergoes no net chemical change. The catalyst lowers the activation energy of the reaction by providing an alternative path that avoids the slow, rate-determining step of the uncatalysed reaction (Fig 20D.5).

Heterogeneous catalysts, which are discussed in Topic 22C, function in a different phase from the reaction mixture. For example, some gas-phase reactions are accelerated in the presence of a solid catalyst. **Homogeneous catalysts** function in the same phase as the reaction mixture. For example, the OH^- ion is a catalyst for a number of organic and inorganic transformations in solution.

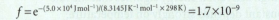

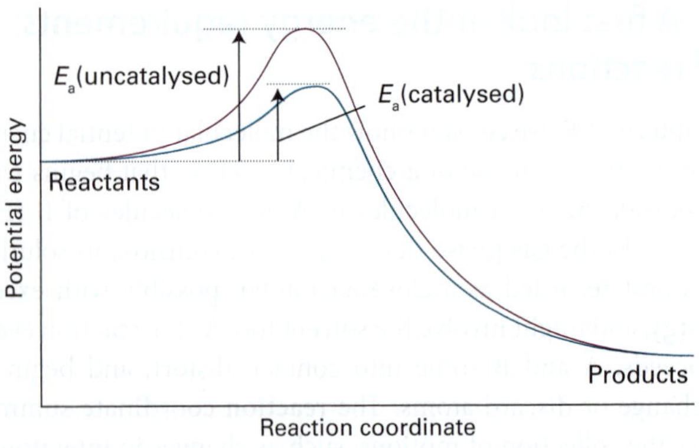

Figure 20D.5 A catalyst provides a different path with a lower activation energy. The result is an increase in the rate of formation of products.

Brief illustration 20D.3 The effect of a catalyst on the rate constant

The enzyme catalase reduces the activation energy for the decomposition of hydrogen peroxide from 76 kJ mol^{-1} to 8 kJ mol^{-1}. From eqn 20D.4 and assuming that the exponential factor is the same in both cases, it follows that the ratio of rate constants is:

$$\frac{k_{r,\text{catalysed}}}{k_{r,\text{uncatalysed}}} = \frac{Ae^{-E_{a,\text{catalysed}}/RT}}{Ae^{-E_{a,\text{uncatalysed}}/RT}} = e^{-(E_{a,\text{catalysed}}-E_{a,\text{uncatalysed}})/RT}$$

$$= e^{(68\times10^3\,\text{J}\,\text{mol}^{-1})/(8.3145\,\text{J}\,\text{K}^{-1}\,\text{mol}^{-1})\times(298\,\text{K})} = 8.3\times10^{11}$$

Self-test 20D.4 Consider the decomposition of hydrogen peroxide, which can be catalysed in solution by iodide ion. By how much is the activation energy of the reaction reduced if the rate constant of reaction increases by a factor of 2000 at 298 K upon addition of the catalyst?

Answer: 25 per cent

Checklist of concepts

- ☐ 1. The **activation energy**, the parameter E_a in the **Arrhenius equation**, is the minimum energy of close molecular encounters able to result in reaction.
- ☐ 2. The larger the activation energy, the more sensitive the rate constant is to the temperature.
- ☐ 3. The **pre-exponential factor** is a measure of the rate at which encounters occur irrespective of their energy.
- ☐ 4. A **catalyst** lowers the activation energy of a reaction.

Checklist of equations

Property	Equation	Comment	Equation number
Arrhenius equation	$\ln k_r = \ln A - E_a/RT$		20D.1
Activation energy	$E_a = RT^2(\mathrm{d}\ln k_r/\mathrm{d}T)$	Definition	20D.3

20E Reaction mechanisms

Contents

➤ Why do you need to know this material?

You need to know how to construct the rate law for a reaction that takes place by a sequence of steps partly because that gives insight into the atomic processes going on when reactions take place, but also because it indicates how the yield of desired products can be optimized.

➤ What is the key idea?

Many chemical reactions occur as a sequence of simpler steps, with corresponding rate laws that can be combined together by applying one or more approximations.

➤ What do you need to know already?

You need to be familiar with the concept of rate laws (Topic 20A) and how to integrate them (Topics 20B and 20C). You also need to be familiar with the Arrhenius equation for the effect of temperature on reaction rate (Topic 20D).

The study of reaction rates leads to an understanding of the **mechanisms** of reactions, their analysis into a sequence of elementary steps. Simple elementary steps have simple rate laws, which can be combined together by invoking one or more approximations. These approximations include the concept of the rate-determining step of a reaction, the steady-state concentration of a reaction intermediate, and the existence of a pre-equilibrium.

20E.1 Elementary reactions

Most reactions occur in a sequence of steps called **elementary reactions**, each of which involves only a small number of molecules or ions. A typical elementary reaction is

$$\mathrm{H + Br_2 \rightarrow HBr + Br}$$

Note that the phase of the species is not specified in the chemical equation for an elementary reaction and the equation represents the specific process occurring to individual molecules. This equation, for instance, signifies that an H atom attacks a Br_2 molecule to produce an HBr molecule and a Br atom. The **molecularity** of an elementary reaction is the number of molecules coming together to react in an elementary reaction. In a **unimolecular reaction**, a single molecule shakes itself apart or its atoms into a new arrangement, as in the isomerization of cyclopropane to propene. In a **bimolecular reaction**, a pair of molecules collide and exchange energy, atoms, or groups of atoms, or undergo some other kind of change. It is most important to distinguish molecularity from order:

- *reaction order* is an empirical quantity, and obtained from the experimentally determined rate law;
- *molecularity* refers to an elementary reaction proposed as an individual step in a mechanism.

The rate law of a unimolecular elementary reaction is first-order in the reactant:

$$\mathrm{A \rightarrow P} \qquad \frac{\mathrm{d[A]}}{\mathrm{d}t} = -k_r[\mathrm{A}] \qquad \text{Unimolecular elementary reaction} \qquad (20\text{E}.1)$$

where P denotes products (several different species may be formed). A unimolecular reaction is first order because the number of A molecules that decay in a short interval is proportional to the number available to decay. For instance, ten times as many decay in the same interval when there are initially 1000 A molecules as when there are only 100 present. Therefore, the rate of decomposition of A is proportional to its molar concentration at any moment during the reaction.

An elementary bimolecular reaction has a second-order rate law:

$$\mathrm{A+B\rightarrow P} \qquad \frac{\mathrm{d[A]}}{\mathrm{d}t}=-k_\mathrm{r}\mathrm{[A][B]} \qquad \text{Bimolecular elementary reaction} \tag{20E.2}$$

A bimolecular reaction is second-order because its rate is proportional to the rate at which the reactant species meet, which in turn is proportional to both their concentrations. Therefore, if we have evidence that a reaction is a single-step, bimolecular process, we can write down the rate law (and then go on to test it).

Brief illustration 20E.1 The rate laws of elementary steps

Bimolecular elementary reactions are believed to account for many homogeneous reactions, such as the dimerizations of alkenes and dienes and reactions such as

$$\mathrm{CH_3I(alc)+CH_3CH_2O^-(alc)\rightarrow CH_3OCH_2CH_3(alc)+I^-(alc)}$$

(where 'alc' signifies alcohol solution). There is evidence that the mechanism of this reaction is a single elementary step:

$$\mathrm{CH_3I+CH_3CH_2O^-\rightarrow CH_3OCH_2CH_3+I^-}$$

This mechanism is consistent with the observed rate law

$$v=k_\mathrm{r}\mathrm{[CH_3I][CH_3CH_2O^-]}$$

Self-test 20E.1 The following are elementary processes: (a) the dimerization of NO(g) to form N_2O_2(g), and (b) the decomposition of the N_2O_2(g) dimer into NO(g) molecules. Write the rate laws for these processes.

Answer: (a) bimolecular process: $k_\mathrm{r}\mathrm{[NO]^2}$, (b) unimolecular process: $k_\mathrm{r}\mathrm{[N_2O_2]}$

We shall see in the following sections how to combine a series of simple steps together into a mechanism and how to arrive at the corresponding overall rate law. For the present we emphasize that, *if the reaction is an elementary bimolecular process, then it has second-order kinetics, but if the kinetics is second order, then the reaction might be complex.* The postulated mechanism can be explored only by detailed detective work on the system and by investigating whether side products or intermediates appear during the course of the reaction. Detailed analysis of this kind was one of the ways, for example, in which the reaction $\mathrm{H_2(g)+I_2(g)\rightarrow 2\,HI(g)}$ was shown to proceed by a complex mechanism. For many years the reaction had been accepted on good but insufficiently meticulous evidence as a fine example of a simple bimolecular reaction, $\mathrm{H_2+I_2\rightarrow HI+HI}$, in which atoms exchanged partners during a collision.

20E.2 Consecutive elementary reactions

Some reactions proceed through the formation of an intermediate (I), as in the consecutive unimolecular reactions

$$\mathrm{A}\xrightarrow{k_\mathrm{a}}\mathrm{I}\xrightarrow{k_\mathrm{b}}\mathrm{P}$$

Note that the intermediate occurs in the reaction steps but does not appear in the overall reaction, which in this case is A → P. We are ignoring any reverse reactions, so the reaction proceeds from all A to all P, not to an equilibrium mixture of the two. An example of this type of mechanism is the decay of a radioactive family, such as

$$^{239}\mathrm{U}\xrightarrow{23.5\,\mathrm{min}}{}^{239}\mathrm{Np}\xrightarrow{2.35\,\mathrm{days}}{}^{239}\mathrm{Pu}$$

(The times are half-lives.) The characteristics of this type of reaction are discovered by setting up the rate laws for the net rate of change of the concentration of each substance and then combining them in the appropriate manner.

The rate of unimolecular decomposition of A is

$$\frac{\mathrm{d[A]}}{\mathrm{d}t}=-k_\mathrm{a}\mathrm{[A]} \tag{20E.3a}$$

and A is not replenished. The intermediate I is formed from A (at a rate k_a[A]) but decays to P (at a rate k_b[I]). The net rate of formation of I is therefore

$$\frac{\mathrm{d[I]}}{\mathrm{d}t}=k_\mathrm{a}\mathrm{[A]}-k_\mathrm{b}\mathrm{[I]} \tag{20E.3b}$$

The product P is formed by the unimolecular decay of I:

$$\frac{\mathrm{d[P]}}{\mathrm{d}t}=k_\mathrm{b}\mathrm{[I]} \tag{20E.3c}$$

We suppose that initially only A is present and that its concentration is then $\mathrm{[A]_0}$.

The first of the rate laws, eqn 20E.3a, is an ordinary first-order decay, so we can write

$$\mathrm{[A]=[A]_0\,e}^{-k_\mathrm{a}t} \tag{20E.4a}$$

When this equation is substituted into eqn 20E.3b, we obtain after rearrangement

$$\frac{\mathrm{d[I]}}{\mathrm{d}t}+k_\mathrm{b}\mathrm{[I]}=k_\mathrm{a}\mathrm{[A]_0\,e}^{-k_\mathrm{a}t}$$

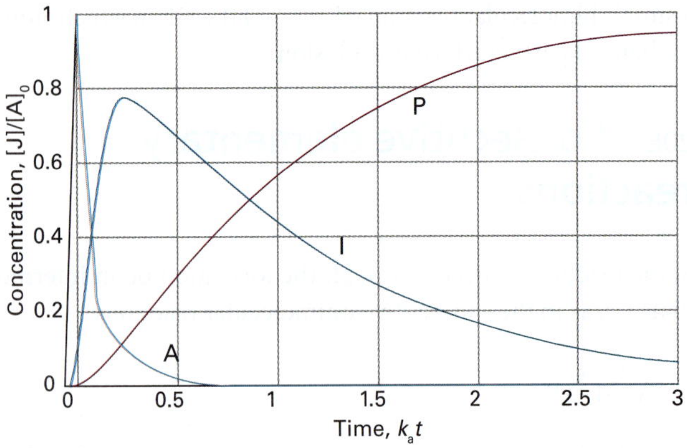

Figure 20E.1 The concentrations of A, I, and P in the consecutive reaction scheme A → I → P. The curves are plots of eqns 20E.4a–c with $k_a = 10k_b$. If the intermediate I is in fact the desired product, it is important to be able to predict when its concentration is greatest; see *Example* 20E.1.

This differential equation has a standard form (see *Mathematical background* 4) and, after setting $[I]_0=0$ (no intermediate present initially), the solution is

$$[I]=\frac{k_a}{k_b-k_a}(e^{-k_at}-e^{-k_bt})[A]_0 \qquad (20E.4b)$$

At all times $[A]+[I]+[P]=[A]_0$, so it follows that

$$[P]=\left\{1+\frac{k_ae^{-k_bt}-k_be^{-k_at}}{k_b-k_a}\right\}[A]_0 \qquad (20E.4c)$$

The concentration of the intermediate I rises to a maximum and then falls to zero (Fig. 20E.1). The concentration of the product P rises from zero towards $[A]_0$, when all A has been converted to P.

Example 20E.1 Analysing consecutive reactions

Suppose that in an industrial batch process a substance A produces the desired compound I which goes on to decay to a worthless product C, each step of the reaction being first order. At what time will I be present in greatest concentration?

Method The time dependence of the concentration of I is given by eqn 20E.4b. We can find the time at which [I] passes through a maximum, t_{max}, by calculating d[I]/dt and setting the resulting rate equal to zero.

Answer It follows from eqn 20E.4b that

$$\frac{d[I]}{dt}=-\frac{k_a(k_ae^{-k_at}-k_be^{-k_bt})[A]_0}{k_b-k_a}$$

This rate is equal to zero when $k_ae^{-k_at}=k_be^{-k_bt}$. Therefore,

$$t_{max}=\frac{1}{k_a-k_b}\ln\frac{k_a}{k_b}$$

For a given value of k_a, as k_b increases both the time at which [I] is a maximum and the yield of I decrease.

Self-test 20E.2 Calculate the maximum concentration of I and justify the last remark.

Answer: $[I]_{max}/[A]_0=(k_a/k_b)^c$, $c=k_b/(k_b-k_a)$

20E.3 The steady-state approximation

One feature of the calculation so far has probably not gone unnoticed: there is a considerable increase in mathematical complexity as soon as the reaction mechanism has more than a couple of steps and reverse reactions are taken into account. A reaction scheme involving many steps is nearly always unsolvable analytically, and alternative methods of solution are necessary. One approach is to integrate the rate laws numerically. An alternative approach, which continues to be widely used because it leads to convenient expressions and more readily digestible results, is to make an approximation.

The **steady-state approximation** (which is also widely called the **quasi-steady-state approximation**, QSSA, to distinguish it from a true steady state) assumes that the intermediate, I, is in a low, constant concentration. More specifically, after an initial **induction period**, an interval during which the concentrations of intermediates rise from zero, and during the major part of the reaction, the rates of change of concentrations of all reaction intermediates are negligibly small (Fig. 20E.2):

$$\frac{d[I]}{dt}\approx 0 \qquad \text{Steady-state approximation} \qquad (20E.5)$$

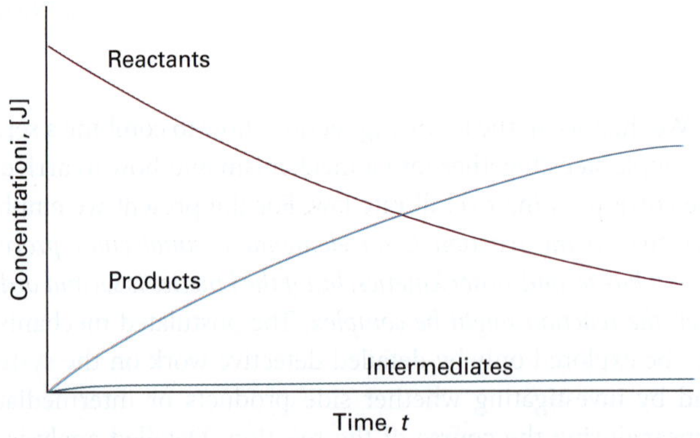

Figure 20E.2 The basis of the steady-state approximation. It is supposed that the concentrations of intermediates remain small and hardly change during most of the course of the reaction.

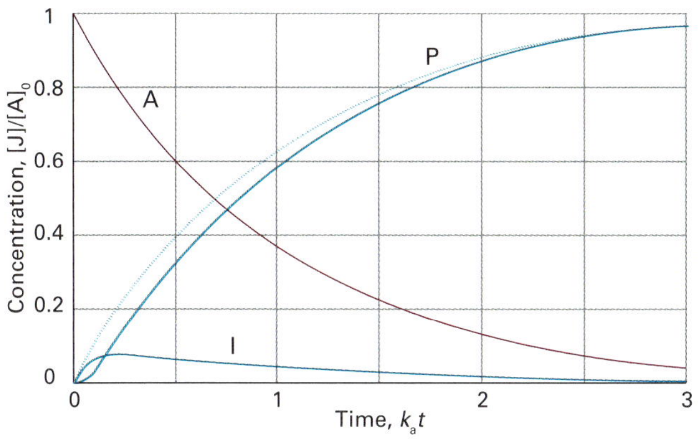

Figure 20E.3 A comparison of the exact result for the concentrations of a consecutive reaction and the concentrations obtained by using the steady-state approximation (dotted lines) for $k_b = 20k_a$. (The curve for [A] is unchanged.)

This approximation greatly simplifies the discussion of reaction schemes. For example, when we apply the approximation to the consecutive first-order mechanism, we set $d[I]/dt = 0$ in eqn 20E.3b, which then becomes $k_a[A] - k_b[I] = 0$. Then

$$[I] = (k_a/k_b)[A] \qquad (20E.6)$$

For this expression to be consistent with eqn 20E.5, we require $k_a/k_b \ll 1$ (so that, even though [A] does depend on the time, the dependence of [I] on the time is negligible). On substituting this value of [I] into eqn 20E.3c, that equation becomes

$$\frac{d[P]}{dt} = k_b[I] \approx k_a[A] \qquad (20E.7)$$

and we see that P is formed by a first-order decay of A, with a rate constant k_a, the rate-constant of the slower, rate-determining, step. We can write down the solution of this equation at once by substituting the solution for [A], eqn 20E.4a, and integrating:

$$[P] = k_a[A]_0 \int_0^t e^{-k_a t} dt = (1 - e^{-k_a t})[A]_0 \qquad (20E.8)$$

This is the same (approximate) result as before, eqn 20E.4c (when $k_b \gg k_a$), but much more quickly obtained. Figure 20E.3 compares the approximate solutions found here with the exact solutions found earlier: k_b does not have to be very much bigger than k_a for the approach to be reasonably accurate.

Example 20E.2 Using the steady-state approximation

Devise the rate law for the decomposition of N_2O_5, $2\,N_2O_5(g) \rightarrow 4\,NO_2(g) + O_2(g)$ on the basis of the following mechanism:

$$N_2O_5 \rightarrow NO_2 + NO_3 \qquad k_a$$
$$NO_2 + NO_3 \rightarrow N_2O_5 \qquad k_a'$$
$$NO_2 + NO_3 \rightarrow NO_2 + O_2 + NO \qquad k_b$$
$$NO + N_2O_5 \rightarrow NO_2 + NO_2 + NO_2 \qquad k_c$$

A note on good practice Note that when writing the equation for an elementary reaction all the species are displayed individually; so we write $A \rightarrow B + B$, for instance, not $A \rightarrow 2\,B$.

Method First identify the intermediates and write expressions for their net rates of formation. Then, all net rates of change of the concentrations of intermediates are set equal to zero and the resulting equations are solved algebraically.

Answer The intermediates are NO and NO_3; the net rates of change of their concentrations are

$$\frac{d[NO]}{dt} = k_b[NO_2][NO_3] - k_c[NO][N_2O_5] \approx 0$$

$$\frac{d[NO_3]}{dt} = k_a[N_2O_5] - k_a'[NO_2][NO_3] - k_b[NO_2][NO_3] \approx 0$$

The solutions of these two simultaneous equations (in blue) are

$$[NO_3] = \frac{k_a[N_2O_5]}{(k_a' + k_b)[NO_2]} \qquad [NO] = \frac{k_b[NO_2][NO_3]}{k_c[N_2O_5]} = \frac{k_a k_b}{(k_a' + k_b)k_c}$$

The net rate of change of concentration of N_2O_5 is then

$$\frac{d[N_2O_5]}{dt} = -k_a[N_2O_5] + k_a'[NO_2][NO_3] - k_c[NO][N_2O_5]$$
$$= -k_a[N_2O_5] + \frac{k_a k_a'[N_2O_5]}{k_a' + k_b} - \frac{k_a k_b}{k_a' + k_b}[N_2O_5]$$
$$= -\frac{2k_a k_b[N_2O_5]}{k_a' + k_b}$$

That is, N_2O_5 decays with a first-order rate law with a rate constant that depends on k_a, k_a' and k_b but not on k_c.

Self-test 20E.3 Derive the rate law for the decomposition of ozone in the reaction $2\,O_3(g) \rightarrow 3\,O_2(g)$ on the basis of the (incomplete) mechanism

$$O_3 \rightarrow O_2 + O \qquad k_a$$
$$O_2 + O \rightarrow O_3 \qquad k_a'$$
$$O + O_3 \rightarrow O_2 + O_2 \qquad k_b$$

Answer: $d[O_3]/dt = -2k_a k_b[O_3]^2/(k_a'[O_2] + k_b[O_3])$

20E.4 The rate-determining step

Equation 20E.8 shows that when $k_b \gg k_a$ the formation of the final product P depends on only the *smaller* of the two rate constants. That is, the rate of formation of P depends on the rate at

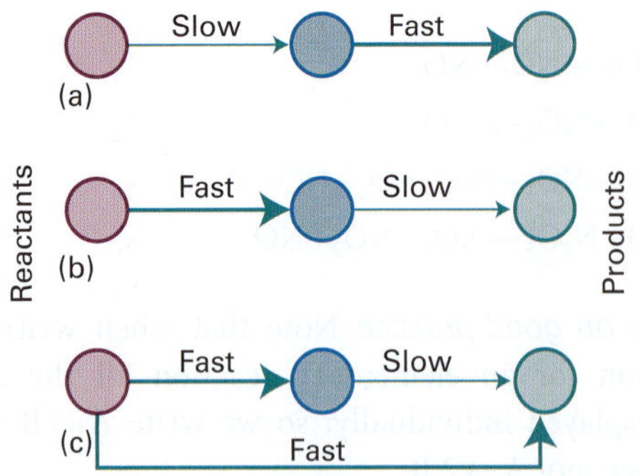

Figure 20E.4 In these diagrams of reaction schemes, heavy arrows represent fast steps and light arrows represent slow steps. (a) The first step is rate-determining; (b) the second step is rate-determining; (c) although one step is slow, it is not rate-determining because there is a fast route that circumvents it.

which I is formed, not on the rate at which I changes into P. For this reason, the step $A \rightarrow I$ is called the 'rate-determining step' of the reaction. Its existence has been likened to building a six-lane highway up to a single-lane bridge: the traffic flow is governed by the rate of crossing the bridge. Similar remarks apply to more complicated reaction mechanisms. In general, the **rate-determining step** (RDS) is the slowest step in a mechanism and controls the overall rate of the reaction. The rate-determining step is not just the slowest step: it must be slow *and* be a crucial gateway for the formation of products. If a faster reaction can also lead to products, then the slowest step is irrelevant because the slow reaction can then be sidestepped (Fig. 20E.4).

The rate law of a reaction that has a rate-determining step can often—but certainly not always—be written down almost by inspection. If the first step in a mechanism is rate-determining, then the rate of the overall reaction is equal to the rate of the first step because all subsequent steps are so fast that once the first intermediate is formed it results immediately in the formation of products. Figure 20E.5 shows the reaction profile for a mechanism of this kind in which the slowest step is the one with the highest activation energy. Once over the initial barrier, the intermediates cascade into products. However, a rate-determining step may also stem from the low concentration of a crucial reactant and need not correspond to the step with highest activation barrier.

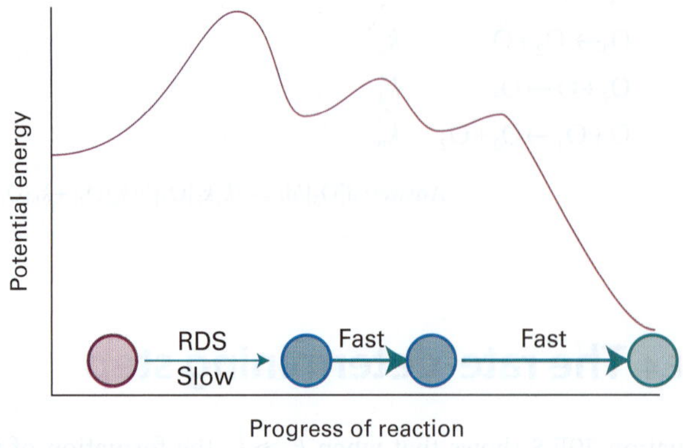

Figure 20E.5 The reaction profile for a mechanism in which the first step (RDS) is rate-determining.

Brief illustration 20E.2 The rate law of a mechanism with a rate-determining step

The oxidation of NO to NO_2, $2\,NO(g) + O_2(g) \rightarrow 2\,NO_2(g)$, proceeds by the following mechanism:

$$NO + NO \rightarrow N_2O_2 \qquad k_a$$

$$N_2O_2 \rightarrow NO + NO \qquad k_a'$$

$$N_2O_2 + O_2 \rightarrow NO_2 + NO_2 \qquad k_b$$

with rate law (see the *Self-test*)

$$\frac{d[NO_2]}{dt} = \frac{2k_a k_b [NO]^2[O_2]}{k_a' + k_b[O_2]}$$

When the concentration of O_2 in the reaction mixture is so large that the third step is very fast in the sense that $[O_2]k_b \gg k_a'$, then the rate law simplifies to

$$\frac{d[NO_2]}{dt} = 2k_a[NO]^2$$

and the formation of N_2O_2 in the first step is rate-determining. We could have written the rate law by inspection of the mechanism, because the rate law for the overall reaction is simply the rate law of that rate-determining step.

Self-test 20E.4 Verify that application of the steady-state approximation to the intermediate N_2O_2 results in the rate law.

20E.5 Pre-equilibria

From a simple sequence of consecutive reactions we now turn to a slightly more complicated mechanism in which an intermediate I reaches an equilibrium with the reactants A and B:

$$A + B \rightleftharpoons I \rightarrow P \qquad \text{Pre-equilibrium} \qquad (20E.9)$$

The rate constants are k_a and k_a' for the forward and reverse reactions of the equilibrium and k_b for the final step. This scheme involves a **pre-equilibrium**, in which an intermediate is in equilibrium with the reactants. A pre-equilibrium can arise when the rate of decay of the intermediate back into reactants is much faster than the rate at which it forms products; thus, the condition is possible when $k_a' \gg k_b$ but not when $k_b \gg k_a'$. Because we assume that A, B, and I are in equilibrium, we can write

$$K = \frac{[I]}{[A][B]} \quad \text{with} \quad K = \frac{k_a}{k_a'} \qquad (20E.10)$$

In writing these equations, we are presuming that the rate of reaction of I to form P is too slow to affect the maintenance of the pre-equilibrium (see the following *Example*). We are also ignoring the fact, as is commonly done, that the standard concentration $c^{\ominus}$ should appear in the expression for K to ensure that it is dimensionless. The rate of formation of P may now be written:

$$\frac{d[P]}{dt}=k_b[I]=k_bK[A][B] \qquad (20E.11)$$

This rate law has the form of a second-order rate law with a composite rate constant:

$$\frac{d[P]}{dt}=k_r[A][B] \quad \text{with} \quad k_r=k_bK=\frac{k_ak_b}{k'_a} \qquad (20E.12)$$

Example 20E.3 Analysing a pre-equilibrium

Repeat the pre-equilibrium calculation but without ignoring the fact that I is slowly leaking away as it forms P.

Method Begin by writing the net rates of change of the concentrations of the substances and then invoke the steady-state approximation for the intermediate I. Use the resulting expression to obtain the rate of change of the concentration of P.

Answer The net rates of change of P and I are

$$\frac{d[P]}{dt}=k_b[I]$$

$$\frac{d[I]}{dt}=k_a[A][B]-k'_a[I]-k_b[I]\approx 0$$

The second equation solves to

$$[I]\approx\frac{k_a[A][B]}{k'_a+k_b}$$

When we substitute this result into the expression for the rate of formation of P, we obtain

$$\frac{d[P]}{dt}\approx k_r[A][B] \quad \text{with} \quad k_r=\frac{k_ak_b}{k'_a+k_b}$$

This expression reduces to that in eqn 20E.12 when the rate constant for the decay of I into products is much smaller than that for its decay into reactants, $k_b \ll k'_a$.

Self-test 20E.5 Show that the pre-equilibrium mechanism in which $2\,A \rightleftharpoons I$ (K) followed by $I+B \rightarrow P$ (k_b) results in an overall third-order reaction.

Answer: $d[P]/dt=k_bK[A]^2[B]$

One feature to note is that although each of the rate constants in eqn 20E.12 increases with temperature, that might not be true of k_r itself. Thus, if the rate constant k'_a increases more rapidly than the product k_ak_b increases, then $k_r=k_ak_b/k'_a$ will decrease with increasing temperature and the reaction will go more slowly as the temperature is raised. Mathematically, we would say that the composite reaction had a 'negative activation energy'. For example, suppose that each rate constant in eqn 20E.12 exhibits an Arrhenius temperature dependence (Topic 20D). It follows from the Arrhenius equation (eqn 20D.4, $k_r=Ae^{-E_a/RT}$) that

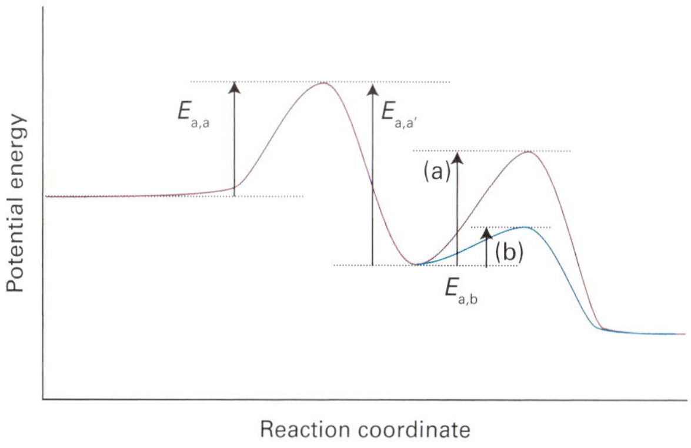

Figure 20E.6 For a reaction with a pre-equilibrium, there are three activation energies to take into account: two referring to the reversible steps of the pre-equilibrium and one for the final step. The relative magnitudes of the activation energies determine whether the overall activation energy is (a) positive or (b) negative.

$$k_r=\frac{(A_ae^{-E_{a,a}/RT})(A_be^{-E_{a,b}/RT})}{A_{a'}e^{-E_{a,a'}/RT}} \overset{e^{x+y}=e^xe^y,\; e^{x+y}=e^x/e^y}{=} \frac{A_aA_b}{A_{a'}}e^{-(E_{a,a}+E_{a,b}-E_{a,a'})/RT}$$

The effective activation energy of the reaction is therefore

$$E_a=E_{a,a}+E_{a,b}-E_{a,a'} \qquad (20E.13)$$

This activation energy is positive if $E_{a,a}+E_{a,b}>E_{a,a'}$ (Fig. 20E.6a) but negative if $E_{a,a'}>E_{a,a}+E_{a,b}$ (Fig. 20E.6b). An important consequence of this discussion is that we have to be very cautious about making predictions about the effect of temperature on reactions that are the outcome of several steps.

20E.6 Kinetic and thermodynamic control of reactions

In some cases reactants can give rise to a variety of products, as in nitrations of mono-substituted benzene, when various proportions of the *ortho*-, *meta*-, and *para*-substituted products are obtained, depending on the directing power of the original substituent. Suppose two products, P_1 and P_2, are produced by the following competing reactions:

$$A+B \rightarrow P_1 \qquad v(P_1)=k_{r,1}[A][B]$$

$$A+B \rightarrow P_2 \qquad v(P_2)=k_{r,2}[A][B]$$

The relative proportion in which the two products have been produced at a given stage of the reaction (before it has reached equilibrium) is given by the ratio of the two rates, and therefore of the two rate constants:

$$\frac{[P_2]}{[P_1]}=\frac{k_{r,2}}{k_{r,1}} \qquad \text{Kinetic control} \qquad (20E.14)$$

This ratio represents the **kinetic control** over the proportions of products, and is a common feature of the reactions encountered in organic chemistry where reactants are chosen that facilitate pathways favouring the formation of a desired product. If a reaction is allowed to reach equilibrium, then the proportion of products is determined by thermodynamic rather than kinetic considerations, and the ratio of concentrations is controlled by considerations of the standard Gibbs energies of all the reactants and products.

Brief illustration 20E.3 The outcome of kinetic control

Consider two products formed from reactant R in reactions for which: (a) product P_1 is thermodynamically more stable than product P_2; and (b) the activation energy E_a for the reaction leading to P_2 is greater than that leading to P_1. It follows from eqn 20E.14 and the Arrhenius equation ($k_r = Ae^{-E_a/RT}$, eqn 20D.4) that the ratio of products is

$$\frac{[P_2]}{[P_1]}=\frac{k_2}{k_1}=\frac{A_2 e^{-E_{a,2}/RT}}{A_1 e^{-E_{a,1}/RT}}=\frac{A_2}{A_1}e^{-(E_{a,2}-E_{a,1})/RT}=\frac{A_2}{A_1}e^{-\Delta E_a/RT}$$

Because $\Delta E_a = E_{a,2} - E_{a,1} > 0$, as T increases,

- the term $\Delta E_a/RT$ decreases, and
- the term $e^{-\Delta E_a/RT}$ increases.

Consequently, the ratio $[P_2]/[P_1]$ increases with increasing temperature before equilibrium is reached.

Self-test 20E.6 Consider the reactions from *Brief illustration* 20E.3. Derive an expression for the ratio $[P_2]/[P_1]$ when the reaction is under thermodynamic control. State your assumptions.

Answer: $[P_2]/[P_1]=e^{-(\Delta_r G_2^{\ominus}-\Delta_r G_1^{\ominus})/RT}$, assuming that activities can be replaced by concentrations

Checklist of concepts

- ☐ 1. The **mechanism** of reaction is the sequence of elementary steps that leads from reactants to products.
- ☐ 2. The **molecularity** of an elementary reaction is the number of molecules coming together to react.
- ☐ 3. An elementary unimolecular reaction has first-order kinetics; an elementary bimolecular reaction has second-order kinetics.
- ☐ 4. The **rate-determining step** is the slowest step in a reaction mechanism that controls the rate of the overall reaction.
- ☐ 5. In the **steady-state approximation**, it is assumed that the concentrations of all reaction intermediates remain constant and small throughout the reaction.
- ☐ 6. **Pre-equilibrium** is a state in which an intermediate is in equilibrium with the reactants and which arises when the rates of formation of the intermediate and its decay back into reactants are much faster than its rate of formation of products.
- ☐ 7. Provided a reaction has not reached equilibrium, the products of competing reactions are controlled by kinetics.

Checklist of equations

Property	Equation	Comment	Equation number
Unimolecular reaction	$d[A]/dt=-k_r[A]$	$A \rightarrow P$	20E.1
Bimolecular reaction	$d[A]/dt=-k_r[A][B]$	$A+B \rightarrow P$	20E.2
Consecutive reactions	$[A]=[A]_0 e^{-k_a t}$ $[I]=(k_a/(k_b-k_a))(e^{-k_a t}-e^{-k_b t})[A]_0$ $[P]=\{1+(k_a e^{-k_b t}-k_b e^{-k_a t})/(k_b-k_a)\}[A]_0$	$A \xrightarrow{k_a} I \xrightarrow{k_b} P$	20E.4
Steady-state approximation	$d[I]/dt \approx 0$	I is an intermediate	20E.5

20F Examples of reaction mechanisms

Contents

➤ Why do you need to know this material?

Some important reactions have complex mechanisms and need special treatment, so you need to see how to make and implement assumptions about the relative rates of the steps in a mechanism.

➤ What is the key idea?

The steady-state approximation can often be used to derive rate laws for proposed mechanisms.

➤ What do you need to know already?

You need to be familiar with the concept of rate laws (Topic 20A) and the steady-state approximation (Topic 20E).

Many reactions take place by mechanisms that involve several elementary steps. We focus here on the kinetic analysis of a special class of reactions in the gas phase and polymerization kinetics. Photochemical processes are treated in Topic 20G and the role of catalysis in Topics 20H and 22C.

20F.1 Unimolecular reactions

A number of gas-phase reactions follow first-order kinetics, as in the isomerization of cyclopropane:

$$cyclo\text{-}C_3H_6(g) \rightarrow CH_3CH{=}CH_2(g) \quad v = k_r[cyclo\text{-}C_3H_6]$$

The problem with the interpretation of first-order rate laws is that presumably a molecule acquires enough energy to react as a result of its collisions with other molecules. However, collisions are simple bimolecular events, so how can they result in a first-order rate law? First-order gas-phase reactions are widely called 'unimolecular reactions' because they also involve an elementary unimolecular step in which the reactant molecule changes into the product. This term must be used with caution, however, because the overall mechanism has bimolecular as well as unimolecular steps.

The first successful explanation of unimolecular reactions was provided by Frederick Lindemann in 1921 and then elaborated by Cyril Hinshelwood. In the **Lindemann–Hinshelwood mechanism** it is supposed that a reactant molecule A becomes energetically excited by collision with another A molecule in a bimolecular step (Fig. 20F.1):

$$A + A \rightarrow A^* + A \qquad \frac{d[A^*]}{dt} = k_a[A]^2 \qquad (20F.1a)$$

The energized molecule (A*) might lose its excess energy by collision with another molecule:

$$A + A^* \rightarrow A + A \qquad \frac{d[A^*]}{dt} = -k_a'[A][A^*] \qquad (20F.1b)$$

Alternatively, the excited molecule might shake itself apart and form products P. That is, it might undergo the unimolecular decay

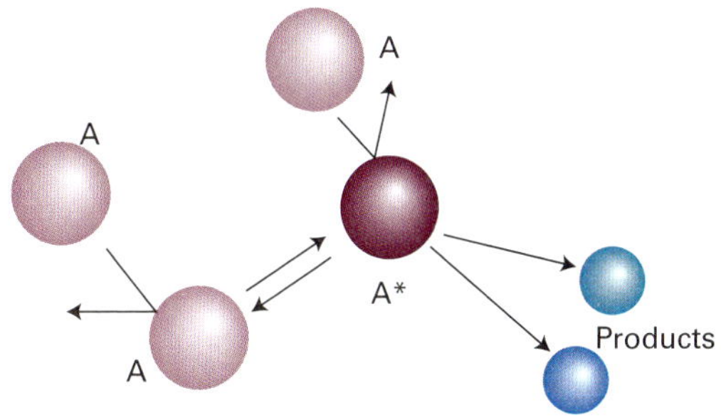

Figure 20F.1 A representation of the Lindemann–Hinshelwood mechanism of unimolecular reactions. The species A is excited by collision with A, and the excited A molecule (A*) may either be deactivated by a collision with A or go on to decay by a unimolecular process to form products.

$$A^* \rightarrow P \qquad \frac{d[A^*]}{dt} = -k_b[A^*] \tag{20F.1c}$$

If the unimolecular step is slow enough to be the rate-determining step, the overall reaction will have first-order kinetics, as observed. This conclusion can be demonstrated explicitly by applying the steady-state approximation to the net rate of formation of A*:

$$\frac{d[A^*]}{dt} = k_a[A]^2 - k_a'[A][A^*] - k_b[A^*] \approx 0 \tag{20F.2}$$

This equation solves to

$$[A^*] = \frac{k_a[A]^2}{k_b + k_a'[A]} \tag{20F.3}$$

so the rate law for the formation of P is

$$\frac{d[P]}{dt} = k_b[A^*] = \frac{k_a k_b [A]^2}{k_b + k_a'[A]} \tag{20F.4}$$

At this stage the rate law is not first-order. However, if the rate of deactivation by (A*,A) collisions is much greater than the rate of unimolecular decay, in the sense that $k_a'[A][A^*] \gg k_b[A^*]$, or (after cancelling the $[A^*]$), $k_a'[A] \gg k_b$, then we can neglect k_b in the denominator and obtain

$$\frac{d[P]}{dt} = k_r[A] \quad \text{with} \quad k_r = \frac{k_a k_b}{k_a'} \qquad \text{Lindemann–Hinshelwood rate law} \tag{20F.5}$$

Equation 20F.5 is a first-order rate law, as we set out to show.

The Lindemann–Hinshelwood mechanism can be tested because it predicts that, as the concentration (and therefore the partial pressure) of A is reduced, the reaction should switch to overall second-order kinetics. Thus, when $k_a'[A] \ll k_b$, the rate law in eqn 20F.4 becomes

$$\frac{d[P]}{dt} = k_a[A]^2 \tag{20F.6}$$

The physical reason for the change of order is that at low pressures the rate-determining step is the bimolecular formation of A*. If we write the full rate law in eqn 20F.4 as

$$\frac{d[P]}{dt} = k_r[A] \quad \text{with} \quad k_r = \frac{k_a k_b [A]}{k_b + k_a'[A]} \tag{20F.7}$$

then the expression for the effective rate constant, k_r, can be rearranged (by inverting each side) to

$$\frac{1}{k_r} = \frac{k_a'}{k_a k_b} + \frac{1}{k_a[A]} \qquad \textit{Lindemann–Hinshelwood mechanism} \qquad \text{Effective rate constant} \tag{20F.8}$$

Hence, a test of the theory is to plot $1/k_r$ against $1/[A]$, and to expect a straight line. This behaviour is observed often at low concentrations but deviations are common at high concentrations. In Topic 21A we develop the description of the mechanism to take into account experimental results over a range of concentrations and pressures.

Example 20F.1 Analysing the Lindemann–Hinshelwood mechanism

At 300 K the effective rate constant for a gaseous reaction A → P, which has a Lindemann–Hinshelwood mechanism, is $k_{r,1} = 2.50 \times 10^{-4}\,\text{s}^{-1}$ at $[A]_1 = 5.21 \times 10^{-4}\,\text{mol dm}^{-3}$ and $k_{r,2} = 2.10 \times 10^{-5}\,\text{s}^{-1}$ at $[A]_2 = 4.81 \times 10^{-6}\,\text{mol dm}^{-3}$. Calculate the rate constant for the activation step in the mechanism.

Method Use eqn 20F.8 to write an expression for the difference $1/k_{r,2} - 1/k_{r,1}$ and then use the data to solve for k_a, the rate constant for the activation step.

Answer It follows from eqn 20F.8 that

$$\frac{1}{k_{r,2}} - \frac{1}{k_{r,1}} = \frac{1}{k_a}\left(\frac{1}{[A]_2} - \frac{1}{[A]_1}\right)$$

and so

$$k_a = \frac{1/[A]_2 - 1/[A]_1}{1/k_{r,2} - 1/k_{r,1}}$$

$$= \frac{1/(4.81\times10^{-6}\,\text{mol dm}^{-3}) - 1/(5.21\times10^{-4}\,\text{mol dm}^{-3})}{1/(2.10\times10^{-5}\,\text{s}^{-1}) - 1/(2.50\times10^{-4}\,\text{s}^{-1})}$$

$$= 4.72\,\text{dm}^3\,\text{mol}^{-1}\,\text{s}^{-1}$$

Self-test 20F.1 The effective rate constants for a gaseous reaction A → P, which has a Lindemann–Hinshelwood mechanism, are $1.70 \times 10^{-3}\,\text{s}^{-1}$ and $2.20 \times 10^{-4}\,\text{s}^{-1}$ at $[A] = 4.37 \times 10^{-4}\,\text{mol dm}^{-3}$ and $1.00 \times 10^{-5}\,\text{mol dm}^{-3}$, respectively. Calculate the rate constant for the activation step in the mechanism.

Answer: $24.6\,\text{dm}^3\,\text{mol}^{-1}\,\text{s}^{-1}$

20F.2 Polymerization kinetics

There are two major classes of polymerization processes and the average molar mass of the product varies with time in distinctive ways. In **stepwise polymerization** any two monomers present in the reaction mixture can link together at any time and growth of the polymer is not confined to chains that are already forming (Fig. 20F.2). As a result, monomers are consumed early in the reaction and, as we shall see, the average molar mass of the product grows linearly with time. In **chain polymerization** an activated monomer, M, attacks another monomer, links to it, then that unit attacks another monomer, and so on. The monomer is used up as it becomes linked to

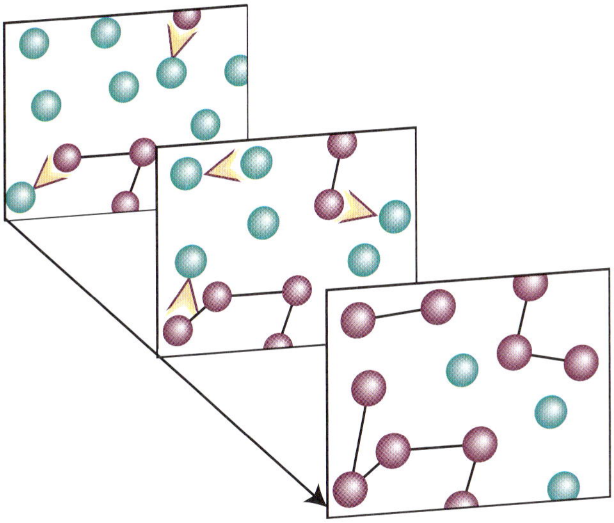

Figure 20F.2 In stepwise polymerization, growth can start at any pair of monomers (in green), and so new chains (in purple) begin to form throughout the reaction.

the growing chains (Fig. 20F.3). High polymers are formed rapidly and only the yield, not the average molar mass, of the polymer is increased by allowing long reaction times.

(a) Stepwise polymerization

Stepwise polymerization commonly proceeds by a **condensation reaction**, in which a small molecule (typically H_2O) is eliminated in each step. Stepwise polymerization is the mechanism of production of polyamides, as in the formation of nylon-66:

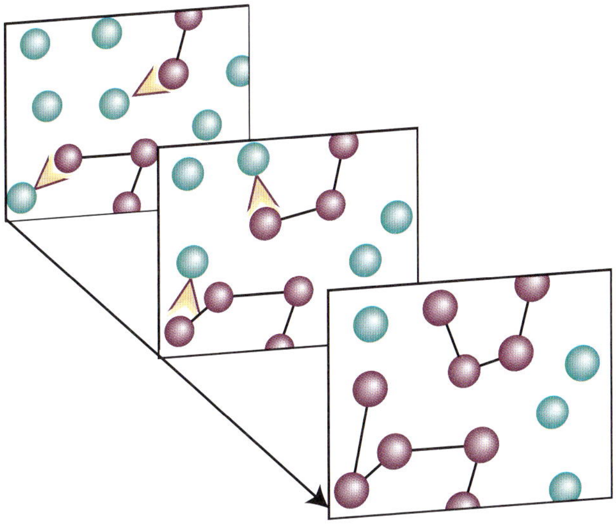

Figure 20F.3 The process of chain polymerization. Chains (in purple) grow as each chain acquires additional monomers (in green).

$$
\begin{aligned}
&H_2N(CH_2)_6NH_2 + HOOC(CH_2)_4COOH \rightarrow \\
&\quad H_2N(CH_2)_6NHCO(CH_2)_4COOH + H_2O \\
&\quad \xrightarrow{\text{continuing to}} H{-}[HN(CH_2)_6NHCO(CH_2)_4CO]_n{-}OH
\end{aligned}
$$

Polyesters and polyurethanes are formed similarly (the latter without elimination). A polyester, for example, can be regarded as the outcome of the stepwise condensation of a hydroxyacid HO—R—COOH. We shall consider the formation of a polyester from such a monomer, and measure its progress in terms of the concentration of the —COOH groups in the sample (which we denote A), for these groups gradually disappear as the condensation proceeds. Because the condensation reaction can occur between molecules containing any number of monomer units, chains of many different lengths can grow in the reaction mixture.

In the absence of a catalyst, we can expect the condensation to be overall second-order in the concentration of the —OH and —COOH (or A) groups, and write

$$\frac{d[A]}{dt} = -k_r[OH][A] \tag{20F.9a}$$

However, because there is one —OH group for each —COOH group, this equation is the same as

$$\frac{d[A]}{dt} = -k_r[A]^2 \tag{20F.9b}$$

If we assume that the rate constant for the condensation is independent of the chain length, then k_r remains constant throughout the reaction. The solution of this rate law is given by eqn 20B.4, and is

$$[A] = \frac{[A]_0}{1 + k_r t[A]_0} \tag{20F.10}$$

The fraction, p, of —COOH groups that have condensed at time t is, after application of eqn 20F.10:

$$p = \frac{[A]_0 - [A]}{[A]_0} = \frac{k_r t[A]_0}{1 + k_r t[A]_0} \quad \textit{Stepwise polymerization} \quad \text{Fraction of condensed groups} \tag{20F.11}$$

Next, we calculate the **degree of polymerization**, which is defined as the average number of monomer residues per polymer molecule. This quantity is the ratio of the initial concentration of A, $[A]_0$, to the concentration of end groups, [A], at the time of interest, because there is one A group per polymer molecule. For example, if there were initially 1000 A groups and there are now only 10, each polymer must be 100 units long on average. Because we can express [A] in terms of p (the first part of eqn 20F.11), the average number of monomers per polymer molecule, $\langle N \rangle$, is

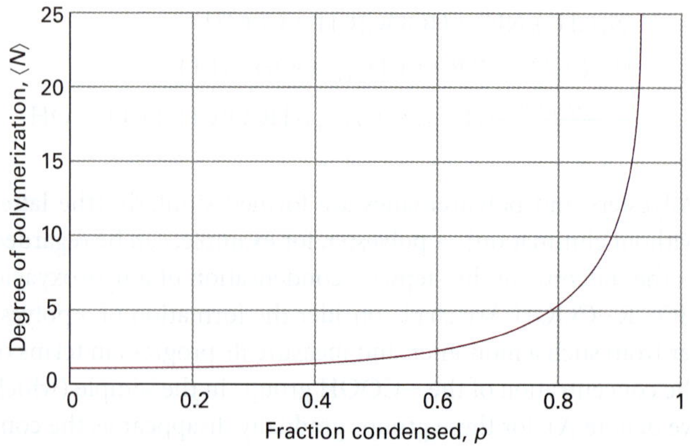

Figure 20F.4 The average chain length of a polymer as a function of the fraction of reacted monomers, p. Note that p must be very close to 1 for the chains to be long.

$$\langle N\rangle=\frac{[A]_0}{[A]}=\frac{1}{1-p} \quad \textit{Stepwise polymerization} \quad \text{Degree of polymerization} \quad (20F.12a)$$

This result is illustrated in Fig. 20F.4. When we express p in terms of the rate constant k_r (the second part of eqn 20F.11), we find

$$\langle N\rangle=1+k_r t[A]_0 \quad \textit{Stepwise polymerization} \quad \text{Degree of polymerization in terms of the rate constant} \quad (20F.12b)$$

The average length grows linearly with time. Therefore, the longer a stepwise polymerization proceeds, the higher the average molar mass of the product.

Brief illustration 20F.1 The degree of polymerization

Consider a polymer formed by a stepwise process with $k_r=1.00\,\mathrm{dm^3\,mol^{-1}\,s^{-1}}$ and an initial monomer concentration of $[A]_0=4.00\times10^{-3}\,\mathrm{mol\,dm^{-3}}$. From eqn 20F.12b, the degree of polymerization at $t=1.5\times10^4\,\mathrm{s}$ is

$$\langle N\rangle=1+(1.00\,\mathrm{dm^3\,mol^{-1}\,s^{-1}})\times(1.5\times10^4\,\mathrm{s})\times(4.00\times10^{-3}\,\mathrm{mol\,dm^{-3}})=61$$

From eqn 20F.12a, the fraction condensed, p, is

$$p=\frac{\langle N\rangle-1}{\langle N\rangle}=\frac{61-1}{61}=0.98$$

Self-test 20F.2 Calculate the fraction condensed and the degree of polymerization at $t=1.0\,\mathrm{h}$ of a polymer formed by a stepwise process with $k_r=1.80\times10^{-2}\,\mathrm{dm^3\,mol^{-1}\,s^{-1}}$ and an initial monomer concentration of $3.00\times10^{-2}\,\mathrm{mol\,dm^{-3}}$.

Answer: $\langle N\rangle=2.9$; $p=0.66$

(b) Chain polymerization

Many gas-phase reactions and liquid-phase polymerization reactions are **chain reactions**. In a chain reaction, a reaction intermediate produced in one step generates an intermediate in a subsequent step, then that intermediate generates another intermediate, and so on. The intermediates in a chain reaction are called **chain carriers**. In a **radical chain reaction** the chain carriers are radicals (species with unpaired electrons).

Chain polymerization occurs by addition of monomers to a growing polymer, often by a radical chain process. It results in the rapid growth of an individual polymer chain for each activated monomer. Examples include the addition polymerizations of ethene, methyl methacrylate, and styrene, as in

$$-\mathrm{CH_2CH_2X}\cdot+\mathrm{CH_2{=}CHX}\rightarrow-\mathrm{CH_2CHXCH_2CHX}\cdot$$

and subsequent reactions. The central feature of the kinetic analysis (which is summarized in the following *Justification*) is that the rate of polymerization is proportional to the square root of the initiator, In, concentration:

$$v=k_r[\mathrm{In}]^{1/2}[\mathrm{M}] \quad \textit{Chain polymerization} \quad \text{Rate of polymerization} \quad (20F.13)$$

Justification 20F.1 The rate of chain polymerization

There are three basic types of reaction step in a chain polymerization process:

(a) Initiation:

$$\mathrm{In}\rightarrow\mathrm{R}\cdot+\mathrm{R}\cdot \qquad v_i=k_i[\mathrm{In}]$$
$$\mathrm{M}+\mathrm{R}\cdot\rightarrow\cdot\mathrm{M}_1 \qquad \text{(fast)}$$

where In is the initiator, R· the radical that In forms, and $\cdot M_1$ is a monomer radical. We have shown a reaction in which a radical is produced, but in some polymerizations the initiation step leads to the formation of an ionic chain carrier. The rate-determining step is the formation of the radicals R· by homolysis of the initiator, so the rate of initiation is equal to the v_i given above.

(b) Propagation:

$$\mathrm{M}+\cdot\mathrm{M}_1\rightarrow\cdot\mathrm{M}_2$$
$$\mathrm{M}+\cdot\mathrm{M}_2\rightarrow\cdot\mathrm{M}_3$$
$$\vdots$$
$$\mathrm{M}+\cdot\mathrm{M}_{n-1}\rightarrow\cdot\mathrm{M}_n \qquad v_p=k_p[\mathrm{M}][\cdot\mathrm{M}]$$

If we assume that the rate of propagation is independent of chain size for sufficiently large chains, then we can use only the equation given above to describe the propagation process. Consequently, for sufficiently large chains, the rate of propagation is equal to the overall rate of polymerization.

Because this chain of reactions propagates quickly, the rate at which the total concentration of radicals grows is equal to the rate of the rate-determining initiation step. It follows that

$$\left(\frac{d[\cdot M]}{dt}\right)_{\text{production}} = 2f k_i[\text{In}]$$

where f is the fraction of radicals R· that successfully initiate a chain.

(c) Termination:

mutual termination: $\cdot M_n + \cdot M_m \rightarrow M_{n+m}$

disproportionation: $\cdot M_n + \cdot M_m \rightarrow M_n + M_m$

chain transfer: $M + \cdot M_n \rightarrow \cdot M + M_n$

In **mutual termination** two growing radical chains combine. In termination by **disproportionation** a hydrogen atom transfers from one chain to another, corresponding to the oxidation of the donor and the reduction of the acceptor. In **chain transfer**, a new chain initiates at the expense of the one currently growing.

Here we suppose that only mutual termination occurs. If we assume that the rate of termination is independent of the length of the chain, the rate law for termination is

$$v_t = k_t[\cdot M]^2$$

and the rate of change of radical concentration by this process is

$$\left(\frac{d[\cdot M]}{dt}\right)_{\text{depletion}} = -2k_t[\cdot M]^2$$

The steady-state approximation gives:

$$\frac{d[\cdot M]}{dt} = 2f k_i[\text{In}] - 2k_t[\cdot M]^2 \approx 0$$

The steady-state concentration of radical chains is therefore

$$[\cdot M] = \left(\frac{f k_i}{k_t}\right)^{1/2}[\text{In}]^{1/2}$$

Because the rate of propagation of the chains is the negative of the rate at which the monomer is consumed, we can write $v_p = -d[M]/dt$ and

$$v_p = k_p[\cdot M][M] = k_p\left(\frac{f k_i}{k_t}\right)^{1/2}[\text{In}]^{1/2}[M]$$

This rate is also the rate of polymerization, which has the form of eqn 20F.13.

The **kinetic chain length**, λ, is the ratio of the number of monomer units consumed per activated centre produced in the initiation step:

$$\lambda = \frac{\text{number of monomer units consumed}}{\text{number of activated centres produced}}$$

Definition Kinetic chain length (20F.14a)

The kinetic chain length can be expressed in terms of the rate expressions in *Justification* 20F.1. To do so, we recognize that monomers are consumed at the rate that chains propagate. Then,

$$\lambda = \frac{\text{rate of propagation of chains}}{\text{rate of production of radicals}}$$

Kinetic chain length in terms of reaction rates (20F.14b)

By making the steady-state approximation, we set the rate of production of radicals equal to the termination rate. Therefore, we can write the expression for the kinetic chain length as

$$\lambda = \frac{k_p[\cdot M][M]}{2k_t[\cdot M]^2} = \frac{k_p[M]}{2k_t[\cdot M]}$$

When we substitute the steady-state expression, $[\cdot M] = (fk_i/k_t)^{1/2}[\text{In}]^{1/2}$, for the radical concentration, we obtain

$$\lambda = k_r[M][\text{In}]^{-1/2} \quad \text{with } k_r = k_p(fk_ik_t)^{-1/2}$$

Chain polymerization Kinetic chain length (20F.15)

Consider a polymer produced by a chain mechanism with mutual termination. In this case, the average number of monomers in a polymer molecule, $\langle N\rangle$, produced by the reaction is the sum of the numbers in the two combining polymer chains. The average number of units in each chain is λ. Therefore,

$$\langle N\rangle = 2\lambda = 2k_r[M][\text{In}]^{-1/2}$$

Chain polymerization Degree of polymerization (20F.16)

with k_r given in eqn 20F.15. We see that, the slower the initiation of the chain (the smaller the initiator concentration and the smaller the initiation rate constant), the greater is the kinetic chain length, and therefore the higher is the average molar mass of the polymer. Some of the consequences of molar mass for polymers are explored in Topic 17D: here we have seen how we can exercise kinetic control over them.

Checklist of concepts

☐ 1. The **Lindemann–Hinshelwood mechanism** of 'unimolecular' reactions account for the first-order kinetics of some gas-phase reactions.

☐ 2. In **stepwise polymerization** any two monomers in the reaction mixture can link together at any time.

☐ 3. The longer a stepwise polymerization proceeds, the higher the average molar mass of the product.

☐ 4. In **chain polymerization** an activated monomer attacks another monomer and links to it.

☐ 5. The slower the initiation of the chain, the higher the average molar mass of the polymer.

☐ 6. The **kinetic chain length** is the ratio of the number of monomer units consumed per activated centre produced in the initiation step.

Checklist of equations

Property	Equation	Comment	Equation number
Lindemann–Hinshelwood rate law	$d[P]/dt=k_r[A]$ with $k_r=k_ak_b/k_a'$	$k_a'[A]\gg k_b$	20F.5
Effective rate constant	$1/k_r=k_a'/k_ak_b+1/k_a[A]$	Lindemann–Hinshelwood mechanism	20F.8
Fraction of condensed groups	$p=k_rt[A]_0/(1+k_rt[A]_0)$	Stepwise polymerization	20F.11
Degree of polymerization	$\langle N\rangle=1/(1-p)=1+k_rt[A]_0$	Stepwise polymerization	20F.12
Rate of polymerization	$v=k_r[In]^{1/2}[M]$	Chain polymerization	20F.13
Kinetic chain length	$\lambda=k_r[M][In]^{-1/2}$, $k_r=k_p(fk_ik_t)^{-1/2}$	Chain polymerization	20F.15
Degree of polymerization	$\langle N\rangle=2k_r[M][In]^{-1/2}$	Chain polymerization	20F.16

20G Photochemistry

Contents

➤ Why do you need to know this material?

Many chemical and biological processes, including photosynthesis and vision, can be initiated by the absorption of electromagnetic radiation, so you need to know how to include its effect in rate laws. You also need to see how to obtain insight into these processes by the quantitative analysis of their mechanisms.

➤ What is the key idea?

The mechanisms of many photochemical reactions lead to relatively simple rate laws that yield rate constants and quantitative measures of the efficiency with which radiant energy induces reactions.

➤ What do you need to know already?

You need to be familiar with the concepts of singlet and triplet states (Topics 9B and 13B), modes of radiative decay (fluorescence and phosphorescence, Topic 13B), concepts of electronic spectroscopy (Topic 13A), and the formulation of a rate law from a proposed mechanism (Topic 20E).

Photochemical processes are initiated by the absorption of electromagnetic radiation. Among the most important of these processes are those that capture the radiant energy of the Sun. Some of these reactions lead to the heating of the atmosphere during the daytime by absorption of ultraviolet radiation. Others include the absorption of visible radiation during photosynthesis. Without photochemical processes, the Earth would be simply a warm, sterile, rock.

20G.1 Photochemical processes

Table 20G.1 summarizes common photochemical reactions. Photochemical processes are initiated by the absorption of radiation by at least one component of a reaction mixture. In a **primary process**, products are formed directly from the excited state of a reactant. Examples include fluorescence (Topic 13B) and the *cis–trans* photoisomerization of retinal. Products of a **secondary process** originate from intermediates that are

Table 20G.1 Examples of photochemical processes

Process	General form	Example
Ionization	$A^* \rightarrow A^+ + e^-$	$NO^* \xrightarrow{134\,nm} NO^+ + e^-$
Electron transfer	$A^* + B \rightarrow A^+ + B^-$ or $A^- + B^+$	$\left[Ru(bpy)_3^{2+}\right]^* + Fe^{3+} \xrightarrow{452\,nm} Ru(bpy)_3^{3+} + Fe^{2+}$
Dissociation	$A^* \rightarrow B + C$	$O_3^* \xrightarrow{1180\,nm} O^2 + O$
	$A^* + B{-}C \rightarrow A + B + C$	$Hg^* + CH_4 \xrightarrow{254\,nm} Hg + CH_3 + H$
Addition	$2\,A^* \rightarrow B$	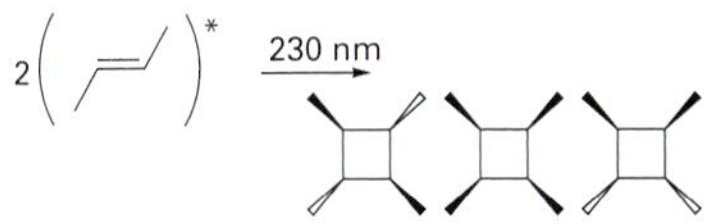
	$A^* + B \rightarrow AB$	
Abstraction	$A^* + B{-}C \rightarrow A{-}B + C$	$Hg^* + H_2 \xrightarrow{254\,nm} HgH + H$
Isomerization or rearrangement	$A^* \rightarrow A'$	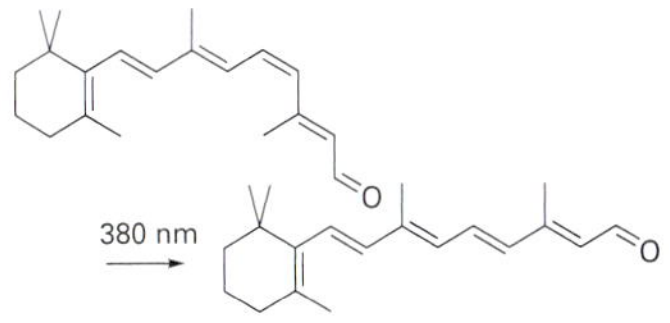

* Excited state.

Table 20G.2 Common photophysical processes

Primary absorption	$S+h\nu \rightarrow S^*$
Excited-state absorption	$S^*+h\nu \rightarrow S^{**}$
	$T^*+h\nu \rightarrow T^{**}$
Fluorescence	$S^* \rightarrow S+h\nu$
Stimulated emission	$S^*+h\nu \rightarrow S+2\,h\nu$
Intersystem crossing (ISC)	$S^* \rightarrow T^*$
Phosphorescence	$T^* \rightarrow S+h\nu$
Internal conversion (IC)	$S^* \rightarrow S$
Collision-induced emission	$S^*+M \rightarrow S+M+h\nu$
Collisional deactivation	$S^*+M \rightarrow S+M$
	$T^*+M \rightarrow S+M$
Electronic energy transfer:	
Singlet–singlet	$S^*+S \rightarrow S+S^*$
Triplet–triplet	$T^*+T \rightarrow T+T^*$
Excimer formation	$S^*+S \rightarrow (SS)^*$
Energy pooling	
Singlet–singlet	$S^*+S^* \rightarrow S^{**}+S$
Triplet–triplet	$T^*+T^* \rightarrow S^{**}+S$

* Denotes an excited state, S a singlet state, T a triplet state, and M is a third-body.

formed directly from the excited state of a reactant, such as oxidative processes initiated by the oxygen atom formed by ozone photodissociation.

Competing with the formation of photochemical products are numerous primary photophysical processes that can deactivate the excited state (Table 20G.2). Therefore, it is important to consider the timescales of the formation and decay of excited states before describing the mechanisms of photochemical reactions.

Electronic transitions caused by absorption of ultraviolet and visible radiation occur within 10^{-16}–10^{-15} s. We expect, then, the upper limit for the rate constant of a first-order photochemical reaction to be about $10^{16}\,s^{-1}$. Fluorescence is slower than absorption, with typical lifetimes of 10^{-12}–10^{-6} s. Therefore, the excited singlet state can initiate very fast photochemical reactions in the femtosecond (10^{-15} s) to picosecond (10^{-12} s) range. Examples of such ultrafast reactions are the initial events of vision and of photosynthesis. Typical intersystem crossing (ISC, Topic 13B) and phosphorescence times for large organic molecules are 10^{-12}–10^{-4} s and 10^{-6}–10^{-1} s, respectively. As a consequence of their long lifetimes, excited triplet states are photochemically important. Indeed, because phosphorescence decay is several orders of magnitude slower than most typical reactions, species in excited triplet states can undergo a very large number of collisions with other reactants before they lose their energy radiatively.

Brief illustration 20G.1 The nature of the excited state

To judge whether the excited singlet or triplet state of the reactant is a suitable product precursor, we compare the emission lifetimes with the time constant for chemical reaction of the reactant, τ (Topic 20B). Consider a unimolecular photochemical reaction with rate constant $k_r = 1.7\times10^4\,s^{-1}$ and therefore time constant $\tau = 1/(1.7\times10^4\,s^{-1}) = 59\,\mu s$ that involves a reactant with an observed fluorescence lifetime of 1.0 ns and an observed phosphorescence lifetime of 1.0 ms. The excited singlet state is too short-lived to be a major source of product in this reaction. On the other hand, the relatively long-lived excited triplet state is a good candidate for a precursor.

Self-test 20G.1 Consider a molecule with a fluorescence lifetime of 10.0 ns that undergoes unimolecular photoisomerization. What approximate value of the half-life would be consistent with the excited singlet state being the product precursor?

Answer: The value of $t_{1/2}$ should be less than about 7 ns

20G.2 The primary quantum yield

The rates of deactivation of the excited state by radiative, non-radiative, and chemical processes determine the yield of product in a photochemical reaction. The **primary quantum yield**, ϕ, is defined as the number of photophysical or photochemical events that lead to primary products divided by the number of photons absorbed by the molecule in the same interval:

$$\phi = \frac{\text{number of events}}{\text{number of photons absorbed}} \quad \textit{Definition} \quad \text{Primary quantum yield} \quad (20G.1a)$$

When both the numerator and denominator of this expression are divided by the time interval over which the events occur, we see that the primary quantum yield is also the rate of radiation-induced primary events divided by the rate of photon absorption, I_{abs}:

$$\phi = \frac{\text{rate of process}}{\text{rate of photon absorption}} = \frac{v}{I_{abs}} \quad \text{Primary quantum yield in terms of rates of processes} \quad (20G.1b)$$

Example 20G.1 Calculating a primary quantum yield

In an experiment to determine the quantum yield of a photochemical reaction, the absorbing substance was exposed to 490 nm light from a 100 W source for 2700 s, with 60 per cent of the incident light being absorbed. As a result of irradiation,

0.344 mol of the absorbing substance decomposed. Determine the primary quantum yield.

Method We need to calculate the terms used in eqn 20G.1a. To calculate the number of absorbed photons N_{abs}, which is the denominator of the expression on the right-hand side of eqn 20G.1a, we note that:

- The energy absorbed by the substance is $E_{abs}=fPt$, where P is the incident power, t is the time of exposure, and the factor f (in this case $f=0.60$) is the proportion of incident light that is absorbed.
- E_{abs} is also related to the number N_{abs} of absorbed photons through $E_{abs}=N_{abs}hc/\lambda$, where hc/λ is the energy of a single photon of wavelength λ (eqn 7A.5).

By combining both expressions for the absorbed energy, the value of N_{abs} follows readily. The number of photochemical events, and hence the numerator of the expression on the right-hand side of eqn 20G.1a, is simply the number of decomposed molecules $N_{decomposed}$. The primary quantum yield follows from $\phi=N_{decomposed}/N_{abs}$.

Answer From the expressions for the absorbed energy, it follows that

$$E_{abs}=fPt=N_{abs}\left(\frac{hc}{\lambda}\right)$$

and therefore that $N_{abs}=fPt\lambda/hc$. Now we use eqn 20G.1a to write

$$\phi=\frac{N_{decomposed}}{N_{abs}}=\frac{N_{decomposed}hc}{fPt\lambda}$$

With $N_{decomposed}=(0.344\,\text{mol})\times(6.022\times10^{23}\,\text{mol}^{-1})$, $P=100\,\text{W}=100\,\text{J s}^{-1}$, $t=2700\,\text{s}$, $\lambda=490\,\text{nm}=4.90\times10^{-7}\,\text{m}$, and $f=0.60$ it follows that

$$\phi=\frac{(0.344\,\text{mol})\times(6.022\times10^{23}\,\text{mol}^{-1})\times(6.626\times10^{-34}\,\text{J s})\times(2.998\times10^{8}\,\text{m s}^{-1})}{0.60\times(100\,\text{J s}^{-1})\times(2700\,\text{s})\times(4.90\times10^{-7}\,\text{m})}$$
$$=0.52$$

Self-test 20G.2 In an experiment to measure the quantum yield of a photochemical reaction, the absorbing substance was exposed to 320 nm radiation from a 87.5 W source for 38 min. The intensity of the transmitted light was 0.35 that of the incident light. As a result of irradiation, 0.324 mol of the absorbing substance decomposed. Determine the primary quantum yield.

Answer: $\phi=0.93$

A molecule in an excited state must either decay to the ground state or form a photochemical product. Therefore, the total number of molecules deactivated by radiative processes, non-radiative processes, and photochemical reactions must be equal to the number of excited species produced by absorption of the incident radiation. We conclude that the sum of primary quantum yields ϕ_i for *all* photophysical and photochemical events i must be equal to 1, regardless of the number of reactions involving the excited state:

$$\sum_i \phi_i=\sum_i \frac{v_i}{I_{abs}}=1 \qquad (20G.2)$$

It follows that for an excited singlet state that decays to the ground state only by the photophysical processes described in Section 20G.1 (and without reacting), we write

$$\phi_F+\phi_{IC}+\phi_P=1$$

where ϕ_F, ϕ_{IC}, and ϕ_P are the quantum yields of fluorescence, internal conversion, and phosphorescence, respectively (intersystem crossing from the singlet to the triplet state is taken into account by the presence of ϕ_P). The quantum yield of photon emission by fluorescence and phosphorescence is $\phi_{emission}=\phi_F+\phi_P$, which is less than 1. If the excited singlet state also participates in a primary photochemical reaction with quantum yield ϕ_r, we write

$$\phi_F+\phi_{IC}+\phi_P+\phi_r=1$$

We can now strengthen the link between reaction rates and primary quantum yield already established by eqns 20G.1 and 20G.2. By taking the constant I_{abs} out of the sum in eqn 20G.2 and rearranging, we obtain $I_{abs}=\Sigma_i v_i$. Substituting this result into eqn 20G.2 gives the general result

$$\phi_i=\frac{v_i}{\sum_i v_i} \qquad (20G.3)$$

Therefore, the primary quantum yield may be determined directly from the experimental rates of *all* photophysical and photochemical processes that deactivate the excited state.

20G.3 Mechanism of decay of excited singlet states

Consider the formation and decay of an excited singlet state in the absence of a chemical reaction:

Absorption:	$S+h\nu_i\rightarrow S^*$	$v_{abs}=I_{abs}$
Fluorescence:	$S^*\rightarrow S+h\nu_f$	$v_F=k_F[S^*]$
Internal conversion:	$S^*\rightarrow S$	$v_{IC}=k_{IC}[S^*]$
Intersystem crossing:	$S^*\rightarrow T^*$	$v_{ISC}=k_{ISC}[S^*]$

in which S is an absorbing singlet-state species, S^* an excited singlet state, T^* an excited triplet state, and $h\nu_i$ and $h\nu_f$ are the energies of the incident and fluorescent photons, respectively. From the methods presented in Topic 20E and the rates of the steps that form and destroy the excited singlet state S^*, we write the rate of formation and decay of S^* as:

$$\text{Rate of formation of } S^* = I_{abs}$$
$$\text{Rate of disappearance of } S^* = k_F[S^*]+k_{ISC}[S^*]+k_{IC}[S^*] = (k_F+k_{ISC}+k_{IC})[S^*]$$

It follows that the excited state decays by a first-order process so, when the light is turned off, the concentration of S^* varies with time t as:

$$[S^*](t)=[S^*]_0 e^{-t/\tau_0} \quad (20G.4)$$

where the **observed lifetime**, τ_0, of the first excited singlet state is defined as:

$$\tau_0=\frac{1}{k_F+k_{ISC}+k_{IC}} \quad \textit{Definition} \quad \text{Observed lifetime of the excited singlet state} \quad (20G.5)$$

We show in the following *Justification* that the quantum yield of fluorescence is

$$\phi_{F,0}=\frac{k_F}{k_F+k_{ISC}+k_{IC}} \quad \text{Quantum yield of fluorescence} \quad (20G.6)$$

Justification 20G.1 The quantum yield of fluorescence

Most fluorescence measurements are conducted by illuminating a relatively dilute sample with a continuous and intense beam of light or ultraviolet radiation. It follows that $[S^*]$ is small and constant, so we may invoke the steady-state approximation (Topic 20E) and write:

$$\frac{d[S^*]}{dt}=I_{abs}-k_F[S^*]-k_{ISC}[S^*]-k_{IC}[S^*] = I_{abs}-(k_F+k_{ISC}+k_{IC})[S^*]\approx 0$$

Consequently,

$$I_{abs}=(k_F+k_{ISC}+k_{IC})[S^*]$$

By using this expression and eqn 20G.1b, we write the quantum yield of fluorescence as:

$$\phi_{F,0}=\frac{v_F}{I_{abs}}=\frac{k_F[S^*]}{(k_F+k_{ISC}+k_{IC})[S^*]}$$

which, by cancelling the $[S^*]$, simplifies to eqn 20G.6.

The observed fluorescence lifetime can be measured by using a pulsed laser technique. First, the sample is excited with a short light pulse from a laser using a wavelength at which S absorbs strongly. Then, the exponential decay of the fluorescence intensity after the pulse is monitored. From eqns 20G.5 and 20G.6, it follows that

$$\tau_0=\frac{1}{k_F+k_{ISC}+k_{IC}}=\frac{k_F}{k_F+k_{ISC}+k_{IC}}\times\frac{1}{k_F}=\frac{\phi_{F,0}}{k_F} \quad (20G.7)$$

Brief illustration 20G.2 The fluorescence rate constant

The fluorescence quantum yield and observed fluorescence lifetime of tryptophan in water are $\phi_{F,0}=0.20$ and $\tau_0=2.6$ ns, respectively. It follows from eqn 20G.7 that the fluorescence rate constant k_F is

$$k_F=\frac{\phi_{F,0}}{\tau_0}=\frac{0.20}{2.6\times10^{-9}\,\text{s}}=7.7\times10^7\,\text{s}^{-1}$$

Self-test 20G.3 A substance has a fluorescence quantum yield of $\phi_{F,0}=0.35$. In an experiment to measure the fluorescence lifetime of this substance, it was observed that the fluorescence emission decayed with a half-life of 5.6 ns. Determine the fluorescence rate constant of this substance.

Answer: $k_F=4.3\times10^7\,\text{s}^{-1}$

20G.4 Quenching

The shortening of the lifetime of the excited state by the presence of another species is called **quenching**. Quenching may be either a desired process, such as in energy or electron transfer, or an undesired side reaction that can decrease the quantum yield of a desired photochemical process. Quenching effects may be studied by monitoring the emission from the excited state that is involved in the photochemical reaction.

The addition of a quencher, Q, opens an additional channel for deactivation of S^*:

$$\text{Quenching: } S^*+Q\rightarrow S+Q \qquad v_Q=k_Q[Q][S^*]$$

The **Stern–Volmer equation**, which is derived in the following *Justification*, relates the fluorescence quantum yields $\phi_{F,0}$ and ϕ_F measured in the absence and presence, respectively, of a quencher Q at a molar concentration [Q]:

$$\frac{\phi_{F,0}}{\phi_F}=1+\tau_0 k_Q[Q] \quad \text{Stern–Volmer equation} \quad (20G.8)$$

This equation tells us that a plot of $\phi_{F,0}/\phi_F$ against [Q] should be a straight line with slope $\tau_0 k_Q$. Such a plot is called a **Stern–Volmer plot** (Fig. 20G.1). The method may also be applied to the quenching of phosphorescence.

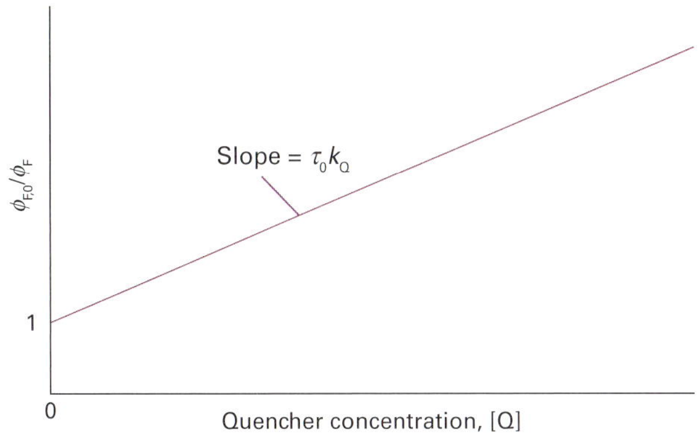

Figure 20G.1 The format of a Stern–Volmer plot and the interpretation of the slope in terms of the rate constant for quenching and the observed fluorescence lifetime in the absence of quenching.

Justification 20G.2 The Stern–Volmer equation

With the addition of quenching, the steady-state approximation for $[S^*]$ now gives:

$$\frac{d[S^*]}{dt} = I_{abs} - (k_F + k_{ISC} + k_{IC} + k_Q[Q])[S^*] \approx 0$$

and the fluorescence quantum yield in the presence of the quencher is:

$$\phi_F = \frac{k_F}{k_F + k_{ISC} + k_{IC} + k_Q[Q]}$$

It follows that

$$\frac{\phi_{F,0}}{\phi_F} = \frac{k_F}{k_F + k_{ISC} + k_{IC}} \times \frac{k_F + k_{ISC} + k_{IC} + k_Q[Q]}{k_F} = \frac{k_F + k_{ISC} + k_{IC} + k_Q[Q]}{k_F + k_{ISC} + k_{IC}} = 1 + \frac{k_Q}{k_F + k_{ISC} + k_{IC}}[Q]$$

By using eqn 20G.7, this expression simplifies to eqn 20G.8.

Because the fluorescence intensity and lifetime are both proportional to the fluorescence quantum yield (specifically, from eqn 20G.7, $\tau = \phi_F/k_F$), plots of $I_{F,0}/I_F$ and τ_0/τ (where the subscript 0 indicates a measurement in the absence of quencher) against [Q] should also be linear with the same slope and intercept as those shown for eqn 20G.8.

Example 20G.2 Determining the quenching rate constant

The molecule 2,2′-bipyridine (**1**, bpy) forms a complex with the Ru^{2+} ion. Tris-(2,2′-bipyridyl)ruthenium(II), $Ru(bpy)_3^{2+}$ (**2**), has a strong metal-to-ligand charge transfer (MLCT) transition (Topic 13A) at 450 nm.

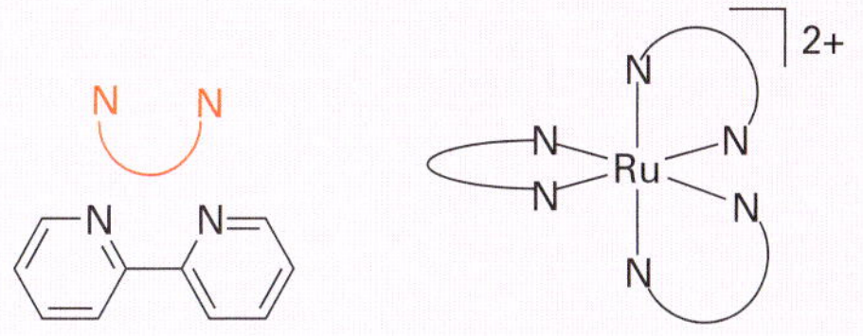

1 2,2′-Bipyridine (bpy) **2** $[Ru(bpy)_3]^{2+}$

The quenching of the $^*Ru(bpy)_3^{2+}$ excited state by Fe^{3+} (present as the complex ion $Fe(OH_2)_6^{3+}$) in acidic solution was monitored by measuring emission lifetimes at 600 nm. Determine the quenching rate constant for this reaction from the following data:

$[Fe(OH_2)_6^{3+}]/(10^{-4}\,mol\,dm^{-3})$	0	1.6	4.7	7	9.4
$\tau/(10^{-7}\,s)$	6	4.05	3.37	2.96	2.17

Method Rewrite the Stern–Volmer equation (eqn 20G.8) for use with lifetime data; then fit the data to a straight-line.

Answer Upon substitution of τ_0/τ for $\phi_{F,0}/\phi_F$ in eqn 20G.8 and after rearrangement, we obtain:

$$\frac{1}{\tau} = \frac{1}{\tau_0} + k_Q[Q]$$

Figure 20G.2 shows a plot of $1/\tau$ against $[Fe^{3+}]$ and the results of a fit to this expression. The slope of the line is 2.8×10^9, so $k_Q = 2.8 \times 10^9\,dm^3\,mol^{-1}\,s^{-1}$. This example shows that measurements of emission lifetimes are preferred because they yield the value of k_Q directly. To determine the value of k_Q from intensity or quantum yield measurements, it is necessary to make an independent measurement of τ_0.

Self-test 20G.4 The quenching of tryptophan fluorescence by dissolved O_2 gas was monitored by measuring

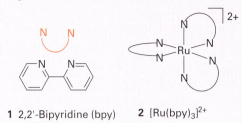

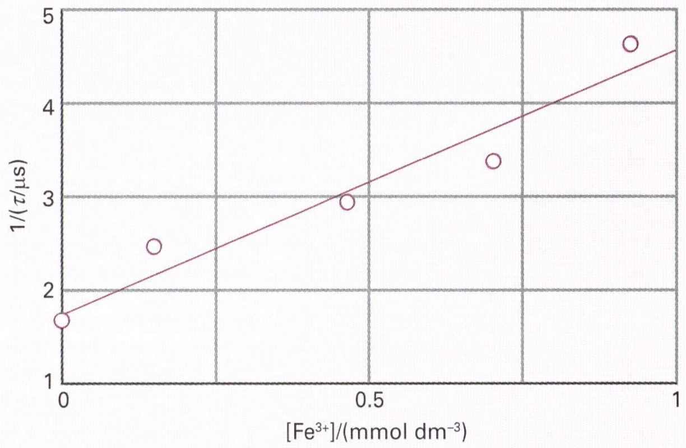

Figure 20G.2 The Stern–Volmer plot of the data for *Example* 20G.2.

emission lifetimes at 348 nm in aqueous solutions. Determine the quenching rate constant for this process from the following data:

$[O_2]/(10^{-2}\,mol\,dm^{-3})$	0	2.3	5.5	8	10.8
$\tau/(10^{-9}\,s)$	2.6	1.5	0.92	0.71	0.57

Answer: $1.3\times10^{10}\,dm^3\,mol^{-1}\,s^{-1}$

Three common mechanisms for bimolecular quenching of an excited singlet (or triplet) state are:

Collisional deactivation:	$S^*+Q\rightarrow S+Q$
Resonance energy transfer:	$S^*+Q\rightarrow S+Q^*$
Electron ransfer:	$S^*+Q\rightarrow S^{+/-}+Q^{-/+}$

The quenching rate constant itself does not give much insight into the mechanism of quenching. For the system of *Example* 20G.2, it is known that the quenching of the excited state of $Ru(bpy)_3^{2+}$ is a result of electron transfer to Fe^{3+}, but the quenching data do not allow us to prove the mechanism.

There are, however, some criteria that govern the relative efficiencies of collisional quenching, resonance energy transfer, and electron transfer. Collisional quenching is particularly efficient when Q is a species, such as iodide ion, which receives energy from S* and then decays to the ground state primarily by releasing energy as heat. As we show in detail in Topic 21E, according to the **Marcus theory** of electron transfer, which was proposed by R.A. Marcus in 1965, the rates of electron transfer (from ground or excited states) depend on:

- The distance between the donor and acceptor, with electron transfer becoming more efficient as the distance between donor and acceptor decreases.
- The reaction Gibbs energy, $\Delta_r G$, with electron transfer becoming more efficient as the reaction becomes more exergonic. For example, it follows from the thermodynamic principles that lead to the electrochemical series (Topic 6D) that efficient photo-oxidation of S requires that the reduction potential of S* be lower than the reduction potential of Q.
- The reorganization energy, the energy cost incurred by molecular rearrangements of donor, acceptor, and solvent medium during electron transfer. The electron transfer rate is predicted to increase as this reorganization energy is matched closely by the reaction Gibbs energy.

Electron transfer can also be studied by time-resolved spectroscopy (Topic 13C). The oxidized and reduced products often have electronic absorption spectra distinct from those of their neutral parent compounds. Therefore, the rapid appearance of such known features in the absorption spectrum after excitation by a laser pulse may be taken as indication of quenching by electron transfer. In the following section we explore resonance energy transfer in detail.

20G.5 Resonance energy transfer

We visualize the process $S^*+Q\rightarrow S+Q^*$ as follows. The oscillating electric field of the incoming electromagnetic radiation induces an oscillating electric dipole moment in S. Energy is absorbed by S if the frequency of the incident radiation, ν, is such that $\nu=\Delta E_S/h$, where ΔE_S is the energy separation between the ground and excited electronic states of S and h is Planck's constant. This is the 'resonance condition' for absorption of radiation (essentially the Bohr frequency condition, eqn 7A.12). The oscillating dipole on S can now affect electrons bound to a nearby Q molecule by inducing an oscillating dipole moment in them. If the frequency of oscillation of the electric dipole moment in S is such that $\nu=\Delta E_Q/h$ then Q will absorb energy from S.

The efficiency, η_T, of resonance energy transfer is defined as

$$\eta_T=1-\frac{\phi_F}{\phi_{F,0}} \quad \textit{Definition} \quad \text{Efficiency of resonance energy transfer} \quad (20G.9)$$

According to the **Förster theory** of resonance energy transfer, energy transfer is efficient when:

- The energy donor and acceptor are separated by a short distance (of the order of nanometres).
- Photons emitted by the excited state of the donor can be absorbed directly by the acceptor.

For donor–acceptor systems held rigidly either by covalent bonds or by a protein 'scaffold', η_T increases with decreasing distance, R, according to

$$\eta_T=\frac{R_0^6}{R_0^6+R^6} \quad \text{Efficiency of energy transfer in terms of the donor–acceptor distance} \quad (20G.10)$$

where R_0 is a parameter (with dimensions of distance) that is characteristic of each donor–acceptor pair. It can be regarded as the distance at which energy transfer is 50 per cent efficient for a given donor–acceptor pair. (You can verify this assertion by using $R=R_0$ in eqn 20G.10.) Equation 20G.10 has been verified experimentally and values of R_0 are available for a number of donor–acceptor pairs (Table 20G.3).

The emission and absorption spectra of molecules span a range of wavelengths, so the second requirement of the Förster theory is met when the emission spectrum of the donor molecule overlaps significantly with the absorption spectrum of the

Table 20G.3 Values of R_0 for some donor–acceptor pairs*

Donor‡	Acceptor‡	R_0/nm
Naphthalene	Dansyl	2.2
Dansyl	ODR	4.3
Pyrene	Coumarin	3.9
1.5-I AEDANS	FITC	4.9
Tryptophan	1.5-I AEDANS	2.2
Tryptophan	Haem (heme)	2.9

*Additional values may be found in J.R. Lacowicz *Principles of fluorescence spectroscopy*, Kluwer Academic/Plenum, New York (1999).

‡Abbreviations:

Dansyl: 5-dimethylamino-1-naphthalenesulfonic acid

FITC: fluorescein 5-isothiocyanate

1.5-I AEDANS: 5-((((2-iodoacetyl)amino)ethyl)amino)naphthalene-1-sulfonic acid

ODR: octadecyl-rhodamine

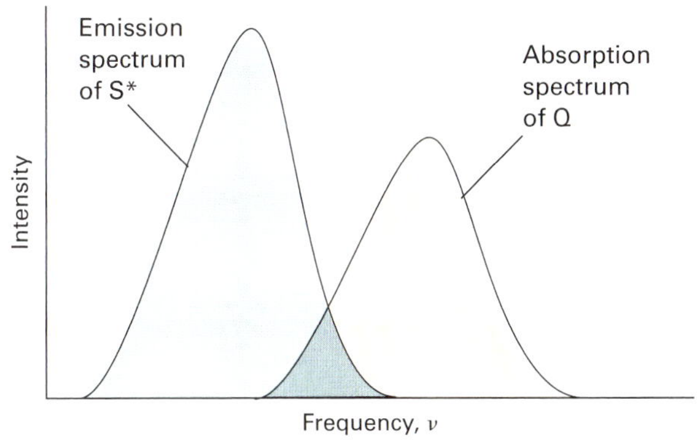

Figure 20G.3 According to the Förster theory, the rate of energy transfer from a molecule S^* in an excited state to a quencher molecule Q is optimized at radiation frequencies in which the emission spectrum of S^* overlaps with the absorption spectrum of Q, as shown in the (dark green) shaded region.

acceptor. In the overlap region, photons emitted by the donor have the appropriate energy to be absorbed by the acceptor (Fig. 20G.3).

Equation 20G.10 forms the basis of **fluorescence resonance energy transfer** (FRET), in which the dependence of the energy transfer efficiency, η_T, on the distance, R, between energy donor and acceptor is used to measure distances in biological systems. In a typical FRET experiment, a site on a biopolymer or membrane is labelled covalently with an energy donor and another site is labelled covalently with an energy acceptor. In certain cases, the donor or acceptor may be natural constituents of the system, such as amino acid groups, cofactors, or enzyme substrates. The distance between the labels is then calculated from the known value of R_0 and eqn 20G.10. Several tests have shown that the FRET technique is useful for measuring distances ranging from 1 to 9 nm.

Brief illustration 20G.3 The FRET technique

As an illustration of the FRET technique, consider a study of the protein rhodopsin. When an amino acid on the surface of rhodopsin was labelled covalently with the energy donor 1.5-I AEDANS (3), the fluorescence quantum yield of the label decreased from 0.75 to 0.68 due to quenching by the visual pigment 11-*cis*-retinal (4). From eqn 20G.10, we calculate $\eta_T = 1 - 0.68/0.75 = 0.093$ and from eqn 20G.10 and the known value of $R_0 = 5.4$ nm for the 1.5-I AEDANS/11-*cis*-retinal pair we calculate $R = 7.9$ nm. Therefore, we take 7.9 nm to be the distance between the surface of the protein and 11-*cis*-retinal.

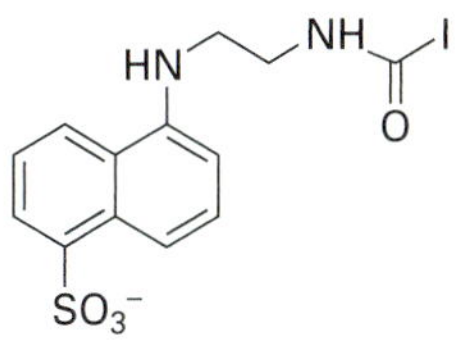

3 1.5-I AEDANS

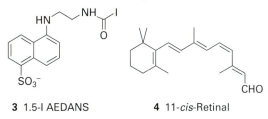

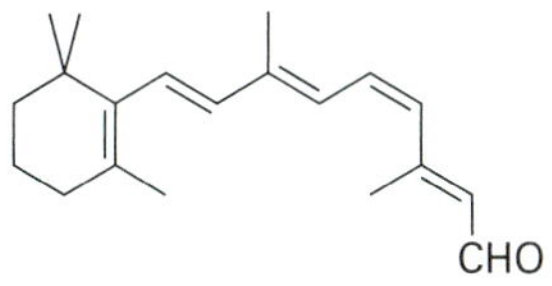

4 11-*cis*-Retinal

Self-test 20G.5 An amino acid on the surface of a protein was labelled covalently with 1.5-I AEDANS and another was labelled covalently with FITC. The fluorescence quantum yield of 1.5-I AEDANS decreased by 10 per cent due to quenching by FITC. What is the distance between the amino acids?

Answer: 7.1 nm

If donor and acceptor molecules diffuse in solution or in the gas phase, Förster theory predicts that the efficiency of quenching by energy transfer increases as the average distance travelled between collisions of donor and acceptor decreases. That is, the quenching efficiency increases with concentration of quencher, as predicted by the Stern–Volmer equation.

Checklist of concepts

- ☐ 1. The **primary quantum yield** of a photochemical reaction is the number of reactant molecules producing specified primary products for each photon absorbed.
- ☐ 2. The **observed lifetime** of an excited state is related to the quantum yield and rate constant of emission.
- ☐ 3. A **Stern–Volmer plot** is used to analyse the kinetics of fluorescence quenching in solution.
- ☐ 4. Collisional deactivation, electron transfer, and resonance energy transfer are common fluorescence quenching processes.

☐ 5. The efficiency of resonance energy transfer decreases with increasing separation between donor and acceptor molecules.

Checklist of equations

Property	Equation	Comment	Equation number
Primary quantum yield	$\phi = v/I_{abs}$		20G.1b
Excited state lifetime	$\tau_0 = 1/(k_F + k_{ISC} + k_{IC})$	No quencher present	20G.5
Quantum yield of fluorescence	$\phi_{F,0} = k_F/(k_F + k_{ISC} + k_{IC})$	Without quencher present	20G.6
Observed excited state lifetime	$\tau_0 = \phi_{F,0}/k_F$		20G.7
Stern–Volmer equation	$\phi_{F,0}/\phi_F = 1 + \tau_0 k_Q[Q]$		20G.8
Efficiency of resonance energy transfer	$\eta_T = 1 - \phi_F/\phi_{F,0}$	Definition	20G.9
	$\eta_T = R_0^6/(R_0^6 + R^6)$	Förster theory	20G.10

20H Enzymes

Contents

➤ Why do you need to know this material?

The role of enzymes in controlling chemical reactions is central to biology and the maintenance of life. It is at the centre of attention of much of the application of physical chemistry to biology.

➤ What is the key idea?

Enzymes are homogeneous catalysts that can have a dramatic effect on the rates of the reactions they control but are subject to inhibition.

➤ What do you need to know already?

You need to be familiar with the analysis of reaction mechanisms in terms of the steady-state approximation (Topic 20E).

A catalyst is a substance that accelerates a reaction but undergoes no net chemical change (Topic 20D): the catalyst lowers the activation energy of the reaction by providing an alternative path that avoids the slow, rate-determining step of the uncatalysed reaction (Fig. 20H.1). **Enzymes**, which are homogeneous biological catalysts, are very specific and can have a dramatic effect on the reactions they control. The enzyme catalase reduces the activation energy from 76 kJ mol^{-1} to 8 kJ mol^{-1}, corresponding to an acceleration of the reaction by a factor of 10^{15} at 298 K.

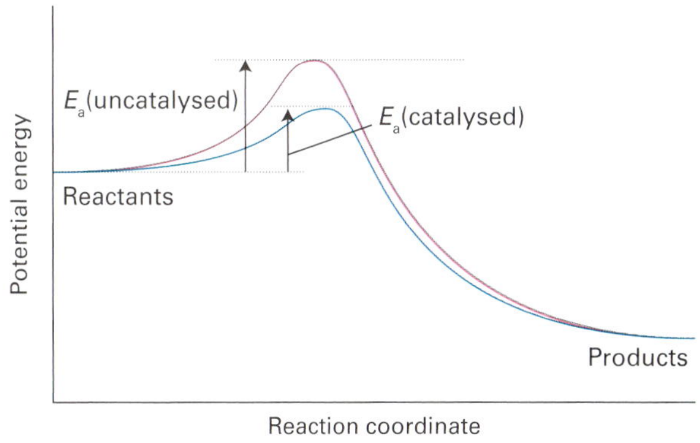

Figure 20H.1 A catalyst provides a different path with a lower activation energy. The result is an increase in the rate of formation of product.

20H.1 Features of enzymes

Enzymes act in the aqueous environment of cells. These biologically ubiquitous compounds are special proteins or nucleic acids that contain an **active site**, which is responsible for binding the **substrates**, the reactants, and processing them into products. As is true of any catalyst, the active site returns to its original state after the products are released. Many enzymes consist primarily of proteins, some featuring organic or inorganic co-factors in their active sites. However, certain RNA molecules can also be biological catalysts, forming *ribozymes*. A very important example of a ribozyme is the *ribosome*, a large assembly of proteins and catalytically active RNA molecules responsible for the synthesis of proteins in the cell.

The structure of the active site is specific to the reaction that it catalyses, with groups in the substrate interacting with groups in the active site by intermolecular interactions, such as hydrogen bonding, electrostatic forces, and van der Waals interactions. Figure 20H.2 shows two models that explain the binding of a substrate to the active site of an enzyme. In the **lock-and-key model**, the active site and substrate have complementary three-dimensional structures and dock without the need for major atomic rearrangements. Experimental evidence favours the **induced fit model**, in which binding of the substrate induces a conformational change in the active site. Only after the change does the substrate fit snugly in the active site.

Enzyme-catalysed reactions are prone to inhibition by molecules that interfere with the formation of product. Many drugs for the treatment of disease function by inhibiting

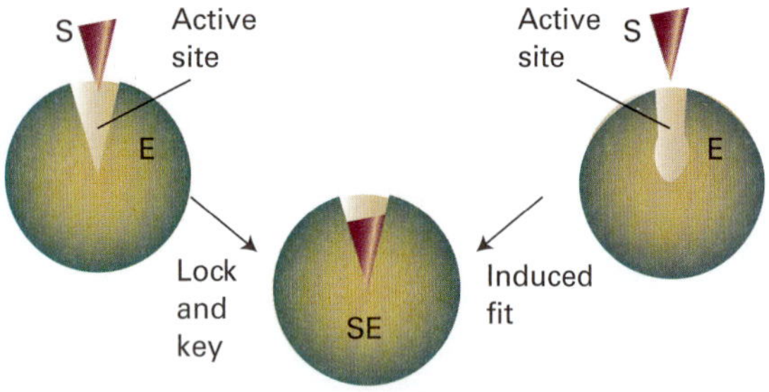

Figure 20H.2 Two models that explain the binding of a substrate to the active site of an enzyme. In the lock-and-key model, the active site and substrate have complementary three-dimensional structures and dock without the need for major atomic rearrangements. In the induced fit model, binding of the substrate induces a conformational change in the active site. The substrate fits well in the active site after the conformational change has taken place.

enzymes. For example, an important strategy in the treatment of acquired immune deficiency syndrome (AIDS) involves the steady administration of a specially designed protease inhibitor. The drug inhibits an enzyme that is key to the formation of the protein envelope surrounding the genetic material of the human immunodeficiency virus (HIV). Without a properly formed envelope, HIV cannot replicate in the host organism.

20H.2 The Michaelis–Menten mechanism

Experimental studies of enzyme kinetics are typically conducted by monitoring the initial rate of product formation in a solution in which the enzyme is present at very low concentration. Indeed, enzymes are such efficient catalysts that significant accelerations may be observed even when their concentration is more than three orders of magnitude smaller than that of the substrate.

The principal features of many enzyme-catalysed reactions are as follows:

- For a given initial concentration of substrate, $[S]_0$, the initial rate of product formation is proportional to the total concentration of enzyme, $[E]_0$.
- For a given $[E]_0$ and low values of $[S]_0$, the rate of product formation is proportional to $[S]_0$.
- For a given $[E]_0$ and high values of $[S]_0$, the rate of product formation becomes independent of $[S]_0$, reaching a maximum value known as the **maximum velocity**, v_{max}.

The **Michaelis–Menten mechanism** accounts for these features. According to this mechanism, an enzyme–substrate complex is formed in the first step and either the substrate is released unchanged or after modification to form products:

$$\mathrm{E+S \rightleftarrows ES} \quad k_a, k_a'$$
$$\mathrm{ES \rightarrow P+E} \quad k_b$$

Michaelis–Menten mechanism

We show in the following *Justification* that this mechanism leads to the **Michaelis–Menten equation** for the rate of product formation

$$v = \frac{k_b[\mathrm{E}]_0}{1+K_M/[\mathrm{S}]_0} \qquad \text{Michaelis–Menten equation} \qquad (20\text{H}.1)$$

where $K_M = (k_a' + k_b)/k_a$ is the **Michaelis constant**, characteristic of a given enzyme acting on a given substrate and having the dimensions of a molar concentration.

Justification 20H.1 The Michaelis–Menten equation

The rate of product formation according to the Michaelis–Menten mechanism is

$$v = k_b[\mathrm{ES}]$$

We can obtain the concentration of the enzyme–substrate complex by invoking the steady-state approximation and writing

$$\frac{d[\mathrm{ES}]}{dt} = k_a[\mathrm{E}][\mathrm{S}] - k_a'[\mathrm{ES}] - k_b[\mathrm{ES}] \approx 0$$

It follows that

$$[\mathrm{ES}] = \frac{k_a[\mathrm{E}][\mathrm{S}]}{k_a' + k_b}$$

where [E] and [S] are the concentrations of *free* enzyme and substrate, respectively. Now we define the Michaelis constant as

$$K_M = \frac{k_a' + k_b}{k_a}$$

To express the rate law in terms of the concentrations of enzyme and substrate added, we note that $[\mathrm{E}]_0 = [\mathrm{E}] + [\mathrm{ES}]$ and

$$[\mathrm{E}]_0 = \overbrace{\frac{K_M[\mathrm{ES}]}{[\mathrm{S}]}}^{[\mathrm{E}]} + [\mathrm{ES}] = [\mathrm{ES}]\{1 + K_M/[\mathrm{S}]\}$$

Moreover, because the substrate is typically in large excess relative to the enzyme, the free substrate concentration is approximately equal to the initial substrate concentration and we can write $[\mathrm{S}] \approx [\mathrm{S}]_0$. It then follows that:

$$[\mathrm{ES}] = \frac{[\mathrm{E}]_0}{1+K_M/[\mathrm{S}]_0}$$

Equation 20H.1 is obtained when this expression for [ES] is substituted into that for the rate of product formation ($v = k_b[\mathrm{ES}]$).

Equation 20H.1 shows that, in accord with experimental observations (Fig. 20H.3):

- When $[S]_0 \ll K_M$, the rate is proportional to $[S]_0$:

$$v = \frac{k_b}{K_M}[S]_0[E]_0 \tag{20H.2a}$$

- When, $[S]_0 \gg K_M$, the rate reaches its maximum value and is independent of $[S]_0$:

$$v = v_{max} = k_b[E]_0 \tag{20H.2b}$$

Substitution of the definition of v_{max} into eqn 20H.1 gives

$$v = \frac{v_{max}}{1 + K_M/[S]_0} \tag{20H.3a}$$

which can be rearranged into a form suitable for data analysis by linear regression by taking reciprocals of both sides:

$$\frac{1}{v} = \frac{1}{v_{max}} + \left(\frac{K_M}{v_{max}}\right)\frac{1}{[S]_0} \qquad \text{Lineweaver–Burk plot} \tag{20H.3b}$$

A **Lineweaver–Burk plot** is a plot of $1/v$ against $1/[S]_0$, and according to eqn 20H.3b it should yield a straight line with slope of K_M/v_{max}, a y-intercept at $1/v_{max}$, and an x-intercept at $-1/K_M$ (Fig. 20H.4). The value of k_b is then calculated from the y-intercept and eqn 20H.2b. However, the plot cannot give the individual rate constants k_a and k_a' that appear in the expression for K_M. The stopped-flow technique described in Topic 20A can give the additional data needed, because the rate of formation of the enzyme–substrate complex can be found by monitoring the concentration after mixing the enzyme and substrate. This procedure gives a value for k_a, and k_a' is then found by combining this result with the values of k_b and K_M.

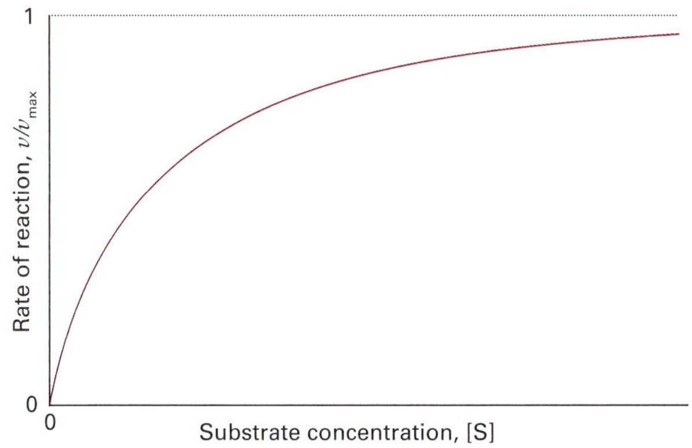

Figure 20H.3 The variation of the rate of an enzyme-catalysed reaction with substrate concentration. The approach to a maximum rate, v_{max}, for large [S] is explained by the Michaelis–Menten mechanism.

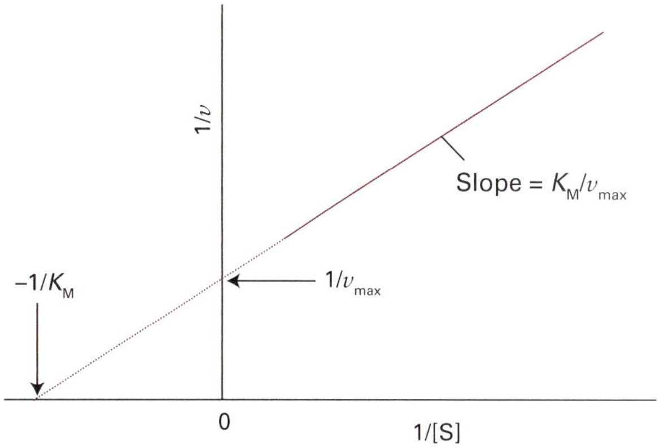

Figure 20H.4 A Lineweaver–Burk plot for the analysis of an enzyme-catalysed reaction that proceeds by a Michaelis–Menten mechanism and the significance of the intercepts and the slope.

Example 20H.1 Analysing a Lineweaver–Burk plot

The enzyme carbonic anhydrase catalyses the hydration of CO_2 in red blood cells to give bicarbonate (hydrogencarbonate) ion: $CO_2(g) + H_2O(l) \rightarrow HCO_3^-(aq) + H^+(aq)$. The following data were obtained for the reaction at pH = 7.1, 273.5 K, and an enzyme concentration of 2.3 nmol dm^{-3}:

$[CO_2]$/(mmol dm^{-3})	1.25	2.5	5	20
v/(mmol dm^{-3} s^{-1})	2.78×10^{-2}	5.00×10^{-2}	8.33×10^{-2}	1.67×10^{-1}

Determine the maximum velocity and the Michaelis constant for the reaction at 273.5 K.

Method Prepare a Lineweaver–Burk plot and determine the values of K_M and v_{max} by linear regression analysis.

Answer We draw up the following table:

$1/([CO_2]$/(mmol dm^{-3}))	0.800	0.400	0.200	0.0500
$1/(v$/(mmol dm^{-3} s^{-1}))	36.0	20.0	12.0	6.0

Figure 20H.5 shows the Lineweaver–Burk plot for the data. The slope is 40.0 and the y-intercept is 4.00. Hence,

$$v_{max}/(\text{mmol dm}^{-3}\,\text{s}^{-1}) = \frac{1}{\text{intercept}} = \frac{1}{4.00} = 0.250$$

and

$$K_M/(\text{mmol dm}^{-3}) = \frac{\text{slope}}{\text{intercept}} = \frac{40.00}{4.00} = 10.0$$

A note on good practice The slope and the intercept are unit-less: all graphs should be plotted as pure numbers.

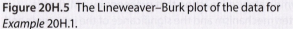

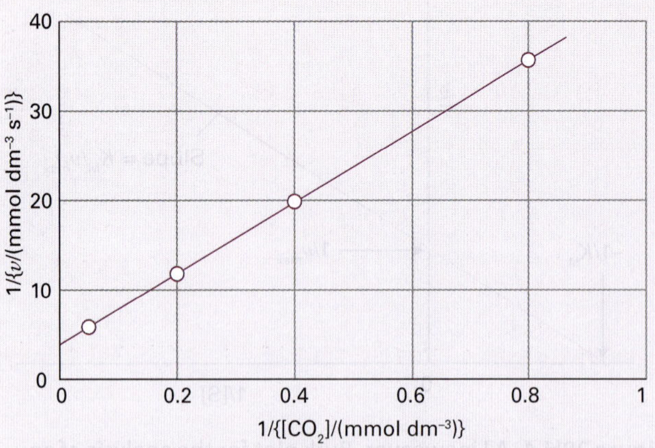

Figure 20H.5 The Lineweaver–Burk plot of the data for *Example* 20H.1.

Self-test 20H.1 The enzyme α-chymotrypsin is secreted in the pancreas of mammals and cleaves peptide bonds made between certain amino acids. Several solutions containing the small peptide *N*-glutaryl-l-phenylalanine-*p*-nitroanilide at different concentrations were prepared and the same small amount of α-chymotrypsin was added to each one. The following data were obtained on the initial rates of the formation of product:

[S]/(mmol dm^{-3})	0.334	0.450	0.667	1.00	1.33	1.67
v/(mmol dm^{-3} s^{-1})	0.152	0.201	0.269	0.417	0.505	0.667

Determine the maximum velocity and the Michaelis constant for the reaction.

Answer: $v_{max}=2.80$ mmol dm^{-3} s^{-1}, $K_M=5.89$ mmol dm^{-3}

20H.3 The catalytic efficiency of enzymes

The **turnover frequency**, or **catalytic constant**, of an enzyme, k_{cat}, is the number of catalytic cycles (turnovers) performed by the active site in a given interval divided by the duration of the interval. This quantity has units of a first-order rate constant and, in terms of the Michaelis–Menten mechanism, is numerically equivalent to k_b, the rate constant for release of product from the enzyme–substrate complex. It follows from the identification of k_{cat} with k_b and from eqn 20H.2b that

$$k_{cat}=k_b=\frac{v_{max}}{[E]_0} \quad \text{Turnover frequency} \quad (20H.4)$$

The **catalytic efficiency**, η (eta), of an enzyme is the ratio k_{cat}/K_M. The higher the value of η, the more efficient is the enzyme. We can think of the catalytic efficiency as the effective rate constant of the enzymatic reaction. From $K_M=(k_a'+k_b)/k_a$ and eqn 20H.4, it follows that

$$\eta=\frac{k_{cat}}{K_M}=\frac{k_a k_b}{k_a'+k_b} \quad \text{Catalytic efficiency} \quad (20H.5)$$

The efficiency reaches its maximum value of k_a when $k_b \gg k_a'$. Because k_a is the rate constant for the formation of a complex from two species that are diffusing freely in solution, the maximum efficiency is related to the maximum rate of diffusion of E and S in solution. This limit (which is discussed further in Topic 21B) leads to rate constants of about 10^8–10^9 dm^3 mol^{-1} s^{-1} for molecules as large as enzymes at room temperature. The enzyme catalase has $\eta=4.0\times10^8$ dm^3 mol^{-1} s^{-1} and is said to have attained 'catalytic perfection', in the sense that the rate of the reaction it catalyses is controlled only by diffusion: it acts as soon as a substrate makes contact.

Brief illustration 20H.1 The catalytic efficiency of an enzyme

To determine the catalytic efficiency of carbonic anhydrase at 273.5 K from the results from *Example* 20H.1, we begin by using eqn 20H.4 to calculate k_{cat}:

$$k_{cat}=\frac{v_{max}}{[E]_0}=\frac{2.5\times10^{-4}\ \text{mol dm}^{-3}\ \text{s}^{-1}}{2.3\times10^{-9}\ \text{mol dm}^{-3}}=1.1\times10^5\ \text{s}^{-1}$$

The catalytic efficiency follows from eqn 20H.5:

$$\eta=\frac{k_{cat}}{K_M}=\frac{1.1\times10^5\ \text{s}^{-1}}{2.3\times10^{-9}\ \text{mol dm}^{-3}}=1.1\times10^7\ \text{dm}^3\ \text{mol}^{-1}\ \text{s}^{-1}$$

Self-test 20H.2 The enzyme-catalysed conversion of a substrate at 298 K has $K_M=0.032$ mol dm^{-3} and $v_{max}=4.25\times10^{-4}$ mol dm^{-3} s^{-1} when the enzyme concentration is 3.60×10^{-9} mol dm^{-3}. Calculate k_{cat} and η. Is the enzyme 'catalytically perfect'?

Answer: $k_{cat}=1.18\times10^5$ s^{-1}, $\eta=7.9\times10^6$ dm^3 mol^{-1} s^{-1}; the enzyme is not 'catalytically perfect'

20H.4 Mechanisms of enzyme inhibition

An inhibitor, In, decreases the rate of product formation from the substrate by binding to the enzyme, to the ES complex, or to the enzyme and ES complex simultaneously. The most general kinetic scheme for enzyme inhibition is then:

$$\text{E}+\text{S}\rightleftarrows\text{ES} \quad k_a, k_a'$$
$$\text{ES}\rightarrow\text{P}+\text{E} \quad k_b$$

$$\mathrm{EIn} \rightleftharpoons \mathrm{E} + \mathrm{In} \qquad K_\mathrm{I} = \frac{[\mathrm{E}][\mathrm{In}]}{[\mathrm{EI}]} \tag{20H.6a}$$

$$\mathrm{ESIn} \rightleftharpoons \mathrm{ES} + \mathrm{In} \qquad K'_\mathrm{I} = \frac{[\mathrm{ES}][\mathrm{In}]}{[\mathrm{ESIn}]} \tag{20H.6b}$$

The lower the values of K_I and K'_I the more efficient are the inhibitors. The rate of product formation is always given by $v = k_\mathrm{b}[\mathrm{ES}]$, because only ES leads to product. As shown in the following *Justification*, the rate of reaction in the presence of an inhibitor is

$$v = \frac{v_\mathrm{max}}{\alpha' + \alpha K_\mathrm{M}/[\mathrm{S}]_0} \qquad \text{Effect of inhibition on the rate} \tag{20H.7}$$

where $\alpha = 1 + [\mathrm{In}]/K_\mathrm{I}$ and $\alpha' = 1 + [\mathrm{In}]/K'_\mathrm{I}$. This equation is very similar to the Michaelis–Menten equation for the uninhibited enzyme (eqn 20H.1) and is also amenable to analysis by a Lineweaver–Burk plot:

$$\frac{1}{v} = \frac{\alpha'}{v_\mathrm{max}} + \left(\frac{\alpha K_\mathrm{M}}{v_\mathrm{max}}\right)\frac{1}{[\mathrm{S}]_0} \tag{20H.8}$$

Justification 20H.2 Enzyme inhibition

By mass balance, the total concentration of enzyme is:

$$[\mathrm{E}]_0 = [\mathrm{E}] + [\mathrm{EIn}] + [\mathrm{ES}] + [\mathrm{ESIn}]$$

By using eqns 20H.6a and 20H.6b and the definitions

$$\alpha = 1 + \frac{[\mathrm{In}]}{K_\mathrm{I}} \quad \text{and} \quad \alpha' = 1 + \frac{[\mathrm{In}]}{K'_\mathrm{I}}$$

it follows that

$$[\mathrm{E}]_0 = [\mathrm{E}]\alpha + [\mathrm{ES}]\alpha'$$

By using $K_\mathrm{M} = [\mathrm{E}][\mathrm{S}]/[\mathrm{ES}]$ and replacing [S] with $[\mathrm{S}]_0$ we can write

$$[\mathrm{E}]_0 = \frac{K_\mathrm{M}[\mathrm{ES}]}{[\mathrm{S}]_0}\alpha + [\mathrm{ES}]\alpha' = [\mathrm{ES}]\left(\frac{\alpha K_\mathrm{M}}{[\mathrm{S}]_0} + \alpha'\right)$$

The expression for the rate of product formation is then:

$$v = k_\mathrm{b}[\mathrm{ES}] = \frac{k_\mathrm{b}[\mathrm{E}]_0}{\alpha K_\mathrm{M}/[\mathrm{S}]_0 + \alpha'}$$

which, upon replacement of $k_\mathrm{b}[\mathrm{E}]_0$ with v_max, gives eqn 20H.7.

There are three major modes of inhibition that give rise to distinctly different kinetic behaviour (Fig. 20H.6). In **competitive inhibition** the inhibitor binds only to the active site of the enzyme and thereby inhibits the attachment of the substrate.

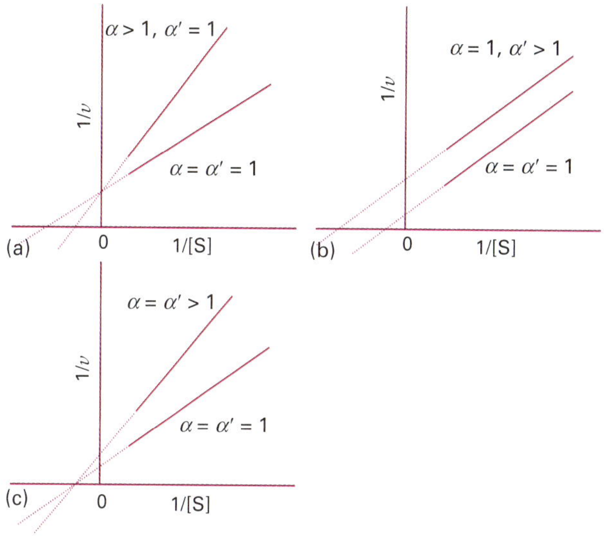

Figure 20H.6 Lineweaver–Burk plots characteristic of the three major modes of enzyme inhibition: (a) competitive inhibition, (b) uncompetitive inhibition, and (c) non-competitive inhibition, showing the special case $\alpha = \alpha' > 1$.

This condition corresponds to $\alpha > 1$ and $\alpha' = 1$ (because ESIn does not form). In this limit, eqn 20H.8 becomes

$$\frac{1}{v} = \frac{1}{v_\mathrm{max}} + \left(\frac{\alpha K_\mathrm{M}}{v_\mathrm{max}}\right)\frac{1}{[\mathrm{S}]_0} \qquad \text{Competitive inhibition}$$

The y-intercept is unchanged but the slope of the Lineweaver–Burk plot increases by a factor of α relative to the slope for data on the uninhibited enzyme (Fig. 20H.6a).

In **uncompetitive inhibition** the inhibitor binds to a site of the enzyme that is removed from the active site, but only if the substrate is already present. The inhibition occurs because ESI reduces the concentration of ES, the active type of complex. In this case $\alpha = 1$ (because EI does not form) and $\alpha' > 1$ and eqn 20H.8 becomes

$$\frac{1}{v} = \frac{\alpha'}{v_\mathrm{max}} + \left(\frac{K_\mathrm{M}}{v_\mathrm{max}}\right)\frac{1}{[\mathrm{S}]_0} \qquad \text{Uncompetitive inhibition}$$

The y-intercept of the Lineweaver–Burk plot increases by a factor of α' relative to the y-intercept for data on the uninhibited enzyme but the slope does not change (Fig. 20H.6b).

In **non-competitive inhibition** (also called **mixed inhibition**) the inhibitor binds to a site other than the active site, and its presence reduces the ability of the substrate to bind to the active site. Inhibition occurs at both the E and ES sites. This condition corresponds to $\alpha > 1$ and $\alpha' > 1$. Both the slope and y-intercept of the Lineweaver–Burk plot increase upon addition of the inhibitor. Figure 20H.6c shows the special case of $K_\mathrm{I} = K'_\mathrm{I}$ and $\alpha = \alpha'$, which results in intersection of the lines at the x-axis.

In all cases, the efficiency of the inhibitor may be obtained by determining K_M and v_{max} from a control experiment with uninhibited enzyme and then repeating the experiment with a known concentration of inhibitor. From the slope and *y*-intercept of the Lineweaver–Burk plot for the inhibited enzyme, the mode of inhibition, the values of α or α', and the values of K_I and K_I' may be obtained.

Example 20H.2 Distinguishing between types of inhibition

Five solutions of a substrate, S, were prepared with the concentrations given in the first column below and each one was divided into five equal volumes. The same concentration of enzyme was present in each one. An inhibitor, In, was then added in four different concentrations to the samples, and the initial rate of formation of product was determined with the results given below. Does the inhibitor act competitively or noncompetitively? Determine K_I and K_M.

$[S]_0$/(mmol dm^{-3})	[In]/(mmol dm^{-3})				
	0	0.20	0.40	0.60	0.80
	v/(μmol dm^{-3} s^{-1})				
0.050	0.033	0.026	0.021	0.018	0.016
0.10	0.055	0.045	0.038	0.033	0.029
0.20	0.083	0.071	0.062	0.055	0.050
0.40	0.111	0.100	0.091	0.084	0.077
0.60	0.116	0.116	0.108	0.101	0.094

Method Draw a series of Lineweaver–Burk plots for different inhibitor concentrations. If the plots resemble those in Fig. 20H.6a, then the inhibition is competitive. On the other hand, if the plots resemble those in Fig. 20H.6b, then the inhibition is non-competitive. To find K_I, we need to determine the slope at each value of [In], which is equal to $\alpha K_M/v_{max}$, or $K_M/v_{max}+K_M[\mathrm{In}]/K_I v_{max}$, then plot this slope against [In]: the intercept at [In] = 0 is the value of K_M/v_{max} and the slope is $K_M/K_I v_{max}$.

Answer First we draw up a table of $1/[S]_0$ and $1/v$ for each value of [I]:

$1/([S]_0$/(mmol dm^{-3}))	[In]/(mmol dm^{-3})				
	0	0.20	0.40	0.60	0.80
	$1/(v$/(μmol dm^{-3} s^{-1}))				
20	30	38	48	56	62
10	18	22	26	30	34
5.0	12	14	16	18	20
2.5	9.01	10.0	11.0	11.9	13.0
1.7	7.94	8.62	9.26	9.90	10.6

The five plots (one for each [In]) are given in Fig. 20H.7. We see that they pass through the same intercept on the vertical axis, so the inhibition is competitive.

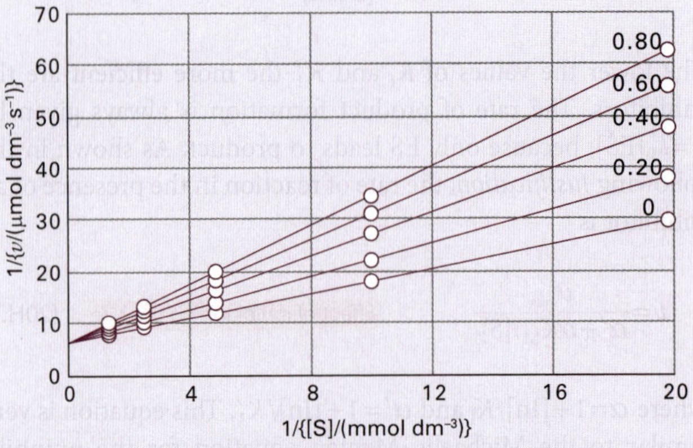

Figure 20H.7 Lineweaver–Burk plots for the data in *Example* 20H.2. Each line corresponds to a different concentration of inhibitor.

The mean of the (least squares) intercepts is 5.83, so $v_{max} = 0.172\,\mu\mathrm{mol\,dm^{-3}\,s^{-1}}$ (note how it picks up the units for v in the data). The (least squares) slopes of the lines are as follows:

[In]/(mmol dm^{-3})	0	0.20	0.40	0.60	0.80
Slope	1.219	1.627	2.090	2.489	2.832

These values are plotted in Fig. 20H.8. The intercept at [In] = 0 is 1.234, so $K_M = 0.212\,\mathrm{mmol\,dm^{-3}}$. The (least squares) slope of the line is 2.045, so

$$K_I/(\mathrm{mmol\,dm^{-3}}) = \frac{K_M}{\mathrm{slope} \times v_{max}} = \frac{0.212}{2.045 \times 0.172}$$

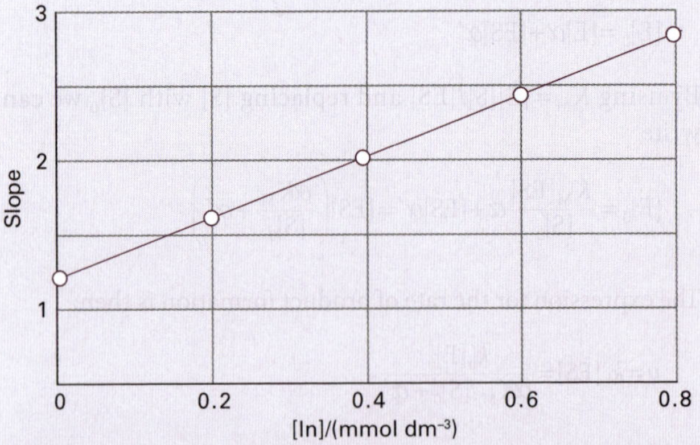

Figure 20H.8 Plot of the slopes of the plots in Fig. 20H.7 against [In] based on the data in *Example* 20H.2.

Self-test 20H.3 Repeat the question using the following data:

$[S]_0/(mmol\ dm^{-3})$	$[In]/(mmol\ dm^{-3})$				
	0	0.20	0.40	0.60	0.80
0.050	0.020	0.015	0.012	0.0098	0.0084
0.10	0.035	0.026	0.021	0.017	0.015
0.20	0.056	0.042	0.033	0.028	0.024
0.40	0.080	0.059	0.047	0.039	0.034
0.60	0.093	0.069	0.055	0.046	0.039

$v/(\mu mol\ dm^{-3}\ s^{-1})$

Answer: Non-competitive, $K_M = 0.30\ mmol\ dm^{-3}$, $K_I = 0.57\ mmol\ dm^{-3}$

Checklist of concepts

- ☐ 1. **Enzymes** are homogeneous biological catalysts.
- ☐ 2. The **Michaelis–Menten mechanism** of enzyme kinetics accounts for the dependence of rate on the concentration of the substrate and the enzyme.
- ☐ 3. A **Lineweaver–Burk plot** is used to determine the parameters that occur in the Michaelis–Menten mechanism.
- ☐ 4. In **competitive inhibition** of an enzyme, the inhibitor binds only to the active site of the enzyme.
- ☐ 5. In **uncompetitive inhibition** the inhibitor binds to a site of the enzyme that is removed from the active site, but only if the substrate is already present.
- ☐ 6. In **non-competitive inhibition**, the inhibitor binds to a site other than the active site.

Checklist of equations

Property	Equation	Comment	Equation number
Michaelis–Menten equation	$v = v_{max}/(1 + K_M/[S]_0)$		20H.3a
Lineweaver–Burk plot	$1/v = 1/v_{max} + (K_M/v_{max})(1/[S]_0)$		20H.3b
Turnover frequency	$k_{cat} = v_{max}/[E]_0$	Definition	20H.4
Catalytic efficiency	$\eta = k_{cat}/K_M$	Definition	20H.5
Effect of inhibition	$v = v_{max}/(\alpha' + \alpha K_M/[S]_0)$	Assumes Michaelis–Menten mechanism	20H.7

CHAPTER 20 Chemical kinetics

TOPIC 20A The rates of chemical reactions

Discussion question

20A.1 Summarize the characteristic of zeroth-order, first-order, second-order, and pseudofirst-order reactions.

20A.2 When can a reaction order not be ascribed?

20A.3 What are the advantages of ascribing an order to a reaction?

20A.4 Summarize the experimental procedures that can be used to monitor the composition of a reaction system.

Exercises

20A.1(a) Predict how the total pressure varies during the gas-phase reaction $2\,ICl(g)+H_2(g)\rightarrow I_2(g)+2\,HCl(g)$ in a constant-volume container.
20A.1(b) Predict how the total pressure varies during the gas-phase reaction $N_2(g)+3\,H_2(g)\rightarrow 2\,NH_3(g)$ in a constant-volume container.

20A.2(a) The rate of the reaction $A+2\,B\rightarrow 3\,C+D$ was reported as $2.7\,mol\,dm^{-3}\,s^{-1}$. State the rates of formation and consumption of the participants.
20A.2(b) The rate of the reaction $A+3\,B\rightarrow C+2\,D$ was reported as $2.7\,mol\,dm^{-3}\,s^{-1}$. State the rates of formation and consumption of the participants.

20A.3(a) The rate of formation of C in the reaction $2\,A+B\rightarrow 2\,C+3\,D$ is $2.7\,mol\,dm^{-3}\,s^{-1}$. State the reaction rate, and the rates of formation or consumption of A, B, and D.
20A.3(b) The rate of consumption of B in the reaction $A+3\,B\rightarrow C+2\,D$ is $2.7\,mol\,dm^{-3}\,s^{-1}$. State the reaction rate, and the rates of formation or consumption of A, C, and D.

20A.4(a) The rate law for the reaction in Exercise 20A.2(a) was found to be $v=k_r[A][B]$. What are the units of k_r when the concentrations are in moles per cubic decimetre? Express the rate law in terms of the rates of formation and consumption of (i) A, (ii) C.
20A.4(b) The rate law for the reaction in Exercise 20A.2(b) was found to be $v=k_r[A][B]^2$. What are the units of k_r when the concentrations are in moles per cubic decimetre? Express the rate law in terms of the rates of formation and consumption of (i) A, (ii) C.

20A.5(a) The rate law for the reaction in Exercise 20A.3(a) was reported as $d[C]/dt=k_r[A][B][C]$. Express the rate law in terms of the reaction rate v; what are the units for k_r in each case when the concentrations are in moles per cubic decimetre?
20A.5(b) The rate law for the reaction in Exercise 20A.3(b) was reported as $d[C]/dt=k_r[A][B][C]^{-1}$. Express the rate law in terms of the reaction rate v; what are the units for k_r in each case when the concentrations are in moles per cubic decimetre?

20A.6(a) If the rate laws are expressed with (i) concentrations in moles per cubic decimetre, (ii) pressures in kilopascals, what are the units of the second-order and third-order rate constants?
20A.6(b) If the rate laws are expressed with (i) concentrations in molecules per cubic metre, (ii) pressures in pascals, what are the units of the second-order and third-order rate constants?

Problems

20A.1 At 400 K, the rate of decomposition of a gaseous compound initially at a pressure of 12.6 kPa, was $9.71\,Pa\,s^{-1}$ when 10.0 per cent had reacted and $7.67\,Pa\,s^{-1}$ when 20.0 per cent had reacted. Determine the order of the reaction.

20A.2 The following initial-rate data were obtained on the rate of binding of glucose with the enzyme hexokinase present at a concentration of $1.34\,mmol\,dm^{-3}$. What is (a) the order of reaction with respect to glucose, (b) the rate constant?

$[C_6H_{12}O_6]/(mmol\,dm^{-3})$	1.00	1.54	3.12	4.02
$v_0/(mol\,dm^{-3}\,s^{-1})$	5.0	7.6	15.5	20.0

20A.3 The following data were obtained on the initial rates of a reaction of a d-metal complex with a reactant Y in aqueous solution. What is (a) the order of reaction with respect to the complex and Y, (b) the rate constant? For the experiments (i), $[Y]=2.7\,mmol\,dm^{-3}$ and for experiments (ii) $[Y]=6.1\,mmol\,dm^{-3}$.

$[complex]/(mmol\,dm^{-3})$		8.01	9.22	12.11
$v_0/(mol\,dm^{-3}\,s^{-1})$	(i)	125	144	190
	(ii)	640	730	960

20A.4 The following kinetic data (v_0 is the initial rate) were obtained for the reaction $2\,ICl(g)+H_2(g)\rightarrow I_2(g)+2\,HCl(g)$:

Experiment	$[ICl]_0/(mmol\,dm^{-3})$	$[H_2]_0/(mmol\,dm^{-3})$	$v_0/(mol\,dm^{-3}\,s^{-1})$
1	1.5	1.5	3.7×10^{-7}
2	3.0	1.5	7.4×10^{-7}
3	3.0	4.5	22×10^{-7}
4	4.7	2.7	?

(a) Write the rate law for the reaction. (b) From the data, determine the value of the rate constant. (c) Use the data to predict the reaction rate for experiment 4.

TOPIC 20B Integrated rate laws

Discussion questions

20B.1 Describe the main features, including advantages and disadvantages, of the following experimental methods for determining the rate law of a reaction: the isolation method, the method of initial rates, and fitting data to integrated rate law expressions.

20B.2 Write the rate law that corresponds to each of the following expressions: (a) $[A]=[A]_0-k_rt$, (b) $\ln([A]/[A]_0)=-k_rt$, and (c) $[A]=[A]_0/(1+k_rt[A]_0)$.

Exercises

20B.1(a) At 518 °C, the half-life for the decomposition of a sample of gaseous acetaldehyde (ethanal) initially at 363 Torr was 410 s. When the pressure was 169 Torr, the half-life was 880 s. Determine the order of the reaction.

20B.1(b) At 400 K, the half-life for the decomposition of a sample of a gaseous compound initially at 55.5 kPa was 340 s. When the pressure was 28.9 kPa, the half-life was 178 s. Determine the order of the reaction.

20B.2(a) The rate constant for the first-order decomposition of N_2O_5 in the reaction $2\,N_2O_5(g)\rightarrow 4\,NO_2(g)+O_2(g)$ is $k_r=3.38\times10^{-5}\,s^{-1}$ at 25 °C. What is the half-life of N_2O_5? What will be the pressure, initially 500 Torr, after (i) 50 s, (ii) 20 min after initiation of the reaction?

20B.2(b) The rate constant for the first-order decomposition of a compound A in the reaction $2\,A\rightarrow P$ is $k_r=3.56\times10^{-7}\,s^{-1}$ at 25 °C. What is the half-life of A? What will be the pressure, initially 33.0 kPa after (i) 50 s, (ii) 20 min after initiation of the reaction?

20B.3(a) The second-order rate constant for the reaction $CH_3COOC_2H_5(aq)+OH^-(aq)\rightarrow CH_3CO_2^-(aq)+CH_3CH_2OH(aq)$ is $0.11\,dm^3\,mol^{-1}\,s^{-1}$. What is the concentration of ester ($CH_3COOC_2H_5$) after (i) 20 s, (ii) 15 min when ethyl acetate is added to sodium hydroxide so that the initial concentrations are $[NaOH]=0.060\,mol\,dm^{-3}$ and $[CH_3COOC_2H_5]=0.110\,mol\,dm^{-3}$?

20B.3(b) The second-order rate constant for the reaction $A+2\,B\rightarrow C+D$ is $0.34\,dm^3\,mol^{-1}\,s^{-1}$. What is the concentration of C after (i) 20 s, (ii) 15 min when the reactants are mixed with initial concentrations of $[A]=0.027\,mol\,dm^{-1}$ and $[B]=0.130\,mol\,dm^{-3}$?

20B.4(a) A reaction $2\,A\rightarrow P$ has a second-order rate law with $k_r=4.30\times10^{-4}\,dm^3\,mol^{-1}\,s^{-1}$. Calculate the time required for the concentration of A to change from $0.210\,mol\,dm^{-3}$ to $0.010\,mol\,dm^{-3}$.

20B.4(b) A reaction $2\,A\rightarrow P$ has a third-order rate law with $k_r=6.50\times10^{-4}\,dm^6\,mol^{-2}\,s^{-1}$. Calculate the time required for the concentration of A to change from $0.067\,mol\,dm^{-3}$ to $0.015\,mol\,dm^{-3}$.

Problems

20B.1 For a first-order reaction of the form $A\rightarrow n\,B$ (with n possibly fractional), the concentration of the product varies with time as $[B]=n[B]_0(1-e^{-k_rt})$. Plot the time dependence of [A] and [B] for the cases $n=\frac{1}{2}$, 1, and 2.

20B.2 For a second-order reaction of the form $A\rightarrow n\,B$ (with n possibly fractional), the concentration of the product varies with time as $[B]=nk_rt[A]_0^2/(1+k_rt[A]_0)$. Plot the time dependence of [A] and [B] for the cases $n=\frac{1}{2}$, 1, and 2.

20B.3 The data below apply to the formation of urea from ammonium cyanate, $NH_4CNO\rightarrow NH_2CONH_2$. Initially 22.9 g of ammonium cyanate was dissolved in enough water to prepare 1.00 dm³ of solution. Determine the order of the reaction, the rate constant, and the mass of ammonium cyanate left after 300 min.

t/min	0	20.0	50.0	65.0	150
m(urea)/g	0	7.0	12.1	13.8	17.7

20B.4 The data below apply to the reaction, $(CH_3)_3CBr+H_2O\rightarrow(CH_3)_3COH+HBr$. Determine the order of the reaction, the rate constant, and the molar concentration of $(CH_3)_3CBr$ after 43.8 h.

t/h	0	3.15	6.20	10.00	18.30	30.80
$[(CH_3)_3CBr]/(10^{-2}\,mol\,dm^{-3})$	10.39	8.96	7.76	6.39	3.53	2.07

20B.5 The thermal decomposition of an organic nitrile produced the following data:

$t/(10^3\,s)$	0	2.00	4.00	6.00	8.00	10.00	12.00	∞
[nitrile]/(mol dm⁻³)	1.50	1.26	1.07	0.92	0.81	0.72	0.65	0.40

Determine the order of the reaction and the rate constant.

20B.6 A second-order reaction of the type $A+2\,B\rightarrow P$ was carried out in a solution that was initially $0.050\,mol\,dm^{-3}$ in A and $0.030\,mol\,dm^{-3}$ in B. After 1.0 h the concentration of A had fallen to $0.010\,mol\,dm^{-3}$. (a) Calculate the rate constant. (b) What is the half-life of the reactants?

20B.7‡ The oxidation of HSO_3^- by O_2 in aqueous solution is a reaction of importance to the processes of acid rain formation and flue gas desulfurization. R.E. Connick et al. (*Inorg. Chem.* **34**, 4543 (1995)) report that the reaction $2\,HSO_3^-(aq)+O_2(g)\rightarrow 2\,SO_4^{2-}(aq)+2\,H^+(aq)$ follows the rate law $v=k_r[HSO_4^-]^2[H^+]^2$. Given pH = 5.6 and an oxygen molar concentration of $0.24\,mmol\,dm^{-3}$ (both presumed constant), an initial HSO_3^- molar concentration of $50\,\mu mol\,dm^{-3}$, and a rate constant of $3.6\times10^6\,dm^9\,mol^{-3}\,s^{-1}$, what is the initial rate of reaction? How long would it take for HSO_3^- to reach half its initial concentration?

20B.8 Pharmacokinetics is the study of the rates of absorption and elimination of drugs by organisms. In most cases, elimination is slower than absorption and is a more important determinant of availability of a drug for binding to its target. A drug can be eliminated by many mechanisms, such as metabolism in the liver, intestine, or kidney followed by excretion of breakdown products through urine or faeces. As an example of pharmacokinetic analysis, consider the elimination of beta adrenergic blocking agents (beta blockers), drugs used in the treatment of hypertension. After intravenous administration of a beta blocker, the blood plasma of a patient was analysed for remaining drug and the data are shown below, where c is the drug concentration measured at a time t after the injection.

t/min	30	60	120	150	240	360	480
c/(ng cm⁻³)	699	622	413	292	152	60	24

(a) Is removal of the drug a first- or second-order process? (b) Calculate the rate constant and half-life of the process. *Comment*: An essential

‡ These problems were supplied by Charles Trapp and Carmen Giunta.

aspect of drug development is the optimization of the half-life of elimination, which needs to be long enough to allow the drug to find and act on its target organ but not so long that harmful side effects become important.

20B.9 The following data have been obtained for the decomposition of $N_2O_5(g)$ at 67 °C according to the reaction $2\,N_2O_5(g) \rightarrow 4\,NO_2(g) + O_2(g)$. Determine the order of the reaction, the rate constant, and the half-life. It is not necessary to obtain the result graphically; you may do a calculation using estimates of the rates of change of concentration.

t/min	0	1	2	3	4	5
$[N_2O_5]/(\text{mol dm}^{-3})$	1.000	0.705	0.497	0.349	0.246	0.173

20B.10 The gas phase decomposition of acetic acid at 1189 K proceeds by way of two parallel reactions:

(1) $CH_3COOH \rightarrow CH_4 + CO_2$ $\quad k_1 = 3.74\,s^{-1}$

(2) $CH_3COOH \rightarrow CH_2CO + H_2O$ $\quad k_2 = 4.65\,s^{-1}$

What is the maximum percentage yield of the ketene CH_2CO obtainable at this temperature?

20B.11 Sucrose is readily hydrolysed to glucose and fructose in acidic solution. The hydrolysis is often monitored by measuring the angle of rotation of plane-polarized light passing through the solution. From the angle of rotation the concentration of sucrose can be determined. An experiment on the hydrolysis of sucrose in 0.50 M HCl(aq) produced the following data:

t/min	0	14	39	60	80	110	140	170	210
[sucrose]/(mol dm^{-3})	0.316	0.300	0.274	0.256	0.238	0.211	0.190	0.170	0.146

Determine the rate constant of the reaction and the half-life of a sucrose molecule.

20B.12 The composition of a liquid phase reaction $2\,A \rightarrow B$ was followed by a spectrophotometric method with the following results:

t/min	0	10	20	30	40	∞
$[B]/(\text{mol dm}^{-3})$	0	0.089	0.153	0.200	0.230	0.312

Determine the order of the reaction and its rate constant.

20B.13 The ClO radical decays rapidly by way of the reaction, $2\,ClO \rightarrow Cl_2 + O_2$. The following data have been obtained:

t/ms	0.12	0.62	0.96	1.60	3.20	4.00	5.75
[ClO]/(μmol dm^{-3})	8.49	8.09	7.10	5.79	5.20	4.77	3.95

Determine the rate constant of the reaction and the half-life of a ClO radical.

20B.14 Cyclopropane isomerizes into propene when heated to 500 °C in the gas phase. The extent of conversion for various initial pressures has been followed by gas chromatography by allowing the reaction to proceed for a time with various initial pressures:

p_0/Torr	200	200	400	400	600	600
t/s	100	200	100	200	100	200
p/Torr	186	173	373	347	559	520

where p_0 is the initial pressure and p is the final pressure of cyclopropane. What is the order and rate constant for the reaction under these conditions?

20B.15 The addition of hydrogen halides to alkenes has played a fundamental role in the investigation of organic reaction mechanisms. In one study (M.J. Haugh and D.R. Dalton, *J. Amer. Chem. Soc.* 97, 5674 (1975)), high pressures of hydrogen chloride (up to 25 atm) and propene (up to 5 atm) were examined over a range of temperatures and the amount of 2-chloropropane formed was determined by NMR. Show that if the reaction $A + B \rightarrow P$ proceeds for a short time δt, the concentration of product follows $[P]/[A] = k_r[A]^{m-1}[B]^n\delta t$ if the reaction is mth-order in A and nth-order in B. In a series of runs the ratio of [chloropropane] to [propene] was independent of [propene] but the ratio of [chloropropane] to [HCl] for constant amounts of propene depended on [HCl]. For $\delta t \approx 100$ h (which is short on the time scale of the reaction) the latter ratio rose from zero to 0.05, 0.03, 0.01 for $p(HCl) = 10$ atm, 7.5 atm, 5.0 atm. What are the orders of the reaction with respect to each reactant?

20B.16 Show that $t_{1/2}$ is given by eqn 20B.6 for a reaction that is nth order in A. Then deduce an expression for the time it takes for the concentration of a substance to fall to one-third the initial value in an nth-order reaction.

20B.17 Derive an integrated expression for a second-order rate law $v = k_r[A][B]$ for a reaction of stoichiometry $2\,A + 3\,B \rightarrow P$.

20B.18 Derive the integrated form of a third-order rate law $v = k_r[A]^2[B]$ in which the stoichiometry is $2\,A + B \rightarrow P$ and the reactants are initially present in (a) their stoichiometric proportions, (b) with B present initially in twice the amount.

20B.19 Show that the ratio $t_{1/2}/t_{3/4}$, where $t_{1/2}$ is the half-life and $t_{3/4}$ is the time for the concentration of A to decrease to $^3/_4$ of its initial value (implying that $t_{3/4} < t_{1/2}$), can be written as a function of n alone, and can therefore be used as a rapid assessment of the order of a reaction.

TOPIC 20C Reactions approaching equilibrium

Discussion questions

20C.1 Describe the strategy of a temperature-jump experiment. What parameters of a reaction are accessible from this technique?

20C.2 What feature of a reaction would ensure that its rate would respond to a pressure jump?

Exercises

20C.1(a) The equilibrium $NH_3(aq) + H_2O(l) \rightleftharpoons NH_4^+(aq) + OH^-(aq)$ at 25 °C is subjected to a temperature jump which slightly increased the concentration of $NH_4^+(aq)$ and $OH^-(aq)$. The measured relaxation time is 7.61 ns. The equilibrium constant for the system is 1.78×10^{-5} at 25 °C, and the equilibrium concentration of $NH_3(aq)$ is 0.15 mol dm^{-3}. Calculate the rate constants for the forward and reverse steps.

20C.1(b) The equilibrium $A \rightleftharpoons B + C$ at 25 °C is subjected to a temperature jump which slightly increases the concentrations of B and C. The measured relaxation time is 3.0 μs. The equilibrium constant for the system is 2.0×10^{-16} at 25 °C, and the equilibrium concentrations of B and C at 25 °C are both 0.20 mmol dm^{-3}. Calculate the rate constants for the forward and reverse steps.

Problems

20C.1 Show by differentiation that eqn 20C.4 is a solution of eqn 20C.3.

20C.2 Set up the rate equations and plot the corresponding graphs for the approach to an equilibrium of the form $A \rightleftharpoons 2B$.

20C.3 The reaction $A \rightleftharpoons 2B$ is first-order in both directions. Derive an expression for the concentration of A as a function of time when the initial molar concentrations of A and B are $[A]_0$ and $[B]_0$. What is the final composition of the system?

20C.4 Show that eqn 20C.8 is an expression for the overall equilibrium constant in terms of the rate constants for the intermediate steps of a reaction mechanism. To facilitate the task, begin with a mechanism containing three steps, and then argue that your expression may be generalized for any number of steps.

20C.5 Consider the dimerization $2A \rightleftarrows A_2$, with forward rate constant k_a and backward rate constant k_a'. (a) Derive the following expression for the relaxation time in terms of the total concentration of protein, $[A]_{tot} = [A] + 2[A_2]$:

$$\frac{1}{\tau^2} = k_a'^2 + 8k_a k_a'[A]_{tot}$$

(b) Describe the computational procedures that lead to the determination of the rate constants k_a and k_a' from measurements of τ for different values of $[A]_{tot}$. (c) Use the data provided below and the procedure you outlined in part (b) to calculate the rate constants k_a and k_a', and the equilibrium constant K for formation of hydrogen-bonded dimers of 2-pyridone:

$[P]/(\text{mol dm}^{-3})$	0.500	0.352	0.251	0.151	0.101
τ/ns	2.3	2.7	3.3	4.0	5.3

20C.6 Consider the dimerization $2A \rightleftarrows A_2$ with forward rate constant k_r and backward rate constant k_r'. Show that the relaxation time is $\tau = 1/(k_r' + 4k_r[A]_{eq})$.

TOPIC 20D The Arrhenius equation

Discussion question

20D.1 Define the terms in and discuss the validity of the expression $\ln k_r = \ln A - E_a/RT$.

20D.2 What might account for the failure of the Arrhenius equation at low temperatures?

Exercises

20D.1(a) The rate constant for the decomposition of a certain substance is 3.80×10^{-3} dm^3 mol^{-1} s^{-1} at 35 °C and 2.67×10^{-2} dm^3 mol^{-1} s^{-1} at 50 °C. Evaluate the Arrhenius parameters of the reaction.

20D.1(b) The rate constant for the decomposition of a certain substance is 2.25×10^{-2} dm^3 mol^{-1} s^{-1} at 29 °C and 4.01×10^{-2} dm^3 mol^{-1} s^{-1} at 37 °C. Evaluate the Arrhenius parameters of the reaction.

20D.2(a) The rate of a chemical reaction is found to triple when the temperature is raised from 24 °C to 49 °C. Determine the activation energy.

20D.2(b) The rate of a chemical reaction is found to double when the temperature is raised from 25 °C to 35 °C. Determine the activation energy.

Problems

20D.1 Show that the definition of E_a given in eqn 20D.3 reduces to eqn 20D.1 for a temperature-independent activation energy.

20D.2 A first-order decomposition reaction is observed to have the following rate constants at the indicated temperatures. Estimate the activation energy.

$k_r/(10^{-3}\ \text{s}^{-1})$	2.46	45.1	576
θ/°C	0	20.0	40.0

20D.3 The second-order rate constants for the reaction of oxygen atoms with aromatic hydrocarbons have been measured (R. Atkinson and J.N. Pitts, *J. Phys. Chem.* **79**, 295 (1975)). In the reaction with benzene the rate constants are 1.44×10^{7} dm^3 mol^{-1} s^{-1} at 300.3 K, 3.03×10^{7} dm^3 mol^{-1} s^{-1} at 341.2 K, and 6.9×10^{7} dm^3 mol^{-1} s^{-1} at 392.2 K. Find the pre-exponential factor and activation energy of the reaction.

20D.4‡ Methane is a by-product of a number of natural processes (such as digestion of cellulose in ruminant animals, anaerobic decomposition of organic waste matter), and industrial processes (such as food production and fossil fuel use). Reaction with the hydroxyl radical OH is the main path by which CH_4 is removed from the lower atmosphere. T. Gierczak et al. (*J. Phys. Chem. A* **101**, 3125 (1997)) measured the rate constants for the elementary bimolecular gas-phase reaction of methane with the hydroxyl radical over a range of temperatures of importance to atmospheric chemistry. Deduce the Arrhenius parameters A and E_a from the following measurements:

T/K	295	223	218	213	206	200	195
$k_r/(10^{6}\ \text{dm}^3\ \text{mol}^{-1}\ \text{s}^{-1})$	3.55	0.494	0.452	0.379	0.295	0.241	0.217

20D.5‡ As we saw in Problem 20D.4, reaction with the hydroxyl radical OH is the main path by which CH_4, a by-product of many natural and industrial processes, is removed from the lower atmosphere. T. Gierczak et al. (*J. Phys. Chem. A* **101**, 3125 (1997)) measured the rate constants for the bimolecular gas-phase reaction $CH_4(g) + OH(g) \rightarrow CH_3(g) + H_2O(g)$ and found $A = 1.13\times10^9\,dm^3\,mol^{-1}\,s^{-1}$ and $E_a = 14.1\,kJ\,mol^{-1}$ for the Arrhenius parameters. (a) Estimate the rate of consumption of CH_4. Take the average OH concentration to be $1.5\times10^{-21}\,mol\,dm^{-3}$, that of CH_4 to be 40 nmol dm^{-3}, and the temperature to be −10 °C. (b) Estimate the global annual mass of CH_4 consumed by this reaction (which is slightly less than the amount introduced to the atmosphere) given an effective volume for the Earth's lower atmosphere of $4\times10^{21}\,dm^3$.

TOPIC 20E Reaction mechanisms

Discussion questions

20E.1 Distinguish between reaction order and molecularity.

20E.2 Assess the validity of the statement that the rate-determining step is the slowest step in a reaction mechanism.

20E.3 Distinguish between a pre-equilibrium approximation and a steady-state approximation. Why might they lead to different conclusions?

20E.4 Explain and illustrate how reaction orders may change under different circumstances.

20E.5 Distinguish between kinetic and thermodynamic control of a reaction. Suggest criteria for expecting one rather than the other.

20E.6 Explain how it is possible for the activation energy of a reaction to be negative.

Exercises

20E.1(a) The reaction mechanism for the decomposition of A_2

$$A_2 \rightleftharpoons A + A\ (\text{fast})$$
$$A + B \rightarrow P\ (\text{slow})$$

involves an intermediate A. Deduce the rate law for the reaction in two ways by (i) assuming a pre-equilibrium and (ii) making a steady-state approximation.

20E.1(b) The reaction mechanism for renaturation of a double helix from its strands A and B:

$$A + B \rightleftharpoons \text{unstable helix}\ (\text{fast})$$
$$\text{Unstable helix} \rightarrow \text{stable double helix}\ (\text{slow})$$

involves an intermediate. Deduce the rate law for the reaction in two ways by (i) assuming a pre-equilibrium and (ii) making a steady-state approximation.

20E.2(a) The mechanism of a composite reaction consists of a fast pre-equilibrium step with forward and reverse activation energies of 25 kJ mol^{-1} and 38 kJ mol^{-1}, respectively, followed by an elementary step of activation energy 10 kJ mol^{-1}. What is the activation energy of the composite reaction?

20E.2(b) The mechanism of a composite reaction consists of a fast pre-equilibrium step with forward and reverse activation energies of 27 kJ mol^{-1} and 35 kJ mol^{-1}, respectively, followed by an elementary step of activation energy 15 kJ mol^{-1}. What is the activation energy of the composite reaction?

Problems

20E.1 Use mathematical software or a spreadsheet to examine the time dependence of [I] in the reaction mechanism $A \xrightarrow{k_a} I \xrightarrow{k_b} P$. In all of the following calculations, use $[A]_0 = 1\,mol\,dm^{-3}$ and a time range of 0 to 5 s. (a) Plot [In] against t for $k_a = 10\,s^{-1}$ and $k_b = 1\,s^{-1}$. (b) Increase the ratio k_b/k_a steadily by decreasing the value of k_a and examine the plot of [I] against t at each turn. What approximation about d[I]/dt becomes increasingly valid?

20E.2 Use mathematical software or a spreadsheet to investigate the effects on [A], [I], [P], and t_{max} of decreasing the ratio k_a/k_b from 10 (as in Fig. 20E.1) to 0.01. Compare your results with those shown in Fig. 20E.3.

20E.3 Set up the rate equations for the reaction mechanism:

$$A \underset{k_a'}{\overset{k_a}{\rightleftarrows}} B \underset{k_b'}{\overset{k_b}{\rightleftarrows}} C$$

Show that, under specific circumstances, the mechanism is equivalent to

$$A \underset{k_r'}{\overset{k_r}{\rightleftarrows}} C$$

20E.4 Derive an equation for the steady state rate of the sequence of reactions $A \rightleftarrows B \rightleftarrows C \rightleftarrows D$, with [A] maintained at a fixed value and the product D removed as soon as it is formed.

20E.5 Show that the following mechanism can account for the rate law of the reaction in Problem 20B.15:

$HCl + HCl \rightleftharpoons (HCl)_2$	K_1
$HCl + CH_3CH{=}CH_2 \rightleftharpoons \text{complex}$	K_2
$(HCl)_2 + \text{complex} \rightarrow CH_3CHClCH_3 + HCl + HCl$	k_r (slow)

What further tests could you apply to verify this mechanism?

20E.6 Polypeptides are polymers of amino acids. Suppose that a long polypeptide chain can undergo a transition from a helical conformation to a random coil. Consider a mechanism for a helix–coil transition that begins in the middle of the chain:

hhhh... ⇄ *hchh*...
hchh... ⇄ *cccc*...

in which h and c label, respectively, an amino acid in a helical or coil part of the chain. The first conversion from h to c, also called a nucleation step, is relatively slow, so neither step may be rate determining. (a) Set up the rate equations for this mechanism. (b) Apply the steady-state approximation and show that, under these circumstances, the mechanism is equivalent to *hhhh*... ⇄ *cccc*...

TOPIC 20F Examples of reaction mechanisms

Discussion questions

20F.1 Discuss the range of validity of the expression $k_r = k_a k_b[A]/(k_b + k_a'[A])$ for the effective rate constant of a unimolecular reaction according to the Lindemann–Hinshelwood mechanism.

20F.2 Bearing in mind distinctions between the mechanisms of stepwise and chain polymerization, describe ways in which it is possible to control the molar mass of a polymer by manipulating the kinetic parameters of polymerization.

Exercises

20F.1(a) The effective rate constant for a gaseous reaction which has a Lindemann–Hinshelwood mechanism is $2.50\times10^{-4}\,s^{-1}$ at 1.30 kPa and $2.10\times10^{-5}\,s^{-1}$ at 12 Pa. Calculate the rate constant for the activation step in the mechanism.
20F.1(b) The effective rate constant for a gaseous reaction which has a Lindemann–Hinshelwood mechanism is $1.7\times10^{-3}\,s^{-1}$ at 1.09 kPa and $2.2\times10^{-4}\,s^{-1}$ at 25 Pa. Calculate the rate constant for the activation step in the mechanism.

20F.2(a) Calculate the fraction condensed and the degree of polymerization at $t = 5.00$ h of a polymer formed by a stepwise process with $k_r = 1.39\,dm^3\,mol^{-1}\,s^{-1}$ and an initial monomer concentration of $10.0\,mmol\,dm^{-3}$.

20F.2(b) Calculate the fraction condensed and the degree of polymerization at $t = 10.00$ h of a polymer formed by a stepwise process with $k_r = 2.80\times10^{-2}\,dm^3\,mol^{-1}\,s^{-1}$ and an initial monomer concentration of $50.0\,mmol\,dm^{-3}$.

20F.3(a) Consider a polymer formed by a chain process. By how much does the kinetic chain length change if the concentration of initiator increases by a factor of 3.6 and the concentration of monomer decreases by a factor of 4.2?
20F.3(b) Consider a polymer formed by a chain process. By how much does the kinetic chain length change if the concentration of initiator decreases by a factor of 10.0 and the concentration of increases by a factor of 5.0?

Problems

20F.1 The isomerization of cyclopropane over a limited pressure range was examined in Problem 20B.14. If the Lindemann mechanism of unimolecular reactions is to be tested we also need data at low pressures. These have been obtained (H.O. Pritchard et al., *Proc. R. Soc. A* **217**, 563 (1953)):

p/Torr	84.1	11.0	2.89	0.569	0.120	0.067
$10^4\,k_r/s^{-1}$	2.98	2.23	1.54	0.857	0.392	0.303

Test the Lindemann–Hinshelwood theory with these data.

20F.2 Calculate the average polymer length in a polymer produced by a chain mechanism in which termination occurs by a disproportionation reaction of the form $M\cdot + \cdot M \rightarrow M + {:}M$.

20F.3 Derive an expression for the time dependence of the degree of polymerization for a stepwise polymerization in which the reaction is acid-catalysed by the –COOH acid functional group. The rate law is $d[A]/dt = -k_r[A]^2[OH]$.

TOPIC 20G Photochemistry

Discussion questions

20G.1 Consult literature sources and list the observed ranges of timescales during which the following processes occur: radiative decay of excited electronic states, molecular rotational motion, molecular vibrational motion, proton transfer reactions, energy transfer between fluorescent molecules used in FRET analysis, electron transfer events between complex ions in solution, and collisions in liquids.

20G.2 Discuss experimental procedures that make it possible to differentiate between quenching by energy transfer, collisions, and electron transfer.

Exercises

20G.1(a) In a photochemical reaction $A \rightarrow 2\,B + C$, the quantum yield with 500 nm light is $210\,mol\,einstein^{-1}$ (1 einstein = 1 mol photons). After exposure of 300 mmol of A to the light, 2.28 mmol of B is formed. How many photons were absorbed by A?
20G.1(b) In a photochemical reaction $A \rightarrow B + C$, the quantum yield with 500 nm light is $120\,mol\,einstein^{-1}$ (1 einstein = 1 mol photons). After exposure of 200 mmol A to the light, 1.77 mmol B is formed. How many photons were absorbed by A?

20G.2(a) Consider the quenching of an organic fluorescent species with $\tau_0 = 6.0$ ns by a d-metal ion with $k_Q = 3.0\times10^8\,dm^3\,mol^{-1}\,s^{-1}$. Predict the concentration of quencher required to decrease the fluorescence intensity of the organic species to 50 per cent of the unquenched value.

20G.2(b) Consider the quenching of an organic fluorescent species with $\tau_0 = 3.5$ ns by a d-metal ion with $k_Q = 2.5\times10^9\,dm^3\,mol^{-1}\,s^{-1}$. Predict the concentration of quencher required to decrease the fluorescence intensity of the organic species to 75 per cent of the unquenched value.

Problems

20G.1 In an experiment to measure the quantum yield of a photochemical reaction, the absorbing substance was exposed to 320 nm radiation from a 87.5 W source for 28.0 min. The intensity of the transmitted radiation was 0.257 that of the incident radiation. As a result of irradiation, 0.324 mol of the absorbing substance decomposed. Determine the quantum yield.

20G.2‡ Ultraviolet radiation photolyses O_3 to O_2 and O. Determine the rate at which ozone is consumed by 305 nm radiation in a layer of the stratosphere of thickness 1.0 km. The quantum yield is 0.94 at 220 K, the concentration about 8 nmol dm^{-3}, the molar absorption coefficient 260 dm^3 mol^{-1} cm^{-1}, and the flux of 305 nm radiation about 1×10^{14} photons cm^{-2} s^{-1}. Data from W.B. DeMore et al., *Chemical kinetics and photochemical data for use in stratospheric modeling: Evaluation Number 11*, JPL Publication 94–26 (1994).

20G.3 Dansyl chloride, which absorbs maximally at 330 nm and fluoresces maximally at 510 nm, can be used to label amino acids in fluorescence microscopy and FRET studies. Tabulated below is the variation of the fluorescence intensity of an aqueous solution of dansyl chloride with time after excitation by a short laser pulse (with I_0 the initial fluorescence intensity). The ratio of intensities is equal to the ratio of the rates of photon emission.

t/ns	5.0	10.0	15.0	20.0
I_f/I_0	0.45	0.21	0.11	0.05

(a) Calculate the observed fluorescence lifetime of dansyl chloride in water.
(b) The fluorescence quantum yield of dansyl chloride in water is 0.70. What is the fluorescence rate constant?

20G.4 When benzophenone is exposed to ultraviolet radiation it is excited into a singlet state. This singlet changes rapidly into a triplet, which phosphoresces. Triethylamine acts as a quencher for the triplet. In an experiment in the solvent methanol, the phosphorescence intensity varied with amine concentration as shown below. A time-resolved laser spectroscopy experiment had also shown that the half-life of the fluorescence in the absence of quencher is 29 μs. What is the value of k_Q?

[Q]/(mmol dm^{-3})	1.0	5.0	10.0
I_f/(arbitrary units)	0.41	0.25	0.16

20G.5 An electronically excited state of Hg can be quenched by N_2 according to $Hg^*(g) + N_2(g, v=0) \rightarrow Hg(g) + N_2(g, v=1)$ in which energy transfer from Hg* excites N_2 vibrationally. Fluorescence lifetime measurements of samples of Hg with and without N_2 present are summarized below (for $T = 300$ K):

$p_{N_2} = 0.0$ atm

Relative fluorescence intensity	1.000	0.606	0.360	0.22	0.135
t/μs	0.0	5.0	10.0	15.0	20.0

$p_{N_2} = 9.74\times10^{-4}$ atm

Relative fluorescence intensity	1.000	0.585	0.342	0.200	0.117
t/μs	0.0	3.0	6.0	9.0	12.0

You may assume that all gases are perfect. Determine the rate constant for the energy transfer process.

20G.6 An amino acid on the surface of an enzyme was labelled covalently with 1.5-I AEDANS and it is known that the active site contains a tryptophan residue. The fluorescence quantum yield of tryptophan decreased by 15 per cent due to quenching by 1.5-I AEDANS. What is the distance between the active site and the surface of the enzyme?

20G.7 The Förster theory of resonance energy transfer and the basis for the FRET technique can be tested by performing fluorescence measurements on a series of compounds in which an energy donor and an energy acceptor are covalently linked by a rigid molecular linker of variable and known length. L. Stryer and R.P. Haugland (*Proc. Natl. Acad. Sci. USA* **58**, 719 (1967)) collected the following data on a family of compounds with the general composition dansyl-(l-prolyl)$_n$-naphthyl, in which the distance R between the naphthyl donor and the dansyl acceptor was varied from 1.2 nm to 4.6 nm by increasing the number of prolyl units in the linker:

R/nm	1.2	1.5	1.8	2.8	3.1	3.4	3.7	4.0	4.3	4.6
η_T	0.99	0.94	0.97	0.82	0.74	0.65	0.40	0.28	0.24	0.16

Are the data described adequately by eqn 20G.10? If so, what is the value of R_0 for the naphthyl–dansyl pair?

20G.8 The first step in plant photosynthesis is absorption of light by chlorophyll molecules bound to proteins known as 'light-harvesting complexes', where the fluorescence of a chlorophyll molecule is quenched by nearby chlorophyll molecules. Given that for a pair of chlorophyll *a* molecules $R_0 = 5.6$ nm, by what distance should two chlorophyll *a* molecules be separated to shorten the fluorescence lifetime from 1 ns (a typical value for monomeric chlorophyll *a* in organic solvents) to 10 ps?

TOPIC 20H Enzymes

Discussion questions

20H.1 Discuss the features, advantages, and limitations of the Michaelis–Menten mechanism of enzyme action.

20H.2 A plot of the rate of an enzyme-catalysed reaction against temperature has a maximum, in an apparent deviation from the behaviour predicted by the Arrhenius equation (Topic 20D). Suggest an interpretation.

20H.3 Distinguish between competitive, non-competitive, and uncompetitive inhibition of enzymes. Discuss how these modes of inhibition may be detected experimentally.

20H.4 Some enzymes are inhibited by high concentrations of their own products. Sketch a plot of reaction rate against concentration of substrate for an enzyme that is prone to product inhibition.

Exercises

20H.1(a) Consider the base-catalysed reaction

$$(1)\ \mathrm{AH}+B \underset{k_a'}{\overset{k_a}{\rightleftarrows}} \mathrm{BH}^+ + A^- \quad \text{(both fast)}$$

$$(2)\ \mathrm{A}^- + \mathrm{AH} \xrightarrow{k_b} \text{product} \quad \text{(slow)}$$

Deduce the rate law.

20H.1(b) Consider the acid-catalysed reaction

$$(1)\ \mathrm{HA}+\mathrm{H}^+ \underset{k_a'}{\overset{k_a}{\rightleftarrows}} \mathrm{HAH}^+ \quad \text{(both fast)}$$

$$(2)\ \mathrm{HAH}^+ + \mathrm{B} \xrightarrow{k_b} \mathrm{BH}^+ + \mathrm{AH} \quad \text{(slow)}$$

Deduce the rate law.

20H.2(a) The enzyme-catalysed conversion of a substrate at 25 °C has a Michaelis constant of 0.046 mol dm^{-3}. The rate of the reaction is 1.04 mmol dm^{-3} s^{-1} when the substrate concentration is 0.105 mol dm^{-3}. What is the maximum velocity of this reaction?

20H.2(b) The enzyme-catalysed conversion of a substrate at 25 °C has a Michaelis constant of 0.032 mol dm^{-3}. The rate of the reaction is 0.205 mmol dm^{-3} s^{-1} when the substrate concentration is 0.875 mol dm^{-3}. What is the maximum velocity of this reaction?

20H.3(a) Consider an enzyme-catalysed reaction that follows Michaelis–Menten kinetics with $K_M = 3.0$ mmol dm^{-3}. What concentration of a competitive inhibitor characterized by $K_I = 20$ μmol dm^{-3} will reduce the rate of formation of product by 50 per cent when the substrate concentration is held at 0.10 mmol dm^{-3}?

20H.3(b) Consider an enzyme-catalysed reaction that follows Michaelis–Menten kinetics with $K_M = 0.75$ mmol dm^{-3}. What concentration of a competitive inhibitor characterized by $K_I = 0.56$ mmol dm^{-3} will reduce the rate of formation of product by 75 per cent when the substrate concentration is held at 0.10 mmol dm^{-3}?

Problems

20H.1 Michaelis and Menten derived their rate law by assuming a rapid pre-equilibrium of E, S, and ES. Derive the rate law in this manner, and identify the conditions under which it becomes the same as that based on the steady-state approximation (eqn 20H.1).

20H.2 (a) Use the Michaelis–Menten equation (eqn 20H.1) to generate two families of curves showing the dependence of v on [S]: one in which K_M varies but v_{max} is constant, and another in which v_{max} varies but K_M is constant. (b) Use eqn 20H.7 to explore the effect of competitive, uncompetitive, and non-competitive inhibition on the shapes of the plots of v against [S] for constant K_M and v_{max}. Use mathematical software, a spreadsheet, or the *Living graphs* on the web site of this book.

20H.3 For many enzymes, the mechanism of action involves the formation of two intermediates:

$\mathrm{E+S \rightarrow ES}$	$v = k_a[\mathrm{E}][\mathrm{S}]$
$\mathrm{ES \rightarrow E+S}$	$v = k_a'[\mathrm{ES}]$
$\mathrm{ES \rightarrow ES'}$	$v = k_b[\mathrm{ES}]$
$\mathrm{ES' \rightarrow E+P}$	$v = k_c[\mathrm{ES'}]$

Show that the rate of formation of product has the same form as that shown in eqn 20H.1, but with v_{max} and K_M given by

$$v_{max} = \frac{k_b k_c [\mathrm{E}]_0}{k_b + k_c} \quad \text{and} \quad K_M = \frac{k_c(k_a' + k_b)}{k_a(k_b + k_c)}$$

20H.4 The enzyme-catalysed conversion of a substrate at 25 °C has a Michaelis constant of 90 μmol dm^{-3} and a maximum velocity of 22.4 μmol dm^{-3} s^{-1} when the enzyme concentration is 1.60 nmol dm^{-3}. (a) Calculate k_{cat} and η. (b) Is the enzyme 'catalytically perfect'?

20H.5 The following results were obtained for the action of an ATPase on ATP at 20 °C, when the concentration of the ATPase was 20 nmol dm^{-3}:

[ATP]/(μmol dm^{-3})	0.60	0.80	1.4	2.0	3.0
v/(μmol dm^{-3} s^{-1})	0.81	0.97	1.30	1.47	1.69

Determine the Michaelis constant, the maximum velocity of the reaction, the turnover number, and the catalytic efficiency of the enzyme.

20H.6 Some enzymes are inhibited by high concentrations of their own substrates. (a) Show that when substrate inhibition is important the reaction rate v is given by

$$v = \frac{v_{max}}{1 + K_M/[\mathrm{S}]_0 + [\mathrm{S}]_0/K_I}$$

where K_I is the equilibrium constant for dissociation of the inhibited enzyme–substrate complex. (b) What effect does substrate inhibition have on a plot of $1/v$ against $1/[\mathrm{S}]_0$?

Integrated activities

20.1 Autocatalysis is the catalysis of a reaction by the products. For example, for a reaction A→P it may be found that the rate law is $v = k_r[\mathrm{A}][\mathrm{P}]$ and the reaction rate is proportional to the concentration of P. The reaction gets started because there are usually other reaction routes for the formation of some P initially, which then takes part in the autocatalytic reaction proper. (a) Integrate the rate equation for an autocatalytic reaction of the form A→P, with rate law $v = k_r[\mathrm{A}][\mathrm{P}]$, and show that

$$\frac{[\mathrm{P}]}{[\mathrm{P}]_0} = \frac{(1+b)\mathrm{e}^{at}}{1+b\mathrm{e}^{at}}$$

where $a = ([\mathrm{A}]_0 + [\mathrm{P}]_0)k_r$ and $b = [\mathrm{P}]_0/[\mathrm{A}]_0$. *Hint*: Starting with the expression $v = -\mathrm{d}[\mathrm{A}]/\mathrm{d}t = k_r[\mathrm{A}][\mathrm{P}]$, write $[\mathrm{A}] = [\mathrm{A}]_0 - x$, $[\mathrm{P}] = [\mathrm{P}]_0 + x$ and then write the expression for the rate of change of either species in terms of x. To integrate the resulting expression, use integration by the method of partial fractions

(*The chemist's toolbox* 20B.1). (b) Plot $[P]/[P]_0$ against at for several values of b. Discuss the effect of autocatalysis on the shape of a plot of $[P]/[P]_0$ against t by comparing your results with those for a first-order process, in which $[P]/[P]_0 = 1 - e^{-k_r t}$. (c) Show that for the autocatalytic process discussed in parts (a) and (b), the reaction rate reaches a maximum at $t_{max} = -(1/a)\ln b$. (d) An autocatalytic reaction $A \to P$ is observed to have the rate law $d[P]/dt = k_r[A]^2[P]$. Solve the rate law for initial concentrations $[A]_0$ and $[P]_0$. Calculate the time at which the rate reaches a maximum. (e) Another reaction with the stoichiometry $A \to P$ has the rate law $d[P]/dt = k_r[A][P]^2$; integrate the rate law for initial concentrations $[A]_0$ and $[P]_0$. Calculate the time at which the rate reaches a maximum.

20.2 Many biological and biochemical processes involve autocatalytic steps (Problem 20.1). In the SIR model of the spread and decline of infectious diseases the population is divided into three classes; the 'susceptibles', S, who can catch the disease, the 'infectives', I, who have the disease and can transmit it, and the 'removed class', R, who have either had the disease and recovered, are dead, are immune, or isolated. The model mechanism for this process implies the following rate laws:

$$\frac{dS}{dt} = -rSI \qquad \frac{dI}{dt} = rSI - aI \qquad \frac{dR}{dt} = aI$$

Which are the autocatalytic steps of this mechanism? Find the conditions on the ratio a/r that decide whether the disease will spread (an epidemic) or die out. Show that a constant population is built into this system, namely that $S + I + R = N$, meaning that the time scales of births, deaths by other causes, and migration are assumed large compared to that of the spread of the disease.

20.3‡ J. Czarnowski and H.J. Schuhmacher (*Chem. Phys. Lett.* **17**, 235 (1972)) suggested the following mechanism for the thermal decomposition of F_2O in the reaction $2\,F_2O(g) \to 2\,F_2(g) + O_2(g)$:

(1) $F_2O + F_2O \to F + OF + F_2O$	k_a
(2) $F + F_2O \to F_2 + OF$	k_b
(3) $OF + OF \to O_2 + F + F$	k_c
(4) $F + F + F_2O \to F_2 + F_2O$	k_d

(a) Using the steady-state approximation, show that this mechanism is consistent with the experimental rate law $-d[F_2O]/dt = k_r[F_2O]^2 + k_r'[F_2O]^{3/2}$. (b) The experimentally determined Arrhenius parameters in the range 501–583 K are $A = 7.8 \times 10^{13}\,dm^3\,mol^{-1}\,s^{-1}$, $E_a/R = 1.935 \times 10^4\,K$ for k_r and $A = 2.3 \times 10^{10}\,dm^3\,mol^{-1}\,s^{-1}$, $E_a/R = 1.691 \times 10^4\,K$ for k_r'. At 540 K, $\Delta_f H^{\ominus}(F_2O) = +24.41\,kJ\,mol^{-1}$, $D(F\text{–}F) = 160.6\,kJ\,mol^{-1}$, and $D(O\text{–}O) = 498.2\,kJ\,mol^{-1}$. Estimate the bond dissociation energies of the first and second F–O bonds in F_2O and the Arrhenius activation energy of reaction 2.

20.4 Express the root mean square deviation $\{\langle M^2\rangle - \langle M\rangle^2\}^{1/2}$ of the molar mass of a condensation polymer in terms of the fraction p, and deduce its time-dependence.

20.5 Calculate the ratio of the mean cube molar mass to the mean square molar mass in terms of (a) the fraction p, (b) the chain length.

20.6 Conventional equilibrium considerations do not apply when a reaction is being driven by light absorption. Thus the steady-state concentration of products and reactants might differ significantly from equilibrium values. For instance, suppose the reaction $A \to B$ is driven by light absorption, and that its rate is I_a, but that the reverse reaction $B \to A$ is bimolecular and second order with a rate $k_r[B]^2$. What is the stationary state concentration of B? Why does this 'photostationary state' differ from the equilibrium state?

20.7 The photochemical chlorination of chloroform in the gas phase has been found to follow the rate law $d[CCl_4]/dt = k_r[Cl_2]^{1/2}I_a^{1/2}$. Devise a mechanism that leads to this rate law when the chlorine pressure is high.

CHAPTER 21

Reaction dynamics

Now we are at the heart of chemistry. In this chapter we examine the details of what happens to molecules at the climax of reactions. Extensive changes of structure are taking place and energies the size of dissociation energies are being redistributed among bonds: old bonds are being ripped apart and new bonds are being formed.

As may be imagined, the calculation of the rates of such processes from first principles is very difficult. Nevertheless, like so many intricate problems, the broad features can be established quite simply. Only when we enquire more deeply do the complications emerge. Here we look at several approaches to the calculation of a rate constant for elementary bimolecular processes, ranging from electron transfer to chemical reactions involving bond breakage and formation. Although a great deal of information can be obtained from gas-phase reactions, many reactions of interest take place in condensed phases, and we also see to what extent their rates can be predicted.

21A Collision theory

This Topic explores 'collision theory', the simplest quantitative account of reaction rates. The treatment can be used only for the discussion of reactions between simple species in the gas phase.

21B Diffusion-controlled reactions

In this Topic we see that reactions in solution are classified into two types: 'diffusion-controlled' and 'activation-controlled'. The former can be expressed quantitatively in terms of the diffusion equation.

21C Transition-state theory

This Topic discusses 'transition-state theory', in which it is assumed that the reactant molecules form a complex that can be discussed in terms of the population of its energy levels. The theory inspires a thermodynamic approach to reaction rates, in which the rate constant is expressed in terms of thermodynamic parameters. This approach is useful for parameterizing the rates of reactions in solution.

21D The dynamics of molecular collisions

The highest level of sophistication in the theoretical study of chemical reactions is in terms of potential energy surfaces and the motion of molecules on these surfaces. As we see in this Topic, such an approach gives an intimate picture of the events that occur when reactions occur and is open to experimental study.

21E Electron transfer in homogeneous systems

In this Topic we use transition-state theory to examine the transfer of electrons in homogeneous systems, including those involving proteins.

21F Processes at electrodes

Electron transfer processes on the surface of electrodes are difficult to describe theoretically, but in this Topic we develop a useful phenomenological approach that lends insight into useful experimental techniques and technological applications of electrochemistry.

What is the impact of this material?

The economic consequences of electron transfer reactions are almost incalculable. Most of the modern methods of generating electricity are inefficient, and in *Impact* I21.1 we see how the development of special electrochemical cells known as 'fuel cells' could revolutionize our production and deployment of energy.

To read more about the impact of this material, scan the QR code, or go to bcs.whfreeman.com/webpub/chemistry/pchem10e/impact/pchem-21-1.html

21A Collision theory

Contents

➤ Why do you need to know this material?

A major component of chemistry is the study of the mechanisms of chemical reactions. One of the earliest approaches, which continues to give insight into the details of mechanisms, is collision theory.

➤ What is the key idea?

According to collision theory, in a bimolecular gas-phase reaction, a reaction takes place on the collision of reactants provided their relative kinetic energy exceeds a threshold value and certain steric requirements are fulfilled.

➤ What do you need to know already?

This Topic draws on the kinetic theory of gases (Topic 1B) and extends the account of unimolecular reactions (Topic 20F). The latter uses combinatorial arguments like those described in Topic 15A.

In this Topic we consider the bimolecular elementary reaction

$$\mathrm{A}+\mathrm{B}\rightarrow\mathrm{P} \qquad v=k_\mathrm{r}[\mathrm{A}][\mathrm{B}] \tag{21A.1a}$$

where P denotes products. Our aim is to calculate the second-order rate constant k_r and to justify the form of the Arrhenius expression (Topic 20D):

$$k_\mathrm{r}=A\mathrm{e}^{-E_\mathrm{a}/RT} \qquad \text{Arrhenius expression} \tag{21A.1b}$$

where A is the 'pre-exponential factor' and E_a is the 'activation energy'. The model is then improved by examining how the energy of a collision is distributed over all the bonds in the reactant molecule. This improvement helps to account for the value of the rate constant k_b that appears in the Lindemann theory of unimolecular reactions (Topic 20F).

21A.1 Reactive encounters

We can anticipate the general form of the expression for k_r in eqn 21A.1a by considering the physical requirements for reaction. We can expect the rate v to be proportional to the rate of collisions, and therefore to the mean speed of the molecules, $v_\mathrm{mean}\propto(T/M)^{1/2}$ where M is some combination of the molar masses of A and B; we also expect the rate to be proportional to their collision cross-section, σ, (Topic 1B) and to the number densities $\mathcal{N}_\mathrm{A}$ and $\mathcal{N}_\mathrm{B}$ of A and B:

$$v\propto\sigma(T/M)^{1/2}\mathcal{N}_\mathrm{A}\mathcal{N}_\mathrm{B}\propto\sigma(T/M)^{1/2}[\mathrm{A}][\mathrm{B}]$$

However, a collision will be successful only if the kinetic energy exceeds a minimum value which we denote E'. This requirement suggests that the rate should also be proportional to a Boltzmann factor of the form $\mathrm{e}^{-E'/RT}$ representing the fraction of collisions with at least the minimum required energy E'. Therefore,

$$v\propto\sigma(T/M)^{1/2}\mathrm{e}^{-E'/RT}[\mathrm{A}][\mathrm{B}]$$

and we can anticipate, by writing the reaction rate in the form given in eqn 21A.1, that

$$k_\mathrm{r}\propto\sigma(T/M)^{1/2}\mathrm{e}^{-E'/RT}$$

At this point, we begin to recognize the form of the Arrhenius equation, eqn 21A.1b, and identify the minimum kinetic energy E' with the activation energy E_a of the reaction. This identification, however, should not be regarded as precise, since collision theory is only a rudimentary model of chemical reactivity.

Not every collision will lead to reaction even if the energy requirement is satisfied, because the reactants may need to collide in a certain relative orientation. This 'steric requirement' suggests that a further factor, P, should be introduced, and that

$$k_r \propto P\sigma(T/M)^{1/2}e^{-E'/RT} \qquad (21A.2)$$

As we shall see in detail below, this expression has the form predicted by collision theory. It reflects three aspects of a successful collision:

$$k_r \propto \overbrace{P}^{\text{Steric requirement}}\ \overbrace{\sigma(T/M)^{1/2}}^{\text{Encounter rate}}\ \overbrace{e^{-E'/RT}}^{\text{Minimum energy requirement}}$$

(a) Collision rates in gases

We have anticipated that the reaction rate, and hence k_r, depends on the frequency with which molecules collide. The **collision density**, Z_{AB}, is the number of (A,B) collisions in a region of the sample in an interval of time divided by the volume of the region and the duration of the interval. The frequency of collisions of a single molecule in a gas was calculated in Topic 1B (eqn 1B.11a, $z=\sigma v \mathcal{N}_A$). As shown in the following *Justification*, that result can be adapted to deduce that

$$Z_{AB}=\sigma\left(\frac{8kT}{\pi\mu}\right)^{1/2}N_A^2[A][B] \quad KMT \quad \text{Collision density} \qquad (21A.3a)$$

where σ is the collision cross-section (Fig. 21A.1)

$$\sigma=\pi d^2 \qquad d=\tfrac{1}{2}(d_A+d_B) \quad \text{Collision cross-section} \qquad (21A.3b)$$

d_A and d_B are the diameters of A and B, respectively, and μ is the reduced mass,

$$\mu=\frac{m_A m_B}{m_A+m_B} \quad \text{Reduced mass} \qquad (21A.3c)$$

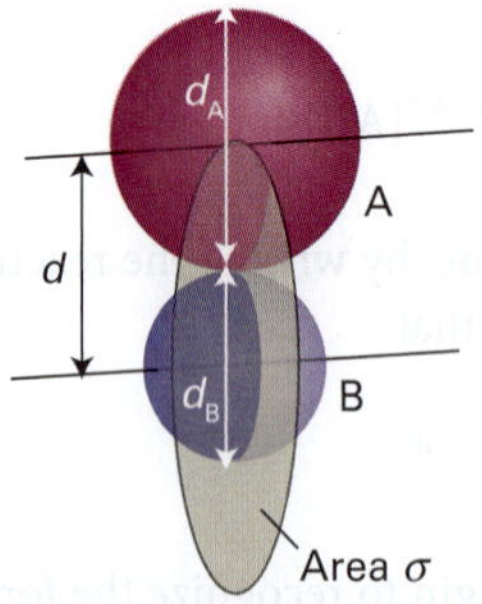

Figure 21A.1 The collision cross-section for two molecules can be regarded to be the area within which the projectile molecule (A) must enter around the target molecule (B) in order for a collision to occur. If the diameters of the two molecules are d_A and d_B, the radius of the target area is $d=\frac{1}{2}(d_A+d_B)$ and the cross-section is πd^2.

For like molecules $\mu=\frac{1}{2}m_A$ and at a molar concentration [A]

$$Z_{AA}=\tfrac{1}{2}\sigma\left(\frac{16kT}{\pi m_A}\right)^{1/2}N_A^2[A]^2=\sigma\left(\frac{4kT}{\pi m_A}\right)^{1/2}N_A^2[A]^2 \qquad (21A.3d)$$

The (blue) factor of $\frac{1}{2}$ is included to avoid double counting of collisions in this instance. If the collision density is required in terms of the pressure of each gas J, then we use $[J]=n_J/V=p_J/RT$.

Brief illustration 21A.1 Collision density

Collision densities may be very large. For example, in nitrogen at 25 °C and 1.0 bar, when $[N_2]\approx 40\,\text{mol m}^{-3}$, with $\sigma=0.43\,\text{nm}^2$ and $m_{N_2}=28.02m_u$ the collision density is

$$Z_{N_2N_2}=(4.3\times10^{-19}\,\text{m}^2)\times\left(\frac{4\times(1.381\times10^{-23}\,\text{J K}^{-1})\times(298\,\text{K})}{\pi\times28.02\times(1.661\times10^{-27}\,\text{kg})}\right)^{1/2}$$
$$\times(6.022\times10^{23}\,\text{mol}^{-1})^2\times(40\,\text{mol m}^{-3})^2=8.4\times10^{34}\,\text{m}^{-3}\,\text{s}^{-1}$$

Even in 1 cm^3, there are over 8×10^{16} collisions in each picosecond.

Self-test 21A.1 Calculate the collision density in molecular hydrogen under the same conditions.

Answer: $Z_{H_2H_2}=2.0\times10^{35}\,\text{m}^{-3}\,\text{s}^{-1}$

Justification 21A.1 The collision density

It follows from Topic 1B that the collision frequency, z, for a single A molecule of mass m_A in a gas of other A molecules is $z=\sigma v_{rel}\mathcal{N}_A$, where $\mathcal{N}_A$ is the number density of A molecules and v_{rel} is their relative mean speed. As indicated in Topic 1B, $v_{rel}=2^{1/2}v_{mean}$ with $v_{mean}=(8kT/\pi m)^{1/2}$. For future convenience, it is sensible to introduce $\mu=\frac{1}{2}m$ (for like molecules of mass m), and then to write $v_{rel}=(8kT/\pi\mu)^{1/2}$. This expression also applies to the mean relative speed of dissimilar molecules provided that μ is interpreted as their reduced mass.

The total collision density is the collision frequency multiplied by the number density of A molecules:

$$Z_{AA}=\tfrac{1}{2}z\mathcal{N}_A=\tfrac{1}{2}\sigma v_{rel}\mathcal{N}_A^2$$

The factor of $\frac{1}{2}$ has been introduced to avoid double counting of the collisions (so one A molecule colliding with another A molecule is counted as one collision regardless of their actual identities). For collisions of A and B molecules present at number densities $\mathcal{N}_A$ and $\mathcal{N}_B$, the collision density is

$$Z_{AB}=\sigma v_{rel}\mathcal{N}_A\mathcal{N}_B$$

The factor of $\frac{1}{2}$ has been discarded because now we are considering an A molecule colliding with any of the B molecules as a collision. The number density of a species J is $\mathcal{N}_J=N_A[J]$, where [J] is their molar concentration and N_A is Avogadro's constant. Equation 21A.3 then follows.

(b) The energy requirement

According to collision theory, the rate of change in the number density, $\mathcal{N}_A$, of A molecules is the product of the collision density and the probability that a collision occurs with sufficient energy. The latter condition can be incorporated by writing the collision cross-section σ as a function of the kinetic energy ε of approach of the two colliding species, and setting the cross-section, $\sigma(\varepsilon)$, equal to zero if the kinetic energy of approach is below a certain threshold value, ε_a. Later, we shall identify $N_A\varepsilon_a$ as E_a, the (molar) activation energy of the reaction. Then, for a collision between A and B with a specific relative speed of approach v_{rel} (not, at this stage, a mean value),

$$\frac{d\mathcal{N}_A}{dt}=-\sigma(\varepsilon)v_{rel}\mathcal{N}_A\mathcal{N}_B \qquad (21A.4a)$$

or, in terms of molar concentrations,

$$\frac{d[A]}{dt}=-\sigma(\varepsilon)v_{rel}N_A[A][B] \qquad (21A.4b)$$

The kinetic energy associated with the relative motion of the two particles takes the form $\varepsilon=\frac{1}{2}\mu v_{rel}^2$ when the centre-of-mass coordinates are separated from the internal coordinates of each particle. Therefore the relative speed is given by $v_{rel}=(2\varepsilon/\mu)^{1/2}$. At this point we recognize that a wide range of approach energies ε is present in a sample, so we should average the expression just derived over a Boltzmann distribution of energies $f(\varepsilon)$, and write

$$\frac{d[A]}{dt}=-\left\{\int_0^\infty \sigma(\varepsilon)v_{rel}f(\varepsilon)d\varepsilon\right\}N_A[A][B] \qquad (21A.5)$$

and hence recognize the rate constant as

$$k_r=N_A\int_0^\infty \sigma(\varepsilon)v_{rel}f(\varepsilon)d\varepsilon \qquad \text{Rate constant} \qquad (21A.6)$$

Now suppose that the reactive collision cross-section is zero below ε_a. We show in the following *Justification* that, above ε_a, $\sigma(\varepsilon)$ varies as

$$\sigma(\varepsilon)=\left(1-\frac{\varepsilon_a}{\varepsilon}\right)\sigma \qquad \text{Energy dependence of } \sigma \qquad (21A.7)$$

with the energy-independent σ given by eqn 21A.3b. This form of the energy-dependence for $\sigma(\varepsilon)$ is broadly consistent with experimental determinations of the reaction between H and D_2 as determined by molecular beam measurements of the kind described in Topic 21D (Fig. 21A.2).

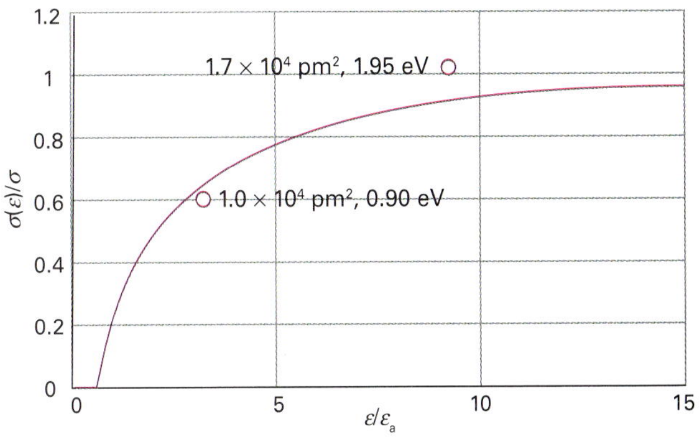

Figure 21A.2 The variation of the reactive cross-section with energy as expressed by eqn 21A.7. The data points are from experiments on the reaction $H+D_2\rightarrow HD+D$ (K. Tsukiyama et al., *J. Chem. Phys.* **84**, 1934 (1986)).

Justification 21A.2 The collision cross-section

Consider two colliding molecules A and B with relative speed v_{rel} and relative kinetic energy $\varepsilon=\frac{1}{2}\mu v_{rel}^2$ (Fig. 21A.3). Intuitively, we expect that a head-on collision between A and B will be most effective in bringing about a chemical reaction. Therefore, $v_{rel,A-B}$, the magnitude of the relative velocity component parallel to an axis that contains the vector connecting the centres of A and B, must be large. From trigonometry and the definitions of the distances a and d and the angle θ given in Fig. 21A.3, it follows that

$$v_{rel,A-B}=v_{rel}\cos\theta=v_{rel}\left(\frac{d^2-a^2}{d^2}\right)^{1/2}$$

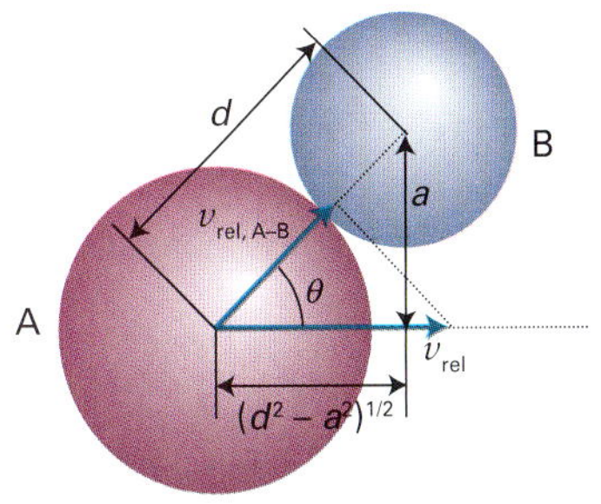

Figure 21A.3 The parameters used in the calculation of the dependence of the collision cross-section on the relative kinetic energy of two molecules A and B.

We assume that only the kinetic energy associated with the head-on component of the collision, ε_{A-B}, can lead to a chemical reaction. After squaring both sides of this equation and multiplying by $\frac{1}{2}\mu$, it follows that

$$\varepsilon_{A-B}=\varepsilon\times\frac{d^2-a^2}{d^2}$$

The existence of an energy threshold, ε_a, for the formation of products implies that there is a maximum value of a, a_{max}, above which reaction does not occur. Setting $a = a_{max}$ and $\varepsilon_{A-B} = \varepsilon_a$ gives

$$a_{max}^2 = \left(1 - \frac{\varepsilon_a}{\varepsilon}\right) d^2$$

Substitution of $\sigma(\varepsilon)$ for πa_{max}^2 and σ for πd^2 in the equation above gives eqn 21A.7. Note that the equation can be used only when $\varepsilon > \varepsilon_a$.

With the energy dependence of the collision cross-section established, we can evaluate the integral in eqn 21A.6. In the following *Justification* we show that

$$k_r = \sigma N_A v_{rel} e^{-E_a/RT} \qquad \textit{Collision theory} \quad \text{Rate constant} \qquad (21A.8)$$

Justification 21A.3 The rate constant

The Maxwell–Boltzmann distribution of molecular speeds is eqn 1B.4 of Topic 1B:

$$f(v)dv = 4\pi \left(\frac{\mu}{2\pi kT}\right)^{3/2} v^2 e^{-\mu v^2/2kT} dv$$

(We have replaced M/R by μ/k.) This expression may be written in terms of the kinetic energy, ε, by writing $\varepsilon = \frac{1}{2}\mu v^2$; then $dv = d\varepsilon/(2\mu\varepsilon)^{1/2}$, when it becomes

$$f(v)dv = 4\pi \left(\frac{\mu}{2\pi kT}\right)^{3/2} \left(\frac{2\varepsilon}{\mu}\right) e^{-\varepsilon/kT} \frac{d\varepsilon}{(2\mu\varepsilon)^{1/2}}$$

$$= 2\pi \left(\frac{1}{\pi kT}\right)^{3/2} \varepsilon^{1/2} e^{-\varepsilon/kT} d\varepsilon = f(\varepsilon) d\varepsilon$$

The integral we need to evaluate is therefore

$$\int_0^\infty \sigma(\varepsilon) \overbrace{v_{rel}}^{(2\varepsilon/\mu)^{1/2}} f(\varepsilon) d\varepsilon = 2\pi \left(\frac{1}{\pi kT}\right)^{3/2} \int_0^\infty \sigma(\varepsilon) \left(\frac{2\varepsilon}{\mu}\right)^{1/2} \varepsilon^{1/2} e^{-\varepsilon/kT} d\varepsilon$$

$$= \left(\frac{8}{\pi \mu kT}\right)^{1/2} \left(\frac{1}{kT}\right) \int_0^\infty \varepsilon \sigma(\varepsilon) e^{-\varepsilon/kT} d\varepsilon$$

To proceed, we introduce the approximation for $\sigma(\varepsilon)$ in eqn 21A.7 and evaluate

$$\int_0^\infty \varepsilon \sigma(\varepsilon) e^{-\varepsilon/kT} d\varepsilon \overset{\sigma=0 \text{ for } \varepsilon<\varepsilon_a}{=} \sigma \int_{\varepsilon_a}^\infty \varepsilon \left(1 - \frac{\varepsilon_a}{\varepsilon}\right) e^{-\varepsilon/kT} d\varepsilon$$

$$= \sigma \left\{ \int_{\varepsilon_a}^\infty \varepsilon e^{-\varepsilon/kT} d\varepsilon - \int_{\varepsilon_a}^\infty \varepsilon_a e^{-\varepsilon/kT} d\varepsilon \right\}$$

$$\overset{\text{Integral E.1}}{=} (kT)^2 \sigma e^{-\varepsilon_a/kT}$$

It follows that

$$\int_0^\infty \sigma(\varepsilon) v_{rel} f(\varepsilon) d\varepsilon = \sigma \left(\frac{8kT}{\pi\mu}\right)^{1/2} e^{-\varepsilon_a/kT}$$

as in eqn 21A.8 (with $\varepsilon_a/kT = E_a/RT$).

Equation 21A.8 has the Arrhenius form $k_r = A e^{-E_a/RT}$ provided the exponential temperature dependence dominates the weak square-root temperature dependence of the pre-exponential factor. It follows that we can identify (within the constraints of collision theory) the activation energy, E_a, with the minimum kinetic energy along the line of approach that is needed for reaction, and that the pre-exponential factor is a measure of the rate at which collisions occur in the gas.

The simplest procedure for calculating k_r is to use for σ the values obtained for non-reactive collisions (for example, typically those obtained from viscosity measurements) or from tables of molecular radii. If the collision cross-sections of A and B are $\sigma_A = \pi d_A^2$ and $\sigma_B = \pi d_B^2$, then an approximate value of the AB cross-section can be estimated from $\sigma = \pi d^2$, with $d = \frac{1}{2}(d_A + d_B)$. That is,

$$\sigma \approx \tfrac{1}{4}(\sigma_A^{1/2} + \sigma_B^{1/2})^2$$

Brief illustration 21A.2 The rate constant

To estimate the rate constant for the reaction $H_2 + C_2H_4 \rightarrow C_2H_6$ at 628 K we first calculate the reduced mass using $m(H_2) = 2.016 m_u$ and $m(C_2H_4) = 28.05 m_u$. A straightforward calculation gives $\mu = 3.123 \times 10^{-27}$ kg. It then follows that

$$\left(\frac{8kT}{\pi\mu}\right)^{1/2} = \left(\frac{8 \times (1.381 \times 10^{-23}\ \text{J K}^{-1}) \times (628\ \text{K})}{\pi \times (3.123 \times 10^{-27}\ \text{kg})}\right)^{1/2} = 2.65\ldots \text{km s}^{-1}$$

From Table 1B.1, $\sigma(H_2) = 0.27\ \text{nm}^2$ and $\sigma(C_2H_4) = 0.64\ \text{nm}^2$, giving $\sigma(H_2, C_2H_4) \approx 0.44\ \text{nm}^2$. The activation energy, Table 20D.1, is large: 180 kJ mol^{-1}. Therefore,

$$k_r = (4.4 \times 10^{-19}\ \text{m}^2) \times (2.65\ldots \times 10^3\ \text{m s}^{-1}) \times (6.022 \times 10^{23}\ \text{mol}^{-1}) \times e^{-(1.80 \times 10^5\ \text{J mol}^{-1})/(8.3145\ \text{J K}^{-1}\ \text{mol}^{-1}) \times (628\ \text{K})}$$

$$= \overbrace{7.04\ldots \times 10^8\ \text{m}^3\ \text{mol}^{-1}\ \text{s}^{-1}}^{A} \times e^{-34.4\ldots} = 7.5 \times 10^{-7}\ \text{m}^3\ \text{mol}^{-1}\ \text{s}^{-1}$$

or $7.5 \times 10^{-4}\ \text{dm}^3\ \text{mol}^{-1}\ \text{s}^{-1}$.

Self-test 21A.2 Evaluate the rate constant for the reaction $NO + Cl_2 \rightarrow NOCl + Cl$ at 298 K from $\sigma(NO) = 0.42\ \text{nm}^2$ and $\sigma(Cl_2) = 0.93\ \text{nm}^2$ and data in Table 1B.1.

Answer: $2.7 \times 10^{-4}\ \text{dm}^3\ \text{mol}^{-1}\ \text{s}^{-1}$

(c) The steric requirement

Table 21A.1 compares some values of the pre-exponential factor calculated from the collisional data in Table 1B.1 with values obtained from Arrhenius plots. One of the reactions shows fair agreement between theory and experiment, but for others there are major discrepancies. In some cases the experimental values are orders of magnitude smaller than those calculated, which suggests that the collision energy is not the only criterion for reaction and that some other feature, such as the relative orientation of the colliding species, is important. Moreover, one reaction in the table has a pre-exponential factor larger than theory, which seems to indicate that the reaction occurs more quickly than the particles collide!

The disagreement between experiment and theory can be eliminated by introducing a **steric factor**, P, and expressing the **reactive cross-section**, σ^*, as a multiple of the collision cross-section, $\sigma^* = P\sigma$ (Fig. 21A.4). Then the rate constant becomes

$$k_r = P\sigma N_A \left(\frac{8kT}{\pi\mu}\right)^{1/2} e^{-E_a/RT} \quad (21A.9)$$

This expression has the form we anticipated in eqn 21A.2. The steric factor is normally found to be several orders of magnitude smaller than 1.

Table 21A.1* Arrhenius parameters for gas-phase reactions

	$A/(\text{dm}^3\,\text{mol}^{-1}\,\text{s}^{-1})$		$E_a/(\text{kJ mol}^{-1})$	P
	Experiment	Theory		
$2\,NOCl \rightarrow 2\,NO + 2\,Cl$	9.4×10^9	5.9×10^{10}	102	0.16
$2\,ClO \rightarrow Cl_2 + O_2$	6.3×10^7	2.5×10^{10}	0	2.5×10^{-3}
$H_2 + C_2H_4 \rightarrow C_2H_6$	1.24×10^6	7.4×10^{11}	180	1.7×10^{-6}
$K + Br_2 \rightarrow KBr + Br$	1.0×10^{12}	2.1×10^{11}	0	4.8

* More values are given in the *Resource section*.

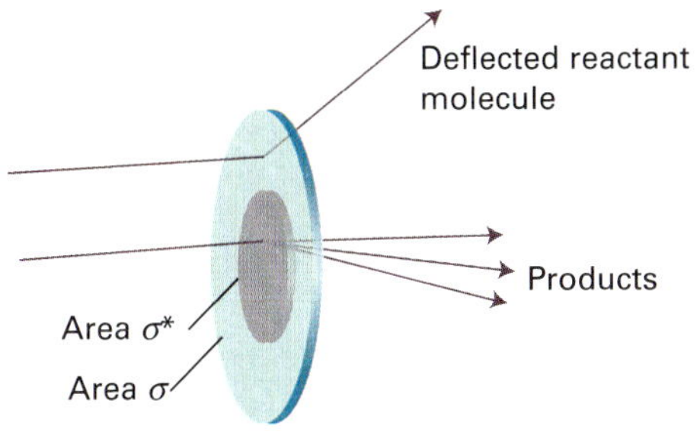

Figure 21A.4 The collision cross-section is the target area that results in simple deflection of the projectile molecule; the reactive cross-section is the corresponding area for chemical change to occur on collision.

Brief illustration 21A.3 The steric factor

It is found experimentally that the pre-exponential factor for the reaction $H_2 + C_2H_4 \rightarrow C_2H_6$ at 628 K is $1.24\times10^6\,\text{dm}^3\,\text{mol}^{-3}\,\text{s}^{-1}$. In *Brief illustration* 21A.2 we calculated the result that can be expressed as $A = 7.04\ldots\times10^{11}\,\text{dm}^3\,\text{mol}^{-1}\,\text{s}^{-1}$. It follows that the steric factor for this reaction is

$$P = \frac{A_{\text{experimental}}}{A_{\text{calculated}}} = \frac{1.24\times10^6\,\text{dm}^3\,\text{mol}^{-1}\,\text{s}^{-1}}{7.04\ldots\times10^{11}\,\text{dm}^3\,\text{mol}^{-1}\,\text{s}^{-1}} \approx 1.8\times10^{-6}$$

The very small value of P is one reason why catalysts are needed to bring this reaction about at a reasonable rate. As a general guide, the more complex the reactant molecules, the smaller the value of P.

Self-test 21A.3 It is found for the reaction $NO + Cl_2 \rightarrow NOCl + Cl$ that $A = 4.0\times10^9\,\text{dm}^3\,\text{mol}^{-1}\,\text{s}^{-1}$ at 298 K. Estimate the P factor for the reaction (see *Self–test* 21A.2).

Answer: 0.019

An example of a reaction for which it is possible to estimate the steric factor is $K + Br_2 \rightarrow KBr + Br$, with the experimental value $P = 4.8$. In this reaction, the distance of approach at which reaction occurs appears to be considerably larger than the distance needed for deflection of the path of the approaching molecules in a non-reactive collision. It has been proposed that the reaction proceeds by a **harpoon mechanism**. This brilliant name is based on a model of the reaction which pictures the K atom as approaching a Br_2 molecule, and when the two are close enough an electron (the harpoon) flips across from K to Br_2. In place of two neutral particles there are now two ions, so there is a Coulombic attraction between them: this attraction is the line on the harpoon. Under its influence the ions move together (the line is wound in), the reaction takes place, and KBr + Br emerge. The harpoon extends the cross-section for the reactive encounter, and the reaction rate is significantly underestimated by taking for the collision cross-section the value for simple mechanical contact between K and Br_2.

Example 21A.1 Estimating a steric factor

Estimate the value of P for the harpoon mechanism by calculating the distance at which it becomes energetically favourable for the electron to leap from K to Br_2. Take the sum of the radii of the reactants (treating them as spherical) to be 400 pm.

Method Begin by identifying all the contributions to the energy of interaction between the colliding species. There are three contributions to the energy of the process $K + Br_2 \rightarrow K^+ + Br_2^-$. The first is the ionization energy, I, of K. The second is the electron affinity, E_{ea}, of Br_2. The third is the

Coulombic interaction energy between the ions when they have been formed: when their separation is R, this energy is $-e^2/4\pi\varepsilon_0 R$. The electron flips across when the sum of these three contributions changes from positive to negative (that is, when the sum is zero) and becomes energetically favourable.

Answer The net change in energy when the transfer occurs at a separation R is

$$E = I - E_{ea} - \frac{e^2}{4\pi\varepsilon_0 R}$$

The ionization energy I is larger than E_{ea}, so E becomes negative only when R has decreased to less than some critical value R^* given by

$$R = \frac{e^2}{4\pi\varepsilon_0(I - E_{ea})}$$

When the particles are at this separation, the harpoon shoots across from K to Br_2, so we can identify the reactive cross-section as $\sigma^* = \pi R^{*2}$. This value of σ^* implies that the steric factor is

$$P = \frac{\sigma^*}{\sigma} = \frac{R^{*2}}{d^2} = \left(\frac{e^2}{4\pi\varepsilon_0 d(I - E_{ea})}\right)^2$$

where $d = R(\mathrm{K}) + R(\mathrm{Br_2})$, the sum of the radii of the spherical reactants. With $I = 420\,\mathrm{kJ\,mol^{-1}}$ (corresponding to 0.70 aJ), $E_{ea} \approx 250\,\mathrm{kJ\,mol^{-1}}$ (corresponding to 0.42 aJ), and $d = 400$ pm, we find $P = 4.2$, in good agreement with the experimental value (4.8).

Self-test 21A.4 Estimate the value of P for the harpoon reaction between Na and Cl_2 for which $d \approx 350$ pm; take $E_{ea} \approx 230\,\mathrm{kJ\,mol^{-1}}$.

Answer: 2.2

Example 21A.1 illustrates two points about steric factors. First, the concept of a steric factor is not wholly useless because in some cases its numerical value can be estimated. Second, and more pessimistically, most reactions are much more complex than $K + Br_2$, and we cannot expect to obtain P so easily.

21A.2 The RRK model

The steric factor P can also be estimated for unimolecular gas-phase reactions (Topic 20F), by a calculation based on the **Rice–Ramsperger–Kassel model** (RRK model), which was proposed in 1926 by O.K. Rice and H.C. Ramsperger and almost simultaneously by L.S. Kassel. The model has been elaborated, largely by R.A. Marcus, into the 'RRKM model'. Here we outline Kassel's original approach to the RRK model; the details are set out in the following *Justification*. The essential feature of the model is that although a molecule might have enough energy to react, that energy is distributed over all the modes of motion of the molecule, and reaction will occur only when enough of that energy has migrated into a particular location (such as a particular bond) in the molecule. This distribution effect leads to a P factor of the form

$$P = \left(1 - \frac{E^*}{E}\right)^{s-1} \qquad \text{RRK theory} \qquad (21A.10a)$$

where s is the number of modes of motion over which the energy E may be dissipated and E^* is the energy required for the bond of interest to break. The resulting **Kassel form** of the unimolecular rate constant for the decay of A^* to products is

$$k_b(E) = \left(1 - \frac{E^*}{E}\right)^{s-1} k_b \qquad \text{for } E \geq E^* \qquad \text{Kassel form} \qquad (21A.10b)$$

where k_b is the rate constant used in the original Lindemann theory for the decomposition of the activated intermediate (eqn 20F.8 of Topic 20F).

Justification 21A.4 The RRK model of unimolecular reactions

To set up the RRK model, we suppose that a molecule consists of s identical harmonic oscillators, each of which has frequency ν. In practice, of course, the vibrational modes of a molecule have different frequencies, but assuming that they are all the same is a reasonable first approximation. Next, we suppose that the vibrations are excited to a total energy $E = nh\nu$ and then set out to calculate the number of ways N in which the energy can be distributed over the oscillators.

We can represent the n quanta as follows:

□□□□□□□□□□□□□□□□□□□□□□□□□□
□□□□□□...□□□

These quanta must be put in s containers (the s oscillators), which can be represented by inserting $s-1$ walls, denoted by . One such distribution is

□□|□□□□|□□||□□□|□□□□□□□□|□□□□|||
□□□□□|□□□□□...□|□□

The total number of arrangements of each quantum and wall (of which there are $n+s-1$ in all) is $(n+s-1)!$ where, as usual, $x! = x(x-1)\ldots1$. However, the $n!$ arrangements of the n quanta are indistinguishable, as are the $(s-1)!$ arrangements of the $s-1$ walls. Therefore, to find N we must divide $(n+s-1)!$ by these two factorials. It follows that

$$N = \frac{(n+s-1)!}{n!(s-1)!}$$

The distribution of the energy throughout the molecule means that it is too sparsely spread over all the modes for any particular bond to be sufficiently highly excited to undergo dissociation. We suppose that a bond will break only if it is excited to at least an energy $E^* = n^*h\nu$. Therefore, we isolate one critical oscillator as the one that undergoes dissociation if it has *at least* n^* of the quanta, leaving up to $n - n^*$ quanta to be accommodated in the remaining $s - 1$ oscillators (and therefore with $s - 2$ walls in the partition in place of the $s - 1$ walls we used above). For example, consider 28 quanta distributed over six oscillators, with excitation by at least six quanta required for dissociation. Then all the following partitions will result in dissociation:

□□□□□□|□□□□□|□□□□□□□□|□□□□||□□□□□
□□□□□□□|□□□□|□□□□□□□□|□□□□||□□□□□
□□□□□□□□|□□□|□□□□□□□□|□□□□||□□□□□
⋮   ⋮   ⋮   ⋮   ⋮

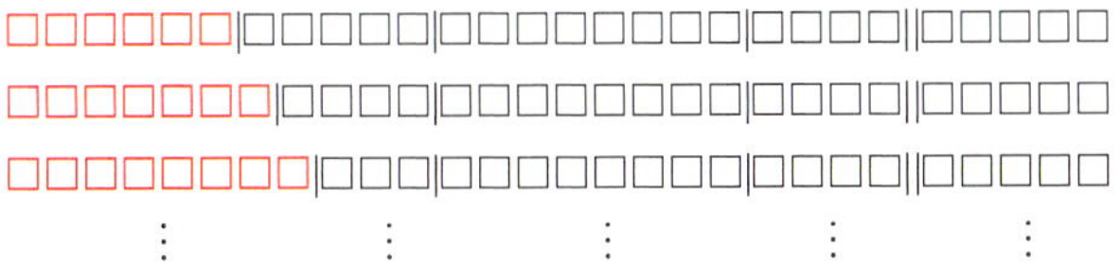

(The leftmost partition is the critical oscillator.) However, these partitions are equivalent to

□□□□□□ |□□□□□|□□□□□□□□□□|□□□□||□□□□□
□□□□□□ □|□□□□|□□□□□□□□□□|□□□□||□□□□□
□□□□□□ □□|□□□|□□□□□□□□□□|□□□□||□□□□□
⋮   ⋮   ⋮   ⋮   ⋮

and we see that we have the problem of permuting $28 - 6 = 22$ (in general, $n - n^*$) quanta and five (in general, $s - 1$) walls, and therefore a total of 27 (in general, $n - n^* + s - 1$ objects). Therefore, the calculation is exactly like the one above for N, except that we have to find the number of distinguishable permutations of $n - n^*$ quanta in s containers (and therefore $s - 1$ walls). The number N^* is therefore obtained from the expression for N by replacing n by $n - n^*$ and is

$$N^* = \frac{(n - n^* + s - 1)!}{(n - n^*)!(s - 1)!}$$

From the preceding discussion we conclude that the probability that one specific oscillator will have undergone sufficient excitation to dissociate is the ratio N^*/N, which is

$$P = \frac{N^*}{N} = \frac{n!(n - n^* + s - 1)!}{(n - n^*)!(n + s - 1)!}$$

This equation is still awkward to use, even when written out in terms of its factors:

$$P = \frac{n(n-1)(n-2)\ldots 1}{(n - n^*)(n - n^* - 1)\ldots 1} \times \frac{(n - n^* + s - 1)(n - n^* + s - 2)\ldots 1}{(n + s - 1)(n + s - 2)\ldots 1}$$

$$= \frac{(n - n^* + s - 1)(n - n^* + s - 2)\ldots(n - n^* + 1)}{(n + s - 1)(n + s - 2)\ldots(n + 2)(n + 1)}$$

However, because $s - 1$ is small (in the sense $s - 1 \ll n - n^*$), we can approximate this expression by

$$P = \frac{\overbrace{(n - n^*)(n - n^*)\ldots(n - n^*)}^{s-1 \text{ factors}}}{\underbrace{(n)(n)\ldots(n)}_{s-1 \text{ factors}}} = \left(\frac{n - n^*}{n}\right)^{s-1}$$

An alternative derivation of this expression for P is developed in Problem 21A.7. Because the energy of the excited molecule is $E = nh\nu$ and the critical energy is $E^* = n^*h\nu$, this expression may be written

$$P = \left(1 - \frac{E^*}{E}\right)^{s-1}$$

as in eqn 21A.10.

The energy dependence of the rate constant given by eqn 21A.10b is shown in Fig. 21A.5 for various values of s. We see that the rate constant is smaller at a given excitation energy if s is large, as it takes longer for the excitation energy to migrate through all the oscillators of a large molecule and accumulate in the critical mode. As E becomes very large, however, the term in parentheses approaches 1, and $k_b(E)$ becomes independent of the energy and the number of oscillators in the molecule, as there is now enough energy to accumulate immediately in the critical mode regardless of the size of the molecule.

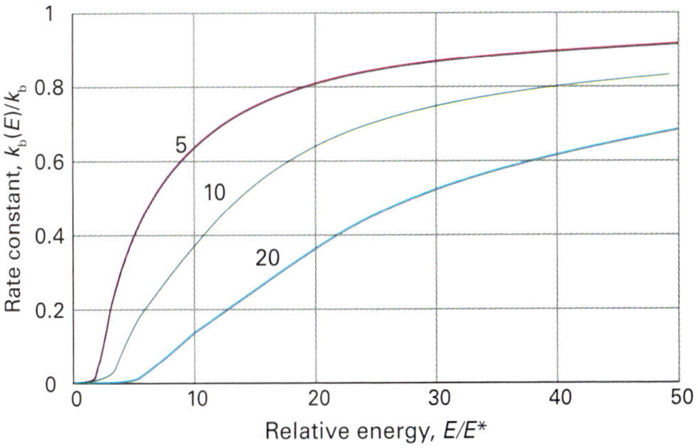

Figure 21A.5 The energy dependence of the rate constant given by eqn 21A.10b for three values of s.

Brief illustration 21A.4 The RRK model

In *Brief illustration* 21A.3 we calculated a value of $P = 1.8 \times 10^{-6}$ for the reaction $H_2 + C_2H_4 \rightarrow C_2H_6$. Although this is not a unimolecular process, it is interesting to analyse it on the basis of the RRK theory because in some sense the collision energy must accumulate in a region where bonds are broken and formed. Thus, C_2H_4 has six atoms and therefore $s=12$ vibrational modes. We can estimate the ratio E^*/E by solving

$$\left(1-\frac{E^*}{E}\right)^{11}=1.8\times10^{-6} \quad \text{or} \quad \frac{E^*}{E}=1-(1.8\times10^{-6})^{1/11}=0.70$$

This result suggests in one interpretation that the energy needed to proceed in the reaction (identified here with the energy to break the carbon-carbon bond in C_2H_4) is typically 70 per cent of the energy of a typical collision. If all eight atoms are taken to be involved in sharing the energy of the collision, the ratio works out as 0.54.

Self-test 21A.5 Apply the same analysis to the reaction in Self-test 21A.3, where it is found that $P=0.019$ for $NO+Cl_2 \rightarrow NOCl+Cl$. Take the number of atoms in the complex to be 4, so $s=6$.

Answer: 0.55

Checklist of concepts

☐ 1. In **collision theory**, it is supposed that the rate is proportional to the collision frequency, a steric factor, and the fraction of collisions that occur with at least the kinetic energy E_a along their lines of centres.

☐ 2. The **collision density** is the number of collisions in a region of the sample in an interval of time divided by the volume of the region and the duration of the interval.

☐ 3. The **activation energy** is the minimum kinetic energy along the line of approach of reactant molecules that is required for reaction.

☐ 4. The **steric factor** is an adjustment that takes into account the orientational requirements for a successful collision.

☐ 5. For unimolecular reactions, the steric factor can be computed by using the **RRK model**.

Checklist of equations

Property	Equation	Comment	Equation number
Collision density	$Z_{AB}=\sigma(8kT/\pi\mu)^{1/2}N_A^2[A][B]$	Unlike molecules, KMT (kinetic molecular theory)	21A.3a
	$Z_{AA}=\sigma(4kT/\pi m_A)^{1/2}N_A^2[A]^2$	Like molecules, KMT	21A.3d
Energy-dependence of σ	$\sigma(\varepsilon)=(1-\varepsilon_a/\varepsilon)\sigma$	$\varepsilon\geq\varepsilon_a$, 0 otherwise	21A.7
Rate constant	$k_r=P\sigma N_A(8kT/\pi\mu)^{1/2}e^{-E_a/RT}$	KMT, collision theory	21A.9
Steric factor	$P=(1-E^*/E)^{s-1}$	RRK theory	21A.10a

21B Diffusion-controlled reactions

Contents

➤ Why do you need to know this material?

Most chemical reactions take place in solution and for a thorough grasp of chemistry it is important to understand what controls their rates and how those rates can be modified.

➤ What is the key idea?

There are two limiting types of chemical reaction mechanism in solution: diffusion control and activation control.

➤ What do you need to know already?

This Topic makes use of the steady-state approximation (Topic 20E) and draws on the formulation and solution of the diffusion equation (Topic 19C). At one point it uses the Stokes–Einstein relation (Topic 19B).

To consider reactions in solution we have to imagine processes that are entirely different from those in gases. No longer are there collisions of molecules hurtling through space; now there is the jostling of one molecule through a dense but mobile collection of molecules making up the fluid environment.

21B.1 Reactions in solution

Encounters between reactants in solution occur in a very different manner from encounters in gases. The encounters of reactant molecules dissolved in solvent are considerably less frequent than in a gas. However, because a molecule also migrates only slowly away from a location, two reactant molecules that encounter each other stay near each other for much longer than in a gas. This lingering of one molecule near another on account of the hindering presence of solvent molecules is called the **cage effect**. Such an **encounter pair** may accumulate enough energy to react even though it does not have enough energy to do so when it first forms. The activation energy of a reaction is a much more complicated quantity in solution than in a gas because the encounter pair is surrounded by solvent and we need to consider the energy of the entire local assembly of reactant and solvent molecules.

(a) Classes of reaction

The complicated overall process can be divided into simpler parts by setting up a simple kinetic scheme. We suppose that the rate of formation of an encounter pair AB is first order in each of the reactants A and B:

$$\mathrm{A+B \rightarrow AB} \qquad v=k_\mathrm{d}[\mathrm{A}][\mathrm{B}]$$

As we shall see, k_d (where the d signifies diffusion) is determined by the diffusional characteristics of A and B. The encounter pair can break up without reaction or it can go on to form products P. If we suppose that both processes are pseudofirst-order reactions (with the solvent perhaps playing a role), then we can write

$$\mathrm{AB \rightarrow A+B} \qquad v=k_\mathrm{d}'[\mathrm{AB}]$$

$$\mathrm{AB \rightarrow P} \qquad v=k_\mathrm{a}[\mathrm{AB}]$$

The concentration of AB can now be found from the equation for the net rate of change of concentration of AB:

$$\frac{\mathrm{d[AB]}}{\mathrm{d}t}=k_\mathrm{d}[\mathrm{A}][\mathrm{B}]-k_\mathrm{d}'[\mathrm{AB}]-k_\mathrm{a}[\mathrm{AB}]=0$$

where we have applied the steady-state approximation (Topic 20E). This expression solves to

$$[\mathrm{AB}]=\frac{k_\mathrm{d}[\mathrm{A}][\mathrm{B}]}{k_\mathrm{a}+k_\mathrm{d}'}$$

The rate of formation of products is therefore

$$\frac{\mathrm{d[P]}}{\mathrm{d}t}=k_\mathrm{a}[\mathrm{AB}]=k_\mathrm{r}[\mathrm{A}][\mathrm{B}] \qquad k_\mathrm{r}=\frac{k_\mathrm{a}k_\mathrm{d}}{k_\mathrm{a}+k_\mathrm{d}'} \qquad (21\mathrm{B}.1)$$

Two limits can now be distinguished. If the rate of separation of the unreacted encounter pair is much slower than the rate at which it forms products, then $k_d' \ll k_a$ and the effective rate constant is

$$k_r \approx \frac{k_a k_d}{k_a} = k_d \quad \text{Diffusion-controlled limit} \quad (21B.2a)$$

In this **diffusion-controlled limit**, the rate of reaction is governed by the rate at which the reactant molecules diffuse through the solvent. Because the combination of radicals involves very little activation energy, radical and atom recombination reactions are often diffusion-controlled. An **activation-controlled reaction** arises when a substantial activation energy is involved in the reaction AB→P. Then $k_a \ll k_d'$ and

$$k_r \approx \frac{k_a k_d}{k_d'} = k_a K \quad \text{Activation-controlled limit} \quad (21B.2b)$$

where K is the equilibrium constant for A+B⇌AB. In this limit, the reaction proceeds at the rate at which energy accumulates in the encounter pair from the surrounding solvent. Some experimental data are given in Table 21B.1.

Table 21B.1* Arrhenius parameters for solvolysis reactions in solution

	Solvent	$A/(\text{dm}^3\,\text{mol}^{-1}\,\text{s}^{-1})$	$E_a/(\text{kJ mol}^{-1})$
$(CH_3)_3CCl$	Water	7.1×10^{16}	100
	Ethanol	3.0×10^{13}	112
	Chloroform	1.4×10^{4}	45
CH_3CH_2Br	Ethanol	4.3×10^{11}	90

* More values are given in the *Resource section*.

(b) Diffusion and reaction

The rate of a diffusion-controlled reaction is calculated by considering the rate at which the reactants diffuse together. As shown in the following *Justification*, the rate constant for a reaction in which the two molecules react if they come within a distance R^* of one another is

$$k_d = 4\pi R^* D N_A \quad (21B.3)$$

where D is the sum of the diffusion coefficients of the two reactant species in the solution.

Brief illustration 21B.1 Diffusion control 1

The order of magnitude of R^* is 10^{-7} m (100 nm) and that of D for a species in water is $10^{-9}\,\text{m}^2\,\text{s}^{-1}$. It follows from eqn 21B.3 that

$$k_d \approx 4\pi\times(10^{-7}\,\text{m})\times(10^{-9}\,\text{m}^2\,\text{s}^{-1})\times(6.022\times10^{23}) \approx 10^{9}\,\text{m}^3\text{mol}^{-1}\text{s}^{-1}$$

which corresponds to $10^{12}\,\text{dm}^3\,\text{mol}^{-1}\,\text{s}^{-1}$. An indication that a reaction is diffusion-controlled is therefore that its rate constant is of the order of $10^{12}\,\text{dm}^3\,\text{mol}^{-1}\,\text{s}^{-1}$.

Self-test 21B.1 Estimate the rate constant for a diffusion-controlled reaction in benzene ($D \approx 2\times10^{-9}\,\text{m}^2\,\text{s}^{-1}$), taking $R^* \approx 100$ nm.

Answer: $1.5\times10^{12}\,\text{dm}^3\,\text{mol}^{-1}\,\text{s}^{-1}$

Justification 21B.1 Solution of the radial diffusion equation

The general form of the diffusion equation (Topic 19A) corresponding to motion in three dimensions is $D_B\nabla^2[\text{B}](r,t) = \partial[\text{B}](r,t)/\partial t$; therefore, the concentration of B when the system has reached a steady state ($\partial[\text{B}](r,t)/\partial t = 0$) satisfies $\nabla^2[\text{B}](r) = 0$, with the concentration of B now depending only on location not time. For a spherically symmetrical system, ∇^2 can be replaced by radial derivatives alone (see Table 7B.1), so the equation satisfied by $[\text{B}](r)$, as $[\text{B}](r)$ can now be written, is

$$\frac{d^2[\text{B}](r)}{dr^2} + \frac{2}{r}\frac{d[\text{B}](r)}{dr} = 0$$

The general solution of this equation is

$$[\text{B}](r) = a + \frac{b}{r}$$

as may be verified by substitution. We need two boundary conditions to pin down the values of the two constants (a and b). One condition is that $[\text{B}](r)$ has its bulk value [B] as $r\to\infty$. The second condition is that the concentration of B is zero at $r = R^*$, the distance at which reaction occurs. It follows that $a = [\text{B}]$ and $b = -R^*[\text{B}]$, and hence that (for $r \geq R^*$)

$$[\text{B}](r) = \left(1 - \frac{R^*}{r}\right)[\text{B}]$$

Figure 21B.1 illustrates the variation of concentration expressed by this equation.

The rate of reaction is the (molar) flux, J, of the reactant B towards A multiplied by the area of the spherical surface of radius R^* through which B must pass:

$$\text{Rate of reaction} = 4\pi R^{*2} J$$

From Fick's first law (eqn 19C.3 of Topic 19C, $J = -D\partial[\text{J}]/\partial x$), the flux of B towards A is proportional to the concentration gradient, so at a radius R^*:

$$J = D_B\left(\frac{d[\text{B}](r)}{dr}\right)_{r=R^*} = -D_B[\text{B}]R^*\left(-\frac{1}{r^2}\right)_{r=R^*} = \frac{D_B[\text{B}]}{R^*}$$

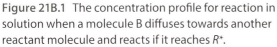

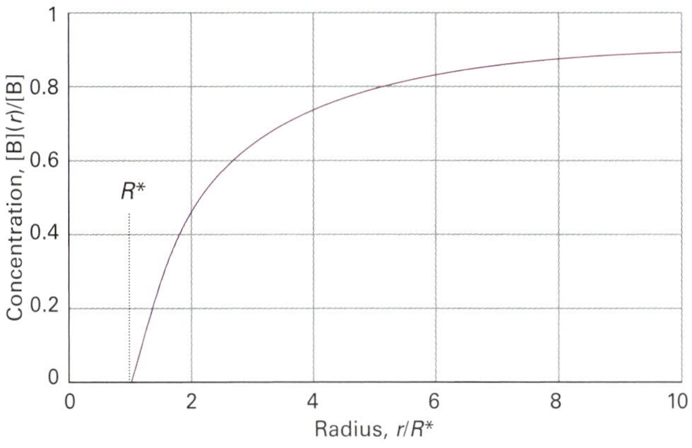

Figure 21B.1 The concentration profile for reaction in solution when a molecule B diffuses towards another reactant molecule and reacts if it reaches R^*.

(A sign change has been introduced because we are interested in the flux towards decreasing values of r.) It follows that

$$\text{Rate of reaction} = 4\pi R^* D_B[B]$$

The rate of the diffusion-controlled reaction is equal to the average flow of B molecules to all the A molecules in the sample. If the bulk concentration of A is [A], the number of A molecules in the sample of volume V is $N_A[A]V$; the global flow of all B to all A is therefore $4\pi R^* D_B N_A[A][B]V$. Because it is unrealistic to suppose that all A molecules are stationary; we replace D_B by the sum of the diffusion coefficients of the two species and write $D = D_A + D_B$. Then the rate of change of concentration of AB is

$$\frac{d[AB]}{dt} = 4\pi R^* D N_A[A][B]$$

Hence, the diffusion-controlled rate constant is as given in eqn 21B.3.

We can take eqn 21B.3 further by incorporating the Stokes–Einstein equation (eqn 19B.19 of Topic 19B, $D_J = kT/6\pi\eta R_J$) relating the diffusion constant and the hydrodynamic radius R_A and R_B of each molecule in a medium of viscosity η. As this relation is approximate, little extra error is introduced if we write $R_A = R_B = \frac{1}{2}R^*$, which leads to

$$k_d = \frac{8RT}{3\eta} \quad \text{Diffusion-controlled rate constant} \quad (21B.4)$$

(The R in this equation is the gas constant.) The radii have cancelled because, although the diffusion constants are smaller when the radii are large, the reactive collision radius is larger and the particles need travel a shorter distance to meet. In this approximation, the rate constant is independent of the identities of the reactants, and depends only on the temperature and the viscosity of the solvent.

Brief illustration 21B.2 Diffusion control 2

The rate constant for the recombination of I atoms in hexane at 298 K, when the viscosity of the solvent is 0.326 cP (with $1\,\text{P} = 10^{-1}\,\text{kg m}^{-1}\,\text{s}^{-1}$) is

$$k_d = \frac{8\times(8.3145\,\text{J K}^{-1}\,\text{mol}^{-1})\times(298\,\text{K})}{3\times(3.26\times10^{-4}\,\text{kg m}^{-1}\,\text{s}^{-1})} = 2.0\times10^{7}\,\text{m}^3\,\text{mol}^{-1}\,\text{s}^{-1}$$

where we have used $1\,\text{J} = 1\,\text{kg m}^2\,\text{s}^{-2}$. Because $1\,\text{m}^3 = 10^3\,\text{dm}^3$, this result corresponds to $2.0\times10^{10}\,\text{dm}^3\,\text{mol}^{-1}\,\text{s}^{-1}$. The experimental value is $1.3\times10^{10}\,\text{dm}^3\,\text{mol}^{-1}\,\text{s}^{-1}$, so the agreement is very good considering the approximations involved.

Self-test 21B.2 Evaluate a typical rate constant for a reaction taking place in ethanol at 20 °C, for which the viscosity is 1.06 cP.

Answer: $6.1\times10^{9}\,\text{dm}^3\,\text{mol}^{-1}\,\text{s}^{-1}$

21B.2 The material-balance equation

The diffusion of reactants plays an important role in many chemical processes, such as the diffusion of O_2 molecules into red blood corpuscles and the diffusion of a gas towards a catalyst. We can catch a glimpse of the kinds of calculations involved by considering the diffusion equation (Topic 19C) generalized to take into account the possibility that the diffusing, convecting molecules are also reacting.

(a) The formulation of the equation

Consider a small volume element in a chemical reactor (or a biological cell). The net rate at which J molecules enter the region by diffusion and convection is given by eqn 19C.10 of Topic 19C, which we repeat here:

$$\frac{\partial[J]}{\partial t} = D\frac{\partial^2[J]}{\partial x^2} - v\frac{\partial[J]}{\partial x} \quad \text{Diffusion equation} \quad (21B.5)$$

where v is the velocity of the convective flow of J and [J] in general depends on both position and time. The net rate of change of molar concentration due to chemical reaction is

$$\frac{\partial[J]}{\partial t} = -k_r[J]$$

if we suppose that J disappears by a pseudofirst-order reaction. Therefore, the overall rate of change of the concentration of J is

$$\frac{\partial[J]}{\partial t} = D\overbrace{\frac{\partial^2[J]}{\partial x^2}}^{\text{Spread due to non-uniform distribution}} - v\overbrace{\frac{\partial[J]}{\partial x}}^{\text{Change due to convection}} - \overbrace{k_r[J]}^{\text{Loss due to reaction}} \quad \text{Material-balance equation} \quad (21B.6)$$

Equation 21B.6 is called the **material-balance equation**. If the rate constant is large, then [J] will decline rapidly. However, if the diffusion constant is large, then the decline can be replenished as J diffuses rapidly into the region. The convection term, which may represent the effects of stirring, can sweep material either into or out of the region according to the signs of v and the concentration gradient $\partial[\mathrm{J}]/\partial x$.

(b) Solutions of the equation

The material-balance equation is a second-order partial differential equation and is far from easy to solve in general. Some idea of how it is solved can be obtained by considering the special case in which there is no convective motion (as in an unstirred reaction vessel):

$$\frac{\partial[\mathrm{J}]}{\partial t}=D\frac{\partial^2[\mathrm{J}]}{\partial x^2}-k_\mathrm{r}[\mathrm{J}] \quad (21\mathrm{B}.7)$$

As may be verified by substitution (Problem 21B.1), if the solution of this equation in the absence of reaction (that is, for $k_\mathrm{r}=0$) is [J], then the solution $[\mathrm{J}]^*$ in the presence of reaction ($k_\mathrm{r}>0$) is

$$[\mathrm{J}]^*=[\mathrm{J}]\mathrm{e}^{-k_\mathrm{r}t} \quad \text{Diffusion with reaction} \quad (21\mathrm{B}.8)$$

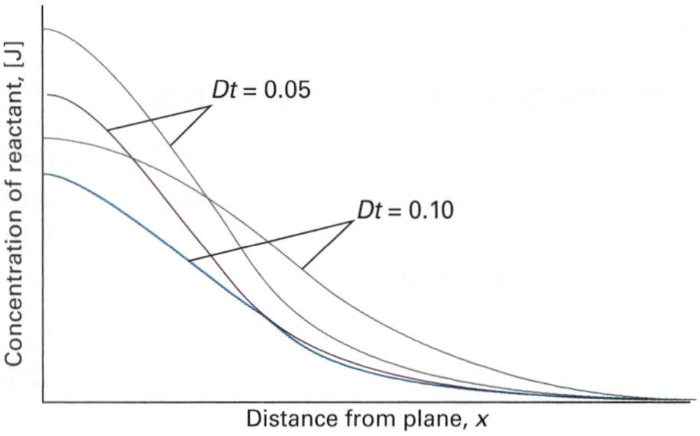

Figure 21B.2 The concentration profiles for a diffusing, reacting system (for example, a column of solution) in which one reactant is initially in a layer at $x=0$. In the absence of reaction (grey lines), the concentration profiles are the same as in Fig. 19C.3.

An example of a solution of the diffusion equation in the absence of reaction is that given in Topic 19C for a system in which initially a layer of n_0N_A molecules is spread over a plane of area A:

$$[\mathrm{J}]=\frac{n_0\mathrm{e}^{-x^2/4Dt}}{A(\pi Dt)^{1/2}} \quad (21\mathrm{B}.9)$$

When this expression is substituted into eqn 21B.8, we obtain the concentration of J as it diffuses away from its initial surface layer and undergoes reaction in the overlying solution (Fig. 21B.2).

Brief illustration 21B.3 Reaction with diffusion

Suppose 1.0 g of iodine (3.9 mmol I_2) is spread over a surface of area 5.0 cm² under a column of hexane ($D=4.1\times10^{-9}\,\mathrm{m^2\,s^{-1}}$). As it diffuses upwards it reacts with a pseudofirst-order rate constant $k_\mathrm{r}=4.0\times10^{-5}\,\mathrm{s^{-1}}$. By substituting these values into

$$[\mathrm{J}]^*=\frac{n_0\mathrm{e}^{-x^2/4Dt-k_\mathrm{r}t}}{A(\pi Dt)^{1/2}}$$

we can construct the following table:

$[\mathrm{J}]^*/(\mathrm{mmol\,dm^{-3}})$		x	
T	1 mm	5 mm	1 cm
100 s	3.72	0	0
1000 s	1.96	0.45	0.005
10 000 s	0.46	0.40	0.25

Self-test 21B.3 What is the value of [J] at 15 000 s at the same three locations?

Answer: 0.31, 0.28, 0.21 mmol dm⁻³

Even this relatively simple example has led to an equation that is difficult to solve, and only in some special cases can the full material balance equation be solved analytically. Most modern work on reactor design and cell kinetics uses numerical methods to solve the equation, and detailed solutions for realistic environments, such as vessels of different shapes (which influence the boundary conditions on the solutions) and with a variety of inhomogeneously distributed reactants, can be obtained reasonably easily.

Checklist of concepts

- ☐ 1. A reaction in solution may be **diffusion-controlled** if its rate is controlled by the rate at which reactant molecules encounter each other in solution.
- ☐ 2. The rate of an **activation-controlled reaction** is controlled by the rate at which the encounter pair accumulates sufficient energy.

- ☐ 3. The **material-balance equation** relates the overall rate of change of the concentration of a species to its rates of diffusion, convection and reaction.
- ☐ 4. The **cage effect**, the lingering of one reactant molecule near another due to the hindering presence of solvent molecules, results in the formation of an **encounter pair** of reactant molecules.

Checklist of equations

Property	Equation	Comment	Equation number
Diffusion-controlled limit	$k_r = k_d$	$v = k_d[A][B]$ for the encounter rate	21B.2a
Activation-controlled limit	$k_r = k_d K$	K for $A + B \rightleftharpoons AB$, k_a for the decomposition of AB	21B.2b
Diffusion-controlled rate constant	$k_d = 4\pi R^* D N_A$	$D = D_A + D_B$	21B.3
	$k_d = 8RT/3\eta$	Assumes Stokes–Einstein relation	21B.4
Material-balance equation	$\partial[J]/\partial t = D\partial^2[J]/\partial x^2 - v\partial[J]/\partial x - k_r[J]$	First-order reaction	21B.6

21C Transition-state theory

Contents

➤ Why do you need to know this material?

Transition-state theory provides a way to relate the rate constant of reactions to models of the cluster of atoms that is proposed to be formed when reactants come together. It provides a link between information about the structures of reactants and the rate constant for their reaction.

➤ What is the key idea?

Reactants come together to form an activated complex that decays into products.

➤ What do you need to know already?

This Topic makes use of two strands: one is the relation between equilibrium constants and partition functions (Topic 15F); the other is the relation between equilibrium constants and thermodynamic functions, such as the Gibbs energy, enthalpy, and entropy of reaction (Topic 6A). You need to be aware of the Arrhenius equation for the temperature dependence of the rate constant (Topic 20D).

In **transition-state theory** (which is also widely referred to as *activated complex theory*), the notion of the transition state is used in conjunction with concepts of statistical thermodynamics to provide a more detailed calculation of rate constants than that presented by collision theory (Topic 21A). Transition-state theory has the advantage that a quantity corresponding to the steric factor appears automatically, and P does not need to be grafted on to an equation as an afterthought; it is an attempt to identify the principal features governing the size of a rate constant in terms of a model of the events that take place during the reaction. There are several approaches to the formulation of transition-state theory; here we present the simplest.

21C.1 The Eyring equation

In the course of a chemical reaction that begins with a collision between molecules of A and molecules of B, the potential energy of the system typically changes in a manner shown in Fig. 21C.1. Although the illustration displays an exothermic reaction, a potential barrier is also common for endothermic reactions. As the reaction event proceeds, A and B come into contact, distort, and begin to exchange or discard atoms.

(a) The formulation of the equation

The **reaction coordinate** is a representation of the atomic displacements, such as changes in interatomic distances and bond angles, that are directly involved in the formation of products from reactants. The potential energy rises to a maximum and the cluster of atoms that corresponds to the region close to the maximum is called the **activated complex**. After the maximum, the potential energy falls as the atoms rearrange in the cluster and reaches a value characteristic of the products. The climax of the reaction is at the peak of the potential energy, which can be identified with the activation energy E_a; however, as in collision theory, this identification should be regarded as approximate and we clarify it later. Here two reactant molecules have come to such a degree of closeness and distortion that a small further distortion will send them in the direction of products. This crucial configuration is called the **transition state** of the reaction. Although some molecules entering the transition state might revert to reactants, if they pass through this configuration then it is inevitable that products will emerge from the encounter.

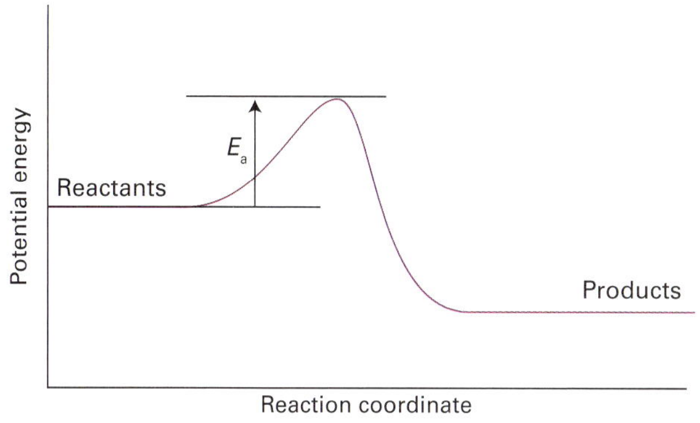

Figure 21C.1 A potential energy profile for an exothermic reaction. The height of the barrier between the reactants and products is the activation energy of the reaction.

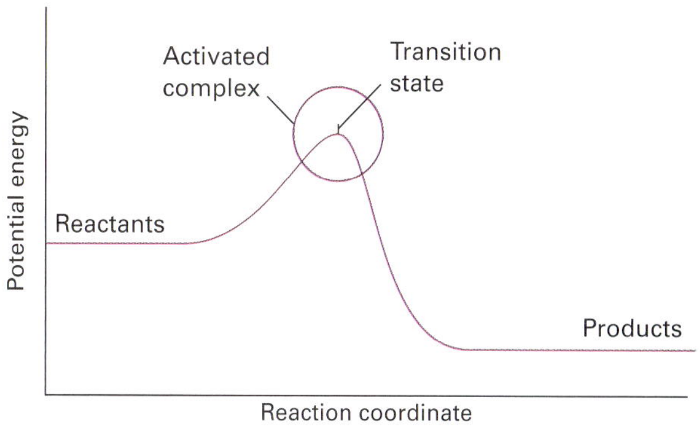

Figure 21C.2 A reaction profile (for an exothermic reaction). The horizontal axis is the reaction coordinate, and the vertical axis is potential energy. The activated complex is the region near the potential maximum, and the transition state corresponds to the maximum itself.

A note on good practice The terms *activated complex* and *transition state* are often used as synonyms; however, it is best to preserve the distinction, with the former referring to the cluster of atoms and the latter to their critical configuration.

Transition state theory pictures a reaction between A and B as proceeding through the formation of an activated complex, $C^{\ddagger}$, in a rapid pre-equilibrium (Fig. 21C.2):

$$A+B \rightleftharpoons C^{\ddagger} \qquad K^{\ddagger}=\frac{p_{C^{\ddagger}}p^{\ominus}}{p_A p_B} \tag{21C.1}$$

where we have replaced the activity of each species by $p/p^{\ominus}$. When we express the partial pressures, p_J, in terms of the molar concentrations, [J], by using $p_J=RT[J]$, the concentration of activated complex is related to the (dimensionless) equilibrium constant by

$$[C^{\ddagger}]=\frac{RT}{p^{\ominus}}K^{\ddagger}[A][B] \tag{21C.2}$$

The activated complex falls apart by unimolecular decay into products, P, with a rate constant $k^{\ddagger}$:

$$C^{\ddagger} \rightarrow P \qquad v=k^{\ddagger}[C^{\ddagger}] \tag{21C.3}$$

It follows that

$$v=k_r[A][B] \qquad k_r=\frac{RT}{p^{\ominus}}k^{\ddagger}K^{\ddagger} \tag{21C.4}$$

Our task is to calculate the unimolecular rate constant $k^{\ddagger}$ and the equilibrium constant $K^{\ddagger}$.

(b) The rate of decay of the activated complex

An activated complex can form products if it passes through the transition state. As the reactant molecules approach the activated complex region, some bonds are forming and shortening while others are lengthening and breaking; therefore, along the reaction coordinate, there is a vibration-like motion of the atoms in the activated complex. If this motion occurs with a frequency $\nu^{\ddagger}$, then the frequency with which the cluster of atoms forming the complex approaches the transition state is also $\nu^{\ddagger}$. However, it is possible that not every oscillation along the reaction coordinate takes the complex through the transition state. For instance, the centrifugal effect of rotations might also be an important contribution to the break-up of the complex, and in some cases the complex might be rotating too slowly or rotating rapidly but about the wrong axis. Therefore, we suppose that the rate of passage of the complex through the transition state is proportional to the vibrational frequency along the reaction coordinate, and write

$$k^{\ddagger}=\kappa\nu^{\ddagger} \tag{21C.5}$$

where κ (kappa) is the **transmission coefficient**. In the absence of information to the contrary, κ is assumed to be about 1.

Brief illustration 21C.1 The transmission coefficient

Typical molecular vibration wavenumbers of small molecules occur at wavenumbers of the order of $10^3\,cm^{-1}$ (C–H bends, for example, occur in the range 1340–1465 cm^{-1}) and therefore occur at frequencies of the order of 10^{13} Hz. If we suppose that the loosely bound cluster vibrates at one or two orders of magnitude lower frequency, then $\nu^{\ddagger}\approx 10^{11}$–$10^{12}$ Hz. These figures suggest that $\nu^{\ddagger}\approx 10^{11}$–$10^{12}\,s^{-1}$, with κ perhaps reducing that value further.

Self-test 21C.1 Estimate the change in $\nu^{\ddagger}$ that would occur if ^1H is replaced by ^2H in a C–H group at the site of reaction. Assume that the C atom is immobile.

Answer: $\nu^{\ddagger}\rightarrow\nu^{\ddagger}/2^{1/2}$

(c) The concentration of the activated complex

Topic 15F explains how to calculate equilibrium constants from structural data. Equation 15F.10 of that Topic (K in terms of the standard molar partition functions $q_J^{\ominus}$) can be used directly, which in this case gives

$$K^{\ddagger}=\frac{N_A q_{C^{\ddagger}}^{\ominus}}{q_A^{\ominus} q_B^{\ominus}}\mathrm{e}^{-\Delta E_0/RT} \quad (21C.6)$$

with

$$\Delta E_0=E_0(C^{\ddagger})-E_0(A)-E_0(B) \quad (21C.7)$$

Note that the units of N_A and the $q_J^{\ominus}$ are mol^{-1}, so $K^{\ddagger}$ is dimensionless (as is appropriate for an equilibrium constant).

In the final step of this part of the calculation, we focus attention on the partition function of the activated complex. We have already assumed that a vibration of the activated complex $C^{\ddagger}$ tips it through the transition state. The partition function for this vibration is (see eqn 15B.15 of Topic 15B, which is essentially the following):

$$q=\frac{1}{1-\mathrm{e}^{-h\nu^{\ddagger}/kT}}$$

where $\nu^{\ddagger}$ is its frequency (the same frequency that determines $k^{\ddagger}$). This frequency is much lower than for an ordinary molecular vibration because the oscillation corresponds to the complex falling apart (Fig. 21C.3), so the force constant is very low.

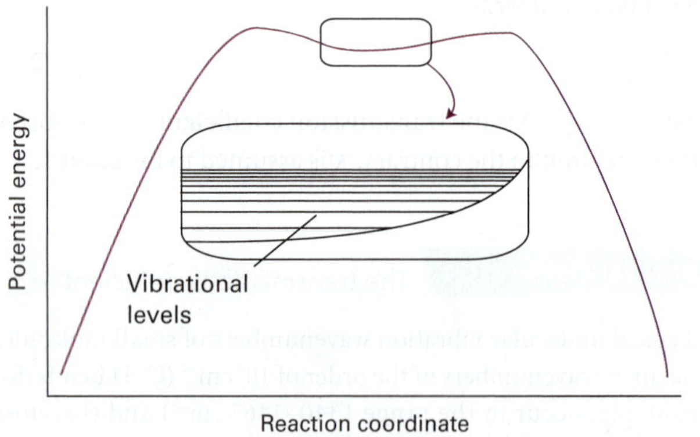

Figure 21C.3 In an elementary depiction of the activated complex close to the transition state, there is a broad, shallow dip in the potential energy surface along the reaction coordinate. The complex vibrates harmonically and almost classically in this well. However, this depiction is an oversimplification, for in many cases there is no dip at the top of the barrier, and the curvature of the potential energy, and therefore the force constant, is negative. Formally, the vibrational frequency is then imaginary. We ignore this problem here.

Therefore, provided that $h\nu^{\ddagger}/kT \ll 1$, the exponential may be expanded and the partition function reduces to

$$q=\frac{1}{1-(1-h\nu^{\ddagger}/kT+\cdots)}\approx\frac{kT}{h\nu^{\ddagger}}$$

We can therefore write

$$q_{C^{\ddagger}}^{\ominus}=\frac{kT}{h\nu^{\ddagger}}\bar{q}_{C^{\ddagger}} \quad (21C.8)$$

where $\bar{q}_{C^{\ddagger}}$ denotes the partition function for all the other modes of the complex. The constant $K^{\ddagger}$ is therefore

$$K^{\ddagger}=\frac{kT}{h\nu^{\ddagger}}\bar{K}^{\ddagger} \qquad \bar{K}^{\ddagger}=\frac{N_A \bar{q}_{C^{\ddagger}}^{\ominus}}{q_A^{\ominus} q_B^{\ominus}}\mathrm{e}^{-\Delta E_0/RT} \quad (21C.9)$$

with $\bar{K}^{\ddagger}$ a kind of equilibrium constant, but with one vibrational mode of $C^{\ddagger}$ discarded.

Brief illustration 21C.2 The discarded mode

Consider the case of two structureless particles A and B colliding to give an activated complex that resembles a diatomic molecule. The activated complex is a diatomic cluster. It has one vibrational mode, but that mode corresponds to motion along the reaction coordinate and therefore does not appear in $\bar{q}_{C^{\ddagger}}^{\ominus}$. It follows that the standard molar partition function of the activated complex has only rotational and translational contributions.

Self-test 21C.2 Which mode would be discarded for a reaction in which the activated complex is modelled as a linear triatomic cluster?

Answer: Antisymmetric stretch

(d) The rate constant

We can now combine all the parts of the calculation into

$$k_r=\frac{RT}{p^{\ominus}}k^{\ddagger}K^{\ddagger}=\kappa\nu^{\ddagger}\frac{kT}{h\nu^{\ddagger}}\frac{RT}{p^{\ominus}}\bar{K}^{\ddagger}$$

At this stage the unknown frequencies $\nu^{\ddagger}$ (in blue) cancel, and after writing $\bar{K}_c^{\ddagger}=(RT/p^{\ominus})\bar{K}^{\ddagger}$, we obtain the **Eyring equation**:

$$k_r=\kappa\frac{kT}{h}\bar{K}_c^{\ddagger} \quad \text{Eyring equation} \quad (21C.10)$$

The factor $\bar{K}_c^{\ddagger}$ is given by eqn 21C.9 and the definition $\bar{K}_c^{\ddagger}=(RT/p^{\ominus})\bar{K}^{\ddagger}$ in terms of the partition functions of A, B, and $C^{\ddagger}$, so in principle we now have an explicit expression for calculating the second-order rate constant for a bimolecular reaction in terms of the molecular parameters for the reactants and the activated complex and the quantity κ.

The partition functions for the reactants can normally be calculated quite readily by using either spectroscopic information about their energy levels or the approximate expressions set out in Table 15C.1. The difficulty with the Eyring equation, however, lies in the calculation of the partition function of the activated complex: $C^‡$ is difficult to investigate spectroscopically (but see the following section), and in general we need to make assumptions about its size, shape, and structure. We shall illustrate what is involved in one simple but significant case.

Example 21C.1 Analysing the collision of structureless particles

Consider the case of two structureless (and different) particles A and B colliding to give an activated complex that resembles a diatomic molecule and deduce an expression for the rate constant of the reaction A + B → Products.

Method Because the reactants J = A, B are structureless 'atoms', the only contributions to their partition functions are the translational terms. The activated complex is a diatomic cluster of mass $m_{C^‡}=m_A+m_B$ and moment of inertia I. It has one vibrational mode but, as explained in *Brief illustration* 21C.2, that mode corresponds to motion along the reaction coordinate. It follows that the standard molar partition function of the activated complex has only rotational and translational contributions. Expressions for the relevant partition functions are given in Topic 15B.

Answer The translational partition functions are

$$q_J^{\ominus}=\frac{V_m^{\ominus}}{\Lambda_J^3}\qquad \Lambda_J=\frac{h}{(2\pi m_J kT)^{1/2}}\qquad V_m^{\ominus}=\frac{RT}{p^{\ominus}}$$

with J = A, B, and $C^‡$ and with $m_{C^‡}=m_A+m_B$. The expression for the partition function of the activated complex is

$$\bar{q}_{C^‡}^{\ominus}=\frac{2IkT}{\hbar^2}\frac{V_m^{\ominus}}{\Lambda_{C^‡}^3}$$

where we have used the high-temperature form of the rotational partition function (Topic 15B). By substituting these expressions into the Eyring equation, we find that the rate constant is

$$k_r=\kappa\frac{kT}{h}\frac{RT}{p^{\ominus}}\left(\frac{N_A\Lambda_A^3\Lambda_B^3}{\Lambda_{C^‡}^3V_m^{\ominus}}\right)\frac{2IkT}{\hbar^2}e^{-\Delta E_0/RT}$$

$$=\kappa\frac{kT}{h}N_A\left(\frac{\Lambda_A\Lambda_B}{\Lambda_{C^‡}}\right)^3\frac{2IkT}{\hbar^2}e^{-\Delta E_0/RT}$$

The moment of inertia of a diatomic molecule of bond length r is μr^2, where $\mu=m_Am_B/(m_A+m_B)$, so after introducing the expressions for the thermal wavelengths and cancelling common terms, we find (Problem 21C.3)

$$k_r=\kappa N_A\left(\frac{8kT}{\pi\mu}\right)^{1/2}\pi r^2e^{-\Delta E_0/RT}$$

Finally, by identifying $\kappa\pi r^2$ as the reactive cross-section σ^*, we arrive at precisely the same expression as that obtained from simple collision theory (eqn 21A.9):

$$k_r=N_A\left(\frac{8kT}{\pi\mu}\right)^{1/2}\sigma^*e^{-\Delta E_0/RT}$$

Self-test 21C.3 What additional factors would be present if the reaction were AB + C → Products through a linear activated complex?

Answer: Rotation and vibration of AB, bends and symmetric stretch of the activated complex

(e) Observation and manipulation of the activated complex

The development of femtosecond pulsed lasers has made it possible to make observations on species that have such short lifetimes that in a number of respects they resemble an activated complex, which often survive for only a few picoseconds. In a typical experiment designed to detect an activated complex, a femtosecond laser pulse is used to excite a molecule to a dissociative state, and then the system is exposed to a second femtosecond pulse at an interval after the dissociating pulse. The frequency of the second pulse is set at an absorption of one of the free fragmentation products, so the intensity of its absorption is a measure of the abundance of the dissociation product. For example, when ICN is dissociated by the first pulse, the emergence of CN from the photoactivated state can be monitored by watching the growth of the free CN absorption (or, more commonly, its laser-induced fluorescence). In this way it has been found that the CN signal remains zero until the fragments have separated by about 600 pm, which takes about 205 fs.

Some sense of the progress that has been made in the study of the intimate mechanism of chemical reactions can be obtained by considering the decay of the ion pair Na^+I^-. As shown in Fig. 21C.4, excitation of the ionic species with a femtosecond laser pulse forms an excited state that corresponds to a covalently bonded NaI molecule. The system can be described with two potential energy surfaces, one largely 'ionic' and another 'covalent', which cross at an internuclear separation of 693 pm. A short laser pulse is composed of a wide range of frequencies, which excite many vibrational states of NaI simultaneously. Consequently, the electronically excited complex exists as a superposition of states, or a localized wavepacket, which oscillates between the 'covalent' and 'ionic' potential energy

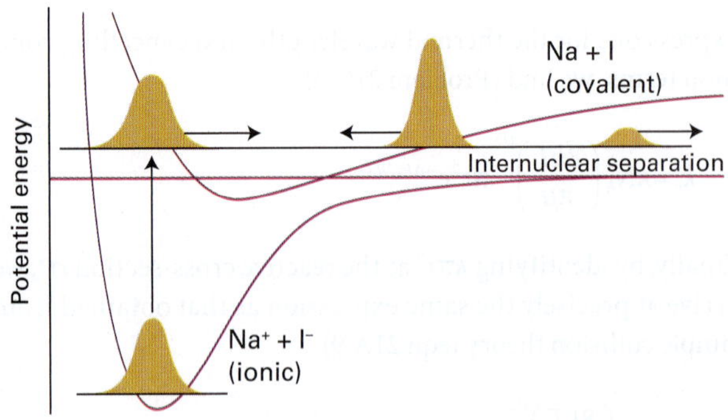

Figure 21C.4 Excitation of the ion pair Na^+I^- forms an excited state with covalent character. Also shown is migration between a 'covalent' surface (upper curve) and an 'ionic' surface (lower curve) of the wavepacket formed by laser excitation.

surfaces, as shown in Fig. 21C.4. The complex can also dissociate, shown as movement of the wavepacket towards very long internuclear separation along the dissociative surface. However, not every outward-going swing leads to dissociation because there is a chance that the I atom can be harpooned again, in which case it fails to make good its escape. The dynamics of the system is probed by a second laser pulse with a frequency that corresponds to the absorption frequency of the free Na product or to the frequency at which Na absorbs when it is a part of the complex. The latter frequency depends on the Na···I distance, so an absorption (in practice, a laser-induced fluorescence) is obtained each time the wavepacket returns to that separation.

Brief illustration 21C.3 Femtosecond analysis

A typical set of results is shown in Fig. 21C.5. The bound Na absorption intensity shows up as a series of pulses that recur in about 1 ps, showing that the wavepacket oscillates with about

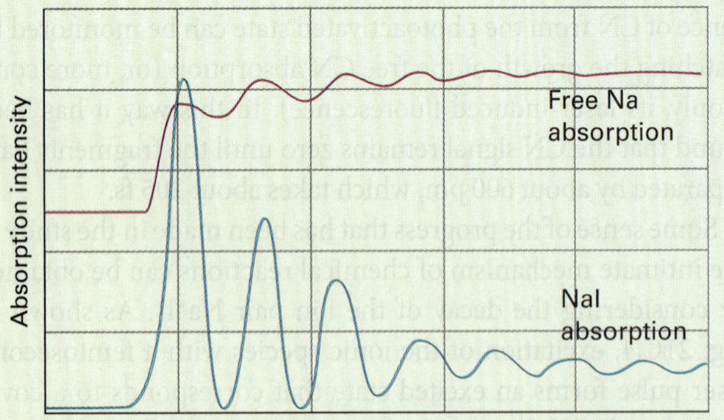

Figure 21C.5 Femtosecond spectroscopic results for the reaction in which sodium iodide separates into Na and I. The lower curve is the absorption of the electronically excited complex and the upper curve is the absorption of free Na atoms. (Adapted from A.H. Zewail, *Science* **242**, 1645 (1988).)

that period. The decline in intensity shows the rate at which the complex can dissociate as the two atoms swing away from each other. The free Na absorption also grows in an oscillating manner, showing the periodicity of wavepacket oscillation, each swing of which gives it a chance to dissociate. The precise period of the oscillation in NaI is 1.25 ps. The complex survives for about ten oscillations. In contrast, although the oscillation frequency of NaBr is similar, it barely survives one oscillation.

Self-test 21C.4 Confirm the assumption in transition-state theory that the vibrational frequency of the dissociative mode of the activated complex is very low by calculating the vibrational wavenumber corresponding to the 1.25 ps period of oscillation in NaI.

Answer: $27\,\text{cm}^{-1}$

Femtosecond spectroscopy has also been used to examine analogues of the activated complex involved in bimolecular reactions. Thus, a molecular beam can be used to produce a complex held together by van der Waals interactions (a 'van der Waals molecule'), such as IH···OCO. The HI bond can be dissociated by a femtosecond pulse, and the H atom is ejected towards the O atom of the neighbouring CO_2 molecule to form HOCO. Hence, the van der Waals molecule is a source of a species that resembles the activated complex of the reaction

$$H + CO_2 \rightarrow [HOCO]^{\ddagger} \rightarrow HO + CO$$

The probe pulse is tuned to the OH radical, which enables the evolution of $[HOCO]^{\ddagger}$ to be studied in real time.

The techniques used for the spectroscopic detection of transition states can also be used to control the outcome of a chemical reaction by direct manipulation of the activated complex. Consider the reaction $I_2 + Xe \rightarrow XeI^* + I$, which occurs by a harpoon mechanism with an activated complex denoted as $[Xe^+\cdots I^-\cdots I]$. The reaction can be initiated by exciting I_2 to an electronic state at least $52\,460\,\text{cm}^{-1}$ above the ground state and then followed by measuring the time dependence of the chemiluminescence of XeI^*. To exert control over the yield of the product, a pair of femtosecond pulses can be used to induce the reaction. The first pulse excites the I_2 molecule to a low energy and unreactive electronic state. We already know that excitation by a femtosecond pulse generates a wavepacket that can be treated as a particle travelling across the potential energy surface. In this case, the wavepacket does not have enough energy to react, but excitation by another laser pulse with the appropriate wavelength can provide the necessary additional energy. It follows that activated complexes with different geometries can be prepared by varying the time delay between the two pulses, as the partially localized wavepacket will be at different locations on the potential energy surface as it evolves after being formed by the first pulse. Because the reaction occurs by the harpoon mechanism, the product yield is expected to be optimal if the second pulse is applied when the wavepacket is at a point where the Xe···I_2 distance is just right for electron transfer from Xe to I_2 to occur. This type of control of the I_2+Xe reaction has been demonstrated.

21C.2 Thermodynamic aspects

The statistical thermodynamic version of transition state theory rapidly runs into difficulties because only in some cases is anything known about the structure of the activated complex. However, the concepts that it introduces, principally that of an equilibrium between the reactants and the activated complex, have motivated a more general, empirical approach in which the activation process is expressed in terms of thermodynamic functions.

(a) Activation parameters

If we accept that $\bar{K}^{\ddagger}$ is an equilibrium constant (despite one mode of $C^{\ddagger}$ having been discarded), we can express it in terms of a **Gibbs energy of activation**, $\Delta^{\ddagger}G$, through the definition

$$\Delta^{\ddagger}G=-RT\ln\bar{K}^{\ddagger}$$ *Definition* Gibbs energy of activation (21C.11)

All the $\Delta^{\ddagger}X$ in this section are *standard* thermodynamic quantities, $\Delta^{\ddagger}X^{\ominus}$, but we shall omit the standard state sign to avoid overburdening the notation. Then the rate constant becomes

$$k_r=\kappa\frac{kT}{h}\frac{RT}{p^{\ominus}}e^{-\Delta^{\ddagger}G/RT} \qquad (21C.12)$$

Because $G=H-TS$, the Gibbs energy of activation can be divided into an **entropy of activation**, $\Delta^{\ddagger}S$, and an **enthalpy of activation**, $\Delta^{\ddagger}H$, by writing

$$\Delta^{\ddagger}G=\Delta^{\ddagger}H-T\Delta^{\ddagger}S$$ *Definition* Entropy and enthalpy of activation (21C.13)

When eqn 21C.13 is used in eqn 21C.12 and κ is absorbed into the entropy term, we obtain

$$k_r=Be^{\Delta^{\ddagger}S/R}e^{-\Delta^{\ddagger}H/RT} \qquad B=\frac{kT}{h}\frac{RT}{p^{\ominus}} \qquad (21C.14)$$

The formal definition of activation energy (eqn 20D.2 of Topic 20D, $E_a=RT^2(\partial\ln k_r/\partial T)$), then gives $E_a=\Delta^{\ddagger}H+2RT$, so[1]

$$k_r=e^2Be^{\Delta^{\ddagger}S/R}e^{-E_a/RT} \qquad (21C.15a)$$

from which it follows that the Arrhenius factor A can be identified as

$$A=e^2Be^{\Delta^{\ddagger}S/R}$$ *Transition-state theory* A-factor (21C.15b)

The entropy of activation is negative because throughout the system reactant species are combining to form reactive pairs. However, if there is a reduction in entropy below what would be expected for the simple encounter of A and B, then the Arrhenius factor A will be smaller than that expected on the basis of simple collision theory. Indeed, we can identify that *additional* reduction in entropy, $\Delta^{\ddagger}S_{\text{steric}}$, as the origin of the steric factor of collision theory, and write

$$P=e^{\Delta^{\ddagger}S_{\text{steric}}/R}$$ *Transition-state theory* P-factor (21C.15c)

Thus, the more complex the steric requirements of the encounter, the more negative the value of $\Delta^{\ddagger}S_{\text{steric}}$, and the smaller the value of P.

Brief illustration 21C.4 Activation parameters

The reaction of propylxanthate ion in acetic acid buffer solutions can be represented by the equation $A^-+H^+\rightarrow P$. Near 30 °C, $A=2.05\times10^{13}\,dm^3\,mol^{-1}\,s^{-1}$. To evaluate the entropy of activation at 30 °C we first note that because the reaction is in solution the e^2 of eqn 21C.15 should be replaced by e (see footnote 1), and then use eqn 21C.15b in the form

$$\Delta^{\ddagger}S=R\ln\frac{A}{eB} \quad \text{with } B=\frac{kT}{h}\frac{RT}{p^{\ominus}}=1.592\times10^{14}\,dm^3\,mol^{-1}\,s^{-1}$$

Therefore,

$$\Delta^{\ddagger}S=R\ln\frac{2.05\times10^{13}\,dm^3\,mol^{-1}\,s^{-1}}{e\times(1.592\times10^{14}\,dm^3\,mol^{-1}\,s^{-1})}=R\ln 0.047\ldots$$
$$=-25.4\,J\,K^{-1}\,mol^{-1}$$

Self-test 21C.5 The reaction $A^-+H^+\rightarrow P$ in solution has $A=6.92\times10^{12}\,dm^3\,mol^{-1}\,s^{-1}$. Evaluate the entropy of activation at 25 °C.

Answer: $-34.1\,J\,K^{-1}\,mol^{-1}$

Gibbs energies, enthalpies, and entropies of activation (and volumes and heat capacities of activation) are widely used to report experimental reaction rates, especially for organic reactions in solution. They are encountered when relationships between equilibrium constants and rates of reaction are explored using **correlation analysis**, in which ln K (which is equal to $-\Delta_rG^{\ominus}/RT$) is plotted against ln k_r (which is proportional to $-\Delta^{\ddagger}G/RT$). In many cases the correlation is linear, signifying that as the reaction becomes thermodynamically more

[1] For reactions of the type $A+B\rightarrow P$ in the gas phase, $E_a=\Delta^{\ddagger}H+2RT$. For such reactions in solution, $E_a=\Delta^{\ddagger}H+RT$.

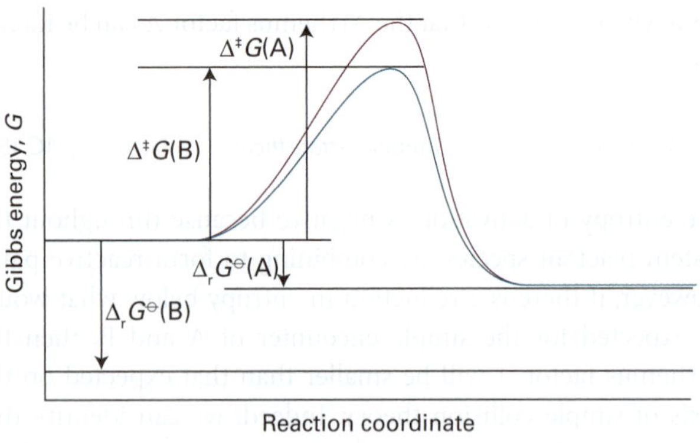

Figure 21C.6 For a related series of reactions, as the magnitude of the standard reaction Gibbs energy increases, so the activation barrier decreases and the rate constant increases. The approximate linear correlation between $\Delta^{\ddagger}G$ and $\Delta_r G^{\ominus}$ is the origin of linear free energy relations.

favourable, its rate constant increases (Fig. 21C.6). This linear correlation is the origin of the alternative name **linear free energy relation** (LFER).

(b) Reactions between ions

The full statistical thermodynamic theory is very complicated to apply because the solvent plays a role in the activated complex. The thermodynamic version of transition-state theory simplifies the discussion of reactions in solution and is applicable to non-ideal systems. In the thermodynamic approach we combine the rate law

$$\frac{d[P]}{dt}=k^{\ddagger}[C^{\ddagger}]$$

with the thermodynamic equilibrium constant (Topic 6A)

$$K=\frac{a_{C^{\ddagger}}}{a_A a_B}=K_{\gamma}\frac{[C^{\ddagger}]c^{\ominus}}{[A][B]} \qquad K_{\gamma}=\frac{\gamma_{C^{\ddagger}}}{\gamma_A\gamma_B}$$

Then

$$\frac{d[P]}{dt}=k_r[A][B] \qquad k_r=\frac{k^{\ddagger}K}{K_{\gamma}c^{\ominus}} \tag{21C.16a}$$

If k_r° is the rate constant when the activity coefficients are 1 (that is, $k_r^{\circ}=k^{\ddagger}K/c^{\ominus}$),we can write

$$k_r=\frac{k_r^{\circ}}{K_{\gamma}} \qquad \log k_r=\log k_r^{\circ}-\log K_{\gamma} \tag{21C.16b}$$

At low concentrations the activity coefficients can be expressed in terms of the ionic strength, I, of the solution by using the Debye–Hückel limiting law (Topic 5F, particularly eqn 5F.8, $\log\gamma_{\pm}=-A|z_+z_-|I^{1/2}$). However, we need the expressions for the individual ions rather than the mean value, and so write $\log \gamma_J=-Az_J^2I^{1/2}$ and

$$\log\gamma_A=-Az_A^2I^{1/2} \qquad \log\gamma_B=-Az_B^2I^{1/2} \tag{21C.17a}$$

with $A=0.509$ in aqueous solution at 298 K and z_A and z_B the (signed) charge numbers of A and B, respectively. Because the activated complex forms from reaction of one of the ions of A with one of the ions of B, the charge number of the activated complex is z_A+z_B where z_J is positive for cations and negative for anions. Therefore

$$\log\gamma_{C^{\ddagger}}=-A(z_A+z_B)^2I^{1/2} \tag{21C.17b}$$

Inserting these relations into eqn 21C.16b results in

$$\begin{aligned}\log k_r&=\log k_r^{\circ}-A\{z_A^2+z_B^2-(z_A+z_B)^2\}I^{1/2}\\&=\log k_r^{\circ}+2Az_Az_BI^{1/2}\end{aligned} \tag{21C.18}$$

Equation 21C.18 expresses the **kinetic salt effect**, the variation of the rate constant of a reaction between ions with the ionic strength of the solution (Fig. 21C.7). If the reactant ions have the same sign (as in a reaction between cations or between anions), then increasing the ionic strength by the addition of inert ions increases the rate constant. The formation of a single, highly charged ionic complex from two less highly charged ions is favoured by a high ionic strength because the new ion has a denser ionic atmosphere and interacts with that atmosphere more strongly. Conversely, ions of opposite charge react more slowly in solutions of high ionic strength. Now the charges cancel and the complex has a less favourable interaction with its atmosphere than the separated ions.

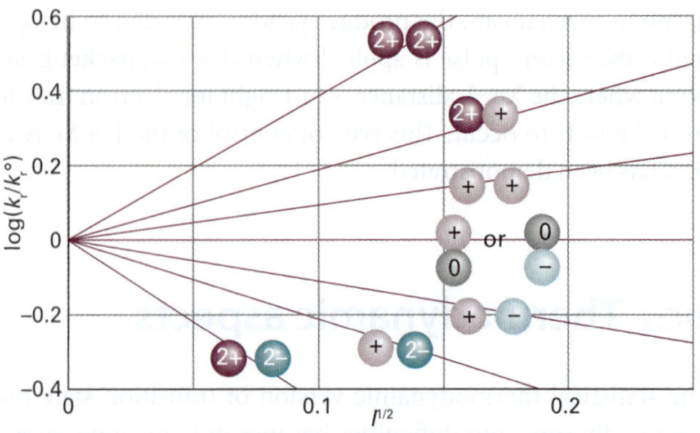

Figure 21C.7 Experimental tests of the kinetic salt effect for reactions in water at 298 K. The ion types are shown as spheres, and the slopes of the lines are those given by the Debye–Hückel limiting law and eqn 21C.18.

Example 21C.2 Analysing the kinetic salt effect

The rate constant for the base (OH^-) hydrolysis of $[CoBr(NH_3)_5]^{2+}$ varies with ionic strength as tabulated below. What can be deduced about the charge of the activated complex in the rate-determining stage? We cannot assume without more evidence that it is a bimolecular process with an activated complex of charge +1.

I	0.0050	0.0100	0.0150	0.0200	0.0250	0.0300
k_r/k_r°	0.718	0.631	0.562	0.515	0.475	0.447

Method According to eqn 21C.18, a plot of $\log(k_r/k_r^\circ)$ against $I^{1/2}$ will have a slope of $1.02z_Az_B$, from which we can infer the charges of the ions involved in the formation of the activated complex.

Answer Form the following table:

I	0.0050	0.0100	0.0150	0.0200	0.0250	0.0300
$I^{1/2}$	0.071	0.100	0.122	0.141	0.158	0.173
$\log(k_r/k_r^\circ)$	−0.14	−0.20	−0.25	−0.29	−0.32	−0.35

These points are plotted in Fig. 21C.8. The slope of the (least squares) straight line is −2.04, indicating that $z_Az_B=-2$. Because $z_A=-1$ for the OH^- ion, if that ion is involved in the formation of the activated complex, then the charge number of the second ion is +2. This analysis suggests that the pentaamminebromidocobalt(III) cation $[CoBr(NH_3)_5]^{2+}$ participates in the formation of the activated complex and that the charge of the activated complex is $-1+2=+1$. Although we do not pursue the point here, you should be aware that the rate constant is also influenced by the relative permittivity of the medium.

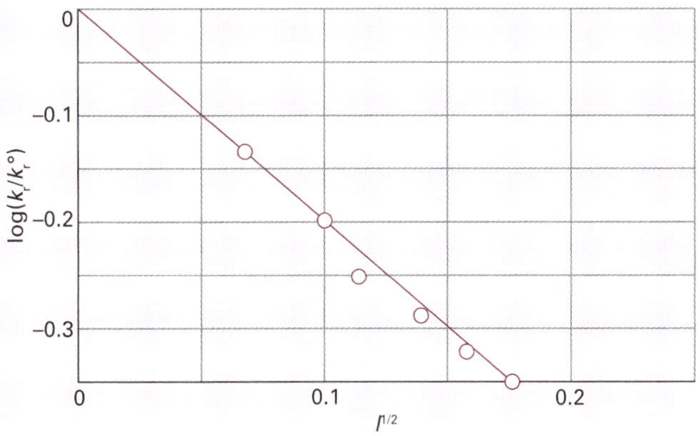

Figure 21C.8 The experimental ionic strength dependence of the rate constant of a hydrolysis reaction: the slope gives information about the charge types involved in the activated complex of the rate determining step. See *Example* 21C.2.

Self-test 21C.6 An ion of charge number +1 is known to be involved in the activated complex of a reaction. Deduce the charge number of the other ion from the following data:

I	0.0050	0.0100	0.0150	0.0200	0.0250	0.0300
k_r/k_r°	0.930	0.902	0.884	0.867	0.853	0.841

Answer: −1

21C.3 The kinetic isotope effect

The postulation of a plausible reaction mechanism requires careful analysis of many experiments designed to determine the fate of atoms during the formation of products. Observation of the **kinetic isotope effect**, a decrease in the rate of a chemical reaction upon replacement of one atom in a reactant by a heavier isotope, facilitates the identification of bond-breaking events in the rate-determining step. A **primary kinetic isotope effect** is observed when the rate-determining step requires the scission of a bond involving the isotope. A **secondary kinetic isotope effect** is the reduction in reaction rate even though the bond involving the isotope is not broken to form product. In both cases, the effect arises from the change in activation energy that accompanies the replacement of an atom by a heavier isotope on account of changes in the zero-point vibrational energies. We now explore the primary kinetic isotope effect in some detail.

Consider a reaction in which a C–H bond is cleaved. If scission of this bond is the rate-determining step (Topic 20E), then the reaction coordinate corresponds to the stretching of the C–H bond and the potential energy profile is shown in Fig. 21C.9. On deuteration, the dominant change is the reduction of the zero-point energy of the bond (because the deuterium atom is heavier). The whole reaction profile is not lowered, however, because the relevant vibration in the activated complex has a very low force constant, so there is little zero-point energy associated with the reaction coordinate in either form of the activated complex. We show in the following *Justification*, that, as a consequence of this reduction, the activation energy change upon deuteration is

$$E_a(\mathrm{C{-}D})-E_a(\mathrm{C{-}H})=\tfrac{1}{2}N_A\hbar\omega(\mathrm{C{-}H})\left\{1-\left(\frac{\mu_{\mathrm{CH}}}{\mu_{\mathrm{CD}}}\right)^{1/2}\right\} \tag{21C.19}$$

where ω is the relevant vibrational frequency (in radians per second), μ is the relevant effective mass, and

$$\frac{k_r(\mathrm{C{-}D})}{k_r(\mathrm{C{-}H})}=e^{-\zeta} \quad \text{with} \quad \zeta=\frac{\hbar\omega(\mathrm{C{-}H})}{2kT}\left\{1-\left(\frac{\mu_{\mathrm{CH}}}{\mu_{\mathrm{CD}}}\right)^{1/2}\right\} \tag{21C.20}$$

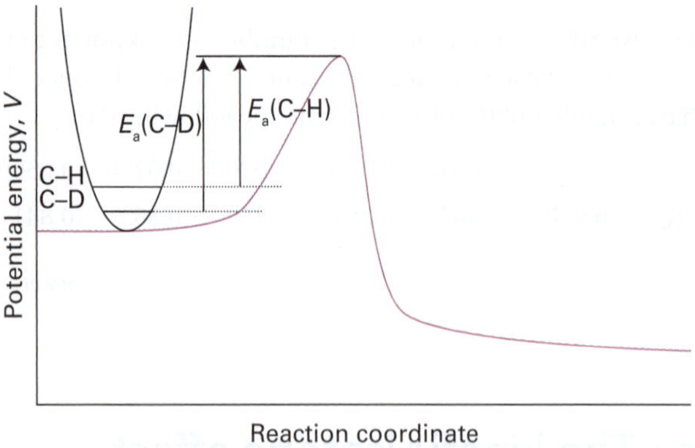

Figure 21C.9 Changes in the reaction profile when a C–H bond undergoing cleavage is deuterated. In this figure the C–H and C–D bonds are modelled as harmonic oscillators. The only significant change is in the zero-point energy of the reactants, which is lower for C–D than for C–H. As a result, the activation energy is greater for C–D cleavage than for C–H cleavage.

Note that $\zeta > 0$ (ζ is zeta) because $\mu_{CD} > \mu_{CH}$ and that $k_r(C\text{–}D)/k_r(C\text{–}H) < 1$, meaning that, as expected from Fig. 21C.9, the rate constant decreases upon deuteration. We also conclude that $k_r(C\text{–}D)/k_r(C\text{–}H)$ decreases with decreasing temperature.

Justification 21C.1 The primary kinetic isotope effect

We assume that, to a good approximation, a change in the activation energy arises only from the change in zero-point energy of the stretching vibration, so, from Fig. 21C.9,

$$E_a(C\text{–}D) - E_a(C\text{–}H) = N_A\{\tfrac{1}{2}\hbar\omega(C\text{–}H) - \tfrac{1}{2}\hbar\omega(C\text{–}D)\}$$
$$= \tfrac{1}{2}N_A\hbar\{\omega(C\text{–}H) - \omega(C\text{–}D)\}$$

where ω is the relevant vibrational frequency. From Topic 12D, we know that $\omega(C\text{–}D) = (\mu_{CH}/\mu_{CD})^{1/2}\omega(C\text{–}H)$, where μ is the relevant effective mass. Making this substitution in the equation above gives eqn 21C.19.

If we assume further that the pre-exponential factor does not change upon deuteration, then the rate constants for the two species should be in the ratio

$$\frac{k_r(C\text{–}D)}{k_r(C\text{–}H)} = e^{-\{E_a(C\text{–}D)-E_a(C\text{–}H)\}/RT} = e^{-\{E_a(C\text{–}D)-E_a(C\text{–}H)\}/N_A kT}$$

where we used $R = N_A k$. Equation 21C.20 follows after using eqn 21C.19 for $E_a(C\text{–}D) - E_a(C\text{–}H)$ in this expression.

Brief illustration 21C.5 The primary kinetic isotope effect

From infrared spectra, the fundamental vibrational wavenumber $\tilde{\nu}$ for stretching of a C–H bond is about 3000 cm^{-1}. To convert this wavenumber to an angular frequency, $\omega = 2\pi\nu$, we use $\omega = 2\pi c\tilde{\nu}$ and it follows that

$$\omega = 2\pi \times (2.998\times10^{10}\,\text{cm s}^{-1}) \times (3000\,\text{cm}^{-1})$$
$$= 5.65\ldots \times 10^{14}\,\text{s}^{-1}$$

The ratio of effective masses is

$$\frac{\mu_{CH}}{\mu_{CD}} = \left(\frac{m_C m_H}{m_C + m_H}\right) \times \left(\frac{m_C + m_D}{m_C m_D}\right)$$
$$= \left(\frac{12.01\times1.0078}{12.01+1.0078}\right) \times \left(\frac{12.01+2.0140}{12.01\times2.0140}\right) = 0.539\ldots$$

Now we can use eqn 21C.20, to calculate

$$\zeta = \frac{(1.055\times10^{-34}\,\text{J s})\times(5.65\ldots\times10^{14}\,\text{s}^{-1})}{2\times(1.381\times10^{-23}\,\text{J K}^{-1})\times(298\,\text{K})} \times (1-0.539\ldots^{1/2})$$
$$= 1.92\ldots$$

and

$$\frac{k_r(C\text{–}D)}{k_r(C\text{–}H)} = e^{-1.92\ldots} = 0.146$$

We conclude that at room temperature C–H cleavage should be about seven times faster than C–D cleavage, other conditions being equal. However, experimental values of $k_r(C\text{–}D)/k_r(C\text{–}H)$ can differ significantly from those predicted by eqn 21C.20 on account of the severity of the assumptions in the model.

Self-test 21C.7 The bromination of a deuterated hydrocarbon at 298 K proceeds 6.4 times more slowly than the bromination of the undeuterated material. What value of the force constant for the cleaved bond can account for this difference?

$k_f = 450\,\text{N m}^{-1}$, which is consistent with k_f(C–H)

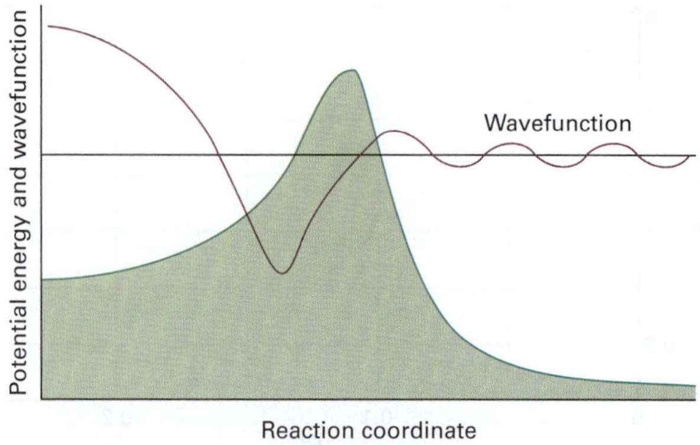

Figure 21C.10 A proton can tunnel through the activation energy barrier that separates reactants from products, so the effective height of the barrier is reduced and the rate of the proton transfer reaction increases. The effect is represented by drawing the wavefunction of the proton near the barrier. Proton tunnelling is important only at low temperatures, when most of the reactants are trapped on the left of the barrier.

In some cases, substitution of deuterium for hydrogen results in values of k_r(C–D)/k_r(C–H) that are too low to be accounted for by eqn 21C.20, even when more complete models are used to predict ratios of rate constants. Such abnormal kinetic isotope effects are evidence for a path in which quantum mechanical tunnelling of hydrogen atoms takes place through the activation barrier (Fig 21C.10). The probability of tunnelling through a barrier decreases as the mass of the particle increases (Topic 8A), so deuterium tunnels less efficiently through a barrier than hydrogen and its reactions are correspondingly slower. Quantum mechanical tunnelling can be the dominant process in reactions involving hydrogen atom or proton transfer when the temperature is so low that very few reactant molecules can overcome the activation energy barrier. We see in Topic 21E that because m_e is so small, tunnelling is also a very important contributor to the rates of electron transfer reactions.

Checklist of concepts

☐ 1. In transition-state theory, it is supposed that an **activated complex** is in equilibrium with the reactants.

☐ 2. The rate at which the activated complex forms products depends on the rate at which it passes through a **transition state**.

☐ 3. The rate constant may be parameterized in terms of the **Gibbs energy, entropy, and enthalpy of activation**.

☐ 4. The **kinetic salt effect** is the effect of an added inert salt on the rate of a reaction between ions.

☐ 5. The **kinetic isotope effect** is the decrease in the rate constant of a chemical reaction upon replacement of one atom in a reactant by a heavier isotope.

Checklist of equations

Property	Equation	Comment	Equation number
'Equilibrium constant' for activated complex formation	$\bar{K}^{\ddagger}=\left(N_A\bar{q}_{C^{\ddagger}}^{\ominus}/q_A^{\ominus}q_B^{\ominus}\right)e^{-\Delta E_0/RT}$	Assume equilibrium; one vibrational mode of $C^{\ddagger}$ discarded	21C.9
Eyring equation	$k_r=\kappa(kT/h)\bar{K}_c^{\ddagger}$	Transition-state theory	21C.10
Gibbs energy of activation	$\Delta^{\ddagger}G=-RT\ln\bar{K}^{\ddagger}$	Definition	21C.11
Enthalpy and entropy of activation	$\Delta^{\ddagger}G=\Delta^{\ddagger}H-T\Delta^{\ddagger}S$	Definition	21C.13
Parameterization	$k_r=e^nBe^{\Delta^{\ddagger}S/R}e^{-E_a/RT}$	$n=2$ for gas-phase reactions; $n=1$ for solution	21C.15a
A-factor	$A=e^nBe^{\Delta^{\ddagger}S/R}$		21C.15b
P-factor	$P=e^{\Delta^{\ddagger}S_{\text{steric}}/R}$		21C.15c
Kinetic salt effect	$\log k_r=\log k_r^{\circ}+2Az_Az_BI^{1/2}$	Assumes Debye–Hückel limiting law valid	21C.18
Primary kinetic isotope effect	$k_r(\text{C–D})/k_r(\text{C–H})=e^{-\zeta}$, $\zeta=(\hbar\omega(\text{C–H})/2kT)\times\{1-(\mu_{CH}/\mu_{CD})^{1/2}\}$	Cleavage of a C–H/D bond in the rate-determining step	21C.20

21D The dynamics of molecular collisions

Contents

➤ **Why do you need to know this material?**

Chemists need to be interested in the details of chemical reactions, and there is no more detailed approach than that involved in the study of the dynamics of reactive encounters, when one molecule collides with another and atoms exchange partners.

➤ **What is the key idea?**

The rates of reactions in the gas phase can be investigated by exploring the trajectories of molecules on potential energy surfaces.

➤ **What do you need to know already?**

This Topic builds on the concept of rate constant (Topic 20A) and in one part of the discussion uses the concept of partition function (Topic 15B). The discussion of potential energy surfaces is qualitative, but the underlying calculations are those of self-consistent field theory (Topic 10E).

The investigation of the dynamics of the collisions between reactant molecules is the most detailed level of the examination of the factors that govern the rates of reactions. There are two approaches: an experimental one that uses molecular beams and a theoretical one that uses the results of computations. In this Topic we describe both approaches and the link between them.

21D.1 Molecular beams

Molecular beams, which consist of collimated, narrow streams of molecules travelling through an evacuated vessel, allow collisions between molecules in preselected states (for example, specific rotational and vibrational states) to be studied, and can be used to identify the states of the products of a reactive collision. Information of this kind is essential if a full picture of the reaction is to be built, because the rate constant is an average over events in which reactants in different initial states evolve into products in their final states.

(a) Techniques

The basic arrangement of a molecular beam experiment is shown in Fig. 21D.1. If the pressure of vapour in the source is increased so that the mean free path of the molecules in the emerging beam is much shorter than the diameter of the pinhole, many collisions take place even outside the source. The net effect of these collisions, which give rise to **hydrodynamic**

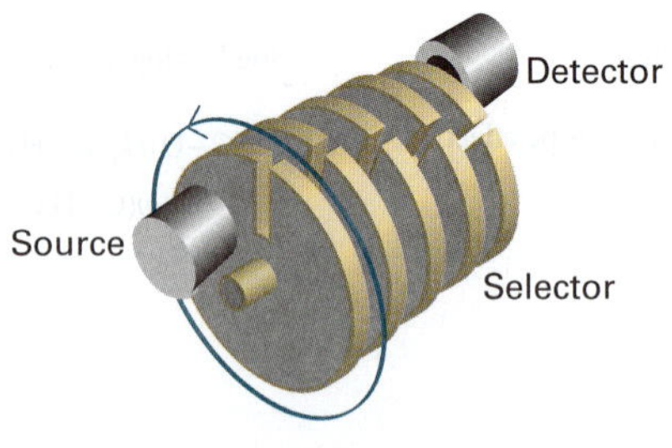

Figure 21D.1 The basic arrangement of a molecular beam apparatus. The atoms or molecules emerge from a heated source, and pass through the velocity selector, a rotating series of slotted discs, such as that discussed in Topic 1B. The scattering occurs from the target gas (which might take the form of another beam), and the flux of particles entering the detector set at some angle is recorded.

flow, is to transfer momentum into the direction of the beam. The molecules in the beam then travel with very similar speeds, so further downstream few collisions take place between them. This condition is called **molecular flow**. Because the spread in speeds is so small, the molecules are effectively in a state of very low translational temperature (Fig. 21D.2). The translational temperature may reach as low as 1 K. Such jets are called **supersonic** because the average speed of the molecules in the jet is much greater than the speed of sound in the jet.

A supersonic jet can be converted into a more parallel **supersonic beam** if it is 'skimmed' in the region of hydrodynamic flow and the excess gas pumped away. A skimmer consists of a conical nozzle shaped to avoid any supersonic shock waves spreading back into the gas and so increasing the translational temperature (Fig. 21D.3). A jet or beam may also be formed by using helium or neon as the principal gas, and injecting molecules of interest into it in the hydrodynamic region of flow.

The low translational temperature of the molecules is reflected in their low rotational and vibrational temperatures. In this context, a rotational or vibrational temperature means the temperature that should be used in the Boltzmann distribution to reproduce the observed populations of the states. However, as rotational states equilibrate more slowly than translational states, and vibrational states equilibrate even more slowly, the rotational and vibrational populations of the species correspond to somewhat higher temperatures, of the order of 10 K for rotation and 100 K for vibrations.

The target gas may be either a bulk sample or another molecular beam. The detectors may consist of a chamber fitted with a sensitive pressure gauge, a bolometer (a detector that responds to the incident energy by making use of the temperature-dependence of resistance), or an ionization detector, in which the incoming molecule is first ionized and then detected electronically. The rotational and vibrational state of the scattered molecules may also be determined spectroscopically.

(b) Experimental results

The primary experimental information from a molecular beam experiment is the fraction of the molecules in the incident beam that are scattered into a particular direction. The fraction is normally expressed in terms of $\mathrm{d}I$, the rate at which molecules are scattered into a cone (described by a solid angle $\mathrm{d}\Omega$) that represents the area covered by the 'eye' of the detector (Fig. 21D.4). This rate is reported as the **differential scattering cross-section**, σ, the constant of proportionality between the value of $\mathrm{d}I$ and the intensity, I, of the incident beam, the number density of target molecules, $\mathcal{N}$, and the infinitesimal path length $\mathrm{d}x$ through the sample:

$$\mathrm{d}I = \sigma I \mathcal{N} \mathrm{d}x \quad \textit{Definition} \quad \text{Differential scattering cross-section} \quad (21\text{D}.1)$$

The value of σ (which has the dimensions of area) depends on the **impact parameter**, b, the initial perpendicular separation of the paths of the colliding molecules (Fig. 21D.5), and the details of the intermolecular potential.

The role of the impact parameter is most easily seen by considering the impact of two hard spheres (Fig. 21D.6). If $b=0$,

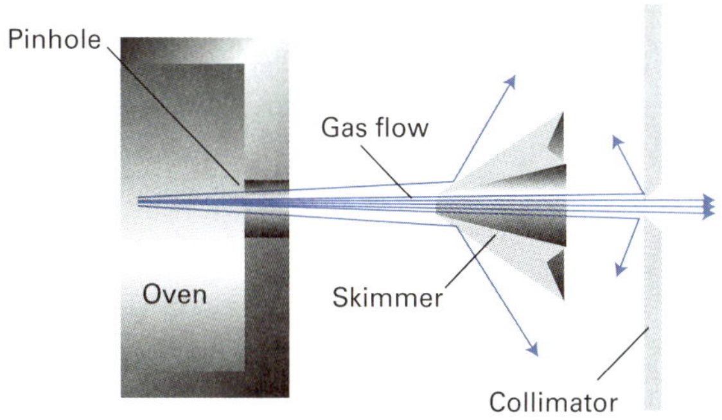

Figure 21D.3 A supersonic nozzle skims off some of the molecules of the beam and leads to a beam with well-defined velocity.

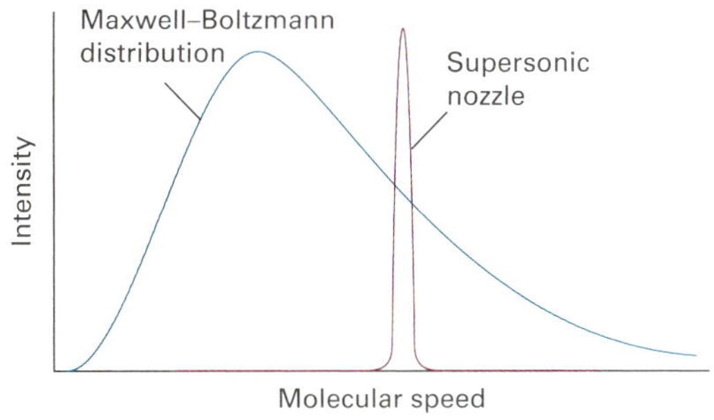

Figure 21D.2 The shift in the mean speed and the width of the distribution brought about by use of a supersonic nozzle.

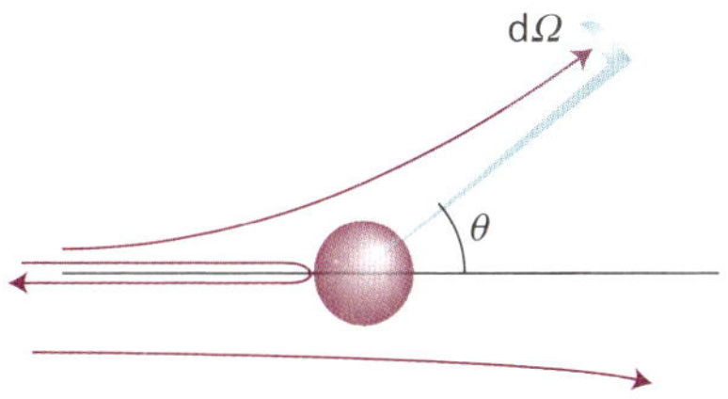

Figure 21D.4 The definition of the solid angle, $\mathrm{d}\Omega$, for scattering.

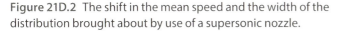

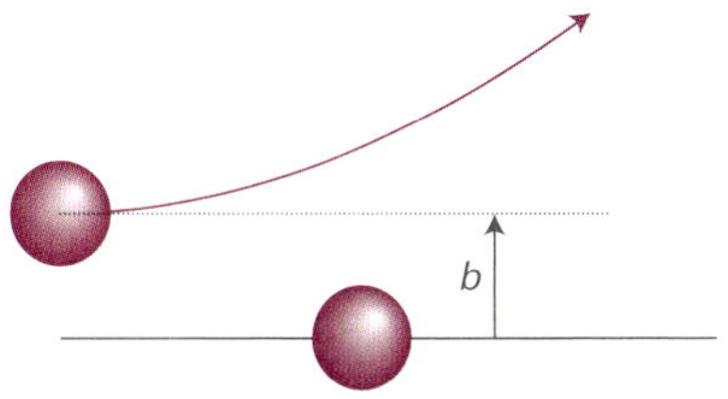

Figure 21D.5 The definition of the impact parameter, b, as the perpendicular separation of the initial paths of the particles.

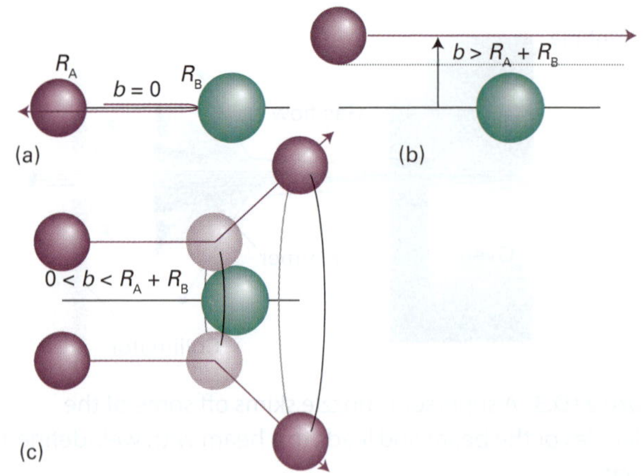

Figure 21D.6 Three typical cases for the collisions of two hard spheres: (a) $b=0$, giving backward scattering; (b) $b>R_A+R_B$, giving forward scattering; (c) $0<b<R_A+R_B$, leading to scattering into one direction on a ring of possibilities. (The target molecule is taken to be so heavy that it remains virtually stationary.)

the projectile is on a trajectory that leads to a head-on collision, so the only scattering intensity is detected when the detector is at $\theta=\pi$. When the impact parameter is so great that the spheres do not make contact ($b>R_A+R_B$), there is no scattering and the scattering cross-section is zero at all angles except $\theta=0$. Glancing blows, with $0<b\leq R_A+R_B$, lead to scattering intensity in cones around the forward direction

The scattering pattern of real molecules, which are not hard spheres, depends on the details of the intermolecular potential, including the anisotropy that is present when the molecules are non-spherical. The scattering also depends on the relative speed of approach of the two particles: a very fast particle might pass through the interaction region without much deflection, whereas a slower one on the same path might be temporarily captured and undergo considerable deflection (Fig. 21D.7). The variation of the scattering cross-section with the relative speed of approach should therefore give information about the strength and range of the intermolecular potential.

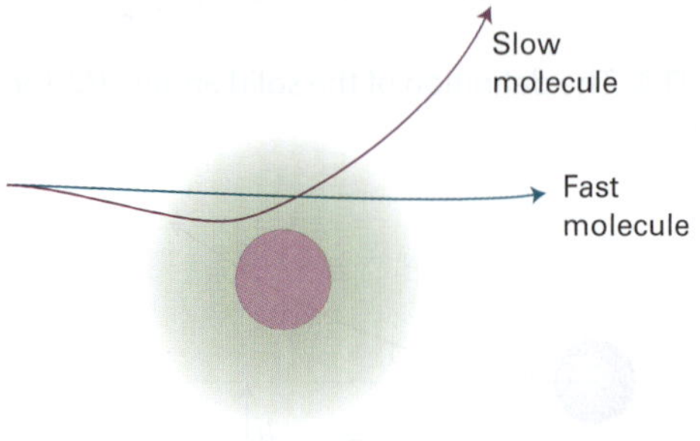

Figure 21D.7 The extent of scattering may depend on the relative speed of approach as well as the impact parameter. The dark central zone represents the repulsive core; the fuzzy outer zone represents the long-range attractive potential.

A further point is that the outcome of collisions is determined by quantum, not classical, mechanics. The wave nature of the particles can be taken into account, at least to some extent, by drawing all classical trajectories that take the projectile particle from source to detector, and then considering the effects of interference between them.

Two quantum mechanical effects are of great importance. A particle with a certain impact parameter might approach the attractive region of the potential in such a way that the particle is deflected towards the repulsive core (Fig. 21D.8), which then repels it out through the attractive region to continue its flight in the forward direction. Some molecules, however, also travel in the forward direction because they have impact parameters so large that they are undeflected. The wavefunctions of the particles that take the two types of path interfere, and the intensity in the forward direction is modified. The effect is called **quantum oscillation**. The same phenomenon accounts for the optical 'glory effect', in which a bright halo can sometimes be seen surrounding an illuminated object. (The coloured rings around the shadow of an aircraft cast on clouds by the Sun, and often seen in flight, are an example of an optical glory.)

The second quantum effect we need consider is the observation of a strongly enhanced scattering in a non-forward direction. This effect is called **rainbow scattering** because the same mechanism accounts for the appearance of an optical rainbow. The origin of the phenomenon is illustrated in Fig. 21D.9. As the impact parameter decreases, there comes a stage at which the scattering angle passes through a maximum and the interference between the paths results in a strongly scattered beam. The **rainbow angle**, θ_r, is the angle for which $d\theta/db=0$ and the scattering is strong.

Another phenomenon that can occur in certain beams is the capturing of one species by another. The vibrational temperature in supersonic beams is so low that **van der Waals molecules** may be formed, which are complexes of the form AB in which A and B are held together by van der Waals forces or hydrogen bonds. Large numbers of such molecules have been studied spectroscopically, including ArHCl, $(HCl)_2$, $ArCO_2$, and $(H_2O)_2$. More recently, van der Waals clusters of water molecules have been pursued as far as $(H_2O)_6$. The study of their

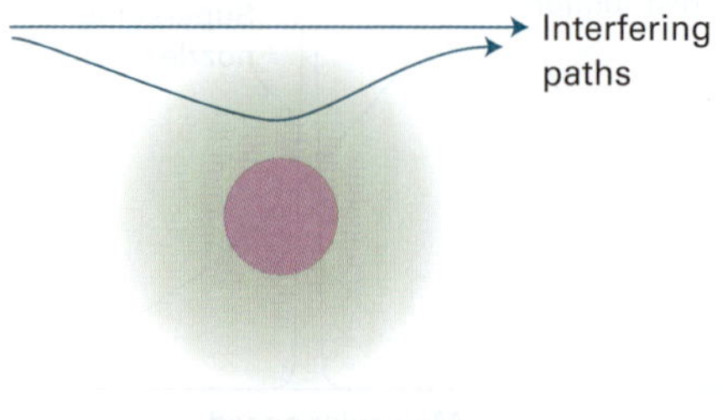

Figure 21D.8 Two paths leading to the same destination will interfere quantum mechanically; in this case they give rise to quantum oscillations in the forward direction.

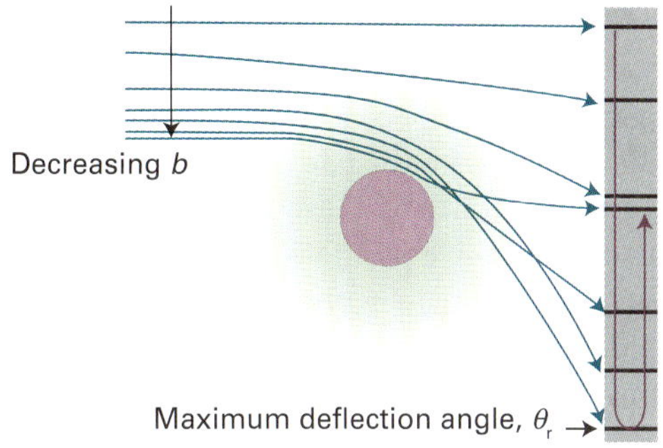

Figure 21D.9 The interference of paths leading to rainbow scattering. The rainbow angle, θ_r, is the maximum scattering angle reached as b is decreased. Interference between the numerous paths at that angle modifies the scattering intensity markedly.

spectroscopic properties gives detailed information about the intermolecular potentials involved.

21D.2 Reactive collisions

Detailed experimental information about the intimate processes that occur during reactive encounters comes from molecular beams, especially **crossed molecular beams** (Fig. 21D.10). The detector for the products of the collision of molecules in the two beams can be moved to different angles to observe so the angular distribution of the products. Because the molecules in the incoming beams can be prepared with different energies (for example, with different translational energies by using rotating sectors and supersonic nozzles, with different vibrational energies by using selective excitation with lasers, and with different orientations by using electric fields), it is possible to study the dependence of the success of collisions on these variables and to study how they affect the properties of the product molecules.

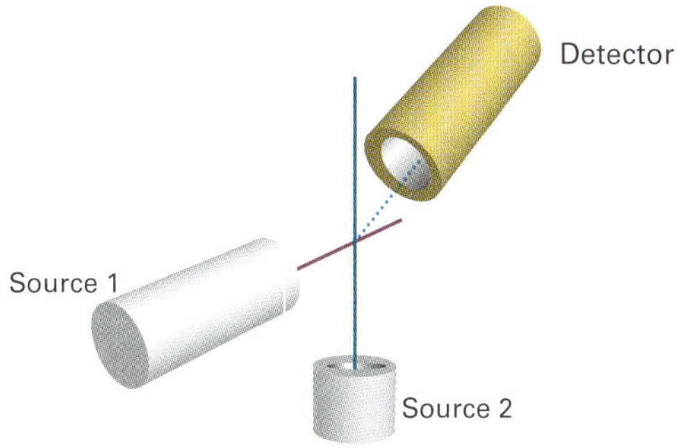

Figure 21D.10 In a crossed-beam experiment, state-selected molecules are generated in two separate sources, and are directed perpendicular to one another. The detector responds to molecules (which may be product molecules if a chemical reaction occurs) scattered into a chosen direction.

(a) Probes of reactive collisions

One method for examining the energy distribution in the products is **infrared chemiluminescence**, in which vibrationally excited molecules emit infrared radiation as they return to their ground states. By studying the intensities of the infrared emission spectrum, the populations of the vibrational states of the products may be determined (Fig. 21D.11). Another method makes use of **laser-induced fluorescence**. In this technique, a laser is used to excite a product molecule from a specific vibration–rotation level; the intensity of the fluorescence from the upper state is monitored and interpreted in terms of the population of the initial vibration–rotation state. When the molecules being studied do not fluoresce efficiently, versions of Raman spectroscopy (Topic 12A) can be used to monitor the progress of reaction. **Multiphoton ionization** (MPI) techniques are also good alternatives for the study of weakly fluorescing molecules. In MPI, the absorption by a molecule of several photons from one or more pulsed lasers results in ionization if the total photon energy is greater than the ionization energy of the molecule.

The angular distribution of products can be determined by **reaction product imaging**. In this technique, product ions are accelerated by an electric field towards a phosphorescent screen and the light emitted from specific spots where the ions struck the screen is imaged by a charge-coupled device (CCD). An important variant of MPI is **resonant multiphoton ionization** (REMPI), in which one or more photons promote a molecule to an electronically excited state and then additional photons are used to generate ions from the excited state. The power of REMPI lies in the fact that the experimenter can choose which reactant or product to study by tuning the laser frequency to the electronic absorption band of a specific molecule.

(b) State-to-state reaction dynamics

The concept of collision cross-section is introduced in connection with collision theory in Topic 21A, where it is shown

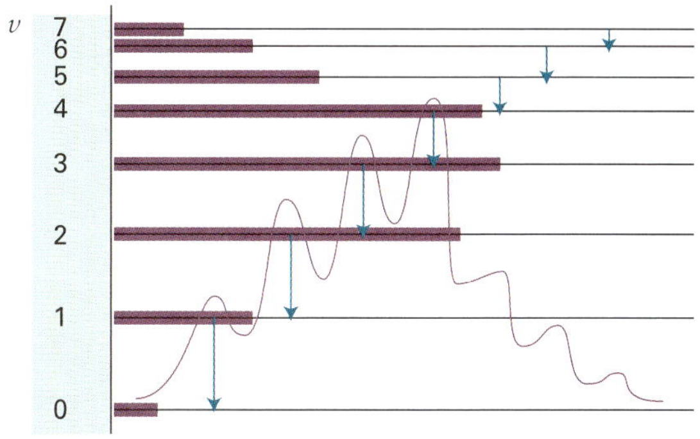

Figure 21D.11 Infrared chemiluminescence from CO produced in the reaction $O + CS \rightarrow CO + S$ arises from the non-equilibrium populations of the vibrational states of CO and the radiative relaxation to equilibrium.

that the second-order rate constant, k_r, can be expressed as a Boltzmann-weighted average of the reactive cross-section and the relative speed of approach of the colliding reactant molecules. We shall write eqn 21A.6 of that Topic ($k_r = N_A\int_0^\infty \sigma(\varepsilon)v_{rel}f(\varepsilon)d\varepsilon$) as

$$k_r = \langle \sigma v_{rel} \rangle N_A \tag{21D.2}$$

where the angle brackets denote a Boltzmann average. Molecular beam studies provide a more sophisticated version of this quantity, for they provide the **state-to-state cross-section**, $\sigma_{nn'}$, and hence the **state-to-state rate constant**, $k_{nn'}$ for the reactive transition from initial state n of the reactants to final state n' of the products:

$$k_{nn'} = \langle \sigma_{nn'} v_{rel} \rangle N_A \quad \text{State-to-state rate constant} \tag{21D.3}$$

The rate constant k_r is the sum of the state-to-state rate constants over all final states (because a reaction is successful whatever the final state of the products) and over a Boltzmann-weighted sum of initial states (because the reactants are initially present with a characteristic distribution of populations at a temperature T):

$$k_r = \sum_{n,n'} k_{nn'}(T) f_n(T) \tag{21D.4}$$

where $f_n(T)$ is the Boltzmann factor at a temperature T. It follows that if we can determine or calculate the state-to-state cross-sections for a wide range of approach speeds and initial and final states, then we have a route to the calculation of the rate constant for the reaction.

Brief illustration 21D.1 The state-to-state rate constant

Suppose a harmonic oscillator collides with another oscillator of the same effective mass and force constant. If the state-to-state rate constant for the excitation of the latter's vibration is $k_{vv'} = k_r^\circ \delta_{vv'}$ for all the states v and v', implying that an excitation can flow only from any level to the same level of the second oscillator, then at a temperature T, when $f_v(T) = e^{-vh\nu/kT}/q$, where q is the molecular vibrational partition function (Topic 15B, $q = 1/(1-e^{-h\nu/kT})$), the overall rate constant is

$$k_r = \frac{k_r^\circ}{q}\sum_{v,v'} \delta_{vv'} e^{-vh\nu/kT} = \frac{k_r^\circ}{q}\overbrace{\sum_{v'} e^{-v'h\nu/kT}}^{q} = k_r^\circ$$

Self-test 21D.1 Now suppose that $k_{vv'} = k_r^\circ \delta_{vv'} e^{-\lambda v}$, implying that the transfer becomes less efficient as the vibrational quantum number increases. Evaluate k_r.

Answer: $k_r = k_r^\circ(1-e^{-h\nu/kT})/(1-e^{-(\lambda+h\nu/kT)})$

21D.3 Potential energy surfaces

One of the most important concepts for discussing beam results and calculating the state-to-state collision cross-section is the **potential energy surface** of a reaction, the potential energy as a function of the relative positions of all the atoms taking part in the reaction. Potential energy surfaces may be constructed from experimental data and from results of quantum chemical calculations (Topic 10E). The theoretical method requires the systematic calculation of the energies of the system in a large number of geometrical arrangements. Special computational techniques, such as those described in Topic 10E, are used to take into account electron correlation, which arises from interactions between electrons as they move closer to and farther from each other in a molecule or molecular cluster. Techniques that incorporate electron correlation accurately are very time consuming and, consequently, only reactions between relatively simple particles, such as the reactions $H+H_2 \rightarrow H_2+H$ and $H+H_2O \rightarrow OH+H_2$, currently are amenable to this type of theoretical treatment. An alternative is to use semi-empirical methods, in which results of calculations and experimental parameters are used to construct the potential energy surface.

To illustrate the features of a potential energy surface, consider the collision between an H atom and an H_2 molecule. Detailed calculations show that the approach of an atom H_A along the H_B–H_C axis requires less energy for reaction than any other approach, so initially we confine our attention to that collinear approach. Two parameters are required to define the nuclear separations: the H_A–H_B separation R_{AB} and the H_B–H_C separation R_{BC}.

At the start of the encounter R_{AB} is effectively infinite and R_{BC} is the H_2 equilibrium bond length. At the end of a successful reactive encounter R_{AB} is equal to the equilibrium bond length and R_{BC} is infinite. The total energy of the three-atom system depends on their relative separations, and can be found by doing an electronic structure calculation. The plot of the total energy of the system against R_{AB} and R_{BC} gives the potential energy surface of this collinear reaction (Fig. 21D.12). This surface is normally depicted as a contour diagram (Fig. 21D.13).

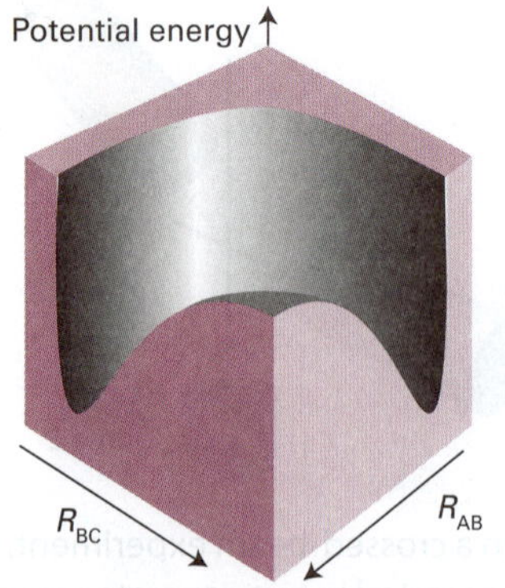

Figure 21D.12 The potential energy surface for the $H+H_2 \rightarrow H_2+H$ reaction when the atoms are constrained to be collinear.

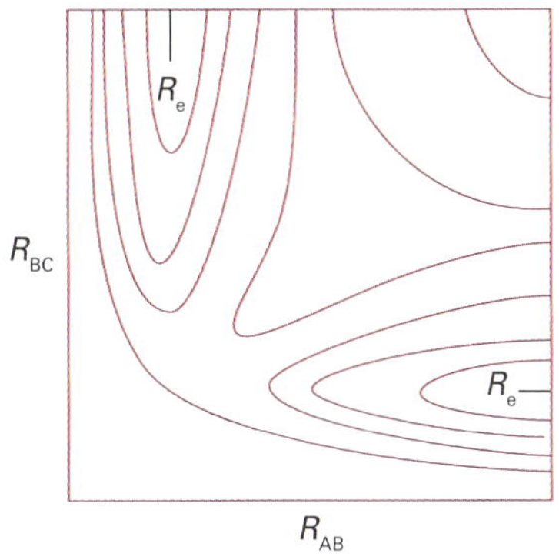

Figure 21D.13 The contour diagram (with contours of equal potential energy) corresponding to the surface in Fig. 21D.12. R_e marks the equilibrium bond length of an H_2 molecule (strictly, it relates to the arrangement when the third atom is at infinity).

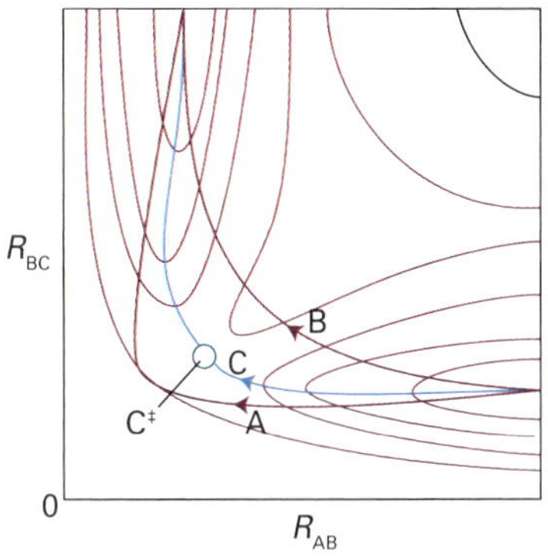

Figure 21D.14 Various trajectories through the potential energy surface shown in Fig. 21D.13. Path A corresponds to a route in which R_{BC} is held constant as H_A approaches; path B corresponds to a route in which R_{BC} lengthens at an early stage during the approach of H_A; path C is the route along the floor of the potential valley.

When R_{AB} is very large, the variation in potential energy represented by the surface as R_{BC} changes is that of an isolated H_2 molecule as its bond length is altered. A section through the surface at $R_{AB}=\infty$, for example, is the same as the H_2 bonding potential energy curve. At the edge of the diagram where R_{BC} is very large, a section through the surface is the molecular potential energy curve of an isolated H_AH_B molecule.

Brief illustration 21D.2 A potential energy surface

The bimolecular reaction $H+O_2 \rightarrow OH+O$ plays an important role in combustion processes. The reaction can be characterized in terms of the HO_2 potential energy surface and the two distances for collinear approach R_{HO_A} and $R_{O_AO_B}$. When R_{HO_A} is very large, the variation of the HO_2 potential energy with $R_{O_AO_B}$ is that of an isolated dioxygen molecule as its bond length is changed. Similarly, when $R_{O_AO_B}$ is very large, a section through the potential energy surface is the molecular potential energy curve of an isolated OH radical.

Self-test 21D.2 Repeat the analysis for $H+OD \rightarrow OH+D$.

Answer: R_{HO} at infinity: OD potential energy curve; R_{OD} at infinity: OH potential energy curve

The actual path of the atoms in the course of the encounter depends on their total energy, the sum of their kinetic and potential energies. However, we can obtain an initial idea of the paths available to the system by identifying paths that correspond to least potential energy. For example, consider the changes in potential energy as H_A approaches H_BH_C. If the H_B–H_C bond length is constant during the initial approach of H_A, then the potential energy of the H_3 cluster rises along the path marked A in Fig. 21D.14. We see that the potential energy reaches a high value as H_A is pushed into the molecule and then decreases sharply as H_C breaks off and separates to a great distance. An alternative reaction path can be imagined (B) in which the H_B–H_C bond length increases while H_A is still far away. Both paths, although feasible if the molecules have sufficient initial kinetic energy, take the three atoms to regions of high potential energy in the course of the encounter.

The path of least potential energy is the one marked C, corresponding to R_{BC} lengthening as H_A approaches and begins to form a bond with H_B. The H_B–H_C bond relaxes at the demand of the incoming atom, and the potential energy climbs only as far as the saddle-shaped region of the surface, to the **saddle point** marked $C^‡$. The encounter of least potential energy is one in which the atoms take route C up the floor of the valley, through the saddle point, and down the floor of the other valley as H_C recedes and the new H_A–H_B bond achieves its equilibrium length. This path is the reaction coordinate.

We can now make contact with the transition-state theory of reaction rates (Topic 21C). In terms of trajectories on potential surfaces with a total energy close to the saddle point energy, the transition state can be identified with a critical geometry such that every trajectory that goes through this geometry goes on to react (Fig. 21D.15). Most trajectories on potential energy

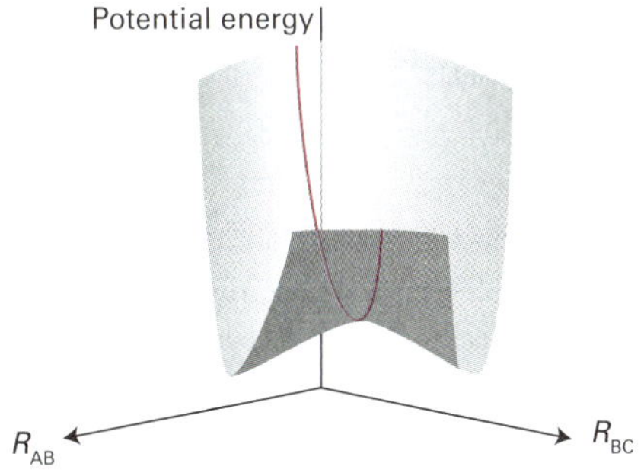

Figure 21D.15 The transition state is a set of configurations (here, marked by the purple line across the saddle point) through which successful reactive trajectories must pass.

surfaces do not go directly over the saddle point and therefore, to result in a reaction, they require a total energy significantly higher than the saddle point energy. As a result, the experimentally determined activation energy is often significantly higher than the calculated saddle-point energy.

21D.4 Some results from experiments and calculations

Although quantum mechanical tunnelling can play an important role in reactivity, particularly in hydrogen atom and electron transfer reactions, initially we can consider the classical trajectories of particles over surfaces. From this viewpoint, to travel successfully from reactants to products, the incoming molecules must possess enough kinetic energy to be able to climb to the saddle point of the potential surface. Therefore, the shape of the surface can be explored experimentally by changing the relative speed of approach (by selecting the beam velocity) and the degree of vibrational excitation and observing whether reaction occurs and whether the products emerge in a vibrationally excited state (Fig. 21D.16). For example, one question that can be answered is whether it is better to smash the reactants together with a lot of translational kinetic energy or to ensure instead that they approach in highly excited vibrational states. Thus, is trajectory C_2*, where the H_BH_C molecule is initially vibrationally excited, more efficient at leading to reaction than the trajectory C_1*, in which the total energy is the same but reactants have a high translational kinetic energy?

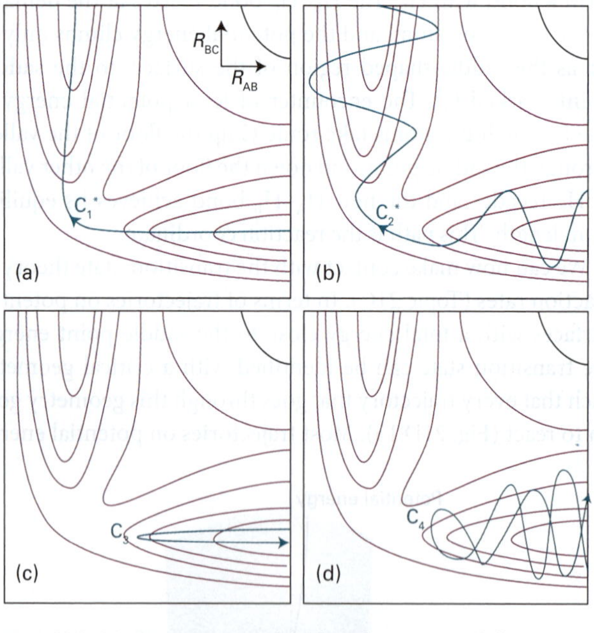

Figure 21D.16 Some successful (*) and unsuccessful encounters. (a) C_1* corresponds to the path along the foot of the valley; (b) C_2* corresponds to an approach of A to a vibrating BC molecule, and the formation of a vibrating AB molecule as C departs. (c) C_3 corresponds to A approaching a non-vibrating BC molecule, but with insufficient translational kinetic energy; (d) C_4 corresponds to A approaching a vibrating BC molecule, but still the energy, and the phase of the vibration, is insufficient for reaction.

(a) The direction of attack and separation

Figure 21D.17 shows the results of a calculation of the potential energy as an H atom approaches an H_2 molecule from different angles, the H_2 bond being allowed to relax to the optimum length in each case. The potential barrier is least for collinear attack, as we assumed earlier. (But we must be aware that other lines of attack are feasible and contribute to the overall rate.) In contrast, Fig. 21D.18 shows the potential energy changes that occur as a Cl atom approaches an HI molecule. The lowest barrier occurs for approaches within a cone of half-angle 30° surrounding the H atom. The relevance of this result to the calculation of the steric factor of collision theory should be noted: not every collision is successful, because they do not all lie within the reactive cone.

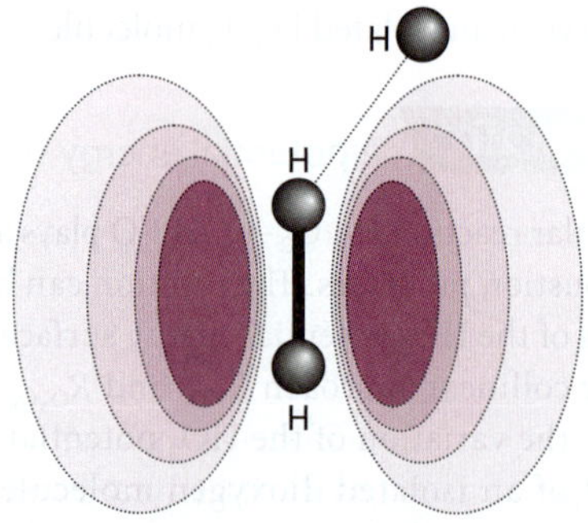

Figure 21D.17 An indication of how the anisotropy of the potential energy changes as H approaches H_2 with different angles of attack. The collinear attack has the lowest potential barrier to reaction. The surface indicates the potential energy profile along the reaction coordinate for each configuration.

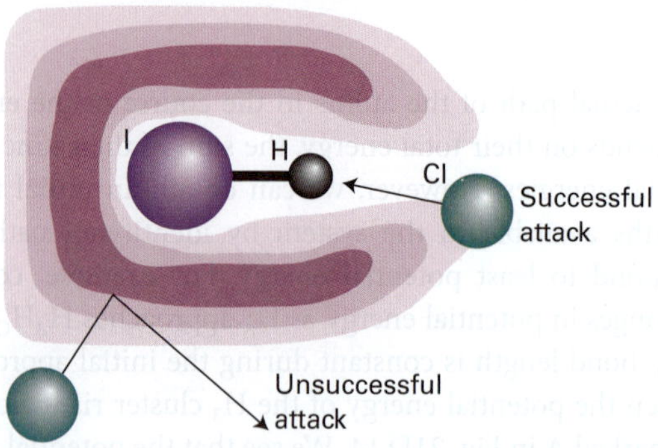

Figure 21D.18 The potential energy barrier for the approach of Cl to HI. In this case, successful encounters occur only when Cl approaches within a cone surrounding the H atom.

If the collision is sticky, so that when the reactants collide they orbit around each other, the products can be expected to emerge in random directions because all memory of the approach direction has been lost. A rotation takes about 1 ps, so if the collision is over in less than that time the complex will not have had time to rotate and the products will be thrown off in a specific direction. In the collision of K and I_2, for example, most of the products are thrown off in the forward direction (forward and backward directions refer to directions in a centre-of-mass coordinate system with the origin at the centre of mass of the colliding reactants and collision occurring when molecules are at the origin.) This product distribution is consistent with the harpoon mechanism (Topic 20A) because the transition takes place at long range. In contrast, the collision of K with CH_3I leads to reaction only if the molecules approach each other very closely. In this mechanism, K effectively bumps into a brick wall, and the KI product bounces out in the backward direction. The detection of this anisotropy in the angular distribution of products gives an indication of the distance and orientation of approach needed for reaction, as well as showing that the event is complete in less than about 1 ps.

(b) Attractive and repulsive surfaces

Some reactions are very sensitive to whether the energy has been pre-digested into a vibrational mode or left as the relative translational kinetic energy of the colliding molecules. For example, if two HI molecules are hurled together with more than twice the activation energy of the reaction, then no reaction occurs if all the energy is solely translational. For $F+HCl \rightarrow Cl+HF$, for example, the reaction is about five times more efficient when the HCl is in its first vibrational excited state than when, although HCl has the same total energy, it is in its vibrational ground state.

The origin of these requirements can be found by examining the potential energy surface. Figure 21D.19 shows an **attractive**

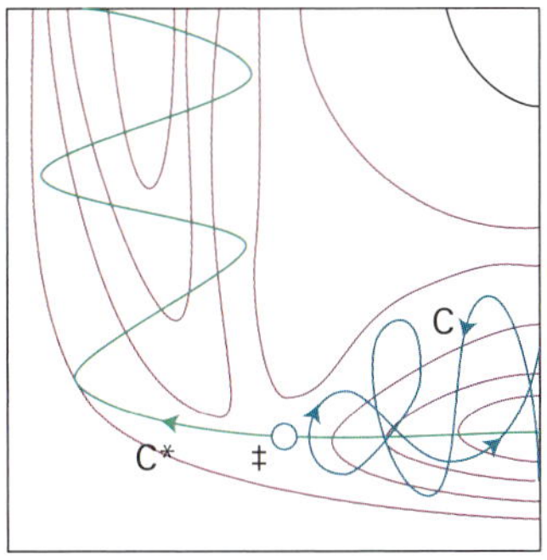

Figure 21D.19 An attractive potential energy surface. A successful encounter (C*) involves high translational kinetic energy and results in a vibrationally excited product.

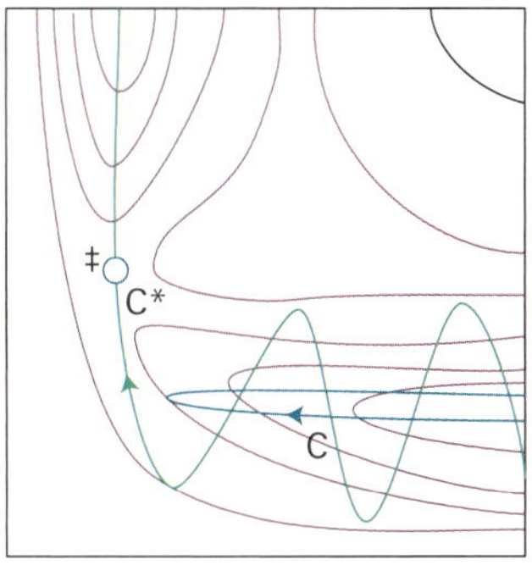

Figure 21D.20 A repulsive potential energy surface. A successful encounter (C*) involves initial vibrational excitation and the products have high translational kinetic energy. A reaction that is attractive in one direction is repulsive in the reverse direction.

surface in which the saddle point occurs early in the reaction coordinate. Figure 21D.20 shows a **repulsive surface** in which the saddle point occurs late. A surface that is attractive in one direction is repulsive in the reverse direction.

Consider first the attractive surface. If the original molecule is vibrationally excited, then a collision with an incoming molecule takes the system along C. This path is bottled up in the region of the reactants, and does not take the system to the saddle point. If, however, the same amount of energy is present solely as translational kinetic energy, then the system moves along C* and travels smoothly over the saddle point into products. We can therefore conclude that reactions with attractive potential energy surfaces proceed more efficiently if the energy is in relative translational motion. Moreover, the potential surface shows that once past the saddle point the trajectory runs up the steep wall of the product valley, and then rolls from side to side as it falls to the foot of the valley as the products separate. In other words, the products emerge in a vibrationally excited state.

Now consider the repulsive surface. On trajectory C the collisional energy is largely in translation. As the reactants approach, the potential energy rises. Their path takes them up the opposing face of the valley, and they are reflected back into the reactant region. This path corresponds to an unsuccessful encounter, even though the energy is sufficient for reaction. On C* some of the energy is in the vibration of the reactant molecule and the motion causes the trajectory to weave from side to side up the valley as it approaches the saddle point. This motion may be sufficient to tip the system round the corner to the saddle point and then on to products. In this case, the product molecule is expected to be in an unexcited vibrational state. Reactions with repulsive potential surfaces can therefore be expected to proceed more efficiently if the excess energy is present as vibrations. This is the case with the $H+Cl_2 \rightarrow HCl+Cl$ reaction, for instance.

Brief illustration 21D.3 Attractive and repulsive surfaces

The reaction $H+Cl_2 \rightarrow HCl+Cl$ has a repulsive potential surface. Of the following four reactive processes, $H+Cl_2(v) \rightarrow HCl(v')+Cl$, which we denote (v,v'), all at the same total energy, (a) (0,0), (b) (2,0), (c) (0,2), (d) (2,2), reaction (b) is most probable with reactants vibrationally excited and products vibrationally unexcited.

Self-test 21D.3 Which of the four reactive processes of the reaction $HCl(v)+Cl \rightarrow H+Cl_2(v')$, all at the same total energy, (a) (0,0), (b) (2,0), (c) (0,2), (d) (2,2), is most probable?

Answer: (0,2); attractive surface

(c) Classical trajectories

A clear picture of the reaction event can be obtained by using classical mechanics to calculate the trajectories of the atoms taking place in a reaction from a set of initial conditions, such as velocities, relative orientations, and internal energies of the reacting particles. The initial values used for the internal energy reflect the quantization of electronic, vibrational, and rotational energies in molecules but the features of quantum mechanics are not used explicitly in the calculation of the trajectory.

Figure 21D.21 shows the result of such a calculation of the positions of the three atoms in the reaction $H+H_2 \rightarrow H_2+H$, the horizontal coordinate now being time and the vertical coordinate the separations. This illustration shows clearly the vibration of the original molecule and the approach of the attacking atom. The reaction itself, the switch of partners, takes place very rapidly and is an example of a **direct mode process**. The newly formed molecule shakes, but quickly settles down to steady, harmonic vibration as the expelled atom departs. In contrast, Fig. 21D.22 shows an example of a **complex mode process**, in which the activated complex survives for an extended period. The reaction in the figure is the exchange reaction $KCl+NaBr \rightarrow KBr+NaCl$. The tetratomic activated complex survives for about 5 ps, during which time the atoms make about 15 oscillations before dissociating into products.

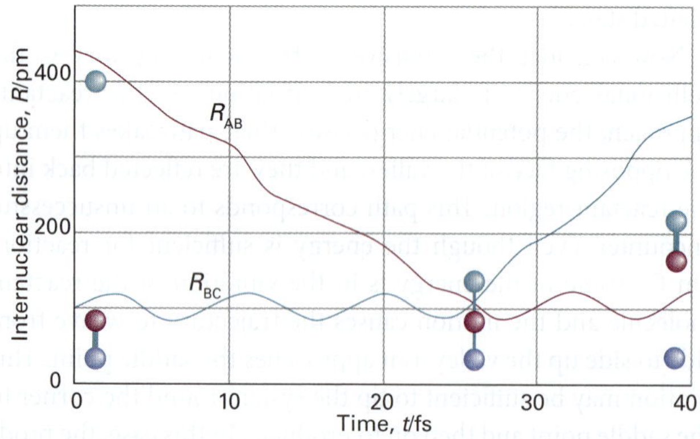

Figure 21D.21 The calculated trajectories for a reactive encounter between A and a vibrating BC molecule leading to the formation of a vibrating AB molecule. This direct-mode reaction is between H and H_2 (M. Karplus et al., *J. Chem. Phys.* **43**, 3258 (1965)).

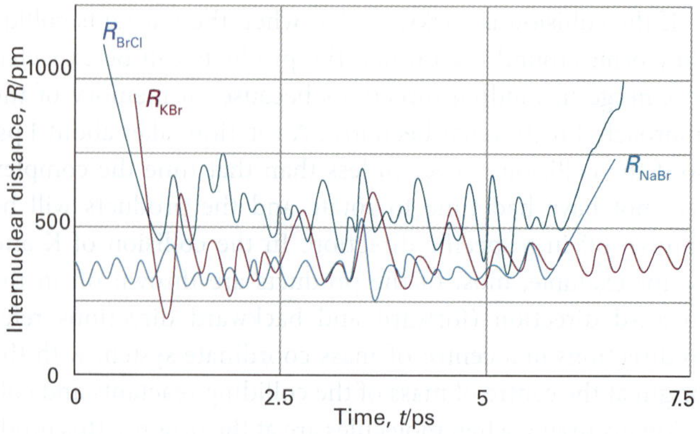

Figure 21D.22 An example of the trajectories calculated for a complex-mode reaction, $KCl+NaBr \rightarrow KBr+NaCl$, in which the collision cluster has a long lifetime (P. Brumer and M. Karplus, *Faraday Disc. Chem. Soc.* **55**, 80 (1973)).

(d) Quantum mechanical scattering theory

Classical trajectory calculations do not recognize the fact that the motion of atoms, electrons, and nuclei is governed by quantum mechanics. The concept of trajectory then fades and is replaced by the unfolding of a wavefunction that represents initially the reactants and finally products.

Complete quantum mechanical calculations of trajectories and rate constants are very onerous because it is necessary to take into account all the allowed electronic, vibrational, and rotational states populated by each atom and molecule in the system at a given temperature. It is common to define a 'channel' as a group of molecules in well-defined quantum mechanically allowed states. Then, at a given temperature, there are many channels that represent the reactants and many channels that represent possible products, with some transitions between channels being allowed but others not allowed. Furthermore, not every transition leads to a chemical reaction. For example, the process $H_2^*+OH \rightarrow H_2+OH^*$, where the asterisk denotes an excited state, amounts to energy transfer between H_2 and OH, whereas the process $H_2^*+OH \rightarrow H_2O+H$ represents a chemical reaction. What complicates a quantum mechanical calculation of rate constants even in this simple four-atom system is that many reacting channels present at a given temperature can lead to the desired products H_2O+H, which themselves may be formed as many distinct channels.

The **cumulative reaction probability**, $\bar{P}(E)$, at a fixed total energy E is then written as

$$\bar{P}(E)=\sum_{i,j} P_{ij}(E) \qquad \text{Cumulative reaction probability} \quad (21D.5)$$

where $P_{i,j}(E)$ is the probability for a transition between a reacting channel i and a product channel j and the summation is over all possible transitions that lead to product. It is then possible to show that the rate constant is given by

$$k_r(T)=\frac{\int_0^\infty \bar{P}(E)e^{-E/kT}dE}{hQ_R(T)} \qquad \text{Rate constant} \quad (21D.6)$$

where $Q_R(T)$ is the partition function density (the partition function divided by the volume) of the reactants at the temperature T. The significance of eqn 21D.6 is that it provides a direct connection between an experimental quantity, the rate constant, and a theoretical quantity, $\bar{P}(E)$.

Checklist of concepts

☐ 1. A **molecular beam** is a collimated, narrow stream of molecules travelling through an evacuated vessel.

☐ 2. In a molecular beam, the scattering pattern of real molecules depends on quantum mechanical effects and the details of the intermolecular potential.

☐ 3. A **van der Waals molecule** is a complex of the form AB in which A and B are held together by van der Waals forces or hydrogen bonds.

☐ 4. Techniques for the study of reactive collisions include **infrared chemiluminescence**, **laser-induced fluorescence**, **multiphoton ionization** (MPI), **reaction product imaging**, and **resonant multiphoton ionization** (REMPI).

☐ 5. A **potential energy surface** maps the potential energy as a function of the relative positions of all the atoms taking part in a reaction.

☐ 6. In an **attractive surface**, the saddle point (the highest point) occurs early on the reaction coordinate.

☐ 7. In a **repulsive surface**, the saddle point occurs late on the reaction coordinate.

Checklist of equations

Property	Equation	Comment	Equation number
Rate of molecular scattering	$dI=\sigma I\mathcal{N}dx$	σ is the differential scattering cross section	21D.1
Rate constant	$k_r=\langle\sigma v_{rel}\rangle N_A$		21D.2
State-to-state rate constant	$k_{nn'}=\langle\sigma_{nn'}v_{rel}\rangle N_A$		21D.3
Overall rate constant	$k_r=\sum_{n,n'} k_{nn'}(T)f_n(T)$		21D.4
Cumulative reaction probability	$\bar{P}(E)=\sum_{i,j} P_{ij}(E)$		21D.5
Rate constant	$k_r(T)=\int_0^\infty \bar{P}(E)e^{-E/kT}dE/hQ_R(T)$	$Q_R(T)$ is the partition function density	21D.6

21E Electron transfer in homogeneous systems

Contents

➤ Why do you need to know this material?

Electron transfer reactions between protein-bound cofactors or between proteins play an important role in a variety of biological processes. Electron transfer is also important in homogeneous, non-biological catalysis (especially biomimetic systems).

➤ What is the key idea?

The rate constant of electron transfer in a donor–acceptor complex depends on the distance between electron donor and acceptor, the standard reaction Gibbs energy, and the energy needed to reach a particular arrangement of atoms.

➤ What do you need to know already?

This Topic makes use of transition-state theory (Topic 21C). It also uses the concept of tunnelling (Topic 8A), the steady-state approximation (Topic 20E), and the Franck–Condon principle (Topic 13A).

Here we apply the concepts of transition state theory and quantum theory to the study of a deceptively simple process, electron transfer between molecules in homogeneous systems. We describe a theoretical approach to the calculation of rate constants and discuss the theory in the light of experimental results on a variety of systems, including protein complexes. We shall see that relatively simple expressions can be used to predict the rates of electron transfer with reasonable accuracy.

21E.1 The electron transfer rate law

Consider electron transfer from a donor species D to an acceptor species A in solution. The overall reaction is

$$\mathrm{D}+\mathrm{A}\rightarrow\mathrm{D}^{+}+\mathrm{A}^{-}\qquad v=k_{\mathrm{r}}[\mathrm{D}][\mathrm{A}] \tag{21E.1}$$

In the first step of the mechanism, D and A must diffuse through the solution and on meeting form a complex DA:

$$\mathrm{D}+\mathrm{A}\underset{k_{\mathrm{a}}'}{\overset{k_{\mathrm{a}}}{\rightleftarrows}}\mathrm{DA} \tag{21E.2a}$$

We suppose that in the complex D and A are separated by d, the distance between their outer surfaces. Next, electron transfer occurs within the DA complex to yield $\mathrm{D}^{+}\mathrm{A}^{-}$:

$$\mathrm{DA}\underset{k_{\mathrm{et}}'}{\overset{k_{\mathrm{et}}}{\rightleftarrows}}\mathrm{D}^{+}\mathrm{A}^{-} \tag{21E.2b}$$

The complex $\mathrm{D}^{+}\mathrm{A}^{-}$ can also break apart and the ions diffuse through the solution:

$$\mathrm{D}^{+}\mathrm{A}^{-}\xrightarrow{k_{\mathrm{d}}}\mathrm{D}^{+}+\mathrm{A}^{-} \tag{21E.2c}$$

We show in the following *Justification* that on the basis of this model

$$\frac{1}{k_{\mathrm{r}}}=\frac{1}{k_{\mathrm{a}}}+\frac{k_{\mathrm{a}}'}{k_{\mathrm{a}}k_{\mathrm{et}}}\left(1+\frac{k_{\mathrm{et}}'}{k_{\mathrm{d}}}\right)\qquad \text{Electron transfer rate constant} \tag{21E.3}$$

Justification 21E.1 The rate constant for electron transfer in solution

We begin by identifying the rate of the overall reaction (eqn 21E.1) with the rate of formation of separated ions:

$$v = k_r[D][A] = k_d[D^+A^-]$$

There are two reaction intermediates, DA and D^+A^-, and we apply the steady-state approximation (Topic 20E) to both. From

$$\frac{d[D^+A^-]}{dt} = k_{et}[DA] - k'_{et}[D^+A^-] - k_d[D^+A^-] = 0$$

it follows that

$$[DA] = \frac{k'_{et} + k_d}{k_{et}}[D^+A^-]$$

and from

$$\frac{d[DA]}{dt} = k_a[D][A] - k'_a[DA] - k_{et}[DA] + k'_{et}[D^+A^-] = 0$$

it follows that

$$k_a[D][A] \overbrace{-k'_a[DA] - k_{et}\overbrace{[DA]}^{((k'_{et}+k_d)/k_{et})[D^+A^-]}}^{-(k_a+k_{et})} + k'_{et}[D^+A^-]$$

$$= k_a[D][A] - \left\{\frac{(k'_a + k_{et})(k'_{et} + k_d)}{k_{et}} - k'_{et}\right\}[D^+A^-]$$

$$= k_a[D][A] - \frac{1}{k_{et}}(k'_a k'_{et} + k'_a k_d + k_d k_{et})[D^+A^-] = 0$$

therefore

$$[D^+A^-] = \frac{k_a k_{et}}{k'_a k'_{et} + k'_a k_d + k_d k_{et}}[D][A]$$

When this expression is multiplied by k_d, the resulting equation has the form of the rate of electron transfer, $v = k_r[D][A]$, with k_r given by

$$k_r = \frac{k_a k_{et} k_d}{k'_a k'_{et} + k'_a k_d + k_d k_{et}}$$

To obtain eqn 21E.3, divide the numerator and denominator on the right-hand side of this expression by $k_d k_{et}$ and solve for the reciprocal of k_r.

To gain insight into eqn 21E.3 and the factors that determine the rate of electron transfer reactions in solution, we assume that the main decay route for D^+A^- is dissociation of the complex into separated ions, and therefore that $k_d \gg k'_{et}$. It follows that

$$\frac{1}{k_r} \approx \frac{1}{k_a}\left(1 + \frac{k'_a}{k_{et}}\right)$$

Physical interpretation

- When $k_{et} \gg k'_a$, $k_r \approx k_a$ and the rate of product formation is controlled by diffusion of D and A in solution, which favours formation of the DA complex.
- When $k_{et} \ll k'_a$, $k_r \approx (k_a/k'_a)k_{et} = Kk_{et}$, where K is the equilibrium constant for the diffusive encounter. The process is controlled by k_{et} and therefore the activation energy of electron transfer in the DA complex.

21E.2 The rate constant

This analysis can be taken further by introducing the implication from transition-state theory (Topic 21C) that, at a given temperature, $k_{et} \propto e^{-\Delta^{\ddagger}G/RT}$, where $\Delta^{\ddagger}G$ is the Gibbs energy of activation. Our remaining task, therefore, is to find expressions for the proportionality constant and $\Delta^{\ddagger}G$.

Our discussion concentrates on the following two key aspects of the theory of electron transfer processes, which was developed independently by R.A. Marcus, N.S. Hush, V.G. Levich, and R.R. Dogonadze:

- Electrons are transferred by tunnelling through a potential energy barrier, the height of which is partly determined by the ionization energies of the DA and D^+A^- complexes. Electron tunnelling influences the magnitude of the proportionality constant in the expression for k_{et}.
- The complex DA and the solvent molecules surrounding it undergo structural rearrangements prior to electron transfer. The energy associated with these rearrangements and the standard reaction Gibbs energy determine $\Delta^{\ddagger}G$.

According to the Franck–Condon principle (Topic 13A), electronic transitions are so fast that they can be regarded as taking place in a stationary nuclear framework. This principle also applies to an electron transfer process in which an electron migrates from one energy surface, representing the dependence of the energy of DA on its geometry, to another representing the energy of D^+A^-. We can represent the potential energy (and the Gibbs energy) surfaces of the two complexes (the reactant complex, DA, and the product complex, D^+A^-) by the parabolas characteristic of harmonic oscillators, with the displacement coordinate corresponding to the changing geometries (Fig. 21E.1). This coordinate represents a collective mode of the donor, acceptor, and solvent.

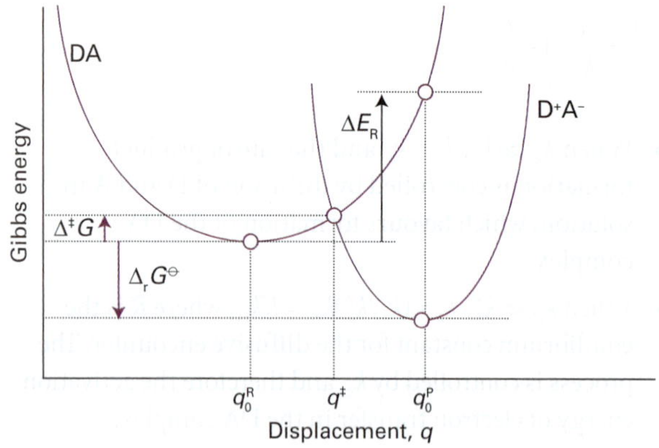

Figure 21E.1 The Gibbs energy surfaces of the complexes DA and D^+A^- involved in an electron transfer process are represented by parabolas characteristic of harmonic oscillators, with the displacement coordinate q corresponding to the changing geometries of the system.

According to the Franck–Condon principle, the nuclei do not have time to move when the system passes from the reactant to the product surface as a result of the transfer of an electron. Therefore, electron transfer can occur only after thermal fluctuations bring the geometry of DA to $q^‡$ in Fig 21E.1, the value of the nuclear coordinate at which the two parabolas intersect.

(a) The role of electron tunnelling

The proportionality constant in the expression for k_{et} is a measure of the rate at which the system will convert from reactants (DA) to products (D^+A^-) at $q^‡$ by electron transfer within the thermally excited DA complex. To understand the process, we must turn our attention to the effect that the rearrangement of nuclear coordinates has on electronic energy levels of DA and D^+A^- for a given distance d between D and A (Fig. 21E.2). Initially, the overall energy of DA is lower than that of D^+A^- (Fig 21E.2a). As the nuclei rearrange to a configuration represented by $q^‡$ in Fig. 21E.2b, the HOMO of DA and the LUMO of D^+A^- become degenerate and electron transfer becomes energetically feasible. Over reasonably short distances d, the main mechanism of electron transfer is tunnelling through the potential energy barrier depicted in Fig 21E.2b. After an electron moves between the two frontier orbitals, the system relaxes to the configuration represented by q_0^P in Fig 21E.2c. As shown in the illustration, now the energy of D^+A^- is lower than that of DA, reflecting the thermodynamic tendency for A to remain reduced (as A^-) and for D to remain oxidized (as D^+).

The tunnelling event responsible for electron transfer is similar to that described in Topic 8A, except that in this case the electron tunnels from an electronic level of DA, with

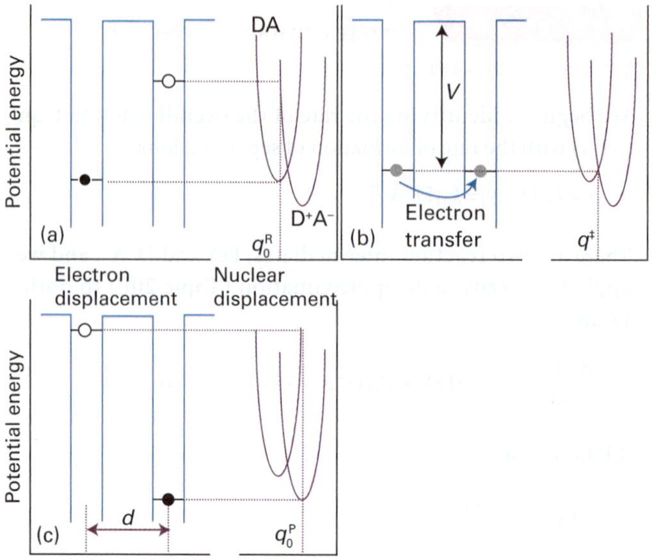

Figure 21E.2 (a) At the nuclear configuration denoted by q_0^R, the electron to be transferred in DA is in an occupied electronic energy level and the lowest unoccupied energy level of $D^+A^-$is of too high an energy to be a good electron acceptor. (b) As the nuclei rearrange to a configuration represented by $q^‡$, DA and D^+A^- become degenerate and electron transfer occurs by tunnelling. (c) The system relaxes to the equilibrium nuclear configuration of D^+A^- denoted by q_0^P, in which the lowest unoccupied electronic level of DA is higher in energy than the highest occupied electronic level of D^+A^-. Adapted from R.A. Marcus and N. Sutin, *Biochim. Biophys. Acta* **811**, 265 (1985).

wavefunction ψ_{DA}, to an electronic level of D^+A^-, with wavefunction $\psi_{D^+A^-}$. The rate of an electronic transition from a level described by the wavefunction ψ_{DA} to a level described by $\psi_{D^+A^-}$ is proportional to the square of the integral

$$H_{et} = \int \psi_{DA} \hat{h} \psi_{D^+A^-} d\tau$$

where $\hat{h}$ is a hamiltonian that describes the coupling of the electronic wavefunctions. The probability of tunnelling through a potential barrier typically has an exponential dependence on the width of the barrier (Topic 8A), which suggests that we should write

$$H_{et}(d)^2 = H_{et}^{\circ 2} e^{-\beta d} \qquad (21E.4)$$

where d is the edge-to-edge distance between D and A, β is a parameter that measures the sensitivity of the electronic coupling matrix element to distance, and H_{et}° is the value of the electronic coupling matrix element when DA and D^+A^- are in contact ($d=0$).

Brief illustration 21E.1 The distance dependence of the coupling

The value of β depends on the medium through which the electron must travel from donor to acceptor. In a vacuum, $28\,\text{nm}^{-1} < \beta < 35\,\text{nm}^{-1}$, whereas $\beta \approx 9\,\text{nm}^{-1}$ when the intervening medium is a molecular link between donor and acceptor. Electron transfer between protein-bound cofactors can occur at distances of up to about 2.0 nm, a long distance on a molecular scale, corresponding to about 20 carbon atoms, with the protein providing an intervening medium between donor and acceptor.

Self-test 21E.1 By how much does H_{et}^2 change when d is increased from 1.0 nm to 2.0 nm, with $\beta \approx 9\,\text{nm}^{-1}$?

Answer: Decrease by a factor of 8100

(b) The reorganization energy

The proportionality constant in $k_{\text{et}} \propto e^{-\Delta^{\ddagger}G/RT}$ is proportional to $H_{\text{et}}(d)^2$, as expressed by eqn 21E.4. Therefore, we can expect the full expression for k_{et} to have the form

$$k_{\text{et}} = CH_{\text{et}}(d)^2 e^{-\Delta^{\ddagger}G/RT} \quad (21\text{E}.5)$$

with C a constant of proportionality and $H_{\text{et}}(d)^2$ given by eqn 21E.5. We show in the following *Justification* that the Gibbs energy of activation $\Delta^{\ddagger}G$ is

$$\Delta^{\ddagger}G = \frac{(\Delta_r G^{\ominus} + \Delta E_R)^2}{4\Delta E_R} \quad \text{Gibbs energy of activation} \quad (21\text{E}.6)$$

where $\Delta_r G^{\ominus}$ is the standard reaction Gibbs energy for the electron transfer process $DA \rightarrow D^+A^-$, and ΔE_R is the **reorganization energy**, the energy change associated with molecular rearrangements that must take place so that DA can take on the equilibrium geometry of D^+A^-. These molecular rearrangements include the relative reorientation of the D and A molecules in DA and the relative reorientation of the solvent molecules surrounding DA. Equation 21E.6 shows that $\Delta^{\ddagger}G = 0$, with the implication that the reaction is not slowed down by an activation barrier, when $\Delta_r G^{\ominus} = -\Delta E_R$, corresponding to the cancellation of the reorganization energy term by the standard reaction Gibbs energy.

Justification 21E.2 The Gibbs energy of activation of electron transfer

The simplest way to derive an expression for the Gibbs energy of activation of electron transfer processes is to construct a model in which the surfaces for DA (the 'reactant complex', denoted R) and for D^+A^- (the 'product complex', denoted P) are described by classical harmonic oscillators with identical reduced masses μ and angular frequencies ω, but displaced minima, as shown in Fig. 21E.3. The molar Gibbs energies $G_{m,R}(q)$ and $G_{m,P}(q)$ of the reactant and product complexes, respectively, may then be written

$$G_{m,R}(q) = \tfrac{1}{2}N_A\mu\omega^2(q - q_0^R)^2 + G_{m,R}(q_0^R)$$
$$G_{m,P}(q) = \tfrac{1}{2}N_A\mu\omega^2(q - q_0^P)^2 + G_{m,P}(q_0^P)$$

where q_0^R and q_0^P are the values of q at which the minima of the reactant and product parabolas occur, respectively. The standard reaction Gibbs energy for the electron transfer process $R \rightarrow P$ is $\Delta_r G^{\ominus} = G_{m,P}(q_0^P) - G_{m,R}(q_0^R)$, the difference in standard molar Gibbs energy between the minima of the parabolas. In Fig. 21E.3, $\Delta_r G^{\ominus} < 0$.

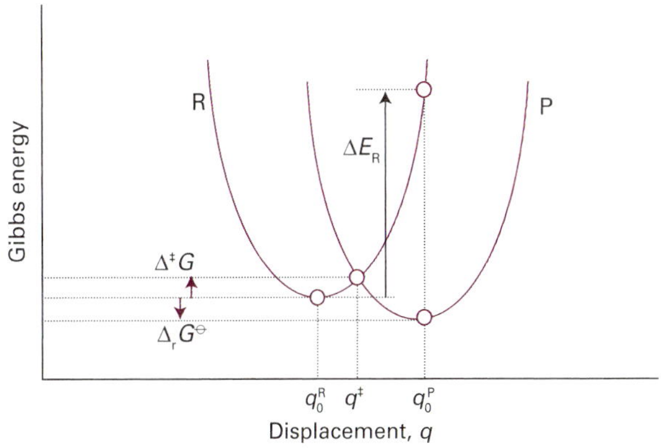

Figure 21E.3 The model system used in *Justification* 21E.2.

The value of q corresponding to the transition state of the complex, $q^{\ddagger}$, may be written in terms of the parameter α, the fractional change in q:

$$q^{\ddagger} = q_0^R + \alpha(q_0^P - q_0^R)$$

We see from Fig. 21E.3 that $\Delta^{\ddagger}G = G_{m,R}(q^{\ddagger}) - G_{m,R}(q_0^R)$. It then follows that

$$\Delta^{\ddagger}G = \tfrac{1}{2}N_A\mu\omega^2(q^{\ddagger} - q_0^R)^2 = \tfrac{1}{2}N_A\mu\omega^2\{\alpha(q_0^P - q_0^R)^2\}$$

We now define the reorganization energy, ΔE_R, as

$$\Delta E_R = \tfrac{1}{2}N_A\mu\omega^2(q_0^P - q_0^R)^2$$

which can be interpreted as $G_{m,R}(q_0^P) - G_{m,R}(q_0^R)$ and, consequently, as the (Gibbs) energy required to deform the equilibrium configuration of R to the equilibrium configuration of P (as shown in Fig. 21E.3). Then $\Delta^{\ddagger}G = \alpha^2\Delta E_R$. Because $G_{m,R}(q^{\ddagger}) = G_{m,P}(q^{\ddagger})$, it follows that

$$\alpha^2\Delta E_R = \tfrac{1}{2}N_A\mu\omega^2\{(\alpha - 1)(q_0^P - q_0^R)\}^2 + \Delta_r G^{\ominus}$$
$$= (\alpha - 1)\Delta E_R + \Delta_r G^{\ominus}$$

which implies that

$$\alpha = \frac{1}{2}\left(\frac{\Delta_r G^{\ominus}}{\Delta E_R} + 1\right)$$

By inserting this equation into $\Delta^{\ddagger}G=\alpha^2\Delta E_R$, we obtain eqn 21E.6. We can obtain an identical relation if we allow the harmonic oscillators to have different angular frequencies and hence different curvatures.

The only missing piece of the expression for k_{et} is the value of the constant of proportionality C. Detailed calculation, which we do not repeat here, gives

$$C=\frac{1}{h}\left(\frac{\pi^3}{RT\Delta E_R}\right)^{1/2} \tag{21E.7}$$

Equation 21E.6 has some limitations as might be expected because perturbation theory arguments have been used. For instance, it describes processes with weak electronic coupling between donor and acceptor. Weak coupling is observed when the electroactive species are sufficiently far apart that the tunnelling is an exponential function of distance. An example of a weakly coupled system is the complex of the proteins cytochrome c and cytochrome b_5, in which the electroactive haem-bound iron ions shuttle between oxidation states Fe(II) and Fe(III) during electron transfer and are about 1.7 nm apart. Strong coupling is observed when the wavefunctions ψ_A and ψ_D overlap very extensively and, as well as other complications, the tunnelling rate is no longer a simple exponential function of distance. Examples of strongly coupled systems are mixed-valence, binuclear d-metal complexes with the general structure L_mM^{n+}–B–$M^{p+}L_m$, in which the electroactive metal ions are separated by a bridging ligand B. In these systems, $d<1.0$ nm. The weak coupling limit applies to a large number of electron transfer reactions, including those between proteins during metabolism.

The most meaningful experimental tests of the dependence of k_{et} on d are those in which the same donor and acceptor are positioned at a variety of distances, perhaps by covalent attachment to molecular linkers (see **1** in *Brief illustration* 21E.2 for an example). Under these conditions, the term $e^{-\Delta^{\ddagger}G/RT}$ becomes a constant and, after taking the natural logarithm of eqn 21E.5 and using eqn 21E.4, we obtain

$$\ln k_{et}=-\beta d+\text{constant} \tag{21E.8}$$

which implies that a plot of ln k_{et} against d should be a straight line of slope $-\beta$.

The dependence of k_{et} on the standard reaction Gibbs energy has been investigated in systems where the edge-to-edge distance and the reorganization energy are constant for a series of reactions. Then, by using eqn 21E.6 for $\Delta^{\ddagger}G$, eqn 21E.5 becomes

$$\ln k_{et}=-\frac{RT}{4\Delta E_R}\left(\frac{\Delta_rG^{\ominus}}{RT}\right)^2-\frac{1}{2}\left(\frac{\Delta_rG^{\ominus}}{RT}\right)+\text{constant} \tag{21E.9}$$

and a plot of ln k_{et} (or log k_{et}=ln k_{et}/ln 10) against $\Delta_rG^{\ominus}$ (or $-\Delta_rG^{\ominus}$) is predicted to be shaped like a downward parabola (Fig. 21E.4). Equation 21E.9 implies that the rate constant increases as $\Delta_rG^{\ominus}$ decreases but only up to $-\Delta_rG^{\ominus}=\Delta E_R$. Beyond that, the reaction enters the **inverted region**, in which the rate constant decreases as the reaction becomes more exergonic ($\Delta_rG^{\ominus}$ becomes more negative). The inverted region has been observed in a series of special compounds in which the electron donor and acceptor are linked covalently to a molecular spacer of known and fixed size (Fig. 21E.5).

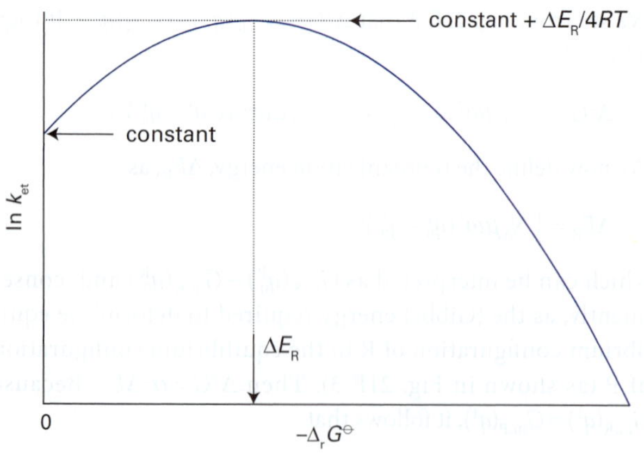

Figure 21E.4 The parabolic dependence of ln k_{et} on $-\Delta_rG^{\ominus}$ predicted by eqn 21E.9.

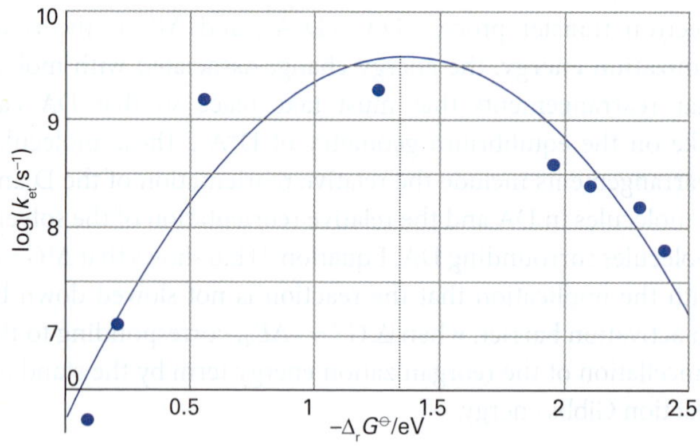

Figure 21E.5 Variation of log k_{et} with $-\Delta_rG^{\ominus}$ for a series of compounds with the structures given in **1** and as described in *Brief illustration* 21E.2. Based on J.R. Miller, et al., *J. Am. Chem. Soc.* **106**, 3047 (1984).

Brief illustration 21E.2 The determination of the reorganization energy

Kinetic measurements were conducted in 2-methyltetrahydrofuran and at 296 K for a series of compounds with the structures given in **1**. The distance between donor (the reduced biphenyl group) and the acceptor is constant for all compounds in the series because the molecular linker remains the same. Each acceptor has a characteristic standard potential, so it follows that the standard Gibbs energy for the electron transfer process is different for each compound in the series. The line in Fig. 21E.5 is a fit to a version of eqn 21E.9 and the maximum of the parabola occurs at $-\Delta_r G^{\ominus} = \Delta E_R = 1.4\,\text{eV} = 1.4\times10^2\,\text{kJ mol}^{-1}$.

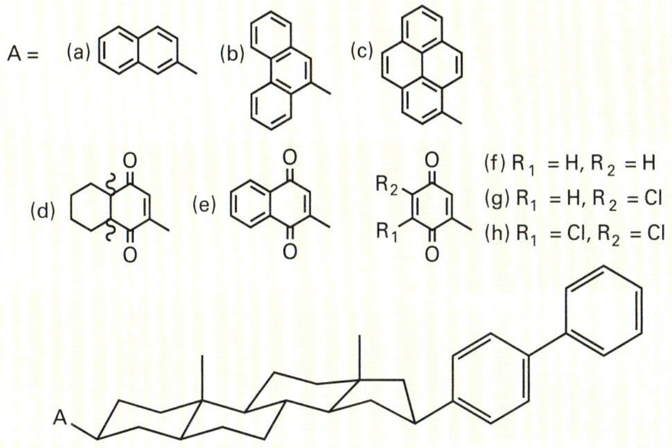

1 An electron donor–acceptor system

Self-test 21E.2 Some (invented) data on a series of complexes are as follows:

$-\Delta_r G^{\ominus}$/eV	0.20	0.60	1.0	1.3	1.6	2.0	2.4
$\log k_{et}$	8.2	9.7	10.2	10.1	9.4	7.7	5.1

Determine the reorganization energy.

Answer: 1.05 eV

Checklist of concepts

- ☐ 1. Electron transfer can occur only after thermal fluctuations bring the nuclear coordinate to the point at which the donor and acceptor have the same configuration.
- ☐ 2. The tunnelling rate is supposed to depend exponentially on the separation of the donor and acceptor.
- ☐ 3. The **reorganization energy** is the energy change associated with molecular rearrangements that must take place so that DA can acquire the equilibrium geometry of D^+A^-.
- ☐ 4. In the **inverted region**, the rate constant k_{et} decreases as the reaction becomes more exergonic ($\Delta_r G^{\ominus}$ becomes more negative).

Checklist of equations

Property	Equation	Comment	Equation number
Electron transfer rate constant	$1/k_r = 1/k_a + (k'_a/k_a k_{et})(1 + k'_{et}/k_d)$	Steady-state assumption	21E.3
Tunnelling probability	$H_{et}(d)^2 = H_{et}^{\circ 2} e^{-\beta d}$	Assumed	21E.4
Rate constant	$k_{et} = CH_{et}(d)^2 e^{-\Delta^{\ddagger}G/RT}$	Transition-state theory	21E.5
Gibbs energy of activation	$\Delta^{\ddagger}G = (\Delta_r G^{\ominus} + \Delta E_R)^2/4\Delta E_R$	Assumes parabolic potential energy	21E.6
Dependence on separation	$\ln k_{et} = -\beta d + \text{constant}$		21E.8
Dependence on $\Delta_r G^{\ominus}$	$\ln k_{et} = a\Delta_r G^{\ominus 2} + b\Delta_r G^{\ominus} + c$	$a = -1/4\Delta E_R RT$, $b = -1/2RT$, c = constant	21E.9

21F Processes at electrodes

Contents

➤ Why do you need to know this material?

A knowledge of the factors that determine the rate of electron transfer at electrodes leads to a better understanding of power production in batteries, and of electron conduction in metals, semiconductors, and nanometre-sized electronic devices, all of which are highly important in modern technology.

➤ What is the key idea?

Transition-state theory can be applied to the description of electron transfer processes at the surface of electrodes.

➤ What do you need to know already?

You need to be familiar with electrochemical cells (Topic 6C), electrode potentials (Topic 6D), and the thermodynamic version of transition-state theory (Topic 21C), particularly the activation Gibbs energy.

As for homogeneous systems (Topic 21E), electron transfer at the surface of an electrode involves electron tunnelling. However, the electrode possesses a nearly infinite number of closely spaced electronic energy levels rather than the small number of discrete levels of a typical complex. Furthermore, specific interactions with the electrode surface give the solute and solvent special properties that can be very different from those observed in the bulk of the solution. For this reason, we begin with a description of the electrode–solution interface. Then we describe the kinetics of electrode processes that draws on the thermodynamic language inspired by transition-state theory.

21F.1 The electrode–solution interface

The most primitive model of the boundary between the solid and liquid phases is as an **electrical double layer**, which consists of a sheet of positive charge at the surface of the electrode and a sheet of negative charge next to it in the solution (or vice versa). We shall see that this arrangement creates an electrical potential difference, called the **Galvani potential difference**, between the bulk of the metal electrode and the bulk of the solution.

More sophisticated models for the electrode–solution interface attempt to describe the gradual changes in the structure of the solution between two extremes, one the charged electrode surface and the other the bulk solution. In the **Helmholtz layer model** of the interface the solvated ions arrange themselves along the surface of the electrode but are held away from it by their hydration spheres (Fig. 21F.1). The location of the sheet

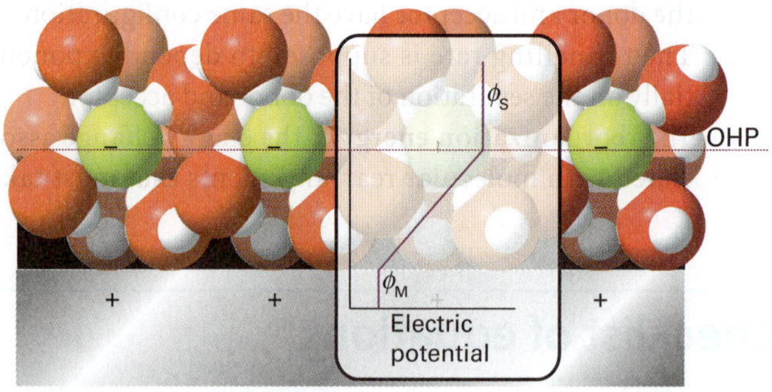

Figure 21F.1 A simple model of the electrode–solution interface treats it as two rigid planes of charge. One plane, the outer Helmholtz plane (OHP), is due to the ions with their solvating molecules and the other plane is that of the electrode itself. The plot shows the dependence of the electric potential with distance from the electrode surface according to this model. Between the electrode surface and the OHP, the potential varies linearly from ϕ_M, the value in the metal, to ϕ_S, the value in the bulk of the solution.

of ionic charge, which is called the **outer Helmholtz plane** (OHP), is identified as the plane running through the solvated ions. In this simple model, the electrical potential changes linearly within the layer bounded by the electrode surface on one side and the OHP on the other. In a refinement of this model, ions that have discarded their solvating molecules and have become attached to the electrode surface by chemical bonds are regarded as forming the **inner Helmholtz plane** (IHP). The Helmholtz layer model ignores the disrupting effect of thermal motion, which tends to break up and disperse the rigid outer plane of charge. In the **Gouy–Chapman model** of the **diffuse double layer**, the disordering effect of thermal motion is taken into account in much the same way as the Debye–Hückel model describes the ionic atmosphere of an ion (Topic 5F) with the latter's single central ion replaced by an infinite, plane electrode.

Figure 21F.2 shows how the local concentrations of cations and anions differ in the Gouy–Chapman model from their bulk concentrations. Ions of opposite charge cluster close to the electrode and ions of the same charge are repelled from it. The modification of the local concentrations near an electrode implies that it might be misleading to use activity coefficients characteristic of the bulk to discuss the thermodynamic properties of ions near the interface. This is one of the reasons why measurements of the dynamics of electrode processes are almost always done using a large excess of supporting electrolyte (for example, a 1 M solution of a salt, an acid, or a base). Under such conditions, the activity coefficients are almost constant because the inert ions dominate the effects of local changes caused by any reactions taking place. The use of a concentrated solution also minimizes ion migration effects.

Neither the Helmholtz nor the Gouy–Chapman model is a very good representation of the structure of the double layer. The former overemphasizes the rigidity of the local solution; the latter underemphasizes its structure. The two are combined in the

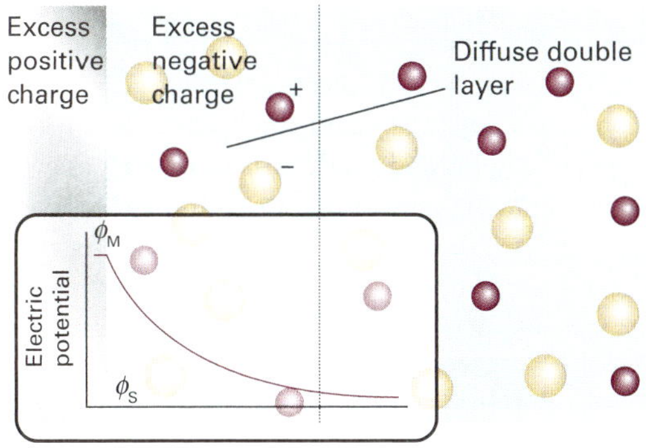

Figure 21F.2 The Gouy–Chapman model of the electrical double layer treats the outer region as an atmosphere of counter-charge, similar to the Debye–Hückel theory of ion atmospheres. The plot of electrical potential against distance from the electrode surface shows the meaning of the diffuse double layer (see text for details).

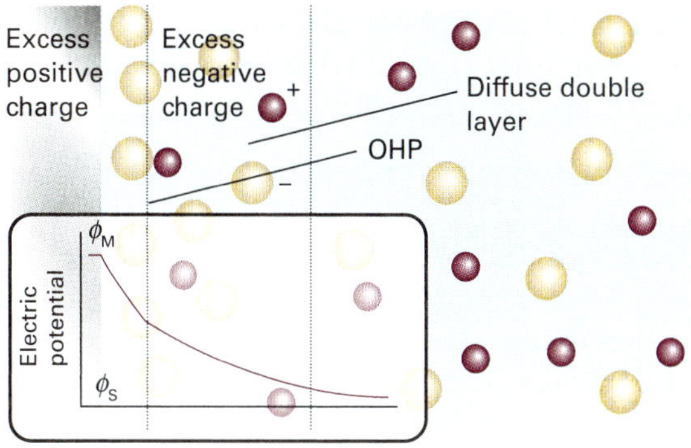

Figure 21F.3 A representation of the Stern model of the electrode–solution interface. The model incorporates the idea of an outer Helmholtz plane near the electrode surface and of a diffuse double layer further away from the surface.

Stern model, in which the ions closest to the electrode are constrained into a rigid Helmholtz plane while outside that plane the ions are dispersed as in the Gouy–Chapman model (Fig. 21F.3). Yet another level of sophistication is found in the **Grahame model**, which adds an inner Helmholtz plane to the Stern model.

The potential difference between points in the bulk metal and the bulk solution is the **Galvani potential difference**, $\Delta\phi$. Apart from a constant, this Galvani potential difference is the electrode potential that was discussed in Topic 6D. We shall ignore the constant, which cannot be measured anyway, and identify changes in $\Delta\phi$ with changes in electrode potential.

21F.2 The rate of electron transfer

We shall consider a reaction at the electrode in which an ion is reduced by the transfer of a single electron in the rate-determining step. We focus on the **current density**, j, the electric current flowing through a region of an electrode divided by the area of the region.

(a) The Butler–Volmer equation

We show in the following *Justification* that an analysis of the effect of the Galvani potential difference at the electrode on the current density leads to the **Butler–Volmer equation**:

$$j = j_0\{e^{(1-\alpha)f\eta} - e^{-\alpha f\eta}\} \qquad \text{Butler–Volmer equation} \qquad (21F.1)$$

where we have written $f = F/RT$, with F as Faraday's constant. The equation contains the following parameters:

- η (eta), the **overpotential:**

$$\eta = E' - E \qquad \text{Definition} \quad \text{Overpotential} \qquad (21F.2)$$

where E is the electrode potential at equilibrium (when there is no net flow of current), and E' is the electrode potential when a current is being drawn from the cell.

- α, the **transfer coefficient**, an indication of where the transition state between the reduced and oxidized forms of the electroactive species in solution is reactant-like ($\alpha=0$) or product-like ($\alpha=1$).
- j_0, the **exchange-current density**, the magnitude of the equal but opposite current densities when the electrode is at equilibrium.

Justification 21F.1 The Butler–Volmer equation

Because an electrode reaction is heterogeneous, we express the rate of charge transfer as the flux of products, the amount of material produced over a region of the electrode surface in an interval of time divided by the area of the region and the duration of the interval.

A first-order heterogeneous rate law has the form

$$\text{Product flux} = k_r[\text{species}]$$

where [species] is the molar concentration of the relevant electroactive species in solution close to the electrode, just outside the double layer. The rate constant has dimensions of length/time (with units, for example, of centimetres per second, cm s^{-1}). If the molar concentrations of the oxidized and reduced materials outside the double layer are [Ox] and [Red], respectively, then the rate of reduction of Ox, v_{Ox}, is $v_{Ox}=k_c[\text{Ox}]$ and the rate of oxidation of Red, v_{Red}, is $v_{Red}=k_a[\text{Red}]$. (The notation k_c and k_a is justified below.)

Now consider a reaction at the electrode in which an ion is reduced by the transfer of a single electron in the rate-determining step. The net current density at the electrode is the difference between the current densities arising from the reduction of Ox and the oxidation of Red. Because the redox processes at the electrode involve the transfer of one electron per reaction event, the current densities, j, arising from the redox processes are the rates v_{Ox} and v_{Red} multiplied by the charge transferred per mole of reaction, which is given by Faraday's constant. Therefore, there is a **cathodic current density** of magnitude

$$j_c = Fk_c[\text{Ox}] \qquad \text{for Ox} + e^- \rightarrow \text{Red}$$

arising from the reduction (because, as defined in Topic 6C, the cathode is the site of reduction). There is also an opposing **anodic current density** of magnitude

$$j_a = Fk_a[\text{Red}] \qquad \text{for Red} \rightarrow \text{Ox} + e^-$$

arising from the oxidation (because the anode is the site of oxidation). The net current density at the electrode is the difference

$$j = j_a - j_c = Fk_a[\text{Red}] - Fk_c[\text{Ox}]$$

Note that, when $j_a > j_c$, so that $j>0$, the current is anodic (Fig. 21F.4a); when $j_c > j_a$, so that $j>0$, the current is cathodic (Fig. 21F.4b).

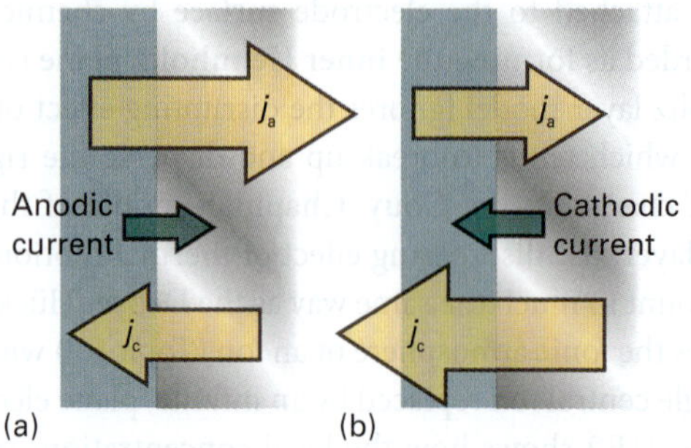

Figure 21F.4 The net current density is defined as the difference between the cathodic and anodic contributions. (a) When $j_a > j_c$, the net current is anodic, and there is a net oxidation of the species in solution. (b) When $j_c > j_a$, the net current is cathodic, and the net process is reduction.

If a species is to participate in reduction or oxidation at an electrode, it must discard any solvating molecules, migrate through the electrode–solution interface, and adjust its hydration sphere as it receives or discards electrons. Likewise, a species already at the inner plane must be detached and migrate into the bulk. Because both processes are activated, we can expect to write their rate constants in the form suggested by transition-state theory (Topic 21C) as

$$k_r = Be^{-\Delta^{\ddagger}G/RT}$$

where $\Delta^{\ddagger}G$ is the activation Gibbs energy and B is a constant with the same dimensions as k_r.

When the expressions for k_r, specifically k_c and k_a, are inserted, we obtain

$$j = FB_a[\text{Red}]e^{-\Delta^{\ddagger}G_a/RT} - FB_c[\text{Ox}]e^{-\Delta^{\ddagger}G_c/RT}$$

This expression allows the activation Gibbs energies to be different for the cathodic and anodic processes. That they are different is the central feature of the remaining discussion.

Next, we relate j to the Galvani potential difference, which varies across the electrode–solution interface as shown schematically in Fig. 21F.5. Consider the reduction reaction, Ox + e$^-$ → Red, and the corresponding reaction profile. If the transition state of the activated complex is product-like (as represented by the peak of the reaction profile being close to the electrode in Fig. 21F.6), the activation Gibbs energy is changed from $\Delta^{\ddagger}G_c(0)$, the value it has in the absence of a potential difference across the double layer, to $\Delta^{\ddagger}G_c = \Delta^{\ddagger}G_c(0) + F\Delta\phi$. Thus, if the electrode is more positive than the solution, $\Delta\phi > 0$, then more work has to be done to form an activated complex from Ox; in this case the activation Gibbs energy is increased.

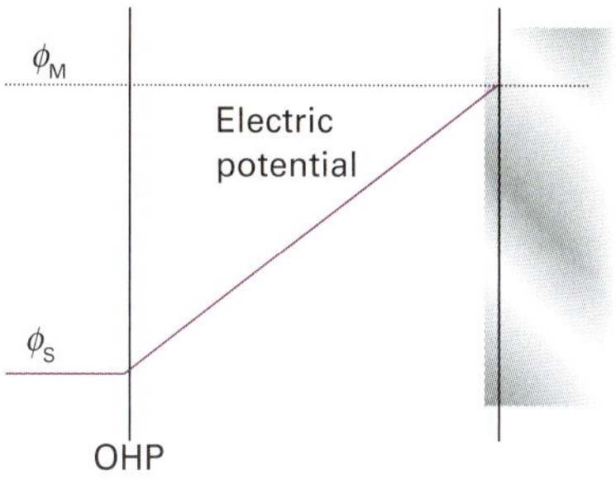

Figure 21F.5 The potential, ϕ, varies linearly between two plane parallel sheets of charge, and its effect on the Gibbs energy of the transition state depends on the extent to which the latter resembles the species at the inner or outer planes.

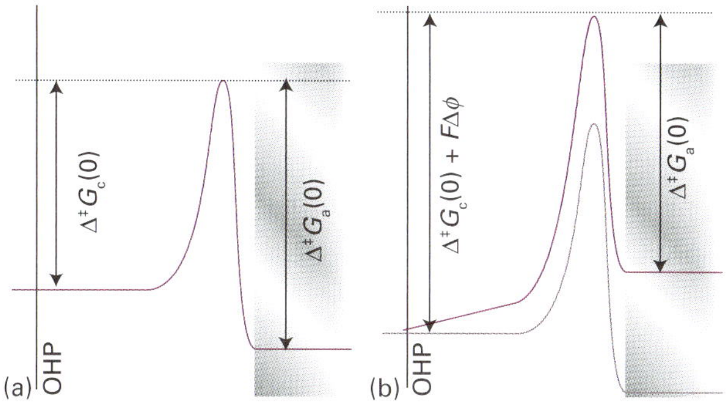

Figure 21F.6 When the transition state resembles a species that has undergone reduction, the activation Gibbs energy for the anodic current is almost unchanged, but the full effect applies to the cathodic current. (a) Zero potential difference; (b) nonzero potential difference.

If the transition state is reactant-like (represented by the peak of the reaction profile being close to the outer plane of the double-layer in Fig. 21F.7), then $\Delta^{\ddagger}G_c$ is independent of $\Delta\phi$. In a real system, the transition state has an intermediate resemblance to these extremes (Fig. 21F.8) and the activation Gibbs energy for reduction may be written as

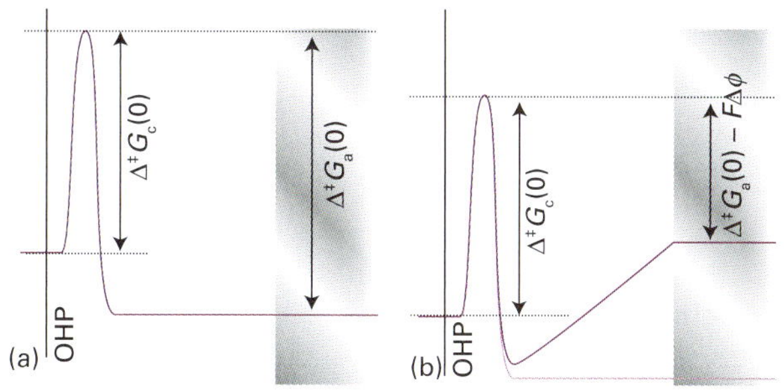

Figure 21F.7 When the transition state resembles a species that has undergone oxidation, the activation Gibbs energy for the cathodic current is almost unchanged but the activation Gibbs energy for the anodic current is strongly affected. (a) Zero potential difference; (b) nonzero potential difference.

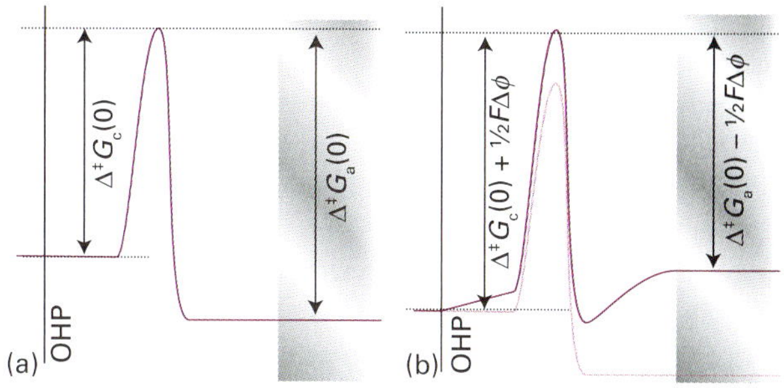

Figure 21F.8 When the transition state is intermediate in its resemblance to reduced and oxidized species, as represented here by a peak located at an intermediate position as measured by α (with $0<\alpha<1$), both activation Gibbs energies are affected; here, $\alpha \approx 0.5$. (a) Zero potential difference; (b) nonzero potential difference.

$$\Delta^{\ddagger}G_c = \Delta^{\ddagger}G_c(0) + \alpha F\Delta\phi$$

The parameter α lies in the range 0 to 1. Experimentally, α is often found to be about 0.5.

Now consider the oxidation reaction, $\text{Red} + e^- \rightarrow \text{Ox}$ and its reaction profile. Similar remarks apply. In this case, Red discards an electron to the electrode, so the extra work is zero if the transition state is reactant-like (represented by a peak close to the electrode). The extra work is the full $-F\Delta\phi$ if it resembles the product (the peak close to the outer plane). In general, the activation Gibbs energy for this anodic process is

$$\Delta^{\ddagger}G_a = \Delta^{\ddagger}G_a(0) - (1-\alpha)F\Delta\phi$$

The two activation Gibbs energies can now be inserted in the expression for j, with the result that

$$j = FB_a[\text{Red}]e^{-\Delta^{\ddagger}G_a(0)/RT}e^{(1-\alpha)F\Delta\phi/RT} - FB_c[\text{Ox}]e^{-\Delta^{\ddagger}G_c(0)/RT}e^{-\alpha F\Delta\phi/RT}$$

This is an explicit, if complicated, expression for the net current density in terms of the potential difference.

The appearance of the new expression for j can be simplified. First, in a purely cosmetic step we write $f = F/RT$. Next, we identify the individual cathodic and anodic current densities:

$$j = \overbrace{FB_a[\text{Red}]e^{-\Delta^{\ddagger}G_a(0)/RT}e^{(1-\alpha)f\Delta\phi}}^{j_a} - \overbrace{FB_c[\text{Ox}]e^{-\Delta^{\ddagger}G_c(0)/RT}e^{-\alpha f\Delta\phi}}^{j_c}$$

If the cell is balanced against an external source, the Galvani potential difference, $\Delta\phi$, can be identified as the (zero-current) electrode potential, E, and we can write

$$j_a = FB_a[\text{Red}]e^{-\Delta^{\ddagger}G_a(0)/RT}e^{(1-\alpha)fE}$$
$$j_c = FB_c[\text{Ox}]e^{-\Delta^{\ddagger}G_c(0)/RT}e^{-\alpha fE}$$

When these equations apply, there is no net current at the electrode (as the cell is balanced), so the two current densities must be equal. From now on we denote them both as j_0.

When the cell is producing current (that is, when a load is connected between the electrode being studied and a second counter electrode) the electrode potential changes from its zero-current value, E, to a new value, E', and the difference is the electrode's overpotential, $\eta = E' - E$. Hence, $\Delta\phi$ changes from E to $E + \eta$ and the two current densities become

$$j_a = j_0 e^{(1-\alpha)f\eta} \qquad j_c = j_0 e^{-\alpha f\eta}$$

Then from $j = j_a - j_c$ we obtain the Butler–Volmer equation, eqn 21F.1.

Figure 21F.9 shows how eqn 21F.1 predicts the current density to depend on the overpotential for different values of the transfer coefficient. When the overpotential is so small that $f\eta \ll 1$ (in practice, η less than about 10 mV) the exponentials in eqn 21F.1 can be expanded by using $e^x = 1 + x + \cdots$ to give

$$j = j_0\{\overbrace{1 + (1-\alpha)f\eta + \cdots}^{e^{(1-\alpha)f\eta}} - \overbrace{(1 - \alpha f\eta + \cdots)}^{e^{-\alpha f\eta}}\} \approx j_0 f\eta \qquad (21F.3)$$

This equation shows that the current density is proportional to the overpotential, so at low overpotentials the interface behaves like a conductor that obeys Ohm's law. When there is a small positive overpotential the current is anodic ($j > 0$ when $\eta > 0$), and when the overpotential is small and negative the current is cathodic ($j < 0$ when $\eta < 0$). The relation can also be reversed to calculate the potential difference that must exist if a current density j has been established by some external circuit:

$$\eta = \frac{RTj}{Fj_0} \qquad (21F.4)$$

The importance of this interpretation will become clear shortly.

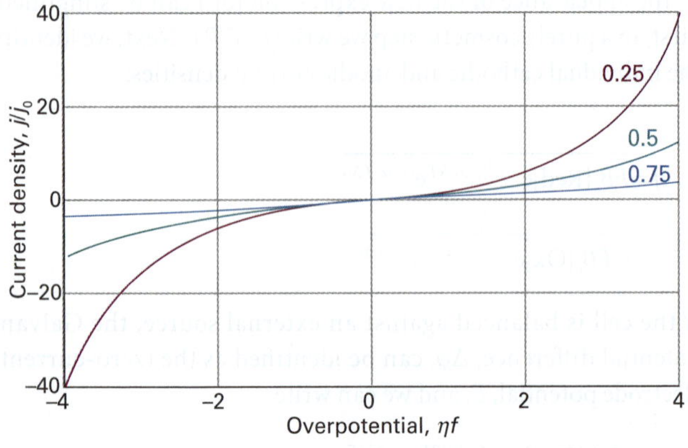

Figure 21F.9 The dependence of the current density on the overpotential for different values of the transfer coefficient.

Brief illustration 21F.1 The current density

The exchange current density of a Pt(s)|H_2(g)|H^+(aq) electrode at 298 K is 0.79 mA cm^{-2}. Therefore, the current density when the overpotential is +5.0 mV is obtained by using eqn 21F.4 and $f = F/RT = 1/(25.69\,\text{mV})$:

$$j = j_0 f\eta = \frac{(0.79\,\text{mA cm}^{-2}) \times (5.0\,\text{mV})}{25.69\,\text{mV}} = 0.15\,\text{mA cm}^{-2}$$

The current through an electrode of total area 5.0 cm^2 is therefore 0.75 mA.

Self-test 21F.1 What would be the current at pH = 2.0, the other conditions being the same?

Answer: −18 mA (cathodic)

Some experimental values for the Butler–Volmer parameters are given in Table 21F.1. From them we can see that exchange current densities vary over a very wide range. Exchange currents are generally large when the redox process involves no bond breaking (as in the $[Fe(CN)_6]^{3-}$, $[Fe(CN)_6]^{4-}$ couple) or if only weak bonds are broken (as in Cl_2,Cl^-). They are generally small when more than one electron needs to be transferred, or when multiple or strong bonds are broken, as in the N_2,N_3^- couple and in redox reactions of organic compounds.

(b) Tafel plots

When the overpotential is large and positive (in practice, $\eta \geq 0.12$ V), corresponding to the electrode being the anode in electrolysis, the second exponential in eqn 21F.1 is much smaller than the first, and may be neglected. Then

$$j = j_0 e^{(1-\alpha)f\eta} \qquad (21F.5a)$$

so

$$\ln j = \ln j_0 + (1-\alpha)f\eta \qquad (21F.5b)$$

The plot of the logarithm of the current density against the overpotential is called a **Tafel plot**. The slope, which is equal to

Table 21F.1* Exchange current densities and transfer coefficients at 298 K

Reaction	Electrode	j_0/(A cm^{-2})	α
$2\,H^+ + 2\,e^- \rightarrow H_2$	Pt	7.9×10^{-4}	
	Ni	6.3×10^{-6}	0.58
	Pb	5.0×10^{-12}	
$Fe^{3+} + e^- \rightarrow Fe^{2+}$	Pt	2.5×10^{-3}	0.58

* More values are given in the *Resource section*.

$(1-\alpha)f$, gives the value of α and the intercept at $\eta=0$ gives the exchange-current density. If instead the overpotential is large but negative (in practice, $\eta \leq -0.12\,\text{V}$), the first exponential in eqn 21F.1 may be neglected. Then

$$j = j_0 e^{-\alpha f \eta} \qquad \text{(21F.6a)}$$

so

$$\ln j = \ln j_0 - \alpha f \eta \qquad \text{(21F.6b)}$$

In this case the slope of the Tafel plot is $-\alpha f$.

Example 21F.1 Interpreting a Tafel plot

The data below refer to the anodic current through a platinum electrode of area 2.0 cm^2 in contact with an Fe^{3+}, Fe^{2+} aqueous solution at 298 K. Calculate the exchange-current density and the transfer coefficient for the electrode process.

η/mV	50	100	150	200	250
I/mA	8.8	25.0	58.0	131	298

Method The anodic process is the oxidation $Fe^{2+}(aq) \rightarrow Fe^{3+}(aq) + e^-$. To analyse the data, we make a Tafel plot (of $\ln j$ against η) using the anodic form (eqn 21F.5b). The intercept at $\eta=0$ is $\ln j_0$ and the slope is $(1-\alpha)f$.

Answer Draw up the following table:

η/mV	50	100	150	200	250
j/(mA cm^{-2})	4.4	12.5	29.0	65.5	149
ln(j/(mA cm^{-2}))	1.48	2.53	3.37	4.18	5.00

The points are plotted in Fig. 21F.10. The high overpotential region gives a straight line of intercept 0.88 and slope 0.0165. From the former it follows that $\ln(j_0/(\text{mA cm}^{-2}))=0.88$, so $j_0 = 2.4\,\text{mA cm}^{-2}$. From the latter,

$$(1-\alpha)f = \overbrace{0.0165}^{\text{Slope}}\ \overbrace{\text{mV}^{-1}}^{\text{Units of }\ln j/\eta\ (1/\text{mV})}$$

so

$$\alpha = 1 - \frac{\overbrace{0.0165\,\text{mV}^{-1}}^{16.5\,\text{V}^{-1}}}{\underbrace{f}_{38.9\,\text{V}^{-1}}} = 1 - 0.42\ldots = 0.58$$

Note that the Tafel plot is nonlinear for $\eta < 100$ mV; in this region $\alpha f \eta = 2.3$ and the approximation that $\alpha f \eta \gg 1$ fails.

Self-test 21F.2 Repeat the analysis using the following cathodic current data:

η/mV	−50	−100	−150	−200	−250	−300
I/mA	0.3	1.5	6.4	27.6	118.6	510

Answer: $\alpha=0.75$, $j_0=0.041\,\text{mA cm}^{-2}$

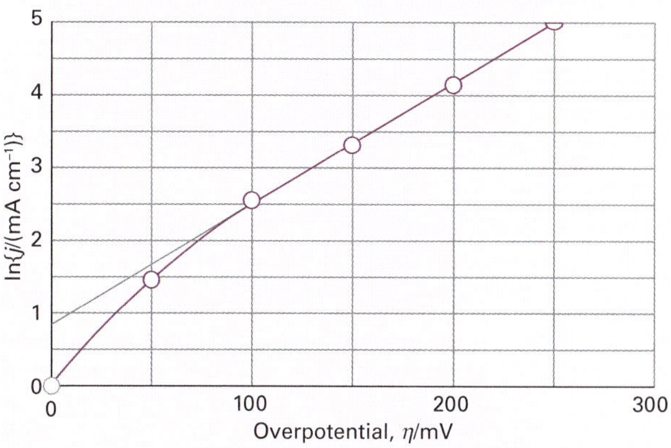

Figure 21F.10 A Tafel plot is used to measure the exchange current density (given by the extrapolated intercept at $\eta=0$) and the transfer coefficient (from the slope). The data are from *Example* 21F.1.

21F.3 Voltammetry

One of the assumptions in the derivation of the Butler–Volmer equation is the negligible conversion of the electroactive species at low current densities, resulting in uniformity of concentration near the electrode. This assumption fails at high current densities because the consumption of electroactive species close to the electrode results in a concentration gradient. The diffusion of the species towards the electrode from the bulk is slow and may become rate determining; a larger overpotential is then needed to produce a given current. This effect is called **concentration polarization**. Concentration polarization is important in the interpretation of **voltammetry**, the study of the current through an electrode as a function of the applied potential difference.

The kind of output from **linear-sweep voltammetry** is illustrated in Fig. 21F.11. Initially, the absolute value of the potential

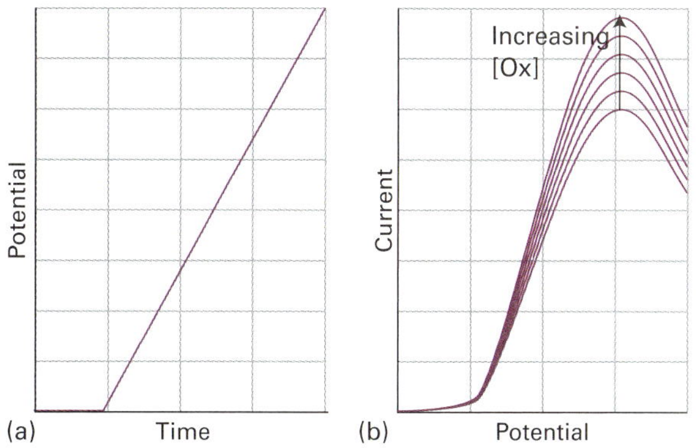

Figure 21F.11 (a) The change of potential with time and (b) the resulting current/potential curve in a voltammetry experiment. The peak value of the current density is proportional to the concentration of electroactive species (for instance, [Ox]) in solution.

is low, and the current is due to the migration of ions in the solution. However, as the potential approaches the reduction potential of the reducible solute, the current grows. Soon after the potential exceeds the reduction potential the current rises and reaches a maximum value. This maximum current is proportional to the molar concentration of the species, so that concentration can be determined from the peak height after subtraction of an extrapolated baseline.

In **cyclic voltammetry** the potential is applied with a triangular waveform (linearly up, then linearly down) and the current is monitored. A typical cyclic voltammogram is shown in Fig. 21F.12. The shape of the curve is initially like that of a linear-sweep experiment, but after reversal of the sweep there is a rapid change in current on account of the high concentration of oxidizable species close to the electrode that was generated on the reductive sweep. When the potential is close to the value required to oxidize the reduced species, there is a substantial current until all the oxidation is complete, and the current returns to zero. Cyclic voltammetry data are obtained at scan rates of about 50 mV s^{-1}, so a scan over a range of 2 V takes about 80 s.

When the reduction reaction at the electrode can be reversed, as in the case of the $[Fe(CN)_6]^{3-}/[Fe(CN)_6]^{4-}$ couple, the cyclic voltammogram is broadly symmetric about the standard potential of the couple (as in Fig. 21F.12). The scan is initiated with $[Fe(CN)_6]^{3-}$ present in solution, and as the potential approaches $E^\ominus$ for the couple, the $[Fe(CN)_6]^{3-}$ near the electrode is reduced and current begins to flow. As the potential continues to change, the current begins to decline again because all the $[Fe(CN)_6]^{3-}$ near the electrode has been reduced and the current reaches its limiting value. The potential is now returned linearly to its initial value, and the reverse series of events occurs with the $[Fe(CN)_6]^{4-}$ produced during the forward scan now undergoing oxidation. The peak of current lies on the other side of $E^\ominus$, so the species present and its standard potential can be identified, as indicated in the illustration, by noting the locations of the two peaks.

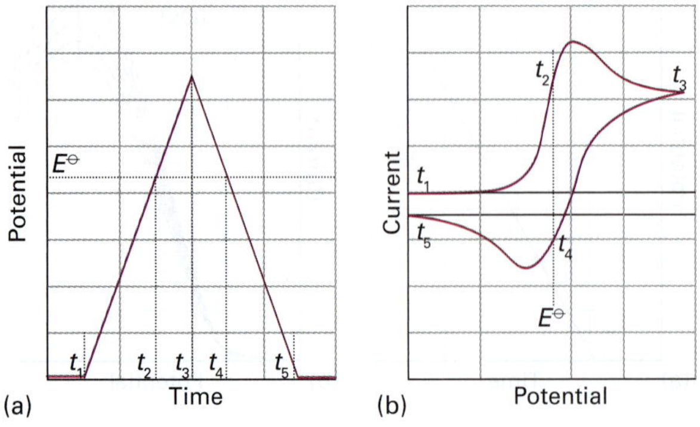

Figure 21F.12 (a) The change of potential with time and (b) the resulting current/potential curve in a cyclic voltammetry experiment.

The overall shape of the curve gives details of the kinetics of the electrode process and the change in shape as the rate of change of potential is altered gives information on the rates of the processes involved. For example, the matching peak on the return phase of the potential sweep may be missing, which indicates that the oxidation (or reduction) is irreversible. The appearance of the curve may also depend on the timescale of the sweep, for if the sweep is too fast some processes might not have time to occur. This style of analysis is illustrated in the following example.

Example 21F.2 Analysing a cyclic voltammetry experiment

The electroreduction of *p*-bromonitrobenzene in liquid ammonia is believed to occur by the following mechanism:

$$BrC_6H_4NO_2 + e^- \rightarrow BrC_6H_4NO_2^-$$
$$BrC_6H_4NO_2^- \rightarrow {\cdot}C_6H_4NO_2 + Br^-$$
$${\cdot}C_6H_4NO_2 + e^- \rightarrow C_6H_4NO_2^-$$
$$C_6H_4NO_2^- + H^+ \rightarrow C_6H_5NO_2$$

Suggest the likely form of the cyclic voltammogram expected on the basis of this mechanism.

Method Decide which steps are likely to be reversible on the timescale of the potential sweep: such processes will give symmetrical voltammograms. Irreversible processes will give unsymmetrical shapes as reduction (or oxidation) might not occur. However, at fast sweep rates, an intermediate might not have time to react, and a reversible shape will be observed.

Answer At slow sweep rates, the second reaction has time to occur, and a curve typical of a two-electron reduction will be

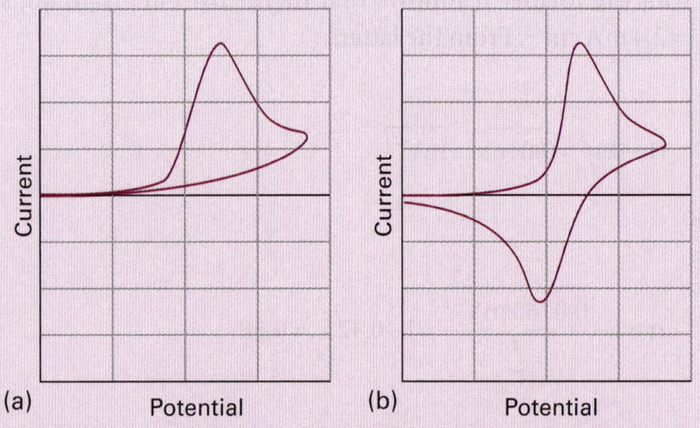

Figure 21F.13 (a) When a non-reversible step in a reaction mechanism has time to occur, the cyclic voltammogram may not show the reverse oxidation or reduction peak. (b) However, if the rate of sweep is increased, the return step may be caused to occur before the irreversible step has had time to intervene, and a typical 'reversible' voltammogram is obtained.

observed, but there will be no oxidation peak on the second half of the cycle because the product, $C_6H_5NO_2$, cannot be oxidized (Fig. 21F.13a). At fast sweep rates, the second reaction does not have time to take place before oxidation of the $BrC_6H_4NO_2^-$ intermediate starts to occur during the reverse scan, so the voltammogram will be typical of a reversible one-electron reduction (Fig. 21F.13b).

Self-test 21F.3 Suggest an interpretation of the cyclic voltammogram shown in Fig. 21F.14.The electroactive material is ClC_6H_4CN in acid solution; after reduction to $ClC_6H_4CN^-$ the radical anion may form C_6H_5CN irreversibly.

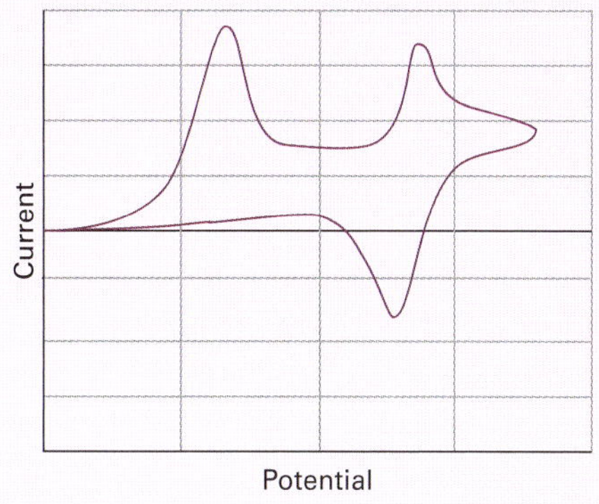

Figure 21F.14 The cyclic voltammogram referred to in *Self-test* 21F.3.

Answer: $ClC_6H_4CN + e^- \rightleftharpoons ClC_6H_4CN^-$, $ClC_6H_4CN^- + H^+ + e^- \rightarrow C_6H_5CN + Cl^-$, $C_6H_5CN + e^- \rightleftharpoons C_6H_5CN^-$

21F.4 Electrolysis

To induce current to flow through an electrolytic cell and bring about a nonspontaneous cell reaction, the applied potential difference must exceed the zero-current potential by at least the **cell overpotential**, the sum of the overpotentials at the two electrodes and the ohmic drop (IR_s, where R_s is the internal resistance of the cell) due to the current through the electrolyte. The additional potential needed to achieve a detectable rate of reaction may need to be large when the exchange current density at the electrodes is small. For similar reasons, a working galvanic cell generates a smaller potential than under zero-current conditions. In this section we see how to cope with both aspects of the overpotential.

The relative rates of gas evolution or metal deposition during electrolysis can be estimated from the Butler–Volmer equation and tables of exchange current densities. From eqn 21F.6a and assuming equal transfer coefficients, we write the ratio of the cathodic currents as

$$\frac{j'}{j} = \frac{j_0'}{j_0} e^{(\eta - \eta')\alpha f} \qquad (21F.7)$$

where j' is the current density for electrodeposition and j is that for gas evolution, and j_0' and j_0 are the corresponding exchange current densities. This equation shows that metal deposition is favoured by a large exchange current density and relatively high gas evolution overpotential (so $\eta - \eta'$ is positive and large). Note that $\eta < 0$ for a cathodic process, so $-\eta' > 0$. The exchange current density depends strongly on the nature of the electrode surface, and changes in the course of the electrodeposition of one metal on another. A very crude criterion is that significant evolution or deposition occurs only if the overpotential exceeds about 0.6 V.

A glance at Table 21F.1 shows the wide range of exchange current densities for a metal/hydrogen electrode. The most sluggish exchange currents occur for lead and mercury: $1\,\text{pA cm}^{-2}$ corresponds to a monolayer of atoms being replaced in about 5 years. For such systems, a high overpotential is needed to induce significant hydrogen evolution. In contrast, the value for platinum ($1\,\text{mA cm}^{-2}$) corresponds to a monolayer being replaced in 0.1 s, so significant gas evolution occurs for a much lower overpotential.

The exchange current density also depends on the crystal face exposed. For the deposition of copper on copper, the (100) face has $j_0 = 1\,\text{mA cm}^{-2}$, so for the same overpotential the (100) face grows at 2.5 times the rate of the (111) face, for which $j_0 = 0.4\,\text{mA cm}^{-2}$.

21F.5 Working galvanic cells

In working galvanic cells (those not balanced against an external potential), the overpotential leads to a smaller potential than under zero-current conditions. Furthermore, we expect the cell potential to decrease as current is generated because it is then no longer working reversibly and can therefore do less than maximum work.

We shall consider the cell $M|M^+(aq)||M'^+(aq)|M'$ and ignore all the complications arising from liquid junctions. The potential of the cell is $E' = \Delta\phi_R - \Delta\phi_L$. Because the cell potential differences differ from their zero-current values by overpotentials, we can write $\Delta\phi_X = E_X + \eta_X$ where X is L or R for the left or right electrode, respectively. The cell potential is therefore

$$E' = E + \eta_R - \eta_L \qquad (21F.8a)$$

To avoid confusion about signs (η_R is negative, η_L is positive) and to emphasize that a working cell has a lower potential than a zero-current cell, we shall write this expression as

$$E' = E - |\eta_R| - |\eta_L| \tag{21F.8b}$$

with E the cell potential. We should also subtract the ohmic potential difference IR_s, where R_s is the cell's internal resistance:

$$E' = E - |\eta_R| - |h_L| - IR_s \tag{21F.8c}$$

The ohmic term is a contribution to the cell's irreversibility—it is a thermal dissipation term—so the sign of IR_s is always such as to reduce the potential in the direction of zero.

The overpotentials in eqn 21F.8 can be calculated from the Butler–Volmer equation for a given current, I, being drawn. We shall simplify the equations by supposing that the areas, A, of the electrodes are the same, that only one electron is transferred in the rate-determining steps at the electrodes, that the transfer coefficients are both $\frac{1}{2}$, and that the high-overpotential limit of the Butler–Volmer equation may be used. Then from eqns 21F.6a and 21F.8c we find

$$E' = E - IR_s - \frac{4RT}{F}\ln\left(\frac{I}{A\bar{j}}\right) \qquad \bar{j} = (j_{0L}j_{0R})^{1/2} \tag{21F.9}$$

where j_{0L} and j_{0R} are the exchange current densities for the two electrodes.

Brief illustration 21F.2 The working potential

Suppose that a cell consists of two electrodes each of area $10\,\mathrm{cm}^2$ with exchange current densities $5\,\mu\mathrm{A\,cm}^{-2}$ and has internal resistance $10\,\Omega$. At 298 K $RT/F = 25.7$ mV. The zero-current cell potential is 1.5 V. If the cell is producing a current of 10 mA, its working potential will be

$$E' = 1.5\,\mathrm{V} - \overbrace{(10\,\mathrm{mA})\times(10\,\Omega)}^{0.10\,V} - \overbrace{4(25.7\,\mathrm{mV})\ln\left(\frac{10\,\mathrm{mA}}{(10\,\mathrm{cm}^2)\times(5\,\mu\mathrm{A\,cm}^{-2})}\right)}^{0.54\,V\ldots} = 0.9\,\mathrm{V}$$

We have used $1\,\mathrm{A}\,\Omega = 1\,\mathrm{V}$. Note that we have ignored various other factors that reduce the cell potential, such as the inability of reactants to diffuse rapidly enough to the electrodes.

Self-test 21F.4 What is the effective resistance at 25 °C of an electrode interface when the overpotential is small? Evaluate it for a $\mathrm{Pt,H_2|H^+}$ electrode with a surface area of $1.0\,\mathrm{cm}^2$.

Answer: 33 Ω

Electric storage cells operate as galvanic cells while they are producing electricity but as electrolytic cells while they are being charged by an external supply. The lead–acid battery is an old device, but one well suited to the job of starting cars (and the only one available). During charging the cathode reaction is the reduction of Pb^{2+} and its deposition as lead on the lead electrode. Deposition occurs instead of the reduction of the acid to hydrogen because the latter has a low exchange current density on lead. The anode reaction during charging is the oxidation of Pb(II) to Pb(IV), which is deposited as the oxide PbO_2. On discharge, the two reactions run in reverse. Because they have such high exchange current densities the discharge can occur rapidly, which is why the lead battery can produce large currents on demand.

Checklist of concepts

- ☐ 1. An **electrical double layer** consists of sheets of opposite charge at the surface of the electrode and next to it in the solution.
- ☐ 2. Models of the double layer include the **Helmholtz layer model** and the **Gouy-Chapman** model.
- ☐ 3. The **Galvani potential difference** is the potential difference between the bulk of the metal electrode and the bulk of the solution.
- ☐ 4. The current density at an electrode is expressed by the **Butler–Volmer equation**.
- ☐ 5. A **Tafel plot** is the plot of the logarithm of the current density against the overpotential (see below).
- ☐ 6. **Voltammetry** is the study of the current through an electrode as a function of the applied potential difference.
- ☐ 7. To induce current to flow through an electrolytic cell and bring about a nonspontaneous cell reaction, the applied potential difference must exceed the cell potential by at least the **cell overpotential**.
- ☐ 8. In working galvanic cells the overpotential leads to a smaller potential than under zero-current conditions and the cell potential decreases as current is generated.

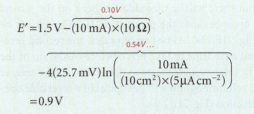

Checklist of equations

Property	Equation	Comment	Equation number
Butler–Volmer equation	$j = j_0\{e^{(1-\alpha)f\eta} - e^{-\alpha f\eta}\}$		21F.1
Tafel plots	$\ln j = \ln j_0 + (1-\alpha)f\eta$	Anodic current density	21F.5b
	$\ln j = \ln j_0 - \alpha f\eta$	Cathodic current density	21F.6b
Potential of a working galvanic cell	$E' = E - IR_s - (4RT/F)\ln(I/A\bar{j})$	$\bar{j} = (j_{0L}j_{0R})^{1/2}$	21F.9

CHAPTER 21 Reaction dynamics

TOPIC 21A Collision theory

Discussion questions

21A.1 Discuss how the collision theory of gases builds on the kinetic–molecular theory.

21A.2 How might collision theory change for real gases?

21A.3 Describe the essential features of the harpoon mechanism.

21A.4 Discuss the significance of the steric *P*-factor in the RRK model.

Exercises

21A.1(a) Calculate the collision frequency, z, and the collision density, Z, in ammonia, $R = 190$ pm, at 30 °C and 120 kPa. What is the percentage increase when the temperature is raised by 10 K at constant volume?
21A.1(b) Calculate the collision frequency, z, and the collision density, Z, in carbon monoxide, $R = 180$ pm at 30 °C and 120 kPa. What is the percentage increase when the temperature is raised by 10 K at constant volume?

21A.2(a) Collision theory depends on knowing the fraction of molecular collisions having at least the kinetic energy E_a along the line of flight. What is this fraction when (i) $E_a = 20\,\text{kJ mol}^{-1}$, (ii) $E_a = 100\,\text{kJ mol}^{-1}$ at (1) 350 K and (2) 900 K?
21A.2(b) Collision theory depends on knowing the fraction of molecular collisions having at least the kinetic energy E_a along the line of flight. What is this fraction when (i) $E_a = 15\,\text{kJ mol}^{-1}$, (ii) $E_a = 150\,\text{kJ mol}^{-1}$ at (1) 300 K and (2) 800 K?

21A.3(a) Calculate the percentage increase in the fractions in Exercise 21A.2(a) when the temperature is raised by 10 K.
21A.3(b) Calculate the percentage increase in the fractions in Exercise 21A.2(b) when the temperature is raised by 10 K.

21A.4(a) Use the collision theory of gas-phase reactions to calculate the theoretical value of the second-order rate constant for the reaction $H_2(g) + I_2(g) \rightarrow 2\,HI(g)$ at 650 K, assuming that it is elementary and bimolecular. The collision cross section is 0.36 nm^2, the reduced mass is 3.32×10^{-27} kg, and the activation energy is 171 kJ mol^{-1}. (Assume a steric factor of 1.)
21A.4(b) Use the collision theory of gas-phase reactions to calculate the theoretical value of the second-order rate constant for the reaction $D_2(g) + Br_2(g) \rightarrow 2\,HI(g)$ at 450 K, assuming that it is elementary and bimolecular. Take the collision cross section as 0.30 nm^2, the reduced mass as $3.930m_u$, and the activation energy as 200 kJ mol^{-1}. (Assume a steric factor of 1.)

21A.5(a) For the gaseous reaction $A + B \rightarrow P$, the reactive cross-section obtained from the experimental value of the pre-exponential factor is 9.2×10^{-22} m^2. The collision cross-sections of A and B estimated from the transport properties are 0.95 and 0.65 nm^2 respectively. Calculate the *P*-factor for the reaction.
21A.5(b) For the gaseous reaction $A + B \rightarrow P$, the reactive cross-section obtained from the experimental value of the pre-exponential factor is 8.7×10^{-22} m^2. The collision cross-sections of A and B estimated from the transport properties are 0.88 and 0.40 nm^2, respectively. Calculate the *P*-factor for the reaction.

21A.6(a) Consider the unimolecular decomposition of a nonlinear molecule containing five atoms according to RRK theory. If $P = 3.0 \times 10^{-5}$, what is the value of E^*/E?
21A.6(b) Consider the unimolecular decomposition of a linear molecule containing four atoms according to RRK theory. If $P = 0.025$, what is the value of E^*/E?

21A.7(a) Suppose that an energy of 250 kJ mol^{-1} is available in a collision but 200 kJ mol^{-1} is needed to break a particular bond in a molecule with $s = 10$. Use the RRK model to calculate the steric *P*-factor.
21.A7(b) Suppose that an energy of 500 kJ mol^{-1} is available in a collision but 300 kJ mol^{-1} is needed to break a particular bond in a molecule with $s = 12$. Use the RRK model to calculate the steric *P*-factor.

Problems

21A.1 In the dimerization of methyl radicals at 25 °C, the experimental pre-exponential factor is 2.4×10^{10} dm^3 mol^{-1} s^{-1}. What are (a) the reactive cross-section, (b) the *P*-factor for the reaction if the C–H bond length is 154 pm?

21A.2 Nitrogen dioxide reacts bimolecularly in the gas phase: $NO_2 + NO_2 \rightarrow NO + NO + O_2$. The temperature dependence of the second-order rate constant for the rate law $d[P]/dt = k_r[NO_2]^2$ is given in the following table. What are the *P*-factor and the reactive cross-section for the reaction?

T/K	600	700	800	1000
k_r/(cm^3 mol^{-1} s^{-1})	4.6×10^2	9.7×10^3	1.3×10^5	3.1×10^6

Take $\sigma = 0.60$ nm^2.

21A.3 The diameter of the methyl radical is about 308 pm. What is the maximum rate constant in the expression $d[C_2H_6]/dt = k_r[CH_3]^2$ for second-order recombination of radicals at room temperature? 10 per cent of a sample of ethane of volume 1.0 dm^3 at 298 K and 100 kPa is dissociated into methyl radicals. What is the minimum time for 90 per cent recombination?

21A.4 The total cross-sections for reactions between alkali metal atoms and halogen molecules are given in the following table (R.D. Levine and R.B. Bernstein, *Molecular reaction dynamics*, Clarendon Press, Oxford, 72 (1974)). Assess the data in terms of the harpoon mechanism.

σ^*/nm^2	Cl_2	Br_2	I_2
Na	1.24	1.16	0.97
K	1.54	1.51	1.27
Rb	1.90	1.97	1.67
Cs	1.96	2.04	1.95

Electron affinities are approximately 1.3 eV (Cl_2), 1.2 eV (Br_2), and 1.7 eV (I_2), and ionization energies are 5.1 eV (Na), 4.3 eV (K), 4.2 eV (Rb), and 3.9 eV (Cs).

21A.5‡ One of the most historically significant studies of chemical reaction rates was that by M. Bodenstein (*Z. physik. Chem.* **29**, 295 (1899)) of the gas-phase reaction 2 HI(g)→H_2(g)+I_2(g) and its reverse, with rate constants k_r and k_r', respectively. The measured rate constants as a function of temperature are

T/K	647	666	683	700	716	781
k_r/(22.4 dm^3 mol^{-1} min^{-1})	0.230	0.588	1.37	3.10	6.70	105.9
k_r'/(22.4 dm^3 mol^{-1} min^{-1})	0.0140	0.0379	0.0659	0.172	0.375	3.58

Demonstrate that these data are consistent with the collision theory of bimolecular gas-phase reactions.

21A.6‡ R. Atkinson (*J. Phys. Chem. Ref. Data* **26**, 215 (1997)) has reviewed a large set of rate constants relevant to the atmospheric chemistry of volatile organic compounds. The recommended rate constant for the bimolecular association of O_2 with an alkyl radical R at 298 K is 4.7×10^9 dm^3 mol^{-1} s^{-1} for R = C_2H_5 and 8.4×10^9 dm^3 mol^{-1} s^{-1} for R = cyclohexyl. Assuming no energy barrier, compute the steric factor, P, for each reaction. *Hint*: Obtain collision diameters from collision cross-sections of similar molecules in the *Resource section*.

21A.7 According to the RRK model (see *Justification* 21A.1)

$$P=\frac{n!(n-n^*+s-1)!}{(n-n^*)!(n+s-1)!}$$

Use Stirling's approximation of the form $\ln x!\approx x\ln x - x$ to deduce that $P\approx(n-n^*/n)^{s-1}$ when $s-1\ll n-n^*$. *Hint*: replace terms of the form $n-n^*+s-1$ by $n-n^*$ inside logarithms but retain $n-n^*+s-1$ when it is a factor of a logarithm.

TOPIC 21B Diffusion-controlled reactions

Discussion questions

21B.1 Distinguish between a diffusion-controlled reaction and an activation-controlled reaction. Do both have activation energies?

21B.2 Describe the role of the encounter pair in the cage effect.

Exercises

21B.1(a) A typical diffusion coefficient for small molecules in aqueous solution at 25 °C is 6×10^{-9} m^2 s^{-1}. If the critical reaction distance is 0.5 nm, what value is expected for the second-order rate constant for a diffusion-controlled reaction?

21B.1(b) Suppose that the typical diffusion coefficient for a reactant in aqueous solution at 25 °C is 5.2×10^{-9} m^2 s^{-1}. If the critical reaction distance is 0.4 nm, what value is expected for the second-order rate constant for the diffusion-controlled reaction?

21B.2(a) Calculate the magnitude of the diffusion-controlled rate constant at 298 K for a species in (i) water, (ii) pentane. The viscosities are 1.00×10^{-3} kg m^{-1} s^{-1}, and 2.2×10^{-4} kg m^{-1} s^{-1}, respectively.

21B.2(b) Calculate the magnitude of the diffusion-controlled rate constant at 298 K for a species in (i) decylbenzene, (ii) concentrated sulfuric acid. The viscosities are 3.36 cP and 27 cP, respectively.

21B.3(a) Calculate the magnitude of the diffusion-controlled rate constant at 320 K for the recombination of two atoms in water, for which η = 0.89 cP. Assuming the concentration of the reacting species is 1.5 mmol dm^{-3} initially, how long does it take for the concentration of the atoms to fall to half that value? Assume the reaction is elementary.

21B.3(b) Calculate the magnitude of the diffusion-controlled rate constant at 320 K for the recombination of two atoms in benzene, for which η = 0.601 cP. Assuming the concentration of the reacting species is 2.0 mmol dm^{-3} initially, how long does it take for the concentration of the atoms to fall to half that value? Assume the reaction is elementary.

21B.4(a) Two neutral species, A and B, with diameters 655 pm and 1820 pm, respectively, undergo the diffusion-controlled reaction A + B→P in a solvent of viscosity 2.93×10^{-3} kg m^{-1} s^{-1} at 40 °C. Calculate the initial rate d[P]/dt if the initial concentrations of A and B are 0.170 mol dm^{-3} and 0.350 mol dm^{-3}, respectively.

21B.4(b) Two neutral species, A and B, with diameters 421 pm and 945 pm, respectively, undergo the diffusion-controlled reaction A + B→P in a solvent of viscosity 1.35 cP at 20 °C. Calculate the initial rate d[P]/dt if the initial concentrations of A and B are 0.155 mol dm^{-3} and 0.195 mol dm^{-3}, respectively.

Problems

21B.1 Confirm that eqn 21B.8 is a solution of eqn 21B.7, where [J] is a solution of the same equation but with k_r = 0 and for the same initial conditions.

21B.2 Use mathematical software, a spreadsheet, or the *Living graphs* on the web site of this book to explore the effect of varying the value of the rate constant k_r on the spatial variation of [J]* (see eqn 21B.8 with [J] given in eqn 21B.9) for a constant value of the diffusion constant D.

21B.3 Confirm that if the initial condition is [J] = 0 at t = 0 everywhere, and the boundary condition is [J] = $[J]_0$ at t > 0 at all points on a surface, then the solutions [J]* in the presence of a first-order reaction that removed J are related to those in the absence of reaction, [J], by

$$[J]^*=k_r\int_0^t [J]e^{-k_rt}dt+[J]e^{-k_rt}$$

Base your answer on eqn 21B.5.

21B.4‡ The compound α-tocopherol, a form of vitamin E, is a powerful antioxidant that may help to maintain the integrity of biological membranes. R.H. Bisby and A.W. Parker (*J. Amer. Chem. Soc.* **117**, 5664 (1995)) studied

‡ These problems were supplied by Charles Trapp and Carmen Giunta.

the reaction of photochemically excited duroquinone with the antioxidant in ethanol. Once the duroquinone was photochemically excited, a bimolecular reaction took place at a rate described as diffusion limited. (a) Estimate the rate constant for a diffusion-limited reaction in ethanol. (b) The reported rate constant was $2.77\times10^9\,dm^3\,mol^{-1}\,s^{-1}$; estimate the critical reaction distance if the sum of diffusion constants is $1\times10^{-9}\,m^2\,s^{-1}$.

TOPIC 21C Transition-state theory

Discussion questions

21C.1 Describe in outline the formulation of the Eyring equation.

21C.2 How is femtosecond spectroscopy used to examine the structures of activated complexes?

21C.3 Explain the physical origin of the kinetic salt effect. What might be the effect of the relative permittivity of the medium?

21C.4 How do kinetic isotope effects provide insight into the mechanism of a reaction?

Exercises

21C.1(a) The reaction of propylxanthate ion in acetic acid buffer solutions has the mechanism $A^- + H^+ \rightarrow P$. Near 30 °C the rate constant is given by the empirical expression $k_r = (2.05\times10^{13})\,e^{-(8681\,K)/T}\,dm^3\,mol^{-1}\,s^{-1}$. Evaluate the energy and entropy of activation at 30 °C.

21C.1(b) The reaction $A^- + H^+ \rightarrow P$ has a rate constant given by the empirical expression $k_r = (6.92\times10^{12})e^{-(5925\,K)/T}\,dm^3\,mol^{-1}\,s^{-1}$. Evaluate the energy and entropy of activation at 25 °C.

21C.2(a) When the reaction in Exercise 21C.1(a) occurs in a dioxane/water mixture which is 30 per cent dioxane by mass, the rate constant fits $k_r = (7.78\times10^{14})e^{-(9134\,K)/T}\,dm^3\,mol^{-1}\,s^{-1}$ near 30 °C. Calculate $\Delta^{\ddagger}G$ for the reaction at 30 °C.

21C.2(b) A rate constant is found to fit the expression $k_r = (4.98\times10^{13})e^{-(4972\,K)/T}\,dm^3\,mol^{-1}\,s^{-1}$ near 25 °C. Calculate $\Delta^{\ddagger}G$ for the reaction at 25 °C.

21C.3(a) The gas phase association reaction between F_2 and IF_5 is first order in each of the reactants. The energy of activation for the reaction is 58.6 kJ mol^{-1}. At 65 °C the rate constant is $7.84\times10^{-3}\,kPa^{-1}\,s^{-1}$. Calculate the entropy of activation at 65 °C.

21C.3(b) A gas-phase recombination reaction is first order in each of the reactants. The energy of activation for the reaction is 39.7 kJ mol^{-1}. At 65 °C the rate constant is $0.35\,m^3\,s^{-1}$. Calculate the entropy of activation at 65 °C.

21C.4(a) Calculate the entropy of activation for a collision between two structureless particles at 300 K, taking $M = 65\,g\,mol^{-1}$ and $\sigma = 0.35\,nm^2$.

21C.4(b) Calculate the entropy of activation for a collision between two structureless particles at 450 K, taking $M = 92\,g\,mol^{-1}$ and $\sigma = 0.45\,nm^2$.

21C.5(a) The pre-exponential factor for the gas-phase decomposition of ozone at low pressures is $4.6\times10^{12}\,dm^3\,mol^{-1}\,s^{-1}$ and its activation energy is 10.0 kJ mol^{-1}. What are (i) the entropy of activation, (ii) the enthalpy of activation, (iii) the Gibbs energy of activation at 298 K?

21C.5(b) The pre-exponential factor for a gas-phase decomposition of a gas at low pressures is $2.3\times10^{13}\,dm^3\,mol^{-1}\,s^{-1}$ and its activation energy is 30.0 kJ mol^{-1}. What are (i) the entropy of activation, (ii) the enthalpy of activation, (iii) the Gibbs energy of activation at 298 K?

21C.6(a) The rate constant of the reaction $H_2O_2(aq) + I^-(aq) + H^+(aq) \rightarrow H_2O(l) + HIO(aq)$ is sensitive to the ionic strength of the aqueous solution in which the reaction occurs. At 25 °C, $k_r = 12.2\,dm^6\,mol^{-2}\,min^{-1}$ at an ionic strength of 0.0525. Use the Debye–Hückel limiting law to estimate the rate constant at zero ionic strength.

21C.6(b) At 25 °C, $k_r = 1.55\,dm^6\,mol^{-2}\,min^{-1}$ at an ionic strength of 0.0241 for a reaction in which the rate-determining step involves the encounter of two singly charged cations. Use the Debye–Hückel limiting law to estimate the rate constant at zero ionic strength.

Problems

21C.1 The rates of thermolysis of a variety of *cis*- and *trans*-azoalkanes have been measured over a range of temperatures in order to settle a controversy concerning the mechanism of the reaction. In ethanol an unstable *cis*-azoalkane decomposed at a rate that was followed by observing the N_2 evolution, and this led to the rate constants given in the following table (P.S. Engel and D.J. Bishop, *J. Amer. Chem. Soc.* **97**, 6754 (1975)). Calculate the enthalpy, entropy, energy, and Gibbs energy of activation at −20 °C.

θ/°C	−24.82	−20.73	−17.02	−13.00	−8.95
$10^4\times k_r/s^{-1}$	1.22	2.31	4.39	8.50	14.3

21C.2 In an experimental study of a bimolecular reaction in aqueous solution, the second-order rate constant was measured at 25 °C and at a variety of ionic strengths and the results are tabulated in the following table. It is known that a singly charged ion is involved in the rate-determining step. What is the charge on the other ion involved?

I/(mol kg^{-1})	0.0025	0.0037	0.0045	0.0065	0.0085
k_r/(dm^3 mol^{-1} s^{-1})	1.05	1.12	1.16	1.18	1.26

21C.3 Derive the expression for k_r given in *Example* 21C.1 by introducing the equations for the thermal wavelengths.

21C.4 The rate constant of the reaction $I^-(aq) + H_2O_2(aq) \rightarrow H_2O(l) + IO^-(aq)$ varies slowly with ionic strength, even though the Debye–Hückel limiting law predicts no effect. Use the following data from 25 °C to find the dependence of $\log k_r$ on the ionic strength:

I/(mol kg^{-1})	0.0207	0.0525	0.0925	0.1575
k_r/(dm^3 mol^{-1} min^{-1})	0.663	0.670	0.679	0.694

Evaluate the limiting value of k_r at zero ionic strength. What does the result suggest for the dependence of $\log \gamma$ on ionic strength for a neutral molecule in an electrolyte solution?

21C.5‡ For the gas phase reaction $A + A \rightarrow A_2$, the experimental rate constant, k_r, has been fitted to the Arrhenius equation with the pre-exponential factor $A = 4.07 \times 10^5\,dm^3\,mol^{-1}\,s^{-1}$ at 300 K and an activation energy of $65.43\,kJ\,mol^{-1}$. Calculate $\Delta^{\ddagger}S$, $\Delta^{\ddagger}H$, $\Delta^{\ddagger}U$, and $\Delta^{\ddagger}G$ for the reaction.

21C.6 Use the Debye–Hückel limiting law to show that changes in ionic strength can affect the rate of reaction catalysed by H^+ from the deprotonation of a weak acid. Consider the mechanism: $H^+ + B \rightarrow P$, where H^+ comes from the deprotonation of the weak acid, HA. The weak acid has a fixed concentration. First show that log $[H^+]$, derived from the ionization of HA, depends on the activity coefficients of ions and thus depends on the ionic strength. Then find the relationship between log(rate) and log $[H^+]$ to show that the rate also depends on the ionic strength.

21C.7‡ Show that bimolecular reactions between nonlinear molecules are much slower than between atoms even when the activation energies of both reactions are equal. Use transition-state theory and make the following assumptions. (1) All vibrational partition functions are close to unity; (2) all rotational partition functions are approximately $1 \times 10^{1.5}$, which is a reasonable order of magnitude number; (3) the translational partition function for each species is 1×10^{26}.

21C.8 This exercise gives some familiarity with the difficulties involved in predicting the structure of activated complexes. It also demonstrates the importance of femtosecond spectroscopy to our understanding of chemical dynamics because direct experimental observation of the activated complex removes much of the ambiguity of theoretical predictions. Consider the attack of H on D_2, which is one step in the $H_2 + D_2$ reaction. (a) Suppose that the H approaches D_2 from the side and forms a complex in the form of an isosceles triangle. Take the H–D distance as 30 per cent greater than in H_2 (74 pm) and the D–D distance as 20 per cent greater than in H_2. Let the critical coordinate be the antisymmetric stretching vibration in which one H–D bond stretches as the other shortens. Let all the vibrations be at about $1000\,cm^{-1}$. Estimate k_r for this reaction at 400 K using the experimental activation energy of about $35\,kJ\,mol^{-1}$. (b) Now change the model of the activated complex in part (a) and make it linear. Use the same estimated molecular bond lengths and vibrational frequencies to calculate k_r for this choice of model. (c) Clearly, there is much scope for modifying the parameters of the models of the activated complex. Use mathematical software or write and run a program that allows you to vary the structure of the complex and the parameters in a plausible way, and look for a model (or more than one model) that gives a value of k_r close to the experimental value, $4 \times 10^5\,dm^3\,mol^{-1}\,s^{-1}$.

21C.9‡ M. Cyfert et al. (*Int. J. Chem. Kinet.* **28**, 103 (1996)) examined the oxidation of tris(1,10-phenanthroline)iron(II) by periodate in aqueous solution, a reaction which shows autocatalytic behaviour. To assess the kinetic salt effect, they measured rate constants at a variety of concentrations of Na_2SO_4 far in excess of reactant concentrations and reported the following data:

$[Na_2SO_4]/(mol\,kg^{-1})$	0.2	0.15	0.1	0.05	0.025	0.0125	0.005
$k_r/(dm^{3/2}\,mol^{-1/2}\,s^{-1})$	0.462	0.430	0.390	0.321	0.283	0.252	0.224

What can be inferred about the charge of the activated complex of the rate-determining step?

21C.10 The study of conditions that optimize the association of proteins in solution guides the design of protocols for formation of large crystals that are amenable to analysis by X-ray diffraction techniques. It is important to characterize protein dimerization because the process is considered to be the rate-determining step in the growth of crystals of many proteins. Consider the variation with ionic strength of the rate constant of dimerization in aqueous solution of a cationic protein P:

$I/(mol\,kg^{-1})$	0.0100	0.0150	0.0200	0.0250	0.0300	0.0350
k_r/k_r'	8.10	13.30	20.50	27.80	38.10	52.00

What can be deduced about the charge of P?

21C.11 Predict the order of magnitude of the primary isotope effect on the relative rates of displacement of (a) ^1H and ^3H in a C–H bond, (b) ^{16}O and ^{18}O in a C–O bond. Will raising the temperature enhance the difference? Take $k_f(C–H) = 450\,N\,m^{-1}$, $k_f(C–O) = 1750\,N\,m^{-1}$.

TOPIC 21D The dynamics of molecular collisions

Discussion questions

21D.1 Describe how the following techniques are used in the study of chemical dynamics: infrared chemiluminescence, laser-induced fluorescence, multiphoton ionization, resonant multiphoton ionization, and reaction product imaging.

21D.2 Discuss the relationship between the saddle-point energy and the activation energy of a reaction.

21D.3 A method for directing the outcome of a chemical reaction consists of using molecular beams to control the relative orientations of reactants during a collision. Consider the reaction $Rb + CH_3I \rightarrow RbI + CH_3$. How should CH_3I molecules and Rb atoms be oriented to maximize the production of RbI?

21D.4 Consider a reaction with an attractive potential energy surface. Discuss how the initial distribution of reactant energy affects how efficiently the reaction proceeds. Repeat for a repulsive potential energy surface.

21D.5 Describe how molecular beams are used to investigate intermolecular potentials.

Exercises

21D.1(a) The interaction between two diatomic molecules is described by an attractive potential energy surface. What distribution of vibrational and translational energies among reactants and products is most likely to lead to a successful reaction?

21D.1(b) The interaction between two diatomic molecules has a repulsive potential energy surface. What distribution of vibrational and translational energies among reactants and products is most likely to lead to a successful reaction?

21D.2(a) If the cumulative reaction probability were independent of energy, what is the temperature dependence of the rate constant predicted by the numerator of eqn 21D.6?

21D.2(b) If the cumulative reaction probability equalled 1 for energies less than a barrier height V and vanished for higher energies, what is the temperature dependence of the rate constant predicted by the numerator of eqn 21D.6?

Problems

21D.1 Show that the intensities of a molecular beam before and after passing through a chamber of length L containing inert scattering atoms are related by $I = I_0 e^{-\mathcal{N}\sigma L}$, where σ is the collision cross-section and $\mathcal{N}$ the number density of scattering atoms.

21D.2 In a molecular beam experiment to measure collision cross-sections it was found that the intensity of a CsCl beam was reduced to 60 per cent of its intensity on passage through CH_2F_2 at 10 μTorr, but that when the target was Ar at the same pressure the intensity was reduced only by 10 per cent. What are the relative cross-sections of the two types of collision? Why is one much larger than the other?

21D.3 Consider the collision between a hard-sphere molecule of radius R_1 and mass m, and an infinitely massive impenetrable sphere of radius R_2. Plot the scattering angle θ as a function of the impact parameter b. Carry out the calculation using simple geometrical considerations.

21D.4 The dependence of the scattering characteristics of atoms on the energy of the collision can be modelled as follows. We suppose that the two colliding atoms behave as impenetrable spheres, as in Problem 21D.3, but that the effective radius of the heavy atoms depends on the speed v of the light atom. Suppose its effective radius depends on v as $R_2 e^{-v/v^*}$, where v^* is a constant. Take $R_1 = \frac{1}{2}R_2$ for simplicity and an impact parameter $b = \frac{1}{2}R_2$, and plot the scattering angle as a function of (a) speed, (b) kinetic energy of approach.

TOPIC 21E Electron transfer in homogeneous systems

Discussion questions

21E.1 Discuss how the following factors determine the rate of electron transfer in homogeneous systems: the distance between electron donor and acceptor, the standard Gibbs energy of the process, and the reorganization energy of the redox active species and the surrounding medium.

21E.2 What role does tunnelling play in electron transfer?

21E.3 Explain why the rate constant decreases as the reaction becomes more exergonic in the inverted region.

Exercises

21E.1(a) For a pair of electron donor and acceptor at 298 K, $H_{et}(d) = 0.04\,cm^{-1}$, $\Delta_r G^{\ominus} = -0.185$ eV and $k_{et} = 37.5\,s^{-1}$. Estimate the value of the reorganization energy.

21E.1(b) For a pair of electron donor and acceptor at 298 K, $k_{et} = 2.02\times10^5\,s^{-1}$ for $\Delta_r G^{\ominus} = -0.665$ eV. The standard reaction Gibbs energy changes to $\Delta_r G^{\ominus} = -0.975$ eV when a substituent is added to the electron acceptor and the rate constant for electron transfer changes to $k_{et} = 3.33\times10^6\,s^{-1}$. Assuming that the distance between donor and acceptor is the same in both experiments, estimate the values of $H_{et}(d)$ and ΔE_R.

21E.2(a) For a pair of electron donor and acceptor, $k_{et} = 2.02\times10^5\,s^{-1}$ when $d = 1.11$ nm and $k_{et} = 4.51\times10^4\,s^{-1}$ when $r = 1.23$ nm. Assuming that $\Delta_r G^{\ominus}$ and ΔE_R are the same in both experiments, estimate the value of β.

21E.2(b) Refer to Exercise 21E.2(a). Estimate the value of k_{et} when $d = 1.59$ nm.

Problems

21E.1 Consider the reaction $D + A \rightarrow D^+ + A^-$. The rate constant k_r may be determined experimentally or may be predicted by the *Marcus cross-relation* $k_r = (k_{DD}k_{AA}K)^{1/2} f$, where k_{DD} and k_{AA} are the experimental rate constants for the electron self-exchange processes $^*D + D^+ \rightarrow {}^*D^+ + D$ and $^*A + A^+ \rightarrow {}^*A^+ + A$, respectively, and f is a function of $K = [D^+][A^-]/[D][A]$, k_{DD}, k_{AA}, and the collision frequencies. Derive the approximate form of the Marcus cross-relation by following these steps. (a) Use eqn 21E.7 to write expressions for $\Delta^\ddagger G$, $\Delta^\ddagger G_{DD}$, and $\Delta^\ddagger G_{AA}$, keeping in mind that $\Delta_r G^{\ominus} = 0$ for the electron self-exchange reactions. (b) Assume that the reorganization energy $\Delta E_{R,DA}$ for the reaction $D + A \rightarrow D^+ + A^-$ is the average of the reorganization energies $\Delta E_{R,DD}$ and $\Delta E_{R,AA}$ of the electron self-exchange reactions. Then show that in the limit of small magnitude of $\Delta_r G^{\ominus}$, or $|\Delta_r G^{\ominus}| \ll \Delta E_{R,DA}$, $\Delta^\ddagger G = \frac{1}{2}(\Delta^\ddagger G_{DD} + \Delta^\ddagger G_{AA} + \Delta_r G^{\ominus})$, where $\Delta_r G^{\ominus}$ is the standard Gibbs energy for the reaction $D + A \rightarrow D^+ + A^-$. (c) Use an equation of the form of eqn 21E.4 to write expressions for k_{DD} and k_{AA}. (d) Use eqn 21E.4 and the result above to write an expression for k_r. (e) Complete the derivation by using the results from part (c), the relation $K = e^{-\Delta_r G^{\ominus}/RT}$, and assuming that all $\kappa\nu^\ddagger$ terms, which may be interpreted as collision frequencies, are identical.

21E.2 Consider the reaction $D + A \rightarrow D^+ + A^-$. The rate constant k_r may be determined experimentally or may be predicted by the Marcus cross-relation (see Problem 21E.1). It is common to make the assumption that $f \approx 1$. Use the approximate form of the Marcus relation to estimate the rate constant for the reaction $Ru(bpy)_3^{3+} + Fe(H_2O)_6^{2+} \rightarrow Ru(bpy)_3^{2+} + Fe(H_2O)_6^{3+}$, where bpy stands for 4,4′-bipyridine. The following data will be useful:

$Ru(bpy)_3^{3+} + e^- \rightarrow Ru(bpy)_3^{2+}$	$E^{\ominus} = 1.26$ V
$Fe(H_2O)_6^{3+} + e^- \rightarrow Fe(H_2O)_6^{2+}$	$E^{\ominus} = 0.77$ V
$^*Ru(bpy)_3^{3+} + Ru(bpy)_3^{2+} \rightarrow {}^*Ru(bpy)_3^{2+} + Ru(bpy)_3^{3+}$	$k_{Ru} = 4.0\times10^8\,dm^3\,mol^{-1}\,s^{-1}$
$^*Fe(H_2O)_6^{3+} + Fe(H_2O)_6^{2+} \rightarrow {}^*Fe(H_2O)_6^{2+} + Fe(H_2O)_6^{3+}$	$k_{Fe} = 4.2\,dm^3\,mol^{-1}\,s^{-1}$

21E.3 A useful strategy for the study of electron transfer in proteins consists of attaching an electroactive species to the protein's surface and then measuring k_{et} between the attached species and an electroactive protein cofactor. J.W. Winkler and H.B. Gray (*Chem. Rev.* **92**, 369 (1992)) summarize data for cytochrome *c* modified by replacement of the haem iron by a zinc ion, resulting in a zinc–porphyrin (ZnP) group in the interior of the protein, and by attachment of a ruthenium ion complex to a surface histidine amino acid. The edge-to-edge distance between the electroactive species was thus fixed at 1.23 nm. A variety of ruthenium ion complexes with different standard potentials was used. For each ruthenium-modified protein, either the $Ru^{2+} \rightarrow ZnP^+$ or the $ZnP^* \rightarrow Ru^{3+}$, in which the electron donor is an electronically excited state of the zinc–porphyrin group formed by laser excitation, was monitored. This arrangement leads to different standard reaction Gibbs energies because the redox couples ZnP^+/ZnP and ZnP^+/ZnP^* have different standard potentials, with the electronically excited porphyrin being a more powerful reductant. Use the following data to estimate the reorganization energy for this system:

$\Delta_r G^{\ominus}/\text{eV}$	0.665	0.705	0.745	0.975	1.015	1.055
$k_{et}/(10^6\,\text{s}^{-1})$	0.657	1.52	1.12	8.99	5.76	10.1

21E.4 The photosynthetic reaction centre of the purple photosynthetic bacterium *Rhodopseudomonas viridis* contains a number of bound co-factors that participate in electron transfer reactions. The following table shows data compiled by Moser et al. (*Nature* **355**, 796 (1992)) on the rate constants for electron transfer between different co-factors and their edge-to-edge distances:

Reaction	$BChl^- \rightarrow BPh$	$BPh^- \rightarrow BChl_2^+$	$BPh^- \rightarrow Q_A$	$cyt\ c_{559} \rightarrow BChl_2$	$Q_A^- \rightarrow Q_B$	$Q_A^- \rightarrow BChl_2^+$
d/nm	0.48	0.95	0.96	1.23	1.35	2.24
k_{et}/s^{-1}	1.58×10^{12}	3.98×10^9	1.00×10^9	1.58×10^8	3.98×10^7	63.1

(BChl, bacteriochlorophyll; $BChl_2$, bacteriochlorophyll dimer, functionally distinct from BChl; BPh, bacteriophaeophytin; Q_A and Q_B, quinone molecules bound to two distinct sites; cyt c_{559}, a cytochrome bound to the reaction centre complex). Are these data in agreement with the behaviour predicted by eqn 21E.9? If so, evaluate the value of β.

21E.5 The rate constant for electron transfer between a cytochrome *c* and the bacteriochlorophyll dimer of the reaction centre of the purple bacterium *Rhodobacter sphaeroides* (Problem 21E.4) decreases with decreasing temperature in the range 300 K to 130 K. Below 130 K, the rate constant becomes independent of temperature. Account for these results.

TOPIC 21F Processes at electrodes

Discussion questions

21F.1 Describe the various models of the electrode–electrolyte interface.

21F.2 In what sense is electron transfer at an electrode an activated process?

21F.3 Discuss the technique of cyclic voltammetry and account for the characteristic shape of a cyclic voltammogram, such as those shown in Figs. 21F.13 and 21F.14.

Exercises

21F.1(a) The transfer coefficient of a certain electrode in contact with M^{3+} and M^{4+} in aqueous solution at 25 °C is 0.39. The current density is found to be 55.0 mA cm^{-2} when the overpotential is 125 mV. What is the overpotential required for a current density of 75 mA cm^{-2}?
21F.1(b) The transfer coefficient of a certain electrode in contact with M^{2+} and M^{3+} in aqueous solution at 25°C is 0.42. The current density is found to be 17.0 mA cm^{-2} when the overpotential is 105 mV. What is the overpotential required for a current density of 72 mA cm^{-2}?

21F.2(a) Determine the exchange current density from the information given in Exercise 21F.1(a).
21F.2(b) Determine the exchange current density from the information given in Exercise 21F.1(b).

21F.3(a) To a first approximation, significant evolution or deposition occurs in electrolysis only if the overpotential exceeds about 0.6 V. To illustrate this criterion determine the effect that increasing the overpotential from 0.40 V to 0.60 V has on the current density in the electrolysis of 1.0 M NaOH(aq), which is 1.0 mA cm^{-2} at 0.4 V and 25 °C. Take $\alpha = 0.5$.
21F.3(b) Determine the effect that increasing the overpotential from 0.50 V to 0.60 V has on the current density in the electrolysis of 1.0 M NaOH(aq), which is 1.22 mA cm^{-2} at 0.50 V and 25 °C. Take $\alpha = 0.50$.

21F.4(a) Use the data in Table 21F.1 for the exchange current density and transfer coefficient for the reaction $2\,H^+ + 2\,e^- \rightarrow H_2$ on nickel at 25 °C to determine what current density would be needed to obtain an overpotential of 0.20 V as calculated from (i) the Butler–Volmer equation, and (ii) the Tafel equation. Is the validity of the Tafel approximation affected at higher overpotentials (of 0.4 V and more)?
21F.4(b) Use the data in Table 21F.1 for the exchange current density and transfer coefficient for the reaction $Fe^{3+} + e^- \rightarrow Fe^{2+}$ on platinum at 25 °C to determine what current density would be needed to obtain an overpotential of 0.30 V as calculated from (i) the Butler–Volmer equation, and (ii) the Tafel equation. Is the validity of the Tafel approximation affected at higher overpotentials (of 0.4 V and more)?

21F.5(a) A typical exchange current density, that for H^+ discharge at platinum, is 0.79 mA cm^{-2} at 25 °C. What is the current density at an electrode when its overpotential is (i) 10 mV, (ii) 100 mV, (iii) −5.0 V? Take $\alpha = 0.5$.
21F.5(b) The exchange current density for a $Pt|Fe^{3+},Fe^{2+}$ electrode is 2.5 mA cm^{-2}. The standard potential of the electrode is +0.77 V. Calculate the current flowing through an electrode of surface area 1.0 cm^2 as a function of the potential of the electrode. Take unit activity for both ions.

21F.6(a) How many electrons or protons are transported through the double layer in each second when the $Pt,H_2|H^+$, $Pt|Fe^{3+},Fe^{2+}$, and $Pb,H_2|H^+$ electrodes are at equilibrium at 25 °C? Take the area as 1.0 cm^2 in each case. Estimate the number of times each second a single atom on the surface takes part in an electron transfer event, assuming an electrode atom occupies about (280 pm)^2 of the surface.
21F.6(b) How many electrons or protons are transported through the double layer in each second when the $Cu,H_2|H^+$ and $Pt|Ce^{4+},Ce^{3+}$ electrodes are at equilibrium at 25 °C? Take the area as 1.0 cm^2 in each case. Estimate the number of times each second a single atom on the surface takes part in an electron transfer event, assuming an electrode atom occupies about (260 pm)^2 of the surface.

21F.7(a) What is the effective resistance at 25 °C of an electrode interface when the overpotential is small? Evaluate it for 1.0 cm² (i) $Pt,H_2|H^+$, (ii) $Hg,H_2|H^+$ electrodes.

21F.7(b) Evaluate the effective resistance at 25 °C of an electrode interface for 1.0 cm² (i) $Pb,H_2|H^+$, (ii) $Pt|Fe^{2+},Fe^{3+}$ electrodes.

21F.8(a) The exchange current density for H^+ discharge at zinc is about 50 pA cm^{-2}. Can zinc be deposited from a unit activity aqueous solution of a zinc salt?

21F.8(b) The standard potential of the $Zn^{2+}|Zn$ electrode is −0.76 V at 25 °C. The exchange current density for H^+ discharge at platinum is 0.79 mA cm^{-2}. Can zinc be plated on to platinum at that temperature? (Take unit activities.)

Problems

21F.1 In an experiment on the $Pt|H_2|H^+$ electrode in dilute H_2SO_4 the following current densities were observed at 25 °C. Evaluate α and j_0 for the electrode.

η/mV	50	100	150	200	250
j/(mA cm^{-2})	2.66	8.91	29.9	100	335

How would the current density at this electrode depend on the overpotential of the same set of magnitudes but of opposite sign?

21F.2 The standard potentials of lead and tin are −126 mV and −136 mV, respectively, at 25 °C, and the overpotential for their deposition are close to zero. What should their relative activities be in order to ensure simultaneous deposition from a mixture?

21F.3‡ The rate of deposition of iron, v, on the surface of an iron electrode from an aqueous solution of Fe^{2+} has been studied as a function of potential, E, relative to the standard hydrogen electrode, by J. Kanya (*J. Electroanal. Chem.* **84**, 83 (1977)). The values in the table below are based on the data obtained with an electrode of surface area 9.1 cm² in contact with a solution of concentration 1.70 μmol dm^{-3} in Fe^{2+}. (a) Assuming unit activity coefficients, calculate the zero current potential of the Fe^{2+}/Fe cathode and the overpotential at each value of the working potential. (b) Calculate the cathodic current density, j_c, from the rate of deposition of Fe^{2+} for each value of E. (c) Examine the extent to which the data fit the Tafel equation and calculate the exchange current density.

v/(pmol s^{-1})	1.47	2.18	3.11	7.26
$-E$/mV	702	727	752	812

21F.4‡ V.V. Losev and A.P. Pchel'nikov (*Soviet Electrochem.* **6**, 34 (1970)) obtained the following current–voltage data for an indium anode relative to a standard hydrogen electrode at 293 K:

$-E$/V	0.388	0.365	0.350	0.335
j/(A m^{-2})	0	0.590	1.438	3.507

Use these data to calculate the transfer coefficient and the exchange current density. What is the cathodic current density when the potential is 0.365 V?

21F.5‡ An early study of the hydrogen overpotential is that of H. Bowden and T. Rideal (*Proc. Roy. Soc.* **A120**, 59 (1928)), who measured the overpotential for H_2 evolution with a mercury electrode in dilute aqueous solutions of H_2SO_4 at 25 °C. Determine the exchange current density and transfer coefficient, α, from their data:

j/(mA m^{-2})	2.9	6.3	28	100	250	630	1650	3300
η/V	0.60	0.65	0.73	0.79	0.84	0.89	0.93	0.96

Explain any deviations from the result expected from the Tafel equation.

21F.6 If $\alpha=\frac{1}{2}$, an electrode interface is unable to rectify alternating current because the current density curve is symmetrical about $\eta=0$. When $\alpha\neq\frac{1}{2}$, the magnitude of the current density depends on the sign of the overpotential, and so some degree of 'faradaic rectification' may be obtained. Suppose that the overpotential varies as $\eta=\eta_0\cos\omega t$. Derive an expression for the mean flow of current (averaged over a cycle) for general α, and confirm that the mean current is zero when $\alpha=\frac{1}{2}$. In each case work in the limit of small η_0 but to second order in η_0F/RT. Calculate the mean direct current at 25 °C for a 1.0 cm² hydrogen–platinum electrode with $\alpha=0.38$ when the overpotential varies between ±10 mV at 50 Hz.

21F.7 Now suppose that the overpotential is in the high overpotential region at all times even though it is oscillating. What waveform will the current across the interface show if it varies linearly and periodically (as a sawtooth waveform) between η_- and η_+ around η_0? Take $\alpha=\frac{1}{2}$.

21F.8 Figure 21.1 shows four different examples of voltammograms. Identify the processes occurring in each system. In each case the vertical axis is the current and the horizontal axis is the (negative) electrode potential.

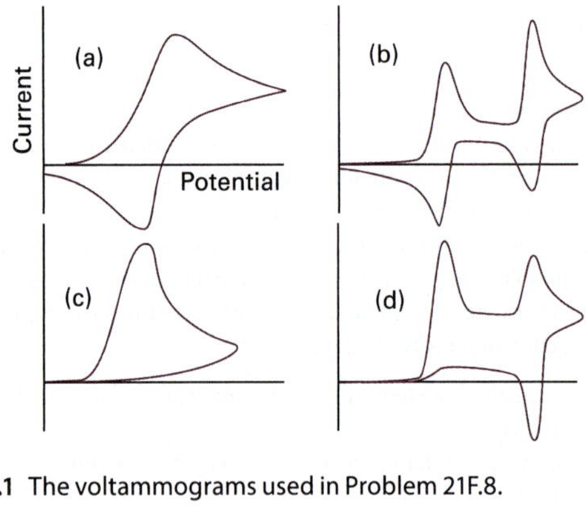

Figure 21.1 The voltammograms used in Problem 21F.8.

Integrated activities

21.1 Estimate the orders of magnitude of the partition functions involved in a rate expression. State the order of magnitude of q_m^T/N_A, q^R, q^V, q^E for typical molecules. Check that in the collision of two structureless molecules the order of magnitude of the pre-exponential factor is of the same order as that predicted by collision theory. Go on to estimate the *P*-factor for a reaction in which A + B→ P, and A and B are nonlinear triatomic molecules.

21.2 Discuss the factors that govern the rates of photo-induced electron transfer according to Marcus theory and that govern the rates of resonance energy transfer according to Förster theory (Topic 20G). Can you find similarities between the two theories?

21.3 Calculate the thermodynamic limit to the zero-current potential of fuel cells operating on (a) hydrogen and oxygen, (b) methane and air, and (c) propane and air. Use the Gibbs energy information in the *Resource section*, and take the species to be in their standard states at 25 °C.

CHAPTER 22

Processes on solid surfaces

Processes at solid surfaces govern the viability of industry constructively, as in catalysis, and the permanence of its products destructively, as in corrosion. Chemical reactions at solid surfaces may differ sharply from reactions in the bulk, for reaction pathways of much lower activation energy may be provided by the surface, and hence result in catalysis. This chapter extends the material introduced in Chapters 20 and 21 by showing how to deal with processes on solid surfaces.

22A An introduction to solid surfaces

We begin by exploring the structure of solid surfaces. This Topic also describes a number of experimental techniques commonly used in surface science.

22B Adsorption and desorption

Although we began with a discussion of clean surfaces, for chemists the important aspects of a surface are the attachment of substances to it and the reactions that take place there. In this Topic we discuss the extent to which a solid surface is covered and the variation of the extent of coverage with pressure and temperature.

22C Heterogeneous catalysis

This Topic discusses chemical reactions on solid surfaces. We focus on how surfaces affect the rate and course of chemical change by acting as the site of catalysis.

What is the impact of this material?

Almost the whole of modern chemical industry depends on the development, selection, and application of catalysts, with heterogeneous catalysts being particularly important. All we can hope to do in *Impact* I22.1 is to give a brief indication of some of the problems involved. Other than the ones we consider, these problems include the danger of the catalyst being poisoned by by-products or impurities, and economic considerations relating to cost and lifetime.

To read more about the impact of this material, scan the QR code, or go to bcs.whfreeman.com/webpub/chemistry/pchem10e/impact/pchem-22-1.html

22A An introduction to solid surfaces

Contents

➤ Why do you need to know this material?

To understand the thermodynamics and kinetics of chemical reactions occurring on solid surfaces, which underlie much of catalysis and therefore the chemical industry, you need to understand surface structure, composition, and growth.

➤ What is the key idea?

Structural features, including defects, play important roles in physical and chemical processes occurring on solid surfaces.

➤ What do you need to know already?

You need to be aware of the structure of solids (Topic 18A), but not in detail. This Topic draws on results from the kinetic theory of gases (Topic 1B).

A great deal of chemistry occurs at solid surfaces. Heterogeneous catalysis (Topic 22C) is just one example, with the surface providing reactive sites where reactants can attach, be torn apart, and react with other reactants. Even as simple an act as dissolving is intrinsically a surface phenomenon, with the solid gradually escaping into the solvent from sites on the surface. Surface deposition, in which atoms are laid down on a surface to create layers, is crucial to the semiconductor industry, as it is the way in which integrated circuits are created. Electrodes are essentially surfaces at which electron transfer occurs, and their efficiency depends crucially on an understanding of the events there (Topic 21F).

22A.1 Surface growth

Adsorption is the attachment of particles to a solid surface; **desorption** is the reverse process. The substance that adsorbs is the **adsorbate** and the material to which it adsorbs is the **adsorbent** or **substrate**.

A simple picture of a perfect crystal surface is as a tray of oranges in a grocery store (Fig. 22A.1). A gas molecule that collides with the surface can be imagined as a ping-pong ball bouncing erratically over the oranges. The molecule loses energy as it bounces, but it is likely to escape from the surface before it has lost enough kinetic energy to be trapped. The same is true, to some extent, of an ionic crystal in contact with a solution. There is little energy advantage for an ion in solution to discard some of its solvating molecules and stick at an exposed position on the surface.

The picture changes when the surface has defects, for then there are ridges of incomplete layers of atoms or ions. A common type of surface defect is a **step** between two otherwise flat layers of atoms called **terraces** (Fig. 22A.2). A step defect might itself have defects, for it might have kinks. When an atom settles on a terrace it bounces across it under the influence of the intermolecular potential, and might come to a step or a corner formed by a kink. Instead of interacting with a single terrace

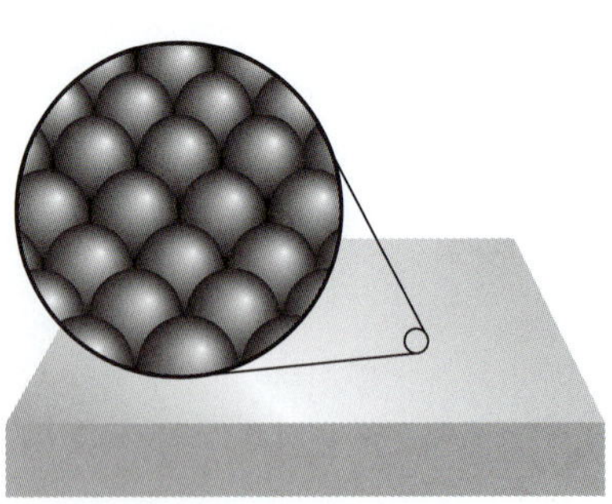

Figure 22A.1 A schematic diagram of the flat surface of a solid. This primitive model is largely supported by scanning tunnelling microscope images.

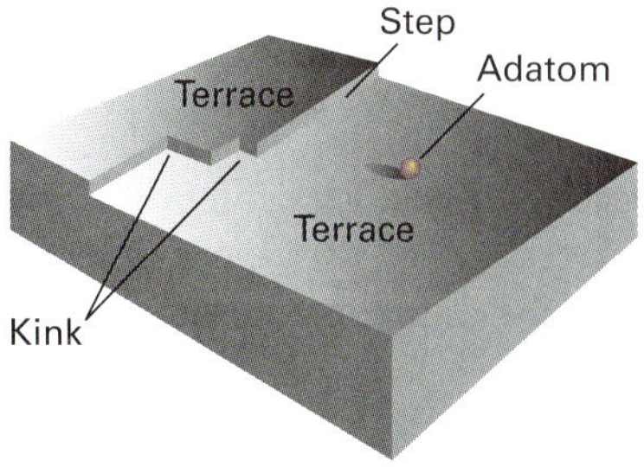

Figure 22A.2 Some of the kinds of defects that may occur on otherwise perfect terraces. Defects play an important role in surface growth and catalysis.

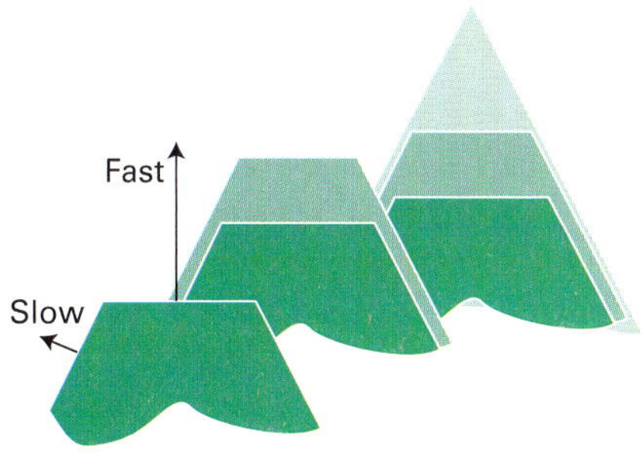

Figure 22A.3 The slower-growing faces of a crystal dominate its final external appearance. Three successive stages of the growth are shown.

atom, the molecule now interacts with several, and the interaction may be strong enough to trap it. Likewise, when ions deposit from solution, the loss of the solvation interaction is offset by a strong Coulombic interaction between the arriving ions and several ions at the surface defect.

The rapidity of growth depends on the crystal plane concerned, and the slowest growing faces dominate the appearance of the crystal. This feature is explained in Fig. 22A.3, where we see that, although the horizontal face grows forward most rapidly, it grows itself out of existence, and the slower-growing faces survive.

Under normal conditions, a surface exposed to a gas is constantly bombarded with molecules and a freshly prepared surface is covered very quickly. Just how quickly can be estimated using the kinetic model of gases and the following expression for the collision flux (eqn 19A.6):

$$Z_W = \frac{p}{(2\pi mkT)^{1/2}} \quad \text{Collision flux} \quad (22A.1)$$

Brief illustration 22A.1 The collision flux

If we write $m = M/N_A$, where M is the molar mass of the gas, eqn 22A.1 becomes

$$Z_W = \frac{\overbrace{(N_A/2\pi k)^{1/2}}^{Z_0} p}{(TM)^{1/2}}$$

After inserting numerical values for the constants and selecting units for the variables, the practical form of this expression is:

$$Z_W = \frac{Z_0(p/\text{Pa})}{\{(T/\text{K})(M/(\text{g mol}^{-1}))\}^{1/2}} \text{ with } Z_0 = 2.63\times10^{24}\,\text{m}^{-2}\,\text{s}^{-1}$$

For air, with $M \approx 29\,\text{g mol}^{-1}$, at $p = 1\,\text{atm} = 1.013\,25\times10^5\,\text{Pa}$ and $T = 298\,\text{K}$, we obtain $Z_W = 2.9\times10^{27}\,\text{m}^{-2}\,\text{s}^{-1}$. Because $1\,\text{m}^2$ of metal surface consists of about 10^{19} atoms, each atom is struck about 10^8 times each second. Even if only a few collisions leave a molecule adsorbed to the surface, the time for which a freshly prepared surface remains clean is very short.

Self-test 22A.1 Calculate the collision flux with a surface of a vessel containing propane at 25 °C when the pressure is 100 Pa.

Answer: $Z_W = 2.30\times10^{20}\,\text{cm}^{-2}\,\text{s}^{-1}$

22A.2 Physisorption and chemisorption

Molecules and atoms can attach to surfaces in two ways. In **physisorption** (a contraction of 'physical adsorption'), there is a van der Waals interaction (for example, a dispersion or a dipolar interaction, Topic 16B) between the adsorbate and the substrate; van der Waals interactions have a long range but are weak, and the energy released when a particle is physisorbed is of the same order of magnitude as the enthalpy of condensation. Such small energies can be absorbed as vibrations of the lattice and dissipated as thermal motion, and a molecule bouncing across the surface will gradually lose its energy and finally adsorb to it in the process called **accommodation**.

The enthalpy of physisorption can be measured by monitoring the rise in temperature of a sample of known heat capacity, and typical values are in the region of $-20\,\text{kJ mol}^{-1}$ (Table 22A.1). This small enthalpy change is insufficient to lead to bond breaking, so a physisorbed molecule retains its identity, although it might be distorted by the presence of the surface.

In **chemisorption** (a contraction of 'chemical adsorption'), the molecules (or atoms) stick to the surface by forming a chemical (usually covalent) bond, and tend to find sites that maximize their coordination number with the substrate. The

Table 22A.1* Maximum observed standard enthalpies of physisorption, $\Delta_{ad}H^\ominus/(\text{kJ mol}^{-1})$, at 298 K

Adsorbate	$\Delta_{ad}H^\ominus/(\text{kJ mol}^{-1})$
CH_4	−21
H_2	−84
H_2O	−59
N_2	−21

* More values are given in the *Resource section*.

Table 22A.2* Standard enthalpies of chemisorption, $\Delta_{ad}H^{\ominus}$/(kJ mol^{-1}), at 298 K

Adsorbate	Adsorbent (substrate)		
	Cr	Fe	Ni
C_2H_4	−427	−285	−243
CO		−192	
H_2	−188	−134	
NH_3		−188	−155

* More values are given in the *Resource section*.

enthalpy of chemisorption is very much greater than that for physisorption, and typical values are in the region of −200 kJ mol^{-1} (Table 22A.2). The distance between the surface and the closest adsorbate atom is also typically shorter for chemisorption than for physisorption. A chemisorbed molecule may be torn apart at the demand of the unsatisfied valencies of the surface atoms, and the existence of molecular fragments on the surface as a result of chemisorption is one reason why solid surfaces catalyse reactions (Topic 22C).

Except in special cases, chemisorption must be exothermic. A spontaneous process requires $\Delta G<0$ at constant pressure and temperature. Because the translational freedom of the adsorbate is reduced when it is adsorbed, ΔS is negative. Therefore, in order for $\Delta G=\Delta H-T\Delta S$ to be negative, ΔH must be negative (that is, the process must be exothermic). Exceptions may occur if the adsorbate dissociates and has high translational mobility on the surface. For example, H_2 adsorbs endothermically on glass because there is a large increase of translational entropy accompanying the dissociation of the molecules into atoms that move quite freely over the surface. In this case, the entropy change in the process $H_2(g) \rightarrow 2\,H(glass)$ is sufficiently positive to overcome the small positive enthalpy change.

The enthalpy of adsorption depends on the extent of surface coverage, mainly because the adsorbate particles interact with each other. If the particles repel each other (as for CO on palladium) the adsorption becomes less exothermic (the enthalpy of adsorption less negative) as coverage increases. Moreover, studies show that such species settle on the surface in a disordered way until packing requirements demand order. If the adsorbate particles attract one another (as for O_2 on tungsten), then they tend to cluster together in islands, and growth occurs at the borders. These adsorbates also show order–disorder transitions when they are heated enough for thermal motion to overcome the particle–particle interactions, but not so much that they are desorbed.

Whether a result of physisorption or chemisorption, the extent of surface coverage is normally expressed as the **fractional coverage**, θ:

$$\theta=\frac{\text{number of adsorption sites occupied}}{\text{number of adsorption sites available}}$$

Definition Fractional coverage (22A.2)

The fractional coverage is often expressed in terms of the volume of adsorbate adsorbed by $\theta=V/V_\infty$, where V_∞ is the volume of adsorbate corresponding to complete monolayer coverage. In each case, the volumes in the definition of θ are those of the free gas measured under the same conditions of temperature and pressure, not the volume the adsorbed gas occupies when attached to the surface.

Brief illustration 22A.2 Fractional coverage

For the adsorption of CO on charcoal at 273 K, $V_\infty=111\,\text{cm}^3$, a value corrected to 1 atm. When the partial pressure of CO is 80.0 kPa, the value of V (also corrected to 1 atm) is 41.6 cm^3, so it follows that $\theta=(41.6\,\text{cm}^3)/(111\,\text{cm}^3)=0.375$.

Self-test 22A.2 It is commonly observed that θ increases sharply with the partial pressure of adsorbate at low pressures, but becomes increasingly less dependent on partial pressure at high pressures. Explain this behaviour.

Answer: See Topic 22B

22A.3 Experimental techniques

A vast array of experimental techniques are used to study the composition and structure of solid surfaces at the atomic level. Many of the arrangements allow for direct visualization of changes in the surface as adsorption and chemical reactions take place there.

Experimental procedures must begin with a clean surface. The obvious way to retain cleanliness of a surface is to reduce the pressure and reduce the number of impacts on the surface. When the pressure is reduced to 0.1 mPa (as in a simple vacuum system) the collision flux falls to about $10^{18}\,\text{m}^{-2}\,\text{s}^{-1}$, corresponding to one hit per surface atom in each 0.1 s. Even that is too frequent in most experiments, and in **ultrahigh vacuum** (UHV) techniques pressures as low as 0.1 μPa (when $Z_W=10^{15}\,\text{m}^{-2}\,\text{s}^{-1}$) are reached on a routine basis and as low as 1 nPa (when $Z_W=10^{13}\,\text{m}^{-2}\,\text{s}^{-1}$) are reached with special care. These collision fluxes correspond to each surface atom being hit once every 10^5 to 10^6 s, or about once a day.

(a) Microscopy

The basic approach of illuminating a small area of a sample and collecting light with a microscope has been used for many years to image small specimens. However, the resolution of a microscope, the minimum distance between two objects that leads to two distinct images, is on the order of the wavelength of the light being used. Therefore, conventional microscopes employing visible light have resolutions in the micrometre range and are blind to features on a scale of nanometres.

One technique that is often used to image nanometre-sized objects is **electron microscopy**, in which a beam of electrons with a well-defined de Broglie wavelength (Topic 7A) replaces the lamp found in traditional light microscopes. Instead of glass or quartz lenses, magnetic fields are used to focus the beam. In **transmission electron microscopy** (TEM), the electron beam passes through the specimen and the image is collected on a screen. In **scanning electron microscopy** (SEM), electrons scattered back from a small irradiated area of the sample are detected and the electrical signal is sent to a video screen. An image of the surface is then obtained by scanning the electron beam across the sample.

As in traditional light microscopy, the wavelength of and the ability to focus the incident beam—in this case a beam of electrons focused by magnetic fields—govern the resolution. It is now possible to achieve atomic resolution with TEM instruments. Resolution on the order of a few nanometres is possible with SEM instruments.

Scanning probe microscopy (SPM) is a collection of techniques that can be used to make visible and manipulate objects as small as atoms on surfaces. One version is **scanning tunnelling microscopy** (STM), in which a platinum–rhodium or tungsten needle is scanned across the surface of a conducting solid. When the tip of the needle is brought very close to the surface, electrons tunnel across the intervening space (Fig. 22A.4). In the 'constant-current mode' of operation, the stylus moves up and down according to the form of the surface, and the topography of the surface, including any adsorbates, can be mapped on an atomic scale. The vertical motion of the stylus is achieved by fixing it to a piezoelectric cylinder, which contracts or expands according to the potential difference it experiences. In the 'constant-z mode', the vertical position of the stylus is held constant and the current is monitored. Because the tunnelling probability is very sensitive to the size of the gap, the microscope can detect tiny, atom-scale variations in the height of the surface.

Figure 22A.5 shows an example of the kind of image obtained with a surface, in this case of gallium arsenide that has been modified by addition of caesium atoms. Each 'bump' on the surface corresponds to an atom. In a further variation of the STM technique, the tip may be used to nudge single atoms around on the surface, making possible the fabrication of complex and yet very tiny nanometre-sized materials and devices.

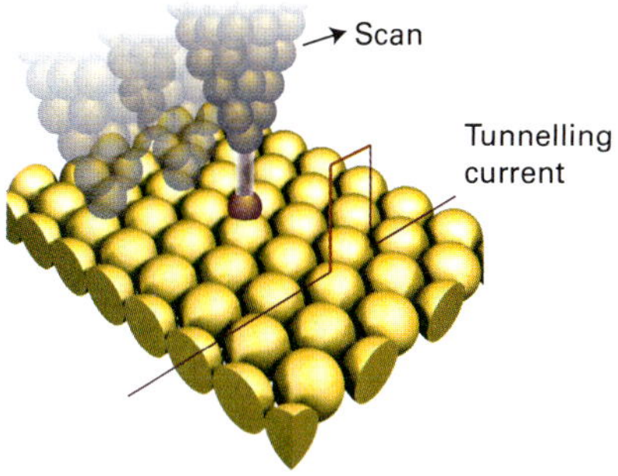

Figure 22A.4 A scanning tunnelling microscope makes use of the current of electrons that tunnel between the surface and the tip. That current is very sensitive to the distance of the tip above the surface.

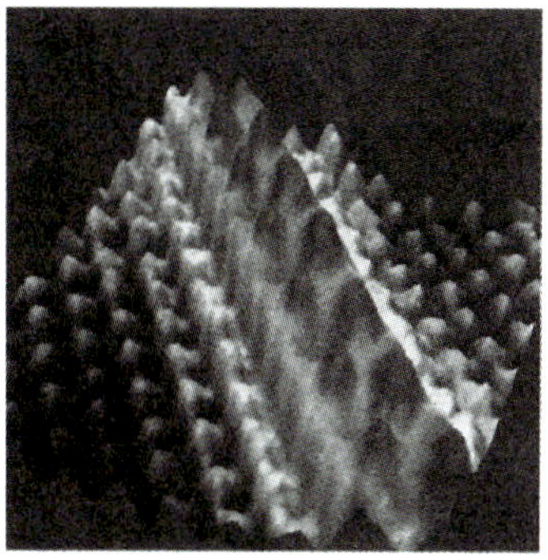

Figure 22A.5 An STM image of caesium atoms on a gallium arsenide surface.

Diffusion characteristics of an adsorbate can be examined by using STM to follow the change in surface characteristics. An adsorbed atom makes a random walk across the surface, and the diffusion coefficient, D, can be inferred from the mean distance, d, travelled in an interval τ by using the two-dimensional random walk expression $d=(D\tau)^{1/2}$. The value of D for different crystal planes at different temperatures can be determined directly in this way, and the activation energy for migration over each plane obtained from the Arrhenius-like expression

$$D=D_0 e^{-E_{a,diff}/RT} \quad \text{Temperature dependence of the diffusion coefficient} \quad (22A.3)$$

where $E_{a,diff}$ is the activation energy for diffusion and D_0 is the diffusion coefficient in the limit of infinite temperature.

Brief illustration 22A.3 Diffusion coefficients

Typical values for W atoms on tungsten have $E_{a,diff}$ in the range 57–87 kJ mol^{-1} and $D_0 \approx 3.8\times10^{-11}$ m^2 s^{-1}. It follows from eqn 22A.3 that at 800 K the diffusion coefficient varies approximately from

$$D=(3.8\times10^{-11}\,\text{m}^2\,\text{s}^{-1})\times e^{-5.7\times10^{-2}\,\text{J}\,\text{mol}^{-1}/(8.3145\,\text{J}\,\text{K}^{-1}\,\text{mol}^{-1}\times800\,\text{K})}$$
$$=7.2\times10^{-15}\,\text{m}^2\,\text{s}^{-1}$$

to

$$D=(3.8\times10^{-11}\,\text{m}^2\,\text{s}^{-1})\times e^{-8.7\times10^{-2}\,\text{J}\,\text{mol}^{-1}/(8.3145\,\text{J}\,\text{K}^{-1}\,\text{mol}^{-1}\times800\,\text{K})}$$
$$=7.9\times10^{-17}\,\text{m}^2\,\text{s}^{-1}$$

Self-test 22A.3 For CO on tungsten, the activation energy falls from 144 kJ mol^{-1} at low surface coverage to 88 kJ mol^{-1} when the coverage is high. Calculate the ratio D_{high}/D_{low} of diffusion coefficients at 800 K.

Answer: 4.5×10^3

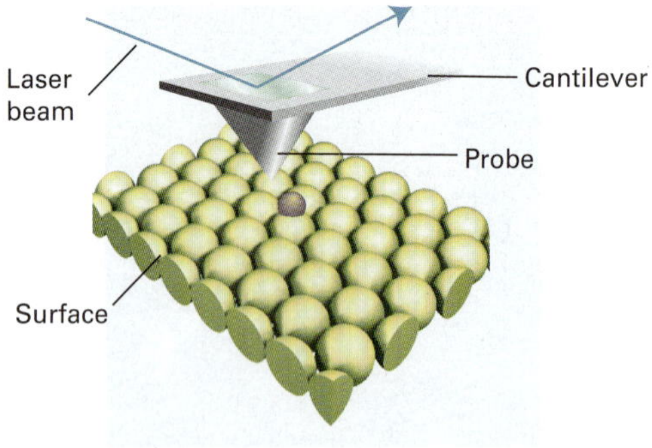

Figure 22A.6 In atomic force microscopy, a laser beam is used to monitor the tiny changes in position of a probe as it is attracted to or repelled by atoms on a surface.

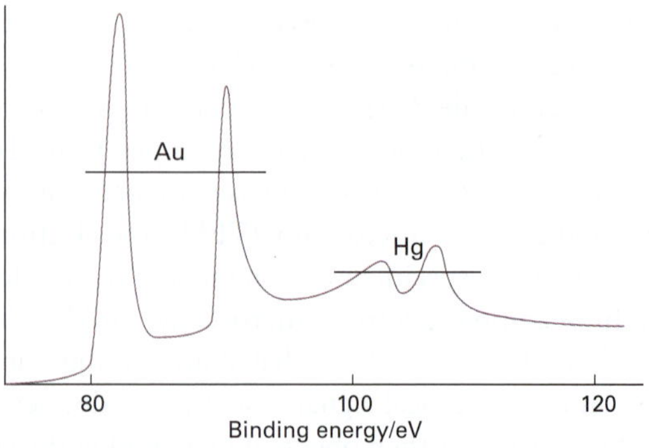

Figure 22A.7 The X-ray photoelectron emission spectrum of a sample of gold contaminated with a surface layer of mercury. (M.W. Roberts and C.S. McKee, *Chemistry of the metal–gas interface*, Oxford (1978).)

In **atomic force microscopy** (AFM), a sharpened tip attached to a cantilever is scanned across the surface. The force exerted by the surface and any molecules attached to it pushes or pulls on the tip and deflects the cantilever (Fig. 22A.6). The deflection is monitored by using a laser beam. Because no current needs to pass between the sample and the probe, the technique can be applied to non-conducting surfaces and to liquid samples.

Two modes of operation of AFM are common. In 'contact mode', or 'constant-force mode', the force between the tip and surface is held constant and the tip makes contact with the surface. This mode of operation can damage fragile samples on the surface. In 'non-contact', or 'tapping mode', the tip bounces up and down with a specified frequency and never quite touches the surface. The amplitude of the tip's oscillation changes when it passes over a species adsorbed on the surface.

(b) Ionization techniques

The chemical composition of a surface can be determined by a variety of ionization techniques. The same techniques can be used to detect any remaining contamination after cleaning and to detect layers of material adsorbed later in the experiment.

One technique is **photoemission spectroscopy**, a derivative of the photoelectric effect (Topic 7A), in which X-rays (for XPS) or hard (short wavelength) ultraviolet (for UPS) ionizing radiation is used, giving rise to ejected electrons from adsorbed species. The kinetic energies of the electrons ejected from their orbitals are measured and the pattern of energies is a fingerprint of the material present (Fig. 22A.7). UPS, which examines electrons ejected from valence shells, is also used to establish the bonding characteristics and the details of valence shell electronic structures of substances on the surface. Its usefulness is its ability to reveal which orbitals of the adsorbate are involved in the bond to the substrate.

Brief illustration 22A.4 A UPS spectrum

The principal difference between the photoemission results on free benzene and benzene adsorbed on palladium is in the energies of the π electrons. This difference is interpreted as meaning that the C_6H_6 molecules lie parallel to the surface and are attached to it by their π orbitals.

Self-test 22A.4 When adsorbed to palladium, pyridine (C_6H_5N) stands almost perpendicular to the surface. Suggest a mode of attachment of the molecule to palladium atoms on the surface.

Answer: Data are consistent with a σ bond formed by the nitrogen lone pair.

A very important technique, which is widely used in the microelectronics industry, is **Auger electron spectroscopy** (AES). The **Auger effect** (pronounced oh-zhey) is the emission of a second electron after high energy radiation has expelled another. The first electron to depart leaves a hole in a low-lying orbital, and an upper electron falls into it. The energy this transition releases may result either in the generation of radiation, which is called **X-ray fluorescence** (Fig. 22A.8a) or in the ejection of another electron (Fig. 22A.8b). The latter is the 'secondary electron' of the Auger effect. The energies of the secondary electrons are characteristic of the material present, so the Auger effect effectively takes a fingerprint of the sample. In practice, the Auger spectrum is normally obtained by irradiating the sample with an electron beam of energy in the range 1–5 keV rather than electromagnetic radiation. In **scanning Auger electron microscopy** (SAM), the finely focused electron beam is scanned over the surface and a map of composition is compiled; the resolution can reach below about 50 nm.

(c) Diffraction techniques

A useful technique for determining the arrangement of the atoms close to the surface is **low energy electron diffraction**

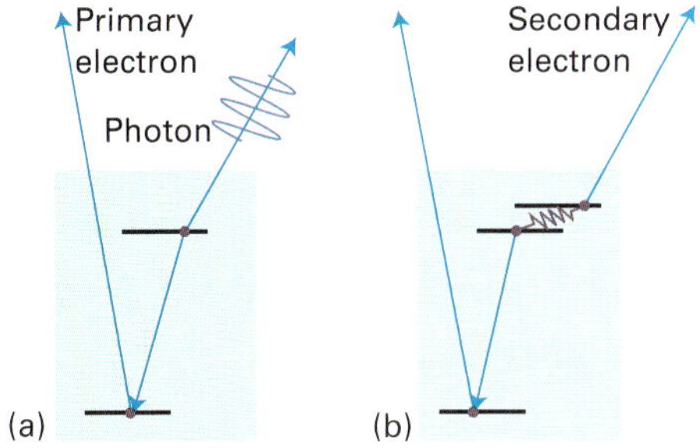

Figure 22A.8 When an electron is expelled from a solid (a) an electron of higher energy may fall into the vacated orbital and emit an X-ray photon to produce X-ray fluorescence. Alternatively (b) the electron falling into the orbital may give up its energy to another electron, which is ejected in the Auger effect.

(LEED). This technique is like X-ray diffraction (Topic 18A) but uses the wave character of electrons, and the sample is now the surface of a solid. The use of low energy electrons (with energies in the range 10–200 eV, corresponding to wavelengths in the range 100–400 pm) ensures that the diffraction is caused only by atoms on and close to the surface. The experimental arrangement is shown in Fig. 22A.9, and typical LEED patterns, obtained by photographing the fluorescent screen through the viewing port, are shown in Fig. 22A.10.

Observations using LEED show that the surface of a crystal rarely has exactly the same form as a slice through the bulk because surface and bulk atoms experience different forces. **Reconstruction** refers to processes by which atoms on the surface achieve their equilibrium structures. As a general rule, it is found that metal surfaces are simply truncations of the bulk lattice, but the distance between the top layer of atoms and the one below is contracted by around 5 per cent. Semiconductors generally have surfaces reconstructed to a depth of several layers. Reconstruction occurs in ionic solids. For example, in

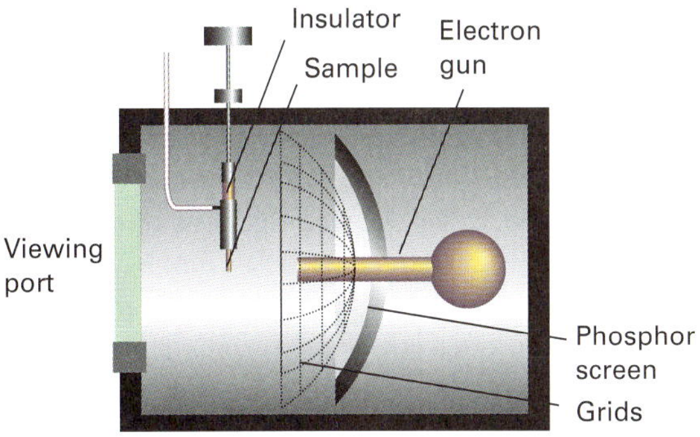

Figure 22A.9 A schematic diagram of the apparatus used for a LEED experiment. The electrons diffracted by the surface layers are detected by the fluorescence they cause on the phosphor screen.

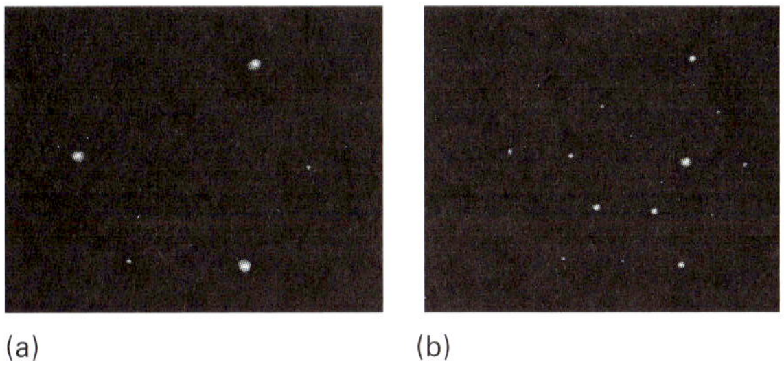

Figure 22A.10 LEED photographs of (a) a clean platinum surface and (b) after its exposure to propyne, $CH_3C{\equiv}CH$. (Photographs provided by Professor G.A. Somorjai.)

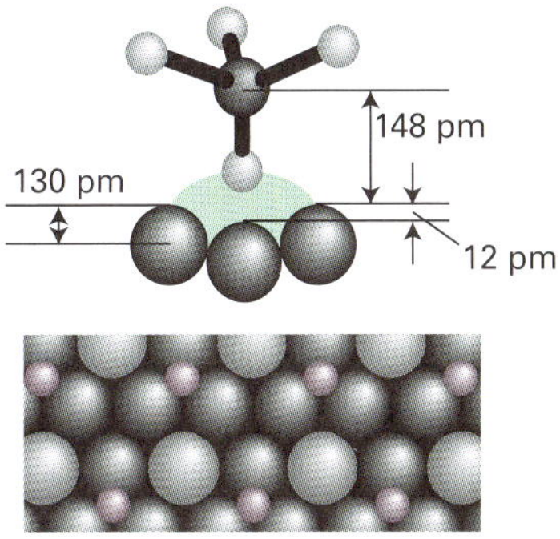

Figure 22A.11 The structure of a surface close to the point of attachment of CH_3C– to the (110) surface of rhodium at 300 K and the changes in positions of the metal atoms that accompany chemisorption.

lithium fluoride the Li^+ and F^- ions close to the surface apparently lie on slightly different planes. An actual example of the detail that can now be obtained from refined LEED techniques is shown in Fig. 22A.11 for CH_3C– adsorbed on a (111) plane of rhodium.

Example 22A.1 Interpreting a LEED pattern

The LEED pattern from a clean (110) face of palladium is shown in (a) below. The reconstructed surface gives a LEED pattern shown as (b). What can be inferred about the structure of the reconstructed surface?

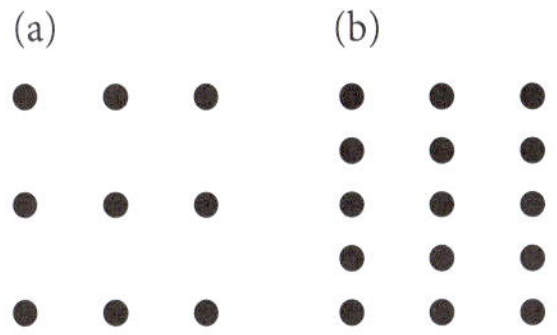

Method Recall from Bragg's law (Topic 18A), $\lambda = 2d \sin \theta$, that for a given wavelength, the greater the separation d of the layers, the smaller is the scattering angle (so that $2d \sin \theta$

remains constant). It follows that, in terms of the LEED pattern, the farther apart the atoms responsible for the pattern, the closer the spots appear in the pattern. Twice the separation between the atoms corresponds to half the separation between the spots, and vice versa. Therefore, inspect the two patterns and identify how the new pattern relates to the old.

Answer The horizontal separation between spots is unchanged, which indicates that the atoms remain in the same position in that dimension when reconstruction occurs. However, the vertical spacing is halved, which suggests that the atoms are twice as far apart in that direction as they are in the unreconstructed surface.

Self-test 22A.5 Sketch the LEED pattern for a surface that differs from that shown in (a) above by tripling the vertical separation.

Answer:

The presence of terraces, steps, and kinks in a surface shows up in LEED patterns, and their surface density (the number of defects in a region divided by the area of the region) can be estimated. The importance of this type of measurement will emerge later. Three examples of how steps and kinks affect the pattern are shown in Fig. 22A.12. The samples used were obtained by cleaving a crystal at different angles to a plane of atoms. Only terraces are produced when the cut is parallel to the plane, and the density of steps increases as the angle of the cut increases. The observation of additional structure in the LEED patterns, rather than blurring, shows that the steps are arrayed regularly.

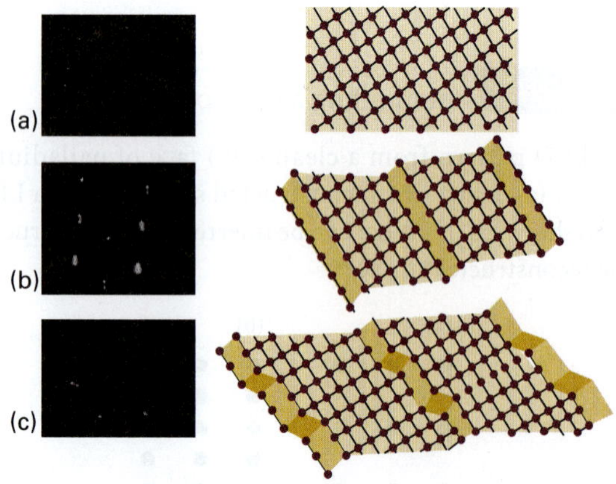

Figure 22A.12 LEED patterns may be used to assess the defect density of a surface. The photographs correspond to a platinum surface with (a) low defect density, (b) regular steps separated by about six atoms, and (c) regular steps with kinks. (Photographs provided by Professor G.A. Samorjai.)

(d) Determination of the extent and rates of adsorption and desorption

A common technique for measuring rates of processes on surfaces is to monitor the rates of flow of gas into and out of the system: the difference is the rate of gas uptake by the sample. Integration of this rate then gives the fractional coverage at any stage.

- **Gravimetry**, in which the sample is weighed on a microbalance during the experiment, is used to determine the extent and kinetics of adsorption and desorption.

The technique commonly uses a **quartz crystal microbalance** (QCM), in which the mass of a sample adsorbed on the surface of a quartz crystal is related to changes in the characteristic vibrational frequency of the crystal. In this way, masses as small as a few nanograms can be measured reliably.

- **Second harmonic generation** (SHG), the conversion of an intense, pulsed laser beam to radiation with twice its initial frequency is very important for the study of all types of surfaces, including thin films and liquid–gas interfaces.

For example, adsorption of gas molecules on to a surface alters the intensity of the SHG signal, allowing for determination of the rates of surface processes and the fractional coverage. Because pulsed lasers are the excitation sources, time-resolved measurements of the kinetics and dynamics of surface processes are possible on timescales as short as femtoseconds.

- **Surface plasmon resonance** (SPR), the absorption of energy from an incident beam of electromagnetic radiation by surface 'plasmons', is a very sensitive technique now used routinely in the study of adsorption and desorption.

To understand the technique we need to examine the terms 'surface plasmon' and 'resonance' in its name.

The mobile delocalized valence electrons of metals form a **plasma**, a dense gas of charged particles. Bombardment of this plasma by light or an electron beam can cause transient changes in the distribution of electrons, with some regions becoming slightly denser than others. Coulomb repulsion in the regions of high density causes electrons to move away from each other, so lowering their density. The resulting oscillations in electron density, the **plasmons**, can be excited both in the bulk and on the surface of a metal. A surface plasmon propagates away from the surface, but the amplitude of the wave, also called an

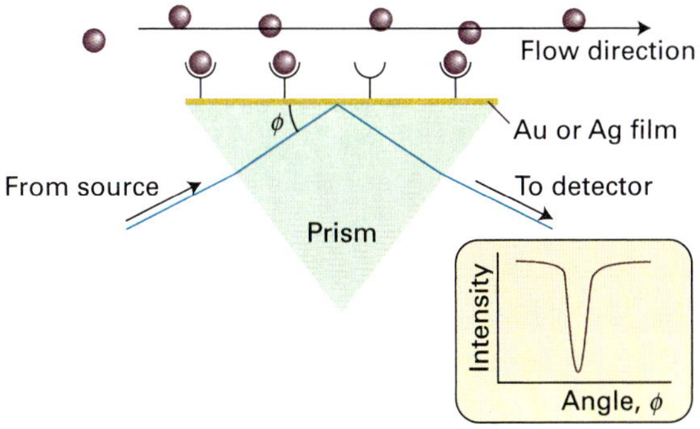

Figure 22A.13 The experimental arrangement for the observation of surface plasmon resonance, as explained in the text.

evanescent wave, decreases sharply with distance from the surface. The 'resonance' in the name refers to the absorption that can be observed with appropriate choice of the wavelength and angle of incidence of the excitation beam.

It is common practice to use a monochromatic beam and to vary the angle of incidence (the ϕ in Fig. 22A.13). The beam passes through a prism that strikes one side of a thin film of gold or silver. Because the evanescent wave interacts with material a short distance away from the surface, the angle at which resonant absorption occurs depends on the refractive index of the medium on the opposite side of the metallic film. Thus, changing the identity and quantity of material on the surface changes the resonance angle.

The SPR technique can be used in the study of the binding of molecules to a surface or binding of ligands to a biopolymer attached to the surface; this interaction mimics the biological recognition processes that occur in cells. Examples of complexes amenable to analysis include antibody–antigen and protein–DNA interactions. The most important advantage of SPR is its sensitivity: it is possible to measure the deposition of nanograms of material on to a surface. The main disadvantage of the technique is its requirement for immobilization of at least one of the components of the system under study.

Checklist of concepts

- ☐ 1. **Adsorption** is the attachment of molecules to a surface; the substance that adsorbs is the adsorbate and the underlying material is the adsorbent or substrate. The reverse of adsorption is desorption.
- ☐ 2. Surface defects play an important role in surface growth and catalysis.
- ☐ 3. **Reconstruction** refers to processes by which atoms on the surface achieve their equilibrium structures.
- ☐ 4. Techniques for studying surfaces include **scanning electron microscopy** (SEM), **transmission electron microscopy** (TEM), **scanning probe microscopy** (SPM), **photoemission spectroscopy**, **Auger electron spectroscopy** (AES), **low energy electron diffraction** (LEED), **gravimetry**, **second harmonic generation** (SHG), and **surface plasmon resonance** (SPR).

Checklist of equations

Property	Equation	Comment	Equation number
Collision flux	$Z_W = p/(2\pi mkT)^{1/2}$	KMT	22A.1
Fractional coverage	θ=(number of adsorption sites occupied)/(number of adsorption sites available)	Definition	22A.2

22B Adsorption and desorption

Contents

➤ Why do you need to know this material?

To understand how surfaces can affect the rates of chemical reactions, you need to know how to assess the extent of surface coverage and the factors that determine the rates at which molecules attach to and detach from solid surfaces.

➤ What is the key idea?

The extent of surface coverage can be expressed in terms of isotherms derived on the basis of dynamic equilibria between adsorbed and free material.

➤ What do you need to know already?

This Topic extends the discussion of adsorption in Topic 22A. You need to be familiar with the basic ideas of chemical kinetics (Topics 20A–20C), the Arrhenius equation (Topic 20D), and the expression of reaction mechanisms as rate laws (Topic 20E).

Here we consider the extent to which a solid surface is covered and the variation of the extent of coverage with pressure and temperature. For simplicity, we consider only gas/solid systems. We use this material in Topic 22C to discuss how surfaces affect the rate and course of chemical change by acting as the site of catalysis.

22B.1 Adsorption isotherms

In adsorption (Topic 22A) the free gas and the adsorbed gas are in dynamic equilibrium, and the fractional coverage, θ, of the surface (eqn 22A.2) depends on the pressure of the overlying gas. The variation of θ with pressure at a chosen temperature is called the **adsorption isotherm**.

Many of the techniques discussed in Topic 22A can be used to measure θ. Another is **flash desorption**, in which the sample is suddenly heated (electrically) and the resulting rise of pressure is interpreted in terms of the amount of adsorbate originally on the sample.

(a) The Langmuir isotherm

The simplest physically plausible isotherm is based on three assumptions:

- Adsorption cannot proceed beyond monolayer coverage.
- All sites are equivalent.
- The ability of a molecule to adsorb at a given site is independent of the occupation of neighbouring sites (that is, there are no interactions between adsorbed molecules).

The dynamic equilibrium is

$$\mathrm{A(g)+M(surface)} \rightleftharpoons \mathrm{AM(surface)}$$

with rate constants k_a for adsorption and k_d for desorption. The rate of change of the surface coverage, $d\theta/dt$, due to adsorption is proportional to the partial pressure p of A and the number of vacant sites $N(1-\theta)$, where N is the total number of sites:

$$\frac{d\theta}{dt}=k_a pN(1-\theta) \qquad \text{Rate of adsorption} \qquad (22B.1a)$$

The rate of change of θ due to desorption is proportional to the number of adsorbed species, $N\theta$:

$$\frac{d\theta}{dt}=-k_d N\theta \qquad \text{Rate of desorption} \qquad (22B.1b)$$

At equilibrium there is no net change (that is, the sum of these two rates is zero), and solving $k_a pN(1-\theta)-k_d N\theta=0$ for θ gives the **Langmuir isotherm**:

$$\theta=\frac{\alpha p}{1+\alpha p} \qquad \alpha=\frac{k_a}{k_d} \qquad \text{Langmuir isotherm} \qquad (22B.2)$$

The dimensions of α are 1/pressure.

Example 22B.1 Using the Langmuir isotherm

The data given below are for the adsorption of CO on charcoal at 273 K. Confirm that they fit the Langmuir isotherm, and find the constant α and the volume corresponding to complete coverage. In each case V has been corrected to 1 atm (101.325 kPa).

p/kPa	13.3	26.7	40.0	53.3	66.7	80.0	93.3
V/cm^3	10.2	18.6	25.5	31.5	36.9	41.6	46.1

Method From eqn 22B.2, $\alpha p\theta+\theta=\alpha p$. With $\theta=V/V_\infty$ (eqn 22A.2), where V_∞ is the volume corresponding to complete coverage, this expression can be rearranged into

$$\frac{p}{V}=\frac{p}{V_\infty}+\frac{1}{\alpha V_\infty}$$

Hence, a plot of p/V against p should give a straight line of slope $1/V_\infty$ and intercept $1/\alpha V_\infty$.

Answer The data for the plot are as follows:

p/kPa	13.3	26.7	40.0	53.3	66.7	80.0	93.3
(p/kPa)/(V/cm^3)	1.30	1.44	1.57	1.69	1.81	1.92	2.02

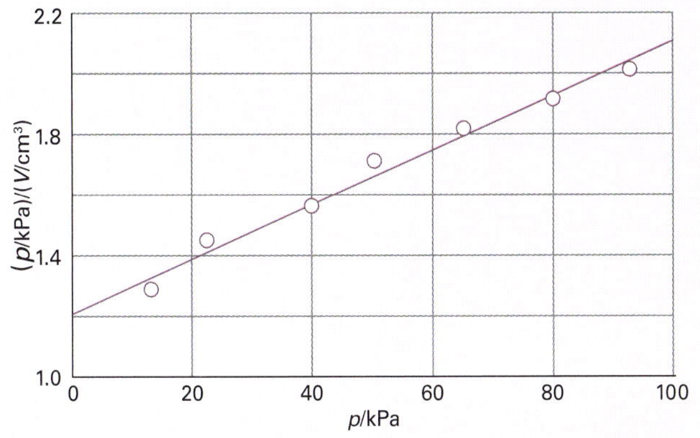

Figure 22B.1 The plot of the data in Example 22B.1. As illustrated here, the Langmuir isotherm predicts that a straight line should be obtained when p/V is plotted against p.

The points are plotted in Fig. 22B.1.The (least squares) slope is 0.009 00, so $V_\infty=111\,\text{cm}^3$. The intercept at $p=0$ is 1.20, so

$$\alpha=\frac{1}{(111\,\text{cm}^3)\times(1.20\,\text{kPa cm}^3)}=7.51\times10^{-3}\,\text{kPa}^{-1}$$

Self-test 22B.1 Repeat the calculation for the following data:

p/kPa	13.3	26.7	40.0	53.3	66.7	80.0	93.3
V/cm^3	10.3	19.3	27.3	34.1	40.0	45.5	48.0

Answer: 128 cm^3, 6.69×10^{-3} kPa^{-1}

For adsorption with dissociation, when A_2 adsorbs as 2 A, the rate of adsorption is proportional to the pressure and to the probability that both fragments A will find sites. The latter is now proportional to the *square* of the number of vacant sites:

$$\frac{d\theta}{dt}=k_a p\{N(1-\theta)\}^2 \qquad (22B.3a)$$

The rate of desorption is proportional to the frequency of encounters of the fragments on the surface, and is therefore second order in the number of fragments present:

$$\frac{d\theta}{dt}=-k_d(N\theta)^2 \qquad (22B.3b)$$

The condition for no net change leads to the isotherm

$$\theta=\frac{(\alpha p)^{1/2}}{1+(\alpha p)^{1/2}} \qquad \text{Langmuir isotherm for adsorption with dissociation} \qquad (22B.4)$$

The surface coverage now depends more weakly on pressure than for non-dissociative adsorption.

The shapes of the Langmuir isotherms with and without dissociation are shown in Figs. 22B.2 and 22B.3. The fractional

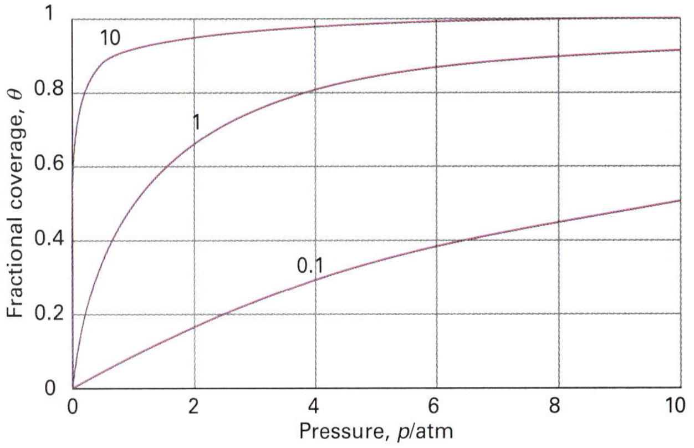

Figure 22B.2 The Langmuir isotherm for dissociative adsorption, $A_2(g)\rightarrow 2\,A(\text{surface})$, for different values of α.

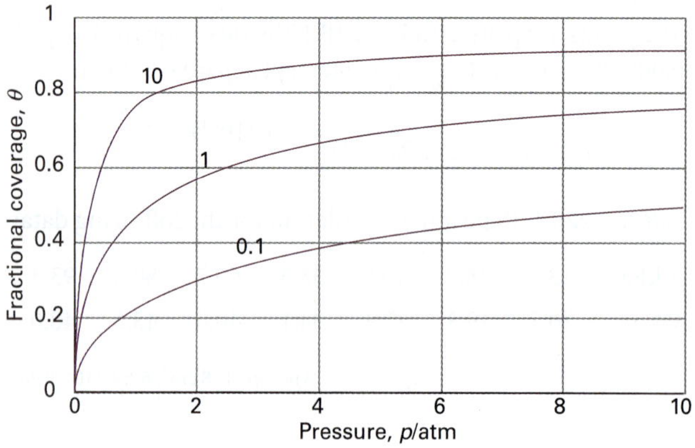

Figure 22B.3 The Langmuir isotherm for non-dissociative adsorption for different values of α.

coverage increases with increasing pressure, and approaches 1 only at very high pressure, when the gas is forced on to every available site of the surface.

(b) The isosteric enthalpy of adsorption

The Langmuir isotherm depends on the value of $\alpha = k_a/k_d$, which depends on the temperature. As we show in the following *Justification*, the temperature dependence of α can be used to determine the **isosteric enthalpy of adsorption**, $\Delta_{ad}H^{\ominus}$, the standard enthalpy of adsorption at a fixed surface coverage, by using the relation

$$\left(\frac{\partial \ln(\alpha p^{\ominus})}{\partial T}\right)_{\theta} = \frac{\Delta_{ad}H^{\ominus}}{RT^2} \quad \text{Isosteric enthalpy of adsorption} \quad (22B.5)$$

Justification 22B.1 The isosteric enthalpy of adsorption

It follows from the treatment in Topic 20C that the quantity $\alpha p^{\ominus} = (k_a/k_d) \times p^{\ominus}$ is an equilibrium constant for the process $A(g) + M(surface) \rightleftharpoons AM(surface)$, and can therefore be expressed in terms of the standard Gibbs energy of adsorption, $\Delta_{ad}G^{\ominus}$ through eqn 6A.14 ($\Delta G^{\ominus} = -RT \ln K$) as

$$\ln(\alpha p^{\ominus}) = -\frac{\Delta_{ad}G^{\ominus}}{RT}$$

We can then infer from the Gibbs–Helmholtz equation (eqn 6B.2, $d((\Delta G/T)/dT = -\Delta H/RT^2)$ that

$$\frac{d\ln(\alpha p^{\ominus})}{dT} = \frac{\Delta_{ad}H^{\ominus}}{RT^2}$$

There is the possibility that the enthalpy of adsorption depends on the fractional coverage, so this expression is confined to constant θ, which implies the partial derivative form in eqn 22B.5.

Example 22B.2 Measuring the isosteric enthalpy of adsorption

The data below show the pressures of CO needed for the volume of adsorption (corrected to 1 atm and 0 °C) to be 10.0 cm³ using the same sample as in *Example* 22B.1. In this case, there is no dissociation. Calculate the adsorption enthalpy at this surface coverage.

T/K	200	210	220	230	240	250
p/kPa	4.00	4.95	6.03	7.20	8.47	9.85

Method The Langmuir isotherm for adsorption without dissociation (eqn 22B.2), can be rearranged to $\alpha p = \theta/(1-\theta)$, a constant when θ is constant. We need to guard against problems with units as we manipulate expressions, and in this case it will prove useful to write αp = constant as $(\alpha p^{\ominus}) \times (p/p^{\ominus})$ = constant. It then follows that

$$\ln\{(\alpha p^{\ominus})(p/p^{\ominus})\} = \ln(\alpha p^{\ominus}) + \ln(p/p^{\ominus}) = \text{constant}$$

and from eqn 22B.5 that

$$\left(\frac{\partial \ln(p/p^{\ominus})}{\partial T}\right)_{\theta} = -\left(\frac{\partial \ln(\alpha p^{\ominus})}{\partial T}\right)_{\theta} = -\frac{\Delta_{ad}H^{\ominus}}{RT^2}$$

With $d(1/T)/dT = -1/T^2$, and therefore $dT = -T^2 d(1/T)$, this expression becomes

$$\left(\frac{\partial \ln(p/p^{\ominus})}{\partial(1/T)}\right)_{\theta} = \frac{\Delta_{ad}H^{\ominus}}{R}$$

Therefore, a plot of $\ln(p/p^{\ominus})$ against $1/T$ should be a straight line of slope $\Delta_{ad}H^{\ominus}/R$.

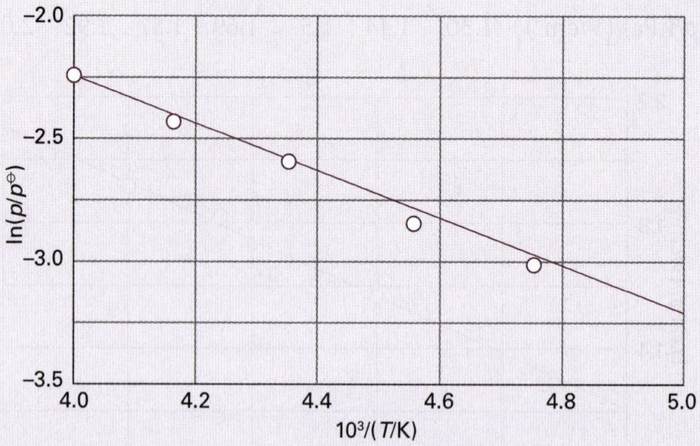

Figure 22B.4 The isosteric enthalpy of adsorption can be obtained from the slope of the plot of $\ln(p/p^{\ominus})$ against $1/T$, where p is the pressure needed to achieve the specified surface coverage. The data used are from *Example* 22B.2.

Answer With $p^{\ominus} = 1\ \text{bar} = 10^2\ \text{kPa}$, we draw up the following table:

T/K	200	210	220	230	240	250
$10^3/(T/\text{K})$	5.00	4.76	4.55	4.35	4.17	4.00
$(p/p^{\ominus})\times 10^2$	4.00	4.95	6.03	7.20	8.47	9.85
$\ln(p/p^{\ominus})$	−3.22	−3.01	−2.81	−2.63	−2.47	−2.32

The points are plotted in Fig. 22B.4. The slope (of the least squares fitted line) is −0.904, so

$$\Delta_{\text{ad}}H^{\ominus} = -(0.904\times 10^3\ \text{K})\times R = -7.52\ \text{kJ mol}^{-1}$$

Self-test 22B.2 Repeat the calculation using the following data:

T/K	200	210	220	230	240	250
p/kPa	4.32	5.59	7.07	8.80	10.67	12.80

Answer: $-9.0\ \text{kJ mol}^{-1}$

Two assumptions of the Langmuir isotherm are the independence and equivalence of the adsorption sites. Deviations from the isotherm can often be traced to the failure of these assumptions. For example, the enthalpy of adsorption often becomes less negative as θ increases, which suggests that the energetically most favourable sites are occupied first. Also, substrate–substrate interactions on the surface can be important. A number of isotherms have been developed to deal with cases where deviations from the Langmuir isotherm are important.

(c) The BET isotherm

If the initial adsorbed layer can act as a substrate for further (for example, physical) adsorption, then, instead of the isotherm levelling off to some saturated value at high pressures, it can be expected to rise indefinitely. The most widely used isotherm dealing with multilayer adsorption was derived by Stephen Brunauer, Paul Emmett, and Edward Teller and is called the **BET isotherm**:

$$\frac{V}{V_{\text{mon}}} = \frac{cz}{(1-z)\{1-(1-c)z\}} \quad \text{with} \quad z = \frac{p}{p^*} \qquad \text{BET isotherm} \qquad (22\text{B}.6)$$

In this expression, which is obtained in the following *Justification*, p^* is the vapour pressure above a layer of adsorbate that is more than one molecule thick and which resembles a pure bulk liquid, V_{mon} is the volume corresponding to monolayer coverage, and c is a constant which is large when the enthalpy of desorption from a monolayer is large compared with the enthalpy of vaporization of the liquid adsorbate:

$$c = e^{(\Delta_{\text{des}}H^{\ominus} - \Delta_{\text{vap}}H^{\ominus})/RT} \qquad (22\text{B}.7)$$

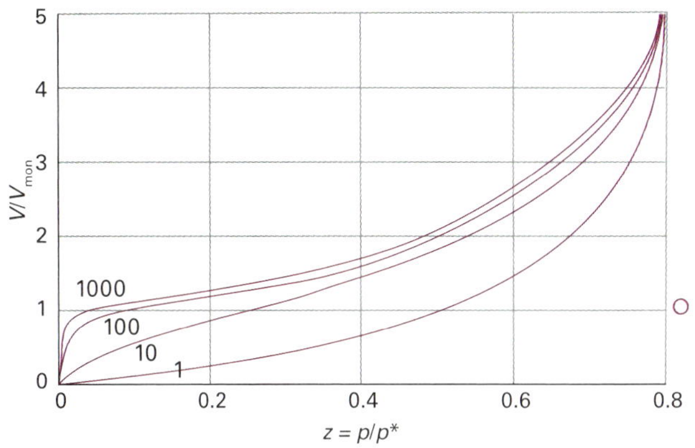

Figure 22B.5 Plots of the BET isotherm for different values of c. The value of V/V_{mon} rises indefinitely because the adsorbate may condense on the covered substrate surface.

Figure 22B.5 illustrates the shapes of BET isotherms. They rise indefinitely as the pressure is increased because there is no limit to the amount of material that may condense when multilayer coverage is possible. A BET isotherm is not accurate at all pressures, but it is widely used in industry to determine the surface areas of solids.

Justification 22B.2 The BET isotherm

We suppose that at equilibrium a fraction θ_0 of the surface sites are unoccupied, a fraction θ_1 is covered by a monolayer, a fraction θ_2 is covered by a bilayer, and so on. The number of adsorbed molecules is therefore

$$N = N_{\text{sites}}(\theta_1 + 2\theta_2 + 3\theta_3 + \cdots)$$

where N_{sites} is the total number of sites. We now follow the derivation that led to the Langmuir isotherm (eqn 22B.2) but allow for different rates of desorption from the substrate and the various layers:

First layer:	Rate of adsorption $= Nk_{\text{a},0}p\theta_0$
	Rate of desorption $= Nk_{\text{d},0}\theta_1$
	At equilibrium, $k_{\text{a},0}p\theta_0 = k_{\text{d},0}\theta_1$
Second layer:	Rate of adsorption $= Nk_{\text{a},1}p\theta_1$
	Rate of desorption $= Nk_{\text{d},1}\theta_2$
	At equilibrium, $k_{\text{a},1}p\theta_1 = k_{\text{d},1}\theta_2$
Third layer:	Rate of adsorption $= Nk_{\text{a},2}p\theta_2$
	Rate of desorption $= Nk_{\text{d},2}\theta_3$
	At equilibrium, $k_{\text{a},2}p\theta_2 = k_{\text{d},2}\theta_3$

and so on. We now suppose that once a monolayer has been formed, all the rate constants involving adsorption and desorption from the physisorbed layers are the same, and write these equations as

$$k_{a,0}p\theta_0 = k_{d,0}\theta_1, \text{ so}$$
$$\theta_1 = (k_{a,0}/k_{d,0})p\theta_0 = \alpha_0 p\theta_0$$
$$k_{a,1}p\theta_1 = k_{d,1}\theta_2, \text{ so}$$
$$\theta_2 = (k_{a,1}/k_{d,1})p\theta_1 = (k_{a,0}/k_{d,0})(k_{a,1}/k_{d,1})p^2\theta_0 = \alpha_0\alpha_1 p^2\theta_0$$
$$k_{a,1}p\theta_2 = k_{d,1}\theta_3, \text{ so}$$
$$\theta_3 = (k_{a,1}/k_{d,1})p\theta_2 = (k_{a,0}/k_{d,0})(k_{a,1}/k_{d,1})^2 p^3\theta_0 = \alpha_0\alpha_1^2 p^3\theta_0$$

and so on, with $\alpha_0 = k_{a,0}/k_{d,0}$ and $\alpha_1 = k_{a,1}/k_{d,1}$ the ratios of rate constants for adsorption to the substrate and an overlayer, respectively. Now, because $\theta_0 + \theta_1 + \theta_2 + \cdots = 1$, it follows that

$$\theta_0 + \alpha_0 p\theta_0 + \alpha_0\alpha_1 p^2\theta_0 + \alpha_0\alpha_1^2 p^3\theta_0 + \cdots$$
$$= \theta_0 + \alpha_0 p\theta_0\{1 + \alpha_1 p + \alpha_1^2 p^2 + \cdots\}$$
$$\overset{1+x+x^2+\cdots=1/(1-x)}{\equiv} \left\{1 + \frac{\alpha_0 p}{1-\alpha_1 p}\right\}\theta_0 = \left\{\frac{1-\alpha_1 p + \alpha_0 p}{1-\alpha_1 p}\right\}\theta_0$$

Then, because this expression is equal to 1,

$$\theta_0 = \frac{1-\alpha_1 p}{1-(\alpha_1-\alpha_0)p}$$

In a similar way, we can write the number of adsorbed species as

$$N = N_{\text{sites}}\alpha_0 p\theta_0 + 2N_{\text{sites}}\alpha_0\alpha_1 p^2\theta_0 + \cdots$$
$$= N_{\text{sites}}\alpha_0 p\theta_0(1 + 2\alpha_1 p + 3\alpha_1^2 p^2 + \cdots)$$
$$\overset{1+2x+3x^2+\cdots=1/(1-x)^2}{\equiv} \frac{N_{\text{sites}}\alpha_0 p\theta_0}{(1-\alpha_1 p)^2}$$

By combining the last two expressions, we obtain

$$N = \frac{N_{\text{sites}}\alpha_0 p}{(1-\alpha_1 p)^2} \times \frac{1-\alpha_1 p}{1-(\alpha_1-\alpha_0)p} = \frac{N_{\text{sites}}\alpha_0 p}{(1-\alpha_1 p)\{1-(\alpha_1-\alpha_0)p\}}$$

The ratio N/N_{sites} is equal to the ratio V/V_{mon}, where V is the total volume adsorbed and V_{mon} the volume adsorbed had there been complete monolayer coverage. The equilibrium of the adsorption and desorption from the overlayers is equivalent to the vaporization $\text{A(l)} \rightleftharpoons \text{A(g)}$ of the pure adsorbate, with matching forward and reverse rates: $k_d = k_a p^*$, where p^* is the vapour pressure of the liquid adsorbate. Therefore, $\alpha_1 = k_{a,1}/k_{d,1} = 1/p^*$. Then, with $z = p/p^*$ and $c = \alpha_0/\alpha_1$, the last equation becomes

$$\frac{V}{V_{\text{mon}}} = \frac{\alpha_0 p}{(1-p/p^*)\{1-(1-\alpha_0/\alpha_1)p/p^*\}} = \frac{cz}{(1-z)\{1-(1-c)z\}}$$

as in eqn 22B.6.

As in *Justification* 22B.1, α_0 and α_1 are related to the Gibbs energy changes accompanying adsorption to the substrate and condensation on the adsorbed layers, $\Delta_{\text{ad}}G^{\ominus}$ and $\Delta_{\text{con}}G^{\ominus}$, which in turn can be related to the Gibbs energies for the opposite processes, desorption from the substrate and vaporization from the overlayer, by $\Delta_{\text{des}}G^{\ominus} = -\Delta_{\text{ad}}G^{\ominus}$ and $\Delta_{\text{vap}}G^{\ominus} = -\Delta_{\text{con}}G^{\ominus}$. Therefore, from $\ln(\alpha p^{\ominus}) = -\Delta G^{\ominus}/RT$ in each case,

$$\alpha_0 p^{\ominus} = e^{-\Delta_{\text{ad}}G^{\ominus}/RT} = e^{\Delta_{\text{des}}G^{\ominus}/RT} \text{ and } \alpha_1 p^{\ominus} = e^{-\Delta_{\text{con}}G^{\ominus}/RT} = e^{\Delta_{\text{vap}}G^{\ominus}/RT}$$

The ratio c then becomes (after cancelling the $p^{\ominus}$ and writing $\Delta G^{\ominus} = \Delta H^{\ominus} - T\Delta S^{\ominus}$ in each case)

$$c = \frac{\alpha_0}{\alpha_1} = \frac{e^{\Delta_{\text{des}}G^{\ominus}/RT}}{e^{\Delta_{\text{vap}}G^{\ominus}/RT}} = \frac{e^{\Delta_{\text{des}}H^{\ominus}/RT}e^{-\Delta_{\text{des}}S^{\ominus}/R}}{e^{\Delta_{\text{vap}}H/RT}e^{-\Delta_{\text{vap}}S^{\ominus}/R}}$$

If the entropies of desorption and vaporization are assumed to be the same because they correspond to very similar processes in terms of the escape of the condensed adsorbate to the gas phase, this ratio becomes

$$c = \frac{e^{\Delta_{\text{des}}H^{\ominus}/RT}}{e^{\Delta_{\text{vap}}H^{\ominus}/RT}} = e^{(\Delta_{\text{des}}H^{\ominus} - \Delta_{\text{vap}}H^{\ominus})/RT}$$

as in eqn 21B.7.

Example 22B.3 Using the BET isotherm

The data below relate to the adsorption of N_2 on rutile (TiO_2) at 75 K. Confirm that they fit a BET isotherm in the range of pressures reported, and determine V_{mon} and c.

p/kPa	0.160	1.87	6.11	11.67	17.02	21.92	27.29
V/mm^3	601	720	822	935	1046	1146	1254

At 75 K, $p^* = 76.0$ kPa. The volumes have been corrected to 1.00 atm and 273 K and refer to 1.00 g of substrate.

Method Equation 22B.6 can be reorganized into

$$\frac{z}{(1-z)V} = \frac{1}{cV_{\text{mon}}} + \frac{(c-1)z}{cV_{\text{mon}}}$$

It follows that $(c-1)/cV_{\text{mon}}$ can be obtained from the slope of a plot of the expression on the left against z, and cV_{mon} can be found from the intercept at $z=0$. The results can then be combined to give c and V_{mon}.

Answer We draw up the following table:

p/kPa	0.160	1.87	6.11	11.67	17.02	21.92	27.29
$10^3 z$	2.11	24.6	80.4	154	224	288	359
$10^4 z/(1-z)(V/\text{mm}^3)$	0.035	0.350	1.06	1.95	2.76	3.53	4.47

These points are plotted in Fig. 22B.6. The least squares best line has an intercept at 0.0398, so

$$\frac{1}{cV_{\text{mon}}} = 3.98 \times 10^{-6}\ \text{mm}^{-3}$$

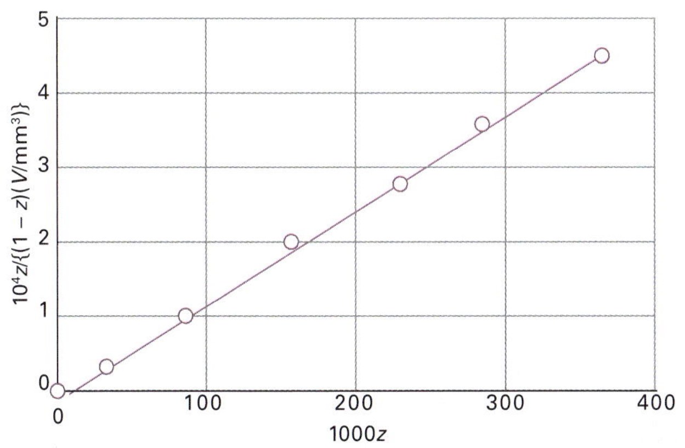

Figure 22B.6 The BET isotherm can be tested, and the parameters determined, by plotting $z/(1-z)V$ against $z = p/p^*$. The data are from *Example* 22B.3.

The slope of the line is 1.23×10^{-2}, so

$$\frac{c-1}{cV_{\text{mon}}} = (1.23\times10^{-2})\times10^3\times10^{-4}\ \text{mm}^{-3} = 1.23\times10^{-3}\ \text{mm}^{-3}$$

The solutions of these equations are $c = 310$ and $V_{\text{mon}} = 811\ \text{mm}^3$. At 1.00 atm and 273 K, $811\ \text{mm}^3$ corresponds to 3.6×10^{-5} mol, or 2.2×10^{19} atoms. Because each atom occupies an area of about $0.16\ \text{nm}^2$, the surface area of the sample is about $3.5\ \text{m}^2$.

Self-test 22B.3 Repeat the calculation for the following data:

p/kPa	0.160	1.87	6.11	11.67	17.02	21.92	27.29
V/cm^3	235	559	649	719	790	860	950

Answer: 370, 615 cm^3

When $c \gg 1$, the BET isotherm takes the simpler form

$$\frac{V}{V_{\text{mon}}} = \frac{1}{1-z} \qquad \text{BET isotherm when } c\gg 1 \qquad (22\text{B}.8)$$

This expression is applicable to unreactive gases on polar surfaces, for which $c \approx 10^2$ because $\Delta_{\text{des}}H^\ominus$ is then significantly greater than $\Delta_{\text{vap}}H^\ominus$ (eqn 22B.7). The BET isotherm fits experimental observations moderately well over restricted pressure ranges, but it errs by underestimating the extent of adsorption at low pressures and by overestimating it at high pressures.

(d) The Temkin and Freundlich isotherms

An assumption of the Langmuir isotherm is the independence and equivalence of the adsorption sites. Deviations from the isotherm can often be traced to the failure of these assumptions. For example, the enthalpy of adsorption often becomes less negative as θ increases, which suggests that the energetically most favourable sites are occupied first. Various attempts have been made to take these variations into account. The **Temkin isotherm**,

$$\theta = c_1 \ln(c_2 p) \qquad \text{Temkin isotherm} \qquad (22\text{B}.9)$$

where c_1 and c_2 are constants, corresponds to supposing that the adsorption enthalpy changes linearly with pressure. The **Freundlich isotherm**

$$\theta = c_1 p^{1/c_2} \qquad \text{Freundlich isotherm} \qquad (22\text{B}.10)$$

corresponds to a logarithmic change. This isotherm attempts to incorporate the role of substrate–substrate interactions on the surface.

Different isotherms agree with experiment more or less well over restricted ranges of pressure, but they remain largely empirical. Empirical, however, does not mean useless for, if the parameters of a reasonably reliable isotherm are known, reasonably reliable results can be obtained for the extent of surface coverage under various conditions. This kind of information is essential for any discussion of heterogeneous catalysis (Topic 22C).

22B.2 The rates of adsorption and desorption

We have noted that adsorption and desorption are activated processes, in the sense that they have an activation energy and follow Arrhenius behaviour. Now we are ready to look more closely at the origin of the activation energy in these processes, with a special focus on chemisorption.

(a) The precursor state

Figure 22B.7 shows how the potential energy of a molecule varies with its distance from the substrate surface. As the molecule approaches the surface its energy falls as it becomes physisorbed into the **precursor state** for chemisorption (see Topic 22A). Dissociation into fragments often takes place as a molecule moves into its chemisorbed state, and after an initial increase of energy as the bonds stretch there is a sharp decrease as the adsorbate–substrate bonds reach their full strength. Even if the molecule does not fragment, there is likely to be an initial increase of potential energy as the molecule approaches the surface and the bonds adjust.

In most cases, therefore, we can expect there to be a potential energy barrier separating the precursor and chemisorbed states. This barrier, though, might be low, and might not rise above the energy of a distant, stationary particle (as in Fig. 22B.7a). In this case, chemisorption is not an activated process and can be expected to be rapid. Many gas adsorptions on clean metals appear to be non-activated. In some cases, however, the

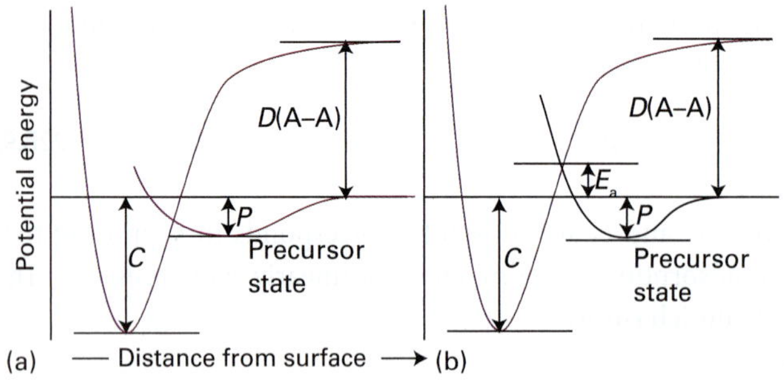

Figure 22B.7 The potential energy profiles for the dissociative chemisorption of an A_2 molecule. In each case, P is the enthalpy of (non-dissociative) physisorption and C that for chemisorption (at $T=0$). The relative locations of the curves determine whether the chemisorption is (a) not activated or (b) activated.

barrier rises above the zero axis (as in Fig. 22B.7b); such chemisorptions are activated and slower than the non-activated kind. An example is H_2 on copper, which has an activation energy in the region of 20–40 kJ mol^{-1}.

One point that emerges from this discussion is that rates are not good criteria for distinguishing between physisorption and chemisorption. Chemisorption can be fast if the activation energy is small or zero, but it may be slow if the activation energy is large. Physisorption is usually fast, but it can appear to be slow if adsorption is taking place on a porous medium.

Brief illustration 22B.1 The rate of activated adsorption

Consider two adsorption experiments for hydrogen on different faces of a copper crystal. In one, Face 1, the activation energy is 28 kJ mol^{-1} and on the other, Face 2, the activation energy is 33 kJ mol^{-1}. The ratio of the rates of chemisorption on equal areas of the two faces at 250 K is

$$\frac{\text{Rate}(1)}{\text{Rate}(2)} = \frac{Ae^{-E_{a,ads}(1)/RT}}{Ae^{-E_{a,ads}(2)/RT}} = e^{-\{E_{a,ads}(1)-E_{a,ads}(2)\}/RT}$$

$$= e^{5\times10^3\,\text{J mol}^{-1}/(8.3145\,\text{J K}^{-1}\,\text{mol}^{-1})\times(250\,\text{K})} = 11$$

We have assumed that the A factor is the same for each face.

Self-test 22B.4 What are the relative rates when the temperature is increased to 300 K?

Answer: 7

(b) Adsorption and desorption at the molecular level

The rate at which a surface is covered by adsorbate depends on the ability of the substrate to dissipate the energy of the incoming particle as thermal motion as it crashes on to the surface. If the energy is not dissipated quickly, the particle migrates over the surface until a vibration expels it into the overlying gas or it reaches an edge. The proportion of collisions with the surface that successfully lead to adsorption is called the **sticking probability**, s:

$$s = \frac{\text{rate of adsorption of particles by the surface}}{\text{rate of collision of particles with the surface}}$$

Definition Sticking probability (22B.11)

The denominator can be calculated from the kinetic model (from Z_W, Topic 22A), and the numerator can be measured by observing the rate of change of pressure.

Values of s vary widely. For example, at room temperature CO has s in the range 0.1–1.0 for several d-metal surfaces, but for N_2 on rhenium $s<10^{-2}$, indicating that more than a hundred collisions are needed before one molecule sticks successfully. Beam studies on specific crystal planes show a pronounced specificity: for N_2 on tungsten, s ranges from 0.74 on the (320) faces down to less than 0.01 on the (110) faces at room temperature. The sticking probability decreases as the surface coverage increases (Fig. 22B.8). A simple assumption is that s is proportional to $1-\theta$, the fraction uncovered, and it is common to write

$$s=(1-\theta)s_0$$

Commonly used form of the sticking probability (22B.12)

where s_0 is the sticking probability on a perfectly clean surface. The results in the illustration do not fit this expression because they show that s remains close to s_0 until the coverage has risen to about 6×10^{13} molecules cm^{-2}, and then falls steeply. The explanation is probably that the colliding molecule does not enter the chemisorbed state at once, but moves over the surface until it encounters an empty site.

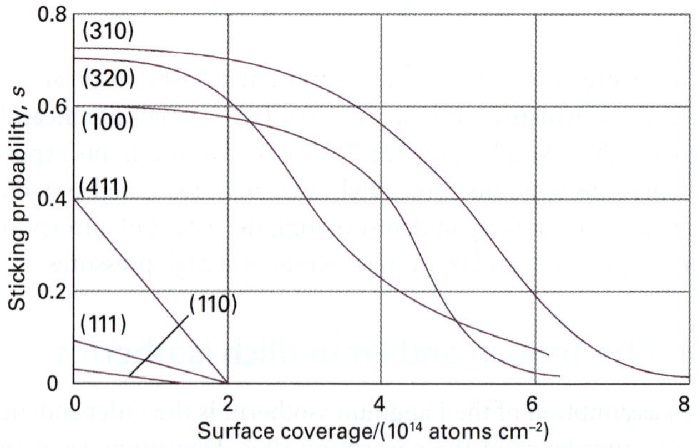

Figure 22B.8 The sticking probability of N_2 on various faces of a tungsten crystal and its dependence on surface coverage. Note the very low sticking probability for the (110) and (111) faces. (Data provided by Professor D.A. King.)

Desorption is always activated because the particles have to be lifted from the foot of a potential well. A physisorbed particle vibrates in its shallow potential well, and might shake itself off the surface after a short time. If the temperature dependence of the first-order rate of departure follows Arrhenius behaviour, then $k_d = Ae^{-E_{a,des}/RT}$, with $E_{a,des}$ the activation energy for desorption. Therefore, the half-life for remaining on the surface has a temperature dependence

$$t_{1/2} = \frac{\ln 2}{k_d} = \tau_0 e^{E_{a,des}/RT} \qquad \tau_0 = \frac{\ln 2}{A} \qquad \text{Residence half-life} \qquad (22B.13)$$

Note the positive sign in the exponent: the greater the activation energy for desorption, the larger the residence half-life.

Brief illustration 22B.2 Residence half-lives

If we suppose that $1/\tau_0$ is approximately the same as the vibrational frequency of the weak particle–surface bond (about 10^{12} Hz) and $E_d \approx 25$ kJ mol^{-1}, then residence half-lives of around 10 ns are predicted at room temperature. Lifetimes close to 1 s are obtained only by lowering the temperature to about 100 K. For chemisorption, with $E_d = 100$ kJ mol^{-1} and guessing that $\tau_0 = 10^{-14}$ s (because the adsorbate–substrate bond is quite stiff), we expect a residence half-life of about 3×10^3 s (about an hour) at room temperature, decreasing to 1 s at about 350 K.

Self-test 22B.5 For how long on average would an atom remain on a surface at 800 K if its desorption activation energy is 200 kJ mol^{-1}? Take $\tau_0 = 0.10$ ps.

Answer: $t_{1/2} = 1.3$ s

The desorption activation energy can be measured in several ways. However, we must be guarded in its interpretation because it often depends on the fractional coverage, and so might change as desorption proceeds. Moreover, the transfer of concepts such as 'reaction order' and 'rate constant' from bulk studies to surfaces is hazardous, and there are few examples of strictly first-order or second-order desorption kinetics (just as there are few integral-order reactions in the gas phase too).

If we disregard these complications, one way of measuring the desorption activation energy is to monitor the rate of increase in pressure when the sample is maintained at a series of temperatures, and to attempt to make an Arrhenius plot. A more sophisticated technique is **temperature programmed desorption** (TPD) or **thermal desorption spectroscopy** (TDS). The basic observation is a surge in desorption rate (as monitored by a mass spectrometer) when the temperature is raised linearly to the temperature at which desorption occurs rapidly, but once the desorption has occurred there is no more adsorbate to escape from the surface, so the desorption flux falls again

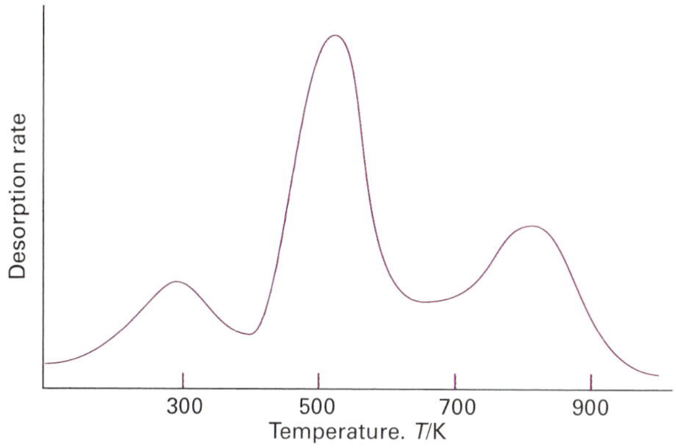

Figure 22B.9 The TPD spectrum of H_2 on the (100) face of tungsten. The three peaks indicate the presence of three sites with different adsorption enthalpies and therefore different desorption activation energies (P.W. Tamm and L.D. Schmidt, *J. Chem. Phys.* **51**, 5352 (1969)).

as the temperature continues to rise. The TPD spectrum, the plot of desorption flux against temperature, therefore shows a peak, the location of which depends on the desorption activation energy. There are three maxima in the example shown in Fig. 22B.9, indicating the presence of three sites with different activation energies.

In many cases only a single activation energy (and a single peak in the TPD spectrum) is observed. When several peaks are observed they might correspond to adsorption on different crystal planes or to multilayer adsorption. For instance, Cd atoms on tungsten show two activation energies, one of 18 kJ mol^{-1} and the other of 90 kJ mol^{-1}. The explanation is that the more tightly bound Cd atoms are attached directly to the substrate, and the less strongly bound are in a layer (or layers) above the primary overlayer. Another example of a system showing two desorption activation energies is CO on tungsten, the values being 120 kJ mol^{-1} and 300 kJ mol^{-1}. The explanation is believed to be the existence of two types of metal–adsorbate binding site, one involving a simple M-CO bond, the other adsorption with dissociation into individually adsorbed C and O atoms.

(c) Mobility on surfaces

A further aspect of the strength of the interactions between adsorbate and substrate is the mobility of the adsorbate. Mobility is often a vital feature of a catalyst's activity, because a catalyst might be impotent if the reactant molecules adsorb so strongly that they cannot migrate.

The activation energy for diffusion over a surface need not be the same as for desorption because the particles may be able to move through valleys between potential peaks without leaving the surface completely. In general, the activation energy for migration is about 10–20 per cent of the energy of the surface–adsorbate bond, but the actual value depends on the extent of coverage. The defect structure of the sample (which depends on the temperature) may also play a dominant role because the adsorbed molecules might find it easier to skip across a terrace than to roll along the foot of a step, and these molecules might become trapped in vacancies in an otherwise flat terrace. Diffusion may also be easier across one crystal face than another, and so the surface mobility depends on which lattice planes are exposed.

Checklist of concepts

☐ 1. An **adsorption isotherm** is the variation of the surface coverage θ with pressure at a chosen temperature.

☐ 2. **Flash desorption** is a technique in which the sample is suddenly heated (electrically) and the resulting rise of pressure is interpreted in terms of the amount of adsorbate originally on the substrate.

☐ 3. Examples of adsorption isotherms include the **Langmuir, BET, Temkin,** and **Freundlich** isotherms.

☐ 4. The **sticking probability** is the proportion of collisions with the surface that successfully lead to adsorption.

☐ 5. Desorption is an activated process; the desorption activation energy is measured by **temperature-programmed desorption** or **thermal desorption spectroscopy**.

☐ 6. The mobility of adsorbates on a surface is dominated by diffusion.

Checklist of equations

Property	Equation	Comment	Equation number
Langmuir isotherm:		Independent and equivalent sites, monolayer coverage	
(a) without dissociation	$\theta=\alpha p/(1+\alpha p)$		22B.2
(b) with dissociation	$\theta=(\alpha p)^{1/2}/\{1+(\alpha p)^{1/2}\}$		22B.4
Isosteric enthalpy of adsorption	$(\partial \ln(\alpha p^{\ominus})/\partial T)_\theta=\Delta_{ad}H^{\ominus}/RT^2$		22B.5
BET isotherm	$V/V_{mon}=cz/(1-z)\{1-(1-c)z\}$, $z=p/p^*$, $c=e^{(\Delta_{des}H^{\ominus}-\Delta_{vap}H^{\ominus})/RT}$	Multilayer adsorption	22B.6–7
Temkin isotherm	$\theta=c_1\ln(c_2p)$	Enthalpy of adsorption varies with θ	22B.9
Freundlich isotherm	$\theta=c_1p^{1/c_2}$	Substrate–substrate interactions	22B.10
Sticking probability	$s=(1-\theta)s_0$	Approximate form	22B.12

22C Heterogeneous catalysis

Contents

➤ Why do we need to know this material?

Catalysis is at the heart of the chemical industry, and an understanding of the concepts is essential for developing new catalysts.

➤ What is the key idea?

In heterogeneous catalysis, the pathway for lowering the activation energy of a reaction commonly involves chemisorption of one or more reactants.

➤ What do we need to know already?

Catalysis is introduced in Topic 20H. This Topic builds on the discussion of reaction mechanisms (Topic 20E), the Arrhenius equation (Topic 20D), and adsorption isotherms (Topic 22B).

A **heterogeneous catalyst** is a catalyst in a different phase from the reaction mixture. For example, the hydrogenation of ethene to ethane, a gas-phase reaction, is accelerated in the presence of a solid catalyst such as palladium, platinum, or nickel. The metal provides a surface to which the reactants bind; this binding facilitates encounters between reactants and increases the rate of the reaction. This Topic is an exploration of catalytic activity on surfaces, building on the concepts developed in Topic 22B.

22C.1 Mechanisms of heterogeneous catalysis

Many catalysts depend on **co-adsorption**, the adsorption of two or more species. One consequence of the presence of a second species may be the modification of the electronic structure at the surface of a metal. For instance, partial coverage of d-metal surfaces by alkali metals has a pronounced effect on the electron distribution at the surface and reduces the work function of the metal (the energy needed to remove an electron; see Topic 7A). Such modifiers can act as promoters (to enhance the action of catalysts) or as poisons (to inhibit catalytic action).

Figure 22C.1 shows the potential energy curve for a reaction influenced by the action of a heterogeneous catalyst. Differences between Fig. 22C.1 and 20H.1 arise from the fact that heterogeneous catalysis normally depends on at least one reactant being adsorbed (usually chemisorbed) and modified to an **active phase** in which it readily undergoes reaction, and desorption of products. Modification of the reactant often takes the form of a fragmentation of the reactant molecules. In practice, the active phase is dispersed as very small particles of linear dimension less than 2 nm on a porous oxide support. **Shape-selective catalysts**, such as the zeolites, which have a pore size that can distinguish shapes and sizes at a molecular scale, have high internal specific surface areas, in the range of 100–500 $m^2\,g^{-1}$.

Mechanisms of reactions catalysed by surfaces can be treated quantitatively by using the techniques of Topic 20E (on the

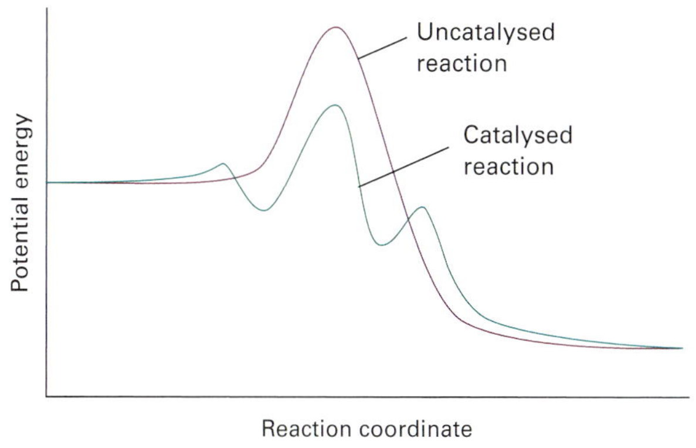

Figure 22C.1 The reaction profile for catalysed and uncatalysed reactions. The catalysed reaction path includes activation energies for adsorption and desorption as well as an overall lower activation energy for the process.

development of rate laws based on proposed reaction mechanisms) and the adsorption isotherms developed in Topic 22B. Here we explore some simple mechanisms that can give significant insight into surface-catalysed reactions.

(a) Unimolecular reactions

The rate law of a surface-catalysed unimolecular reaction, such as the decomposition of a substance on a surface, can be written in terms of an adsorption isotherm if the rate is supposed to be proportional to the extent of surface coverage. For example, if the fractional coverage θ is given by the Langmuir isotherm (eqn 22B.2, $\theta = \alpha p/(1+\alpha p)$), we would write

$$v = k_r\theta = \frac{k_r \alpha p}{1+\alpha p} \qquad (22C.1)$$

where p is the pressure of the adsorbing substance.

Brief illustration 22C.1 Surface-catalysed unimolecular decomposition

Consider the decomposition of phosphine (PH_3) on tungsten, which is first order at low pressures. We can use eqn 22C.1 to account for this observation. When the pressure is so low that $\alpha p \ll 1$, we can neglect αp in the denominator of eqn 22C.1 and obtain $v = k_r \alpha p$. The decomposition is predicted to be first order, as observed experimentally.

Self-test 22C.1 Write a rate law for the decomposition of PH_3 on tungsten at high pressures

Answer: $v = k_r$; the reaction is zeroth order at high pressures

(b) The Langmuir–Hinshelwood mechanism

In the **Langmuir–Hinshelwood mechanism** (LH mechanism) of surface-catalysed reactions, the reaction takes place by encounters between molecular fragments and atoms adsorbed on the surface. We therefore expect the rate law to be second order in the extent of surface coverage:

$$A + B \rightarrow P \quad v = k_r\theta_A\theta_B \qquad \text{Langmuir–Hinshelwood rate law} \qquad (22C.2)$$

Insertion of the appropriate isotherms for A and B then gives the reaction rate in terms of the partial pressures of the reactants.

Example 22C.1 Writing a rate law based on the Langmuir–Hinshelwood mechanism

Consider a reaction $A + B \rightarrow P$ in which A and B follow Langmuir isotherms and adsorb without dissociation. Devise a rate law that is consistent with the Langmuir–Hinshelwood mechanism.

Method Begin by following the procedures outlined in Topic 22B for the derivation of the Langmuir isotherm to write expressions for θ_A and θ_B, the fractional coverages of A and B, respectively. However, note that, unlike the simple situation in Topic 22B, two species compete for the same sites on the surface. Then, use eqn 22C.2 to express the rate law.

Answer Because two species compete for sites on the surface, the number of vacant sites is equal to $N(1-\theta_A-\theta_B)$, where N is the total number of sites. It follows from eqns 22B.1a and 22B.1b that the rates of adsorption and desorption are given by

Rate of adsorption of A $= k_{a,A}p_A N(1-\theta_A-\theta_B)$	Rate of desorption of A $= k_{d,A}N\theta_A$
Rate of adsorption of B $= k_{a,B}p_B N(1-\theta_A-\theta_B)$	Rate of desorption of B $= k_{d,B}N\theta_B$

At equilibrium, the rates of adsorption and desorption for each species are equal, and, with $\alpha_A = k_{a,A}/k_{d,A}$ and $\alpha_B = k_{a,B}/k_{d,B}$, it follows that

$$\alpha_A p_A(1-\theta_A-\theta_B) = \theta_A$$
$$\alpha_B p_B(1-\theta_A-\theta_B) = \theta_B$$

The solutions of this pair of simultaneous equations (see *Self-test* 22C.2) are

$$\theta_A = \frac{\alpha_A p_A}{1+\alpha_A p_A+\alpha_B p_B} \qquad \theta_B = \frac{\alpha_B p_B}{1+\alpha_A p_A+\alpha_B p_B}$$

It follows from eqn 22C.2 that the rate law is

$$v = \frac{k_r\alpha_A\alpha_B p_A p_B}{(1+\alpha_A p_A+\alpha_B p_B)^2}$$

The parameters α in the isotherms and the rate constant k_r are all temperature-dependent, so the overall temperature dependence of the rate may be strongly non-Arrhenius (in the sense that the reaction rate is unlikely to be proportional to $e^{-E_a/RT}$). The LH mechanism is dominant for the catalytic oxidation of CO to CO_2.

Self-test 22C.2 Provide the missing steps in the derivation of the expression for v.

(c) The Eley–Rideal mechanism

In the **Eley–Rideal mechanism** (ER mechanism) of a surface-catalysed reaction, a gas-phase molecule collides with another molecule already adsorbed on the surface. The rate of formation of product is expected to be proportional to the partial pressure, p_B, of the non-adsorbed gas B and the extent of surface coverage, θ_A, of the adsorbed gas A. It follows that the rate law should be

$$A + B \rightarrow P \quad v = k_r p_B\theta_A \qquad \text{Eley–Rideal rate law} \qquad (22C.3)$$

The rate constant, k_r, might be much larger than for the uncatalysed gas-phase reaction because the reaction on the surface has a low activation energy and the adsorption itself is often not activated.

If we know the adsorption isotherm for A, we can express the rate law in terms of its partial pressure, p_A. For example, the adsorption of A follows a Langmuir isotherm in the pressure range of interest, then the rate law would be

$$v = \frac{k_r \alpha p_A p_B}{1 + \alpha p_A} \qquad (22C.4)$$

Brief illustration 22C.2 The Eley–Rideal mechanism

According to eqn 22C.4, when the partial pressure of A is high (in the sense $\alpha p_A \gg 1$) there is almost complete surface coverage, and the rate is equal to $k_r p_B$. Now the rate-determining step is the collision of B with the adsorbed fragments. When the pressure of A is low ($\alpha p_A \ll 1$), perhaps because of its reaction, the rate is equal to $k_r \alpha p_A p_B$; now the extent of surface coverage is important in the determination of the rate.

Self-test 22C.3 Rewrite eqn 22C.4 for cases where A is a diatomic molecule that adsorbs as atoms.

Answer: $v = k_r p_B (\alpha p_A)^{1/2} / (1 + (\alpha p_A)^{1/2})$

Almost all thermal surface-catalysed reactions are thought to take place by the LH mechanism but a number of reactions with an ER mechanism have also been identified from molecular beam investigations. For example, the reaction between H(g) and D(ad) to form HD(g) is thought to be by an ER mechanism involving the direct collision and pick-up of the adsorbed D atom by the incident H atom. However, the two mechanisms should really be thought of as ideal limits with all reactions lying somewhere between the two and showing features of each one.

22C.2 Catalytic activity at surfaces

It has become possible to investigate how the catalytic activity of a surface depends on its structure as well as its composition. For instance, the cleavage of C–H and H–H bonds appears to depend on the presence of steps and kinks, and a terrace often has only minimal catalytic activity.

The reaction $H_2 + D_2 \rightarrow 2HD$ has been studied in detail. For this reaction, terrace sites are inactive but one molecule in ten reacts when it strikes a step. Although the step itself might be the important feature, it may be that the presence of the step merely exposes a more reactive crystal face (the step face itself). Likewise, the dehydrogenation of hexane to hexene depends strongly on the kink density, and it appears that kinks are needed to cleave C–C bonds. These observations suggest a reason why even small amounts of impurities may poison a catalyst: they are likely to attach to step and kink sites, and so impair the activity of the catalyst entirely. A constructive outcome is that the extent of dehydrogenation may be controlled relative to other types of reactions by seeking impurities that adsorb at kinks and act as specific poisons.

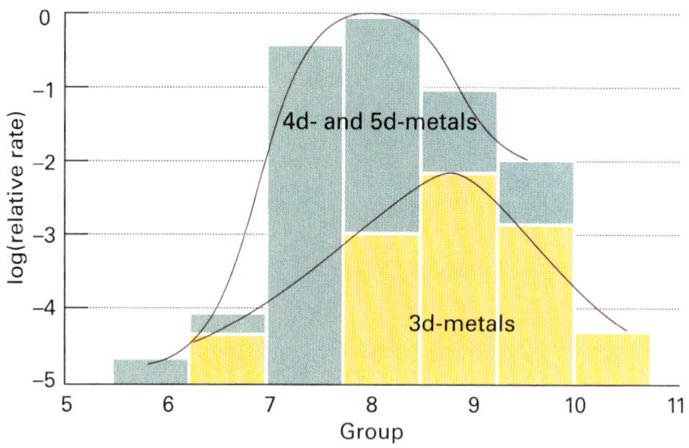

Figure 22C.2 A volcano curve of catalytic activity arises because although the reactants must adsorb reasonably strongly, they must not adsorb so strongly that they are immobilized. The lower curve refers to the first series of d-block metals, the upper curve to the second and third series of d-block metals. The group numbers relate to the periodic table inside the back cover.

The activity of a catalyst depends on the strength of chemisorption as indicated by the 'volcano' curve in Fig. 22C.2 (which is so-called on account of its general shape). To be active, the catalyst should be extensively covered by adsorbate, which is the case if chemisorption is strong. On the other hand, if the strength of the substrate–adsorbate bond becomes too great, the activity declines either because the other reactant molecules cannot react with the adsorbate or because the adsorbate molecules are immobilized on the surface. This pattern of behaviour suggests that the activity of a catalyst should initially increase with strength of adsorption (as measured, for instance, by the enthalpy of adsorption) and then decline, and that the most active catalysts should be those lying near the summit of the volcano. Most active metals are those that lie close to the middle of the d block. Many metals are suitable for adsorbing gases, and some trends are summarized in Table 22C.1.

Brief illustration 22C.3 Trends in chemisorption abilities

We see from Table 22C.1 that for a number of metals the general order of adsorption strengths decreases along the series O_2, C_2H_2, C_2H_4, CO, H_2, CO_2, N_2. Some of these molecules adsorb dissociatively (for example, H_2). Elements from the d block, such as iron, titanium, and chromium, show a strong activity towards all these gases, but manganese and copper are unable to adsorb N_2 and CO_2. Metals towards the left of the periodic table (for example, magnesium) can adsorb (and, in fact, react with) only the most active gas (O_2).

Self-test 22C.4 Why is iron a good catalyst for the formation of ammonia from N_2(g) and H_2(g)?

Answer: See Fig. 22C.2 and Table 22C.1

Table 22C.1 Chemisorption abilities*

	O_2	C_2H_2	C_2H_4	CO	H_2	CO_2	N_2
Ti, Cr, Mo, Fe	+	+	+	+	+	+	+
Ni, Co	+	+	+	+	+	+	–
Pd, Pt	+	+	+	+	+	–	–
Mn, Cu	+	+	+	+	±	–	–
Al, Au	+	+	+	–	–	–	–
Li, Na, K	+	+	–	–	–	–	–
Mg, Ag, Zn, Pb	+	–	–	–	–	–	–

* +, Strong chemisorption; ±, chemisorption; –, no chemisorption.

Checklist of concepts

- ☐ 1. A **catalyst** is a substance that accelerates a reaction but undergoes no net chemical change.
- ☐ 2. A **heterogeneous catalyst** is a catalyst in a different phase from the reaction mixture.
- ☐ 3. In the **Langmuir–Hinshelwood mechanism** of surface-catalysed reactions, the reaction takes place by encounters between molecular fragments and atoms adsorbed on the surface.
- ☐ 4. In the **Eley–Rideal mechanism** of a surface-catalysed reaction, a gas-phase molecule collides with another molecule already adsorbed on the surface.
- ☐ 5. The activity of a catalyst depends on the strength of chemisorption.

Checklist of equations

Property	Equation	Comment	Equation number
Langmuir–Hinshelwood mechanism	$v=k_r\theta_A\theta_B$	Competitive adsorption	22C.2
Eley–Rideal mechanism	$v=k_r p_B\theta_A$	Adsorption of A	22C.3

CHAPTER 22 Processes on solid surfaces

TOPIC 22A An introduction to solid surfaces

Discussion questions

22A.1 (a) What topographical features are found on clean surfaces? (b) Describe how steps and terraces can be formed by dislocations.

22A.2 Drawing from knowledge you have acquired through the text, describe the advantages and limitations of each of the microscopy, diffraction, and ionizations techniques designated by the acronyms AFM, LEED, SAM, SEM, STM, and TEM.

Exercises

22A.1(a) Calculate the frequency of molecular collisions per square centimetre of surface in a vessel containing (i) hydrogen, (ii) propane at 25 °C when the pressure is 0.10 μTorr.

22A.1(b) Calculate the frequency of molecular collisions per square centimetre of surface in a vessel containing (i) nitrogen, (ii) methane at 25 °C when the pressure is 10.0 Pa. Repeat the calculations for a pressure of 0.150 μTorr.

22A.2(a) What pressure of argon gas is required to produce a collision rate of $4.5\times10^{20}\,s^{-1}$ at 425 K on a circular patch of surface of diameter 1.5 mm?

22A.2(b) What pressure of nitrogen gas is required to produce a collision rate of $5.00\times10^{19}\,s^{-1}$ at 525 K on a circular patch of surface of diameter 2.0 mm?

Problems

22A.1 The movement of atoms and ions on a surface depends on their ability to leave one position and stick to another, and therefore on the energy changes that occur. As an illustration, consider a two-dimensional square lattice of singly charged positive and negative ions separated by 200 pm, and consider a cation on the upper terrace of this array. Calculate, by direct summation, its Coulombic interaction when it is in an empty lattice point directly above an anion. Now consider a high step in the same lattice, and let the cation move into the corner formed by the step and the terrace. Calculate the Coulombic energy for this position, and decide on the likely settling point for the cation.

22A.2 In a study of the catalytic properties of a titanium surface it was necessary to maintain the surface free from contamination. Calculate the collision frequency per square centimetre of surface made by O_2 molecules at (a) 100 kPa, (b) 1.00 Pa and 300 K. Estimate the number of collisions made with a single surface atom in each second. The conclusions underline the importance of working at very low pressures (much lower than 1 Pa, in fact) in order to study the properties of uncontaminated surfaces. Take the nearest neighbour distance as 291 pm.

22A.3 Nickel is face-centred cubic with a unit cell of side 352 pm. What is the number of atoms per square centimetre exposed on a surface formed by (a) (100), (b) (110), (c) (111) planes? Calculate the frequency of molecular collisions with a single atom in a vessel containing (i) hydrogen, (ii) propane at 25 °C when the pressure is 100 Pa and 0.10 μTorr.

22A.4 The LEED pattern from a clean unreconstructed (110) face of a metal is shown below. Sketch the LEED pattern for a surface that was reconstructed by tripling the horizontal separation between the atoms.

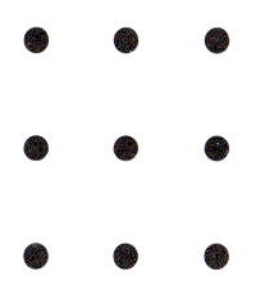

TOPIC 22B Adsorption and desorption

Discussion questions

22B.1 Distinguish between the following adsorption isotherms: Langmuir, BET, Temkin, and Freundlich. Indicate when and why each is likely to be appropriate.

22B.2 What approximations underlie the formulation of the Langmuir and BET isotherms?

Exercises

22B.1(a) The volume of oxygen gas at 0 °C and 104 kPa adsorbed on the surface of 1.00 g of a sample of silica at 0 °C was 0.286 cm^3 at 145.4 Torr and 1.443 cm^3 at 760 Torr. What is the value of V_{mon}?

22B.1(b) The volume of gas at 20 °C and 1.00 bar adsorbed on the surface of 1.50 g of a sample of silica at 0 °C was 1.52 cm^3 at 56.4 kPa and 2.77 cm^3 at 108 kPa. What is the value of V_{mon}?

22B.2(a) The enthalpy of adsorption of CO on a surface is found to be −120 kJ mol^{-1}. Estimate the mean lifetime of a CO molecule on the surface at 400 K.

22B.2(b) The enthalpy of adsorption of ammonia on a nickel surface is found to be −155 kJ mol^{-1}. Estimate the mean lifetime of an NH_3 molecule on the surface at 500 K.

22B.3(a) A certain solid sample adsorbs 0.44 mg of CO when the pressure of the gas is 26.0 kPa and the temperature is 300 K. The mass of gas adsorbed when the pressure is 3.0 kPa and the temperature is 300 K is 0.19 mg. The Langmuir isotherm is known to describe the adsorption. Find the fractional coverage of the surface at the two pressures.

22B.3(b) A certain solid sample adsorbs 0.63 mg of CO when the pressure of the gas is 36.0 kPa and the temperature is 300 K. The mass of gas adsorbed when the pressure is 4.0 kPa and the temperature is 300 K is 0.21 mg. The Langmuir isotherm is known to describe the adsorption. Find the fractional coverage of the surface at the two pressures.

22B.4(a) The adsorption of a gas is described by the Langmuir isotherm with $\alpha = 0.75\,\mathrm{kPa^{-1}}$ at 25 °C. Calculate the pressure at which the fractional surface coverage is (i) 0.15, (ii) 0.95.

22B.4(b) The adsorption of a gas is described by the Langmuir isotherm with $\alpha = 0.548\,\mathrm{kPa^{-1}}$ at 25 °C. Calculate the pressure at which the fractional surface coverage is (i) 0.20, (ii) 0.75.

22B.5(a) A solid in contact with a gas at 12 kPa and 25 °C adsorbs 2.5 mg of the gas and obeys the Langmuir isotherm. The enthalpy change when 1.00 mmol of the adsorbed gas is desorbed is +10.2 J. What is the equilibrium pressure for the adsorption of 2.5 mg of gas at 40 °C?

22B.5(b) A solid in contact with a gas at 8.86 kPa and 25 °C adsorbs 4.67 mg of the gas and obeys the Langmuir isotherm. The enthalpy change when 1.00 mmol of the adsorbed gas is desorbed is +12.2 J. What is the equilibrium pressure for the adsorption of the same mass of gas at 45 °C?

22B.6(a) Nitrogen gas adsorbed on charcoal to the extent of $0.921\,\mathrm{cm^3\,g^{-1}}$ at 490 kPa and 190 K, but at 250 K the same amount of adsorption was achieved only when the pressure was increased to 3.2 MPa. What is the enthalpy of adsorption of nitrogen on charcoal?

22B.6(b) Nitrogen gas adsorbed on a surface to the extent of $1.242\,\mathrm{cm^3\,g^{-1}}$ at 350 kPa and 180 K, but at 240 K the same amount of adsorption was achieved only when the pressure was increased to 1.02 MPa. What is the enthalpy of adsorption of nitrogen on the surface?

22B.7(a) In an experiment on the adsorption of oxygen on tungsten it was found that the same volume of oxygen was desorbed in 27 min at 1856 K and 2.0 min at 1978 K. What is the activation energy of desorption? How long would it take for the same amount to desorb at (i) 298 K, (ii) 3000 K?

22B.7(b) In an experiment on the adsorption of ethene on iron it was found that the same volume of the gas was desorbed in 1856 s at 873 K and 8.44 s at 1012 K. What is the activation energy of desorption? How long would it take for the same amount of ethene to desorb at (i) 298 K, (ii) 1500 K?

22B.8(a) The average time for which an oxygen atom remains adsorbed to a tungsten surface is 0.36 s at 2548 K and 3.49 s at 2362 K. What is the activation energy for chemisorption?

22B.8(b) The average time for which a hydrogen atom remains adsorbed on a manganese surface is 35 per cent shorter at 1000 K than at 600 K. What is the activation energy for chemisorption?

22B.9(a) For how long on average would an H atom remain on a surface at 400 K if its desorption activation energy is (i) $15\,\mathrm{kJ\,mol^{-1}}$, (ii) $150\,\mathrm{kJ\,mol^{-1}}$? Take $\tau_0 = 0.10$ ps. For how long on average would the same atoms remain at 1000 K?

22B.9(b) For how long on average would an atom remain on a surface at 298 K if its desorption activation energy is (i) $20\,\mathrm{kJ\,mol^{-1}}$, (ii) $200\,\mathrm{kJ\,mol^{-1}}$? Take $\tau_0 = 0.12$ ps. For how long on average would the same atoms remain at 800 K?

22B.10(a) Hydrogen iodide is very strongly adsorbed on gold but only slightly adsorbed on platinum. Assume the adsorption follows the Langmuir isotherm and predict the order of the HI decomposition reaction on each of the two metal surfaces.

22B.10(b) Suppose it is known that ozone adsorbs on a particular surface in accord with a Langmuir isotherm. How could you use the pressure dependence of the fractional coverage to distinguish between adsorption (i) without dissociation, (ii) with dissociation into $O + O_2$, (iii) with dissociation into $O + O + O$?

Problems

22B.1 Use mathematical software, a spreadsheet, or the *Living graphs* on the web site of this book to perform the following calculations: (a) Use eqn 22B.2 to generate a family of curves showing the dependence of $1/\theta$ on $1/p$ for several values of α. (b) Use eqn 22B.4 to generate a family of curves showing the dependence of $1/\theta$ on $1/p$ for several values of α. On the basis of your results from parts (a) and (b), discuss how plots of $1/\theta$ against $1/p$ can be used to distinguish between adsorption with and without dissociation. (c) Use eqn 22B.6 to generate a family of curves showing the dependence of $zV_{\mathrm{mon}}/(1-z)V$ on z for different values of c.

22B.2 The following data are for the chemisorption of hydrogen on copper powder at 25 °C. Confirm that they fit the Langmuir isotherm at low coverages. Then find the value of α for the adsorption equilibrium and the adsorption volume corresponding to complete coverage.

p/Pa	25	129	253	540	1000	1593
V/cm^3	0.042	0.163	0.221	0.321	0.411	0.471

22B.3 The data for the adsorption of ammonia on barium fluoride are reported in the following tables. Confirm that they fit a BET isotherm and find values of c and V_{mon}.

(a) $\theta = 0\,°\mathrm{C}$, $p^* = 429.6$ kPa:

p/kPa	14.0	37.6	65.6	79.2	82.7	100.7	106.4
V/cm^3	11.1	13.5	14.9	16.0	15.5	17.3	16.5

(b) $\theta = 18.6\,°\mathrm{C}$, $p^* = 819.7$ kPa:

p/kPa	5.3	8.4	14.4	29.2	62.1	74.0	80.1	102.0
V/cm^3	9.2	9.8	10.3	11.3	12.9	13.1	13.4	14.1

22B.4 The following data have been obtained for the adsorption of H_2 on the surface of 1.00 g of copper at 0 °C. The volume of H_2 below is the volume that the gas would occupy at STP (0 °C and 1 atm).

p/atm	0.050	0.100	0.150	0.200	0.250
V/cm^3	23.8	13.3	8.70	6.80	5.71

Determine the volume of H_2 necessary to form a monolayer and estimate the surface area of the copper sample. The density of liquid hydrogen is $0.708\,\mathrm{g\,cm^{-3}}$.

22B.5‡ M.-G. Olivier and R. Jadot (*J. Chem. Eng. Data* **42**, 230 (1997)) studied the adsorption of butane on silica gel. They report the following amounts of absorption (in moles per kilogram of silica gel) at 303 K:

p/kPa	31.00	38.22	53.03	76.38	101.97
n/(mol kg^{-1})	1.00	1.17	1.54	2.04	2.49

p/kPa	130.47	165.06	182.41	205.75	219.91
n/(mol kg^{-1})	2.90	3.22	3.30	3.35	3.36

Fit these data to a Langmuir isotherm, and determine the value of n that corresponds to complete coverage and the constant K.

‡ These problems were supplied by Charles Trapp and Carmen Giunta.

22B.6 The designers of a new industrial plant wanted to use a catalyst code-named CR-1 in a step involving the fluorination of butadiene. As a first step in the investigation they determined the form of the adsorption isotherm. The volume of butadiene adsorbed per gram of CR-1 at 15 °C varied with pressure as given below. Is the Langmuir isotherm suitable at this pressure?

p/kPa	13.3	26.7	40.0	53.3	66.7	80.0
V/cm^3	17.9	33.0	47.0	60.8	75.3	91.3

Investigate whether the BET isotherm gives a better description of the adsorption of butadiene on CR-1. At 15 °C, p^*(butadiene) = 200 kPa. Find V_{mon} and c.

22B.7‡ C. Huang and W.P. Cheng (*J. Colloid Interface Sci.* **188**, 270 (1997)) examined the adsorption of the hexacyanoferrate(III) ion, $[Fe(CN)_6]^{3-}$, on γ-Al_2O_3 from aqueous solution. They modelled the adsorption with a modified Langmuir isotherm, obtaining the following values of α at pH = 6.5:

T/K	283	298	308	318
$10^{-11}\alpha$	2.642	2.078	1.286	1.085

Determine the isosteric enthalpy of adsorption, $\Delta_{ads}H^\ominus$, at this pH. The researchers also reported $\Delta_{ads}S^\ominus = +146\,J\,mol^{-1}\,K^{-1}$ under these conditions. Determine $\Delta_{ads}G^\ominus$.

22B.8‡ In a study relevant to automobile catalytic converters, C.E. Wartnaby et al. (*J. Phys. Chem.* **100**, 12483 (1996)) measured the enthalpy of adsorption of CO, NO, and O_2 on initially clean platinum (110) surfaces. They report $\Delta_{ads}H^\ominus$ for NO to be −160 kJ mol^{-1}. How much more strongly adsorbed is NO at 500 °C than at 400 °C?

22B.9‡ The removal or recovery of volatile organic compounds (VOCs) from exhaust gas streams is an important process in environmental engineering. Activated carbon has long been used as an adsorbent in this process, but the presence of moisture in the stream reduces its effectiveness. M.-S. Chou and J.-H. Chiou (*J. Envir. Engrg. ASCE* **123**, 437(1997)) have studied the effect of moisture content on the adsorption capacities of granular activated carbon (GAC) for normal hexane and cyclohexane in air streams. From their data for dry streams containing cyclohexane, shown in the following table, they conclude that GAC obeys a Langmuir type model in which $q_{VOC,RH=0} = abc_{VOC}/(1 + bc_{VOC})$, where $q = m_{VOC}/m_{GAC}$, RH denotes relative humidity, a the maximum adsorption capacity, b is an affinity parameter, and p is the abundance in parts per million (ppm). The following table gives values of $q_{VOC,RH=0}$ for cyclohexane:

c/ppm	33.6 °C	41.5 °C	57.4 °C	76.4 °C	99 °C
200	0.080	0.069	0.052	0.042	0.027
500	0.093	0.083	0.072	0.056	0.042
1000	0.101	0.088	0.076	0.063	0.045
2000	0.105	0.092	0.083	0.068	0.052
3000	0.112	0.102	0.087	0.072	0.058

(a) By linear regression of $1/q_{VOC,RH=0}$ against $1/c_{VOC}$, test the goodness of fit and determine values of a and b. (b) The parameters a and b can be related to $\Delta_{ads}H$, the enthalpy of adsorption, and Δ_bH, the difference in activation energy for adsorption and desorption of the VOC molecules, through Arrhenius type equations of the form $a = k_a e^{-\Delta_{ads}H/RT}$ and $b = k_b e^{-\Delta_b H/RT}$. Test the goodness of fit of the data to these equations and obtain values for k_a, k_b, $\Delta_{ads}H$, and Δ_bH. (c) What interpretation might you give to k_a and k_b?

22B.10 The adsorption of solutes on solids from liquids often follows a Freundlich isotherm. Check the applicability of this isotherm to the following data for the adsorption of acetic acid on charcoal at 25 °C and find the values of the parameters c_1 and c_2.

[acid]/(mol dm^{-3})	0.05	0.10	0.50	1.0	1.5
w_a/g	0.04	0.06	0.12	0.16	0.19

where w_a is the mass adsorbed per gram of charcoal.

22B.11‡ A. Akgerman and M. Zardkoohi (*J. Chem. Eng. Data* **41**, 185 (1996)) examined the adsorption of phenol from aqueous solution on to fly ash at 20 °C. They fitted their observations to a Freundlich isotherm of the form $c_{ads} = Kc_{sol}^{1/n}$, where c_{ads} is the concentration of adsorbed phenol and c_{sol} is the concentration of aqueous phenol. Among the data reported are the following:

c_{sol}/(mg g^{-1})	8.26	15.65	25.43	31.74	40.00
c_{ads}/(mg g^{-1})	4.41	9.2	35.2	52.0	67.2

Determine the constants K and n. What further information would be necessary in order to express the data in terms of fractional coverage, θ?

22B.12‡ The following data were obtained for the extent of adsorption, s, of acetone on charcoal from an aqueous solution of molar concentration, c, at 18 °C:

c/(mmol dm^{-3})	15.0	23.0	42.0	84.0	165	390	800
s/(mmol acetone/g charcoal)	0.60	0.75	1.05	1.50	2.15	3.50	5.10

Which isotherm fits this data best, Langmuir, Freundlich, or Temkin?

22B.13‡ M.-S. Chou and J.-H. Chiou (*J. Envir. Engrg. ASCE* **123**, 437(1997)) have studied the effect of moisture content on the adsorption capacities of granular activated carbon (GAC, Norit PK 1-3) for the volatile organic compounds (VOCs) normal hexane and cyclohexane in air streams. The following table shows the adsorption capacities ($q_{water} = m_{water}/m_{GAC}$) of GAC for pure water from moist air streams as a function of relative humidity (RH) in the absence of VOCs at 41.5 °C:

RH	0.00	0.26	0.49	0.57	0.80	1.00
q_{water}	0.00	0.026	0.072	0.091	0.161	0.229

The authors conclude that the data at this and other temperatures obey a Freundlich type isotherm, $q_{water} = k(\text{RH})^{1/n}$. (a) Test this hypothesis for their data at 41.5 °C and determine the constants k and n. (b) Why might VOCs obey the Langmuir model, but water the Freundlich model? (c) When both water vapour and cyclohexane were present in the stream the values given in the table below were determined for the ratio $r_{VOC} = q_{VOC}/q_{VOC,RH=0}$ at 41.5 °C:

The authors propose that these data fit the equation $r_{VOC} = 1 - q_{water}$. Test their proposal and determine values for k and n and compare to those obtained in part (b) for pure water. Suggest reasons for any differences.

RH	0.00	0.10	0.25	0.40	0.53	0.76	0.81
r_{VOC}	1.00	0.98	0.91	0.84	0.79	0.67	0.61

22B.14‡ The release of petroleum products by leaky underground storage tanks is a serious threat to clean ground water. BTEX compounds (benzene, toluene, ethylbenzene, and xylenes) are of primary concern due to their ability to cause health problems at low concentrations. D.S. Kershaw et al. (*J. Geotech. Geoenvir. Engrg.* **123**, 324 (1997)) have studied the ability of ground tyre rubber to sorb (adsorb and absorb) benzene and o-xylene. Though sorption involves more than surface interactions, sorption data is usually found to fit one of the adsorption isotherms. In this study, the authors have tested how well their data fit the linear ($q = Kc_{eq}$), Freundlich ($q = K_F c_{eq}^{1/n}$), and Langmuir ($q = K_L M c_{eq}/(1 + K_L c_{eq})$) type isotherms, where q is the mass of solvent sorbed per gram of ground rubber (in milligrams per gram), the Ks and M are empirical constants, and c_{eq} the equilibrium concentration of contaminant in solution (in milligrams per litre). (a) Determine the units of the empirical constants. (b) Determine which of the isotherms best fits the data in the following table for the sorption of benzene on ground rubber:

c_{eq}/(mg dm^{-3})	97.10	36.10	10.40	6.51	6.21	2.48
q/(mg g^{-1})	7.13	4.60	1.80	1.10	0.55	0.31

(c) Compare the sorption efficiency of ground rubber to that of granulated activated charcoal which for benzene has been shown to obey the Freundlich isotherm in the form $q = 1.0c_{eq}^{1.6}$ with coefficient of determination $R^2 = 0.94$.

TOPIC 22C Heterogeneous catalysis

Discussion questions

22C.1 Describe the essential features of the Langmuir–Hinshelwood and Eley–Rideal mechanisms for surface-catalysed reactions.

22C.2 Account for the dependence of catalytic activity of a surface on the strength of chemisorption, as shown in Fig. 22B.8.

Exercises

22C.1(a) A monolayer of N_2 molecules is adsorbed on the surface of 1.00 g of an Fe/Al_2O_3 catalyst at 77 K, the boiling point of liquid nitrogen. Upon warming, the nitrogen occupies 3.86 cm^3 at 0 °C and 760 Torr. What is the surface area of the catalyst?

22C.1(b) A monolayer of CO molecules is adsorbed on the surface of 1.00 g of an Fe/Al_2O_3 catalyst at 77 K, the boiling point of liquid nitrogen. Upon warming, the carbon monoxide occupies 3.75 cm^3 at 0 °C and 1.00 bar. What is the surface area of the catalyst?

Problem

22C.1 In some catalytic reactions the products adsorb more strongly than the reacting gas. This is the case, for instance, in the catalytic decomposition of ammonia on platinum at 1000 °C. As a first step in examining the kinetics of this type of process, show that the rate of ammonia decomposition should follow

$$\frac{dp_{NH_3}}{dt}=-k_c\frac{p_{NH_3}}{p_{H_2}}$$

in the limit of very strong adsorption of hydrogen. Start by showing that when a gas J adsorbs very strongly, and its pressure is p_J, that the fraction of uncovered sites is approximately $1/Kp_J$. Solve the rate equation for the catalytic decomposition of NH_3 on platinum and show that a plot of $F(t)=(1/t)\times\ln(p/p_0)$ against $G(t)=(p-p_0)/t$, where p is the pressure of ammonia, should give a straight line from which k_c can be determined. Check the rate law on the basis of the following data, and find k_c for the reaction.

t/s	0	30	60	100	160	200	250
p/kPa	13.3	11.7	11.2	10.7	10.3	9.9	9.6

Integrated activities

22.1 Although the attractive van der Waals interaction between individual molecules varies as R^{-6} the interaction of a molecule with a nearby solid (a homogeneous collection of molecules) varies as R^{-3}, where R is its vertical distance above the surface. Confirm this assertion. Calculate the interaction energy between an Ar atom and the surface of solid argon on the basis of a Lennard-Jones (6,12)-potential. Estimate the equilibrium distance of an atom above the surface.

22.2 Electron microscopes can obtain images with much higher resolution than optical microscopes because of the short wavelength obtainable from a beam of electrons. For electrons moving at speeds close to c, the speed of light, the expression for the de Broglie wavelength (eqn 7A.14, $\lambda=h/p$) needs to be corrected for relativistic effects:

$$\lambda=\frac{h}{\left\{2m_e e\Delta\phi\left(1+\dfrac{e\Delta\phi}{2m_e c^2}\right)\right\}^{1/2}}$$

where c is the speed of light in vacuum and $\Delta\phi$ is the potential difference through which the electrons are accelerated. (a) Use this expression to calculate the de Broglie wavelength of electrons accelerated through 50 kV. (b) Is the relativistic correction important?

22.3 The forces measured by AFM arise primarily from interactions between electrons of the stylus and on the surface. To get an idea of the magnitudes of these forces, calculate the force acting between two electrons separated by 2.0 nm. To calculate the force between the electrons, use $F=-dV/dr$ where V is their mutual Coulombic potential energy and r is their separation.

22.4 To appreciate the distance dependence of the tunnelling current in scanning tunnelling microscopy, suppose that the electron in the gap between sample and needle has an energy 2.0 eV smaller than the barrier height. By what factor would the current drop if the needle is moved from $L_1=0.50$ nm to $L_2=0.60$ nm from the surface?

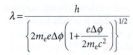

RESOURCE SECTION

Contents

PART 1 Common integrals

Algebraic functions

A.1 $\int x^n \mathrm{d}x = \dfrac{x^{n+1}}{n+1} + \text{constant},\ n \neq -1$

A.2 $\int \dfrac{1}{x} \mathrm{d}x = \ln x + \text{constant}$

Exponential functions

E.1 $\int_0^\infty x^n \mathrm{e}^{-ax} \mathrm{d}x = \dfrac{n!}{a^{n+1}},\quad n! = n(n-1)\ldots 1;\ 0! \equiv 1$

E.2 $\int_0^\infty \dfrac{x^4 \mathrm{e}^x}{(\mathrm{e}^x - 1)^2} \mathrm{d}x = \dfrac{\pi^4}{15}$

Gaussian functions

G.1 $\int_0^\infty \mathrm{e}^{-ax^2} \mathrm{d}x = \dfrac{1}{2}\left(\dfrac{\pi}{a}\right)^{1/2}$

G.2 $\int_0^\infty x\mathrm{e}^{-ax^2} \mathrm{d}x = \dfrac{1}{2a}$

G.3 $\int_0^\infty x^2\mathrm{e}^{-ax^2} \mathrm{d}x = \dfrac{1}{4}\left(\dfrac{\pi}{a^3}\right)^{1/2}$

G.4 $\int_0^\infty x^3\mathrm{e}^{-ax^2} \mathrm{d}x = \dfrac{1}{2a^2}$

G.5 $\int_0^\infty x^4\mathrm{e}^{-ax^2} \mathrm{d}x = \dfrac{3}{8a^2}\left(\dfrac{\pi}{a}\right)^{1/2}$

G.6 $\operatorname{erf} z = \dfrac{2}{\pi^{1/2}} \int_0^z \mathrm{e}^{-x^2} \mathrm{d}x \qquad \operatorname{erfc} z = 1 - \operatorname{erf} z$

G.7 $\int_0^\infty x^{2m+1}\mathrm{e}^{-ax^2} \mathrm{d}x = \dfrac{m!}{2a^{m+1}}$

G.8 $\int_0^\infty x^{2m}\mathrm{e}^{-ax^2} \mathrm{d}x = \dfrac{(2m-1)!!}{2^{m+1}a^m}\left(\dfrac{\pi}{a}\right)^{1/2}$

$(2m-1)!! = 1 \times 3 \times 5 \cdots \times (2m-1)$

Trigonometric functions

T.1 $\int \sin ax\, \mathrm{d}x = -\dfrac{1}{a}\cos ax + \text{constant}$

T.2 $\int \sin^2 ax\, \mathrm{d}x = \dfrac{1}{2}x - \dfrac{\sin 2ax}{4a} + \text{constant}$

T.3 $\int \sin^3 ax\, \mathrm{d}x = -\dfrac{(\sin^2 ax + 2)\cos ax}{3a} + \text{constant}$

T.4 $\int \sin^4 ax\, \mathrm{d}x = \dfrac{3x}{8} - \dfrac{3}{8a}\sin ax \cos ax - \dfrac{1}{4a}\sin^3 ax \cos ax + \text{constant}$

T.5 $\int \sin ax \sin bx\, \mathrm{d}x = \dfrac{\sin(a-b)x}{2(a-b)} - \dfrac{\sin(a+b)x}{2(a+b)} + \text{constant},\ a^2 \neq b^2$

T.6 $\int_0^L \sin nax \sin^2 ax\, \mathrm{d}x = -\dfrac{1}{2a}\left\{\dfrac{1}{n} - \dfrac{1}{2(n+2)} - \dfrac{1}{2(n-2)}\right\} \times \{(-1)^n - 1\}$

T.7 $\int \sin ax \cos ax\, \mathrm{d}x = \dfrac{1}{2a}\sin^2 ax + \text{constant}$

T.8 $\int \sin bx \cos ax\, \mathrm{d}x = \dfrac{\cos(a-b)x}{2(a-b)} - \dfrac{\cos(a+b)x}{2(a+b)} + \text{constant},\ a^2 \neq b^2$

T.9 $\int x \sin ax \sin bx\, \mathrm{d}x = -\dfrac{\mathrm{d}}{\mathrm{d}a}\int \sin bx \cos ax\, \mathrm{d}x$

T.10 $\int \cos^2 ax \sin ax\, \mathrm{d}x = -\dfrac{1}{3a}\cos^3 ax + \text{constant}$

T.11 $\int x \sin^2 ax\, \mathrm{d}x = \dfrac{x^2}{4} - \dfrac{x \sin 2ax}{4a} - \dfrac{\cos 2ax}{8a^2} + \text{constant}$

T.12 $\int x^2 \sin^2 ax\, \mathrm{d}x = \dfrac{x^3}{6} - \left(\dfrac{x^2}{4a} - \dfrac{1}{8a^3}\right)\sin 2ax - \dfrac{x \cos 2ax}{4a^2} + \text{constant}$

T.13 $\int x \cos ax\, \mathrm{d}x = \dfrac{1}{a^2}\cos ax + \dfrac{x}{a}\sin ax + \text{constant}$

PART 2 Units

Table A.1 Some common units

Physical quantity	Name of unit	Symbol for unit	Value*
Time	minute	min	60 s
	hour	h	3600 s
	day	d	86 400 s
	year	a	31 556 952 s
Length	ångström	Å	10^{-10} m
Volume	litre	L, l	1 dm^3
Mass	tonne	t	10^3 kg
Pressure	bar	bar	10^5 Pa
	atmosphere	atm	101.325 kPa
Energy	electronvolt	eV	$1.602\ 177\ 33 \times 10^{-19}$ J 96.485 31 kJ mol^{-1}

* All values are exact, except for the definition of 1 eV, which depends on the measured value of *e*, and the year, which is not a constant and depends on a variety of astronomical assumptions.

Table A.2 Common SI prefixes

Prefix	y	z	a	f	p	n	μ	m	c	d
Name	yocto	zepto	atto	femto	pico	nano	micro	milli	centi	deci
Factor	10^{-24}	10^{-21}	10^{-18}	10^{-15}	10^{-12}	10^{-9}	10^{-6}	10^{-3}	10^{-2}	10^{-1}
Prefix	da	h	k	M	G	T	P	E	Z	Y
Name	deca	hecto	kilo	mega	giga	tera	peta	exa	zeta	yotta
Factor	10	10^2	10^3	10^6	10^9	10^{12}	10^{15}	10^{18}	10^{21}	10^{24}

Table A.3 The SI base units

Physical quantity	Symbol for quantity	Base unit
Length	l	metre, m
Mass	m	kilogram, kg
Time	t	second, s
Electric current	I	ampere, A
Thermodynamic temperature	T	kelvin, K
Amount of substance	n	mole, mol
Luminous intensity	I_v	candela, cd

Table A.4 A selection of derived units

Physical quantity	Derived unit*	Name of derived unit
Force	1 kg m s^{-2}	newton, N
Pressure	1 kg m^{-1} s^{-2} 1 N m^{-2}	pascal, Pa
Energy	1 kg m^2 s^{-2} 1 N m 1 Pa m^3	joule, J
Power	1 kg m^2 s^{-3} 1 J s^{-1}	watt, W

* Equivalent definitions in terms of derived units are given following the definition in terms of base units.

PART 3 Data

The following is a directory of all tables in the text; those included in this *Resource section* are marked with an asterisk. The remainder will be found on the pages indicated. These tables reproduce and expand the data given in the short tables in the text, and follow their numbering. Standard states refer to a pressure of $p^{\ominus} = 1$ bar. The general references are as follows:

AIP: D.E. Gray (ed.), *American Institute of Physics handbook*. McGraw-Hill, New York (1972).

E: J. Emsley, *The elements*. Oxford University Press, Oxford (1991).

HCP: D.R. Lide (ed.), *Handbook of chemistry and physics*. CRC Press, Boca Raton (2000).

JL: A.M. James and M.P. Lord, *Macmillan's chemical and physical data*. Macmillan, London (1992).

KL: G.W.C. Kaye and T.H. Laby (ed.), *Tables of physical and chemical constants*. Longman, London (1973).

LR: G.N. Lewis and M. Randall, revised by K.S. Pitzer and L. Brewer, *Thermodynamics*. McGraw-Hill, New York (1961).

NBS: *NBS tables of chemical thermodynamic properties*, published as *J. Phys. Chem. Reference Data*, **11**, Supplement 2 (1982).

RS: R.A. Robinson and R.H. Stokes, *Electrolyte solutions*, Butterworth, London (1959).

TDOC: J.B. Pedley, J.D. Naylor, and S.P. Kirby, *Thermochemical data of organic compounds*. Chapman & Hall, London (1986).

Table 0.1 Physical properties of selected materials

	$\rho/(\text{g cm}^{-3})$ at 293 K†	T_f/K	T_b/K		$\rho/(\text{g cm}^{-3})$ at 293 K†	T_f/K	T_b/K
Elements				**Inorganic compounds**			
Aluminium(s)	2.698	933.5	2740	$CaCO_3$(s, calcite)	2.71	1612	1171[d]
Argon(g)	1.381	83.8	87.3	$CuSO_4\cdot5H_2O$(s)	2.284	383($-H_2O$)	423($-5H_2O$)
Boron(s)	2.340	2573	3931	HBr(g)	2.77	184.3	206.4
Bromine(l)	3.123	265.9	331.9	HCl(g)	1.187	159.0	191.1
Carbon(s, gr)	2.260	3700[s]		HI(g)	2.85	222.4	237.8
Carbon(s, d)	3.513			H_2O(l)	0.997	273.2	373.2

(Continued)

Table 0.1 (Continued)

	ρ/(g cm^{-3}) at 293 K†	T_f/K	T_b/K		ρ/(g cm^{-3}) at 293 K†	T_f/K	T_b/K
Elements (Continued)				**Inorganic compounds (continued)**			
Chlorine(g)	1.507	172.2	239.2	D_2O(l)	1.104	277.0	374.6
Copper(s)	8.960	1357	2840	NH_3(g)	0.817	195.4	238.8
Fluorine(g)	1.108	53.5	85.0	KBr(s)	2.750	1003	1708
Gold(s)	19.320	1338	3080	KCl(s)	1.984	1049	1773^s
Helium(g)	0.125		4.22	NaCl(s)	2.165	1074	1686
Hydrogen(g)	0.071	14.0	20.3	H_2SO_4(l)	1.841	283.5	611.2
Iodine(s)	4.930	386.7	457.5				
Iron(s)	7.874	1808	3023	**Organic compounds**			
Krypton(g)	2.413	116.6	120.8	Acetaldehyde, CH_3CHO(l)	0.788	152	293
Lead(s)	11.350	600.6	2013	Acetic acid, CH_3COOH(l)	1.049	289.8	391
Lithium(s)	0.534	453.7	1620	Acetone, $(CH_3)_2CO$(l)	0.787	178	329
Magnesium(s)	1.738	922.0	1363	Aniline, $C_6H_5NH_2$(l)	1.026	267	457
Mercury(l)	13.546	234.3	629.7	Anthracene, $C_{14}H_{10}$(s)	1.243	490	615
Neon(g)	1.207	24.5	27.1	Benzene, C_6H_6(l)	0.879	278.6	353.2
Nitrogen(g)	0.880	63.3	77.4	Carbon tetrachloride, CCl_4(l)	1.63	250	349.9
Oxygen(g)	1.140	54.8	90.2	Chloroform, $CHCl_3$(l)	1.499	209.6	334
Phosphorus(s, wh)	1.820	317.3	553	Ethanol, C_2H_5OH(l)	0.789	156	351.4
Potassium(s)	0.862	336.8	1047	Formaldehyde, HCHO(g)		181	254.0
Silver(s)	10.500	1235	2485	Glucose, $C_6H_{12}O_6$(s)	1.544	415	
Sodium(s)	0.971	371.0	1156	Methane, CH_4(g)		90.6	111.6
Sulfur(s, α)	2.070	386.0	717.8	Methanol, CH_3OH(l)	0.791	179.2	337.6
Uranium(s)	18.950	1406	4018	Naphthalene, $C_{10}H_8$(s)	1.145	353.4	491
Xenon(g)	2.939	161.3	166.1	Octane, C_8H_{18}(l)	0.703	216.4	398.8
Zinc(s)	7.133	692.7	1180	Phenol, C_6H_5OH(s)	1.073	314.1	455.0
				Sucrose, $C_{12}H_{22}O_{11}$(s)	1.588	457^d	

d: decomposes; s: sublimes; Data: AIP, E, HCP, KL. † For gases, at their boiling points.

Table 0.2 Masses and natural abundances of selected nuclides

Nuclide		m/m_u	Abundance/%
H	^{1}H	1.0078	99.985
	^{2}H	2.0140	0.015
He	^{3}He	3.0160	0.000 13
	^{4}He	4.0026	100
Li	^{6}Li	6.0151	7.42
	^{7}Li	7.0160	92.58
B	^{10}B	10.0129	19.78
	^{11}B	11.0093	80.22
C	^{12}C	12*	98.89
	^{13}C	13.0034	1.11
N	^{14}N	14.0031	99.63
	^{15}N	15.0001	0.37
O	^{16}O	15.9949	99.76
	^{17}O	16.9991	0.037
	^{18}O	17.9992	0.204
F	^{19}F	18.9984	100
P	^{31}P	30.9738	100
S	^{32}S	31.9721	95.0
	^{33}S	32.9715	0.76

Table 0.2 (Continued)

	Nuclide	m/m_u	Abundance/%
	^{34}S	33.9679	4.22
Cl	^{35}Cl	34.9688	75.53
	^{37}Cl	36.9651	24.4
Br	^{79}Br	78.9183	50.54
	^{81}Br	80.9163	49.46
I	^{127}I	126.9045	100

* Exact value.

Table 1B.1 Collision cross-sections, σ/nm^2

Ar	0.36
C_2H_4	0.64
C_6H_6	0.88
CH_4	0.46
Cl_2	0.93
CO_2	0.52
H_2	0.27
He	0.21
N_2	0.43
Ne	0.24
O_2	0.40
SO_2	0.58

Data: KL.

Table 1C.1 Second virial coefficients, B/(cm^3 mol^{-1})

	100 K	273 K	373 K	600 K
Air	−167.3	−13.5	3.4	19.0
Ar	−187.0	−21.7	−4.2	11.9
CH_4		−53.6	−21.2	8.1
CO_2		−142	−72.2	−12.4
H_2	−2.0	13.7	15.6	
He	11.4	12.0	11.3	10.4
Kr		−62.9	−28.7	1.7
N_2	−160.0	−10.5	6.2	21.7
Ne	−6.0	10.4	12.3	13.8
O_2	−197.5	−22.0	−3.7	12.9
Xe		−153.7	−81.7	−19.6

Data: AIP, JL. The values relate to the expansion in eqn 1C.3 of Topic 1C; convert to eqn 1C.3 using $B' = B/RT$.
For Ar at 273 K, C = 1200 cm^6 mol^{-1}.

Table 1C.2 Critical constants of gases

	p_c/atm	V_c/(cm^3 mol^{-1})	T_c/K	Z_c	T_B/K
Ar	48.0	75.3	150.7	0.292	411.5
Br_2	102	135	584	0.287	
C_2H_4	50.50	124	283.1	0.270	
C_2H_6	48.20	148	305.4	0.285	
C_6H_6	48.6	260	562.7	0.274	
CH_4	45.6	98.7	190.6	0.288	510.0
Cl_2	76.1	124	417.2	0.276	
CO_2	72.9	94.0	304.2	0.274	714.8
F_2	55	144			
H_2	12.8	34.99	33.23	0.305	110.0
H_2O	218.3	55.3	647.4	0.227	
HBr	84.0	363.0			
HCl	81.5	81.0	324.7	0.248	
He	2.26	57.8	5.2	0.305	22.64
HI	80.8	423.2			
Kr	54.27	92.24	209.39	0.291	575.0
N_2	33.54	90.10	126.3	0.292	327.2
Ne	26.86	41.74	44.44	0.307	122.1

(Continued)

Table 1C.2 (Continued)

	p_c/atm	V_c/(cm^3 mol^{-1})	T_c/K	Z_c	T_B/K
NH_3	111.3	72.5	405.5	0.242	
O_2	50.14	78.0	154.8	0.308	405.9
Xe	58.0	118.8	289.75	0.290	768.0

Data: AIP, KL.

Table 1C.3 van der Waals coefficients

	a/(atm dm^6 mol^{-2})	b/(10^{-2} dm^3 mol^{-1})		a/(atm dm^6 mol^{-2})	b/(10^{-2} dm^3 mol^{-1})
Ar	1.337	3.20	H_2S	4.484	4.34
C_2H_4	4.552	5.82	He	0.0341	2.38
C_2H_6	5.507	6.51	Kr	5.125	1.06
C_6H_6	18.57	11.93	N_2	1.352	3.87
CH_4	14.61	4.31	Ne	0.205	1.67
Cl_2	6.260	5.42	NH_3	4.169	3.71
CO	1.453	3.95	O_2	1.364	3.19
CO_2	3.610	4.29	SO_2	6.775	5.68
H_2	0.2420	2.65	Xe	4.137	5.16
H_2O	5.464	3.05			

Data: HCP.

Table 2B.1 Temperature variation of molar heat capacities, $C_{p,m}$/(J K^{-1} mol^{-1}) = $a + bT + c/T^2$

	a	b/(10^{-3} K^{-1})	c/(10^5 K^2)
Monatomic gases			
	20.78	0	0
Other gases			
Br_2	37.32	0.50	−1.26
Cl_2	37.03	0.67	−2.85
CO_2	44.22	8.79	−8.62
F_2	34.56	2.51	−3.51
H_2	27.28	3.26	0.50
I_2	37.40	0.59	−0.71
N_2	28.58	3.77	−0.50
NH_3	29.75	25.1	−1.55
O_2	29.96	4.18	−1.67
Liquids (from melting to boiling)			
$C_{10}H_8$, naphthalene	79.5	0.4075	0
I_2	80.33	0	0
H_2O	75.29	0	0
Solids			
Al	20.67	12.38	0
C (graphite)	16.86	4.77	−8.54
$C_{10}H_8$, naphthalene	−110	936	0
Cu	22.64	6.28	0
I_2	40.12	49.79	0
NaCl	45.94	16.32	0
Pb	22.13	11.72	0.96

Source: Mostly LR.

Table 2C.1 Standard enthalpies of fusion and vaporization at the transition temperature, $\Delta_{trs}H^{\ominus}/(kJ\,mol^{-1})$

	T_f/K	Fusion	T_b/K	Vaporization		T_f/K	Fusion	T_b/K	Vaporization
Elements					**Inorganic compounds**				
Ag	1234	11.30	2436	250.6	CO_2	217.0	8.33	194.6	25.23[s]
Ar	83.81	1.188	87.29	6.506	CS_2	161.2	4.39	319.4	26.74
Br_2	265.9	10.57	332.4	29.45	H_2O	273.15	6.008	373.15	40.656
Cl_2	172.1	6.41	239.1	20.41					44.016 at 298 K
F_2	53.6	0.26	85.0	3.16	H_2S	187.6	2.377	212.8	18.67
H_2	13.96	0.117	20.38	0.916	H_2SO_4	283.5	2.56		
He	3.5	0.021	4.22	0.084	NH_3	195.4	5.652	239.7	23.35
Hg	234.3	2.292	629.7	59.30	**Organic compounds**				
I_2	386.8	15.52	458.4	41.80	CH_4	90.68	0.941	111.7	8.18
N_2	63.15	0.719	77.35	5.586	CCl_4	250.3	2.47	349.9	30.00
Na	371.0	2.601	1156	98.01	C_2H_6	89.85	2.86	184.6	14.7
O_2	54.36	0.444	90.18	6.820	C_6H_6	278.61	10.59	353.2	30.8
Xe	161	2.30	165	12.6	C_6H_{14}	178	13.08	342.1	28.85
K	336.4	2.35	1031	80.23	$C_{10}H_8$	354	18.80	490.9	51.51
					CH_3OH	175.2	3.16	337.2	35.27
									37.99 at 298 K
					C_2H_5OH	158.7	4.60	352	43.5

Data: AIP; s denotes sublimation.

Table 2C.3 Lattice enthalpies at 298 K, $\Delta H_L/(kJ\,mol^{-1})$. See Table 18B.4.

Table 2C.4 Thermodynamic data for organic compounds at 298 K

	$M/(g\,mol^{-1})$	$\Delta_f H^{\ominus}/(kJ\,mol^{-1})$	$\Delta_f G^{\ominus}/(kJ\,mol^{-1})$	$S_m^{\ominus}/(J\,K^{-1}\,mol^{-1})$	$C_{p,m}^{\ominus}/(J\,K^{-1}\,mol^{-1})$	$\Delta_c H^{\ominus}/(kJ\,mol^{-1})$
C(s) (graphite)	12.011	0	0	5.740	8.527	−393.51
C(s) (diamond)	12.011	+1.895	+2.900	2.377	6.113	−395.40
CO_2(g)	44.040	−393.51	−394.36	213.74	37.11	
Hydrocarbons						
CH_4(g), methane	16.04	−74.81	−50.72	186.26	35.31	−890
CH_3(g), methyl	15.04	+145.69	+147.92	194.2	38.70	
C_2H_2(g), ethyne	26.04	+226.73	+209.20	200.94	43.93	−1300
C_2H_4(g), ethene	28.05	+52.26	+68.15	219.56	43.56	−1411
C_2H_6(g), ethane	30.07	−84.68	−32.82	229.60	52.63	−1560
C_3H_6(g), propene	42.08	+20.42	+62.78	267.05	63.89	−2058
C_3H_6(g), cyclopropane	42.08	+53.30	+104.45	237.55	55.94	−2091
C_3H_8(g), propane	44.10	−103.85	−23.49	269.91	73.5	−2220
C_4H_8(g), 1-butene	56.11	−0.13	+71.39	305.71	85.65	−2717
C_4H_8(g), *cis*-2-butene	56.11	−6.99	+65.95	300.94	78.91	−2710
C_4H_8(g), *trans*-2-butene	56.11	−11.17	+63.06	296.59	87.82	−2707
C_4H_{10}(g), butane	58.13	−126.15	−17.03	310.23	97.45	−2878

(Continued)

Table 2C.4 (Continued)

	M/(g mol^{-1})	$\Delta_f H^\ominus$/(kJ mol^{-1})	$\Delta_f G^\ominus$/(kJ mol^{-1})	$S_m^\ominus$/(J K^{-1} mol^{-1})†	$C_{p,m}^\ominus$/(J K^{-1} mol^{-1})	$\Delta_c H^\ominus$/(kJ mol^{-1})
C_5H_{12}(g), pentane	72.15	−146.44	−8.20	348.40	120.2	−3537
C_5H_{12}(l)	72.15	−173.1				
C_6H_6(l), benzene	78.12	+49.0	+124.3	173.3	136.1	−3268
C_6H_6(g)	78.12	+82.93	+129.72	269.31	81.67	−3302
C_6H_{12}(l), cyclohexane	84.16	−156	+26.8	204.4	156.5	−3920
C_6H_{14}(l), hexane	86.18	−198.7		204.3		−4163
$C_6H_5CH_3$(g), methylbenzene (toluene)	92.14	+50.0	+122.0	320.7	103.6	−3953
C_7H_{16}(l), heptane	100.21	−224.4	+1.0	328.6	224.3	
C_8H_{18}(l), octane	114.23	−249.9	+6.4	361.1		−5471
C_8H_{18}(l), iso-octane	114.23	−255.1				−5461
$C_{10}H_8$(s), naphthalene	128.18	+78.53				−5157
Alcohols and phenols						
CH_3OH(l), methanol	32.04	−238.66	−166.27	126.8	81.6	−726
CH_3OH(g)	32.04	−200.66	−161.96	239.81	43.89	−764
C_2H_5OH(l), ethanol	46.07	−277.69	−174.78	160.7	111.46	−1368
C_2H_5OH(g)	46.07	−235.10	−168.49	282.70	65.44	−1409
C_6H_5OH(s), phenol	94.12	−165.0	−50.9	146.0		−3054
Carboxylic acids, hydroxy acids, and esters						
HCOOH(l), formic	46.03	−424.72	−361.35	128.95	99.04	−255
CH_3COOH(l), acetic	60.05	−484.5	−389.9	159.8	124.3	−875
CH_3COOH(aq)	60.05	−485.76	−396.46	178.7		
$CH_3CO_2^-$(aq)	59.05	−486.01	−369.31	+86.6	−6.3	
$(COOH)_2$(s), oxalic	90.04	−827.2			117	−254
C_6H_5COOH(s), benzoic	122.13	−385.1	−245.3	167.6	146.8	−3227
$CH_3CH(OH)COOH$(s), lactic	90.08	−694.0				−1344
$CH_3COOC_2H_5$(l), ethyl acetate	88.11	−479.0	−332.7	259.4	170.1	−2231
Alkanals and alkanones						
HCHO(g), methanal	30.03	−108.57	−102.53	218.77	35.40	−571
CH_3CHO(l), ethanal	44.05	−192.30	−128.12	160.2		−1166
CH_3CHO(g)	44.05	−166.19	−128.86	250.3	57.3	−1192
CH_3COCH_3(l), propanone	58.08	−248.1	−155.4	200.4	124.7	−1790
Sugars						
$C_6H_{12}O_6$(s), α-D-glucose	180.16	−1274				−2808
$C_6H_{12}O_6$(s), β-D-glucose	180.16	−1268	−910	212		
$C_6H_{12}O_6$(s), β-D-fructose	180.16	−1266				−2810
$C_{12}H_{22}O_{11}$(s), sucrose	342.30	−2222	−1543	360.2		−5645
Nitrogen compounds						
$CO(NH_2)_2$(s), urea	60.06	−333.51	−197.33	104.60	93.14	−632
CH_3NH_2(g), methylamine	31.06	−22.97	+32.16	243.41	53.1	−1085
$C_6H_5NH_2$(l), aniline	93.13	+31.1				−3393
$CH_2(NH_2)COOH$(s), glycine	75.07	−532.9	−373.4	103.5	99.2	−969

Data: NBS, TDOC. † Standard entropies of ions may be either positive or negative because the values are relative to the entropy of the hydrogen ion.

Table 2C.5 Thermodynamic data for elements and inorganic compounds at 298 K

	$M/(\text{g mol}^{-1})$	$\Delta_f H^\ominus/(\text{kJ mol}^{-1})$	$\Delta_f G^\ominus/(\text{kJ mol}^{-1})$	$S_m^\ominus/(\text{J K}^{-1}\text{ mol}^{-1})^\dagger$	$C_{p,m}^\ominus/(\text{J K}^{-1}\text{ mol}^{-1})$
Aluminium (aluminum)					
Al(s)	26.98	0	0	28.33	24.35
Al(l)	26.98	+10.56	+7.20	39.55	24.21
Al(g)	26.98	+326.4	+285.7	164.54	21.38
Al^{3+}(g)	26.98	+5483.17			
Al^{3+}(aq)	26.98	−531	−485	−321.7	
Al_2O_3(s, α)	101.96	−1675.7	−1582.3	50.92	79.04
$AlCl_3$(s)	133.24	−704.2	−628.8	110.67	91.84
Argon					
Ar(g)	39.95	0	0	154.84	20.786
Antimony					
Sb(s)	121.75	0	0	45.69	25.23
SbH_3(g)	124.77	+145.11	+147.75	232.78	41.05
Arsenic					
As(s, α)	74.92	0	0	35.1	24.64
As(g)	74.92	+302.5	+261.0	174.21	20.79
As_4(g)	299.69	+143.9	+92.4	314	
AsH_3(g)	77.95	+66.44	+68.93	222.78	38.07
Barium					
Ba(s)	137.34	0	0	62.8	28.07
Ba(g)	137.34	+180	+146	170.24	20.79
Ba^{2+}(aq)	137.34	−537.64	−560.77	+9.6	
BaO(s)	153.34	−553.5	−525.1	70.43	47.78
$BaCl_2$(s)	208.25	−858.6	−810.4	123.68	75.14
Beryllium					
Be(s)	9.01	0	0	9.50	16.44
Be(g)	9.01	+324.3	+286.6	136.27	20.79
Bismuth					
Bi(s)	208.98	0	0	56.74	25.52
Bi(g)	208.98	+207.1	+168.2	187.00	20.79
Bromine					
Br_2(l)	159.82	0	0	152.23	75.689
Br_2(g)	159.82	+30.907	+3.110	245.46	36.02
Br(g)	79.91	+111.88	+82.396	175.02	20.786
Br^-(g)	79.91	−219.07			
Br^-(aq)	79.91	−121.55	−103.96	+82.4	−141.8
HBr(g)	90.92	−36.40	−53.45	198.70	29.142
Cadmium					
Cd(s, γ)	112.40	0	0	51.76	25.98
Cd(g)	112.40	+112.01	+77.41	167.75	20.79
Cd^{2+}(aq)	112.40	−75.90	−77.612	−73.2	
CdO(s)	128.40	−258.2	−228.4	54.8	43.43
$CdCO_3$(s)	172.41	−750.6	−669.4	92.5	

(Continued)

Table 2C.5 (Continued)

	$M/(g\ mol^{-1})$	$\Delta_f H^\ominus/(kJ\ mol^{-1})$	$\Delta_f G^\ominus/(kJ\ mol^{-1})$	$S_m^\ominus/(J\ K^{-1}\ mol^{-1})^\dagger$	$C_{p,m}^\ominus/(J\ K^{-1}\ mol^{-1})$
Caesium (cesium)					
Cs(s)	132.91	0	0	85.23	32.17
Cs(g)	132.91	+76.06	+49.12	175.60	20.79
Cs^+(aq)	132.91	−258.28	−292.02	+133.05	−10.5
Calcium					
Ca(s)	40.08	0	0	41.42	25.31
Ca(g)	40.08	+178.2	+144.3	154.88	20.786
Ca^{2+}(aq)	40.08	−542.83	−553.58	−53.1	
CaO(s)	56.08	−635.09	−604.03	39.75	42.80
$CaCO_3$(s) (calcite)	100.09	−1206.9	−1128.8	92.9	81.88
$CaCO_3$(s) (aragonite)	100.09	−1207.1	−1127.8	88.7	81.25
CaF_2(s)	78.08	−1219.6	−1167.3	68.87	67.03
$CaCl_2$(s)	110.99	−795.8	−748.1	104.6	72.59
$CaBr_2$(s)	199.90	−682.8	−663.6	130	
Carbon (for 'organic' compounds of carbon, see Table 2C.4)					
C(s) (graphite)	12.011	0	0	5.740	8.527
C(s) (diamond)	12.011	+1.895	+2.900	2.377	6.113
C(g)	12.011	+716.68	+671.26	158.10	20.838
C_2(g)	24.022	+831.90	+775.89	199.42	43.21
CO(g)	28.011	−110.53	−137.17	197.67	29.14
CO_2(g)	44.010	−393.51	−394.36	213.74	37.11
CO_2(aq)	44.010	−413.80	−385.98	117.6	
H_2CO_3(aq)	62.03	−699.65	−623.08	187.4	
HCO_3^-(aq)	61.02	−691.99	−586.77	+91.2	
CO_3^{2-}(aq)	60.01	−677.14	−527.81	−56.9	
CCl_4(l)	153.82	−135.44	−65.21	216.40	131.75
CS_2(l)	76.14	+89.70	+65.27	151.34	75.7
HCN(g)	27.03	+135.1	+124.7	201.78	35.86
HCN(l)	27.03	+108.87	+124.97	112.84	70.63
CN^-(aq)	26.02	+150.6	+172.4	+94.1	
Chlorine					
Cl_2(g)	70.91	0	0	223.07	33.91
Cl(g)	35.45	+121.68	+105.68	165.20	21.840
Cl^-(g)	34.45	−233.13			
Cl^-(aq)	35.45	−167.16	−131.23	+56.5	−136.4
HCl(g)	36.46	−92.31	−95.30	186.91	29.12
HCl(aq)	36.46	−167.16	−131.23	56.5	−136.4
Chromium					
Cr(s)	52.00	0	0	23.77	23.35
Cr(g)	52.00	+396.6	+351.8	174.50	20.79
CrO_4^{2-}(aq)	115.99	−881.15	−727.75	+50.21	
$Cr_2O_7^{2-}$(aq)	215.99	−1490.3	−1301.1	+261.9	

Table 2C.5 (Continued)

	$M/(\text{g mol}^{-1})$	$\Delta_f H^\ominus/(\text{kJ mol}^{-1})$	$\Delta_f G^\ominus/(\text{kJ mol}^{-1})$	$S_m^\ominus/(\text{J K}^{-1}\text{ mol}^{-1})^\dagger$	$C_{p,m}^\ominus/(\text{J K}^{-1}\text{ mol}^{-1})$
Copper					
Cu(s)	63.54	0	0	33.150	24.44
Cu(g)	63.54	+338.32	+298.58	166.38	20.79
Cu^+(aq)	63.54	+71.67	+49.98	+40.6	
Cu^{2+}(aq)	63.54	+64.77	+65.49	−99.6	
Cu_2O(s)	143.08	−168.6	−146.0	93.14	63.64
CuO(s)	79.54	−157.3	−129.7	42.63	42.30
$CuSO_4$(s)	159.60	−771.36	−661.8	109	100.0
$CuSO_4\cdot H_2O$(s)	177.62	−1085.8	−918.11	146.0	134
$CuSO_4\cdot 5H_2O$(s)	249.68	−2279.7	−1879.7	300.4	280
Deuterium					
D_2(g)	4.028	0	0	144.96	29.20
HD(g)	3.022	+0.318	−1.464	143.80	29.196
D_2O(g)	20.028	−249.20	−234.54	198.34	34.27
D_2O(l)	20.028	−294.60	−243.44	75.94	84.35
HDO(g)	19.022	−245.30	−233.11	199.51	33.81
HDO(l)	19.022	−289.89	−241.86	79.29	
Fluorine					
F_2(g)	38.00	0	0	202.78	31.30
F(g)	19.00	+78.99	+61.91	158.75	22.74
F^-(aq)	19.00	−332.63	−278.79	−13.8	−106.7
HF(g)	20.01	−271.1	−273.2	173.78	29.13
Gold					
Au(s)	196.97	0	0	47.40	25.42
Au(g)	196.97	+366.1	+326.3	180.50	20.79
Helium					
He(g)	4.003	0	0	126.15	20.786
Hydrogen (see also deuterium)					
H_2(g)	2.016	0	0	130.684	28.824
H(g)	1.008	+217.97	+203.25	114.71	20.784
H^+(aq)	1.008	0	0	0	0
H^+(g)	1.008	+1536.20			
H_2O(s)	18.015			37.99	
H_2O(l)	18.015	−285.83	−237.13	69.91	75.291
H_2O(g)	18.015	−241.82	−228.57	188.83	33.58
H_2O_2(l)	34.015	−187.78	−120.35	109.6	89.1
Iodine					
I_2(s)	253.81	0	0	116.135	54.44
I_2(g)	253.81	+62.44	+19.33	260.69	36.90
I(g)	126.90	+106.84	+70.25	180.79	20.786
I^-(aq)	126.90	−55.19	−51.57	+111.3	−142.3
HI(g)	127.91	+26.48	+1.70	206.59	29.158

(Continued)

Table 2C.5 (Continued)

	M/(g mol^{-1})	$\Delta_f H^\ominus$/(kJ mol^{-1})	$\Delta_f G^\ominus$/(kJ mol^{-1})	$S_m^\ominus$/(J K^{-1} mol^{-1})†	$C_{p,m}^\ominus$/(J K^{-1} mol^{-1})
Iron					
Fe(s)	55.85	0	0	27.28	25.10
Fe(g)	55.85	+416.3	+370.7	180.49	25.68
Fe^{2+}(aq)	55.85	−89.1	−78.90	−137.7	
Fe^{3+}(aq)	55.85	−48.5	−4.7	−315.9	
Fe_3O_4(s) (magnetite)	231.54	−1118.4	−1015.4	146.4	143.43
Fe_2O_3(s) (haematite)	159.69	−824.2	−742.2	87.40	103.85
FeS(s, α)	87.91	−100.0	−100.4	60.29	50.54
FeS_2(s)	119.98	−178.2	−166.9	52.93	62.17
Krypton					
Kr(g)	83.80	0	0	164.08	20.786
Lead					
Pb(s)	207.19	0	0	64.81	26.44
Pb(g)	207.19	+195.0	+161.9	175.37	20.79
Pb^{2+}(aq)	207.19	−1.7	−24.43	+10.5	
PbO(s, yellow)	223.19	−217.32	−187.89	68.70	45.77
PbO(s, red)	223.19	−218.99	−188.93	66.5	45.81
PbO_2(s)	239.19	−277.4	−217.33	68.6	64.64
Lithium					
Li(s)	6.94	0	0	29.12	24.77
Li(g)	6.94	+159.37	+126.66	138.77	20.79
Li^+(aq)	6.94	−278.49	−293.31	+13.4	68.6
Magnesium					
Mg(s)	24.31	0	0	32.68	24.89
Mg(g)	24.31	+147.70	+113.10	148.65	20.786
Mg^{2+}(aq)	24.31	−466.85	−454.8	−138.1	
MgO(s)	40.31	−601.70	−569.43	26.94	37.15
$MgCO_3$(s)	84.32	−1095.8	−1012.1	65.7	75.52
$MgCl_2$(s)	95.22	−641.32	−591.79	89.62	71.38
Mercury					
Hg(l)	200.59	0	0	76.02	27.983
Hg(g)	200.59	+61.32	+31.82	174.96	20.786
Hg^{2+}(aq)	200.59	+171.1	+164.40	−32.2	
Hg_2^{2+}(aq)	401.18	+172.4	+153.52	+84.5	
HgO(s)	216.59	−90.83	−58.54	70.29	44.06
Hg_2Cl_2(s)	472.09	−265.22	−210.75	192.5	102
$HgCl_2$(s)	271.50	−224.3	−178.6	146.0	
HgS(s, black)	232.65	−53.6	−47.7	88.3	
Neon					
Ne(g)	20.18	0	0	146.33	20.786
Nitrogen					
N_2(g)	28.013	0	0	191.61	29.125
N(g)	14.007	+472.70	+455.56	153.30	20.786

Table 2C.5 (Continued)

	M/(g mol^{-1})	$\Delta_f H^\ominus$/(kJ mol^{-1})	$\Delta_f G^\ominus$/(kJ mol^{-1})	$S_m^\ominus$ /(J K^{-1} mol^{-1})†	$C_{p,m}^\ominus$ /(J K^{-1} mol^{-1})
NO(g)	30.01	+90.25	+86.55	210.76	29.844
N_2O(g)	44.01	+82.05	+104.20	219.85	38.45
NO_2(g)	46.01	+33.18	+51.31	240.06	37.20
N_2O_4(g)	92.1	+9.16	+97.89	304.29	77.28
N_2O_5(s)	108.01	−43.1	+113.9	178.2	143.1
N_2O_5(g)	108.01	+11.3	+115.1	355.7	84.5
HNO_3(l)	63.01	−174.10	−80.71	155.60	109.87
HNO_3(aq)	63.01	−207.36	−111.25	146.4	−86.6
NO_3^-(aq)	62.01	−205.0	−108.74	+146.4	−86.6
NH_3(g)	17.03	−46.11	−16.45	192.45	35.06
NH_3(aq)	17.03	−80.29	−26.50	111.3	
NH_4^+(aq)	18.04	−132.51	−79.31	+113.4	79.9
NH_2OH(s)	33.03	−114.2			
HN_3(l)	43.03	+264.0	+327.3	140.6	43.68
HN_3(g)	43.03	+294.1	+328.1	238.97	98.87
N_2H_4(l)	32.05	+50.63	+149.43	121.21	139.3
NH_4NO_3(s)	80.04	−365.56	−183.87	151.08	84.1
NH_4Cl(s)	53.49	−314.43	−202.87	94.6	
Oxygen					
O_2(g)	31.999	0	0	205.138	29.355
O(g)	15.999	+249.17	+231.73	161.06	21.912
O_3(g)	47.998	+142.7	+163.2	238.93	39.20
OH^-(aq)	17.007	−229.99	−157.24	−10.75	−148.5
Phosphorus					
P(s, wh)	30.97	0	0	41.09	23.840
P(g)	30.97	+314.64	+278.25	163.19	20.786
P_2(g)	61.95	+144.3	+103.7	218.13	32.05
P_4(g)	123.90	+58.91	+24.44	279.98	67.15
PH_3(g)	34.00	+5.4	+13.4	210.23	37.11
PCl_3(g)	137.33	−287.0	−267.8	311.78	71.84
PCl_3(l)	137.33	−319.7	−272.3	217.1	
PCl_5(g)	208.24	−374.9	−305.0	364.6	112.8
PCl_5(s)	208.24	−443.5			
H_3PO_3(s)	82.00	−964.4			
H_3PO_3(aq)	82.00	−964.8			
H_3PO_4(s)	94.97	−1279.0	−1119.1	110.50	106.06
H_3PO_4(l)	94.97	−1266.9			
H_3PO_4(aq)	94.97	−1277.4	−1018.7	−222	
PO_4^{3-}(aq)	94.97	−1277.4	−1018.7	−221.8	
P_4O_{10}(s)	283.89	−2984.0	−2697.0	228.86	211.71
P_4O_6(s)	219.89	−1640.1			

(Continued)

Table 2C.5 (Continued)

	M/(g mol^{-1})	$\Delta_f H^\ominus$/(kJ mol^{-1})	$\Delta_f G^\ominus$/(kJ mol^{-1})	$S_m^\ominus$/(J K^{-1} mol^{-1})†	$C_{p,m}^\ominus$/(J K^{-1} mol^{-1})
Potassium					
K(s)	39.10	0	0	64.18	29.58
K(g)	39.10	+89.24	+60.59	160.336	20.786
K^+(g)	39.10	+514.26			
K^+(aq)	39.10	−252.38	−283.27	+102.5	21.8
KOH(s)	56.11	−424.76	−379.08	78.9	64.9
KF(s)	58.10	−576.27	−537.75	66.57	49.04
KCl(s)	74.56	−436.75	−409.14	82.59	51.30
KBr(s)	119.01	−393.80	−380.66	95.90	52.30
KI(s)	166.01	−327.90	−324.89	106.32	52.93
Silicon					
Si(s)	28.09	0	0	18.83	20.00
Si(g)	28.09	+455.6	+411.3	167.97	22.25
SiO_2(s, α)	60.09	−910.94	−856.64	41.84	44.43
Silver					
Ag(s)	107.87	0	0	42.55	25.351
Ag(g)	107.87	+284.55	+245.65	173.00	20.79
Ag^+(aq)	107.87	+105.58	+77.11	+72.68	21.8
AgBr(s)	187.78	−100.37	−96.90	107.1	52.38
AgCl(s)	143.32	−127.07	−109.79	96.2	50.79
Ag_2O(s)	231.74	−31.05	−11.20	121.3	65.86
$AgNO_3$(s)	169.88	−129.39	−33.41	140.92	93.05
Sodium					
Na(s)	22.99	0	0	51.21	28.24
Na(g)	22.99	+107.32	+76.76	153.71	20.79
Na^+(aq)	22.99	−240.12	−261.91	+59.0	46.4
NaOH(s)	40.00	−425.61	−379.49	64.46	59.54
NaCl(s)	58.44	−411.15	−384.14	72.13	50.50
NaBr(s)	102.90	−361.06	−348.98	86.82	51.38
NaI(s)	149.89	−287.78	−286.06	98.53	52.09
Sulfur					
S(s, α) (rhombic)	32.06	0	0	31.80	22.64
S(s, β) (monoclinic)	32.06	+0.33	+0.1	32.6	23.6
S(g)	32.06	+278.81	+238.25	167.82	23.673
S_2(g)	64.13	+128.37	+79.30	228.18	32.47
S^{2-}(aq)	32.06	+33.1	+85.8	−14.6	
SO_2(g)	64.06	−296.83	−300.19	248.22	39.87
SO_3(g)	80.06	−395.72	−371.06	256.76	50.67
H_2SO_4(l)	98.08	−813.99	−690.00	156.90	138.9
H_2SO_4(aq)	98.08	−909.27	−744.53	20.1	−293
SO_4^{2-}(aq)	96.06	−909.27	−744.53	+20.1	−293
HSO_4^-(aq)	97.07	−887.34	−755.91	+131.8	−84
H_2S(g)	34.08	−20.63	−33.56	205.79	34.23

Table 2C.5 (Continued)

	M/(g mol^{-1})	$\Delta_f H^\ominus$/(kJ mol^{-1})	$\Delta_f G^\ominus$/(kJ mol^{-1})	$S_m^\ominus$/(J K^{-1} mol^{-1})†	$C_{p,m}^\ominus$/(J K^{-1} mol^{-1})
H_2S(aq)	34.08	−39.7	−27.83	121	
HS^-(aq)	33.072	−17.6	+12.08	+62.08	
SF_6(g)	146.05	−1209	−1105.3	291.82	97.28
Tin					
Sn(s, β)	118.69	0	0	51.55	26.99
Sn(g)	118.69	+302.1	+267.3	168.49	20.26
Sn^{2+}(aq)	118.69	−8.8	−27.2	−17	
SnO(s)	134.69	−285.8	−256.9	56.5	44.31
SnO_2(s)	150.69	−580.7	−519.6	52.3	52.59
Xenon					
Xe(g)	131.30	0	0	169.68	20.786
Zinc					
Zn(s)	65.37	0	0	41.63	25.40
Zn(g)	65.37	+130.73	+95.14	160.98	20.79
Zn^{2+}(aq)	65.37	−153.89	−147.06	−112.1	46
ZnO(s)	81.37	−348.28	−318.30	43.64	40.25

Source: NBS. † Standard entropies of ions may be either positive or negative because the values are relative to the entropy of the hydrogen ion.

Table 2C.6 Standard enthalpies of formation of organic compounds at 298 K, $\Delta_f H^\ominus$/(kJ mol^{-1}). See Table 2C.4.

Table 2D.1 Expansion coefficients (α) and isothermal compressibilities (κ_T) at 298 K

	α/(10^{-4} K^{-1})	κ_T/(10^{-6} atm^{-1})
Liquids		
Benzene	12.4	92.1
Carbon tetrachloride	12.4	90.5
Ethanol	11.2	76.8
Mercury	1.82	38.7
Water	2.1	49.6
Solids		
Copper	0.501	0.735
Diamond	0.030	0.187
Iron	0.354	0.589
Lead	0.861	2.21

The values refer to 20 °C.
Data: AIP(α), KL(κ_T).

Table 2D.2 Inversion temperatures (T_I), normal freezing (T_f) and boiling points (T_b), and Joule–Thomson coefficients (μ) at 1 atm and 298 K

	T_I/K	T_f/K	T_b/K	μ/(K atm^{-1})
Air	603			0.189 at 50 °C
Argon	723	83.8	87.3	
Carbon dioxide	1500	194.7s		1.11 at 300 K
Helium	40		4.22	−0.062
Hydrogen	202	14.0	20.3	−0.03
Krypton	1090	116.6	120.8	
Methane	968	90.6	111.6	
Neon	231	24.5	27.1	
Nitrogen	621	63.3	77.4	0.27
Oxygen	764	54.8	90.2	0.31

s: sublimes.
Data: AIP, JL, and M.W. Zemansky, *Heat and thermodynamics*. McGraw-Hill, New York (1957).

Table 3A.1 Standard entropies (and temperatures) of phase transitions, $\Delta_{trs}S^{\ominus}/(J\,K^{-1}\,mol^{-1})$

	Fusion (at T_f)	Vaporization (at T_b)
Ar	14.17 (at 83.8 K)	74.53 (at 87.3 K)
Br_2	39.76 (at 265.9 K)	88.61 (at 332.4 K)
C_6H_6	38.00 (at 278.6 K)	87.19 (at 353.2 K)
CH_3COOH	40.4 (at 289.8 K)	61.9 (at 391.4 K)
CH_3OH	18.03 (at 175.2 K)	104.6 (at 337.2 K)
Cl_2	37.22 (at 172.1 K)	85.38 (at 239.0 K)
H_2	8.38 (at 14.0 K)	44.96 (at 20.38 K)
H_2O	22.00 (at 273.2 K)	109.1 (at 373.2 K)
H_2S	12.67 (at 187.6 K)	87.75 (at 212.0 K)
He	4.8 (at 1.8 K and 30 bar)	19.9 (at 4.22 K)
N_2	11.39 (at 63.2 K)	75.22 (at 77.4 K)
NH_3	28.93 (at 195.4 K)	97.41 (at 239.73 K)
O_2	8.17 (at 54.4 K)	75.63 (at 90.2 K)

Data: AIP.

Table 3A.2 The standard enthalpies and entropies of vaporization of liquids at their normal boiling point

	$\Delta_{vap}H^{\ominus}/(kJ\,mol^{-1})$	$\theta_b/°C$	$\Delta_{vap}S^{\ominus}/(J\,K^{-1}\,mol^{-1})$
Benzene	30.8	80.1	+87.2
Carbon disulfide	26.74	46.25	+83.7
Carbon tetrachloride	30.00	76.7	+85.8
Cyclohexane	30.1	80.7	+85.1
Decane	38.75	174	+86.7
Dimethyl ether	21.51	−23	+86
Ethanol	38.6	78.3	+110.0
Hydrogen sulfide	18.7	−60.4	+87.9
Mercury	59.3	356.6	+94.2
Methane	8.18	−161.5	+73.2
Methanol	35.21	65.0	+104.1
Water	40.7	100.0	+109.1

Data: JL.

Table 3B.1 Standard Third-Law entropies at 298 K, $S_m^{\ominus}/(J\,K^{-1}\,mol^{-1})$. See Tables 2C.4 and 2C.5

Table 3C.1 Standard Gibbs energies of formation at 298 K, $\Delta_fG^{\ominus}/(kJ\,mol^{-1})$. See Tables 2C.4 and 2C.5

Table 3D.2 The fugacity coefficients of nitrogen at 273 K, ϕ

p/atm	ϕ	p/atm	ϕ
1	0.999 55	300	1.0055
10	0.9956	400	1.062
50	0.9912	600	1.239
100	0.9703	800	1.495
150	0.9672	1000	1.839
200	0.9721		

To convert to fugacities, use $f=\phi p$
Data: LR.

Table 5A.1 Henry's law constants for gases at 298 K, $K/(kPa\,kg\,mol^{-1})$

	Water	Benzene
CH_4	7.55×10^4	44.4×10^3
CO_2	3.01×10^3	8.90×10^2
H_2	1.28×10^5	2.79×10^4
N_2	1.56×10^5	1.87×10^4
O_2	7.92×10^4	

Data: converted from R.J. Silbey and R.A. Alberty, *Physical chemistry*. Wiley, New York (2001).

Table 5B.1 Freezing-point (K_f) and boiling-point (K_b) constants

	$K_f/(K\,kg\,mol^{-1})$	$K_b/(K\,kg\,mol^{-1})$
Acetic acid	3.90	3.07
Benzene	5.12	2.53
Camphor	40	
Carbon disulfide	3.8	2.37
Carbon tetrachloride	30	4.95
Naphthalene	6.94	5.8
Phenol	7.27	3.04
Water	1.86	0.51

Data: KL.

Table 5F.2 Mean activity coefficients in water at 298 K

$b/b^{\ominus}$	HCl	KCl	$CaCl_2$	H_2SO_4	$LaCl_3$	$In_2(SO_4)_3$
0.001	0.966	0.966	0.888	0.830	0.790	
0.005	0.929	0.927	0.789	0.639	0.636	0.16
0.01	0.905	0.902	0.732	0.544	0.560	0.11
0.05	0.830	0.816	0.584	0.340	0.388	0.035
0.10	0.798	0.770	0.524	0.266	0.356	0.025
0.50	0.769	0.652	0.510	0.155	0.303	0.014
1.00	0.811	0.607	0.725	0.131	0.387	
2.00	1.011	0.577	1.554	0.125	0.954	

Data: RS, HCP, and S. Glasstone, *Introduction to electrochemistry*. Van Nostrand (1942).

Table 6D.1 Standard potentials at 298 K, $E^{\ominus}$/V. (a) In electrochemical order

Reduction half-reaction	$E^{\ominus}$/V	Reduction half-reaction	$E^{\ominus}$/V
Strongly oxidizing		$Cu^+ + e^- \rightarrow Cu$	+0.52
$H_4XeO_6 + 2 H^+ + 2 e^- \rightarrow XeO_3 + 3 H_2O$	+3.0	$NiOOH + H_2O + e^- \rightarrow Ni(OH)_2 + OH^-$	+0.49
$F_2 + 2 e^- \rightarrow 2 F^-$	+2.87	$Ag_2CrO_4 + 2e^- \rightarrow 2Ag + CrO_4^{2-}$	+0.45
$O_3 + 2 H^+ + 2 e^- \rightarrow O_2 + H_2O$	+2.07	$O_2 + 2 H_2O + 4 e^- \rightarrow 4 OH^-$	+0.40
$S_2O_8^{2-} + 2e^- \rightarrow 2 SO_4^{2-}$	+2.05	$ClO_4^- + H_2O + 2e^- \rightarrow ClO_3^- + 2OH^-$	+0.36
$Ag^{2+} + e^- \rightarrow Ag^+$	+1.98	$[Fe(CN)_6]^{3-} + e^- \rightarrow [Fe(CN)_6]^{4-}$	+0.36
$Co^{3+} + e^- \rightarrow Co^{2+}$	+1.81	$Cu^{2+} + 2 e^- \rightarrow Cu$	+0.34
$H_2O_2 + 2 H^+ + 2 e^- \rightarrow 2 H_2O$	+1.78	$Hg_2Cl_2 + 2 e^- \rightarrow 2 Hg + 2 Cl^-$	+0.27
$Au^+ + e^- \rightarrow Au$	+1.69	$AgCl + e^- \rightarrow Ag + Cl^-$	+0.22
$Pb^{4+} + 2 e^- \rightarrow Pb^{2+}$	+1.67	$Bi^{3+} + 3 e^- \rightarrow Bi$	+0.20
$2 HClO + 2 H^+ + 2 e^- \rightarrow Cl_2 + 2 H_2O$	+1.63	$Cu^{2+} + e^- \rightarrow Cu^+$	+0.16
$Ce^{4+} + e^- \rightarrow Ce^{3+}$	+1.61	$Sn^{4+} + 2 e^- \rightarrow Sn^{2+}$	+0.15
$2 HBrO + 2 H^+ + 2 e^- \rightarrow Br_2 + 2 H_2O$	+1.60	$NO_3^- + H_2O + 2e^- \rightarrow NO_2^- + 2OH^-$	+0.10
$MnO_4^- + 8H^+ + 5e^- \rightarrow Mn^{2+} + 4H_2O$	+1.51	$AgBr + e^- \rightarrow Ag + Br^-$	+0.0713
$Mn^{3+} + e^- \rightarrow Mn^{2+}$	+1.51	$Ti^{4+} + e^- \rightarrow Ti^{3+}$	0.00
$Au^{3+} + 3 e^- \rightarrow Au$	+1.40	$2 H^+ + 2 e^- \rightarrow H_2$	0, by definition
$Cl_2 + 2 e^- \rightarrow 2 Cl^-$	+1.36	$Fe^{3+} + 3 e^- \rightarrow Fe$	−0.04
$Cr_2O_7^{2-} + 14H^+ + 6e^- \rightarrow 2Cr^{3+} + 7H_2O$	+1.33	$O_2 + H_2O + 2 e^- \rightarrow HO_2^- + OH^-$	−0.08
$O_3 + H_2O + 2 e^- \rightarrow O_2 + 2 OH^-$	+1.24	$Pb^{2+} + 2 e^- \rightarrow Pb$	−0.13
$O_2 + 4 H^+ + 4 e^- \rightarrow 2 H_2O$	+1.23	$In^+ + e^- \rightarrow In$	−0.14
$ClO_4^- + 2H^+ + 2e^- \rightarrow ClO_3^- + H_2O$	+1.23	$Sn^{2+} + 2 e^- \rightarrow Sn$	−0.14
$MnO_2 + 4 H^+ + 2 e^- \rightarrow Mn^{2+} + 2 H_2O$	+1.23	$AgI + e^- \rightarrow Ag + I^-$	−0.15
$Pt^{2+} + 2 e^- \rightarrow Pt$	+1.20	$Ni^{2+} + 2 e^- \rightarrow Ni$	−0.23
$Br_2 + 2 e^- \rightarrow 2Br^-$	+1.09	$V^{3+} + e^- \rightarrow V^{2+}$	−0.26
$Pu^{4+} + e^- \rightarrow Pu^{3+}$	+0.97	$Co^{2+} + 2 e^- \rightarrow Co$	−0.28
$NO_3^- + 4H^+ + 3e^- \rightarrow NO + 2H_2O$	+0.96	$In^{3+} + 3 e^- \rightarrow In$	−0.34
$2Hg^{2+} + 2e^- \rightarrow Hg_2^{2+}$	+0.92	$Tl^+ + e^- \rightarrow Tl$	−0.34
$ClO^- + H_2O + 2 e^- \rightarrow Cl^- + 2 OH^-$	+0.89	$PbSO_4 + 2 e^- \rightarrow Pb + SO_4^{2-}$	−0.36
$Hg^{2+} + 2 e^- \rightarrow Hg$	+0.86	$Ti^{3+} + e^- \rightarrow Ti^{2+}$	−0.37
$NO_3^- + 2H^+ + e^- \rightarrow NO_2 + H_2O$	+0.80	$Cd^{2+} + 2 e^- \rightarrow Cd$	−0.40
$Ag^+ + e^- \rightarrow Ag$	+0.80	$In^{2+} + e^- \rightarrow In^+$	−0.40
$Hg_2^{2+} + 2e^- \rightarrow 2Hg$	+0.79	$Cr^{3+} + e^- \rightarrow Cr^{2+}$	−0.41
$AgF + e^- \rightarrow Ag + F^-$	+0.78	$Fe^{2+} + 2 e^- \rightarrow Fe$	−0.44
$Fe^{3+} + e^- \rightarrow Fe^{2+}$	+0.77	$In^{3+} + 2 e^- \rightarrow In^+$	−0.44
$BrO^- + H_2O + 2 e^- \rightarrow Br^- + 2 OH^-$	+0.76	$S + 2 e^- \rightarrow S^{2-}$	−0.48
$Hg_2SO_4 + 2e^- \rightarrow 2Hg + SO_4^{2-}$	+0.62	$In^{3+} + e^- \rightarrow In^{2+}$	−0.49
$MnO_4^{2-} + 2H_2O + 2e^- \rightarrow MnO_2 + 4OH^-$	+0.60	$O_2 + e^- \rightarrow O_2^-$	−0.56
$MnO_4^- + e^- \rightarrow MnO_4^{2-}$	+0.56	$U^{4+} + e^- \rightarrow U^{3+}$	−0.61
$I_2 + 2 e^- \rightarrow 2 I^-$	+0.54	$Cr^{3+} + 3 e^- \rightarrow Cr$	−0.74
$I_3^- + 2e^- \rightarrow 3I^-$	+0.53	$Zn^{2+} + 2 e^- \rightarrow Zn$	−0.76

(Continued)

Table 6D.1 (Continued)

Reduction half-reaction	$E^{\ominus}$/V	Reduction half-reaction	$E^{\ominus}$/V
$Cd(OH)_2 + 2e^- \rightarrow Cd + 2OH^-$	−0.81	$Ce^{3+} + 3e^- \rightarrow Ce$	−2.48
$2H_2O + 2e^- \rightarrow H_2 + 2OH^-$	−0.83	$La^{3+} + 3e^- \rightarrow La$	−2.52
$Cr^{2+} + 2e^- \rightarrow Cr$	−0.91	$Na^+ + e^- \rightarrow Na$	−2.71
$Mn^{2+} + 2e^- \rightarrow Mn$	−1.18	$Ca^{2+} + 2e^- \rightarrow Ca$	−2.87
$V^{2+} + 2e^- \rightarrow V$	−1.19	$Sr^{2+} + 2e^- \rightarrow Sr$	−2.89
$Ti^{2+} + 2e^- \rightarrow Ti$	−1.63	$Ba^{2+} + 2e^- \rightarrow Ba$	−2.91
$Al^{3+} + 3e^- \rightarrow Al$	−1.66	$Ra^{2+} + 2e^- \rightarrow Ra$	−2.92
$U^{3+} + 3e^- \rightarrow U$	−1.79	$Cs^+ + e^- \rightarrow Cs$	−2.92
$Be^{2+} + 2e^- \rightarrow Be$	−1.85	$Rb^+ + e^- \rightarrow Rb$	−2.93
$Sc^{3+} + 3e^- \rightarrow Sc$	−2.09	$K^+ + e^- \rightarrow K$	−2.93
$Mg^{2+} + 2e^- \rightarrow Mg$	−2.36	$Li^+ + e^- \rightarrow Li$	−3.05

Table 6D.1 Standard potentials at 298 K, $E^{\ominus}$/V. (b) In alphabetical order

Reduction half-reaction	$E^{\ominus}$/V	Reduction half-reaction	$E^{\ominus}$/V
$Ag^+ + e^- \rightarrow Ag$	+0.80	$Cr^{2+} + 2e^- \rightarrow Cr$	−0.91
$Ag^{2+} + e^- \rightarrow Ag^+$	+1.98	$Cr_2O_7^{2-} + 14H^+ + 6e^- \rightarrow 2Cr^{3+} + 7H_2O$	+1.33
$AgBr + e^- \rightarrow Ag + Br^-$	+0.0713	$Cr^{3+} + 3e^- \rightarrow Cr$	−0.74
$AgCl + e^- \rightarrow Ag + Cl^-$	+0.22	$Cr^{3+} + e^- \rightarrow Cr^{2+}$	−0.41
$Ag_2CrO_4 + 2e^- \rightarrow 2Ag + CrO_4^{2-}$	+0.45	$Cs^+ + e^- \rightarrow Cs$	−2.92
$AgF + e^- \rightarrow Ag + F^-$	+0.78	$Cu^+ + e^- \rightarrow Cu$	+0.52
$AgI + e^- \rightarrow Ag + I^-$	−0.15	$Cu^{2+} + 2e^- \rightarrow Cu$	+0.34
$Al^{3+} + 3e^- \rightarrow Al$	−1.66	$Cu^{2+} + e^- \rightarrow Cu^+$	+0.16
$Au^+ + e^- \rightarrow Au$	+1.69	$F_2 + 2e^- \rightarrow 2F^-$	+2.87
$Au^{3+} + 3e^- \rightarrow Au$	+1.40	$Fe^{2+} + 2e^- \rightarrow Fe$	−0.44
$Ba^{2+} + 2e^- \rightarrow Ba$	−2.91	$Fe^{3+} + 3e^- \rightarrow Fe$	−0.04
$Be^{2+} + 2e^- \rightarrow Be$	−1.85	$Fe^{3+} + e^- \rightarrow Fe^{2+}$	+0.77
$Bi^{3+} + 3e^- \rightarrow Bi$	+0.20	$[Fe(CN)_6]^{3-} + e^- \rightarrow [Fe(CN)_6]^{4-}$	+0.36
$Br_2 + 2e^- \rightarrow 2Br^-$	+1.09	$2H^+ + 2e^- \rightarrow H_2$	0, by definition
$BrO^- + H_2O + 2e^- \rightarrow Br^- + 2OH^-$	+0.76	$2H_2O + 2e^- \rightarrow H_2 + 2OH^-$	−0.83
$Ca^{2+} + 2e^- \rightarrow Ca$	−2.87	$2HBrO + 2H^+ + 2e^- \rightarrow Br_2 + 2H_2O$	+1.60
$Cd(OH)_2 + 2e^- \rightarrow Cd + 2OH^-$	−0.81	$2HClO + 2H^+ + 2e^- \rightarrow Cl_2 + 2H_2O$	+1.63
$Cd^{2+} + 2e^- \rightarrow Cd$	−0.40	$H_2O_2 + 2H^+ + 2e^- \rightarrow 2H_2O$	+1.78
$Ce^{3+} + 3e^- \rightarrow Ce$	−2.48	$H_4XeO_6 + 2H^+ + 2e^- \rightarrow XeO_3 + 3H_2O$	+3.0
$Ce^{4+} + e^- \rightarrow Ce^{3+}$	+1.61	$Hg_2^{2+} + 2e^- \rightarrow 2Hg$	+0.79
$Cl_2 + 2e^- \rightarrow 2Cl^-$	+1.36	$Hg_2Cl_2 + 2e^- \rightarrow 2Hg + 2Cl^-$	+0.27
$ClO^- + H_2O + 2e^- \rightarrow Cl^- + 2OH^-$	+0.89	$Hg^{2+} + 2e^- \rightarrow Hg$	+0.86
$ClO_4^- + 2H^+ + 2e^- \rightarrow ClO_3^- + H_2O$	+1.23	$2Hg^{2+} + 2e^- \rightarrow Hg_2^{2+}$	+0.92
$ClO_4^- + H_2O + 2e^- \rightarrow ClO_3^- + 2OH^-$	+0.36	$Hg_2SO_4 + 2e^- \rightarrow 2Hg + SO_4^{2-}$	+0.62
$Co^{2+} + 2e^- \rightarrow Co$	−0.28	$I_2 + 2e^- \rightarrow 2I^-$	+0.54
$Co^{3+} + e^- \rightarrow Co^{2+}$	+1.81	$I_3^- + 2e^- \rightarrow 3I^-$	+0.53

Table 6D.1a (Continued)

Reduction half-reaction	$E^\ominus$/V	Reduction half-reaction	$E^\ominus$/V
$In^+ + e^- \rightarrow In$	−0.14	$O_3 + 2H^+ + 2e^- \rightarrow O_2 + H_2O$	+2.07
$In^{2+} + e^- \rightarrow In^+$	−0.40	$O_3 + H_2O + 2e^- \rightarrow O_2 + 2OH^-$	+1.24
$In^{3+} + 2e^- \rightarrow In^+$	−0.44	$Pb^{2+} + 2e^- \rightarrow Pb$	−0.13
$In^{3+} + 3e^- \rightarrow In$	−0.34	$Pb^{4+} + 2e^- \rightarrow Pb^{2+}$	+1.67
$In^{3+} + e^- \rightarrow In^{2+}$	−0.49	$PbSO_4 + 2e^- \rightarrow Pb + SO_4^{2-}$	−0.36
$K^+ + e^- \rightarrow K$	−2.93	$Pt^{2+} + 2e^- \rightarrow Pt$	+1.20
$La^{3+} + 3e^- \rightarrow La$	−2.52	$Pu^{4+} + e^- \rightarrow Pu^{3+}$	+0.97
$Li^+ + e^- \rightarrow Li$	−3.05	$Ra^{2+} + 2e^- \rightarrow Ra$	−2.92
$Mg^{2+} + 2e^- \rightarrow Mg$	−2.36	$Rb^+ + e^- \rightarrow Rb$	−2.93
$Mn^{2+} + 2e^- \rightarrow Mn$	−1.18	$S + 2e^- \rightarrow S^{2-}$	−0.48
$Mn^{3+} + e^- \rightarrow Mn^{2+}$	+1.51	$S_2O_8^{2-} + 2e^- \rightarrow 2SO_4^{2-}$	+2.05
$MnO_2 + 4H^+ + 2e^- \rightarrow Mn^{2+} + 2H_2O$	+1.23	$Sc^{3+} + 3e^- \rightarrow Sc$	−2.09
$MnO_4^{2-} + 8H^+ + 5e^- \rightarrow Mn^{2+} + 4H_2O$	+1.51	$Sn^{2+} + 2e^- \rightarrow Sn$	−0.14
$MnO_4^- + e^- \rightarrow MnO_4^{2-}$	+0.56	$Sn^{4+} + 2e^- \rightarrow Sn^{2+}$	+0.15
$MnO_4^{2-} + 2H_2O + 2e^- \rightarrow MnO_2 + 4OH^-$	+0.60	$Sr^{2+} + 2e^- \rightarrow Sr$	−2.89
$Na^+ + e^- \rightarrow Na$	−2.71	$Ti^{2+} + 2e^- \rightarrow Ti$	−1.63
$Ni^{2+} + 2e^- \rightarrow Ni$	−0.23	$Ti^{3+} + e^- \rightarrow Ti^{2+}$	−0.37
$NiOOH + H_2O + e^- \rightarrow Ni(OH)_2 + OH^-$	+0.49	$Ti^{4+} + e^- \rightarrow Ti^{3+}$	0.00
$NO_3^- + 2H^+ + e^- \rightarrow NO_2 + H_2O$	+0.80	$Tl^+ + e^- \rightarrow Tl$	−0.34
$NO_3^- + 4H^+ + 3e^- \rightarrow NO + 2H_2O$	+0.96	$U^{3+} + 3e^- \rightarrow U$	−1.79
$NO_3^- + H_2O + 2e^- \rightarrow + NO_2^- + 2OH^-$	+0.10	$U^{4+} + e^- \rightarrow U^{3+}$	−0.61
$O_2 + 2H_2O + 4e^- \rightarrow 4OH^-$	+0.40	$V^{2+} + 2e^- \rightarrow V$	−1.19
$O_2 + 4H^+ + 4e^- \rightarrow 2H_2O$	+1.23	$V^{3+} + e^- \rightarrow V^{2+}$	−0.26
$O_2 + e^- \rightarrow O_2^-$	−0.56	$Zn^{2+} + 2e^- \rightarrow Zn$	−0.76
$O_2 + H_2O + 2e^- \rightarrow HO_2^- + OH^-$	−0.08		

Table 9B.1 Effective nuclear charge, $Z_{eff} = Z - \sigma$*

	H							He
1s	1							1.6875
	Li	Be	B	C	N	O	F	Ne
1s	2.6906	3.6848	4.6795	5.6727	6.6651	7.6579	8.6501	9.6421
2s	1.2792	1.9120	2.5762	3.2166	3.8474	4.4916	5.1276	5.7584
2p			2.4214	3.1358	3.8340	4.4532	5.1000	5.7584
	Na	Mg	Al	Si	P	S	Cl	Ar
1s	10.6259	11.6089	12.5910	13.5745	14.5578	15.5409	16.5239	17.5075
2s	6.5714	7.3920	8.3736	9.0200	9.8250	10.6288	11.4304	12.2304
2p	6.8018	7.8258	8.9634	9.9450	10.9612	11.9770	12.9932	14.0082
3s	2.5074	3.3075	4.1172	4.9032	5.6418	6.3669	7.0683	7.7568
3p			4.0656	4.2852	4.8864	5.4819	6.1161	6.7641

* The actual charge is $Z_{eff}e$.

Data: E. Clementi and D.L. Raimondi, *Atomic screening constants from SCF functions.* IBM Res. Note NJ-27 (1963). *J. Chem. Phys.* **38**, 2686 (1963).

Table 9B.2 First and subsequent ionization energies, I/(kJ mol^{-1})

H							He
1312.0							2372.3
							5250.4
Li	Be	B	C	N	O	F	Ne
513.3	899.4	800.6	1086.2	1402.3	1313.9	1681	2080.6
7298.0	1757.1	2427	2352	2856.1	3388.2	3374	3952.2
Na	Mg	Al	Si	P	S	Cl	Ar
495.8	737.7	577.4	786.5	1011.7	999.6	1251.1	1520.4
4562.4	1450.7	1816.6	1577.1	1903.2	2251	2297	2665.2
		2744.6		2912			
K	Ca	Ga	Ge	As	Se	Br	Kr
418.8	589.7	578.8	762.1	947.0	940.9	1139.9	1350.7
3051.4	1145	1979	1537	1798	2044	2104	2350
		2963	2735				
Rb	Sr	In	Sn	Sb	Te	I	Xe
403.0	549.5	558.3	708.6	833.7	869.2	1008.4	1170.4
2632	1064.2	1820.6	1411.8	1794	1795	1845.9	2046
		2704	2943.0	2443			
Cs	Ba	Tl	Pb	Bi	Po	At	Rn
375.5	502.8	589.3	715.5	703.2	812	930	1037
2420	965.1	1971.0	1450.4	1610			
		2878	3081.5	2466			

Data: E.

Table 9B.3 Electron affinities, E_{ea}/(kJ mol^{-1})

H							He
72.8							−21
Li	Be	B	C	N	O	F	Ne
59.8	≤0	23	122.5	−7	141	322	−29
					−844		
Na	Mg	Al	Si	P	S	Cl	Ar
52.9	≤0	44	133.6	71.7	200.4	348.7	−35
					−532		
K	Ca	Ga	Ge	As	Se	Br	Kr
48.3	2.37	36	116	77	195.0	324.5	−39
Rb	Sr	In	Sn	Sb	Te	I	Xe
46.9	5.03	34	121	101	190.2	295.3	−41
Cs	Ba	Tl	Pb	Bi	Po	At	Rn
45.5	13.95	30	35.2	101	186	270	−41

Data: E.

Table 10C.1 Bond lengths, R_e/pm

(a) Bond lengths in specific molecules

Br_2	228.3
Cl_2	198.75
CO	112.81
F_2	141.78
H_2^+	106
H_2	74.138
HBr	141.44
HCl	127.45
HF	91.680
HI	160.92
N_2	109.76
O_2	120.75

(b) Mean bond lengths from covalent radii*

H	37						
C	77(1)	N	74(1)	O	66(1)	F	64
	67(2)		65(2)		57(2)		
	60(3)						
Si	118	P	110	S	104(1)	Cl	99
					95(2)		
Ge	122	As	121	Se	104	Br	114
		Sb	141	Te	137	I	133

* Values are for single bonds except where indicated otherwise (values in parentheses). The length of an A–B covalent bond (of given order) is the sum of the corresponding covalent radii.

Table 10C.2a Bond dissociation enthalpies, $\Delta H^{\ominus}$(A–B)/(kJ mol^{-1}) at 298 K*

Diatomic molecules

H–H	436	F–F	155	Cl–Cl	242	Br–Br	193	I–I	151
O=O	497	C=O	1076	N≡N	945				
H–O	428	H–F	565	H–Cl	431	H–Br	366	H–I	299

Polyatomic molecules

H–CH_3	435	H–NH_2	460	H–OH	492	H–C_6H_5	469
H_3C–CH_3	368	H_2C=CH_2	720	HC≡CH	962		
HO–CH_3	377	Cl–CH_3	352	Br–CH_3	293	I–CH_3	237
O=CO	531	HO–OH	213	O_2N–NO_2	54		

* To a good approximation bond dissociation enthalpies and dissociation energies are related by $\Delta H^{\ominus} = D_e + \frac{3}{2}RT$ with $D_e = D_0 + \frac{1}{2}\hbar\omega$. For precise values of D_0 for diatomic molecules, see Table 12D.1.

Data: HCP, KL.

Table 10C.2b Mean bond enthalpies, $\Delta H^{\ominus}$(A–B)/(kJ mol^{-1})*

	H	C	N	O	F	Cl	Br	I	S	P	Si
H	436										
C	412	348(i)									
		612(ii)									
		838(iii)									
		518(a)									
N	388	305(i)	163(i)								
		613(ii)	409(ii)								
		890(iii)	946(iii)								
O	463	360(i)	157	146(i)							
		743(ii)		497(ii)							
F	565	484	270	185	155						
Cl	431	338	200	203	254	242					
Br	366	276				219	193				
I	299	238				210	178	151			
S	338	259			496	250	212		264		
P	322									201	
Si	318		374	466							226

* Mean bond enthalpies are such a crude measure of bond strength that they need not be distinguished from dissociation energies.
(i) Single bond, (ii) double bond, (iii) triple bond, (a) aromatic.
Data: HCP and L. Pauling, *The nature of the chemical bond*. Cornell University Press (1960).

Table 10D.1 Pauling (*italics*) and Mulliken electronegativities

H							He
2.20							
3.06							
Li	Be	B	C	N	O	F	Ne
0.98	*1.57*	*2.04*	*2.55*	*3.04*	*3.44*	*3.98*	
1.28	1.99	1.83	2.67	3.08	3.22	4.43	4.60
Na	Mg	Al	Si	P	S	Cl	Ar
0.93	*1.31*	*1.61*	*1.90*	*2.19*	*2.58*	*3.16*	
1.21	1.63	1.37	2.03	2.39	2.65	3.54	3.36
K	Ca	Ga	Ge	As	Se	Br	Kr
0.82	*1.00*	*1.81*	*2.01*	*2.18*	*2.55*	*2.96*	*3.0*
1.03	1.30	1.34	1.95	2.26	2.51	3.24	2.98
Rb	Sr	In	Sn	Sb	Te	I	Xe
0.82	*0.95*	*1.78*	*1.96*	*2.05*	*2.10*	*2.66*	*2.6*
0.99	1.21	1.30	1.83	2.06	2.34	2.88	2.59
Cs	Ba	Tl	Pb	Bi			
0.79	*0.89*	*2.04*	*2.33*	*2.02*			

Data: Pauling values: A.L. Allred, *J. Inorg. Nucl. Chem.* **17**, 215 (1961); L.C. Allen and J.E. Huheey, ibid., **42**, 1523 (1980). Mulliken values: L.C. Allen, *J. Am. Chem. Soc.* **111**, 9003 (1989). The Mulliken values have been scaled to the range of the Pauling values.

Table 11B.1 The C_{3v} character table; see Part 4

Table 11B.2 The C_{2v} character table; see Part 4

Table 12D.1 Properties of diatomic molecules

	$\tilde{\nu}$/cm^{-1}	θ^V/K	$\tilde{B}$/cm^{-1}	θ^R/K	R_e/pm	k_f/(N m^{-1})	$hc\tilde{D}_0$/(kJ mol^{-1})	σ
$^1H_2^+$	2321.8	3341	29.8	42.9	106	160	255.8	2
$^1H_2$	4400.39	6332	60.864	87.6	74.138	574.9	432.1	2
$^2H_2$	3118.46	4487	30.442	43.8	74.154	577.0	439.6	2
$^1H^{19}F$	4138.32	5955	20.956	30.2	91.680	965.7	564.4	1
$^1H^{35}Cl$	2990.95	4304	10.593	15.2	127.45	516.3	427.7	1
$^1H^{81}Br$	2648.98	3812	8.465	12.2	141.44	411.5	362.7	1
$^1H^{127}I$	2308.09	3321	6.511	9.37	160.92	313.8	294.9	1
$^{14}N_2$	2358.07	3393	1.9987	2.88	109.76	2293.8	941.7	2
$^{16}O_2$	1580.36	2274	1.4457	2.08	120.75	1176.8	493.5	2
$^{19}F_2$	891.8	1283	0.8828	1.27	141.78	445.1	154.4	2
$^{35}Cl_2$	559.71	805	0.2441	0.351	198.75	322.7	239.3	2
$^{12}C^{16}O$	2170.21	3122	1.9313	2.78	112.81	1903.17	1071.8	1
$^{79}Br^{81}Br$	323.2	465	0.0809	10.116	283.3	245.9	190.2	1

Data: AIP.

Table 12E.1 Typical vibrational wavenumbers, $\tilde{\nu}$/cm^{-1}

C–H stretch	2850–2960
C–H bend	1340–1465
C–C stretch, bend	700–1250
C=C stretch	1620–1680
C≡C stretch	2100–2260
O–H stretch	3590–3650
H-bonds	3200–3570
C=O stretch	1640–1780
C≡N stretch	2215–2275
N–H stretch	3200–3500
C–F stretch	1000–1400
C–Cl stretch	600–800
C–Br stretch	500–600
C–I stretch	500
CO_3^{2-}	1410–1450
NO_3^-	1350–1420
NO_2^-	1230–1250
SO_4^{2-}	1080–1130
Silicates	900–1100

Data: L.J. Bellamy, *The infrared spectra of complex molecules* and *Advances in infrared group frequencies*. Chapman and Hall.

Table 13A.1 Colour, wavelength, frequency, and energy of light

Colour	λ/nm	ν/(10^{14} Hz)	$\tilde{\nu}$/(10^4 cm^{-1})	E/eV	E/(kJ mol^{-1})
Infrared	>1000	<3.00	<1.00	<1.24	<120
Red	700	4.28	1.43	1.77	171
Orange	620	4.84	1.61	2.00	193
Yellow	580	5.17	1.72	2.14	206
Green	530	5.66	1.89	2.34	226
Blue	470	6.38	2.13	2.64	254
Violet	420	7.14	2.38	2.95	285
Ultraviolet	<400	>7.5	>2.5	>3.10	>300

Data: J.G. Calvert and J.N. Pitts, *Photochemistry*. Wiley, New York (1966).

Table 13A.2 Absorption characteristics of some groups and molecules

Group	$\tilde{\nu}_{max}/(10^4\ cm^{-1})$	λ_{max}/nm	$\varepsilon_{max}/(dm^3\ mol^{-1}\ cm^{-1})$
C=C ($\pi^* \leftarrow \pi$)	6.10	163	1.5×10^4
	5.73	174	5.5×10^3
C=O ($\pi^* \leftarrow n$)	3.7–3.5	270–290	10–20
–N=N–	2.9	350	15
	>3.9	<260	Strong
$-NO_2$	3.6	280	10
	4.8	210	1.0×10^4
C_6H_5-	3.9	255	200
	5.0	200	6.3×10^3
	5.5	180	1.0×10^5
$[Cu(OH_2)_6]^{2+}(aq)$	1.2	810	10
$[Cu(NH_3)_4]^{2+}(aq)$	1.7	600	50
H_2O ($\pi^* \leftarrow n$)	6.0	167	7.0×10^3

Table 14A.2 Nuclear spin properties

Nuclide	Natural abundance, %	Spin, I	Magnetic Moment, μ/μ_N	g-value	$\gamma/(10^7\ T^{-1}s^{-1})$	NMR frequency at 1 T, ν/MHz
^{1}n*		$\frac{1}{2}$	−1.9130	−3.8260	−18.324	29.164
^{1}H	99.9844	$\frac{1}{2}$	2.792 85	5.5857	26.752	42.576
^{2}H	0.0156	1	0.857 44	0.857 44	4.1067	6.536
^{3}H*		$\frac{1}{2}$	2.978 96	−4.2553	−20.380	45.414
^{10}B	19.6	3	1.8006	0.6002	2.875	4.575
^{11}B	80.4	$\frac{3}{2}$	2.6886	1.7923	8.5841	13.663
^{13}C	1.108	$\frac{1}{2}$	0.7024	1.4046	6.7272	10.708
^{14}N	99.635	1	0.403 56	0.403 56	1.9328	3.078
^{17}O	0.037	$\frac{5}{2}$	−1.893 79	−0.7572	−3.627	5.774
^{19}F	100	$\frac{1}{2}$	2.628 87	5.2567	25.177	40.077
^{31}P	100	$\frac{1}{2}$	1.1316	2.2634	10.840	17.251
^{33}S	0.74	$\frac{3}{2}$	0.6438	0.4289	2.054	3.272
^{35}Cl	75.4	$\frac{3}{2}$	0.8219	0.5479	2.624	4.176
^{37}Cl	24.6	$\frac{3}{2}$	0.6841	0.4561	2.184	3.476

* Radioactive.

μ is the magnetic moment of the spin state with the largest value of m_I: $\mu = g_I\mu_N I$ and μ_N is the nuclear magneton (see inside front cover).

Data: KL and HCP.

Table 14D.1 Hyperfine coupling constants for atoms, a/mT

Nuclide	Spin	Isotropic coupling	Anisotropic coupling
^{1}H	$\frac{1}{2}$	50.8(1s)	
^{2}H	1	7.8(1s)	
^{13}C	$\frac{1}{2}$	113.0(2s)	6.6(2p)
^{14}N	1	55.2(2s)	4.8(2p)
^{19}F	$\frac{1}{2}$	1720(2s)	108.4(2p)
^{31}P	$\frac{1}{2}$	364(3s)	20.6(3p)
^{35}Cl	$\frac{3}{2}$	168(3s)	10.0(3p)
^{37}Cl	$\frac{3}{2}$	140(3s)	8.4(3p)

Data: P.W. Atkins and M.C.R. Symons, *The structure of inorganic radicals*. Elsevier, Amsterdam (1967).

Table 16A.1 Magnitudes of dipole moments (μ), polarizabilities (α), and polarizability volumes (α')

	$\mu/(10^{-30}$ C m)	μ/D	$\alpha'/(10^{-30}$ m^3)	$\alpha/(10^{-40}$ J^{-1} C^2 m^2)
Ar	0	0	1.66	1.85
C_2H_5OH	5.64	1.69		
$C_6H_5CH_3$	1.20	0.36		
C_6H_6	0	0	10.4	11.6
CCl_4	0	0	10.3	11.7
CH_2Cl_2	5.24	1.57	6.80	7.57
CH_3Cl	6.24	1.87	4.53	5.04
CH_3OH	5.70	1.71	3.23	3.59
CH_4	0	0	2.60	2.89
$CHCl_3$	3.37	1.01	8.50	9.46
CO	0.390	0.117	1.98	2.20
CO_2	0	0	2.63	2.93
H_2	0	0	0.819	0.911
H_2O	6.17	1.85	1.48	1.65
HBr	2.67	0.80	3.61	4.01
HCl	3.60	1.08	2.63	2.93
He	0	0	0.20	0.22
HF	6.37	1.91	0.51	0.57
HI	1.40	0.42	5.45	6.06
N_2	0	0	1.77	1.97
NH_3	4.90	1.47	2.22	2.47
1,2-$C_6H_4(CH_3)_2$	2.07	0.62		

Data: HCP and C.J.F. Böttcher and P. Bordewijk, *Theory of electric polarization*. Elsevier, Amsterdam (1978).

Table 16B.2 Lennard-Jones parameters for the (12,6) potential

	(ε/k)/K	r_0/pm
Ar	111.84	362.3
C_2H_2	209.11	463.5
C_2H_4	200.78	458.9
C_2H_6	216.12	478.2
C_6H_6	377.46	617.4
CCl_4	378.86	624.1
Cl_2	296.27	448.5
CO_2	201.71	444.4
F_2	104.29	357.1
Kr	154.87	389.5
N_2	91.85	391.9
O_2	113.27	365.4
Xe	213.96	426.0

Source: F. Cuadros, I. Cachadiña, and W. Ahamuda, *Molec. Engineering* 6, 319 (1996).

Table 16C.1 Surface tensions of liquids at 293 K, γ/(mN m^{-1})

	γ/(mN m^{-})
Benzene	28.88
Carbon tetrachloride	27.0
Ethanol	22.8
Hexane	18.4
Mercury	472
Methanol	22.6
Water	72.75
	72.0 at 25 °C
	58.0 at 100 °C

Data: KL.

Table 17D.1 Radius of gyration

	M/(kg mol^{-1})	R_g/nm
Serum albumin	66	2.98
Myosin	493	46.8
Polystyrene	3.2×10^3	50†
DNA	4×10^3	117
Tobacco mosaic virus	3.9×10^4	92.4

† In a poor solvent.

Table 17D.2 Frictional coefficients and molecular geometry

a/b	Prolate	Oblate
2	1.04	1.04
3	1.18	1.17
4	1.18	1.17
5	1.25	1.22
6	1.31	1.28
7	1.38	1.33
8	1.43	1.37
9	149	1.42
10	1.54	1.46
50	2.95	2.38
100	4.07	2.97

Data: K.E. Van Holde, *Physical biochemistry*. Prentice-Hall, Englewood Cliffs (1971)
Sphere; radius a, $c=af_0$
Prolate ellipsoid; major axis $2a$, minor axis $2b$, $c=(ab)^{1/3}$

$$f=\left\{\frac{(1-b^2/a^2)^{1/2}}{(b/a)^{2/3}\ln\{\left[1+(1-b^2/a^2)^{1/2}\right]/(b/a)\}}\right\}f_0$$

Oblate ellipsoid; major axis $2a$, minor axis $2b$, $c=(a^2b)^{1/3}$

$$f=\left\{\frac{(a^2/b^2-1)^{1/2}}{(a/b)^{2/3}\arctan[(a^2/b^2-1)^{1/2}]}\right\}f_0$$

Long rod; length l, radius a, $c=(3a^2/4)^{1/3}$

$$f=\left\{\frac{(1/2a)^{2/3}}{(3/2)^{1/3}\{2\ln(l/a)-0.11\}}\right\}f_0$$

In each $f_0=6\pi\eta c$ with the appropriate value of c.

Table 17D.3 Intrinsic viscosity

Macromolecule	Solvent	θ/°C	K/(10^{-3} cm^3 g^{-1})	a
Polystyrene	Benzene	25	9.5	0.74
	Cyclobutane	34†	81	0.50
Polyisobutylene	Benzene	23†	83	0.50
	Cyclohexane	30	26	0.70
Amylose	0.33 M KCl(aq)	25†	113	0.50
Various proteins‡	Guanidine hydrochloride + $HSCH_2CH_2OH$		7.16	0.66

† The θ temperature.
‡ Use $[\eta] = KN^a$; N is the number of amino acid residues.
Data: K.E. Van Holde, *Physical biochemistry*. Prentice-Hall, Englewood Cliffs (1971).

Table 18B.2 Ionic radii, r/pm*

Li^+(4)	Be^{2+}(4)	B^{3+}(4)	N^{3-}	O^{2-}(6)	F^-(6)
59	27	12	171	140	133
Na^+(6)	Mg^{2+}(6)	Al^{3+}(6)	P^{3-}	S^{2-}(6)	Cl^-(6)
102	72	53	212	184	181
K^+(6)	Ca^{2+}(6)	Ga^{3+}(6)	As^{3-}(6)	Se^{2-}(6)	Br^-(6)
138	100	62	222	198	196
Rb^+(6)	Sr^{2+}(6)	In^{3+}(6)		Te^{2-}(6)	I^-(6)
149	116	79		221	220
Cs^+(6)	Ba^{2+}(6)	Tl^{3+}(6)			
167	136	88			

d-block elements (high-spin ions)

Sc^{3+}(6)	Ti^{4+}(6)	Cr^{3+}(6)	Mn^{3+}(6)	Fe^{2+}(6)	Co^{3+}(6)	Cu^{2+}(6)	Zn^{2+}(6)
73	60	61	65	63	61	73	75

* Numbers in parentheses are the coordination numbers of the ions. Values for ions without a coordination number stated are estimates.
Data: R.D. Shannon and C.T. Prewitt, *Acta Cryst.* **B25**, 925 (1969).

Table 18B.4 Lattice enthalpies at 298 K, ΔH_L/(kJ mol^{-1})

	F	Cl	Br	I
Halides				
Li	1037	852	815	761
Na	926	787	752	705
K	821	717	689	649
Rb	789	695	668	632
Cs	750	676	654	620
Ag	969	912	900	886
Be		3017		
Mg		2524		
Ca		2255		
Sr		2153		

Oxides							
MgO	3850	CaO	3461	SrO	3283	BaO	3114
Sulfides							
MgS	3406	CaS	3119	SrS	2974	BaS	2832

Entries refer to $MX(s) \rightarrow M^+(g) + X^-(g)$.
Data: Principally D. Cubicciotti et al., *J. Chem. Phys.* **31**, 1646 (1959).

Table 18C.1 Magnetic susceptibilities at 298 K

	$\chi/10^{-6}$	$\chi_m/(10^{-10}\ m^3\ mol^{-1})$
H_2O(l)	−9.02	−1.63
C_6H_6(l)	−8.8	−7.8
C_6H_{12}(l)	−10.2	−11.1
CCl_4(l)	−5.4	−5.2
NaCl(s)	−16	−3.8
Cu(s)	−9.7	−0.69
S(rhombic)	−12.6	−1.95
Hg(l)	−28.4	−4.21
Al(s)	+20.7	+2.07
Pt(s)	+267.3	+24.25
Na(s)	+8.48	+2.01
K(s)	+5.94	+2.61
$CuSO_4 \cdot 5H_2O$(s)	+167	+183
$MnSO_4 \cdot 4H_2O$(s)	+1859	+1835
$NiSO_4 \cdot 7H_2O$(s)	+355	+503
$FeSO_4$(s)	+3743	+1558

Source: Principally HCP, with $\chi_m = \chi V_m = \chi\rho/M$.

Table 19A.1 Transport properties of gases at 1 atm

	$\kappa/(mW\ K^{-1}\ m^{-1})$	$\eta/\mu P$	
	273 K	273 K	293 K
Air	24.1	173	182
Ar	16.3	210	223
C_2H_4	16.4	97	103
CH_4	30.2	103	110
Cl_2	7.9	123	132
CO_2	14.5	136	147
H_2	168.2	84	88
He	144.2	187	196
Kr	8.7	234	250
N_2	24.0	166	176
Ne	46.5	298	313
O_2	24.5	195	204
Xe	5.2	212	228

Data: KL.

Table 19B.1 Viscosities of liquids at 298 K, $\eta/(10^{-3}\ kg\ m^{-1}\ s^{-1})$

Benzene	0.601
Carbon tetrachloride	0.880
Ethanol	1.06
Mercury	1.55
Methanol	0.553
Pentane	0.224
Sulfuric acid	27
Water†	0.891

† The viscosity of water over its entire liquid range is represented with less than 1 per cent error by the expression $\log(\eta_{20}/\eta) = A/B$,
$A = 1.370\ 23(t-20) + 8.36 \times 10^{-4}(t-20)^2$
$B = 109 + t \quad t = \theta/°C$
Convert kg m^{-1} s^{-1} to centipoise (cP) by multiplying by 10^3 (so $\eta \approx 1$ cP for water).
Data: AIP, KL.

Table 19B.2 Ionic mobilities in water at 298 K, $u/(10^{-8}\ m^2\ s^{-1}\ V^{-1})$

Cations		Anions	
Ag^+	6.24	Br^-	8.09
Ca^{2+}	6.17	$CH_3CO_2^-$	4.24
Cu^{2+}	5.56	Cl^-	7.91
H^+	36.23	CO_3^{2-}	7.46
K^+	7.62	F^-	5.70
Li^+	4.01	$[Fe(CN)_6]^{3-}$	10.5
Na^+	5.19	$[Fe(CN)_6]^{4-}$	11.4
NH_4^+	7.63	I^-	7.96
$[N(CH_3)_4]^+$	4.65	NO_3^-	7.40
Rb^+	7.92	OH^-	20.64
Zn^{2+}	5.47	SO_4^{2-}	8.29

Data: Principally Table 19B.2 and $u = \lambda/zF$.

Table 19B.3 Diffusion coefficients at 298 K, $D/(10^{-9}\ m^2\ s^{-1})$

Molecules in liquids				Ions in water			
I_2 in hexane	4.05	H_2 in CCl_4(l)	9.75	K^+	1.96	Br^-	2.08
in benzene	2.13	N_2 in CCl_4(l)	3.42	H^+	9.31	Cl^-	2.03
CCl_4 in heptane	3.17	O_2 in CCl_4(l)	3.82	Li^+	1.03	F^-	1.46
Glycine in water	1.055	Ar in CCl_4(l)	3.63	Na^+	1.33	I^-	2.05
Dextrose in water	0.673	CH_4 in CCl_4(l)	2.89			OH^-	5.03
Sucrose in water	0.5216	H_2O in water	2.26				
		CH_3OH in water	1.58				
		C_2H_5OH in water	1.24				

Data: AIP.

Table 20B.1 Kinetic data for first-order reactions

	Phase	θ/°C	k_r/s^{-1}	$t_{1/2}$
$2\,N_2O_5 \rightarrow 4\,NO_2 + O_2$	g	25	3.38×10^{-5}	5.70 h
	HNO_3(l)	25	1.47×10^{-6}	131 h
	Br_2(l)	25	4.27×10^{-5}	4.51 h
$C_2H_6 \rightarrow 2\,CH_3$	g	700	5.36×10^{-4}	21.6 min
Cyclopropane → propene	g	500	6.71×10^{-4}	17.2 min
$CH_3N_2CH_3 \rightarrow C_2H_6 + N_2$	g	327	3.4×10^{-4}	34 min
Sucrose → glucose + fructose	aq(H^+)	25	6.0×10^{-5}	3.2 h

g: High pressure gas-phase limit.
Data: Principally K.J. Laidler, *Chemical kinetics*. Harper & Row, New York (1987); M.J. Pilling and P.W. Seakins, *Reaction kinetics*. Oxford University Press (1995); J. Nicholas, *Chemical kinetics*. Harper & Row, New York (1976). See also JL.

Table 20B.2 Kinetic data for second-order reactions

	Phase	θ/°C	k_r/(dm^3 mol^{-1} s^{-1})
$2\,NOBr \rightarrow 2\,NO + Br_2$	g	10	0.80
$2\,NO_2 \rightarrow 2\,NO + O_2$	g	300	0.54
$H_2 + I_2 \rightarrow 2\,HI$	g	400	2.42×10^{-2}
$D_2 + HCl \rightarrow DH + DCl$	g	600	0.141
$2\,I \rightarrow I_2$	g	23	7×10^{9}
	hexane	50	1.8×10^{10}
$CH_3Cl + CH_3O^-$	methanol	20	2.29×10^{-6}
$CH_3Br + CH_3O^-$	methanol	20	9.23×10^{-6}
$H^+ + OH^- \rightarrow H_2O$	water	25	1.35×10^{11}
	ice	−10	8.6×10^{12}

Data: Principally K.J. Laidler, *Chemical kinetics*. Harper & Row, New York (1987); M.J. Pilling and P.W. Seakins, *Reaction kinetics*. Oxford University Press, (1995); J. Nicholas, *Chemical kinetics*. Harper & Row, New York (1976).

Table 20D.1 Arrhenius parameters

First-order reactions	A/s^{-1}	E_a/(kJ mol^{-1})
Cyclopropane → propene	1.58×10^{15}	272
$CH_3NC \rightarrow CH_3CN$	3.98×10^{13}	160
cis-CHD=CHD → *trans*-CHD=CHD	3.16×10^{12}	256
Cyclobutane → $2\,C_2H_4$	3.98×10^{13}	261
$C_2H_5I \rightarrow C_2H_4 + HI$	2.51×10^{17}	209
$C_2H_6 \rightarrow 2\,CH_3$	2.51×10^{7}	384
$2\,N_2O_5 \rightarrow 4\,NO_2 + O_2$	4.94×10^{13}	103.4
$N_2O \rightarrow N_2 + O$	7.94×10^{11}	250
$C_2H_5 \rightarrow C_2H_4 + H$	1.0×10^{13}	167

(*Continued*)

Table 20D.1 (Continued)

Second-order, gas-phase	$A/(\mathrm{dm^3\ mol^{-1}\ s^{-1}})$	$E_a/(\mathrm{kJ\ mol^{-1}})$
$O+N_2 \rightarrow NO+N$	1×10^{11}	315
$OH+H_2 \rightarrow H_2O+H$	8×10^{10}	42
$Cl+H_2 \rightarrow HCl+H$	8×10^{10}	23
$2\ CH_3 \rightarrow C_2H_6$	2×10^{10}	*ca.*0
$NO+Cl_2 \rightarrow NOCl+Cl$	4.0×10^{9}	85
$SO+O_2 \rightarrow SO_2+O$	3×10^{8}	27
$CH_3+C_2H_6 \rightarrow CH_4+C_2H_5$	2×10^{8}	44
$C_6H_5+H_2 \rightarrow C_6H_6+H$	1×10^{8}	*ca.*25
Second-order, solution	$A/(\mathrm{dm^3\ mol^{-1}\ s^{-1}})$	$E_a/(\mathrm{kJ\ mol^{-1}})$
$C_2H_5ONa+CH_3I$ in ethanol	2.42×10^{11}	81.6
$C_2H_5Br+OH^-$ in water	4.30×10^{11}	89.5
$C_2H_5I+C_2H_5O^-$ in ethanol	1.49×10^{11}	86.6
$C_2H_5Br+OH^-$ in ethanol	4.30×10^{11}	89.5
CO_2+OH^- in water	1.5×10^{10}	38
$CH_3I+S_2O_3^{2-}$ in water	2.19×10^{12}	78.7
Sucrose$+H_2O$ in acidic water	1.50×10^{15}	107.9
$(CH_3)_3CCl$ solvolysis		
in water	7.1×10^{16}	100
in methanol	2.3×10^{13}	107
in ethanol	3.0×10^{13}	112
in acetic acid	4.3×10^{13}	111
in chloroform	1.4×10^{4}	45
$C_6H_5NH_2+C_6H_5COCH_2Br$ in benzene	91	34

Data: Principally J. Nicholas, *Chemical kinetics*. Harper & Row, New York (1976) and A.A. Frost and R.G. Pearson, *Kinetics and mechanism*. Wiley, New York (1961).

Table 21A.1 Arrhenius parameters for gas-phase reactions

	$A/(\mathrm{dm^3\ mol^{-1}\ s^{-1}})$			
	Experiment	Theory	$E_a/(\mathrm{kJ\ mol^{-1}})$	P
$2\ NOCl \rightarrow 2\ NO+Cl_2$	9.4×10^{9}	5.9×10^{10}	102.0	0.16
$2\ NO_2 \rightarrow 2\ NO+O_2$	2.0×10^{9}	4.0×10^{10}	111.0	5.0×10^{-2}
$2\ ClO \rightarrow Cl_2+O_2$	6.3×10^{7}	2.5×10^{10}	0.0	2.5×10^{-3}
$H_2+C_2H_4 \rightarrow C_2H_6$	1.24×10^{6}	7.4×10^{11}	180	1.7×10^{-6}
$K+Br_2 \rightarrow KBr+Br$	1.0×10^{12}	2.1×10^{11}	0.0	4.8

Data: Principally M.J. Pilling and P.W. Seakins, *Reaction kinetics*. Oxford University Press (1995).

Table 21B.1 Arrhenius parameters for reactions in solution. See Table 20D.1.

Table 21F.1 Exchange current densities (j_0) and transfer coefficients (α) at 298 K

Reaction	Electrode	j_0/(A cm^{-2})	α
$2\,H^+ + 2\,e^- \rightarrow H_2$	Pt	7.9×10^{-4}	
	Cu	1×10^{-6}	
	Ni	6.3×10^{-6}	0.58
	Hg	7.9×10^{-13}	0.50
	Pb	5.0×10^{-12}	
$Fe^{3+} + e^- \rightarrow Fe^{2+}$	Pt	2.5×10^{-3}	0.58
$Ce^{4+} + e^- \rightarrow Ce^{3+}$	Pt	4.0×10^{-5}	0.75

Data: Principally J.O'M. Bockris and A.K.N. Reddy, *Modern electrochemistry*. Pleanum, New York (1970).

Table 22A.1 Maximum observed standard enthalpies of physisorption, $\Delta_{ad}H^{\ominus}$/(kJ mol^{-1}) at 298 K

C_2H_2	−38	H_2	−84
C_2H_4	−34	H_2O	−59
CH_4	−21	N_2	−21
Cl_2	−36	NH_3	−38
CO	−25	O_2	−21
CO_2	−25		

Data: D.O. Haywood and B.M.W. Trapnell, *Chemisorption*. Butterworth (1964).

Table 22A.2 Standard enthalpies of chemisorption, $\Delta_{ad}H^{\ominus}$/(kJ mol^{-1}) at 298 K

Adsorbate	Adsorbent (substrate)											
	Ti	Ta	Nb	W	Cr	Mo	Mn	Fe	Co	Ni	Rh	Pt
H_2		−188			−188	−167	−71	−134			−117	
N_2		−586						−293				
O_2						−720					−494	−293
CO	−640							−192	−176			
CO_2	−682	−703	−552	−456	−339	−372	−222	−225	−146	−184		
NH_3				−301				−188		−155		
C_2H_4		−577		−427	−427			−285		−243	−209	

Data: D.O. Haywood and B.M.W. Trapnell, *Chemisorption*. Butterworth (1964).

PART 4 Character tables

The groups C_1, C_s, C_i

C_1 (1)	E	$h=1$
A	1	

$C_s = C_h$ m	E	σ_h	$h=2$	
A′	1	1	x, y, R_z	x^2, y^2, z^2, xy
A″	1	−1	z, R_x, R_y	yz, zx

$C_i = S_2$ $\bar{1}$	E	i	$h=2$	
A_g	1	1	R_x, R_y, R_z	$x^2, y^2, z^2, xy, yz, zx,$
A_u	1	−1	x, y, z	

The groups C_{nv}

C_{2v}, $2mm$	E	C_2	σ_v	σ'_v	$h=4$	
A_1	1	1	1	1	z, z^2, x^2, y^2	
A_2	1	1	−1	−1	xy	R_z
B_1	1	−1	1	−1	x, zx	R_y
B_2	1	−1	−1	1	y, yz	R_x

C_{3v}, $3m$	E	$2C_3$	$3\sigma_v$	$h=6$	
A_1	1	1	1	z, z^2, x^2+y^2	
A_2	1	1	−1		R_z
E	2	−1	0	$(x, y), (xy, x^2-y^2)$ (yz, zx)	(R_x, R_y)

C_{4v}, $4mm$	E	C_2	$2C_4$	$2\sigma_v$	$2\sigma_d$	$h=8$	
A_1	1	1	1	1	1	z, z^2, x^2+y^2	
A_2	1	1	1	−1	−1		R_z
B_1	1	1	−1	1	−1	x^2-y^2	
B_2	1	1	−1	−1	1	xy	
E	2	−2	0	0	0	$(x, y), (yz, zx)$	(R_x, R_y)

C_{5v}	E	$2C_5$	$2C_5^2$	$5\sigma_v$	$h=10, \alpha=72°$	
A_1	1	1	1	1	z, z^2, x^2+y^2	
A_2	1	1	1	−1		R_z
E_1	2	$2\cos\alpha$	$2\cos 2\alpha$	0	$(x, y), (yz, zx)$	(R_x, R_y)
E_2	2	$2\cos 2\alpha$	$2\cos\alpha$	0	(xy, x^2-y^2)	

C_{6v}, $6mm$	E	C_2	$2C_3$	$2C_6$	$3\sigma_d$	$3\sigma_v$	$h=12$	
A_1	1	1	1	1	1	1	z, z^2, x^2+y^2	
A_2	1	1	1	1	−1	−1		R_z
B_1	1	−1	1	−1	−1	1		
B_2	1	−1	1	−1	1	−1		
E_1	2	−2	−1	1	0	0	$(x, y), (yz, zx)$	(R_x, R_y)
E_2	2	2	−1	−1	0	0	(xy, x^2-y^2)	

$C_{\infty v}$	E	$2C_\phi$†	$\infty\sigma_v$	$h=\infty$	
$A_1(\Sigma^+)$	1	1	1	z, z^2, x^2+y^2	
$A_2(\Sigma^-)$	1	1	−1		R_z
$E_1(\Pi)$	2	$2\cos\phi$	0	$(x, y), (yz, zx)$	(R_x, R_y)
$E_2(\Delta)$	2	$2\cos 2\phi$	0	(xy, x^2-y^2)	

† There is only one member of this class if $\phi=\pi$.

The groups D_n

D_2, 222	E	C_2^z	C_2^y	C_2^x	$h=4$	
A_1	1	1	1	1	x^2, y^2, z^2	
B_1	1	1	−1	−1	z, xy	R_z
B_2	1	−1	1	−1	y, zx	R_y
B_3	1	−1	−1	1	x, yz	R_x

D_3, 32	E	$2C_3$	$3C'_2$	$h=6$	
A_1	1	1	1	z^2, x^2+y^2	
A_2	1	1	−1	z	R_z
E	2	−1	0	$(x, y), (yz, zx), (xy, x^2-y^2)$	(R_x, R_y)

D_4, 422	E	C_2	$2C_4$	$2C'_2$	$2C''_2$	$h=8$	
A_1	1	1	1	1	1	z^2, x^2+y^2	
A_2	1	1	1	−1	−1	z	R_z
B_1	1	1	−1	1	−1	x^2-y^2	
B_2	1	1	−1	−1	1	xy	
E	2	−2	0	0	0	$(x, y), (yz, zx)$	(R_x, R_y)

The groups D_{nh}

D_{3h}, $\bar{6}2m$	E	σ_h	$2C_3$	$2S_3$	$3C_2'$	$3\sigma_v$	$h=12$	
A_1'	1	1	1	1	1	1	z^2, x^2+y^2	
A_2'	1	1	1	1	−1	−1		R_z
A_1''	1	−1	1	−1	1	−1		
A_2''	1	−1	1	−1	−1	1	z	
E'	2	2	−1	−1	0	0	$(x, y), (xy, x^2-y^2)$	
E''	2	−2	−1	1	0	0	(yz, zx)	(R_x, R_y)

D_{4h}, $4/mmm$	E	$2C_4$	C_2	$2C_2'$	$2C_2''$	i	$2S_4$	σ_h	$2\sigma_v$	$2\sigma_d$	$h=16$	
A_{1g}	1	1	1	1	1	1	1	1	1	1	x^2+y^2, z^2	
A_{2g}	1	1	1	−1	−1	1	1	1	−1	−1		R_z
B_{1g}	1	−1	1	1	−1	1	−1	1	1	−1	x^2-y^2	
B_{2g}	1	−1	1	−1	1	1	−1	1	−1	1	xy	
E_g	2	0	−2	0	0	2	0	−2	0	0	(yz, zx)	(R_x, R_y)
A_{1u}	1	1	1	1	1	−1	−1	−1	−1	−1		
A_{2u}	1	1	1	−1	−1	−1	−1	−1	1	1	z	
B_{1u}	1	−1	1	1	−1	−1	1	−1	−1	1		
B_{2u}	1	−1	1	−1	1	−1	1	−1	1	−1		
E_u	2	0	−2	0	0	−2	0	2	0	0	(x, y)	

D_{5h}	E	$2C_5$	$2C_5^2$	$5C_2$	σ_h	$2S_5$	$2S_5^3$	$5\sigma_v$	$h=20$	$\alpha=72°$
A_1'	1	1	1	1	1	1	1	1	x^2+y^2, z^2	
A_2'	1	1	1	−1	1	1	1	−1		R_z
E_1'	2	$2\cos\alpha$	$2\cos 2\alpha$	0	2	$2\cos\alpha$	$2\cos 2\alpha$	0	(x, y)	
E_2'	2	$2\cos 2\alpha$	$2\cos\alpha$	0	2	$2\cos 2\alpha$	$2\cos\alpha$	0	(x^2-y^2, xy)	
A_1''	1	1	1	1	−1	−1	−1	−1		
A_2''	1	1	1	−1	−1	−1	−1	1	z	
E_1''	2	$2\cos\alpha$	$2\cos 2\alpha$	0	−2	$-2\cos\alpha$	$-2\cos 2\alpha$	0	(yz, zx)	(R_x, R_y)
E_2''	2	$2\cos 2\alpha$	$2\cos\alpha$	0	−2	$-2\cos 2\alpha$	$-2\cos\alpha$	0		

$D_{\infty h}$	E	$2C_\phi$	…	$\infty\sigma_v$	i	$2S_\infty$	…	$\infty C_2'$	$h=\infty$	
$A_{1g}(\Sigma_g^+)$	1	1	…	1	1	1	…	1	z^2, x^2+y^2	
$A_{1u}(\Sigma_u^+)$	1	1	…	1	−1	−1	…	−1	z	
$A_{2g}(\Sigma_g^-)$	1	1	…	−1	1	1	…	−1		R_z
$A_{2u}(\Sigma_u^-)$	1	1	…	−1	−1	−1	…	1		
$E_{1g}(\Pi_g)$	2	$2\cos\phi$	…	0	2	$-2\cos\phi$	…	0	(yz, zx)	(R_x, R_y)
$E_{1u}(\Pi_u)$	2	$2\cos\phi$	…	0	−2	$2\cos\phi$	…	0	(x, y)	
$E_{2g}(\Delta_g)$	2	$2\cos 2\phi$	…	0	2	$2\cos 2\phi$	…	0	(xy, x^2-y^2)	
$E_{2u}(\Delta_u)$	2	$2\cos 2\phi$	…	0	−2	$-2\cos 2\phi$	…	0		
⋮	⋮	⋮	…	⋮	⋮	⋮	…	⋮		

The cubic groups

T_d, $\bar{4}3m$	E	$8C_3$	$3C_2$	$6\sigma_d$	$6S_4$	$h=24$	
A_1	1	1	1	1	1	$x^2+y^2+z^2$	
A_2	1	1	1	−1	−1		
E	2	−1	2	0	0	$(3z^2-r^2, x^2-y^2)$	
T_1	3	0	−1	−1	1		(R_x, R_y, R_z)
T_2	3	0	−1	1	−1	(x, y, z), (xy, yz, zx)	

O_h, $m3m$	E	$8C_3$	$6C_2$	$6C_4$	$3C_2(=C_4^2)$	i	$6S_4$	$8S_6$	$3\sigma_h$	$6\sigma_d$	$h=48$	
A_{1g}	1	1	1	1	1	1	1	1	1	1	$x^2+y^2+z^2$	
A_{2g}	1	1	−1	−1	1	1	−1	1	1	−1		
E_g	2	−1	0	0	2	2	0	−1	2	0	$(2z^2-x^2-y^2, x^2-y^2)$	
T_{1g}	3	0	−1	1	−1	3	1	0	−1	−1		(R_x, R_y, R_z)
T_{2g}	3	0	1	−1	−1	3	−1	0	−1	1	(xy, yz, zx)	
A_{1u}	1	1	1	1	1	−1	−1	−1	−1	−1		
A_{2u}	1	1	−1	−1	1	−1	1	−1	−1	1		
E_u	2	−1	0	0	2	−2	0	1	−2	0		
T_{1u}	3	0	−1	1	−1	−3	−1	0	1	1	(x, y, z)	
T_{2u}	3	0	1	−1	−1	−3	1	0	1	−1		

The icosahedral group

I	E	$12C_5$	$12C_5^2$	$20C_3$	$15C_2$	$h=60$	
A	1	1	1	1	1	$x^2+y^2+z^2$	
T_1	3	$\frac{1}{2}(1+5^{1/2})$	$\frac{1}{2}(1-5^{1/2})$	0	−1	(x, y, z)	(R_x, R_y, R_z)
T_2	3	$\frac{1}{2}(1-5^{1/2})$	$\frac{1}{2}(1+5^{1/2})$	0	−1		
G	4	−1	−1	1	0		
H	5	0	0	−1	1	$(2z^2-x^2-y^2, x^2-y^2, xy, yz, zx)$	

Further information: P.W. Atkins, M.S. Child, and C.S.G. Phillips, *Tables for group theory.* Oxford University Press, (1970). In this source, which is available on the web (see p. *x* for more details), other character tables such as D_2, D_4, D_{2d}, D_{3d}, and D_{5d} can be found.

INDEX

D

R

S

T

U